5/24

D0074723

QUATERNARY ENVIRONMENTS

TITLES OF RELATED INTEREST

QE
696
.O3264
1985

QUATERNARY ENVIRONMENTS

EASTERN CANADIAN ARCTIC, BAFFIN BAY AND WESTERN GREENLAND

EDITOR

J. T. ANDREWS

Institute of Arctic and Alpine Research,
University of Colorado

Boston
ALLEN & UNWIN
London   Sydney

WITHDRAWN

Tennessee Tech. Library
Cookeville, Tenn.
357150

© J. T. Andrews and contributors, 1985
This book is copyright under the Berne Convention. No reproduction
without permission. All rights reserved.

This book was typeset, proofed and passed for press by the editor.

Allen & Unwin Inc.,
Fifty Cross Street, Winchester, Mass. 01890, USA

George Allen & Unwin (Publishers) Ltd,
40 Museum Street, London WC1A 1LU, UK

George Allen & Unwin (Publishers) Ltd,
Park Lane, Hemel Hempstead, Herts HP2 4TE, UK

George Allen & Unwin Australia Pty Ltd,
8 Napier Street, North Sydney, NSW 2060, Australia

First published in 1985

Library of Congress Cataloging in Publication Data

Main entry under title:
    Quaternary environments.
Bibliography: p.
1. Geology, Stratigraphic—Quaternary. 2. Geology—
Northwest Territories—Franklin District.
3. Geology—Baffin Bay Region (North Atlantic Ocean)
4. Paleoceanography—Baffin Bay Region (North Atlantic Ocean)
5. Paleoceanography—Labrador Sea. I. Andrews, John T.
QE696.3264 1985   551.7'9'09719   85–6101
ISBN 0–04–551094–6

British Library Cataloguing in Publication Data

    Quaternary environments : the eastern Canadian Arctic, Baffin Bay and
West Greenland.
1. Geology, Stratigraphic—Quaternary
2. Geology—Arctic Regions
I. Andrews, John T.
551.7'9'09998   QE696
ISBN 0–04–551094–6

Printed in Great Britain
by Mackays of Chatham Ltd

Preface

Sometime in the late 1970s at an annual meeting of the Geological
Society of America, Professor A.L. (Art) Bloom of Cornell University
said words to the effect: it would be a good idea to try and
consolidate all the work that has gone on concerning the Quaternary
history of the Eastern Canadian Arctic. Several years later this book
seeks to accomplish that goal. However, as is very clear from the
title and the contents of many chapters the scope of the work has
increased so as to include the important companion research that has
been undertaken in Baffin Bay, on surrounding ice caps, and on the
terrestrial and lacustrine deposits of West Greenland.

This volume represents the combined efforts of scientists from
Denmark, England, Canada, and the U.S.A. The process of compiling the
26 chapters started some 2 to 3 years ago and it is with some relief
that I am now at the moment of writing the Preface and the
Acknowledgements. I have certainly learned a great deal from editing
this volume and the studies reported on will serve as both a testimony
to past concepts and ideas as well as pointing toward new theories.
The ocean and lands surrounding Baffin Bay have a central part to play
in any reconstructions of Quaternary environments at either the
regional or global scales.

John T. Andrews
Boulder, Colorado
14 November 1984

Acknowledgements

Many people have contributed to the successful production of this
volume. Foremost amongst these I must thank Rosella Chavez of the
Institute of Arctic and Alpine Research (INSTAAR), University of
Colorado, Boulder, who has carefully typed and corrected several
versions of each chapter. Her cheerfulness in the face of my edtorial
failings and authors' errors has been remarkable and this book owes
much to her industry. In addition Fatima Al-Rahim assisted with the
word-processing of several chapters during the early stages of the
production of the volume and I wish to thank her for her efforts.

During the final stages of production I was able to "request" the
assistance of several people at INSTAAR who read between one and three
chapters in a final effort to check for typographical errors,
references, and other details. Thus I thank Martha Andrews, Kerstin
Williams, Anne Jennings, Chip Layman, Harvey Thorleifson, Chris
Waythomas, Many Wilson, Lisa Osterman, and Phil Wyatt for their
valuable assistance. Jim Walters did a great deal of the drafting for
those chapters that came from INSTAAR members and, with great care,
undertook the final pasting in of the figures. His work is greatly
appreciated. Roger Jones and Geoffrey Palmer of Allen and Unwin are to
be thanked for their assistance and interest in this volume.

The University of Colorado provided support for the editing of
this volume in the form of a grant which partially covered the costs of
reducing figures. The Institute of Arctic and Alpine Research,
University of Colorado, subsidized the production of the volume in
numerous ways and I would like to thank Pat Webber, Director of
INSTAAR, for his support.

Although the bulk of the material in this volume is original,
there are a number of figures that have been reproduced with the
permission from the publisher or journal. These are noted under the
appropriate figures as "with permission". Here I wish to specifically
outline the source for these permissions.

Figure 3.7   John Wiley and Sons, New York. From "The Last Great Ice
             Sheets". Ed. G.H. Denton & T. Hughes (1980).
Figure 6.2, 6.3, 6.4, and 6.11 Canadian Society of Petroleum
             Geologists (Memoir 7 & 8; Bulletin Canadian Petroleum
             Geology, 29).

Figure 12.4 Academic Press (<u>Quaternary</u> <u>Research</u>, 11 (1979).
Figure 13.9, 13.10, 13.11, and 13.12 Elsevier Science Publishers B.V.
    (<u>Marine Micropaleontology</u>, 5 (1980)).
Figure 20.2, 20.8 <u>Canadian Journal of Earth Sciences</u>, (21, (1984)).
Figure 21.3 John Wiley and Sons, London. From: "Earth Rheology,
    Isostasy and Eustasy". Ed. N-A. Mörner.
Figure 23.2, 23.3, 23.6, 23.7, and 23.14 Meddelelser om Grønland,
    GeoScience (10, (1983)).
Figure 25.1 <u>Arctic and Alpine Research</u> (9, (1977)).
Figure 26.1, 26.2 <u>Arctic and Alpine Research</u> (9, (1977)).

Contents

Part IV -- **Pleistocene Glacial and Nonglacial Stratigraphy**

Part V -- **Holocene Sea Levels and Climate**

List of contributors

John T. Andrews, Institute of Arctic and Alpine Research and Department of Geological Sciences, University of Colorado, Boulder, Colorado, USA

Ali E. Aksu, Department of Geology, University of Dalhousie, Halifax, Nova Scotia, CANADA

Raymond S. Bradley, Department of Geology and Geography, University of Massachusetts, Amherst, Massachusetts, USA

Peter U. Clark, Department of Geological Sciences, University of Illinois, Chicago, Illinois, USA

P. Thompson Davis, Department of Geography and Geology, Mount Holyoke College, South Hadley, Massachusetts, USA

Evelyn K. Dowdeswell, Scott Polar, University of Cambridge, Cambridge, ENGLAND

Rolf W. Feyling-Hanssen, Geologisk Institut, Aarhus Universitet, Universitetsparken, Aarhus C, DENMARK

Richard H. Fillon, Belle W. Baruch Institute of Marine Science, University of South Carolina, Columbia, South Carolina, USA

David A. Fisher, Polar Continental Shelf Project, Department of Energy, Mines and Resources, Ottawa, Ontario, CANADA

Bent Fredskild, Gronlands Botaniske Undersoleglse, Universitetets Botaniske Museum Gothersgade 130, Kobenhavn K, DENMARK

Svend Funder, Universitetets Mineralogiske Museum, Oster Voldgade 5-7, Kobenhavn K, DENMARK

John D. Jacobs, Department of Geography, University of Windsor, Windsor, Ontario, CANADA

Michael Kelly, Department of Environmental Science, University of Lancaster, Lancaster, ENGLAND

Rodney A. Klassen, Terrain Sciences, Geological Survey of Canada, Ottawa, Ontario, CANADA

Roy M. Koerner, Polar Continental Shelf Project, Department of Energy, Mines and Resources, Ottawa, Ontario, CANADA

William W. Locke, III, Department of Geology, Montana State College, Bozeman, Montana, USA

Brian MacLean, Atlantic Geoscience Centre, Geological Survey of Canada, Bedford Institute of Oceanography, Dartmouth, Nova Scotia, CANADA

Gifford H. Miller, Institute of Arctic and Alpine Research and Department of Geological Sciences, University of Colorado, Boulder, Colorado, USA

William N. Mode, Department of Geology, University of Wisconsin, Oshkosh, Wisconsin, USA

Peta J. Mudie, Geological Survey of Canada, Atlantic Geoscience Centre, Dartmouth, Nova Scotia, CANADA

Lisa E. Osterman, Institute of
Arctic and Alpine Research,
University of Colorado,
Boulder, Colorado, USA

Garry Quinlan, Department of
Earth Sciences, Memorial
University, St. John's,
Newfoundland, CANADA

Susan K. Short, Institute of
Arctic and Alpine Research,
University of Colorado, Boulder,
Colorado, USA

Jay A. Stravers, Institute of
Arctic and Alpine Research,
University of Colorado, Boulder,
Colorado, USA

Larry D. Williams, School of
Environmental Sciences, The
University of East Anglia,
Norwich, ENGLAND

Part I
INTRODUCTION TO QUATERNARY STUDIES AND PRESENT ENVIRONMENT

1 Introduction to Quaternary studies

J. T. Andrews and S. Funder

This volume deals with Quaternary history of the areas around Baffin Bay--principally Baffin Island and West Greenland--as well as oceano graphic changes in the Bay itself and in the adjacent Davis Strait. Tucked away at the edge of world maps, to most people it is a remote and unknown region. To Quaternary scientists, however, the region is far from obscure; indeed, the wealth of geological, glaciological, and paleobotanical material in the area have made the Baffin Bay region one of the most important sources of paleoenvironmental data in the entire Northern Hemisphere. Much work has already been accomplished in documenting climatic and environmental changes in the region, but much more remains to be done. It is not a small area, as Figure 1.1 illustrates, and it has a much shorter history of scientific research than regions of comparable size on the other side of the North Atlantic Basin. Nevertheless, studies in and around Baffin Bay have already made important contributions to our understanding of glaciation and deglaciation on a global scale. It is therefore the purpose of this volume to summarize the work accomplished to date, to reflect on the implications of this work, to identify areas of uncertainty and confusion, and to chart research plans for the future. We envisage the book as part of a continuing research program, rather than an end in itself.

Research on the glacial and climatic history of the entire area has been in progress for more than a century, but the pace and volume of research has noticeably increased over the last two decades. It is the purpose of this volume to present a synthesis and an in depth assessment of the Quaternary history of this region. The book is <u>not</u> intended to present a series of detailed reviews of previous observations and theories; the object is to present, in a single volume, the latest thinking and the latest observations. Such an objective is obviously constrained by the "wisdom" of the past, but the

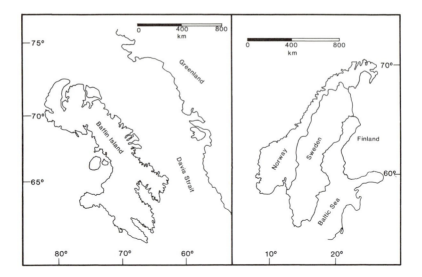

Figure 1.1 Comparison of the size of the area included in studies in this volume with the region of Fennoscandia at the same scale.

pace of research has been so rapid over the last one to two decades that concepts have also been forced to shift and change to keep pace with the new observations. However, it now appears to us that the magnitude of new findings has slowed down--there are still, however, many important questions left unresolved or even unasked. Thus, it is an appropriate moment to take stock of the multifaceted research programs that have been carried out in this region sponsored by different governments, different universities, and from different disciplines. We need to sit back and examine: Where do the different data sets point to a common conclusion?; Where do the data from different areas or disciplines appear contradictory?; Where do the observations from this critical area of the Northern Hemisphere stand in terms of the continuing, and important debate on the question of the necessary conditions for glaciation of Arctic areas?, and the question of the in or out-of phase glacial chronologies of the arctic margins of the great ice sheets?

In a recent book Denton and Hughes (1981) have critically examined the field data from Arctic areas and presented a scenario for glaciation and the timing of that glaciation that is radically different from the interpretation of most field workers (e.g. Funder and Hjort, 1973; England et al., 1981; Klassen, 1981; Miller and Dyke, 1974; Nelson, 1980, 1981; Miller et al., 1977), but which is supported by others (e.g. Blake, 1970, 1975). The major aim of this volume is

to objectively examine the field and laboratory data in order to present whatever model the authors feel explains most facets of the actual Quaternary record. This volume can also be considered as a companion study
of the "Paleoecology of Beringia" (eds. D.M. Hopkins et al. 1982). In that book authors considered the Quaternary events of the Yukon Territory, Canada, Alaska, and eastern Siberia. Together, our two studies include significant new data and interpretations of Quaternary events in two quite different Arctic settings.

Because of the studies on deep-sea marine cores in the north eastern North Atlantic south of 55°N (Ruddiman and McIntyre, 1973; Ruddiman, 1977; Ruddiman and McIntyre, 1981), and the extension of those studies into the Norwegian and Greenland seas (Kellogg, 1976; 1980), the Quaternary oceanographic changes, and their inferred impact on the glacial and climatic history of Northwest Europe has become an accepted tenet of research. In particular, the movement of the oceano graphic "Polar Front" has been mapped by Ruddiman and McIntyre and has been used by several workers to explain the terrestrial Quaternary record. It has been tacitly assumed by many researchers that the Quaternary history of the northwestern North Atlantic simply mimics its larger brethren. The literature to date does not support such an assumption.

The northwestern arm of the North Atlantic Ocean consists of the geographical features Baffin Bay, Davis Strait, and the Labrador Sea (Fig. 1.2). On the east side of these seas is the subcontinent of Greenland which is dominated by the Greenland ice sheet, covering an area of c. $2 \times 10^6 km^2$ (Fig. 1.3). In the north, the ice-sheet meets the ocean for a distance of 100 km in Melville Bugt, the ice-free land being restricted to islands and isolated rocky outliers protruding from under the ice sheet. Calving glaciers, forming major outlets from the Inland Ice, occur in the embayment to the north of the Nugssuaq Peninsula, and in Disko Bugt. To the south the ice-sheet is separated from the Davis Strait by a strip of land, c. 100 kilometers wide, with scattered ice caps and numerous mountain glaciers. The land strip, c. 125,000 km^2 in area, is dissected by long and narrow fjords. It is in this region that the large majority of the Greenland population lives. Across Baffin Bay and Davis Strait is the 450,000 km^2 area of Baffin Island which is largely unglaciated. However, small plateaux ice caps exist on Baffin Island, Bylot Island, and Devon Island and, in addition, all these areas have a variety of valley and cirque glaciers largely concentrated in the mountains and uplands that fringe the northwestern North Atlantic. Relatively warm marine currents sweep northward along the coast of West Greenland, whereas the eastern coast

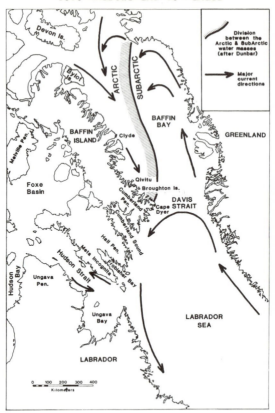

Figure 1.2 Location of the Arctic and Subarctic zoogeographic watermass boundary in Baffin Bay; set of the major surface currents; and
major geographical features.

of Baffin Island is dominated by the cold arctic waters of the Canadian
Current.

Although the physical mileau of our region is outlined in Chapter
2, it is worth noting the gross similarities and differences between
the environment of the northeastern and northwestern arms of the North
Atlantic Ocean. Both consist of large counterclockwise oceanographic
gyres which bring relatively warm water northward along the eastern
coast of each region whereas cold water moves southward along the
western margin (Fig. 1.2). However, this similarity in oceanographic
circulation cannot mask major differences in the glaciological
situation. Thus, on the eastern coast of the northeastern North
Atlantic, the highlands of Norway and Sweden support a few glaciers and
small ice caps. Further north in Spitsbergen and the islands of the
European Arctic, the regional snowline is very low. To the west by

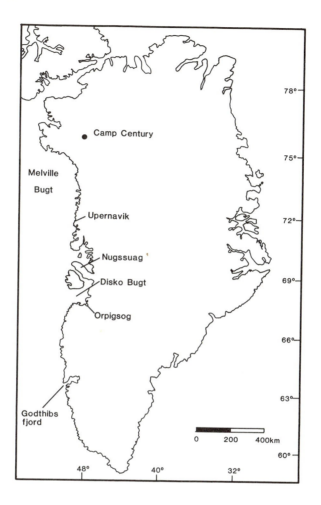

Figure 1.3 Location map of Greenland showing places mentioned in the text.

contrast, the cold East Greenland Current flows southward along the heavily glaciated coast of East Greenland.

The degree of glacierization in the northwestern North Atlantic is, however, the reverse of that noted above. The northward-moving warm West Greenland Current moves off the heavily glaciated coast of South and West Greenland; in contrast, the Canadian Current flows southward past the relatively unglacierized Baffin Island with the glaciation level (or snowline) rising from a minimum of 800 m to about 1200 m in the center of the island (Andrews and Miller, 1972; see Chapter 2).

QUATERNARY RECORDS AND EVENTS

A major problem in many Quaternary studies are the large distances that separated different stratigraphic records. A good example of this are the attempts to correlate the O-18 record from Camp Century, Greenland (Fig. 1.3) with fluctuations of the continental ice-sheet margins in northwestern Europe and the northern USA/southern Canada (Dansgaard et al., 1971). This approach is fraught with dangers because it is not clear what controlled the fluctuations of the ice margins or the variations in the O-18 values. In a similar fashion, the deep-ocean record from the open North Atlantic has been correlated with the glacial   history of the southern margin of the Laurentide Ice Sheet (Sancetta et al., 1973). These correlations associate a variety of stratigraphic records; they involve assumptions about dating; and they associate responses of different systems over distances of 10^3 km. The beauty of the northwestern North Atlantic is the variety of stratigraphic records that exist in close proximity to each other. They can thus provide a framework for scenarios of the interaction between the oceans, the atmosphere, the cryosphere, and the unglaciated land surface.

Figure 1.4 schematically depicts the variety of available stratigraphic records for this region of the North Atlantic. Because of logistical and technical difficulties not all of these records have been exploited to their full potential. However, we feel that this region has perhaps the greatest potential of any region in the Northern Hemisphere for describing and interpreting the complex interplay between process and response that occurs during periods of climatic fluctuation.

One of the challenges we face in this volume is to observe and interpret the various records as objectively as possible and with as few mental constraints as possible. Science is conservative, and rightly so, and hence data from "new" areas are most often fitted into an existing mold. In our region, an example of this problem can be briefly documented (cf. Andrews et al., 1981). The long history of Quaternary studies from the southern margins of the large Northern Hemisphere ice sheets has led to a codification of geological nomenclature. Because of different interests, and the differences in deposits, there is a distinct difference in the northwest European and North American approach to the definition of the geologic-climate unit called "Glaciation", or the opposite "Interglaciation". In Europe these distinctions have been frequently based on pollen assemblages and inferred climates, whereas in North America much more reliance has been placed on the recognition of tills (=Glaciation) and buried or relect

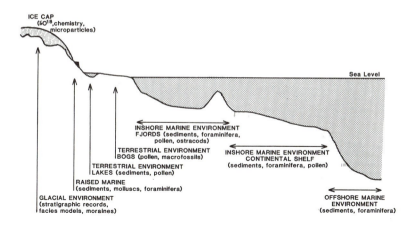

Figure 1.4 Schematic diagram of various environment in the area of study and the types of stratigraphic records available from a number of different sediment sinks (from Church, and Andrews, unpubl).

soils (= Interglaciation). However, both approaches have had in common the assumption that the trend toward Glaciation is governed by a decrease in temperature. In these areas, iceproximal and even distal sedimentary units contain "cold" indicators.

Along the east coast of Baffin Island raised, marine and glacial-marine sediments frequently contain abundant marine fossils (e.g. Feyling-Hanssen, 1976, 1980; Miller et al., 1977; Andrews et al., 1981). However, comparison of the present ranges of key taxa within these raised units, together with an interpretation of their mode of deposition (Nelson, 1978, 1980, 1981; Mode, 1980; Brigham, 1983; Miller et al., 1977), unequivocably indicates that ice-proximal raised glacial marine sediments contain "warm" subarctic molluscs and foraminifera--certainly a "warmer" fauna than exists in these areas today. On the "climatostratigraphic" grounds of European studies these sediments represent Interglacials whereas in reality they represent nearshore glacial marine litho-facies (Nelson, 1981: Mode et al., 1983; Brigham, 1973).

In our view, it is thus very important to separate the basic description of the different Quaternary records (Fig. 1.4) from the interpretation of the inferred relationships and/or climatic significance. Figure 1.5 thus provides a guide to our overall approach of distinguishing between descriptive stratigraphic records (see Hedberg, 1977) and interpreted events; in this context the evens can

NATURE OF THE RECORD	DESCRIPTIVE STRATIGRAPHIC UNITS	DATING	INTERPRETIVE EVENTS (Climatostratigraphy, Glacial-geologic stratigraphy)
Ice Caps (chpt 11)	Litho-, δO^{18}	^{14}C, "Counting"	"Warm" and "Cold" Intervals-"Glaciation" Inferred
Lakes (chpt 21,22)	Litho-, Bio-, (pollen diatoms)	^{210}Pb, ^{14}C	Changes in Sediment Source and Vegetation/Climate
Bogs (chpt 21,22)	Litho-, Bio-, (pollen)	^{14}C	as above
Glacial/Marine Sections (chpt 13,14,15,16,17,18)	Litho-, Bio-, (molluscs,forams,pollen)	Lichens, ^{14}C U-Series, Amino Acids	Glaciation or Interglaciations
Fjords (chpt 4)	Litho-, Bio-, δO^{18} (forams,diatoms,ostracods,pollen)	^{210}Pb, ^{14}C ,Amino Acid,	Changes in Sediment, Changes in Climate/Oceanography inferred
Continental Shelf (chpt 6)	Litho-, Bio-, δO^{18} (forams, etc.)	^{14}C , Amino Acid	as above
Deep Sea (chpt 7,8,9,10)	Litho-, Bio-, δO^{18} (forams,diatoms)	^{14}C , Amino Acid	as above

INTERPRETATIONS CAN INCLUDE :

Warm/Cold, Wet/Dry Climate; Arctic/Subarctic Water Masses; Extent Of Ice Rafting; Extent of Sea Ice; Extent of Glaciers; Changes of Vegetation; Meltwater Influx.

Figure 1.5. Presentation of the descriptive stratigraphic records available from the different environments of Figure 1.4 illustrating what methods of dating control (not correlation) are available and what is the nature of interpreted climate or glacial events that can be deduced from these records. Individual chapters associated with the different records are listed.

be changes in oceanographic circulation, fluctuations of sea level "caused" by glacial loading or unloading, changes in summer temperature inferred from variations in the amount of ice melt (Koerner, 1977) or changes in the percentages of pollen taxa (Fredskild, 1973; Kelly and Funder, 1974; Andrews et al., 1980). This distinction between described units and interpreted events is not trivial; indeed, one can trace much of the current ferment and argument about the glacial history of the Arctic region (e.g. Hughes et al., 1977; Miller et al.,1977; Funder and Hjort, 1973; Blake, 1970; Paterson, 1977; England, 1983; Boulton, 1979; Boulton et al., 1982) to the differences in the interpretation of raised marine sediments. The question is simply: do rasied marine sediments provide necessary and sufficient evidence to argue for the existence of an ice sheet? The argument revolves around different models of Earth's rheology and the geographical response of the Earth to glacial loading. These questions are specifically reviewed by Quinlan in this volume.

When comparing and analysing evidence from a large region, accumulated over long periods of time, it is necessary to take into consideration differences in research histories and traditions--these factors have determined not only which methods have been applied, but also which questions were to be considered. In our region there are two distinct research traditions, one in North America and one in Greenland. Characterized in short, the North American tradition has been one of concentrated efforts often dedicated to the solution of specific problems; conversely, the Greenland research history has centered on systematic and continuous gathering of "standard information". Hence, while the North American development may be described through a series of "landmarks", the Greenland record is characterized by the steady accumulation of knowledge.

Quaternary research in West Greenland

Although information relating to the Quaternary can be traced as far back as the Norse manuscripts from the 12th century, it is appropriate to start the record with the works of Hinrich Rink, who spent three years from 1848 doing a geological survey of the Upernavik District in northern West Greenland (Rink, 1852). Rink's reflections on the nature of the "Indlandsis" (Inland Ice), which he so named, had a decisive impact on the advancement of the glacial theory in Europe, but his survey also set the example for numerous topographical/geological expeditions which were to follow in the period from c. 1865 to c. 1914, relying heavily on the local population not only for logistics, but also for their intimate knowledge of their land. The few scientists involved in these expeditions covered a wide range of scientific disciplines.

Ironically, Rink never became a convinced "glacialist", but his successors, notably the geologist K.J.V. Steenstrup, were in no doubt that the heavily abraded rock surfaces with scattered erratic boulders so frequently met within West Greenland, were due to the work of an ice-sheet that once covered the entire land area. During this period observations on glacial striae, the distribution of erratics, altitudes of raised marine deposits, and the occurrence of moraines accumulated steadily from all parts of western Greenland, and now can be found in the early volumes of "Meddelelser on Gronland" (Communications on Greenland), a periodical published since 1879 by the Commission for Scientific Research in Greenland. These early acounts form a bank of information which it is still essential to consult when new projects are begun.

After 1914, largely through the initiative of Knud Rasmussen and Lauge Koch, geological investigations were concentrated in the more remote and inaccessible areas of northern and eastern Greenland. From his long sledge journey in 1920-1921 Koch brought home the first observations on the Quaternary in northernmost Greenland (Koch, 1928), and later between 1928-1958 he organized large-scale geological expeditions to East Greenland. Unfortunately the Quaternary geological record was rather neglected during this work.

In 1946 the Geological Survey of Greenland was established and systematic mapping of the West Greenland geology, including the Quaternary, was resumed. Since then Quaternary research has been closely associated with this organization, with the emphasis on mapping, and on the construction of stratigraphical models necessary for the mapping. Two Quaternary map sheets at a scale 1:500,000 have been published (Weidick, 1974, 1978), and more are under way. The early work in this period concentrated on the raised marine deposits and their faunas, and a general model of isostatic uplift was worked out by Laursen (1944). Later work has shown this model to be based on faulty observations, and it has been changed radically (e.g. Weidick, 1976a).

In the 1950s A. Weidick began his work for the Survey, and the modern Quaternary research is largely due to his initiatives and inspiration. The most recent fluctuations of glaciers in western Greenland was the first problem to be dealt with, and old written and photographic sources as well as lichenometry were used for the dating of moraines (Weidick, 1959; Beschel, 1961; Beschel and Weidick, 1973).

This aspect of the Quaternary geology has recently gained new impetus (Kelly, 1980a), and is a field of future research in connection with current investigations on the hydro-power potential of glacial melt water, conducted by the Greenland Geological Survey (Weidick and Olesen, 1980).

The introduction, in the late 1960s of C-14 dating as an easily available tool shifted the focus of attention to such topics as Holocene deglaciation history and the history of isostatic uplift. A large amount of information on these topics has been published, mainly in the reports and bulletins of the Survey. This work was summarized by Weidick (1976a); later contributions include those by Kelly (1977, 1979, 1980b), Donner (1978), and Funder (1979).

Although this work continued to be concerned mainly with the Holocene, a new aspect was added to the stratigraphy when, in the 1970s, the Greenland Geological Survey (the Survey) expanded its activities to cover also eastern and, later northern Greenland. In these areas pre-Holocene and interglacial sediments occur frequently. Much recent literature has been devoted to problems concerning the dating of these sediments and their significance for the ice-age regime of the Inland Ice (Funder and Hjort, 1973; Weidick, 1975; Hjort, 1980, 1981; Funder and Hjort, 1980). This aspect of the research is still continuing (see Chapter 16, this volume).

A research tradition different from the Survey's was introduced into northwestern Greenland after the establishment in 1953 of the U.S. air base at Thule. Although the scientific impact of this event lies mainly on the Inland Ice, some Quaternary studies have resulted (Davies, 1963, 1972; Davies et al., 1963; Bendix-Almgreen et al., 1963; Nichols, 1969; Tedrow, 1970). Much attention has recently been focused on this area in connection with a debate over the possible confluence of the North American and Greenland ice sheets (Paterson, 1977; Blake, 1976; 1977a,b, 1978, 1981; Blake and Matthews, 1979; Weidick, 1976b, 1977; Kelly, 1980b; England et al., 1979; England, in press). Individual field workers have reached very different conclusions and it ranks high on the agenda for future research. An important aspect will be to unravel which controversy is due to differences in field workers' scientific backgrounds, and which can be referred to geological circumstances!

Quaternary Biology. Following the traditions in Scandinavia a close cooperation has existed between biologists and geologists in Greenland, providing a wealth of information on environmental change and the long-term stability of modern ecosystems.

Mollusc shells brought back from the raised marine deposits during the early expeditions were identified by the marine biologist Ad.S. Jensen who, from his intimate knowledge of modern faunas, was able to show the general development of climate in the Holocene, linking it to oceanographic changes (Jensen, 1905, 1917, 1942). One site which has been especially prolific is the cliffs at Orpigsoq in the interior of Disko Bugt (Fig. 1.3). The stratigraphy of the cliffs was first worked out by Jensen and the geologist P. Harder in 1906 (Jensen and Harder, 1910; Harder et al., 1949), and later revised by Laursen (1950), who made it a key site in his attempt to correlate the Greenland marine sequence with that in Scandinavia. Later studies and absolute dating have unfortunately shown Laursen's scheme to be wrong (Donner and Jungner, 1975), although, Laursen, continuing Jensen's detailed faunistic work, was able to add many new species to the known subfossil faunas. Subsequently, this line of research was somewhat neglected, for several years during which time mollusc shells were collected mainly for C-14 dating. However, recently detailed studies of interglacial faunas have provided important knowledge about environmental conditions during these periods (Simonarson, 1981; Petersen, 1982), and several sites are currently being studied.

In the 1930s palynology was first introduced into Greenland by the botanist Johs Iversen, who went to the Godthabsford area in order to investigate the enigmatic disappearance of the medieval Norse settlements in this area. Iversen managed not only to provide important information on the ecological adaptations of the settlers and their disappearance, but from an analysis of lake sediment cores, demonstrated the usefulness of palynology as a stratigraphical tool (Iversen, 1934, 1953). Both of these applications of palynology have later been pursued by B. Fredskild (1967, 1973, 1978, Chapter 22, this volume), and a number of lake sediment cores were studied by Kelly and Funder (1974). Palynology is a major source of information on the history and development of terrestrial ecosystems and a number of sites are currently being investigated by this technique.

A controversial issue which is related to Quaternary stratigraphy is a question of survival of organisms in Greenland during one, or several, ice ages. This topic was first raised in a bitter polemic between the botanist E. Warming (1888) and the geologist A. G. Nathorst (1890). Since then a large number of contributions have been dealing especially with the biological aspects of the problem (e.g. Gelting, 1934; Böcher, 1972). Recently the theories have been evaluated in a geological context (Funder, 1980, 1982).

Offshore and Inland Ice Geology. Following oil exploration on the West Greenland shelf in the period from 1968 to 1978, some information on the submarine Quaternary geology has appeared. The evidence includes detailed bathymetric mapping (Henderson, 1975, Sommerhoff, 1975), geophysical survey (Roksandic, 1979, Brett and Zarudski, 1979; Zarudski, 1980), and deep drilling (Henderson, 1979; Risum et al., 1980).

The investigations have shown the shelf, extending up to 230 kilometres from the present coast-line, is covered by a thick mantle of Quaternary sediments, mainly glacigene sediments; glacial landforms such as moraines and glacial troughs have also been identified. Micropaleontological analysis of sediments from drill cores are currently being carried out. An interesting object for future work will be to relate the Quaternary information from the shelf to that from adjacent land areas.

The same may be said about the very large amount of data relating to late Quaternary climates and environments which, in the past decade, has accumulated as a result of the "Greenland Ice Sheet Project," an international scientific project concerning drilling into the Inland Ice, and analysing the ice cores for a variety of parameters (e.g. Dansgaard et al., 1973; Hammer et al., 1978). The project, which was begun after the first successful coring in 1966 of the Inland Ice in its northern part, continued in 1982 with a core to near bedrock in its southern part (Dansgaard et al., 1983). From analysis of the $0^{18}/0^{16}$ content of the ice cores, this project has provided fundamental new knowledge about climatic change and some of its causes (e.g. Hammer et al., 1980). However, the detailed work with the ice cores has shown the importance of obtaining precise information on the subglacial topography, and the dynamics of the Inland Ice, and much work is currently being devoted to these topics (Radock et al., 1982). An important future task will be to combine evidence from these disciplines with that from Quaternary geology in the ice-free land, to construct models of the Inland Ice at various stages in its history.

History of Quaternary research on Baffin Island and in Baffin Bay

In this section we review the development of Quaternary research programs and identify the major landmarks in the process of documenting and understanding the Quaternary history of the region. At the end, we identify those studies that we believe will develop substantially in the next decade and the impact these studies may have on our current status of knowledge. Andrews and Andrews (1980) have published an

annotated bibliography on Quaternary studies from Baffin Island which
provides a useful entry into the literature.

Prior to 1945 AD. The pace of exploration was slow during this
interval. Baffin Island and the islands to the north remained largely
isolated, although important observations were made in the late 19th
century by Bell (1898), Watson (1897), and Tarr (1897). Tarr in
particular commented that Baffin Island was "wonderfully close to
glaciation"--an idea that has been echoed since in several publications
(Ives, 1962; Bradley and Miller, 1972; Locke and Locke, 1977; Williams,
1978). Bell's study of the southern coastline of Baffin Island, where
it fronts Hudson Strait, has yet to be repeated although there is now a
clear need to map the pattern of ice flow(s) throughout the region of
Hudson Strait.

     In the eastern Canadian Arctic the establishment of weather
stations and runways connected with the ferrying of planes between
North America and Europe during World War II led to the establishment
of bases at Fort Chimo, Frobisher Bary, and Clyde River. These weather
stations started to provide important climatic data which complemented
the longterm weather stations established on West Greenland during the
19th century at places such as Upernavik (see Chapter 2) (Fig. 1.3).

1945-1960 AD. The pace of exploration and research quickened during
this interval, although the logistical difficulties involved in
research were still substantial. In Canada, the Geological Survey of
Canada commenced a long-term mapping project on Baffin Island that was
largely associated with R.G. Blackadar (1956, 1961, 1967) and
colleagues. Blackadar noted a variety of glacial and marine phenomena
and was the first to document striations and elevation of the marine
limit over large sections of Baffin Island. A major event, as it
turned out, was the establishment of the Distance Early Warning radar
stations (DEW Lines) at several sites on Baffin Island and Greenland,
and the upgrading of the Frobisher airport to take large SAC refuelling
planes. The positioning of these logistical bases eventually led to
major Quaternary research programs which relied heavily on the use of
these facilities. In the early 1950s the Arctic Institute of North
America organized two major scientific expeditions to Baffin Island;
one to the Barnes Ice Cap (Baird, 1952) and the other to the Penny Ice
Cap (Ward and Baird, 1954) (Fig. 1.2). These two research programs
effectively set the stage for subsequent studies in glaciology and
Quaternary studies (cf. Mercer, 1956a and b).

1960 to Present. The past two decades have seen an increase in research in all aspects of Quaternary science. In the eastern Canadian Arctic, a major research program was begun in 1961 by the former Geographical Branch of the Department of Energy, Mines, and Resources (e.g. Ives, 1962; Ives and Andrews, 1963; Andrews and Webber, 1964; Loken, 1965, 1966; Sagar, 1966). Commencing in 1969, this program was continued by faculty and graduates of the University of Colorado (e.g. Pheasant and Andrews, 1973; Miller, 1973; Miller et al., 1977; Nelson, 1981). The initial work on Baffin Island concentrated on the central and western portions of north-central Baffin Island but then moved eastward to include the fiords and outer coast. Lichenometry and ^{14}C dating were the primary dating tools. The research efforts from the University of Colorado were initially concentrated on nothern Cumberland Peninsula, but then expanded to include the southern part of that peninsula (Dyke, 1979; Birkeland, 1978; Bockheim, 1979) as well as the area of Cape Dyer (Locke 1980) and Clyde River (Miller et al.,1977) (Fig. 1.2). The first studies developed the concept of weathering zones (Boyer and Pheasant, 1975) and considerable work was carried out on the importance of vertical and horizontal variations in bedrock and surface boulder and weathering and soil alterations. Lichenometry and ^{14}C dates were used extensively to develop late glacial and Holocene chronologies, but the presence along the outer coasts of sediments led to the investigation of U-series dating methods on shells and, most significantly, the application of amino-acid racemization studies to the development of a relative sequence of glacial and sea-level events (e.g. Andrews et al., 1975; Miller et al., 1977; Nelson, 1978, 1980; Mode, 1980; Brigham, 1980; Miller et al., 1977; Szabo et al., 1981). In the late 1970s the research effort shifted southward toward Frobisher Bay, Hudson Strait, and nothernmost Labrador (Miller, 1980; Muller, 1980; Osterman, 1982). A continuing theme throughout the two decades of research was the study and interpretation of the pattern of raised marine beaches and sediments (Andrews et al., 1970, Andrews, 1975, 1980; Dyke, 1974, 1979; Clark, 1980), as was the collection and interpretaion of marine shells from raised marine sediments (e.g. Nelson, 1981; Andrews et al., 1981).

The Geological Survey of Canada also continued its work on Baffin Island, on neighbouring Bylot Island, and on the islands at the head of Baffin Bay. These studies emphasised regional mapping, glacial and sea-level events, and chronology (Blake, 1966, 1973; Hodgson and Hazelton, 1974; Klassen, 1981). In addition, there was an increase in other university workers studying facets of the Quaternary geology and morphology, as shown by the work of Sugden and Watts (1977), Mathews, (1967), and Gilbert (1978).

A notable feature of this period was the great increase in the studies of marine geology of Baffin Bay, Davis Strait, the Continental Shelf, and the inshore channels (e.g. Hudson Strait, Lancaster Sound) (e.g. Baker and Friedman, 1973). A great part of this research effort was orchestrated from the Bedford Institute of Oceanography/Atlantic GeoScience Center at Halifax, Nova Scotia and from Dalhousie University in Halifax (Aksu, 1977, 1981; Piper, 1973; MacLean and Falconer, 1979; Fillon and Duplessy, 1980; Grant and Manchester, 1970). The work included fundamental studies on bathymetry, subbottom seismic surveys, cores and grab samples, and the chronology of Quaternary oceanographic events.

During this same interval the Canadian Polar Continental Shelf became involved in a series of ice-core projects on Devon Island (Paterson et al., 1977; Koerner, 1977) which served to complement the earlier studies from Camp Century, Greenland (Dansgaard et al., 1971) (Figs. 1.2 and 1.3). These studies have expanded northwards to include several ice caps in the Queen Elizabeth Islands, north and northwest in northern Baffin Bay. In addition to the ice-core work, significant glaciological studies were carried out on Devon Island.

This brief survey of the nature and tempo of research on and around Baffin Bay/Davis Strait/northern Labrador Sea is meant to highlight the major trends, and the major Quaternary research efforts. There have been significant and interesting differences in research emphasis between the Greenland and Canadian coasts. Two of the significant trends of the past two decades have been (1) the increased diversification of the environments of study (Fig. 1.4); thus, early work was almost totally confined to the glacial geology of land areas, whereas of late, emphasis has shifted to work on continuous stratigraphic records from ice caps and ice sheets, from deep-ocean basins and inshore marine waters, and from lake sediments (Fig. 1.4); and (2) coupled with the increased diversification of the depositional environments has been a substantial increase in our ability to both correlate and date the various records. Great stress has been laid on C-14 dating in all environments, but a large number of techniques and approaches have been developed for both relative and absolute dating. Of these, this book will emphasize lichenometry, amino-acid racemization, and rock and soil weathering characteristics.

ORGANIZATION OF THE VOLUME

The volume is divided into five major parts. Part 1 consists of this chapter and Chapter 2. Chapter 2 is an account of the physical setting of the region, particularly with regard to the area's climatology, oceanography, glaciology, and geology. That chapter is intended to document the present environment against which the various Quaternary descriptive stratigraphies and interpreted climatic and glacial events must be viewed in order to better evaluate the nature and mechanisms of Quaternary climatic variations. Part II consists of a series of chapters concerned with a description and discussion of the origins of glacial landforms and sediments: chapters deal with the distribution of landscapes of glacial erosion on Baffin Island; the morphology and origin of the Baffin Island fiords; the geology and bathymetry of the eastern Baffin Island continental shelf; and finally a survey of the textural characteristics of glacigenic sediments collected from over 600 sites on Baffin Island.

In Part III attention is focused on marine and ice-core stratigraphic records that are available for elucidating Quaternary events within the region; chapters outline various aspects of the deep-sea records from Baffin Bay and the Labrador Sea, and the oxygen isotopic record from local ice caps. A traditional area of study on West Greenland and on Baffin Island has been the record of terrestrial glacial and nonglacial intervals. These sediments are often poorly preserved and there are many problems associated with obtaining accurate age as estimates of the various litho- or biostratigraphic units (e.g. Feyling-Hanssen, 1980; Nelson, 1982; Szabo et al 1981). Part IV consists of chapters that discuss the glacial and nonglacial sediments and associated faunas and floras from records on West Greenland, Bylot Island, and Baffin Island.

Latest Pleistocene and Holocene climatic and glacial (neoglacial events) are discussed in Part V. Many of these chapters use palynology to interpret changes in vegetation and climate during the past 10,000 BP but one chapter presents a study of neoglacial fluctuations of local corrie and valley glaciers from Baffin Island, wheras another focuses on the links between glacial history and glacial isostasy.

REFERENCES

Aksu, A.E., 1977: The Late Quaternary stratigraphy and sedimentation history of Baffin Bay. MSc thesis, Dalhousie University, Halifax, N.S. 170 p.
Aksu, A., 1981: Late Quaternary stratigraphy, paleoenvironmentology and sedimentation history of Baffin and Davis Strait. Ph.D. thesis, Dalhousie University, Halifax. 771 p.
Andrews, M. and Andrews, J.T., 1980: Baffin Island Quaternary Environments. INSTAAR Occasional Paper #33, University of Colorado, Boulder. 123 p.

Andrews, J.T., 1975: Support for a stable late Wisconsin ice margin (14,000 to ca 9000 BP): a test based on glacial rebound. Geology, 4:617-620.

Andrews, J.T., 1980: Progress in relative sea level and ice sheet reconstructions, Baffin Island, N.W.T., for the last 125,000 years. In: Mörner, N-A. (ed.), Earth Rheology, Isostasy and Eustasy. John Wiley, New York. 175-200.

Andrews, J.T., Buckley, J.T., and England, J.G., 1970: Late-glacial chronology and glacio-isostatic recovery, Home Bay, east Baffin Island. Geological Society of America Bulletin,

Andrews, J.T. and Miller, G.H., 1972: Maps of the present Glaciation Limit and lowest equilibrium line altitude for north and south Baffin Island. Arctic and Alpine Research, 4:45-60.

Andrews, J.T., Miller, G.H., Nelson, A.R., Mode, W.N., and Locke, W.W. III, 1981b: Quaternary near-shore environments on eastern Baffin Island, N.W.T. In: W.C. Mahaney (ed.), Quaternary Paleoclimate. Geoabstracts, Norwich, U.K. 13-44.

Andrews, J.T., Mode, W.N., and Davis, P.T., 1980: Holocene climate based on pollen transfer functions, eastern Canadian Arctic. Arctic and Alpine Research, 12:41-64.

Andrews, J.T. and Webber, P.J., 1964: Lichenometrical study of the northwestern margin of the Barnes Ice Cap: a gemorphological technique. Geographical Bulletin, 22:80-104.

Baker, S.R. and Friedman, G.M., 1973: Sedimentation in an Arctic Marine environment: Baffin Bay between Greenland and the Canadian Arctic Archipelago. Geological Survey of Canada, Paper 71-23, 471-498.

Baird, P.D., 1952: The glaciological studies of the Baffin Island Expedition, 1950, part I. Method of nourishment of the Barnes Ice Cap. Journal of Glaciology, 2(11):2-9.

Bell, R., 1898: Report of an exploration on the northern side of Hudson Strait, Canada. Geological Survey of Canada Annual Report, XI, Part M. 33 pp.

Bendix-Almgreen, S.E., Fistrup, B. and Nichols, R.L., 1967: Notes on the geology and geomorphology of the Carey Øer, north-west Greenland. Meddelelser om Grönland, 164, 8, 19 pp.

Beschel, R.E., 1961: Dating rock surfaces by lichen growth and its application to glaciology and physiography (lichenometry). In: Raasch, G.O. (ed.), Geology of the Arctic 2. University of Toronto Press, 1044-1062.

Beschel, R.E. and Weidick, A., 1973: Geobotanical and geomorphological reconnaissance in West Greenland, 1961. Arctic and Alpine Research, 5:311-319.

Birkeland, P.W., 1978: Soil development as an indication of relative age of Quaternary deposits, Baffin Island, N.W.T., Canada. Arctic and Alpine Research, 10:733-747.

Blackadar, R.G., 1956: Geological reconnaissance of Admiralty Inlet, Baffin Island, Arctic Archipelago, Northwest Territories: report, map, and stratigraphic sections. Geological Survey Canada Paper 55-6. 25 pp.

Blackadar, R.G., 1966: Geology, Andrew Gordon Bay - Cory Bay, Baffin Island, District of Franklin. Map sheet with descriptive notes. Geological Survey Canada. Map 5-1962.

Blackadar, R.G., 1967: Geology of Mingo Lake - Macdonald Island map-area, Baffin Island, District of Franklin. Geological Survey Canada Memoir 345. 54 pp.

Blake, W., Jr., 1966: End moraines and deglaciation chronology in northern Canada with special reference to southern Baffin Island. Geological Survey of Canada Paper, 66-26. 31 pp.

Blake, W., Jr., 1970: Studies of glacial history in Arctic Canada. I. Pumice, radiocarbon dates, and differential postglacial uplift in the eastern Queen Elizabeth Islands. Canadian Journal Earth Sciences, 7: 634-664.

Blake, W., Jr., 1973: Former occurence of Mytilus edulis on Coburg Island, Arctic Archipelago. Naturaliste Canadian, 100, 51-58.

Blake, W., Jr., 1975: Radiocarbon age determinations and postglacial emergence at Cape Storm, southern Ellesmere Island. Geografiska Annaler, 62A, 1-71.

Blake, W., Jr., 1976: Sea and land relations during the last 15000 years in the Queen Elizabeth Islands, Arctic Archipelago. Geological Survey of Canada, Paper, 76-1B, 201-207.

Blake W., 1977a: Radiocarbon age determinations from the Carey Islands, Northwest Greenland. Geological Survey of Canada, Paper, 77-1A, 445-453.

Blake, W., 1977b: Glacial sculpture along the east-central coast of Ellesmere Island, Arctic Archipelago. Geological Survey of Canada, Paper, 77-1C, 107-115.

Blake, W., 1978: Aspects of the glacial history, southeastern Ellesmere Island, District of Franklin. Geological Survey of Canada, Paper, 78-1A, 175-182.

Blake, W., 1979: Lake sediment coring along Smith Sound, Ellesmere Island and Greenland. Geological Survey of Canada, Paper, 81-1A, 191-200.

Blake, W. and Matthews, J.V., 1979: New data on an interglacial peat deposit near Makinson Inlet, Ellesmere Island, District of Franklin. Geological Survey of Canada, Paper 79-1A, 157-164.

Bockheim, J.G. 1980: Properties and relative age of soils of south-western Cumberland Peninsula, Baffin Island, N.W.T., Canada. Arctic and Alpine Research, 11:289-306.

Boulton, G.S. 1979: A model of Weischelian glacier variation in the North Atlantic region. Boreas, 8:373-395.

Boulton, G.S. et al. 1982: A glacio-isostatic facies model and amino acid stratigraphy for late Quaternary events in Spitsbergen and the Arctic. Nature, 298:437-441.

Boyer, S.J. and Pheasant, D.R., 1974: Delimination of weathering zones in the fiord area of eastern Baffin Island. Geological Society of America Bulletin, 85:805-810.

Bradley, R.S. and Miller, G.H., 1972: Recent climatic change and increased glacierization in the eastern Canadian Arctic. Nature, 237:385-387.

Brett, C.P., and Zarudski, E.F.K., 1979: Project Westmar, a shallow marine geophysical survey on the West Greenland continental shelf. Grønlands Geologiske Undersögelse, Rapport 87. 27 pp.

Bocher, T.W., 1972: Evolutionary problems in the arctic flora. In: Valentine, D.M. (ed.), Taxonomy, Phytogeography and Evolution, (Academic Press, London), 101-113.

Brigham, J.K., 1983: Stratigraphy, amino acid geochronology, and corre-lation of Quaternary sea-level and glacial events, Broughton Island, arctic Canada. Canadian Journal Earth Sciences, 20:577-598.

Clark, J.A., 1980: A numerical model of world wide sea level changes on a visco-elastic earth. In: Mörner, N-A. (ed.), Earth Rheology, Isos-tasy, and Eustasy. John Wiley, New York, 525-534.

Dansgaard, W., Josnsen, S.J., Clausen, H.B., and Gundestrup, N. 1973: Stable isotope glaciology. Meddelelser om Grönland, 197, 2, 53 pp.

Dansgaard, W., Johnson, S.J., Clausen, H.B., and Langway, C.C.,Jr., 1971: Climatic record revealed by the Camp Century ice core. In: Turekian, K.K. (ed.), Late Cenozoic Ice Ages. Yale University Press, New Haven, 37-56.

Dansgaard. W. et al.   1983 A new Greenland deep ice core. Science.
   218:1273-1277
Davies. W.E.   1963: Glacial geology of northern Greenland. Polar-
   forschung. 5.94-103
Davies, W.E.. Drinsley, D.B.. and Nicol. A.H.. 1963: Geology of the
   North Star Bugt area, Northwest Greenland. Meddelelser om Grönland.
   162. 12. 68 pp.
Denton, G.H. and Hughes, T.J. (eds.). 1981: The Last Great Ice Sheets
   John Wiley, New York. 484 pp.
Donner, J., 1978: Holocene history of the west coast of Disko, central
   West Greenland. Geografiska Annaler. A60, 63-72.
Donner, J. and H. Jungner, 1975: Radiocarbon dating of shells from
   marine Holocene deposits in the Disko Bugt area, West Greenland.
   Boreas, 4:25-45.
Dyke, A.S.. 1974: Deglacial chronology and uplift history: northeastern
   sector, Laurentide Ice Sheet. INSTAAR Occasional Paper #12, Univer-
   sity of Colorado, Boulder. 113 pp.
Dyke, A.S., 1979: Glacial and sea level history of the southwestern
   Cumberland Peninsula, Baffin Island, N.W.T., Canada. Arctic and
   Alpine Research, 11, 179-202.

England, J.H., 1983: Isostatic adjustments in a full glacial sea
   Canadian Journal Earth Sciences, 20:895-917.
England, J., in press: Postglacial emergence along northern Nares
   Strait. In: Dawes, P.R. and Kerr, J.W. (eds.), Nares Strait and the
   drift of Greenland: a conflict in plate tectonics, Meddelelser om
   Grönland, Geoscience, 8.
England, J.H., Bradley, R.S., and Stuckenrath, R. 1981: Multiple glaci-
   ations and marine transgressions, western Kennedy Channel, Northwest
   Territories, Canada Boreas, 10:71-90.

Feyling-Hanssen, R.W., 1976: The Clyde Foreland Formation, a micropale-
   ontological study of Quaternary stratigraphy. Maritime Sediments,
   Special Publication #1, Part B, 315-377.
Feyling-Hanssen, R.W., 1979: A mid-Wisconsin Interstadial on Broughton
   Island, Arctic Canada, and its foraminifera. Arctic and Alpine
   Research, 8, 161-182.
Feyling-Hanssen, R.W., 1980: Microbiostratigraphy of young Cenozoic
   marine deposits at the Qivituq Peninsula, Baffin Island. Marine
   Micropaleontology, 153-184.
Fillon, R.H. and Duplessy, J-C., 1980: Labrador Sea bio-, tephro-,
   oxygen isotope stratigraphy and late Quaternary paleoceanographic
   trends. Canadian Journal Earth Sciences, 17:831-854.
Fredskild, B., 1967: Paleobotanical investigations at Sermermiut,
   Jakobshavn, West Greenland. Meddelelser om Grönland, 178, 4, 54 pp.
Fredskild, B., 1973: Studies in the vegetation history of Greenland.
   Meddelelser om Grönland, Band 198(4), 245 pp.
Fredskild, B., 1978: Palaeobotanical investigations of some peat
   deposits of Norse age at Qagssiarssu, South Greenland. Meddelelser
   om Grönland, 204, 5, 41 pp.
Fredskild, B., in press: The Holocene vegetational development at
   Godthabsfjord, West Greenland. Meddelelser om Grönland, Geoscience.
Funder, S.. 1979: The Quaternary geology of the Narssaq area, South
   Greenland. Grönlands Geologiske Undersögelse, Rapport 86. 14 pp.
Funder, S.. 1980: Ice-age plant refugia in East Greenland. Palaeogeo-
   graphy, Palaeoclimatology, Palaeoecology, 28:279-295.
Funder, S., 1982: Planterefugierne i Grönland. Naturens Verden 241-255.
   Funder, S., and Hjort, C., 1980: A reconnaissance of the Quaternay
   geology of eastern North Greenland. Grönlands Geologiske
   Undersögelse. Rapport 99, 99-106.

Funder, S., and Hjort, C., 1973: Aspects of the Weichselian chronology in central East Greenland. Boreas, 2:69-84.

Gelting, P., 1934: Studies on the vascular flora of East Greenland between Franz Joseph Fjord and Dove Bay. Meddelelser om Grönland, 101, 2, 340 pp.

Gilbert, R., 1978: Observations on oceanography and sedimentation at Pangnirtung Fiord, Baffin Island. Maritime Sediments, 14, 1-9.

Grant, A.C., and Manchester, K.S., 1970: Geophysical investigations in the Ungava Bay-Hudson Strait region of northern Canada. Canadian Journal of Earth Sciences, 7(4):1062-1076.

Hammer, C.U., Clausen, H.B., Dansgaard, W., Gunderstrup, N., Johnsen, S.J., and Reeh, N., 1978: Dating of the Greenland ice cores by flow models, isotopes volcanic debris, and continental dust. Journal of Glaciaology, 20:3-36.

Hammer, C.U., Clausen, G.B., and Dansgaard, W., 1980: Greenland ice sheet evidence of post-glacial volcanism and its climate impact. Nature, 288:230-235.

Harder, P., Jensen, Ad. S., and Laursen, D., 1949: The marine Quaternary sediments in Disko bugt. Meddelelser om Gronland, 149, 1, 85 pp.

Henderson, G., 1975: New bathymetric maps covering Offshore West Greenland 59°-60°30'N. Offshore Technology Conference in Houston, Texas. Paper OTC 2223:761-764.

Henderson, G., 1979: Developments in petroleum exploration offshore West Greenland. Grönlands Geologiske Undersögelse, Rapport 95:42-44.

Hjort, C., 1980: A glacial chronology for East Greenland. Boreas, 10: 259-274.

Hjort, C., 1981: Studies of the Quaternary in Northeast Greenland. University of Lund, Department of Quaternary Geology. Thesis 9, 29 pp.

Hodgson, D.A. and Haselton, G.M., 1974: Reconnaissance glacial geology, northeastern Baffin Island. Geological Survey of Canada, Paper 74-20. 100 pp.

Hughes, T., Denton, G.H. and Grosswald, M.G., 1977: Was there a late-Würm Arctic ice sheet? Nature, 266:596-602.

Iversen, J., 1934: Moorgeologische Untersuchungen auf Grönland. Bulletin of the Geological Survey of Denmark, 8:341-358.

Iversen, J., 1953: Origin of the flora of western Greenland in the light of pollen analysis. Oikos, 4:85-103.

Ives, J.D., 1962: Indications of recent extensive glacierization in north-central Baffin Island, N.W.T. Journal Glaciology, 4:197-205.

Ives, J.D., and Andrews, J.T., 1963: Studies in the physical geography of north central Baffin Island: Geographical Bulletin, 19:5-48.

Jensen, Ad.S., 1905: On the mollusca of East Greenland. I Lamellibranchiata. Meddelelser om Grönland, 29:287-362.

Jensen, Ad.S., 1917: Quaternary fossils collected by the Danmark Expedition. Meddelelser om Grönland, 43:619-632.

Jensen, Ad.S., and P. Harder, 1910: Post-Glacial changes of climate in Arctic regions as revealed by investigations on marine deposits. Postglaziale Klimaveränderungen, 11. International Geological Kongress Stockholm 1910, 399-407.

Kellogg, T.B., 1976: Late Quaternary climatic change: evidence from deep-sea cores of Norwegian and Greenland Seas. In: Cline, R.M. and Hays, J.D. (eds.), Geological Society of America Memoir 125:77-110.

Kellogg, T.C., 1980: Paleoclimatology and paleo-oceanography of the Norwegian and Greenland seas: glacial-interglacial contrasts. Boreas, 9:115-137.

Kelly, M., 1977: Quaternary geology of the Ivigtut-Nunarssuit region, South-West and South Greenland. Grønlands Geologiske Undersøgelse, Rapport 85:6467.

Kelly, M., 1979: Comments on the implications of new radiocarbon dates from the Holsteinsborg region, central West Greenland. Grønlands Geologiske Undersøgelse, Rapport, 95:35-42.

Kelly, M., 1980a: The status of the Neoglacial in Western Greenland. Grønlands Geologiske Undersøgelse, Rapport, 96. 24 pp.

Kelly, M., 1980b: Preliminary investigations of the Quaternary of Melville Bugt and Dundas, North-West Greenland. Grønlands Geologiske Undersøgelse, Rapport, 100:33-38.

Kelly, M. and Funder, S., 1974: The pollen stratigraphy of late Quaternary lake sediments of South-West Greenland. Grønlands Geologiske Undersøgelse, Rapport, 64. 26 pp.

Klassen, R.A., 1981: Aspects of the glacial history of Bylot Island, District of Franklin. Geological Survey of Canada, Paper 81-1A, 317-326.

Koch, L., 1928: Contributions to the glaciology of North Greenland. Meddelelser om Grønland, 65:181-464.

Koerner, R.M., 1977: Devon Island ice cap: core stratigraphy and paleoclimate. Science, 196:15-18.

Laursen, D., 1944: Contributions to the Quaternary geology of northern West Greenland especially the raised marine deposits. Meddelelser om Grønland, 135, 8, 125 pp.

Laursen, D., 1950: The stratigraphy of the marine Quaternary deposits in West Greenland. Meddelelser om Grønland, 151, 1, 152 pp.

Locke, C. and Locke, W.W. III, 1977: Little Ice Age snow-cover extent and paleoglacierization thresholds: north-central Baffin Island, N.W.T., Canada. Arctic and Alpine Research, 9:291-300.

Locke, W.W. III, 1980: The Quaternary geology of the Cape Dyer area, southeasternmost Baffin Island, Canada. PhD thesis, University of Colorado, Boulder. 331 pp.

Loken, O.H., 1965: Postglacial emergence at the south end of Inugsuin Fiord, Baffin Island, N.W.T. Geographical Bulletin, 7:243-258.

Loken, O.H., 1966: Baffin Island refugia older than 54,000 years. Science, 153:1378-1380.

MacLean, B., and Falconer, R.K.H., 1979: Geological geophysical studies in Baffin Bay and Scott Inlet-Buchan Gulf and Cape Dyer-Cumberland Sound areas of the Baffin Island shelf: In: Current Research, Part B, Geological Survey of Canada, Paper 79-1B, 231-244.

Matthews, B., 1967: Late Quaternary marine fossils from Frobisher Bay (Baffin Island, N.W.T., Canada). Palaeogeography, Palaeoclimatology, Palaeoecology. 3:243-263.

Mercer, J.H., 1956a: Geomorphology and glacial history in southernmost Baffin Island. Geological Society of America Bulletin, 67:553-570.

Mercer, J.H., 1956b: The Grinnell and Terra Nivea Ice Caps, Baffin Island. Journal of Glaciology, 2(19):653-656.

Miller, G.H., 1973: Late Quaternary glacial and climatic history of northern Cumberland Peninsula, Baffin Island, N.W.T., Canada. Quaternary Research, 3:561-583.

Miller, G.H., 1980: Late Foxe glaciation of southern Baffin Island, N.W.T., Canada. Geological Society of America Bulletin, Part I, 91, 399-405.

Miller, G.H., Andrews, J.T., and Short, S.K., 1977: The last interglacial/glacial cycle, Clyde Foreland, Baffin Island, N.W.T.: stratigraphy, biostratigraphy, and chronology. Canadian Journal of Earth Sciences, 14:2824-2857.

Miller, G.H. and Dyke, A.S., 1974: Proposed extent of late Wisconsin Laurentide ice on eastern Baffin Island. Geology, 2:125-130.

Mode, W.N., 1980: Quaternary stratigraphy and palynology of the Clyde Foreland, Baffin Island, N.W.T., Canada. Ph.D. dissertation, University of Colorado, Boulder. 219 pp.

Mode, W.N., Nelson, A.R. and Brigham, J.K., 1983: Sedimentologic evidence for Quaternary glaciomarine cyclic sedimentation along eastern Baffin Island, Canada. In: Molnia, B.F. (ed.), Glacial-Marine Sedimentation. Plenum Press, New York, 495-534.

Muller, D.S., 1980: Glacial geology and Quaternary history of southeast Meta Incognita Peninsula, Baffin Island, Canada. M.Sc. thesis, University of Colorado, Boulder. 211 pp.

Nathorst, A.G., 1891: Kritische Bemrkungen über die Geschichte der Vegetation Grönlands. Englers Bot. Jahrb., 14:183-221.

Nelson, A.R., 1978: Quaternary glacial and marine stratigraphy of the Qivitu Peninsula, northern Cumberland Peninsula, Baffin Island, Canada. Ph.D. dissertation, University of Colorado, Boulder. 215 pp.

Nelson, A.R., 1980: Chronology of Quaternary landforms, Qivitu Peninsula, northern Cumberland Peninsula, N.W.T., Canada. Arctic and Alpine Research, 12, 265-286.

Nelson, A.R.: 1981: Quaternary glacial and marine stratigraphy of the Qivitu Peninsula, northern Cumberland Peninsula, Baffin Island: Geological Society of America Bulletin, v. 92, Part I, 512-518, Part II, 1143-1261.

Nichols, R.L., 1969: Geomorphology of Inglefield Land, North Greenland, North Greenland. Meddelelser om Grönland, 188, 1, 109 pp.

Osterman, L.E., 1982: Late Quaternary history of southern Baffin Island, Canada: a study of foraminifera and sediments from Frobisher Bay. Ph.D. dissertation, University of Colorado, Boulder, 380 pp.

Patersen, L, Strand, 1982: Attack by predatory gastropods recongnized in interglacial molluscan fauna from Jameson Land, East Greenland. Malacologia, 22: 721-726.

Paterson, W.S.B., 1977: Extent of the late-Wisconsin Glaciation in Northwest Greenland and northern Ellesmere Island. A review of the glaciological and geological evidence. Quaternary Research, 8:180-190.

Paterson, W.S.G., et al., 1977: An oxygen-isotope climatic record from the Devon Island ice cap. Nature, 266, 508-511.

Pheasant, D.R. and Andrews, J.T., 1973: Wisconsin glacial chronology and relative sea level movements, Narpaing Fiord Broughton Island area, eastern Baffin Island N.W.T. Canadian Journal Earth Science, 10:1621-1641.

Piper, D.J.W., 1973: A late Quaternary stratigraphic marker in the central basin of Baffin Bay. Maritime Sediments, 9:62-63.

Radok, U., Barry, R.G., Jenssen, D., Keen, R.A., Kiladis, G.N., and McInnes, B., 1982: Climatic and physical characteristics of the Greenland Ice Sheet. C.I.R.E.S., University of Colorado, Bouler. 193 pp.

Rink, H., 1852: Om den geographiske Beskaffenhed af de danske Handelsdistrikter i Nordgrönlands tillegemed en Udsigt over Nordgrölands Geognosi. Kgl. danske Vidensk. Selsk. Skrifter, (5), 3, 37-98.

Risum, J.B., Croxton, C.A. and F. Rolle, 1979: Developments in petroleum exploration offshore West Greenland. Grönlands Geologiske Undersögelse, Rapport, 100:51-54.

Roksandic, M.M., 1979: Geology of the Continental Shelf off West Greenland between 61°14'N and 64°N. Grönlands Geologiske Undersögelse, Rapport, 92, 15 pp.

Ruddidman, W.F., 1977: Late Quaternary deposition of ice-rafted sand in
    the sub-polar North Atlantic (40° - 65°N). Geological Society of
    America Bulletin, 88:1813-1827.
Ruddiman, W.F. and McIntyre, I., 1973: Time-transgressive deglacial
    retreat of Polar waters from the North Atlantic. Quaternary Research,
    3:117-130.
Ruddiman, W.F. and McIntyre, A., 1981: The mode and mechanism of the
    last deglaciation: oceanic evidence. Quaternary Research, 16:125-134.

Sager, R.B., 1966: Glaciological and climatological studies on the
    Barnes Ice Cap, 1962-1964. Geography Bulletin, 8(1):3-47.
Sancetta, C., Imbire, J., and Kipp, N.G., 1973: Climatic record of the
    past 130,000 years in the North Atlantic deep-sea core V23-82: Corre-
    lation with the terrestrial record. Quaternary Research, 3:110-116.
Simonarson, L.A., 1981: Upper Pleistocene and Holocene marine deposits
    and faunas on the north coast of Nugssuaq, West Greenland. Grönlands
    Geologiske Undersögelse, Bulletin, 140, 107 pp.
Sommerhogg, G., 1975: Glaziale Gestaltung und marine Überformung der
    Schelfbänke vor SW-Gronland. Polarforschung, 45:22-31.
Sugden, D.E. and Watts, S.H., 1977: Tors, felsenmeer, and glaciation in
    northern Cumberland Peninsula, Baffin Island. Canadian Journal Earth
    Sciences, 3:243-263.
Szabo, B.J., Miller, G.H., Andrews, J.T., and Stuiver, M., 1981:
    Comparison of uranium-series, and amino acid data from marine
    molluscs, Baffin Island, Arctic Canada. Geology, 9:451-457.

Tarr, R.S., 1897: Differences in the climate of the Greenland and Amer-
    ican sides of Davis' and Baffin's Bay. American Journal of
    Science, Series 4, 3, 315-320.
Tedrow, J.C.F., 1970: Soil investigations in Inglefield Land, Green-
    land. Meddeleslser om Grönland, 188, 3, 93 pp.

Ward, H., and Baird, P.D., 1954: Studies in glacier physics on the
    Penny Ice Cap, Baffin Island. Part I. A description of the Penny Ice
    Cap, its accumulation and ablation. Journal of Glaciology, 2(15):342
    -355.
Warming, E., 1888: Om Grönlands Vegetation. Meddelelser om Grönland,
    12. 245 pp.
Watson, T.L., 1897: Evidences of recent elevation of the southern coast
    of Baffin Land. Journal Geology, 5:17-33.
Weidick, A., 1959: glacial variations in West Greenland in historical
    time. Meddelelser om Grönland, 158, 4, 196 pp.
Weidick, A., 1974: Quaternary Map of Greenland 1:500,000 Sondre
    Stromfjord-Nugssuaq. Grönlands Geologiske Undersogelse, Copenhagen.
Weidick, A., 1975: Estimates on the mass balance changes of the Inland
    Ice since Wisconsin-Weichsel. Grönlands Geologiske Undersögelse,
    Rapport, 68. 21 pp.
Weidick, A., 1976a: Glaciation and the Quaternary of Greenland. In:
    Escher, A. and Watt, W.S. (eds.), Geology of Greenland. Geological
    Survey of Greenland, Copenhagen, 430-460.
Weidick, A., 1976b: Glaciations of Northern Greenland. New Evidence.
    Polarforschung, 46:26-37.
Weidick, A., 1977: Comments on radiocarbon dates from northern Green-
    land made during 1977. Grönlands Geologiske Undersögelse, Rapport,
    90, 124-128.
Weidick, A., 1978: Quaternary Map of Greenland Sheet 2. 1:500,000
    Frederikshabs Isblink-Sondre Stromfjord. Grönlands Geologiske
    Undersögelse, Copenhagen.
Weidick, A. and Olesen, O.B., 1980: Hydrological basins in West Green-
    land. Grönlands Geologiske Undersögelse, Rapport, 94. 51 pp.

Williams, L.D., 1978: The Little Ice Age glaciation level on Baffin Island, Arctic Canada. Palaeogeography, Palaeoclimatology, Palaeoecology, 25:199-207.

Zarudski, E.F.K., 1980: Interpretation of shallow seismic profiles over the continental shelf in West Greenland between latitudes 64° and 69°30'N. Grønlands Geologiske Undersøgelse, Rapport, 100:58-60.

2 Environmental background

J. D. Jacobs, J. T. Andrews, and S. Funder

The physical setting of Baffin Bay, the northern Labrador Sea, and the surrounding land plays an important part in the character of the Quaternary glacial and climatic events that are the focus of this book. This chapter has three broad topics: <u>first</u> is a discussion of the bedrock geology of the region and the Cenozoic/Tertiary tectonic evolution which led to the present configuration of land and ocean; <u>second</u> is a brief statement about Cenozoic climates and the timing of glaciation for the region; and <u>third</u> is a discussion and illustration of the present environmental setting, including sections on: glaciology, oceanography, climatology, sea ice, and biogeography.

BEDROCK GEOLOGY AND TECTONIC EVOLUTION

The last decade has seen a fundamental change in our knowledge of the geology of this area. Prior to that geological mapping had resulted in a reasonable picture of the bedrock lithologies and ages or rocks exposed on land, but our knowledge of the submarine geology and structures was virtually nil. Shipboard and overflight geophysical surveys have nowexpanded our knowledge so that we can now portray the distribution of rocks over a significant area (Fig. 2.1). A major impetus for the increased knowledge of the offshore geology has been the search for hydrocarbons. Exploratory wells have been drilled at various sites off Labrador, Baffin Island, and near Devon Island with some promising results.

The great mass of Greenland, Baffin Island and Labrador is composed of Precambrian granite with interbedded, metamorphosed supracrustal rocks. The bedrock units are aligned around an old core of Archaean gneiss, the remnant of a once coherent North Atlantic land mass. Rocks from the Archaean craton occur in areas comprising

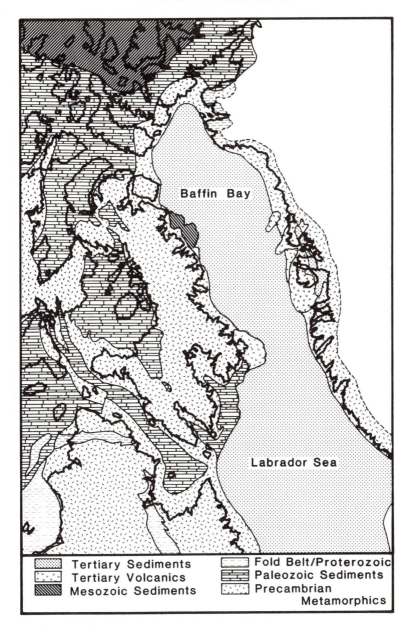

Figure 2.1 Map of the geology of the land and continental margin surrounding Baffin Bay.

south-eastern and western Greenland and eastern Labrador (Bridgwater et al., 1973). The old craton is bordered by Early Proterozoic mobile belts composed of supracrustal, metamorphosed rocks and granites. In

Greenland the Ketilidian mobile belt to the south of the craton, and
the Nagssugtoqidian and Rinkian mobile belts to the north have been
recognised on both sides of the Inland Ice (Kalsbeek, 1982), whereas on
Labrador the Circum-Ungava fold belt runs from the eastern side of
Hudson Bay into the Labrador Trough with rocks frequently consisting of
ironstone formations, high grade metamorphic rocks and volcanites.
Late Proterozoic sandstones and volcanites occur in downfaulted blocks
in southwestern Greenland (Alaart, 1973), in the Milne Inlet Trough
(Kerr, 1980) and in Thule basin and on the adjacent arctic platform of
northwestern Greenland and eastern Ellesmere Island, where they are
overlain by Palaeozoic dolomites, limestones and shales (Dawes et al.,
1982).

Of great significance for the interpretation of Quaternary events
in the Baffin Bay/Labrador Sea region is the distribution of Paleozoic
rocks, largely carbonates. With respect to inferred patterns of
glacial movement, Paleozoic limestones (Fig. 2.1) crop out along the
western and eastern coasts of Foxe Basin and indeed are the major unit
underlying Foxe Basin and Hudson Bay. The Paleozoic limestones extend
southwest-northeast across the Queen Elizabeth Islands to Greenland and
lie within the Franklin Geosyncline. Offshore seismic and magnetic
studies (e.g. Grant, 1975; MacLean and Falconer, 1979) have also
delimited major outcrops of Paleozoic limestones on the floors of
Cumberland Sound, Frobisher Bay, and Hudson Strait as well as extensive
outcrops on the Baffin Island Continental Shelf off Cumberland and Hall
Peninsulas (Grant, 1975; MacLean, et al., 1978; see MacLean, this
volume). Glacial erosion of the Paleozoic outcrop provides important
lithological markers of glacial activity in sediments from Baffin Bay
(Piper, 1973; Aksu, 1981) as well as from Frobisher Bay (Osterman,
1982; Dowdeswell, et al., in press).

Mesozoic sediments are found on either side of Davis Strait, in
Eclipse Channel and on Bylot Island and throughout the Sverdrup Basin
(Fig. 2.1). Sediments of this sequence also have been interpreted to
underlay Lancaster Sound (Kerr, 1980). Seismic studies from ship
platforms have also suggested that Cretaceous to Upper Cretaceous
marine sediments crop-out on the Baffin Island Shelf off Scott Inlet,
Baffin Island, and along the Continental
Shelf in southern Baffin Island (Grant, 1975; MacLean and Falconer,
1979). However, the major group of sediments on the shelves of Baffin
Island, West Greenland and Labrador are Tertiary in age with most
deposition occurring prior to the Miocene.

In the Nugssuaq embayment of western Greenland a thick sequence of
clastic sediments is exposed on land. The sediments comprise deltaic,

prodeltaic, and marine sandstones and shales. They were deposited on a subsiding, peneplaned basement in Upper Cretaceous and Lower Tertiary times. The sediments are believed to be representative of the sequence which is known from geophysical surveys and drillings in the West Greenland basin, a sedimentary basin extending along the entire west coast of Greenland with up to ten thousand meters of mainly Tertiary sediments (Henderson et al., 1981).

The Quaternary sedimentary cover of the structural high regions of Baffin Island and Greenland are remarkably thin (see Chapter 3; Dyke et al., 1982). Only along the outer coastal forelands of eastern Baffin Island and in places along the coast of Greenland do "thick" sequences (10-100 m) of Quaternary sediments occur. Along the shelf of southern Baffin Island, Grant (1975) has estimated the thickness of Quaternary sediments of between 50 and 450 m in places (this might also include late Tertiary sediments as well).

The tectonic history of Baffin Bay and the surrounding lands was responsible to a large measure for the present topographic and bathymetric relationships. Although a good case can be made for an inheritance of structure in some of the larger tectonic features we will simply present a word-model of the development of Baffin Bay and the Labrador Sea since the late Cretaceous (LePinchon et al., 1971; Clark and Upton, 1975; Kerr, 1980). During the late Cretaceous, an offshoot of the Mid-Atlantic ridge led to rifting and spreading. Volcanic rocks associated with early stages of rifting are now located onshore at Cape Dyer, Baffin Island, and on Disko Island, Greenland, but volcanics cover extensive portions of the adjacent Continental shelves. Continued rifting and spreading were also probably associated with isostatic uplift of the margins of the PreCambrian Shield along both the margin of the Eastern Canadian Arctic as well as West Greenland, causing in the former case a pronounced topographic asymmetry which persists today, with the height of land being located in many cases only 50 to 100 km _west_ of the Baffin Bay coast. The uplifted rim of the Shield is the locus of the major alpine glacial forms on Baffin Island and Labrador (Sugden, 1978), whereas the broad, undulating high plateaux at 400-700 m asl was probably the locus of growth for part of the Laurentide Ice Sheet during successive glaciations (Ives et al., 1975).

Although the onset of extensive rifting and faulting cannot be dated with absolute certainty, evidence presented by Kerr (1980) and other offshore geological studies indicate that high-angle faulting, possibly associated with oceanic fracture zones delimit an extensive series of major horsts and grabens. Examples of fault-bounded major

bathymetric lows include: Hudson Strait, Frobisher Bay, Cumberland
Sound, and Lancaster Sound (Grant and Manchester, 1970; Grant, 1975;
MacLean and Falconer, 1979; Kerr, 1980). Dowdeswell and Andrews
(Chapter 4) have also argued that such faulting also controls the
location and shape of many of the eastern Baffin Island fiords.
Geological maps of the offshore part of eastern Baffin Island also
strongly suggest that the major trend of the outer coast is fault
bounded (Chapter 6). Some of the bathymetric highs near the mouths of
some of the major troughs appear to be structural highs, as sedimentary
rocks thin toward them (e.g. Grant, 1975), although this is probably
not universally true.

     With the development of the rifting in the earliest Tertiary time,
models of the development of passive margins indicate that there would
have been a period of isostatic uplift, followed by collapse and
submergence. In addition to these motions, it must be remembered that,
during the late Cretaceous and into the Tertiary, global sea levels
were 550-350 m higher than present. On Baffin Island, uplift of at
least 600 m is indicated by several lines of evidence (Bird, 1967;
MacLean and Falconer, 1979) but, as yet, the offshore Tertiary
sediments appear to imply a continuous marine shelf environment.
Concordant summits across sections of Baffin Island may reflect higher
Tertiary sea levels (Bird, 1967) or possibly exhumed Precambrian
erosional surfaces (Ambrose, 1964; Dyke et al., 1982). However, the
probability of land uplift during an early Tertiary warm and wet
climate (Andrews et al., 1972) followed by subsequent submergence may
be a possible mechanism to explain several aspects of fiord
morphology. Certainly we may expect that the uplift of both the
margins along West Greenland and the eastern Canadian Arctic, combined
with a milder, wetter climate, resulted in extensive subaerial erosion
of the land and shedding of significant clastic sediment to the
continental shelves. This process may have proceeded up to the Miocene
and may have ceased as subsidence caused sediment traps to be developed
in the nearshore zone, possibly in drowned river valleys and canyons.

     Observations on drilling cores from the West Greenland shelf have
suggested that a mature relief existed in adjacent land areas in Eocene
times, while rejuvenation resulting from uplift of coastal Greenland
took place in the Upper Eocene and Oligocene (Henderson et al., 1981).
The uplift caused the formation of very large westward prograding
deltas, a characteristic feature in the morphology of the West
Greenland shelf (Henderson, 1975). The position of these deltas at the
mouths of major fiords show that the drainage pattern which was later
to be inherited by the Inland Ice had already been established in late
Tertiary times.

The next major environmental factor affecting the region was the progressive global cooling that was underway by the Miocene. This was terminated in our area by the growth of the Greenland Ice Sheet, the cyclical growth and decay of the Laurentide Ice Sheet, as well as development of local ice caps and glaciers.

Baffin Bay and the Labrador Sea do not possess an active mid-ocean ridge. It is therefore surprising to note the relatively high seismicity of the region (Basham et al., 1979; MacLean and Falconer, 1979). Figure 2.2 shows the distribution of major recent epicenters. Of particular note are the high concentrations of earthquakes in the vicinity of some of the Baffin Island fiords and bays. Pronounced clusters of epicenters are located close to Scott Inlet in northern Baffin Island and in the vicinity of Home Bay. By contrast the large structural grabens of Cumberland Sound and Frobisher Bay are relatively quiet over the period of record.

ONSET OF CENOZOIC GLACIATION

There is no direct evidence that allows us to attach a firm date to the onset of glaciation on the land surrounding Baffin Bay. On general climatic/oceanographic grounds, Andrews (1974) suggested that glaciation of Greenland may have commenced in the Miocene. However, there has been no supporting evidence uncovered for this. Indeed, the evidence from Tertiary marine faunas in the Labrador Sea (Thunnel and Belyen, 1981) indicates that relatively warm seas persisted throughout our region until well into the Pliocene. In eastern Greenland a distinctly non-arctic marine fauna has tentatively been referred to the Miocene, the "Chlamys beds" of the Kap Brewster Formation (Hassan, 1953; Birkenmajer, 1972). The Deep Sea Drilling Program (DSDP) drilled two sites between Newfoundland and South Greenland that do have information pertaining to when glaciation first affected the seas of the northwestern North Atlantic. Sites #112 and 113 (Fig. 2.3) lie at 54°01'N & 46°36.24'W and 56°47.40'N & 48°19.91'W respectively (DSDP volume XII, 1972). They should be well placed to record, both litho- and biostratigraphically, any major environmental changes than occurred within Baffin Bay and the Labrador Sea (Fig. 2.3). Site #112 was specifically looked at as a site where onset of glaciation could be defined (DSDP volume XII, 1972, p. 165).

Site #112. Core 5, Hole 112A is considered on faunal grounds to be Lower Pliocene. The sediments are interpreted as representing processes operative during a lowering of surface temperature "...prior

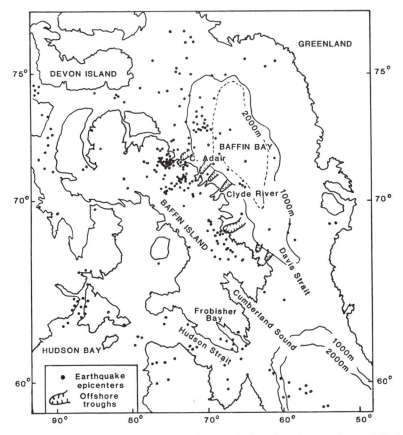

Figure 2.2 Extent of recent seismic activity in the region of Baffin Bay.

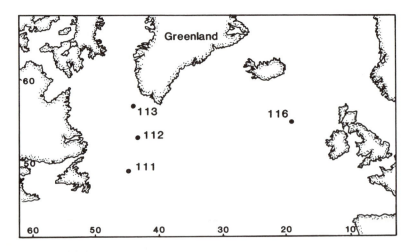

Figure 2.3 Map of the North Atlantic showing the location sites 111, 112, and 116 which are discussed in the text.

to the onset of ice-rafting" (p. 168). Detrital quartz and mica were present in the sand fraction. It is not stated how the sand was emplaced but it was noted that a faint terrestrial influence is also apparent in core 4 which covers the Lower Oligocene to Miocene (p. 168). Mid-Pliocene to Pleistocene sediments were recovered in several cores. They are described thus: "gray, terrigenous, silty to sandy clay with ice-rafted pebbles interbedded with hemipelagic, silty nannofossil clay and marl." (p. 168). In view of the extent of limestones throughout the study area (Fig. 2.1) it is also worth noting that x-ray analysis indicated the presence of 2-6% dolomite. The authors suggest that the dolomite may be detrital and thus may reflect glacial/terrestrial erosion. Sedimentation rates are calculated as between 2.5 to 4.5 cm/1000 yr. The onset of glaciation is placed at a depth of 115 m when there is an increase in the amount of terrestrial material (i.e. quartz, mica, and feldspars). The onset of ice-rafting = glaciation, is dated biostratigraphically at about 3 million years (p. 176).

Site #113. Incomplete core recovery characterized this phase of the DSDP. In addition, sedimentation rates were much higher than at Site #112 with estimates of 10 to 35 cm/1000 yr. Thus the base of the glacially derived sediments can only be approximately fixed as occurring between 550 and 665 m but probably closer to 550 m. This high rate of sedimentation is attributed to gravity-controlled turbidity currents (p. 267) off the Erik Ridge and continental margins. Onset of glaciation is dated at about 3 million years.

Berggren (1972) used the results from Leg 12 of DSDP to discuss the timing of Cenozoic glaciation. In essence, the influence of glaciation on the style of deep sea sedimentation is associated with an increase in detrital minerals and clays which is probably the cause for the increase in the natural gamma count and the preglacial/glacial transition. However, as discussed by Laine (1980) and Andrews (1982) the initial glaciation of northern Canada might be associated with a major increase in detrital carbonates associated with total or partial stripping of some of the Paleozoic limestones (cf Figure2.1). Berggren (1972, p. 954) places a lot of weight in his interpretation on the onset of glaciation at Site #116 which lies in the Rockall-Hatton Basin (57°29.76' & 55.46' W). In particular, the plot of the percent detrital mineral grains (Fig. 2.4) is used to define the onset of glaciation for the northwest North Atlantic at a time which correlates well with the inferred onset of ice-rafting at Sites #111, 112 and 113 (Fig. 2.3). Berggren (1972, p. 956) suggests that peaks in the detrital minerals may define major glaciations at 0.4-0.6, 1.0, 1.3-1.6

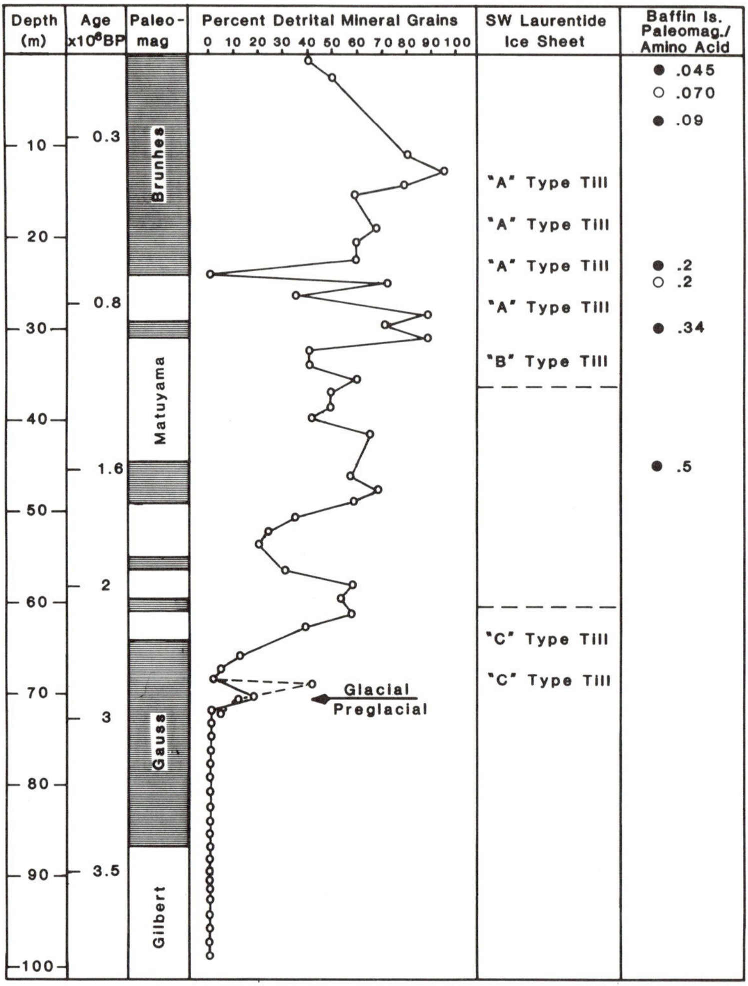

Fig. 2.4.   Onset of detrital mineral influx to site 116 in the North
Atlantic compared to dates for glaciation along the southwestern margin
of the Laurentide Ice Sheet (Berggren, 1972) and paleomagnetic data and
amino acid age estimates from sediments exposed along the west coast of
Baffin Bay near Clyde Inlet.

and 2 million years ago (Fig. 2.4). These dates for inception of
Northern Hemisphere glaciation are in agreement with recent age
determinations on the "Nebraskan" tills of the Midwest, USA (Fig.
2.4). Similarly, on Iceland large-scale glaciation first occurred
about two million years ago (Albertson, 1978). These data may indicate
that glaciation of the Baffin Bay region and the area of the Laurentide
Ice Sheet did not progress through a long minor glaciation mode but
switched abruptly from a nonglacial to a full-scale glacial mode. The

timing of this change in DSDCP Sites #111, 112, 113, and 116 suggests a
link with the changes in oceanographic circulation attendant upon the
erection of a land bridge between North and South Americ. The
correlation, if indeed it is a cause-and-effect relationship, implies
that it was a change in the ocean circulation which may have triggered
glaciation in high latitudes. It can be noted that Gradstein and
Srivastava (1980) argued that large-scale outflow of cold water from
the Arctic Ocean into Baffin Bay first occurred in Pliocene times.
Warnke and Hansen (1977) have suggested, however, that at other DSDP
sites in the North Atlantic glaciation started 7-8 million years ago.

    The possibility that detrital carbonates may better define the
initial continental glaciation of northern Canada (Andrews, 1982) is
not given a great deal of support from Sites #111-113 (Fan and Zemmels,
1972). Dolomite, which may be an indication of erosion of the
widespread Paleozoic platform rocks, is restricted at all sites to the
Pliocene/Pleistocene and occurs in conjunction with ice-rafted
sediments. Dolomite is not recorded in sediments older than 3 million
years.

Glaciation of Baffin Island: oldest sediments

At the base of wave-cut Quaternary outcrops near Clyde River and
Qivitu, shell-bearing marine and glacial marine clastic sediments
record the oldest observed evidence for glaciation (Nelson, 1982; Mode,
1980). Amino-acid geochronological studies give ratios of D
alloisoleucene: L-isoleucene (formerly abreviated to D allo: L-iso
(Miller et al., 1977) but henceforth in this volume abbreviated as
aIle:Ile) of between 0.5 and 0.3 (Total). These sediments lie
stratigraphically below sediments of the Cape Christian Member of the
Clyde Foreland Formation (Feyling-Hanssen, 1976; Mode, 1980). The
aIle:Ile ratios (Total) for the Cape Christian Member are between 0.12
and 0.16 and Szabo et al. (1981), Nelson (1982) and Miller (this
volume) develop arguments to support an age for the Cape Christian
aminozone of between 350,000 to 550,000 BP. Given our uncertain
knowledge about the Pleistocene thermal history of the Baffin Bay
region and the probable non-linear kinetics of racemization at aIle:Ile
values greater than 0.3, we can only suggest that Total ratios close to
0.5 may represent ages of between 1 and 2 million years. Paleomagnetic
studies on several of these older units (Nelson, 1978; Miller, unpubl.)
indicate that most are normally magnetized, although one site near
Clyde River with a total aIle:Ile ratio of 0.07 yielded a reverse
paleomagnetic pole (Fig. 2.4). However, repeated marine transgressions
and regressions along the coastal forelands (Andrews, 1978; Miller et

al., 1977) would result in repeated freezing and thawing of the
fine-grained marine sediments and thus the paleomagnetic character of
the units might have been affected, e.g. the "clock" might be reset
during each thaw (submergence) cycle. However, we cannot ignore the
possibility that ratios of 0.5 might reflect an age which would place
deposition within the Gauss Normal Epoch or for that matter one of the
extensive normal periods of inclination that typify the Matyama Reverse
Epoch (Fig. 2.4). Some paleomagnetic data from Baffin Island are
plotted on Figure 2.4 and indicate that further research might be
warranted on this topic.

In Greenland the earliest known evidence for arctic climate and
glaciation comes from the Lodin Elv Formation in East Greenland. The
diamicton sediments are interpreted as glacial marine and contain a
"modern" high arctic mollusc fauna. The sediments are by
micropaleontological evidence referred to the Pliocene-Pleistocene
transition, and aIle:Ile values of 0.5 have also been obtained here
(Feyling-Hanssen et al., 1983).

Andrews (1972) and Anderson (1978) have used the rate of cirque
erosion on Baffin Island to suggest that the onset of glaciation on
Baffin Island should be measured in millions of years. These estimates
do not conflict with the evidence presented above, but neither are they
sufficiently soundly based to be a major consideration. Mountain and
cirque glaciation may indeed have commenced somewhat earlier than the
first evidence for detrital minerals at DSDP sites #111-113. The flux
of detrital minerals into deep-sea sediments must, however, reflect one
or more of the following states: (1) significant development of sea ice
and transportation of sediment by this mechanism (Andrews and Match,
1983); and (2) development of major outlet glaciers on Greenland and in
Arctic Canada that reached tidewater and calved into Baffin Bay.

PRESENT ENVIRONMENTAL SETTING

Glaciology

Baffin Bay is surrounded by a variety of existing ice masses. Studies
of these ice caps, ice sheets and glaciers have been proceeding for
several decades; however, many of the scientific results consist of
data obtained over a year or two and there are only a few ice masses
which have been studied systematically for many years. The most
detailed and long-term studies have been conducted on the Devon Island
Ice Cap (e.g. Koerner, 1970; Patterson et al., 1977; Koerner, 1977).
In the same general area, the "North Water Project" (Muller et al.,

1975) has amassed a large body of data on glacier climate and mass balance (see Sea Ice Section). Other significant studies have been carried out on the Barnes Ice Cap (Baird, 1952; Sagar, 1966; Holdsworth, 1973; Hooke, 1973); the Decade Glacier (Ostrem et al., 1967); the Penny Ice Cap (Ward and Baird, 1954; Orvig, 1954), and the Boas Glacier (Andrews and Barry, 1972; Jacobs et al., 1973; Weaver, 1975). No systematic studies of any sort have been carried on the icefields and glaciers of Hall Peninsula nor have any mass balance or glacial climatic data been gathered for the two most southerly ice caps of eastern North America, the Terra Nivea and Grinnell ice caps, located close to Frobisher Bay on the northern rim of the Meta Incognita Peninsula. The furthest south that ice bodies exist in eastern North America are in the cirques of the Torngat Mountains.

On West Greenland, studies of the mass balance of the Greenland Ice Sheet have been carried out over many years; however, the sheer size of the ice-sheet has meant that an accurate estimate of the mass balance will be difficult to obtain (e.g. Bauer, 1955; Benson, 1962; Radok et al., 1982). There has been a dearth of glaciological studies in Greenland dealing with the smaller ice bodies, but interest in the hydro-electric power potential may lead to further glacio-climatic studies.

Glaciation level/threshold: Weidick (1975) Andrews and Miller (1972) and Miller et al., (1975) have mapped the Glaciation Level (GL) for the area surrounding Baffin Bay (Fig. 2.5). The GL marks the average elevation between mountains that maintain an ice mass and those that are ice free. The elevation of the Glaciation Level thus represents some notion of the elevation at which, in a regional sense, the net mass balance b = 0. In addition, the elevation of the GL together with the regional topography represents a guide to the sensitivity of a region to a change in mass balance and a subsequent lowering or rising of the GL (e.g. Locke and Locke, 1977; Williams, 1978).

The Glaciation Level around Baffin Bay is lowest along the coasts and rises inland (Fig. 2.5). On Baffin Island and in northern Labrador the GL varies between 1400 and 600 m with the high figure associated with the small cirque glaciers of the Torngat Mountains. Along the outer east coast of Baffin Island the GL is between 600 to 700 m asl but it rises westward to lie between 900 and 1000 m in the vicinity of the Barnes and Penny Ice Caps (Andrews and Miller, 1972). Around northern Baffin Bay the presence of the North Water appears to have a marked effect on precipitation (=snow) and the GL is mapped at 500 m or possibly lower in the vicinity of Coburg Island in northwestern Baffin

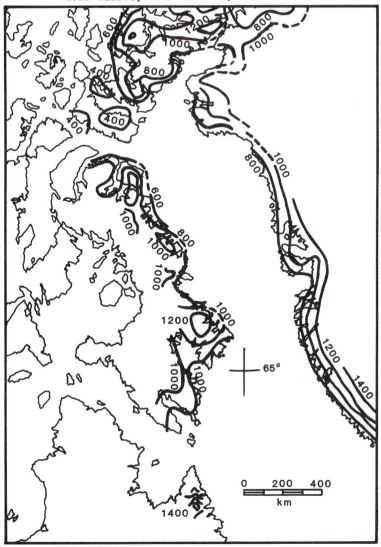

Figure 2.5   Glaciation Level or threshold around Baffin Bay and Davis
Strait (m asl).

Bay (Fig. 2.5). However, the GL rises inland over Ellesmere Island to
heights of 900 m+ before dropping to between 300 and 400 m asl along
the coast facing the Arctic Ocean.

    The Glaciation Level across Greenland (Weidick, 1975) varies
between 500 and 1750 m asl. Along the outer coast of West Greenland
the GL is close to the lower limit but it rises inland to 1500 m over a
distance of 200 km (gradient of 5 m km^{-1} which compares closely with

the gradients established on the west side of Baffin Bay (Andrews and
Miller, 1972).

Mass balance: The annual accumulation over the Greenland Ice Cap
(Hattersley-Smith, 1974) reaches values of 0.9 m H_2O in southern
Greenland and declines northward to 0.15 m over the Ice Sheet in the
vicinity of 78° N. Peak accumulation along the western side of the Ice
Sheet is reached at moderate elevations of 1550-2000 m where
accumulation reaches 0.6 m and it declines in both directions from this
belt of maximum accumulation. An additional belt of high accumulation
is present north of Thule and may be associated with the local effect
of the North Water. Various estimates for the total mass balance of
the Inland Ice range from positive (Bader, 1961; Benson, 1962) to
slightly negative (Bauer, 1955). There is generally agreement that the
total annual gain amounts to c. 500 km^3 water equivalent of snow;
however, the largest mass loss is due to calving but this value is not
well constrained. Weidick (1975) has considered various climatic/
glaciological scenarios associated with an increase in mass and area of
the Greenland Ice Sheet during the late Wisconsin glacial maximum.
This analysis suggested that during the global glacial maximum the
accumulation on the Greenland Ice Sheet was only 0.15 m H_2O at ca
15,000 BP and rose rapidly toward present values between 8000 and 7000
BP. Little is known about the mass balance of the thousands of
individual glaciers on Greenland (e.g. Partl, 1977).

The most intensely investigated ice body in our region is the
Devon Island Ice Cap. Not only have mass balance and glacial-
meteorological studies been conducted on the ice cap for over a decade,
but ice cores taken from the center of the ice cap have a long record
of stable isotope fluctuations (see Koerner and Fisher, this volume)
extending backwards in time to the last interglaciation (Paterson et
al., 1977). The mass balance of the ice cap has been discussed by
Koerner (1970) and Alr (1978). The northwest sector of the ice cap is
estimated to be in balance with the elevation of the equilibrium line
at 920±80 m. The heaviest accumulation occurs along the southeastern
sector of the ice cap, probably associated with enhanced snowfall
resulting from the presence of the North Water just offshore.
Accumulation at 1790 m asl on the ice cap was obtained for the years
1934 to 1968 from snow pit measurements. The results showed long-term
trends with arrange of net accumulation at that elevation of between ca
0.12 to 0.34 m H_2O, with an average of about 0.2 m H_2O. This is
comparable to the accumulation on the main dome of the Greenland Ice
Sheet at the same latitude.

The Barnes Ice Cap has been studied nearly continuously, but with a variety of objectives, since the 1950 Arctic Institute of North America's Expedition (e.g. Baird, 1952). Overall assessments of the mass balance were reported by Sagar (1966) and Löken and Andrews (1966). The ice cap is situated in a marginal environment and the accumulation on the ice cap is largely through the addition of superimposed ice. The ice-sheet is asymmetric with a much lower equilibrium line on the eastern margin of the ice-sheet than on the western. North of 70°N, Sagar (1966) indicated that the ELA for the balance years 1962, 1963, and 1964 varied between 1090 m (1962) and 810 m (1964). For the southern dome of the ice cap, ELAs for the eastern and western margins were, respectively, 595 and 785 m (1964-1965 balance year). The winter balance across the ice cap appears to average ca 0.3 to 0.4 m H_2O.

Further south on Baffin Island there are only a few scattered observations on glacial mass balance and on glacial meteorology (Orvig, 1954; Jacobs, et al., 1973; Weaver, 1975). On the Penny Ice Cap the net accumulation may approach 0.5 m H_2O at the crest of the ice cap (close to 2000 m asl). To the east, the Boas Glacier averaged between 0.4 and 0.32 m H_2O for the winter mass balance for the period 1969 - 1973. Year to year variations in the net mass balance, b, were between +0.4 and -0.54 H_2O.

Little is known about the magnitude of the winter accumulation on the ice bodies on Baffin Island south of Cumberland Sound and in the Labrador mountains. Rogerson (1982) reports that snow depth on the small cirques on the Torngats averaged 1 to 2 m with a maximum depth of 4.25 m. As the snow density was 0.55, the average water equivalent was 1.10 m H_2O.

The overall picture of mass balance determined from around Baffin Bay is that accumulation decreases steadily northward from maximum values in both Greenland and Labrador of ca 1.0 m H_2O to about 0.2 m in the latitude of Devon Island. Detailed glacial meteorological studies have been conducted over the Devon Island Ice Cap, the Barnes and Penny ice caps (Orvig, 1954), and the Boas Glacier (Jacobs, et al., 1973; Jacobs, 1973). These have shown radiation to be the dominant component of the energy inputs that lead to ablation during the summer months.

Paleoglaciological studies around Baffin Bay have involved the question of the changes in climate during the last interglacial/glacial cycle (Andrews et al., 1974; Weidick, 1975; Paterson et al., 1977; Dansgaard et al., 1971) with particular attention on the question: What was the magnitude and importance of changes in accumulation

throughout the region? These problems will be addressed more
specifically in the chapter by Koerner and Fisher. Suffice to note
that the published literature indicates that accumulation was reduced
by 50 to 75% during the main part of the Wisconsin Glaciation (10,000
to 70,000 BP) (Andrews et al., 1974; Weidick, 1975; Williams and
Miller, 1974).

Oceanography

The surface currents in Baffin Bay and the Labrador Sea consist of a
counterclockwise circulation (Fig. 2.6) dominated by a relatively warm
northward flowing current, the West Greenland Current, which flows
toward the head of Baffin Bay along the coast of West Greenland, and a
cold southward current, the Canadian Current. The latter reflects
outflow of water from the Arctic Ocean. The net efflux of water from
the Arctic Ocean across Davis Strait has been estimated as 42,500 to
65,000 km^3 yr^{-1} (Dunbar, 1960; Timofeyev, 1960; Muench, 1971). Surface
current velocities vary throughout the region but are clearly stronger
toward the western side of Baffin Bay and along the Labrador Coast.
Velocities in the last two areas vary between 8 and 15 km d^{-1} (Fig.
2.6) whereas the northward flowing West Greenland Current is generally
much weaker.

    Oceanographers recognize three major water masses within Baffin
Bay. 1. Baffin Bay Surface Water (BBSW) consists of cold, low
salinity water which extends from 0 to ca 250 m. Temperatures vary
between 0° to -1.8°C at depth below the surface layer, subject to
seasonal change (generally within upper 10 m), and with salinities of
31 to 33.7°/oo. In fiords the BBSW is heavily influenced by tidal
pumping and by the influx of freshwater inflows. 2. Baffin Bay
Atlantic Water (BBAW) consists of an intermediate warmer and more
saline layer (0 to 2°C and 34.3 to 34.5°/oo (in Aksu, 1981) that
extends from between 150-300 m to 1200-1300 m. The general depth of
the BBSW indicates that it dominates the physical oceanographic
environment on the Baffin Island Shelf with BBAW lying off the shelf
break. 3. Baffin Bay Bottom Water (BBBW) is cold, saline water with
values of 0° - 0.4°C and 34.4 to 34.5°/oo. The origin of the different
water masses is the subject of debate. BBSW is attributed to cooling
of the surface water in the Arctic Ocean or to cooling and mixing of
water entering Baffin Bay from Davis Strait (Muench, 1971). The
underlying BBAW originates from the Atlantic Ocean via the West
Greenland Current where it is cooled by mixing with surface water. The
origin of the Bottom water is even more unclear about Aksu (1981, p.
46) suggests that the most probable explanation is "...that Baffin Bay

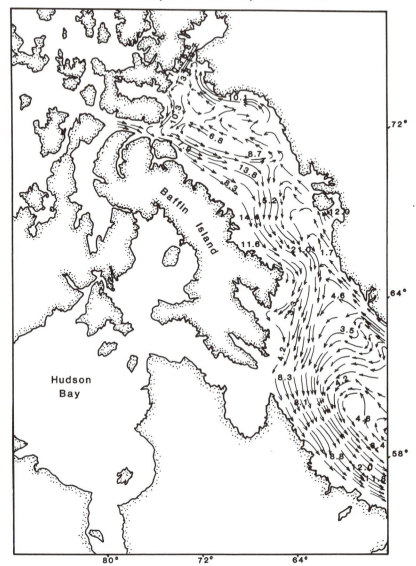

Figure 2.6 Map of surface currents and velocities (km/day) in Baffin
Bay and the Labrador Sea.

Bottom Water originates from the Eurasian Basin of the Arctic Ocean,
was cooled in Nares Strait (between Greenland and Ellesmere Island) to
produce the appropriate temperature and salinity and, consequently,
injected into the deeper basin in irregular pulses or as a slow and
continuous process."

Dunbar (1956) divided Baffin Bay/Labrador Sea into two major
zoogeographic provinces, the Arctic and Subarctic. Lubinsky (1972) and
Andrews et al. (1981), amongst others, have indicated the importance of
these divisions in terms of the present biota and in terms of the
interpretation of paleoenvironmental data from along the east coast of
Baffin Island (e.g. Miller et al., 1977). The zoogeographic boundary
occurs well north of the planktonic/oceanographic Polar Front (Fig.
2.8) which is delimited by the 10°C summer isotherm. The Arctic/
Subarctic boundary (sensus Dunbar) closely parallels the 2 to 3°C
summer surface isotherm across Baffin Bay/Davis Strait and
biogeographically is marked at either end of the boundary by the limit
of the Subarctic blue mussel, Mytilus edulis Linne, and on Baffin
Island by the extreme limit of dwarf birch, Betula nana or/and Betula
glandulosa (Andrews et al., 1980, 1981).

In view of the importance of the Polar Front in the
paleoceanographic literature (Ruddiman and McIntyre, 1973; 1981;

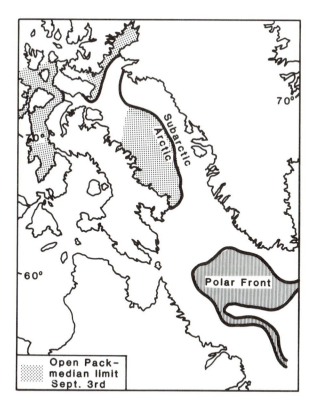

Figure 2.7   Median sea ice extent September 3rd and the location of
Polar Front and arctic subarctic marine boundary.

Kellogg, 1975), it is worth noting the position of the Polar Front in
the northwestern North Atlantic in relation to the extent of "permanent
sea ice" and the numbers of icebergs. This is because the literature
frequently associates the Polar Front with a major control on the limit
of ice-rafted detritus, and areas north of the Polar Front in the
northeastern North Atlantic have been associated by Kellogg (1975;
1980), in a paleoceanographic sense, as being covered by extensive and
permanent sea ice. Such a picture of the northwestern North Atlantic
does not apply to present conditions. By early September to October,
Baffin Bay and the Labrador Sea are largely free of seasonal sea ice.
Icebergs, nearly all of which come from the calving glaciers of West
and Northwest Greenland, track along the western side of the region,
being driven southward by the Canadian and Labrador currents. As shown
in Figure 2.8, iceberg density (number of icebergs per 1000 km^2)
decreases rapidly toward Frobisher Bay. Figure 2.8 illustrates the
numbers of icebergs per 1000 km^2 for the period 1963-1972 AD; the
figure highlights the high concentrations of icebergs in the northern
reaches of Baffin Bay and the rapid decline in numbers/1000 km^2 once
the ice enters the warmer water of the Subarctic zone (compare Figure
2.7 with 2.8). Southward there is an increase associated with a
counterclockwise circulation that sends icebergs westward from the
southern tip of Greenland. These icebergs are largely associated with
calving from East Greenland outlet glaciers. The Polar Front does
serve to mark a major break in iceberg density from the East Greenland
source, and it certainly delimits at the present time a large area of
seasonal sea ice which is, however, not in any sense a "permanent"
sea-ice cover.

A major influence on the nearshore water characteristics and biota
in the area between Cape Dyer and Ungava Bay are the extreme tidal
ranges that prevail throughout the area. The tidal range decreases
north of Cape Dyer to only between 1 to 2 m but in the vicinity of
Pangnirtung, Frobisher Bay, and Ungava Bay the range varies between 8
and 14 m and thus represents one of the world's macrotidal
environments. Gilbert (1978) had demonstrated that tidal pumping in
Pangnirtung Fiord is extremely important to the physical oceanography
of the fiord. Gilbert (1982) has also presented some information on
salinity and temperature in Coronation Fiord, immediately south of
Broughton Island for one day in August 1979. The data indicate that
below 4 km the temperature was 0°C or colder and that the surface
salinity of 15 to 20°/oo increased to 30 to 31°/oo below 6 m. The
fiords of Baffin Island are at the moment poorly studied from the
viewpoint of either physical oceanography or marine geology; however,
The Canadian research vessel Hudson conducted an extensive field

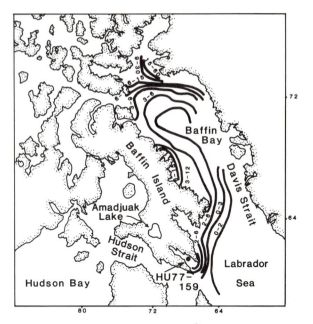

Figure 2.8 Number of icebergs per 1000 km^2 reported in the waters of
Baffin Bay between 1963 and 1972 AD (A.P.O.A. 1973).

sampling program during the 1982 and 1983 ice-free seasons and
publication of the results is awaited with interest.

Present Climate

The earliest meteorological observations in the Baffin region are from
West Greenland stations, several of which have records which extend
back more than one hundred years. By contrast, on Baffin Island, apart
from short-term studies carried out by expeditions such as those of
International Polar Year of 1882-83 (Taylor, 1981), the instrumental
record began only in the early 1900s with the extablishment of
permanent missions, police stations, and trading posts at locations
such as Lake Harour, Pangnirtung, and Pond Inlet (Thomas, 1971;
Bradley, 1973; Fletcher, 1975). Synoptic and upper air observations
began with the establishment of military airdromes during World War II
and the DEW-line radar stations in the late 1950s. At the peak of this
activity, there were more than 30 stations in operation in the region
(Fig. 2.9). Unfortunately for continuity, observations had been
discontinued at some of the older settlements by that time. Also, by
the early 1960s about half of the DEW-line radar sites had been closed,
and only recently have observations again expanded to meet the civil

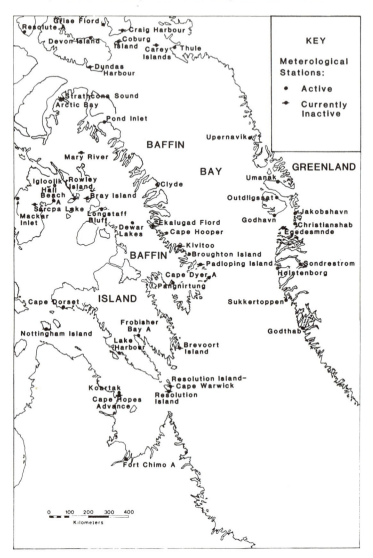

Figure 2.9 Map of weather stations around Baffin Bay (after Maxwell, 1980).

aviation needs of settlements and in support of mining and hydrocarbon exploration activities.

The period of adequate meteorological coverage of the Baffin region is thus scarcely three decades long, barely sufficient for the calculation of climatological normals in the conventional sense; although, as Orvig (1970) has pointed out, a 30 year period may not really be appropriate as an averaging period in the Arctic. Based on this record, supplemented by some special observations, a number of studies have emerged which treat in increasing detail the general climatic conditions of the region, as well as specific problems of local climates, radiation, and energy budgets and short-term climatic variability.

The reviews by Hare (1968) and Barry and Hare (1975) locate the Baffin region within the general Arctic circulation scheme as the locus of a prominent trough in the circumpolar westerly flow. The comprehensive study by Maxwell (1980) places the mean axis of the lower tropospheric (700 mb) trough over northern Baffin Island and Foxe Basin in winter, shifting to central Baffin Bay--Davis strait in summer, with the mean surface low-pressure centers lying several hundred kilometers eastward of these positions. This configuration suggests predominately northwesterly prevailing winds over Foxe Basin and Baffin Island, with more southerly and easterly components along the West Greenland coast. Wind frequency data presented by Maxwell (1980) reveal such a pattern to be the general rule, except for some stations where relief or coastal configuration affect the flow.

Putnins (1970) found that, although surface winds at West Greenland stations tend to either be seasonally variable or dominated by an easterly flow from the ice cap, the winds above about 1 kilometer at these same stations are from the south for a significant amount of time. This flow, with its associated meridional transfers of heat and moisture (as well as pollen and pollutants), contributes both to the milder climate of that coast, compared with east Baffin Island, and to the maintenance of the Greenland Ice Cap.

The average circulation pattern, already indicative of significant links to more southerly regions, does not reveal the true degree of synoptic activity in the region, other than that the Baffin Bay--Davis Strait surface low may be viewed as a statistical reflection of mid-latitude cyclones moving into the region along a more northerly branch of the general northeasterly depression track across the North Atlantic. Synoptic studies in the region (Barry, 1974; Barry, et al., 1975; Keen, 1980) reveal large seasonal and interannual variations in the frequencies of distinctive circulation types which, in many cases, have statistically significant relationships to temperature and precipitation anomalies, as well as to sea-ice conditions and glacier mass balance in the region.

Much of this activity and its variability is associated with the seasonal position of the Arctic Front (Bryson, 1966; Barry, 1967), the

median summer position of which lies just south of Hudson Strait.
Viewed broadly as a boundary between relatively cold, dry Arctic air
and warmer, moister temperate air, the average position of this front
in any given period is obviously climatologically significant. Barry
(1967) found that for the 1961-65 period, the Arctic Frontal zone at
850 mb was located as far north as southern and eastern Baffin Island
up to 10 percent of time in January and over most of Baffin Island and
southwest Greenland 50 percent of the time in July.

Distinguishing air masses entering the region by source region and
trajectory provides another approach to the synoptic climatology of the
region. Maps and interpretation presented by Bryson (1966) and, more
recently, Maxwell (1980) show a high percentage (30 to 60 percent) of
air masses of North Atlantic origin over southern and eastern Baffin
Island in July, compared to central and northern Baffin Island and Foxe
Basin, which are influenced more frequently by air masses originating
over the Arctic Archipelago.

The contrast in synoptic regimes between Foxe Basin and northern
Baffin Island on the one hand and southern and eastern Baffin Island
and West Greenland on the other is strongly reflected in seasonal
temperature and precipitation patterns. Winter gradients are the most
sharply defined: with isotherms running nearly due north-south, mean
January temperatures decrease from -10°C in eastern Davis Strait to
below -30°C in western Foxe Basin. Summer temperatures are very much
controlled by elevation and proximity to the coast. July temperatures
are about 5°C along most of the Baffin Island coast, rising above 7°C
in the interior lowlands of south central Baffin Island, between
Frobisher Bay and Nettiling Lake.

Precipitation follows a somewhat similar SE to NW trend, with
highest values (i.e. mean annual totals around 600 mm) in southeastern
Baffin Island, where high, rugged terrain accentuates the effects of
proximity to Davis Strait and its southerly storm tracks. Rainfall is
at a maximum in July and August at most stations. At Baffin Island
stations facing Davis Strait and Hudson Strait, rain may occur in
winter as late as December, while along the southwest Greenland coast,
rain occurs in all months. Snowfall accounts for about 60 percent of
total precipitation over much of the region, except for the higher
elevations of eastern and central Baffin Island, where it approaches 90
percent. Snowfall is common in summer at higher elevations, amounting
for example to about 83 mm. w.e. or 79 percent of total precipitation
on the Penny Ice Cap (elev. 2050 m) in the summer of 1953 (Orvig, 1954)
and 62 mm w.e. or 78 percent on the Boas Glacier (elev., 1140 m) in
summer, 1970 (Jacobs, et al., 1973). Snowfall has been observed by

Jacobs in mid-July on the smaller ice sheets of Hall Peninsula at
elevations of about 1 km. The locally depressed air temperatures above
these ice bodies constitute a positive feedback mechanism, helping to
maintain them during the summer ablation season.

Regular measurements of solar radiation have been made at some
Baffin Island locations for more than a decade (Jacobs and Andrews,
1983) and intensive surface energy budget studies have been carried out
on glaciers (Ward and Orvig, 1953; Orvig, 1954; Sager, 1966; Jacobs, et
al., 1973), on sea ice (Jacobs et al., 1975; Crane, 1979), and at a
land site near the coast (Thompson and Fahey, 1977).

Regional radiation and energy budget estimates have been presented
by Vowinckel and Orvig (1963), Walmsley (1966), Hare and Hay (1974),
and by Maxwell (1980), who shows a generally south to north decline in
annual average net radiation from about 1.25 MJ m^{-2} d^{-1} (30 ly d^{-1}) in
Hudson Str. to 0.42 MJ m^{-2} d^{-1} (10 ly d^{-1}) along the northeast Baffin
Island coast, northward of which the gradient is fairly flat. Seasonal
values show considerable spatial variability that is, in effect,
averaged out over the year. July net radiation is at a maximum of more
than 10.5 MJ m^{-2} d^{-1} (250 ly d^{-1}) along a broad corridor from northern
Baffin Bay southward through eastern Davis Strait to the Labrador Sea,
with values estimated at about 10 percent less over Baffin Island and
its nearshore waters. The opposite is true in January, when central
and northern Baffin Bay show the lowest values for the region, below
-2.5 MJ m^{-2} d^{-1}.

Because they are based on approximate values for surface albedo
and cloud cover, such estimates belie the true complexity of the
surface energy balance, particularly on the rugged and varied land
areas in summer. Net radiation fluxes measured over snow-free tundra
surfaces in July, 1973, averaged 7.9 and 8.4 MJ m^{-2} d^{-1} at Frobisher
Bay and Broughton Island, respectively (Barry and Jacobs, 1978).

On land, the partitioning of the radiant energy stream in summer
tends to be dominated by the latent heat loss term, although the flux
of sensible heat is positive to the surface most of the time. Thawing
of the ground proceeds slowly, with active layer depths probably being
less than 1 meter over most of the coastal lowland areas. Energy
budget measurements over the sea ice show a wide range of values
between melt-water pools and bare or snow-covered ice. The energy
budget, and thus the progress of melt and breakup of the coastal ice,
is very sensitive to synoptic conditions.

The complexity of terrain in may parts of Baffin Island region
results in a wide range of microclimates under the same regional
synoptic controls. On coastal slopes in summer there is a weak
inversion, with temperature maxima occurring between 50 and 500
m asl on leeward, south-facing slopes. In eastern Cumberland
Peninsula, daily maximum temperatures of as much as 8°C higher than
those on opposite north-facing slopes were recorded (Jacobs and Leung,
1980). Deep snow banks accumulating on these lee slopes in winter
provide a summer moisture reserve, with the result that vegetation is
abundant. Such microclimatic differences and, more generally, the wide
range of topoclimates that exist in the region are reflected strongly
in the flora, with large differences in plant communities occurring
over relatively small vertical and horizontal distances (Beschel and
Weidick, 1973; Andrews et al., 1980a).

The long-term variability of climate in the Baffin Island region
is the subject of other chapters in this volume. It is worth noting,
however, that within the period of observational records, significant
fluctuations are the rule. Changes in relative frequencies of
different synoptic patterns, mentioned previously, are associated with
shifts in the upper tropospheric flow. The most pronounced recent
shift of this kind occurred in the mid-1960s and resulted in more
frequent northerly and northeasterly winds and consequently lower
temperatures in summer in the Baffin Island region (Bradley, 1973;
Barry, et al., 1975; Thomas, 1975; Keen, 1980). Jacobs and Newell
(1979) found that the year-to-year variability in seasonal temperatures
across the Baffin Island region was on the order of 3°C in winter and
1°C in summer, with a high degree of spatial coherence in such
departures throughout the region.

In summary, it is possible to view the Baffin Island and West
Greenland regions as a single climatological unit, from the standpoint
of large-scale circulation patterns. Within this region one can
delimit any number of sub-regions based on such controls as topography,
relief, and marine and sea-ice influences. Below this level, a wide
range of possible microclimates is possible. A useful representation
at the intermediate level has been produced by Maxwell (1981). He
shows the Baffin Island region, his "Eastern" region, to be divided
into seven sub-regions, based on the aforementioned controls and on
regional net radiation estimates (Fig. 2.10). A summary of sub-
regional climatic characteristics taken from Maxwell's work is shown in
Table 2.1.

Although in his analysis of Greenland's climate Putnins (1970)
makes no attempt to draw precise climatic boundaries, one can easily

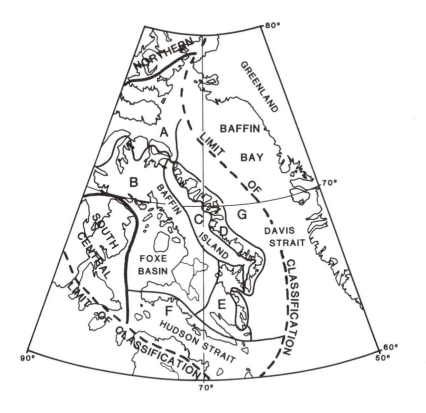

Figure 2.10 Climatic sub-regions of the eastern Canadian Arctic after Maxwell, 1980.

relate his results from the West Greenland coast to Maxwell's for Baffin Island, keeping in mind the limitations on resolution of the sub-regional scale. The coastal area from Disko Island southward, with average temperatures in the range 6 to 8°C in July and -8 to -14°C in January, have no counterpart in Baffin Island. It is only considerably further northward in the vicinity of Upernavik (73°N) that a summer temperature regime comparable to that of the eastern and southeastern coasts of Baffin Island is encountered, and winters are still milder by 2 to 5°C. A sharp climatic gradient occurs northward of this area such that Thule, at 76°N, is characterized by a climate roughly comparable to that of northern Baffin Island. As has already been pointed out, local climate anomalies within these areas may greatly differ from the general pattern.

TABLE 2.1. Characteristics of Climatic Sub-regions of the Baffin
Island Region (Maxwell, 1981) (see Fig. 2.10 for location of
sub-regions)

	Sub-region	Temperature		Annual Precipitation
		Mean Annual Range	Mean Daily Temperature (°C)	
a.	Northern Baffin Bay- Lancaster Sound	33 to 36°C generally: as low as 23°C over "North Open Water"	January: -20 to-33 from northeast to southwest	150-200 mm over water surfaces, up to 300 mm on exposed slopes of Ellesmere and Devon Island
			July: +3 to +5	40-50% falls as liquid precipita- tion 20% at elevation above 400 m
b.	Gulf of Boothia-Foxe Basin- Western Interior Baffin Island	35-37°C	January: -32 to -28 from northwest to southeast	175-250 mm from northwest to southeast
			July: +5-+8 from northwest to southeast	40-50% falls as liquid precipitation
c.	Baffin Island Mountains	33°C	January: -28 to -23 from northwest to southeast	Generally more than 300 mm; 500-600 mm locally on Cumberland Peninsula
			July: +5 to slightly below	20-25% falls as liquid precipita- tion
d.	Baffin Island Eastern Coast	30-26°C from north to south	January: -25 to -20 from north to south	200 mm in north to 600 mm locally in Cape Dyer area to 400 mm in south
			July: +5	15-25% north and 35-45% in south (except lesser at high elevations) falls as liquid precipitation

TABLE 2.1. Characteristics of Climatic Sub-regions of the Baffin
Island Region (Maxwell, 1981) (see Fig. 2.10 for location of
sub-regions)

Sub-region	Temperature Mean Annual Range	Mean Daily Temperature (°C)	Annual Precipitation
e. Southeastern Baffin Island	33-34°C	January: -25	Generally 400 mm except 500 mm at higher elevations north and south of Frobisher Bay
		July: +5 in north to +8 in centre to +5 in south	40-45% falls as liquid precipitation, 20% at elevations above 400 m
f. Hudson Strait	30°C in west to 22°C in east at Resolution Island	January: -25 to -20 from west to east	275-300 mm from west to east
		July: +5 to +8 in west, less than +5 in the east	45-55% falls at liquid precipitation on coasts
g. Baffin Bay Davis Strait	25-30°C	January: -20 to -28 from south to north	Less than 200 mm in north to between 300 and 400 mm in south
		July: between +3 and +5	

Sea Ice

The seas of the Baffin Island-West Greenland region lie entirely within
the seasonal sea-ice zone; that is, the zone in which the ice that
forms in winter melts entirely during the subsequent summer, at least
in most years. Northward and westward of Baffin Island, the straits
and channels of the archipelago are choked with multiyear ice that
rarely clears. By contrast, the waters of southeastern Davis Strait,
influenced by the West Greenland Current, remain largely open
throughout the year, although that area is affected in some years by
"storis", drift ice that rounds Kap Farvel from the Denmark Strait
(Speerschneider, 1931).

Nearly three decades of systematic monitoring using aircraft,
satellites, and ship- and shore-based observations have produced a
comprehensive view of present-day sea-ice conditions in the Canadian

Arctic (Markham, 1981). The extent and thickness of the ice, the
distribution of various age classes, the timing of freeze-up and
break-up, and the year-to-year variations in these characteristics are
well-documented. New satellite technology coupled with improved
predictive models promise better forecasts of seasonal ice patterns
(Dey, et al., 1979; Dey, 1980). Of particular relevance to Quaternary
studies in the Baffin region are the amount and extent of ice and open
water in relation to present climatic conditions and variations in
these patterns from year to year and over longer periods. Figure 2.11
shows median ice conditions based on observations from 1959-1974, as
reported by Markham (1981). The maximum ice extent occurs in late
March but close pack ice fills all of Baffin Bay and Davis Strait west
of 55°W as late as the end of May. The advance of the summer melt is
accelerated in June, and by the end of July much of the region is
clear, although ice usually persists late in Foxe Basin and the Home
Bay area of eastern Baffin Island.

Conspicuous open water areas (polynyas and recurring leads) are
maintained throughout the winter in some places by winds, currents, or
a combination of these. These sources of heat and moisture are of at
least local climatic significance (Muller et al., 1973), and it is from
them that the break up and clearing is seen to advance. The North
Water polynya of northern Baffin Bay and Smith Sound is perhaps the
best known example in the Canadian in seasonal sea-ice conditions
(Jacobs and Newell, 1979); therefore, it may be concluded that severe
ice conditions are a consequence rather than a cause of the
aforementioned shift in atmospheric circulation patterns.

Biogeography

The present flora and terrestrial fauna of Baffin Island may be seen as
a reflection of the insular character of these areas, their recent
glacial history, and the influence of postglacial climatic
fluctuations. The present distribution and relative abundances within
the region are in turn a response to regional and local environmental
conditions.

Vegetation. A broad floristic division of the Arctic region has been
developed by Young (1971). The eastern Canadian Arctic and West
Greenland are included in zones 2, 3, and 4, and 3 and 4 respectively.
Zone 4 represents a Low Arctic floristically richer than comparable
latitudes on Baffin Island across Baffin Bay. This is clearly a
reflection of the combined oceanographic, climatic, and sea-ice effects

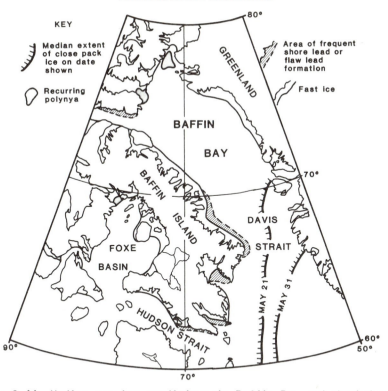

Figure 2.11 Median sea ice conditions in Baffin Bay and the Labrador
Sea region, 1959-1974 AD (after Markham, 1981).

noted earlier in this chapter. Young (1971) characterized his
floristic zones by a term called "summer warmth" which is the sum of
average monthly temperatures above 0°C. Table 2.2 lists this index for
stations within our area. There exist several comprehensive
inventories of arctic plants which encompass Baffin Island (Polunin,
1940-48; Porsild, 1964) and Greenland (Polunin, 1959). Distribution
maps presented by Porsild (1964) and recently updated in Porsild and
Cody (1980) cover Alaska, all of Canada north of the 45th parallel, and
Greenland. These show 340 species of vascular plants to be present in
the Canadian Arctic Archipelago, of which 260 are known to occur on
Baffin Island.

    Most of the Baffin Island species (229) are also found in
Greenland; however, there are some notable exceptions. Fourteen
species of Leguminosae (Lupinus spp., Astragalus spp., and Oxytropis
spp.) are reported by Porsild for the Canadian Arctic islands. Seven
of these are on Baffin Island where several are very abundant in more
southerly areas, yet none are reported to occur in Greenland. Of

TABLE 2.2.   Index of summer warmth for some sites   located on Fig. 2.9
(from Young, 1971)

Site	Index	Zone
Baffin Island		
Pond Inlet	14	3 (2)
Clyde River	12	3 (2)
Padloping Island	16	3
Frobisher Bay	22	4 (3)
Greenland		
Thule	12	3 (2)
Upernavik	18	3
Holsteinburg	32	4
Godhavn	24	4

thirteen arctic willows (Salix spp.), ten are present in Baffin Island,
but only five of these are found in Greenland.

The distribution of birch is of particular interest, since Betula
spp. are recognized as indicators of different floristic and climatic
zones (Young, 1971; Andrews, et al., 1980). Arborescent birch (B.
pubescens) occurs in scattered areas in extreme southern Greenland,
attaining heights of as much as 12 meters; this species has provided
the basis for dendrochronological studies in that region (Kuivinen and
Lawson, 1982). The presence of standing trees might be interpreted
as an indication of subarctic conditions; however, Polunin (1960, p.
390) has argued that the presence of these birch "forests" in southern
Greenland is more a consequence of microhabitat factors than of
regional climate. It should be noted, though, that another subarctic
species, the green alder (Alnus crispa), common in mainland Canada but
not found in the Arctic islands, also occurs in southern Greenland.

No arborescent species of any kind occur on Baffin Island,
although both of the dwarf arctic birches, B. glandulosa and B. nana,
do (Fig. 2.12). The former occurs throughout the arctic and subarctic
mainland of Canada but is restricted in the islands to extreme southern
Baffin Island and a few sites in the western Arctic archipelago. B.
glandulosa occurs in southern Greenland, coinciding approximately with

the distribution of A. crispa (Fig. 2.12). The other dwarf birch, B.
nana, appears to be more of a "true" arctic species in that it is
widespread in western and eastern Greenland, beyond the limits of B.
glandulosa. In Canada, it appears only in the Cumberland Peninsula of
easternmost Baffin Island (Fig. 2.12).

Such anomalies in present-day distributions are intriguing and
raise questions of relative effectiveness in the dispersal and
reestablishment rates of some species following deglaciation, as well
as of the possible survival of relict populations in unglacierized
coastal areas during a glacial maximum (Ives, 1974). Where the
postglacial thermal maximum was succeeded by neoglacial conditions, as
was the case in Baffin Island, some species might not have become
reestablished because the period between deglacierization and the onset
of the neoglacial was too short.

While attempting to extend the mapping of Betula in southern
Baffin Island, Jacobs and Mode (unpublished) found a discontinuity in
the distribution of B. glandulosa at the head of Frobisher Bay which
coincides approximately with the innermost Cockburn moraine, an area
which was deglacierized at about 6750 BP, (Blake, 1966). This
discontinuity cannot be explained by present-day climatic or other
local site conditions. It remains for ongoing palynological studies to
reveal whether or not Betula ever did become established in this inner
zone, retreating during the Neoglacial as happened elsewhere in the
area (Short and Jacobs, 1982), or whether the timing of
deglacierization and emergence was such that terrain conditions did not
become suitable for Betula until after climatic conditions had already
deteriorated.

These observations indicate that the distribution of plant species
in this and other Arctic regions is perhaps as much a function of past
changes in habitat conditions as it is of present-day climate.

Marine Fauna. Reference has been made earlier in this chapter to the
arctic/subarctic boundary in Davis Strait and Baffin Bay as devined by
Dunbar (1954), and its relationship to certain benthic marine fauna.
Setting aside for the present the question of the effects of recent
climatic fluctuations on the marine fauna (discussed by Dunbar, 1954,
1976; Vibe, 1967), we wish here to consider significant aspects of some
present distributions.

Clark (1974) has examined the deep water marine molluscs of Baffin
Bay in relation to North Atlantic distributions and concluded that the

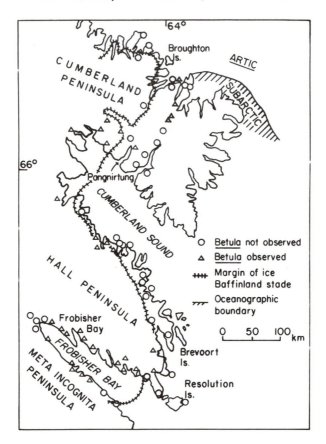

Figure 2.12 Location of _Betula_ spp. in southeastern Baffin Island
(insets from Porsild, 1964).

only pronounced zoogeographic boundary is that between Iceland and East
Greenland. He argues that the arctic/subarctic distinction is an
imprecise one in this context; the molluscan fauna at Baffin Bay being
mainly an extension of that of the Labrador Sea, but with fewer
species. The problem is complicated by the presence of what appears to
Clarke to be "antique" endemic mollusc species (e.g. _Colus Krampi_ and
Acrybia glacialis) at depths in Baffin Bay below the 600 m sill depth
of Davis Strait. Andrews et al. (1980) have noted some ambiguities in
present distributions of subarctic littoral mussels in the Baffin
region which serve to alert us to the combined effects of endemism,
relict populations, and different rates and routes of migration in
compounding the problem of interpreting present environmental
conditions, and of course of interpreting the paleoenvironmental record
as well, since mussels are among the more enduring Quaternary fossils.

In western Greenland it was suggested by Madsen (1940) that the marine subarctic/arctic boundary is defined by the northern limit for shallow water littoral faunas, dominated here by Mytilus edulis, Littorina saxatile, and Balanus balanoides. However, later work has shown that the components of this fauna are subject to short-term fluctuations over several degrees of latitude, as a response to year-to-year fluctuations in the volume of warm Atlantic water brought into Baffin Bay (Theisen, 1973). Hence this important biogeographical boundary must by nature be vaguely defined. Earlier in this chapter, it was pointed out that the northern limit of dwarf birch on Baffin Island is a continuation on land of this boundary. There is, however, in Greenland no correlation between any of the critical marine organisms and the dwarf birches.

A more mobile and thus more responsive climatic indicator than the molluscs is the Atlantic cod (Gadus morhus). Clearly an Atlantic if not subarctic species, this cod, when it appeared in West Greenland waters in harvestable quantities c. 1910, heralded a warmer marine climate (Vibe, 1976), a flourishing fishery in Greenland and hopes of one in Ungava until its retreat with subsequent cooling in the 1960s. A relict population of this species, land-locked in isostatically uplifted Ogak Lake on Frobisher Bay, bears testimony to an earlier "Atlantic" period in southern Baffin Island.

The marine mammals are highly responsive to environmental change, and their present distributions reflect to a large degree the present marine climate. Two seal species serve as an example: the common or ringed seal (Phoca hispada) and the harp seal (Pagophilus groenlandicus). The former is a year-round arctic resident, favouring areas of extensive annual fast ice where it maintains breathing holes during the winter. The harp seal, on the other hand, winters off Newfoundland and in the Gulf of St. Lawrence, migrating northward with the retreating ice in spring into Baffin Bay and the Arctic Archipelago.

Vibe (1967) has studied harvests of both species in Greenland and the pattern that emerges is one in which ringed seal numbers increase in more southerly regions during periods of more severe ice conditions, possibly declining as well in more northerly areas where the ice is heaviest. At the same time, the harp seal numbers decrease in areas of late-lasting ice. The pattern in eastern Baffin Island is similar: severe ice years see an extension of ringed seal habitat and diminution of harp seal numbers (Jacobs, 1975). Looking at both areas together, cne would expect that in the mildest years ringed seals would be present in eastern Baffin Island but not western Greenland, while in

the most severe ice years, they would become more numerous toward the
Greenland side, possibly declining near Baffin Island where the ice
becomes heaviest.

Both Greenlandic and Canadian observations on polar bear numbers
indicate that this marine mammal behaves with respect to ice severity
in a manner similar to the ringed seal, its principal prey. The
densest concentrations of bears in the eastern Canadian Arctic are thus
found off the headlands of Cumberland, Hall, and Meta Incognita
peninsulas of Baffin Island (Stirling et al., 1980), all of which are
areas of converging drift ice and large seal populations.

REFERENCES

Aksu, A., 1981: Late Quaternary stratigraphy, paleoenvironmentology and
    sedimentation history of Baffin Bay and Davis Strait. PhD thesis,
    Dalhousie University, Halifax. 771 pp.
Albertsson, K.J., 1978: Um aldur gajflaga a Tjörnsi. [Some notes on the
    Age of the Tjörnes Strata Sequence, Northern Iceland].
    Natturugraedingurinn 48:1-8.
Allaart, J.H., 1973: Descriptive text to geological map of Greenland
    1:1,000,000 Julianehab 60V2 Nord. Gronlands geol. Unders, Copenhagen.
    41 pp.
Alt, B.T., 1978: Synoptic climate controls of mass-balance variations
    on Devon Island ice cap. Arctic and Alpine Research, 10:61-80.
Ambrose, J.W., 1964: Exhumed paleoplains of the Precambrian shield of
    North Americ. American Journal Science, 262:817-857.
Anderson, L.W., 1978: Cirque glacier erosion rates and characteristics
    of neoglacial tills, Pangnirtung Fiord area, Baffin Island, N.W.T.,
    Canada. Arctic and Alpine Research, 10:749-760.
Andrews, J.T., 1972: Glacial power, mass balances, velocities, and
    erosion potential. Zeitschrift Geomorphologi. Supplement, Bd. 13,
    1-17.
Andrews, J.T., 1974: Cainozoic glaciations and crustal movements of the
    Arctic. In: Ives, J.D. and Barry, J.D. (eds.), Arctic and Alpine
    Environments. Methuen, London, 277-317.
Andrews, J.T., 1978: Sea level history of arctic coasts during the
    Upper Quaternary. Progress in Physical Geography, 2:375-407.
Andrews, J.T., 1982: Comments on "New evidence from beneath the western
    North Atlantic for the depth of glacial erosion in Greenland and
    North America" by E.P. Laine. Quaternary Research, 17:123-124.
Andrews, J.T., et al., 1980. Report on the distribution of dwarf
    birches and present pollen rain, Baffin Island, N.W.T., Canada.
    Arctic, 33:50-58.
Andrews, J.T., et al., 1981. Quaternary near-shore environments on
    eastern Baffin Island, N.W.T. In: Mahaney, W.C. (ed.), Quaternary
    Paleoclimate. GeoAbstracts, Norwich, UK, 13-44.
Andrews, J.T. and Barry, R.G., 1972: Present and paleo-climatic
    influences on the glacierization of Cumberland Peninsula, Baffin
    Island, N.W.T., Canada. INSTAAR Occasional Paper No. 2, University of
    Colorado, Boulder. 245 pp.
Andrews, J.T., Funder, S., Hjort, C. and Imbrie, J., 1974: Comparison
    of the glacial chronology of eastern Baffin Island, East Greenland,
    and the Camp Century accumulation record. Geology, 2:355-358.

Andrews, J.T. Guennel, G.K., Wray, J.L. and Ives, J.D., 1972: An early Tertiary outcrop in north central Baffin Island, Northwest Territories, Canada: environment and significance. Canadian Journal of Earth Sciences, 9:233-238.

Andrews, J.T. and Matsch, C.L., 1983: Glacial marine sediments and sedimentation. An annotated bibliography. GeoAbstracts, Norwich, UK. 227 pp.

Andrews, J.T. and Miller, G.H., 1972: Maps of the present Glaciation Limit and lowest equilibrium line altitude for north and south Baffin Island. Arctic and Alpine Research, 4:45-60.

Bader, H., 1961: The Greenland ice sheet. Rep. Cold Reg. Res. Engng Lab. I-B2. 18 pp.

Baird, P.D., 1952: The glaciological studies of the Baffin Island Expedition, 1950, part I. Method of nourishment of the Barnes Ice Cap. Journal of Glaciology 2(11):2-9.

Banfield, A.W.F., 1954: The role of ice in the distribution of mammals. Journal of Mammalogy, 35(1):104-107.

Banfield, A.W.F., 1961: A revision of the reindeer and caribou, genus Rangifer. National Museums of Canada Bulletin, No. 177.

Barry, R.G., 1967: Seasonal location of the Arctic Front over North Americ. Geographical Bulletin, 9:79-95.

Barry, R.G., 1974: Further climatological studies of Baffin Island. Inland Waters Directorate, Environment Canada, Technical Bulletin, 65. 54 pp.

Barry, R.G., Bradley, R.S., and Jacobs, J.D., 1975: Synoptic climatological studies of the Baffin Island area. In: Weller, G. and Bowling, S (eds.), Climate of the Arctic. Geophysical Institute, University of Alaska, Fairbanks, 82-90.

Barry, R.G. and Hare, F.K., 1975: Arctic climate. In: Ives, J.D. and Barry, R.G. (eds.), Arctic and Alpine Environments. Methuen, London, 17-54.

Barry, R.G. and Jacobs, J.D. (eds.), 1978: Energy budget studies in relation to fast-ice breakup processes in Davis Strait: Climatological Overview. Institute of Arctic and Alpine Research, Occasional Paper, 26. 284 pp.

Barry, R.G., 1980. Meteorology and climatology of the seasonal sea ice zone. Cold Regions Science and Technology, 2:133-150.

Basham, P.W., Forsyth, D.A., and Wetmiller, R.J., 1977: The seismicity of northern Canada. Canadian Journal of Earth Sciences, 14:1646-1667.

Bauer, A. 1955: The balance of the Greenland ice sheet. Journal of Glaciology 2:456-462.

Benson, C., 1962: Stratigraphic studies in the snow and firn of the Greenland ice sheet. Res. Rep. Cold Reg. Res. Engng Lab. 70. 93 pp.

Beschel, R.E. and Weidick, A., 1973: Geobotanical and geomorphological reconnaissance in west Greenland, 1961. Arctic and Alpine Research, 5(4):311-319.

Berggren, W.A., 1972: Late Pliocene/Pleistocene glaciation. Chapter 13, In: Initial Reports of the Deep Sea Drilling Project. Volume XII. Scripps Institute of Oceanography, University of California, 953-964.

Bird, J.B., 1967: The physiography of Arctic Canada. John Hopkins University Press, Baltimore. 336 pp.

Birkenmajer, K., 1972: Report on investigations of Tertiary sediments at Kap Brewster, Scoresby Sund, East Greenland. Rapp. Gronlands geol. Unders. 48:85-91.

Blake, W., Jr. 1966: End moraines and deglaciation chronology in northern Canada, with special reference to southern Baffin Island. Canadian Geological Survey, Paper 66-26, 31 pp.

Boellstorff, J.D., 1978: North American Pleistocene stages reconsidered in light of probably Pliocene/Pleistocene continental glaciation. Science, 202:305-307.

Bradley, R.C., 1973: Seasonal climatic fluctuations on Baffin Island
    during the period of instrumental records. Arctic, 26(3): 230-243.
Bridgewater, D., Watson, J., and Windley, B.F., 1973: The Archaean
    craton of the North Atlantic region. Phil. Trans. R. Soc. Lond. A273:
    493-512.
Bryson, R.A., 1966. Air masses, streamlines, and the boreal forest.
    Geographical Bulletin, 8(3):228-269.

Clarke, A.H., 1974: Molluscs of Baffin Bay and the northern North
    Atlantic Ocean. National Museums of Canada Publications in Biological
    Oceanography No. 7. 23 pp.
Clark, D.B. and Pedersen, A.K., 1976: Tertiary volcanic province of
    West Greenland. In: Escher, A. and Watt, W.S. (eds.), Geology of
    Greenland. Copenhagen, Geological Survey of Greenland, 152-182.
Clarke, D.B. and Upton, B.G., 1971: Tertiary basalts of Baffin Island:
    field relations and tectonic setting. Canadian Journal of Earth
    Sciences, 8:248-258.
Collin, A.E. and Dunbar, M.J., 1964. Physical oceanography in Arctic
    Canada. Oceanography and Marine Biology Annual Review, 2, 45-75.
Crane, R.G. 1978: Seasonal variations of sea ice extent in the Davis
    Strait-Labrador Sea area and relationships with synoptic scale
    atmospheric circulation. Arctic, 31:434-447.
Crane, R.G., 1979. Synoptic controls on the energy budget regime of an
    ablating fast ice surface. Archiv fuer Meteorologie, Geophysik und
    Bioklimatologie, Serie A, 28:53-70.
Barry, R.G., Bradley, R.S., and Jacobs, J.D., 1975: Synoptic
    climatological studies of the Baffin Island area. In: Weller, G. and
    Bowling, S (eds.), Climate of the Arctic. Geophysical Institute,
    University of Alaska, Fairbanks, 82-90.
Barry, R.G. and Hare, F.K., 1975: Arctic climate. In: Ives, J.D. and
    Barry, R.G. (eds.), Arctic and Alpine Environments. Methuen, London,
    17-54.
Barry, R.G. and Jacobs, J.D. (eds.), 1978: Energy budget studies in
    relation to fast-ice breakup processes in Davis Strait:
    Climatological Overview. Institute of Arctic and Alpine Research,
    Occasional Paper, 26. 284 pp.
Barry, R.G., 1980. Meteorology and climatology of the seasonal sea ice
    zone. Cold Regions Science and Technology, 2:133-150.
Basham, P.W., Forsyth, D.A., and Wetmiller, R.J., 1977: The seismicity
    of northern Canada. Canadian Journal of Earth Sciences, 14:1646-1667.
Bauer, A. 1955: The balance of the Greenland ice sheet. Journal of
    Glaciology 2:456-462.
Benson, C., 1962: Stratigraphic studies in the snow and firn of the
    Greenland ice sheet. Res. Rep. Cold Reg. Res. Engng Lab. 70. 93 pp.
Beschel, R.E. and Weidick, A., 1973: Geobotanical and geomorphological
    reconnaissance in west Greenland, 1961. Arctic and Alpine Research,
    5(4):311-319.
Berggren, W.A., 1972: Late Pliocene/Pleistocene glaciation. Chapter 13,
    In: Initial Reports of the Deep Sea Drilling Project. Volume XII.
    Scripps Institute of Oceanography, University of California, 953-964.
Bird, J.B., 1967: The physiography of Arctic Canada. John Hopkins
    University Press, Baltimore. 336 pp.
Birkenmajer, K., 1972: Report on investigations of Tertiary sediments
    at Kap Brewster, Scoresby Sund, East Greenland. Rapp. Gronlands
    geol. Unders. 48:85-91.
Blake, W., Jr. 1966: End moraines and deglaciation chronology in
    northern Canada, with special reference to southern Baffin Island.
    Canadian Geological Survey, Paper 66-26, 31 pp.
Boellstorff, J.D., 1978: North American Pleistocene stages reconsidered
    in light of probably Pliocene/Pleistocene continental glaciation.
    Science, 202:305-307.

Bradley, R.C., 1973: Seasonal climatic fluctuations on Baffin Island during the period of instrumental records. Arctic, 26(3): 230-243.

Bridgewater, D., Watson, J., and Windley, B.F., 1973: The Archaean craton of the North Atlantic region. Phil. Trans. R. Soc. Lond. A273: 493-512.

Bryson, R.A., 1966. Air masses, streamlines, and the boreal forest. Geographical Bulletin, 8(3):228-269.

Clarke, A.H., 1974: Molluscs of Baffin Bay and the northern North Atlantic Ocean. National Museums of Canada Publications in Biological Oceanography No. 7. 23 pp.

Clark, D.B. and Pedersen, A.K., 1976: Tertiary volcanic province of West Greenland. In: Escher, A. and Watt, W.S. (eds.), Geology of Greenland. Copenhagen, Geological Survey of Greenland, 152-182.

Clarke, D.B. and Upton, B.G., 1971: Tertiary basalts of Baffin Island: field relations and tectonic setting. Canadian Journal of Earth Sciences, 8:248-258.

Collin, A.E. and Dunbar, M.J., 1964. Physical oceanography in Arctic Canada. Oceanography and Marine Biology Annual Review, 2, 45-75.

Crane, R.G. 1978: Seasonal variations of sea ice extent in the Davis Strait-Labrador Sea area and relationships with synoptic scale atmospheric circulation. Arctic, 31:434-447.

Crane, R.G., 1979. Synoptic controls on the energy budget regime of an ablating fast ice surface. Archiv fuer Meteorologie, Geophysik und Bioklimatologie, Serie A, 28:53-70.

Dawes, P.R., Frisch, T. and Christie, R.L., 1982: The Proterozoic Thule basin of Greenland and Ellesmere Island: importance to the Nares Strait debate. Meddr Gronland, GeoScience 8:89-104.

Deep Sea Drilling Project, 1972: Initial Reports of the Deep Sea Drilling Project, Volume XII. Scripps Institute of Oceanography, University of California. 1215 pp.

Denham, L.R., 1974: Offshore geology of northern West Greenland (69° to 75°N). Rapp. Gronlands geol. Unders. 63. 24 pp.

Dey, B., 1980a. Applications of satellite thermal infrared images for monitoring North Water during the periods of polar darkness. Journal of Glaciology, 25:425-438.

Dey, B., 1980b. Seasonal and annual variations in ice cover in Baffin Bay and northern Davis Strait. Canadian Geographer. 24(4):368-384.

Dey, B., Moore, H., and Gregory, A.F., 1979. Monitoring and mapping sea-ice breakup and freezeup in Arctic Canada from satellite imagery. Arctic and Alpine Research, 11:229-242.

Dowdeswell, J.A., Osterman, L.E., and Andrews, J.T., in press: SEM and other criteria for distinguishing glacial and non-glacial events in a marine core from Frobisher Bay, N.W.T., Canada. Sedimentology.

Dunbar, M., 1972. Increasing severity of ice conditions in Baffin Bay and Davis Strait. In: Karlsson, T., (ed.), Sea Ice-Proceedings of an International Conference, Rekjavik, Iceland, May 10-13, 1971, Research Council, Reykjavik), 87-93.

Dunbar, M. and Dunbar, M.J., 1972. The history of the North Water. Proc. Royal Soc. Edinburgh, Ser. B., 72: 231-242.

Dunbar, M.J., 1955: The present status of climatic change in the Atlantic sector of Northern seas, with special reference to Canadian eastern arctic waters. Transactions Royal Society of Canada XLIX, Series III: 10-17.

Dunbar, M.J., 1960: Preliminary report on the Bering Strait scheme. NCRC 60-1, Ottawa, Department Norther Affairs and Natural Resources. 14 pp.

Dunbar, M.J., 1976: Climatic change and northern development. Arctic, 29(4):183-193.

Dyke, A.S., Andrews, J.T., and Miller, G.H., 1982: Quaternary geology
    of Cumberland Peninsula, Baffin Island, District of Franklin.
    Geological Survey of Canada Memoir, 403. 32 pp.

Fan, Pow foog and Zemmels, I., 1972: X-ray mineralogy studies.
    Chapter 19, In: Initial Reports of the Deep Sea Drilling Project,
    Volume XII. Scripps Institute of Oceanography, University of
    California, 1127-1154.
Feyling-Hanssen, R.W., 1976: The stratigraphy of the Quaternary Clyde
    Foreland Formation, Baffin Island, illustrated by the distribution of
    benthic foraminifera. Boreas, 5:77-94.
Feyling-Hanssen, R.W., Funder, S., and Petersen, K. Strand, in press:
    The Lodin Elv Formation: a Plio-Pleistocene occurrence on Jameson
    Land, Greenland. Bull. Geol. Soc. Denmark.
Fletcher, R.J., 1975. A guide to current published meteorological data
    for northern lands. Polar Record, 17:495-519.

Gilbert, R., 1978: Observations on oceanography and sedimentation at
    Pangnirtung Fiord, Baffin Island. Maritime Sediments, 14:109.
Gilbert, R., 1982: Contemporary sedimentary environments on Baffin
    Island, N.W.T., Canada: Glaciomarine processes in fiords of eastern
    Cumberland Peninsula. Arctic and Alpine Research, 14:1-12.
Gradstein, F.M. and Srivastave, S.P., 1980: Aspects of the Cenozoic
    stratigraphy and paleoceanography of the Labrador Sea and Baffin
    Bay. Palaeogeogr., palaeoclimatol., palaeoecol. 30:261-295.
Grant, A.C., 1975: Geophysical results from the continental margin off
    southern Baffin Island. In: Yorath, C.J., Parker, E.R., and Glass,
    D.J. (eds.), Canada's continental margins and offshore petroleum
    exploration. Canadian Society of Petroleum Geologists, Memoir 6,
    411-431.
Grant, A.C. and Manchester, K.S., 1970: Geophysical investigation in
    the Ungava Bay - Hudson Strait region of northern Canada. Canadian
    Journal of Earth Sciences, 7:1062-1076.

Hare, F.K. and Hay, J.E., 1974: The climate of Canada and Alaska. In:
    Bryson, R. and Hare, F.K. (eds.), World Survey of Climatology, Vol.
    11. Elsevier, Amsterdam.
Hare, F.K., 1968: The Arctic. Quarterly Journal of the Royal
    Meteorological Society, 94(402):439-459.
Hassan, M.Y., 1953: Tertiary faunas from Kap Brewster, East Greenland.
    Meddr Grønland 111: 5. 42 pp.
Hattersley-Smith, G., 1974: Present arctic ice cover. In: Ives, J.D.
    and Barry, R.G. (eds.), Arctic and Alpine Environments.
Henderson, G., 1975: New bathymetric Maps Covering Offshore West
    Greenland 59°-69°30'N. Offshore Technology Conference Dallas,
    Texas. OTC paper 2223, 761-772.
Henderson, G., Schiener, E.J., Risum, J.B., Croxton, C.A. and Andersen,
    B.B., 1981: In: Kerr, J.W. and Ferguson, A.J. (eds.), Geology of the
    North Atlantic Boderlands. Mem. Can. Soc. Petrol. Geol. 7:399-428.
Holdsworth, G., 1973: Evidence of a surge on Barnes Ice Cap, Baffin
    Island. Canadian Journal of Earth Sciences, 10(10):1565-1574.
Hooke, R.LeB., 1973: Structure and flow in the margin of the Barnes Ice
    Cap, Baffin Island, N.W.T., Canada. Journal of Glaciology,
    12(66):423-438.

Ives, J.D., Andrews, J.T., and Barry, R.G., 1975: Growth and decay of
    the Laurentide Ice Sheet and comparison with Fennoscandia.
    Naturwissenschaften, 62:118-125.

Jacobs, J.D., 1975: Some aspects of the Eskimo community at Broughton Island, N.W.T., Canada, in relation to climatic conditions. Arctic and Alpine Research, 7(1):69-75.

Jacobs, J.D. and Andrews, J.T., 1983: A note on solar radiation measurments on Cumberland Peninsula, Baffin Island, N.W.T., Canada in Summer, 1970-1979. Arctic and Alpine Research, 15(1):91-96.

Jacobs, J.D., Andrews, J.T., Barry, R.G., Weaver, R.L., and Williams, R.D., 1973: Glaciological and meteorological studies on the Boas Glacier for two contrasting seasons (1960-70 and 1970-71). In: The role of snow and ice in hydrology, Proceedings of the Unesco International Symposium, 4:1-12.

Jacobs, J.D., Barry, R.G., and Weaver, R.L., 1975. Fast ice characteristics, with special reference to the esatern Canadian arctic. Polar Record, 17(110):521-536.

Jacobs, J.D. and Leung, C.Y., 1980. Paleoclimatic implications of topoclimatic diversity in Arctic Canada. In: Mahaney, W. (ed.), Quaternary Paleoclimate. Geoabstracts, Norwich, 63-76.

Jacobs, J.D. and Newell, J.P., 1979: Recent year-to-year variations in seasonal temperatures and sea ice conditions in the eastern Canadian Arctic. Arctic, 32(4):345-354.

Kalbeek, F., 1982: The evolution of the Precambrian sheld of Greenland. Geol. Rundschau 71:38-60.

Keen, R.A., 1977. The response of Baffin Bay ice conditions to changes in atmospheric circulation patterns. In: Mahaney, W. (ed.), Fourth International Conference on Port and Ocean Engineering under Arctic Conditions Memorial University, Vol. II. St. Johns, Newfoundland, 963-967.

Kerr, J. W^m., 1980: Structural framework of Lancaster Aulacogen, Arctic Canada: Geological Survey of Canada, Bulletin 319,

Koerner, R.M., 1970: The mass balance of Devon Island ice cap Northwest Territories. Journal Glaciology, 9:325-336.

Koerner, R.M., 1977: Devon Island ice cap: core stratigraphy and paleoclimate. Science 196:15-18.

Kuivinen, K.C. and Lawson, M.P., 1982: Dendroclimatic analysis of birch in South Greenland. Arctic and Alpine Research, 14(3): 243-250.

Laine, E.P., 1980: New evidence from beneath the western North Atlantic for depth of glacial erosion in Greenland and North America. Quaternary Research, 14:188-198.

LeDrew, E.F., 1979. Physical mechanisms responsible for the major synoptic systems in the eastern Canadian Arctic in the winter and summer of 1973. Institute of Arctic and Alpine Research, Occasional Paper, 22. 205 pp.

LePichon, X., Houtz, R.E., Drake, C.L. and Nafe, J.E., 1971: Crustal structure of the mid-ocean ridge. I. Seismic refraction measurements: Journal Geophysical Research, 70:319-339.

Locke, C. and Locke, W.W., III., 1977: Little Ice Age now-cover extent and paleoglacierization thresholds: north-central Baffin Island, N.W.T., Canada. Arctic and Alpine Research, 9:291-300.

Löken, O.H. and Andrews, J.T., 1966. Glaciological and chronology of fluctuations of the ice margin at the south end of the Barnes Ice Cap, Baffin Island, N.W.T., Geographical Bulletin, 8:341-359.

Lubinsky, I., 1972: The marine bivalve molluscs of the Canadian Arctic, Ph.D. thesis, McGill Univeristy, Montreal. 318 pp.

MacLean, B, and Falconer, R.K.H., 1979: Geological/geophysical studies in Baffin Bay and Scott Inlet-Buchan Gulf and Cape Dyer-Cumberland Sound areas of the Baffin Island shelf. In: Current Research, Part B. Geological Survey of Canada, Paper 79-1B, 231-244.

MacLean, B., Jansa, L.F., Falconer, R.K.H., and Srivastava, S.P., 1977: Ordovician strata on the southeastern Baffin Island shelf revealed by shallow drilling. Canadian Journal of Earch Sciences, 15:1925-1939.

Macpherson, E., 1971: The marine molluscs of Arctic Canada: Prosobranch Gastropods, Chitons, and Scaphopods. National Museums of Canada, Publications in Biological Oceanography No. 3. 149 pp.

Madsen, H., 1940: A study of the littoral fauna of northwest Greenland. Meddr Grönland, 124:3, 24 pp.

Markham, W.E., 1981: Ice Atlas: Canadian Arctic Waterways, Atmospheric Environment Service, Toronto, 199 pp.

Maxwell, J.B., 1980. The climate of the Canadian Arctic Islands and Adjacent Waters. Climatological Studies, 30. Atmospheric Environment Service, Toronto. 532 pp.

Maxwell, J.B., 1981. Climatic regions of the Canadian Arctic Islands. Arctic, 34(3):225-240.

Miller, G.H., Andrews, J.T., and Short, S.K., 1977: The last interglacial/glacial cycle, Clyde Foreland, Baffin Island, N.W.T.: stratigraphy, biostratigraphy, and chronology. Canadian Journal of Earth Sciences, 14:2824-2857.

Miller, G.H., Bradley, R.S., and Andrews, J.T., 1975: The Glaciation Level and lowest equilibrium line altitude in the High Canadian Arctic: maps and climatic interpretation. Arctic and Alpine Research, 7:155-168.

Miller, G.H. and Williams, L.D., 1974: Late Wisconsin paleoclimate derived from a snowmelt program and variations in glacier response: eastern Baffin Island. Geological Survey of America, Abstracts, 6:870.

Mode, W.N., 1980: Quaternary stratigraphy and palynology of the Clyde Foreland, Baffin Island, N.W.T., Canada. Ph.D. dissertation, University of Colorado, Boulder. 219 pp.

Muench, R.D., 1971: The Baffin Bay-North Water Project. The physical oceanography of the northern Baffin Bay region. The Baffin Bay-North Water Project, Science Report #1. 152 pp.

Muller, F., Stauffer, B., and Schriber, G., 1975: Isotope measurements and firm stratigraphy on ice caps surounding the North Water polynya. International Association Scientific Hydrology, Publication No. 118, 188-194.

Muller, F., Ohmara, A., and Braithwaite, R., 1973: The North Water Project 9 Canadian-Greenalnd Arctic), Geographica Helvetica, No. 2: 111-117. (Reprinted in Polar Geography, 1 [1977]: 75-85).

Nelson, A.R., 1978: Quaternary glacial and marine stratigraphy of the Qivitu Peninsula, northern Cumberland Peninsula, Baffin Island, Canada. Ph.D. dissertation, University of Colorado, Boulder. 215 pp.

Nelson, A.R., 1982: Aminostratigraphy of Quaternary marine and glaciomarine sediments, Qivitu Peninsula, Baffin Island. Canadian Journal Earth Sciences, 19:945-961.

Orvig, S., 1954. Glacial-meteorological observations on ice caps on Baffin Island. Geografiska Annaler, 36(3):197-318.

Orvig, S., 1970: Climates of the Polar Regions. Vol. 14 of World Survey of Climatology, ed., H.E. Landsberg. Elsevier, Amsterdam, 370 pp.

Osterman, L.E., 1982: Late Quaternary history of southern Baffin Island, Canada: a study of foraminifera and sediments from Frobisher Bay. Ph.D. dissertation, University of Colorado, Boulder. 380 pp.

Ostrem, G., Bridge, C.W., and Rannie, W.F., 1967: Glacio-hydrology, discharge and sediment transport in the Decade Glacier area, Baffin Island N.W.T. Geografiska Annaler, 49A (2-4):268-282.

Partl, R., 1977: Power from glaciers: the hydropower potential of Greenland's glacial waters. International Institute for Applied Systems Analysis, Laxenburg, Austria, Report RR-77-20. 54 pp.

Paterson, W.S.B. et al., 1977: An oxygen-isotope climatic record from the Devon Island ice cap. Nature, 266:508-511.

Piper, D.J.W., 1973: A late Quaternary stratigraphic marker in the central basin of Baffin Bay. Maritime Sediments, 9:62-63.

Polunin, N., 1940-48: Botany of Canadian Arctic Part I (1940), Part II (1947) and Part III (1948). National Museum of Canada, Ottawa.

Polunin, N., 1959: Circumpolar Arctic Flora. Oxford University Press.

Polunin, N., 1960: Introduction to Plant Geography. Longmans, London. 640 pp.

Porsild, A.E., 1964: Illustrated Flora of the Canadian arctic Archipelago (2nd ed., revised). National Museum of Canada, Ottawa. 218 pp.

Putnins, P., 1970. The climate of Greenland. In: Orvig, S. (ed.), Climates of the Polar Regions. Vol. 14 of World Survey of Climatology. Elsevier, Amsterdam. 370 pp.

Radok, U., Barry, R.G., Jenssen, D., Keen, R.A., Kiladis, G.N., and McInnes, B., 1982: Climatic and physical characteristics of the Greenland Ice Sheet. C.I.R.E.S., University of Colorado, Boulder. 193 pp.

Rogerson, R., 1982: personal communication. Lecture given at the 11th Annual Arctic Workshop, Boulder, Colorado.

Ruddiman, W.F. and McIntyre, I., 1973: Time-transgressive deglacial retreat of Polar waters from the North Atlantic. Quaternary Research, 3:117-130.

Ruddiman, W.F. and McIntyre, A., 1981: The mode and mechanism of the last deglaciation: oceanic evidence. Quaternary Research, 16:125-134.

Sager, R.B., 1966: Glaciological and climatological studies on the Barnes Ice Cap, 1962-64. Geogr. Bull. 8 (1):3-47.

Sanford, B.V., Grant, A.C., Wade, J.A., and Barss, M.S., 1979: Geology of Eastern Canada and adjacent areas: Geological Survey of Canada, Map 1401A (4 sheets).

Schledermann, P., 1980. Polynyas and prehistoric settlement patterns. Arctic, 33:292-302.

Short, S.K. and Jacobs, J.D., 1982: A 1100 year paleoclimatic record from Burton Bay-Tarr Inlet, Baffin Island. Canadian Journal of Earth Science, 19(3):398-409.

Smith, T.G., 1973: Population dynamics of the ringed seal in the Canadian Eastern Arctic. Fisheries Research Board of Canada, Bulletin No. 181, 55 p.

Speerschneider, C.I.H., 1931: The state of ice in Davis Strait 1820-1930 Det Danske Meteor. Inst. Medd. Nr. 8, Copenhagen. 51 pp.

Srivastava, S.P., Falconer, R.K.H., and MacLean, B., 1981: Labrador Sea, Davis Strait, Baffin Bay: geology and geophysics - a review. In: Kerr, J. and Ferguson, A.J. (eds.), Geology of the North Atlantic Boderlands. Mem. Can. Soc. Petrol. Geol. 7:333-399.

Stirling, I., 1980: The biological importance of polynyas in the Canadian Arctic. Arctic, 33:303-315.

Stirling, I., Calvert, W., and Andriakshek, D., 1980: Population ecology studies of the polar bear in the area of southeastern Baffin Island. Canadian Wildlife Service, Occasional Paper No 44. 33 pp.

Sugden, D.E., 1978: Glacial erosion by the Laurentide ice sheet. Journal of Glaciology, 20:367-392.

Szabo, B.J., Miller, G.H., Andrews, J.T., and Stuiver, M., 1981: Comparison of uranium-series, radiocarbon, and amino acid data from marine molluscs, Baffin Island, Arctic Canada. Geology, 9:451-457.

Taylor, C.J., 1981. First International Polar Year. Arctic, 34(4): 370-376.

Theisen, B.F., 1973: The growth of Mytilus edulis L. (Bivalvia) from Disko and Thule District, Greenland. Ophekia 12:59-77.

Thomas, M.K., 1971. A brief history of the meteorological services in Canada, Part 1. Atmosphere, 9(1):3-15.

Thomas, M.K., 1975. Recent climatic fluctuations in Canada. Climatological Studies, 28. Atmospheric Environment Service, Toronto, 92 pp.

Thompson R.D. and Fahey, B.D., 1977. Energy balance and ground thermal regime studies, Broughton Island, Northwest Territories, Canada. Archiv fuer Meteorologie, Geophysik und Bioklimatologie, Serie A, 25, 251-272.

Thunell, R.C. and Belyea, P.R., 1981: Neogene planktonic foraminiferal biogeography of the Atlantic Ocean: A synthesis of DSDP legs 1051. Geological Society of America, Abstracts, 13. 567 pp.

Timofeyev, V.T., 1960: Water masses of the Arctic Basin. Leningrad: Hydrometer. Izd., 190 p. Translated by L.K. Coachman, University of Washington, M61-17-1961.

Tooma, S.G., 1978. Summer circulation patterns, northern Smith Sound. Arctic, 31(2):85-92.

Vibe, Christian, 1967. Arctic animals in relation to climaticfluctuations. Medd. om Grönland, 170(5):1-228.

Vowincke, E. and Orvig, S., 1963: Long-wave radiation and total radiation balance at the surface in the Arctic. Arctic Meteorological Research Group, Publications in Meteorology, 62, (McGill University, Montreal), 53 pp.

Walker, E.R., 1977: Aspects of Oceanography in the Archipelago. Institute of Ocean Science, Note, 3. Institute of Ocean Science, Patricial Bay, British Columbia, 186 pp.

Walmsley, J.L., 1966: Ice cover and surface heat fluxes in Baffin Bay. Publications in Meteorology, 84. Department of Meteorology, McGill University, Montreal, 94 pp.

Ward, H. and Baird, P.D., 1954: Studies in glacier physics on the Penny Ice Cap Baffin Island. Part I. A description of the Penny Ice Cap, its accumulation and ablation. Journal of Glaciology, 2(15):342-355.

Warnke, D.A. and Hansen, M.A., 1977: Sediments of glacial origin in the area of D.S.D.P. Leg 38 (Norwegin: Greenland seas): Preliminary results from sites 336 and 344. Naturforschende Gesellschaft zu Freiburg Breisgau, Berichte, 67:371-392.

Weaver, R.L., 1975: Boas Glacier (Baffin Island, N.W.T., Canada Mass Balance for five budget years - 1969 to 1979. Arctic and Alpine Research, 7(3):279-284.

Weidick, A., 1975: Estimates on the mass balance changes of the Inland Ice since Wisxonsin-Welchsel. Grönlands Geologiske Undersögelse, Rapport, 68, 21 pp.

Weidick, A. and Olesen, O.B., 1980: Hydrological basins in West Greenland. Grönlands Geologiske Undersögelse Rapp. 94, 51 pp.

Part II
GLACIAL LANDFORMS AND SEDIMENTS

3 The patterns of glacial erosion across the eastern Canadian Arctic

J. T. Andrews, P. Clark, and J. A. Stravers

Sugden (1974, 1977, 1978) and Gordon (1979) proposed that glacial erosion beneath former ice sheets produced zones of characteristic landscape assemblages. Using the density of small lakes as a measure of the intensity of glacial scour, Sugden (1978) distinguished the following classes of glacial erosion: areas of light, moderate, or intense areal scour and areas of selective linear erosion. This approach has been studied intensively in the Eastern Canadian Arctic by analysis of small-scale topographic maps (1:250,000 to 1:1,000,000) air photographs (1:50,000 or smaller), and LANDSAT imagery (1:1,000,000) (Sugden, 1978; Müller, 1980; Colvill, 1982).

Sugden (1977) argued that regional patterns of glacial erosion in North America were the product of long-term ice sheet stability (i.e. repeated glaciations on time scales of 10^5 yr), and that boundaries between erosional zones reflect changes in basal thermal regimes. In addition, the symmetry of glacial erosional zones centered over Hudson Bay was emphasized, thus echoing statements by White (1972) that the greatest erosion occurred toward the center of the ice sheet, although Sugden (1976) argued against "deep erosion" advocated by White (1972).

In this chapter we expand on the analysis of glacial landforms from the Eastern Canadian Arctic through an examination of LANDSAT imagery and 1:250,000 topographic maps. Our intent is twofold: (1) to extend Sugden's analysis of the relationship between erosional patterns and basal thermal regimes; and (2) examine additional factors which may influence the character of glacial erosion. Additional factors affecting erosion include the following: elevation, bedrock geology, and influence of local and/or regional ice centers (cf. Andrews and Miller, 1979: Dyke et al., 1982a; Quinlan, this volume).

Larger-scale studies, using similar methods, have been carried out on
southern Baffin Island (Muller, 1980; Colvill, 1982).

A major discussion in the literature concerns the configuration of
the Late Wisconsin Laurentide Ice Sheet (Shilts, 1980; Denton and
Hughes, 1981; Andrews, 1982). In part, this argument questions the
time required to create such features as roches moutenees, striae, and
grooves. Proponents of a single-domed ice sheet centered over Hudson
Bay relate erosional features to ephemeral; late-glacial flow regimes
following collapse of major ice centers (Flint, 1943; Denton and
Hughes, 1981). Those supporting a multi-domed ice sheet argue that:
(1) erosional bedforms such as 20-50 m roches moutennees cannot be
ephemeral and reflect primary ice flow; and (2) that shifts in the
pattern of glacial flow should be documented by crossing striae (cf.
Andrews and Falconer, 1969; Andrews, 1970).

As part of our analysis, ice directional indicators from the
Eastern Canadian Arctic have been plotted (Fig. 3.1). Flow patterns
suggested by this plot are summarized as follows: (1) a zone of radial
flow centered on Amadjauk Lake (cf. Dyke et al., 1982a); and (2) a zone
of flow radiating from a point between Longstaff Bluff and Steensby
Inlet along the coast of western Baffin Island (cf. Ives and Andrews,
1963). Of notable interest is the paucity of flow indicators parallel
to Hudson Strait; such features might be expected if an ice stream
draining the major interior portion of the Laurentide Ice Sheet existed
(Denton and Hughes, 1981). These patterns are discussed further in
relation to other features of glacial erosion and proposed ice sheet
configurations.

DATA COLLECTION AND REDUCTION

Using a latitude and longitude grid at a 30' X 30' interval as a
sampling network, our data set is based on 580 sites located across the
Eastern Canadian Arctic. This includes the region between 60° to 72° N
and 60° to 90° W. Because of the decrease in length of a degree of
longitude toward the pole, the actual size of each grid progressively
decreases from south to north. However, our main purpose was to
estimate average conditions within each grid rectangle. Over the 12°
of latitude, the length of a degree of longitude changes from 25 km to
17 km. Within each rectangle, we evaluated the following variables:
elevation, relative relief, distance from the outer coast of Baffin
Island/Labrador, susceptibility of the bedrock to glacial erosion, a
quantitative measure of lake density, and a qualitative assessment of
the intensity of areal scouring.

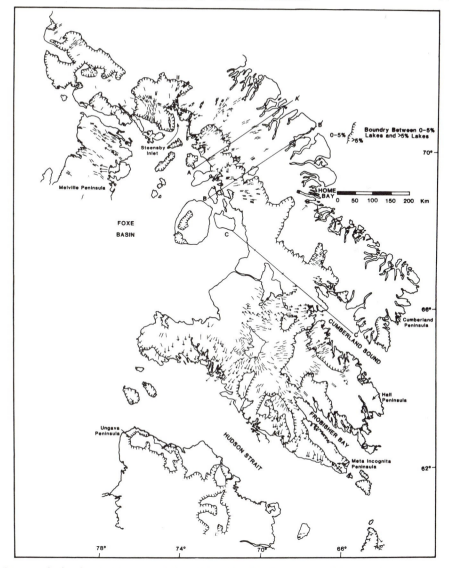

Figure 3.1 Striations and ice flow indicators over the eastern
Canadian Arctic in relationship to the "scoured/non-scoured"
bedrock boundary.

Elevation. Elevations were taken at the center of each grid using
1:1,000,000 scale contour maps (Canadian National Topographic
Series). Contour intervals on this series are 100, 200, 500 and at 500
m above 500 m. The data were reduced to an ordinal series of 4 with
all elevations above 500 m assigned to a single class.

Bedrock. Bedrock geology was abstracted from Douglas (1970) and from
regional maps (e.g. Sanford et al., 1979) and were classified into five
categories according to a subjective assessment of erodibility. These
categories are (from high to low): (1) Paleozoic limestones; (2) low
grade metamorphics; (3) fold belts; (4) amphibolites; and (5) granites
and gneisses. Maps of regional bedrock are presented in Chapters 2 and
6.

Distance. A line was drawn along the outermost coast of eastern Baffin
Island and Labrador and the distance between this line and the center
of each grid was measured. The data were reduced to classes of 100
km. In this analysis, the distance measured is used as a surrogate for
ice thickness. Hollin (1961) concluded that ice sheet profiles are
well approximated by the square root of distance, and perhaps a more
useful category of distance (= ice thickness) would have included
working with transformed data.

Qualitative extent of glacial erosion. The degree and nature of
glacial erosion was assessed from LANDSAT imagery (scale of
1:1,000,000) using the landscape classification scheme proposed by
Sugden (1978). Four units were mapped as areas of: (1) areal scour;
(2) light or no areal scour; (3) selective linear erosion; and (4)
alpine landscapes. Areal scour was further subdivided as heavy or
moderate.

     The lowlands bordering the coast of Foxe Basin west of Amadjuak
and Nettling lakes were excluded from this analysis because of
thermokarst lake development in unconsolidated marine (?) sediments
which mantle the bedrock.

Quantitative measure of areal scour. Lake density was measured within
each grid area using density charts prepared by Terry and Chiligar
(1955). Sugden (1978) restricted lake sizes to diameters of 0.5 to 2
km whereas we have included all lakes visible on black and white
LANDSAT images except for larger lakes such as Amadjuak and Nettling.
One possible problem in using LANDSAT imagery for lake density in the
Arctic is the influence of snow and ice cover on lakes which can result
in an apparent lake density lower than the actual value. This problem
is compensated for by comparing topographic maps (1:250,000) and air
photos with those areas on the imagery obviously affected by snow
cover. Four units of increasing lake density were mapped (Fig. 3.2):
(1) unit A, ranging from 0-5% lake density; (2) unit B (5-10% lake
density); (3) unit C (10-25% lake density); and (4) unit D, the area of
highest lake density with values greater than 25%.

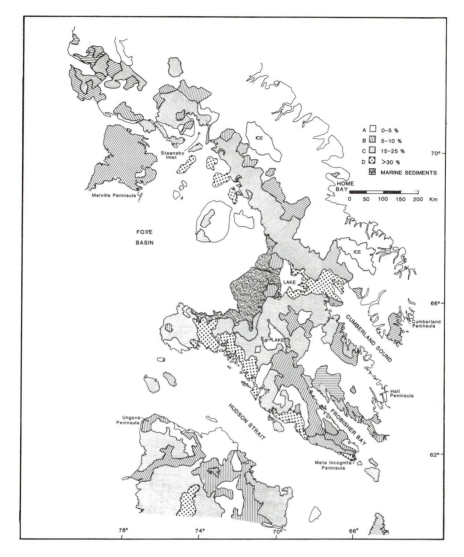

Figure 3.2 Map showing variations in the percentage of lakes across the eastern Canadian Arctic. The map is based on an analysis of LANDSAT imagery and 1:500,000 topography.

Relative Relief. The relative relief for each grid square was measured from 1:500,000 scale topographic maps with contour intervals of 152 m (500 ft.) Several 1:250,000 scale topographic maps were used in areas where higher resolution was needed in order to measure accurately the elevation differences. This was done in order to determine if total relief could be related to such factors as bedrock lithology, lake

density, and basal thermal regime of the reconstructed ice sheet. Five
relief intervals were used to characterize the topography: (1) 0-100
m; (2) 100-300 m; (3) 300-600 m; (4) 600-1000 m; (5) >1000 m. Each
grid square was assigned one of these intervals based on the total
elevation difference between the lowest and highest points within the
square. Small-scale topographic extremes such as very narrow, deeply
incised steam channels in an otherwise low relief plateau surface were
not included in assigning the relief interval. In these cases, the
dominant topographic feature (the low relief plateau surface) was used
to characterize the grid rather than the extreme elevation difference.
It was assumed that such features were not due to primary glacial
erosion and thus their inclusion would bias the data.

DATA ANALYSIS

Sugden (1974, 1977, 1978) argued that intensity of areal scour is
primarily a function of the thermal regime at the base of an ice
sheet. Sugden (1978, p. 387) stated: "Areal scour affects most of the
central ice sheet zone where basal ice conditions change from
warm-melting at the centre, through warm-freezing to a cold-based
zone (but containing debris). A maximum amount of erosion occurs in
association with the warm-freezing zone which affords an effective
means of debris evacuation." The basal temperature of an ice sheet is
a function of several interacting variables (Robin, 1955; Sugden, 1977;
Hooke, 1977) including ice thickness, net mass balance, geothermal heat
flux, amount of internal ice deformation, and amount of basal sliding.
These conditions may be modelled (e.g. Sugden, 1977) or the landscapes
of glacial erosion may be used to control the character of the basal
thermal regime (Denton and Hughes, 1981). Either approach involves a
number of assumptions, some of which are difficult to rigorously define
or defend.

    In order to understand further the reasons for variations in
extent of areal scour, we have concentrated on a basic descriptive
analysis. For example, we can infer that ice thickness, one of the
more important variables, varied across the region as a function of
distance from the former ice margin(s) and existing elevation. Thus,
as a first approach, the distribution of the degree of areal scour as a
function of elevation was examined. The summary figure (Fig. 3.3F)
indicates that areas of low lake density (unit A) rarely occur at
elevations below 200 m but are nearly co-dominant in the elevation
ranges 200-500 m and dominant above 500 m. The opposite applies to

Summary

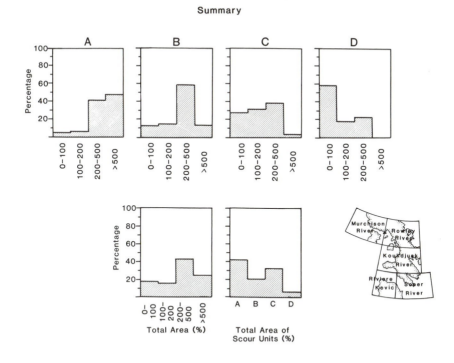

Figure 3.3 Histograms for each of the 1:1,000,000 NTS map areas
(see insert) showing the percentage of land of a specific lake
density against elevation. Fig. 3.3E is a summary for the entire
mapped area.

areas characterized by heavy areal scour (unit D) (Fig. 3.2, 3.3F)
where the largest area occurs between 0-100 m with no areas of heavy
scour above 500 m. Over the 515,254 km^2 area mapped (Fig. 3.2), unit A
occupies the most area (42%) followed by units C (32%), B (20%), and
unit D (heavy scour) which occupies some 29,000 km^2, or 6% of the total
area. Figure 3.3 (A-E) shows, however, that considerable regional
variations in erosional patterns exist across the Eastern Canadian
Arctic. On the Riviere Kovic sheet (Fig. 3.3), for example, the
dominant landform classification is unit C with a total area of 44,138
km^2 compared to 18,202 km^2 for unit A. Further discussion on the
distribution of lake density units will be carried out later in this
chapter.

Association Analysis

We have examined the associations between "dependent variables", i.e.
the qualitative and quantitative assessments of the extent of areal
scour, and "independent variables" that probably reflect glaciological
and bedrock control. In the latter category, elevation, distance from
the outer coast, relative relief, and bedrock erodibility were
measured. These variables were expressed as ordinal data (i.e. ranked)
and the SPSS Program (CROSSTABS) was used to examine the interactions
between the dependent and independent variables (Nie et al., 1975). In
addition, statistically significant associations between pairs of
variables were tested, such as the association between bedrock
erodibility and lake density. The underlying null hypothesis used here
is that the measured parameters are independent of each other and that
measures of association (such as chi square) are no larger than would
arise by chance sampling of two or more independent populations.

Hypothesis 1

We expect some degree of association between the qualitative and
quantitative estimates of aerial scour (see above). However, lake
density estimates did not distinguish areas of selective linear erosion
nor alpine areas. Consequently, there were 530 cells of the total 580
where comparative measurements between the two estimates could be
conducted. The chi-square value of 311.9 is significant at the
0.000001 level, thus indicating substantial agreement between both
methods.

Hypothesis 2

The influence of bedrock on the visual extent of glacial scour is
difficult to assess. However, the association between the qualitative
assessment of glacial erosion and geology was not significant at the $p
= 0.10$ level whereas the percent of lakes versus erodibility of the
bedrock was significant at $p = 0.057$. However, there is a strong
interdependency between rock type, elevation, and distance from the
coast such that the $p = 0.057$ is considered a misleading index of the
strength of the association.

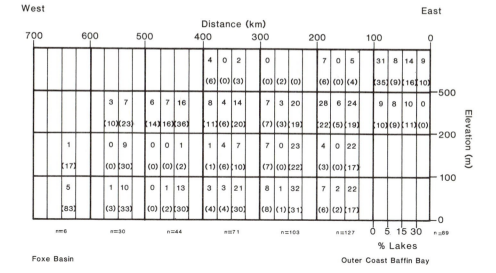

Figure 3.4 This figure shows an analysis which seeks to explain the variation in lake density as a function of (a) distances from the outer coast of Baffin Island; and (b) elevation. Data are from a contingency analysis.

Hypothesis 3

This hypothesis examines the effect of elevation and distance from the outer east coast on lake density variations. Figure 3.4 shows cell counts and percentages (in brackets) for glacial erosion (= lake cover) by elevation controlling for distance. There are seven individual statistical assessments in this design, one for each interval of distance. Because the measurements are ranked (= ordinal data), the significance of the associations was measured by Kendall's tau c (Nie et al., 1975). The association between lake cover and elevation was not significant for the 0-100 km interval. For the remaining distances, the probabilities were $p = 0.0002$, 0.104, 0.0035, 0.0052, and 0.092 for 100-200, 200-300, 300-400, 400-500, and 500-600 km intervals respectively (Fig. 3.4).

These results support the previous suggestion (e.g. Fig. 3.2) that intensity of glacial erosion increases westward away from the eastern coast of Baffin Island. In general, the Pleistocene ice sheet(s) over this region would thicken westward due to increased distance from the terminus and decreasing elevation toward Foxe Basin and Hudson Bay. The analysis under Hypothesis 3 merely confirms that this pattern exists at a statistically useful level in the majority of comparisons (Fig. 3.4).

Spatial Patterns of Lake Density

In general, broad similarities exist between lake densities mapped here
(Fig. 3.2) and those mapped by Sugden (1978). However, a comparison of
the two maps reveals some differences. We examine in detail the
spatial distribution of lake density units across the Eastern Canadian
Arctic (Fig. 3.2).

Some areas of heavy glacial scour (unit D) are associated with
specific bedrock types (Hypothesis 2), particularly Paleozoic
limestones around the eastern rim of Foxe Basin and fold belt rocks
south of Hudson Strait (Fig. 3.2). The other major area of intense
aerial scour occurs in the lowland between the head of Cumberland Sound
and Nettling Lake (Figs 3.2). Areas of medium lake density (unit C)
usually occur on the assumed downstream side of unit D. A broad band
of unit C extends from northeastern Foxe Basin, near Steensby Inlet,
along the west coast of Baffin Island toward Cumberland Sound. Similar
terrain comprises most of Foxe Peninsula and parts of the interior of
Frobisher Bay.

Areas of unit B form protuberances extending toward the outer
coast of Baffin Island or paralleling areas of unit A, such as on the
Ungava Peninsula (Fig. 3.2). A major lobe of unit B extends toward
Home Bay and may represent a major area of ice convergence draining
into this saddle within the Baffin Island uplands (Andrews et al.,
1970; Andrews and Miller, 1979; Dyke et al., 1982a). This suggestion
is supported by the occurrence of limestone erratics in Home Bay, one
of the few areas within the fiord landscapes of Baffin Island, where
such erratics are common (Andrews et al., 1970). Another lobate
pattern occurs near the northwestern margin of the Barnes Ice Cap; to
the north, limestones and shells were carried north to northeast from
Foxe Basin (Fig. 3.1) (Andrews, 1966).

Large sections of Baffin Island and portions of the Ungava
Peninsula exhibit little evidence of glacial erosion. Areas of unit A
form a broad zone extending along the axis of Baffin Island to
Cumberland Peninsula and over parts of Hall and Meta Incognita
peninsulas. These areas are associated with a combination of: (1)
proximity to the outer coast; and (2) elevations frequently above 500 m
(Fig. 3.4; Hypothesis 3). These factors suggest that ice may have been
thin and cold over these high plateaux (Sugden 1977, 1978; Denton and
Hughes, 1981). Other patterns of unit A which suggest former
glaciological conditions include the large wedge of unit A approaching
the eastern coast of Steensby Inlet. This wedge delineates the margins
of two major scour areas to the south and north, respectively (Fig.
3.2).

The mapped boundary between units A and B, if representing a
glacial phenomenon, is not expected to vary significantly in elevation
over moderate distances. The elevation of the contact between units A
and B was estimated by using 1:250,000 topographic maps. From
Cumberland Sound to north of Foxe Basin, the contact is remarkably
consistent and varies between 300-430 m (Fig. 3.5) with a mean
elevation = 369±91 m. Furthermore, the contact between units B and C
varies between 300 and 210 m, and the area of intense scour (unit D) is
primarily below 100 m. The boundaries between major landscape units
appear, therefore, to be a function of elevation, a result implicit in
Hypothesis 3 (Fig. 3.4) and recognized by Sugden (1977) in his
modelling of basal thermal regimes of a maximum Pleistocene ice sheet.

It has often been assumed that Hudson Strait acted as a major
conduit for drainage of the Laurentide Ice Sheet. Denton and Hughes
(1981), for example, proposed that this large bathymetric feature was
the location of a major ice stream. Andrews and Falconer (1969)
suggested that rapid calving through Hudson Strait are led to the
collapse of the interior Laurentide Ice Sheet by 8000 years BP.
Patterns of glacial erosion on adjacent land masses of Hudson Strait,
therefore, are of considerable interest in evaluating these
assumptions.

Combined with evidence reported for ice movement (Fig. 3.1;
Osterman et al., this volume), patterns of glacial erosion around
Hudson Strait are more complex than would be suggested from
unidirectional flow of an ice stream. Areas of intense scour (unit D)
fringe large parts of Hudson Strait along the north shore, but the
distribution is broken by sections of unit C. In contrast, the Ungava

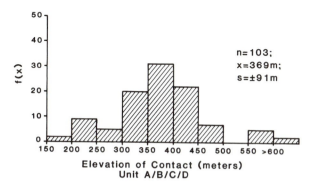

Figure 3.5 Histogram of the elevation of the contact between
light and heavy or moderate areal scour.

Peninsula coastline is only lightly scoured (unit A) with large
intervening areas of unit B. These relationships suggest a more
complex glacial history for Hudson Strait than has been previously
assumed.

GLACIAL EROSION, ICE-FLOW INDICATORS, AND DISTRIBUTION OF ERRATICS

Sugden (1977, 1978) and Boulton (1982) have argued that large scale
glacial features of the Canadian Shield and surrounding areas must
reflect long-term equilibrium conditions of the Laurentide Ice Sheet.
In addition, there are arguments on the significance of erratic trains
and striations as indicators of long-term flow regimes (Boulton et al.,
1977; Shilts, 1980: Denton and Hughes, 1981; Dyke et al., 1982a;
Andrews, 1982). Figure 3.1 illustrates the pattern of glacial
striations and other ice directional features that we have obtained
from the literature and from unpublished data (e.g. Blackadar, 1967;
Ives and Andrews, 1963; Sims, 1964; Blake, 1966; Bell, 1898; Dyke et
al., 1982b; Osterman et al., this volume). We have also included the
boundary between units A and B; the distribution of carbonate tills
overlying Precambrian bedrock is shown on Figure 3.6. Comparison of
Figures 3.1 and 3.2 indicates two major conclusions. (1) Striations
and/or bedrock flow forms are not totally restricted to units C and D.
Distributions of these features (Fig. 3.1) partly reflect areas of
intensive field work (e.g. around the Barnes Ice Cap). It does appear,
however, that although little scouring/plucking occurred within unit A,
ice was able to accomplish limited erosion. (2) The pattern of ice
flow across Baffin Island has been discussed by several workers
(Mercer, 1956; Ives and Andrews, 1963; Andrews and Miller, 1979; Denton
and Hughes, 1981; Dyke et al., 1982a); the data on Figure 3.1 tends to
support an ice sheet configuration similar to that postulated by Dyke
et al., (1982a).

We suggest that a critical test of postulated ice streams is
whether these have left evidence in the form of (1) areas of intense
areal scour, (2) striations or flow forms, and (3) the transportation
of distinctive erratics over large distances. Using evidence presented
in Figures 3.1 and 3.2, the postulated location of major ice streams
around the northeastern margin of the Laurentide Ice Sheet have been
reconstructed (Fig. 3.6). This reconstruction has similarities and
differences with the reconstruction of Dyke et al., (1982a, their Fig.
4), and bears a resemblance to the minimum ice sheet reconstruction of
Denton and Hughes (1981).

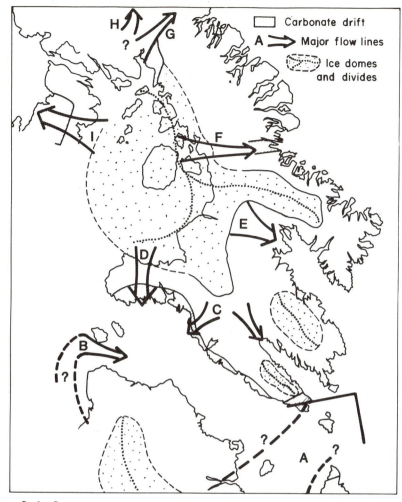

Figure 3.6 Suggested position of major ice streams based on an analysis of the data in Figures 3.1 and 3.2, and areas of carbonate drift (Andrews and Sim, 1964; Andrews and Miller, 1979; Miller, 1982; Dyke et al., 1982a).

The major trajectories that we can identify with confidence are labelled A, B, ... (Fig. 3.6). We do not intend to catalog all the ice streams depicted here, but several deserve some discussion. The major difference between Figure 3.6 and previous publications is the evidence for a major trajectory heading between north and northeast from the Ungava Bay region, overriding eastern Meta Incognita Peninsula on southern Baffin Island (Miller, 1982; Osterman and Miller, this volume). The ice stream (A) is recognized from striations, carbonate content in drift, shelly tills, and erratics that are probably derived from the circum-Labrador fold belt (Andrews et al., in press). This

ice stream indicates that ice drainage down Hudson Strait from Hudson
Bay, Foxe Basin, and the Keewatin dispersal zone was either not as
extensive as previously assumed, or was diachronous with northward flow
off the Ungava or Labrador plateaux. Conversely, as discussed in
Andrews et al., (in press), the NE flow across the tip of SE Baffin
Island may include a series of complex glaciological responses
involving shifts in dominance of Labrador and Keewatin ice. Dyke et
al. (1982a) remarked on the radial pattern of flow centered on Amadjuak
Lake. Blake (1966) considered this pattern to be primarily a
late-glacial phenomenon and only observed limited limestone erratics
eastward of the Paleozoic/Precambrian contact. However, just to the
south and east, Blackadar (1967) reports over 100 m of carbonate-rich
till and notes that limestone till decreases southward. This suggests
that southward flow from the general area of Amadjuak Lake was not a
short-lived event.

On the southern tip of Nottingham Island, Bell (1898) reported the
occurrence of drift dominated by circum-Labrador fold belt erratics.
Ice stream C is thus shown as being forced to turn and flow northward
along the west coast of Ungava Peninsula and along the southern shore
of Hudson Strait. This flow path is consistent with evidence for the
earliest flow on the Ottawa Islands, Hudson Bay, which was directed to
the northeast (Andrews and Falconer, 1969).

Further north, bedforms, distribution of glacial landscapes, and
the relative abundance of limestone erratics in Home Bay (Andrews et
al., 1970) suggests a major ice stream (E) was directed from an ice
center in Foxe Basin toward Baffin Bay (Ives and Andrews, 1963; Dyke et
al., 1982a).

These reconstructions suggest that toward the outer margin of the
Laurentide Ice Sheet, several semi-independent/independent areas of
glaciation existed. Certainly in many areas, local cirque and valley
glaciers responded independently to climatic changes that affected the
main sectors of the Foxe/Baffin Laurentide ice complex (e.g. Miller,
1976; Locke, 1980; Hawkins, 1980; Muller, 1980; Mercer, 1956; Dyke et
al., 1982b; Clark, 1982). Otherwise, the high, eastern plateaux are
not heavily glaciated (Fig. 3.4) with the exception of the Everrett
Mountains fronting Frobisher Bay northwest of the Grinnell and Terra
Nivea ice caps (Fig. 3.4). The Everrett Mountains are scoured (unit B)
(Muller, 1980; Colvill, 1982), but the origin of the ice that caused it
is uncertain. It might represent overriding ice from Labrador crossing
Hudson Strait - an idea suggested by Mercer (1956).

RESULTS OF A GLACIOLOGICAL FLOW MODEL - BASAL TEMPERATURES AND RATES OF
TRANSPORT

Important questions arising from Figures 3.1, 3.2, and 3.6 are the
degree to which erosional patterns reflect long-term conditions at the
base of an ice sheet, and the amount of time required to transport
limestone erratics across distances measured in hundreds of
kilometers. Sugden (1977) calculated basal temperature conditions by
varying input parameters. Hughes (1981) adopted a different approach,
using geological data to imply that units B, C, and D characterized
warm-based ice and unit A represented cold-based ice (Figs. 3.2 & 3.7).

    In order to examine these questions in more detail, we used a
flowline program developed by Jenssen (1977) in conjunction with
research on the Greenland Ice Sheet (Radok et al., 1982). The basic
inputs to the program (called STEMP) are: bedrock elevation, surface
elevation of the ice sheet, ice thickness, net mass balance, estimated
balance velocity, and surface air temperature. The mathematical basis

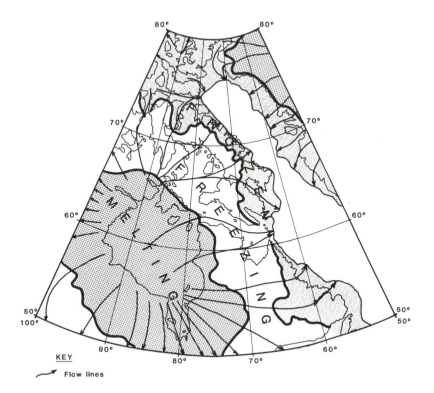

Figure 3.7 Flow directions and basal temperature regimes according to
Denton and Hughes (1981) (maximum model). (Reprinted with permission)

to STEMP are outlined in Radock et al. (1982, P. 57-81). In addition, we used program SPRFIL to calculate particle paths and isochrons for the different models that we developed (see Radock et al., 1982, p. 86-88). The outputs from the program of most interest here are the basal ice temperature, the amount of melting at the base, and the amount of refreezing at the base. SPRFI depicts trajectory paths and isochrons within the ice sheet (developed by B. McInnes in Radock et al., 1982).

There are, of course, an infinite number of ways to vary parameters, but we concentrated our efforts to investigate the following:
(1) What is the effect of glacial isostatic depression (= ice thickening) on the basal temperature;
(2) What is the effect of a change in accumulation rate and air temperature on a glaciological situation similar to the present,   as opposed to a cold, dry model (other combinations were also attempted).

We generalized the topography of Baffin Island (e.g. Fig. 3.8) into two categories: (1) where the ice flowed through a major lowland, such as a flowline heading toward Cumberland Sound (Fig. 3.8, C-C'); and (2) examination of profiles across high plateaux of Baffin Island (Fig. 3.8, A-A', B-B'). Glacial isostatic depression was estimated from the lithospheric bending model of Brotchie and Silvester (1969). Elevations of the ice sheet followed earlier estimates based on

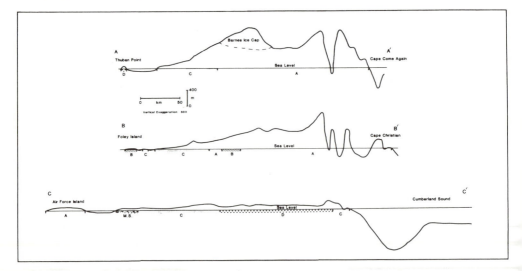

Figure 3.8 Cross-sections on Baffin Island - see Figure 3.1 for location.

Greenland analog arguments (Andrews, 1980), which are generally similar
to elevations used by Denton and Hughes (1981) in their minimum
reconstruction.

Models 1, 2, & 3:

One series of runs represent a flowline across a region such as C-C'
(Figs 3.1 & 3.8) where the landscape is dominated by unit D (Fig.
3.2). Table 3.1 lists the changes in the input variables. Basal
temperature conditions are relatively insensitive to the changes in the
three models; this can readily be seen in Figure 3.9. Basal
temperatures are close to the pressure melting point 40 km from the ice
divide to the margin of the ice sheet. In all three cases, the
computed temperatures were always slightly below the pressure melting
point by 0.1 to 0.3°C. Of the three alternative models, we suggest
that the inclusion of glacial isostatic depression is a necessary
control and is demonstrated by the marine transgressions at the outer
coast of Baffin Island which preceded the encroachment of outlet
glaciers to the mouths of the fiords (Miller et al., 1977; Nelson,
1981). If the temperature estimates of Figure 3.9 are considered
together with a realistic glacier bed, it is possible that bedrock

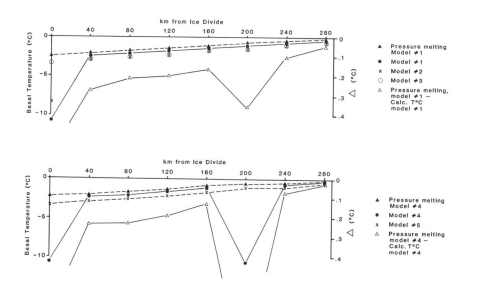

Figure 3.9 Calculated basal temperatures along constrasting
topographic flowlines (A-A and C-C') (Fig. 3.1) using program
STEMP (Radok et al., 1982).

TABLE 3.1. Data input for basal temperature models #1 through #5 (See Figs 3.11, 3.12, & 3.13). Measurements are taken from the ice divide.

Distance (km)	1 Bedrock #1 (m)	2 Bedrock #2 (m)	3 Isostatic bedrock #1 (m)	4 #2	5 Surface elev. (m)	6 Accumulation (m)	7	8 Surface T°C	9
0	-100	-100	-1100	-1000	3100	0.25	0.1	-31	-37
40	150	0	-950	-950	2800	0.23	0.09	-28	-36
80	200	200	-850	-650	2400	0.2	0.08	-26	-34
120	150	400	-650	-250	2000	0.15	0.07	-24	-32
160	100	600	-450	150	1600	0.1	0.05	-22	-30
200	100	100	-350	-200	1200	0	0	-19	-27
240	-100	-400	-250	-600	800	-0.5	-0.5	-15	-21
280	-100	-400	-180	-500	0	-2.0	-1.0	-9	-15
Column #	1	2	3	4	5	6	7	8	9

NB Ice thickness is given by, for example, column 5 - column 3

Model #1 = columns 1, 5, 6, & 9
Model #2 = columns 3, 5, 6, & 9
Model #3 = columns 3, 5, 7, & 10
Model #4 = columns 2, 5, 6, & 9
Model #5 = columns 4, 5, 7 & 10

protuberances on the scale of 1-10 m would cause localized
pressure-melting which would greatly facilitate glacial plucking and
scouring (Boulton, 1974).

Basal temperature estimates for an ice flowline heading from Foxe
Basin toward the fiords (A-A', B-B' Fig. 3.8) are more strongly
affected by changes in the model input parameters (Fig. 3.9). In the
case of ice flowing across a non-isostatically compensated upland, the
temperatures at the base are usually well below the pressure-melting
point for several combinations of net mass balance and surface
temperatures. From Figure 3.9, it can be seen that only the
combination of 0.25 m H_2O at the dome over Foxe Basin with a
temperature of -9°C at the coast resulted in extensive pressure-melting
and even in this case there is a zone of "cold" ice between 120-40 km
from the dome (Fig. 3.10). This zone of cold ice corresponds to
plateaux elevations between 300 and 700 m. With glacial isostatic
compensation, and with less accumulation and colder surface
temperatures, the basal temperatures increase across this "cold" zone
and reach to near the pressure melting point. This situation is
portrayed as model #5 (Table 3.1) with a surface temperature at the
dome of -41°C. However, this model does not mimic the long-term
erosional landforms depicted on the LANDSAT imagery for north central
Baffin Island (Fig. 3.2) where there is a sharp boundary between areas
of glacial scour (units B and C) and units where the bedrock is masked
by felsenmeer/till, much of which shows little evidence of glacial
erosion. However, note that striations and flow forms have been mapped
across this region (Ives and Andrews, 1963; Fig. 3.1). The elevation
of the boundary between unit A and B/C is reflected in model #4 but as

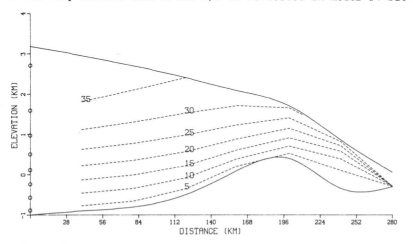

Figure 3.10 Englacial temperatures from STEMP for cross-section across
Baffin Island showing intersection of -5°C isotherm with the uplands.
This inntersection coindices approximately with the 360 m contour.

noted earlier this calculation assumed no glacial isostatic
compensation--which we feel is unrealistic. It is clear from Figure
3.9 that cold basal temperatures reflect a high mass flux associated
with high accumulation rates. The presence of unit A over large areas
of Baffin Island (Fig. 3.2) implies, within the limits of this model,
that the landscapes of glacial erosion reflect atmospheric conditions
that give a maximum accumulation rate at the summit dome of +0.25m and
temperatures > -40°C.

Rates of Transportation

The mapping of glacial erratics across parts of northern Canada
(Andrews and Miller, 1979; Shilts, 1980; Dyke et al., 1982a) has
resulted in a rejection of the notion that the emplacement of these
erratics could be ascribed to very late glacial movements associated
with the final stages of deglaciation. This idea had been espoused by
Flint (1971) and largely accepted by Denton and Hughes (1981). In
contrast, Boulton et al. (1977) estimated that it must have taken
several thousands of years for erratics to be moved from the source to
final deposition within Great Britain. The problem is certainly not
simple, as erratics can be moved through several glaciations and some
erratics might be brought into an area, at least initially, by
ice-rafting through icebergs or sea ice.

The input data for models #1-5 can be used in a program (Radok et
al., 1982) that computes the trajectories and the position of
isochrones within the ice sheet (Fig. 3.11). Calculations of velocity

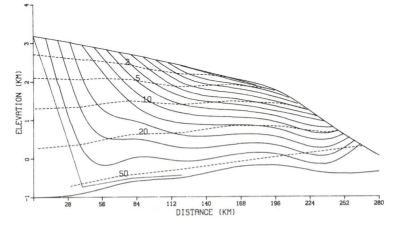

Figure 3.11 Isochrones on reconstructed Pleistocene flowlines
across Baffin Island, using program SPRFIL (Radock et al., 1982) for
models on Table 3.1.

for a 1 m wide strip of the ice sheet were carried out for the 5 different models. These calculations produce <u>average</u> column velocities based on the conservation of mass. For model #1 (Fig. 3.11), velocities increased from 3.1 m yr^{-1} at 40 km to 47 m yr^{-1} at 280 km, whereas, in model #3 (Table 3.1) the velocities at the same distances were 0.6 and 9.3 m yr^{-1}. These calculations do not take into account changes in velocity attributable to either converging or diverging flow. In all cases, the isochrones indicate that the movement of a limestone erratic from the western shore of Foxe Basin into the fiords fronting Baffin Bay would take a minimum of 10,000 years with a realistic estimate closer to >>20,000 years (cf. Haefli, 1961). Thus how far the erratic trains of Fig. 3.6 reflect a series of events, with the erratics being carried forward step by step, is a major research problem and one that will only be answered by detailed analysis of glacial stratigraphic sections where dating methods can delimit the duration of the glacial events. One such area is the lowlands of southern Hudson Bay where extensive exposures of glacial and nonglacial sediments are exposed along the rivers (Skinner, 1973; Andrews et al., 1983).

CONCLUSIONS

LANDSAT mapping of the landscapes of glacial erosion across the Eastern Canadian Arctic indicates that there are systematic changes in the intensity of glacial erosion that can be explained partly by changes in the bedrock geology, but more fully by variations in elevation and distance from the coast of Baffin Island. These last two parameters must, in some way, reflect a combination of glaciological factors that control basal ice regimes and, by inference, the ice/bed interactions (e.g. Boulton, 1974; Sugden, 1977, 1978; Gordon, 1979).

ACKNOWLEDGEMENTS

This study is a contribution to NSF grant EAR-81-21296. We thank S. Forman for his contribution in providing mapped data on glacial landforms. B. McInnes provided us with valuable advice and assistance in running the programs whereas funds for the computer time were provided through the office of Dr. U. Radock, CIRES, University of Colorado.

REFERENCES

Andrews, J.T., 1966: Pattern of coastal uplift and deglacierization, west Baffin Island, N.W.T. Geographical Bulletin, 8:174-193.

Andrews, J.T., 1970: Differential crustal recovery and glacial chronology (6,700- 0 BP), west Baffin Island, N.W.T., Canada. Arctic and Alpine Research, 2:115-134.

Andrews, J.T., 1980: Progress in relative sea level and ice sheet reconstructions, Baffin Island, N.W.T., for the last 125,000 years. In: Mörner, N-A. (ed.), Earth Rheology, Isostasy and Eustasy. New York: John Wiley, 175-200.

Andrews, J.T., 1982: On the reconstruction of Pleistocene ice sheets: a review. Quaternary Science Reviews, 1:1-30.

Andrews, J.T., Buckley, J.T., and England, J.H., 1970: Late-glacial chronology and glacio-isostatic recovery, Home Bay, east Baffin Island. Geological Society of America Bulletin, 81:1123-1148.

Andrews, J.T., and Falconer, G., 1969: Late glacial and postglacial history and emergence of the Ottawa Islands, Hudson Bay, N.W.T., evidence on the deglaciation of Hudson Bay. Canadian Journal Earth Sciences, 6:1263-1276.

Andrews, J.T. and Miller, G.H., 1979: Glacial erosion and ice sheet divides, northeastern Laurentide ice sheet, on the basis of the distribution of limestone erratics. Geology, 7:592-596.

Andrews, J.T., Shilts, W.W., and Miller, G.H., 1983: Multiple deglaciations of the Hudson Bay lowlands, Canada, since deposition of the Missinaibi (Last-Interglacial?) Formation. Quaternary Research, 19:18-37.

Andrews, J.T., Stravers, J.A., and Miller, G.H., Patterns of glacial erosion and deposition around Cumberland Sound, Frobisher Bay, and Hudson Strait, and the location of ice streams in the eastern Canadian Arctic. In: Waldenburg, M. (ed.), Models in Geomorphology. Allen and Unwin. In press.

Bell, R., 1898: Report of an exploration on the northern side of Hudson Strait, Canada. Geological Survey of Canada Annual Report, XI, Part M, 33 pp.

Blackadar, R.G., 1967: Geology of Mingo Lake-MacDonald Island map-area, Baffin Island District of Franklin. Canada Geological Survey Memoir 345. 54 p. Map 1:85A, scale 1:253,440.

Blake, W., Jr., 1966: End moraines and deglaciation chronology in northern Canada with special reference to southern Baffin Island. Geological Survey of Canada Paper, 66-26. 31 pp.

Boulton, G.S., 1982: Sedimentary and geomorphological indicators of ice sheet dynamics. International Association of Sedimentologists 11th International Congress, Abstracts, McMaster University, Ontario, Canada. p. 77.

Boulton, G.S., 1974: Processes and patterns of glacial erosion. In: Coates, D.R. (ed.), Glacial Geomorphology. New York, Publications in Geomorphology, 41-87.

Boulton, G.S., Jones, A.S., Clayton, K.M., and Kenning, M.J., 1977: A British ice sheet model and patterns of glacial erosion and deposition in Britian. In: Shotton, F.W. (ed.), British Quaternary Studies. Oxford, Clarendon Press, 231-246.

Brotchie, J.F. and Sylvester, R., 1979: On crustal flexture. Journal of Geophysical Research, 74:5240-5252.

Clark, P.U., 1982: Late Quaternary history of the Iron Strand region, Torngat Mountains, northernmost Labrador, Canada. Seventh Biennial Meeting, Seattle, Washington, p. 82.

Colvill, A.J., 1982: Glacial landforms at the head of Frobisher Bay, Baffin Island, Canada. MA Thesis, University of Colorado, Boulder. 202 pp.

Denton, G.H. and Hughes, T.H. (eds.), 1981: The Last Great Ice Sheets. New York, John Wiley. 484 pp.

Douglas, R.J.W., 1970: Geological Map of Canada. Geological Survey of Canada, Map 1250A.

Dyke, A.S., Dredge, L.A., Vincent, J-S., 1982a: Configuration and dynamics of the Laurentide ice sheet during the late Wisconsin maximum. Geographie physique et Quaternaire, 36:5-14.

Dyke, A.S., Andrews, J.T., and Miller, G.H., 1982b: Quaternary geology of Cumberland Peninsula, Baffin Island, District of Franklin. Geological Survey of Canada Memoir 403, 32 pp.

Flint, R.F., 1943: Growth of the North American ice sheet during the Wisconsin Age. Geological Society of America Bulletin, 54:325-362.

Flint, R.F., 1971: Glacial and Quaternary Geology. New York, John Wiley and Sons, 892 pp.

Gordon, J.E., 1979: Reconstructed Pleistocene ice-sheet temperatures and glacial erosion in northern Scotland. Journal of Glaciology, 22:331-344.

Haefeli, R., 1961: Contribution to the movement and the form of ice sheets in the Arctic and Antarctic. Journal of Glaciology, 3:1133-1152.

Hawkins, F.F., 1980: Glacial Geology and Late Quaternary paleoenvironment in the Merchants Bay Area, Baffin Island, N.W.T., Canada. M.S. Thesis, University of Colorado, Boulder, 146 pp.

Hollin, J.T., 1962: On the glacial history of Antarctica. Journal of Glaciology, 4:173-195.

Hooke, R.L., 1977: Basal temperatures in polar ice sheets: a qualitative review. Quaternary Research, 7:1-13.

Hughes, T.J., 1981, Numerical reconstruction of paleo-ice sheets. In: Denton, G.H., and Hughes, T.J. (eds.), The Last Great Ice Sheets. New York, John Wiley and Sons, 222-274.

Ives, J.D. and Andrews, J.T., 1963: Studies in the physical geography of north-central Baffin Island, N.W.T. Geographical Bulletin, 19:5-48.

Jenssen, D., 1977: A three-dimensional polar ice-sheet model. Journal Glaciology, 18:373-390.

Locke, W.W. III, 1980: The Quaternary geology of the Cape Dyer area, southeasternmost Baffin Island, Canada. Ph.D. Thesis, University of Colorado, Boulder, 331 pp.

Mercer, J.H., 1956: Geomorphology and glacial history of southernmost Baffin Island. Geological Society of America Bulletin, 67:553-570.

Miller, G.H., 1976: Anomalous local glacier activity, Baffin Island, Canada: Paleoclimatic implications. Geology, 4:502-504.

Miller, G.H., 1982: Dynamics of the Laurentide ice sheet based on field evidence from northeastern Canada. IX INQUA Congress, Moscow, Abstract 1:222.

Miller, G.H., Andrews, J.T. and Short, S.K., 1977: The last interglacial/glacial cycle, Clyde Foreland, Baffin Island, N.W.T.: stratigraphy, biostratigraphy, and chronology. Canadian Journal of Earth Sciences, 14:2824-2857.

Müller, D.S., 1980: Glacial geology and Quaternary history of southeast Meta Incognita Peninsula, Baffin Island, Canada. M.S. Thesis, University of Colorado, Boulder, 211 pp.

Nelson, A.R., 1981: Quaternary glacial and marine stratigraphy of the Qivitu Peninsula, northern Cumberland Peninsula, Baffin Island. Geological Society of America Bulletin, v. 92, Part I, 512-518, Part II, 1143-1261.

Nie, N.H. et al., 1975: Statistical package for the Social Sciences. New York: McGraw-Hill, 675 pp.

Nie, N.H. et al., 1975: Statistical package for the Social Sciences, New York: McGraw-Hill, 675 pp.

Radock, V. Barry, R.G., Jenssen, D., Keen, R.A., Kiladis, G.N., and McInnes, B., 1982: Climatic and physical characteristics of the Greenland Ice Sheet. C.I.R.E.S., University of Colorado, Boulder, 193 pp.

Robin, G. de Q. 1955: Ice movement and temperature distribution in glaciers and ice sheets. Journal Glaciology, 2:523-532.

Sanford, B.V., Grant, A.C., Wade, J.A., and Barss, M.S., 1979: Geology of Eastern Canada and adjacent areas. Canada Geological Survey, Map 1401 A (4 sheets).

Shilts, W.W., 1980: Flow patterns in the central North American ice sheet. Nature, 286, 213-218.

Sim, V.W., 1964: Terrain analysis of west-central Baffin Island. Geographical Bulletin, 21:66-92.

Skinner, R.G., 1973: Quaternary stratigraphy of the Moose River Basin, Ontario. Canada Geological Survey Bulletin 225, 77 pp.

Sugden, D.E., 1974: Landscapes of glacial erosion in Greenland and their relationship to ice, topographic and bedrock conditions. Institute British Geographers, Special Publication 7:177-195.

Sugden, D.E., 1976: A case against deep erosion of shields by ice sheets. Geology, 4:580-582.

Sugden, D.E., 1977: Reconstruction of the morphology, dynamics   and thermal characteristics of the Laurentide ice sheet. Arctic and Alpine Research, 9:21-47.

Sugden, D.E., 1978: Glacial erosion by the Laurentide ice sheet. Journal of Glaciology, 20:367-392.

Terry, R.D., and Chilingar, G.V., 1955: Summary of "Concerning some additional aids in studying sedimentary formations" by M.S. Shvetsov. Journal of Sedimentary Petrology, 25:229-234.

White, W.A., 1972: Deep erosion by continental ice sheets: Geological Society of America Bulletin, 83:1037-1056.

4 The fiords of Baffin Island: description and classification

E. K. Dowdeswell and J. T. Andrews

Arguments on the origin of fiords can be traced back nearly 100 years.
However, very few studies have provided key measurements of fiord
dimensions or other critical parameters (see Pickard and Stanton,
1980). Thus the study of fiords has been largely theoretical, with
little empirical data available. The major objective of this chapter
is to describe the fiords on the eastern coast of Baffin Island and to
attempt a classification based on measured parameters. These data lead
to some commentary and conclusions on the origin of fiords and in
particular on the old argument that fiords are primarily structural in
origin (e.g. Gregory, 1913; Johnson, 1915). Most modern text books
(Flint, 1971; Sugden and John, 1976) subscribe to a predominantly
glacial origin for fiords although Andrews (1974) did express concern
over the easy acceptance of a theory supported with so little data.

Tectonic and Geological Setting

Rifting between West Greenland and Baffin Island occurred in late
Cretaceous/early Tertiary time and is presumed to have been the result
of an offshoot of the Mid-Atlantic Ridge. Many authors (Le Pichon et
al., 1977; Clarke and Upton, 1971; Hay, 1981) have suggested a model
for topographic development during the course of rifting which includes
isostatic uplift along the rifted margin. This accounts for the
topographic asymmetry of Baffin Island, with the highest peaks
occurring generally along the eastern coast. In Baffin Bay, rifting
took place during a time when the climate is inferred to have been
subtropical to warm-temperate (Andrews et al., 1972), resulting in
accelerated fluvial erosion along the steep, seaward-facing flank of
the rift. This led to a series of major east-facing canyons being
incised into the isostatically rising margin. These canyons were, we
suggest, fault-controlled. However, once continental drift ceases,

isostatic submergence occurs (Hay, 1981), and this might explain, in part, the presence of fiords and offshore troughs.

Recent geophysical surveys of the Baffin Island shelf suggest that the outer coastline may be fault-bounded (MacLean and Falconer, 1979; see Chapter 6 this volume). Major faults have been mapped or inferred with strikes NW-SE (Martin, 1973; Douglas, 1969) as well as normal to the coastline (i.e. striking SW-NE). However, the structural history is only roughly depicted in the geologic record. Precambrian rocks of various lithologies crop out along most of the east coast of Baffin Island, whereas Paleozoic to early/middle Tertiary deposits overlain by Quaternary sediments are found on the continental shelf and the floors of major troughs (see Chapters 2 and 6 this volume). Hudson Strait, Frobisher Bay, Cumberland Sound, and the channels around Bylot Island contain Paleozoic limestone and were therefore down-faulted after this time (Grant, 1975; Grant and Manchester, 1970; MacLean et al., 1977; Sanford et al., 1979).The presence of Eocene rocks along the walls of the Scott Inlet trough indicate that graben formation continued until at least the late Eocene. Dyke et al. (1982) suggest that Cumberland Sound was down-dropped during middle Eocene time. An important unresolved question is how far the horst and graben origin of these major structural lows (Kerr, 1980) relate to features the size of fiords. Could many fiords represent grabens that have been glacially modified? Reid and Falconer (1982, p. 1530) note: "Also conjectural at this stage is the relation between the seismic activity and the glaciated valley (Buchan Gulf) that appears to coincide with the fault zone."

In contrast with the offshore bathymetry of Labrador (e.g. Johnson et al., 1975), Baffin Island lacks a marginal trough although the shelf is crossed by a limited number of large transverse depressions (Fig. 4.1) (see Chapter 6 this volume). Troughs do not occur south of Cape Dyer and their location and size is not easily explained with our present knowledge of ice extent and volume during the Quaternary (Andrews, 1974; Gilbert, 1982). In particular, the trough off Broughton Island (Gilbert, 1982) trends roughly N-S whereas the major ice drainage via Maktak, Coronation and north Pangnirtung fiords is directed (Fig. 4.1) E to NE. Finally, we emphasise the present active seismic nature of the Baffin Island offshore zone. Figure 2.2 illustrates the occurrence of major epicenters and the location of the major transverse troughs (MacLean and Falconer, 1979; Reid and Falconer, 1982). Note that Cumberland Sound and, to a lesser extent, Frobisher Bay, are aseismic, whereas major concentrations of earthquake foci occur offshore and onshore in northern Baffin Island (Fig. 2.2).

Glacial geological setting

The fiords of eastern Baffin Island are located within a zone of
selective linear glacial erosion (Sugden, 1974, 1978). By definition,
this zone is characterized by incised troughs bounded by interfluves
unaffected by erosion. This term begs the issue of fiord origin in
that the presence of fiords was used by Sugden (1974, 1978) to argue
for glacial erosion of fiords. Maps of the various landforms on Baffin
Island (Andrews et al., Chapter 3 this volume) confirm Sugden's (1978)
analysis of the landform distribution, but add considerable detail on
the constraints exercised by elevation (and hence ice thickness) and
bedrock geology. The fiords lie generally within the area of alpine
glaciation or in areas where ice was channeled through the uplands.

The spatial and temporal extent of glacial erosion within the
fiords of Baffin Island can now be discussed with more confidence than
for many areas because of the increased glacial geologic data
available. Amino acid and U-series studies suggest (Szabo et al.,
1981; Nelson, 1982; Miller, this volume) that the oldest exposed
sediments along the outer coast of Baffin Island may be as old as 1 to
2 million years. These deposits are marine or glacial marine and their
presence in successive repetitive sequences of marine-dominated clastic
sediments indicates that glaciers have not penetrated seaward of the
outer coast for several hundred thousand years. This finding draws
attention to the origin and chronology of the formation of the offshore
troughs. One of the major anomalies with the glacial argument for the
troughs and sounds (cf. Denton and Hughes, 1981) is seen in the
comparison between the glacial landform mapping around Cumberland Sound
(Andrews et al., Chapter 3 this volume), the chronology of glaciation
(Dyke, 1977), and the offshore bathymetry and geology (Fig. 4.1).
Areas of intense glacial scour cannot be traced much beyond the head of
Cumberland Sound, and on the northern side of the sound aerial scour
stops at the mouth of Pangnirtung Fiord. Dyke (1977), Locke (1979),
Birkeland (1978) and Bockheim (1979) show major glaciers draining
toward Cumberland Sound along Pangnirtung and Kingnait fiords but these
glaciers terminated at the fiord mouths, whereas ice from Foxe Basin is
only mapped as far east as the west shore of Pangnirtung Fiord.

Reconstruction of late Wisconsin fiord glaciers (Pheasant, 1971;
Boyer, 1972; Mears, 1972; Locke, 1980; Mode, 1980) along Baffin Island
have invariably shown basal shear stresses ranging between 0.2 and 0.5
bars. These reconstructions are based on ice thicknesses derived from
glacial geological mapping. They imply that the glaciers which flowed
through the fiords were either sliding on a water film or overriding
deformable muds (cf. Boulton and Jones, 1979). In either case active

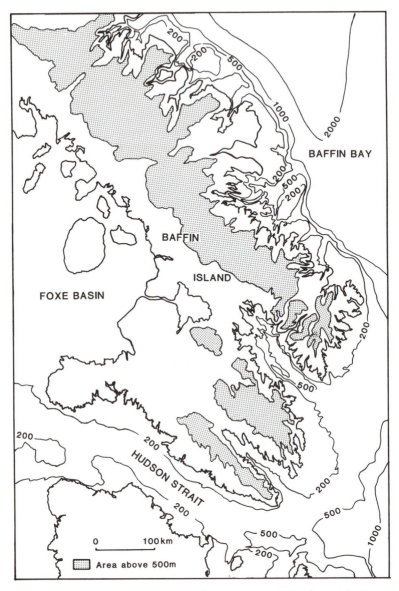

Figure 4.1 Generalized topography (area above 500 m) and bathymetry of
the Baffin Island area showing the asymmetry of topography and the
major troughs off the eastern coast of the island.

crushing and abrasion may not have been taking place at the ice/bedrock
interface (Boulton, 1974).

The remarks above challenge, but do not necessarily disprove, the current ruling hypothesis of fiord formation which is summarized by Flint (1971, p. 131): "Many of the strongly glaciated valleys of high-standing coasts underlain by resistant rocks in high latitudes are partly submerged, and constitute fiords. A fiord is a segment of a glaciated trough partly filled by an arm of the sea. It differs from other strongly glaciated valleys only in the fact of submergence."

Our chapter cannot resolve all of these problems of fiord origin as many answers must lie in detailed geophysical surveys of the fiords. This chapter does present a detailed description of the fiords of Baffin Island, which acts as a basis for subsequent discussion.

BAFFIN ISLAND FIORD MORPHOLOGY

Although it is generally agreed that the location and shape of fiords in part reflects the pre-existing topography, the amount of glacial alteration is uncertain. A statistical analysis of the morphometry of Baffin Island fiords constitutes a quantitative approach to this problem, which has previously been dealt with qualitatively (Gregory, 1913; Linton, 1963; Holtedahl, 1967). Gilbert (1984) has also examined some aspects of Baffin Island fiord morphology.

Measurements of 227 fiords along the northern, eastern, and southern coasts of Baffin Island were taken. Inlets shorter than 5 km were excluded, as were the five larger "fiords"--Admiralty Inlet, Navy Board Inlet, Eclipse Sound, Cumberland Sound, and Frobisher Bay.

Examinations of 29 parameters detailing the size, shape, area, volume, orientation, topography, bathymetry, and spatial distribution of each fiord were made. Information on the nature of access to the interior plateau and the presence or absence of sills or troughs at fiord mouths was also included. Topographic and size/shape data were obtained from the Canadian Department of Energy, Mines, and Resources 1:500,000 topographic maps and bathymetric data were taken from Canadian Hydrographic Service 1:500,000 hydrographic charts, both of which gave elevation in feet. Figures 4.2, 4.3, and 4.4 illustrate a number of the measured parameters (items) and how they are defined. A complete list of variable definitions is presented in the Appendix. The use of precise parameter definitions insures the reproducibility of the raw data and allows comparative studies in other areas to be made.

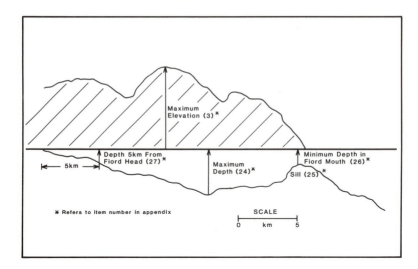

Figure 4.2   Schematic longitudinal cross-section of a fiord.

Descriptive statistics

Baffin Island fiord characteristics are summarized in Table 4.1 for
parameters containing interval or ratio data. Table 4.2 lists
percentage data for those variables containing nominal scale
information. The average Baffin Island fiord is thus described as 28
km long, 6.5 km wide at the mouth, 2.5 km wide at its midpoint, and 1.3
km wide at its head (Fig. 4.4). It encompassses an area of 105 km^2 and
approximates a rectangular plan form as indicated by a Shape Index
value of 0.2 (the Shape Index is calculated as 1.27 x area ÷ $length^2$
and varies from 1 to 0 as a circular object becomes increasingly
elongate and approaches a straight line; Haggett, 1965; Haggett, et
al., 1977). The average fiord makes 2 bends, one with an angle between
91° and 135°, the other with an angle between 136° and 180°.
Subsidiary sediments intersect at a 70° angle. Other parameters are
not easily discussed in terms of mean values and are examined
separately below.

Fiord Bathymetry. Bathymetric data are available for a limited number
of Baffin Island Fiords. However, it should be noted that recent
seismic surveys in ten Baffin Island fiords indicates that they all
contain substantial sediment fills (Gilbert and MacLean, 1983) that
frequently exceed 200 m thickness. Thus our estimates of maximum fiord
depth are minimum estimates of depth to bedrock. The mean values for
the bathymetric parameters define a fiord with a depth of 127 m 5 km

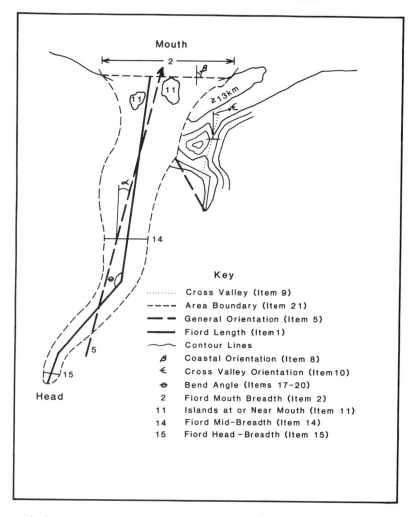

Figure 4.3   Fiord parameter definitions.

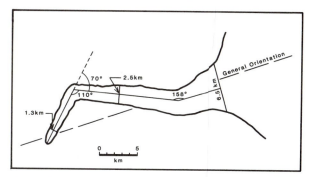

Figure 4.4   The size and shape of the "average" Baffin Island Fiord.
Precise bend angles were not measured; however, the angles given are
necessary to form an average (70°) angle between subsidiary segments.

Table 4.1. Statistics summarizing Baffin Island fiord morphology

ITEM	NUMBER AND PARAMETER TITLE	RANGE	MEAN	STANDARD DEVIATION	SKEWNESS	N*
1	Fiord length	5-121km	27.9	24.7	1.6	227
2	Fiord mouth breadth	1-39 km	6.5	5.8	2.6	227
3	Maximum elevation along fiord	229-1905m	787.1	396.6	0.7	227
4	Number of bends	0-9	2.2	1.8	1.1	227
5	General fiord orientation	0-179°	77.4	52.3	0.3	227
6	Subsidiary segment I orientation	1-179°	86.3	49.5	-0.1	192
7	Subsidiary segment II orientation	2-180°	91.0	55.1	0.1	164
8	Outer coast orientation	0-178°	96.3	50.9	-0.3	227
10	Cross-valley orientation	3-179°	107.4	51.8	-0.6	90
12	Maximum island elevation	76-858m	192.5	199.5	1.8	101
14	Fiord mid-breadth	0.4-8km	2.5	1.4	1.3	227
15	Fiord head-breadth	0.2-5km	1.3	0.9	1.6	227
16	Fiord density down-coast	0-12	4.7	2.5	0.4	227
21	Area	3-953km^2	104.7	158.3	2.8	227
24	Maximum depth within the fiord	20-950m	330.8	196.6	0.8	44
26	Minimum depth in the fiord mouth	2-402m	146.6	110.5	0.3	52
27	Depth 5 km from fiord head	12-512m	127.2	104.0	1.9	32
28	Shape index	0.02-0.6	0.2	0.1	1.4	227
29	Fiord volume	2-1220km^3	179.7	260.5	2.2	44

*Number of fiords with data available for each variable

TABLE 4.2. Fiord summary statistics - percentage data

Item 9	Presence of Cross-Valleys	39% Yes 61% No	N*
Item 11	Presence of islands at or near fiord mouth	57% Yes 43% No	227
Item 13	Access to the Interior Plateau	35% No 37% Yes 8% local access only 19% Indirect access 1% Connected to fiords with access	227
Item 17 Item 18 Item 19 Item 20	Bend Angles: Angles from 0° to 45° Angles from 46° to 90° Angles from 91° to 135% Angles from 136° to 180°	0.2%** 4% 48% 48%	227
Item 22	Presence of trough extending from fiord mouth	7% Yes 5% No 88% No data available	27
Item 23	Trough Applicable†	54% Yes 46% No	227
Item 25	Presence of Sill at the Mouth of a fiord	11% Yes 8% No 81% No data avaiable	43

* Number of fiords with data available for each variable.
* Expressed as a percentage of the total number of bends.
† Indicates the number of fiords situated in a position to have an
  offshore trough, whether or not such data were available.

from the head, a maximum depth of 331 m, a minimum depth of 147 m at
the fiord mouth, and a volume of 180 km^3. These data suggest that the
bottom profile of the fiord approximates a trough of similar depth near
the head and mouth, with a deeper mid-section.

Lôken and Hodgson (1971) found comparable results from soundings
of 26 fiords and major inlets along the eastern coast of Baffin Island,
suggesting that a "characteristic fiord profile" is one with a deep
middle section and a shallower mouth area. They also noted a
correlation between the maximum fiord depth and the highest elevation
along the fiord. This association only appears valid for the general
shape of the topographic and bottom profiles. The area of maximum
fiord depth in Okoa Bay and Coronation Fiord, for example, is located
between 5 and 10 km mouthward of the area of maximum elevation on
land. Lôken and Hodgson (1971) note that Eglington Fiord is also an
exception to their generalization.

Regression between the maximum depth within a fiord and maximum elevation along a fiord showed only low predictive value (r^2 = 0.36). With regard to glacial erosion, the significance of this becomes apparent if an attempt is made to infer ice thickness from the maximum elevation along the fiord (i.e. to associate greater ice thicknesses with higher side-wall elevations and therefore increased depth of erosion), as has been done elsewhere (Holtedahl, 1967). These correlations are not valid for eastern Baffin Island. As indicated by the position of contemporary ice caps, the ice center from which these fiords were fed was located westward of the highest mountains which lie along the eastern coast. Thus the ice thickness and the highest valley-side elevations would not necessarily be related.

Sills and offshore troughs. Data recording the presence of sills at fiord mouths and troughs offshore are also limited. Further, difficulties exist in interpreting sill-like features because their composition is uncertain. Some are believed to be moraines (Løken and Hodgson, 1971), whereas geophysical evidence indicates others are bedrock (MacLean and Falconer, 1979). Where data are available, 58% of the fiords have either a sill or trough. Of the fiords with data for both of these parameters (n = 27), 33% had both a sill and trough, 37% had only a sill, and 30% had only a trough. Further, 57% of the fiords had islands present at or near their mouths, representing emerged sills. Approximately two-thirds of these had offshore troughs.

The presence of sills and troughs at fiord mouths and offshore raises several questions concerning the processes of fiord formation. If, for example, the fiords were eroded largely by ice, the presence of sills must be explained. Where sills are located in the outer fiord areas, the simplest assumption is that they represent areas where glacier ice was no longer confined, permitting spreading and thinning which would decrease the erosive power of the glaciers. However, many sills are found within confined sections of fiord valleys. Here they may represent the glacier grounding line, thus marking the abrupt change from an erosive to non-erosive mode which would also explain the high gradients observed for many of the sills. Even so, it would not explain the presence of offshore troughs beyond the sills. In addition, some of the offshore troughs appear to be offset and do not align with the fiords. For example, Maktak, Coronation, and North Pangnirtung Fiords converge south of Broughton Island though the large offshore trough, which under an assumption of glacial origin would have been eroded in the area of confluence, is situated northeast of Broughton Island.

Access. This variable records whether or not a fiord has access to the interior plateau which represents the largest source area for glacier ice. Interpretation of this parameter is based on the assumptions that the access variable reflects the availability of ice to the valleys, and that the correlation between source or accumulation area and fiord size is valid. Of the 227 fiords measured, 35% did not have access, whereas 37% of the fiord valleys continued onto the interior plateau (Table 4.2). Of the remaining fiords, 8% had access to a local ice body only and hence small accumulation area; 19% may have been supplied ice from tributary valleys. Contingency analysis of the 42 fiords where we could approximate the volume of the fiord and investigate their association with access gave probabilities of the relationship being significant at only $p = .23$.

Haynes (1972) found a positive relationship between the amount of ice in the drainage basin area and the size of glacial valleys around the Sukkertoppen Ice Cap in West Greenland, with 87% of the variance in trough size explained statistically by local drainage basin area. On Baffin Island this correlation is dependent on the method by which the valley/fiord is defined. For example, Løken and Hodgson (1971) studied the relationship between depth and drainage area for several Baffin Island fiords. They found that Inugsuin Fiord had a smaller drainage area than the shallower McBeth and Clyde fiords, thus questioning the link between size and drainage basin area. However, comparing the drainage area of these fiords with fiord volume (V) approximated by the equation: $V = [(\text{maximum elevation} + \text{maximum depth})/2] \times \text{area}$ shows that Inugsuin Fiord has both a smaller drainage area and volume (311 km^3 versus 533 km^3 for McBeth and 817 km^3 for Clyde). In this case fiord volume seems a better measure of size than depth. A study of the relationships between volume and access for the 16 fiords with volumes greater than 150 km^3, shows that 75% have direct access whereas 25% do not. Of the remaining 28 fiords with volume data, 43% have direct access. Though it seems that the correlation between fiord volume and drainage area is valid, we cannot at this point determine the sensitivity of the relationship between access and the amount of ice in the drainage basin area. The importance of the access variable in terms of fiord formational processes is that it shows 35% of the fiords do not have access to the interior plateau. If one assumes that access to an ice body of significant size must be available for a glacial mode of origin, the access parameter also indicates that the fiords of Baffin Island may not all be dominated by the same formational process.

Fiord orientation. Fiord orientation was measured in terms of a general orientation, two subsidiary segment orientations, and a cross-valley orientation where applicable (see Appendix for definitions). Rose diagrams of orientation data are given in Fig. 4.5a. The dominant trend in general orientation is north-northeast; subsidiary segment II shows a N15W mode and a NE mean (Fig. 4.5a). The orientation of subsidiary segment I was essentially random and is not shown. This was not expected as qualitative interpretation indicates that subsidiary segments trend both N-NE and NW, and is due to the interrelated nature of orientation parameters.

A brief discussion of some of the problems evident in quantifying orientation data is needed to aid interpretation and future use of this information. The parameter "general orientation" provides an overall trend which is a composite of the individual segment orientations weighted according to the length of each segment. Subsidiary segment orientations are provided to measure the actual direction of subsidiary segments. However, these variables become interrelated when trying to follow precise parameter definitions. The latitude of the study area must also be considered when examining and interpreting data based on a spatially fixed reference point. The use of a grid centered on the north pole can be problematic when measurements are made over large areas in high latitudes. For example, a straight line drawn through the axis of Baffin Island and measured on a 180° scale relative to north varies by approximately 20°.

Cross-valley orientations (Fig. 4.5(a)C) primarily trend NW, nearly perpendicular to the mean general orientation. The angles between cross-valleys and the general fiord trend have high frequencies for 60° to 70° and 90° to 100° (Fig. 4.5(b)). Subsidiary segment intersections also display a 60° to 70° mode. The trellis pattern exhibited by many fiords in eastern Baffin Island is suggestive of a structural rather than glacial control for the original positioning of these fiords (Fig. 4.6).

Structural influence on fiord morphology

The large-scale tectonic framework of the area is described as a series of transcurrent and tensional faults typical of spreading centers (Wilson, 1963; Keen, et al., 1972; Manchester and Clark, 1973). An examination of the angles between these faults (Martin 1973) shows that the values obtained from the large-scale geology resemble those obtained from the fiord data (Figs 4.5(a) and 4.5(c)). A majority of the faults mapped at the 1:15,000,000scale from the Canadian Geologic Survey tectonic map of Canada (Douglas, 1969) trend northwest (Fig. 4.5(c)B. Those sections of the fiord which trend roughly northwest may also be fault-controlled.

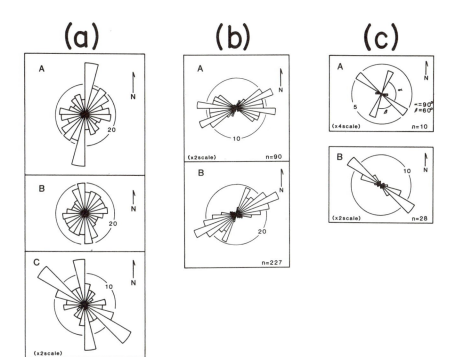

Figure 4.5(a) Rose diagrams of orientation data: (A) Fiord general
orientation; (B) Subsidiary segment orientations based on the secondary
fiord segment which makes the sharpest angle with the general
orientation (subsidiary segment II); and (C) Cross-valley
orientations. Both submerged and emerged valleys were included
provided the floor of an emerged cross-valley did not exceed 152 m (500
ft.). (b) (A) Distribution of intersection angles between
cross-valleys and general fiord orientation. These were calculated as
the absolute value of the difference between the two orientations
measured on a 180° scale. (B) Intersection angles between subsidiary
segments I and II. (c) (A) Rose diagram of Baffin Island fault trends
according to the large-scale geologic framework given by Martin (1973,
Fig. 1). (B) Baffin Island fault orientations compiled from the
Canadian Geologic Survey 1:5,000,000 tectonic map of Canada (Douglas,
1969).

CLASSIFICATION OF BAFFIN ISLAND FIORDS

A statistical approach to classification has long been used to delimit
regions with similar attributes (e.g. Berry, 1960; Stone, 1960) and to

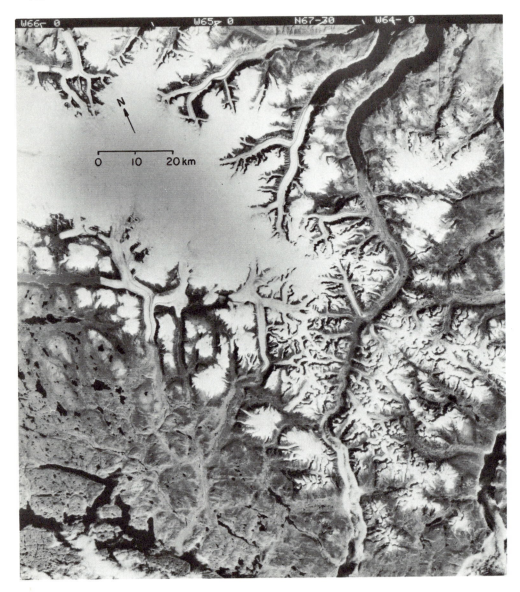

Figure 4.6 Landsat 1:1,000,000 aerial photograph showing a section of
east Bafffin Island centered on the Penny Ice Cap. Note the high angle
fiord bends.

develop taxonomic hierarchies (Sokal and Sneath, 1963). Statistical
grouping methods have recently been applied to problems in Quaternary
Geology (e.g. pollen analysis; Andrews and Nichols, 1981; relative age
dating, Dowdeswell, 1982). Cluster analysis is used here to define

groups of fiords with similar attributes (Krumbein and Graybill, 1969; Davis, 1973). Discriminant analysis is then applied to test the significance level of these groups (King, 1969; Johnston, 1978).

Statistical analyses

Cluster analysis differentiates groups according to a Euclidean distance measure defined as:

$$D_{ij} = \frac{\sum_{k=1}^{m} mX_{ik} - X_{jk}n2}{m}$$

where X_{ik} equals the kth variable measured on case i, X_{jk} equals the kth variable measured on case j, m equals the number of variables in the analysis, and D_{ij} is the distance between case i and case j (Davis, 1973). Two geographically distinct groups were identified: one occupying the northern and southern coasts, the other situated along the eastern coast (Fig. 4.7). Only the variables containing interval or ratio data with a minimum of missing cases were included. The statistical significance of the difference between these two groups can be tested using discriminant analysis, which evaluates the amount of separation between groups and addresses the research hypothesis that distances between group means in n-dimensional space are significantly different (Davis, 1973; Johnston, 1978). Discriminant analysis of the two fiord groups showed that only one discriminant function was significant at the 95% confidence level. This function, representing the maximum elevation along the fiord, explained 89% of the variance. Correct classifications were made in 86% of the cases. The parameter "maximum elevation along the fiord" was the best discriminating variable. The importance of other variables is evaluated using principle component analysis.

R-mode principal component analysis examines variable associations and simplifies the data matrix (Davis, 1973). Principal components are derived to form the best linear combination of variables which explain the highest proportion of variance in the data (Nie, et al., 1975). Only the first component, with an eigenvalue of 4.6 (explaining 32% of the variance in the data) is considered substantively meaningful. Other eigenvalues were near 1 which, for standardized data, represents an average amount of explained variance (Johnston, 1978). The parameters with high loadings on this component are the number of bends a fiord makes, and the length and area of the fiord (i.e. measures of

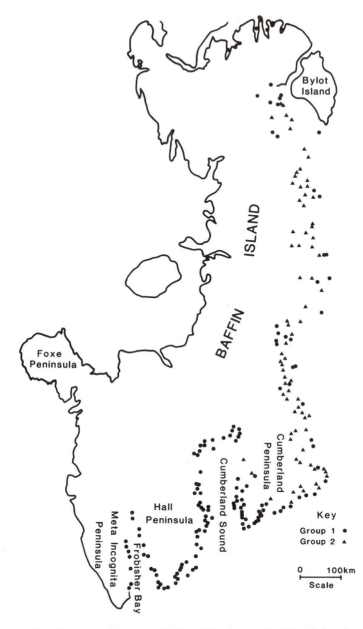

Figure 4.7 Cluster diagram of the fiords on Baffin Island.

fiord size). Mean values for fiords in each group defined through
cluster and discriminant analysis are given in Table 4.3.

TABLE 4.3. Mean parameter values for fiord groups

Item Number and Description		East x	N	North/South x	N
1	Fiord length	36 km	110	20 km	117
2	Fiord mouth-breadth	7.8 km	110	5 km	117
3	Maximum elevation along fiord	106 7m	110	52 4m	117
4	Number of bends	2.5	110	1. 8	117
12	Maximum island elevation	30 1m	47	13 6m	54
14	Fiord mid-breadth	2.8 km	110	2.2 km	117
15	Fiord head-breadth	1.5 km	110	1.1 km	117
21	Area	149 km^2	110	63 km^2	117
24	Maximum depth within the fiord	35 7m	36	21 3m	8
26	Minimum depth in the fiord mouth	15 4m	42	11 6m	10
27	Depth 5km from fiord head	13 8m	28	5 3m	4
28	Shape index	0.1 6	110	0.19	117
29	Fiord volume	211 km^3	36	39 km^3	8

Percentage Data:

9	Presence of cross-valleys	58% No 42% Yes	110	63% No 37% Yes	117
11	Presence of islands at or near fiord mouth	59% No 41% Yes	110	55% No 45% Yes	117
13	Access: No	22%	110	48%	117
	Yes	43%		31%	
	Local	13%		3%	
	Indirect	21%		18%	
	With others	2%		1%	

	Bends:	*	**	*	**
17	0-45°	1.0	1	0.0	0
18	46-90°	1.4	16	1.0	4
19	91-135°	1.7	133	1.5	100
20	136-180°	1.9	126	1.6	112

* Mean number of bends of the specified angle.
* Total number of bends per specified angle.

Fiord group description

The east coast fiord group corresponds with the "Davis Highlands" of Baffin Island and has higher values for the parameter "maximum elevation along the fiord." The mean value for the north/south group is less than half that for the eastern group (Table 4.3). Variations in the maximum elevation along fiords are due to spatial changes in the physiography of Baffin Island and cannot alone be related to the mode of fiord formation. Orientation evidence is examined in conjunction with bend data in order to shed more light on possible modes of fiord formation. The fiords in the eastern group contain a mean of 2.5 bends per fiord and those of the north/south group have an average of 1.8

bends. The mean number of bends of a given angle per fiord also shows
higher values for the eastern group (Table 4.3). The bend parameters
show the eastern group has both higher numbers of bends per fiord and
higher bend angles relative to the north/south group. Further, nearly
two-thirds (63%) of all fiords with more than three bends are in the
eastern group.

General orientation data for the eastern group are shown in Figure
4.8(a). The low frequency of northwest trending fiords and fiord
segments is contrasted with the predominance of northwest trending
cross-valleys (Fig. 4.8(a)C). As with the composite fiord data, the
northwest trending fiord sections and cross-valleys are considered
fault-controlled. The general and subsidiary segment orientations for
the north/south group are shown in Figure 4.9. The northwest component
of this distribution is much more marked here than in the east coast
group, particularly in the northern fiords which may represent a subset
of the north/south group (Fig. 4.8(b)). The northwest trend is again
considered to be related to tectonism. This correlation is especially
probable for the northern fiords because a large number of northwest
trending faults have been mapped on northern Baffin Island (Douglas,
1969).

Variations in length and mouth-width (and the accompanying changes
in area) are in part due to differences in the physiographic setting
between the two groups. Most of the fiords in the eastern group open
into Baffin Bay from a highland plateau some distance inland ($\geq$ 100
km). Many of the fiords in the north/south group open into the
down-dropped grabens of Frobisher Bay and Cumberland Sound, within
relatively short distances (ca. 30 km) of the upland surfaces. Two
subsets are apparent in the north/south group. These are: (1)
relatively short northeast-trending fiords which empty into Frobisher
Bay and Cumberland Sound along their southwestern shores, and (2)
somewhat longer northwest-trending fiords which empty into Davis
Strait. However, these subgroups have not been tested or examined
quantitatively and exceptions do exist (e.g. Kingnait and Pangnirtung
fiords). Combining the size data with the bend data presented
earlier, suggests that a relationship may exist between fiord length and
bend number (recall that the fiords in the eastern group have higher
percentages of fiords with more than 3 bends as well as a larger
size). However, regression analysis shows that there is little
correlation between length and number of bends ($r = 0.5$).
Sugden (1974) also identified two groups of fiords in eastern
Greenland: one group trending roughly east-west, perpendicular to the
coastline, and the other trending north-south, parallel to the coast.

(a)

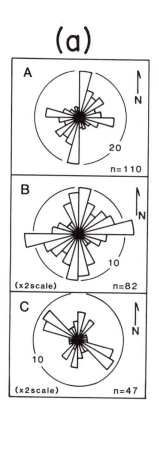

(b)

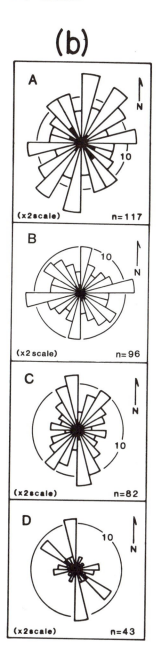

Figure 4.8(a) Orientation data for the fiords in the east coast group:
(A) general orientation; (B) subsidiary segment II orientation; and (C)
cross-valley orientation. Subsidiary segment I orientation was
generally random and is not shown.
     (b) Orientation data for the fiords in the north/south
group: (A) general orientation; (B) subsidiary segment I orientation;
(C) subsidiary segment II orientation; and D) cross-valley
orientation. The northern subset is shaded.

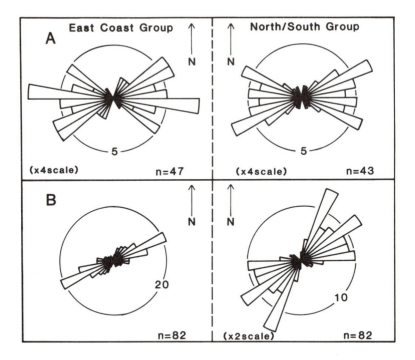

Figure 4.9 Intersection angles between fiord segments. (A) Angles between general and cross-valley orientations. (B) Angles between subsidiary fiord segments.

Sugden (1974) also identified two groups of fiords in eastern Greenland: one group trending roughly east-west, perpendicular to the coastline, and the other trending north-south, parallel to the coast. He notes that the fiords trending north-south were straighter, shallower, and wider than those trending east-west. The north-south trending fiords follow lines of structural weakness. Bretz (1935) had earlier commented on these two fiord groups, concluding that many of the north-south orientated troughs were originally formed by faulting. Similarly, tectonic processes are evident in the locations and positions of the fiords of Baffin Island. The data suggest that the later glacial modification was insufficient to alter substantially the pre-existing structurally and fluvially controlled topography.

Summary of statistical findings

The fiords of Baffin Island fall into two groups: one situated along the eastern coast and the other along the northern and southern coasts. The groups are best distinguished on the basis of the variable "maximum elevation along the fiord" with the group boundaries thus defined coincident with the upland plateau paralleling the eastern

coast of Baffin Island. The fiords in the eastern group trend
generally away from the inland plateau, with a slight radiating pattern
centered around the Barnes Ice Cap. Most cross-valleys to these fiords
trend parallel with the major faults of the island (roughly northwest;
Douglas, 1969). This results in a trellis pattern of fiords (Fig.
4.6). The fiords in the north/south group generally make fewer bends
and are shorter in length than those in the east. The northern subset
of this group trends northwest, also parallel to the dominant fault
pattern. This trend is present, but less pronounced, in the southern
subset. Both groups have many fiords with segments and cross-valleys
intersecting at high angles. The north/south group also has a low
angle component of subsidiary segment intersections, which may reflect
both fluvial and tectonic influences. The north/south fiords are on
average smaller in area than those of the east coast.

DISCUSSION

The assembled data describing the fiords of Baffin Island (Table 4.1)
enable us to make some comparisons with fiords in British Columbia,
Chile, Alaska, and New Zealand (Pickard and Stanton, 1980; Gilbert, in
press). These comparisons are summarized in Table 4.4, and indicate
that the morphology of Baffin Island fiords falls within the range of
shapes observed in other fiord regions. The ratios of some fiord
parameters (Table 4.4) indicate that Baffin Island fiords are on
average (= median) broader relative to length than fiords in other
areas and have low sill depths when compared to maximum fiord depths.
However, although useful, this comparison would be more meaningful if
there were additional data on such properties as the number of bends
and the presence or absence of offshore troughs.

Structural and glaciological differences between these fiord
regions must also be considered. The areas examined by Pickard and
Stanton (1980) are from regions of high snowfall and mass exchange. In
addition, the length of the paleoglacial flowlines in each of these
regions varies dramatically from relatively short distances in New
Zealand and Vancouver Island, to intermediate distances involving
ice-cap sources for British Columbia, Chile and probably many of the
Alaskan fiords. Tectonically these fiords represent coastlines located
along active continental margins (Bostrem, 1970), in contrast to Baffin
Island which is associated with a passive margin (Steckler and Watts,
1981).

The origin of the Baffin Island fiords cannot be resolved by our
data, but we suggest that several of the points discussed earlier must

TABLE 4.4. Fiord parameter ratios from Alaska, British columbia, Chile, New Zealand, and Baffin Island (data from Pickard and Stanton, 1981 and this study).

Medians

	Length (km)	Mean Width (km)	Sill depth (m)	Mx depth (m)
Alaska (16)	95 (20-400)	5.5 (1-14)	80 (10-330)	450 (250-630)
British Columbia (44)	36 (8-140)	1.8 (.8-7)	110 (10-510)	390 (100-720)
Vancouver Island (25)	20 (10-70)	1.2 (.5-2)	50 (10-220)	180 (70-380)
Chile (32)	44 (8-300)	3.2 (.8-15)	140 (10-470)	280 (70-1050)
New Zealand (14)	18 (10-45)	1.2 (1-22)	70 (30-160)	280 (170-450)
BAFFIN ISLAND (227)	18 (5-121)	2.1 (0.4-8)	141 (2-402)	300 (20-950)
(this study)	18 (5-121)	2.1 (0.4-8)	141 (2-402)	300 (20-950)

Ratios

	Length/Width	Rank	Length/Mx depth	Rank	Mx depth/sill depth	Rank
Alaska	17.3	2	211.1	1	5.6	1
British Columbia	20	1	92.3	5	3.5	4
Vancouver Island	16.7	3	111.1	4	3.6	3
Chile	13.8	5	157.1	2	2.0	6
New Zealand	15.0	4	64.3	6	4.0	2
Baffin Island (this study)	9.0	6	141.0	3	2.1	5

* ( ) number of fiords in study.

lead to the rejection of a <u>simple</u> glacial origin. A search for the origin of fiords must start with the pre-Quaternary tectonic, structural and geomorphic history of the region. Mesozoic/early Tertiary rifting on either side of Baffin Bay produced extensive faulting generally trending NW-SE (possibly associated with coastal flexure, Nielson and Brooks, 1981). These trends are parallel to major horst and graben structures as outlined in the geology and bathymetry of Hudson Strait, Frobisher Bay and Cumberland Sound (Kerr, 1980; MacLean and Falconer, 1979). The major east coast fiords also contain sections which follow these structures. Nilson (1973) suggested that certain Norwegian fiords followed major extensional joint sets which he associated with Mesozoic rifting in the North Sea area. The distinction between whether fiords reflect geomorphic processes controlled by structural joint or fault sets, or represent long and narrow grabens is significant. Extensional features of the scale of fiords are reported from the Basin and Range Province of the United States (McGill & Stromquist, 1974; Stromquist, 1976). We propose that drainage would develop and follow these structural trends down the steep east-facing rifted margin. However, a feature of rifted and/or flexing continental margins (Vogt et al., 1969; Menard, 1964) is the presence of fracture zones commonly represented by deep linear troughs cross-cutting the strike of the rift. As streams worked back, these trends would also be exploited and lead to a generally rectangular drainage pattern (Fig. 4.6). Subsidence of the rift margin later on in the process (Hay, 1981) could lead to drowned valleys. These drainage systems would be further exploited with the onset of glaciation.

The relationship between the major offshore troughs (Fig. 2.2) and fiords is unclear. We have discussed some of the problems earlier in this chapter. The absence of an offshore marginal trough at the contact (faulted?) between the Precambrian and Cretaceous sediments (MacLean and Falconer, 1979; MacLean Chapter 6, this volume) suggests a different style of erosion between the Baffin Island and Labrador shelves. As Andrews (1974) noted, the presence of a trough extending off-shore from fiord sills poses a major theoretical problem to the glacial origin of fiords. Löken and Hodgson (1971) and MacLean and Falconer (1974) both describe lateral moraines along trough walls, and the seismic stratigraphy suggests that the transverse troughs are probably erosional. However, the lack of basal tills throughout many sections from the forelands of Baffin Island (Feyling-Hanssen, 1976a; Nelson, 1981; Miller et al., 1977; Mode, 1980) indicates that the outlet glaciers from the interior of Baffin Island have advanced to the outer coast repeatedly throughout the Quaternary but were not grounded beyond it for significant periods (Nelson, 1982; Mode, 1980; Brigham,

1980). This raises the question of when and how the transverse troughs were eroded (cf. Gilbert 1982).

Lastochkin (1978) suggests that troughs on the Eurasian Shelf represent periods of fluvial erosion during a Pliocene regression. If the troughs offshore Baffin Island are eroded by ice, a mechanism for the channeling of ice down these troughs needs to be established. One possibility is the confinement provided by ice shelves. Sanderson (1981) has discussed the requirements for topographically confined glaciers to spread out and form ice shelves. On Baffin Island the angle the glaciers make with the coast is nearly 90° and an ice shelf is unlikely to be stable in this situation unless it can be grounded offshore on shallow banks. However, in Antarctica several glaciers (e.g. Hughes, 1983) extend beyond the mountains and cut through a peripheral ice shelf. In doing so, the glaciers are confined laterally and retain their morphology for considerable distances. England et al. (1978) have presented data in support of an ice shelf along portions of Eastern Ellesmere Island. A peripheral ice shelf may also have existed off of Baffin Island. Such an ice shelf could have formed under extremely cold arctic conditions and represent the consolidation of the fast-ice sheet that extends seasonally from the outer coast to the shelf break. Under severe summer seasons, such as 1972, large segments of the fast-ice remain intact and with time might form a fringing ice shelf. If the climatic deterioration also led to the formation of an ice sheet over Baffin Island/Foxe Basin, advancing fiord glaciers would have to punch through this permanent sea-ice cover. The lateral and frontal buttressing it provided might explain how the glaciers were able to advance down these deep, narrow fiords. The problem with this reconstruction is the limited stratigraphic evidence for such events. MacLean (Chapter 6, this volume) has reported diamictons on the outer Baffin Island Shelf. However, even if the diamictons are tills, they may represent evidence for either: (1) an old, extensive glaciation during which most of the landforms were eroded to their present size and shape; or (2) ice-shelves grounding offshore possibly associated with the thickening of the fast-ice sheet.

A further point concerns the total volume of rock removed from the fiords of Baffin Island (if they are assumed to be erosional). This is of the order of 15×10^{12} m^3, equivalent to about 30×10^{12} m^3 of sediment. Approximately half of this must be stored on the shelves and in the deep basins of Baffin Bay, whereas the more southerly fiords might contribute sediment to the Labrador Sea. If the data can be gathered, a sediment balance approach might assist in resolving some of the unanswered questions on fiord erosion. Vogt (1980) examined

"delta-like excrescences" along the Labrador-Baffin Bay rift. He
argued that these deltaic forms occur at the mouths of troughs and were
formed from deposition by Plio-Pleistocene ice streams. He estimated a
Plio-Pleistocene erosional lowering of the glaciated source region by
200 m based on the sediment volume. Given the (1) glacial landform
distribution (see Chapter 3 this volume), (2) age of glacigenic
sediments on the outer coast of Baffin Island (see Chapter 2), and (3)
reconstructed fiord glacier profiles, it is difficult to see that such
an explanation is correct. Laine's (1980) analysis of deep sea
sediment volume also inferred substantial glacial erosion of the
northeastern Canadian Shield.

This chapter has described the morphology of Baffin Island fiords,
and made a number of inferences on the basis of these data. Several
questions have been raised, but additional evidence is needed before
they can be answered.

In the autumn of 1982 and 1983 the Canadian vessed <u>Hudson</u> cruised
into a large number of Baffin Island fiords. The geophysical and
geological data collected as part of this cruise will undoubtedly shed
much light on the problems raised in our paper.

REFERENCES

Andrews, J.T., 1974: Cainozoic glaciations and crustal movements of the
    Arctic. In: Ives, J.D. and Barry, R.G. (eds.), <u>Arctic and Alpine</u>
    <u>Environments</u>. Methuen, London, 277-318.
Andrews, J.T. and Nichols, H., 1981: Modern pollen deposition and
    Holocene paleotemperature reconstructions, Central northern Canada.
    <u>Arctic and Alpine Research</u>, 13:387-408.
Andrews, J.T., Buennell, G.K., Wray, J.L., and Ives, J.D., 1972: An
    early Tertiary outcrop in north-central Baffin Island, N.W.T.,
    Canada: environment and significance. <u>Canadian Journal of Earth</u>
    <u>Sciences</u>, 9:233-238.

Berry, B.J.L., 1960: The impact of expanding metropolitan communities
    upon the central place hierarchy. Annals, <u>Association of American</u>
    <u>Geographers</u>, 50:112-116.
Birkeland, P.W., 1978: Soil development as an indication of relative
    age of Quaternary deposits, Baffin Island, N.W.T. Canada. <u>Arctic and</u>
    <u>Alpine Research</u>, 10:733-747.
Bockheim, J.G., 1979: Properties and relative age of soil of
    southwestern Cumberland Peninsula, Baffin Island, N.W.T., Canada.
    <u>Arctic and Alpine Research</u>, 11:289-306.
Bostrem, R.C., 1970: Relationship of the world rift system to the fiord
    regions in Chile and British Columbia. <u>Modern Geology</u>, 1:261-273.
Boulton, G.S., 1974: Processes and patterns of glacial erosion. In:
    Coates, D.R. (ed.), <u>Glacial Geomorphology</u>. 5th Annual Geomorphology
    Symposium, State University of New York, Binghampton, 41-87.
Boulton, G.S. and Jones, A.S., 1979: Stability of temperate ice caps
    and ice sheets resting on beds of deformable sediment. <u>Journal of</u>
    <u>Glaciology</u>, 24:29-43.

Boyer, S.J., 1972: Pre-Wisconsin, and Neoglacial ice limits in Maktak
    Fiord, Baffin Island: a statistical analysis. Unpublished M.S.
    thesis, University of Colorado, Boulder. 117 pp.
Bretz, J.H., 1935: Physiographic studies in east Greenland. In: Boyd,
    L.A. (ed.), The Fiord Region of East Greenland. American Geographical
    Society Special Publication 18:159-266.
Brigham, J.K., 1980: Stratigraphy, amino acid geochronology and genesis
    of Quaternary sediments, Broughton Island, east Baffin Island,
    Canada. Unpublished M.S. thesis, University of Colorado, Boulder. 199
    pp.

Clarke, D.B. and Upton, B.G.J., 1971: Tertiary basalts of Baffin
    Island: field relations and tectonic setting. Canadian Journal of
    Earth Sciences, 8:248-258.

Davis, J.C., 1973: Statistics and Data Analysis in Geology. John Wiley
    and Sons, New York. 550 pp.
Denton, G.H. and Hughes, T.J., editors, 1981: The Last Great Ice
    Sheets. John Wiley and Sons, New York. 484 pp.
Douglas, R.J.W., editor, 1969: Geologic and economic minerals of
    Canada, Map 1251A. Geological Survey of Canada Economic Geology
    Report No. 1. 838 pp.
Dowdeswell, J.A., 1982: Relative dating of Late Quaternary deposits
    using cluster and discriminant analysis, Audubon Cirque, Mt. Audubon,
    Colorado Front Range, Boreas, 11:151-161.
Dyke, A.S., 1977: Quaternary geomorphology, glacial chronology and
    climatic and sea-level history of southwestern Cumberland Peninsula,
    Baffin Island, N.W.T., Canada. Unpublished Ph.D. thesis, University
    of Colorado, Boulder. 184 pp.
Dyke, A.S., Andrews, J.T., and Miller, G.H., 1982: Quaternary geology
    of Cumberland Peninsula, Baffin Island, District of Franklin.
    Geological Survey of Canada, Memoir 403, 32 pp.

England, J., Bradley, R.S., and Miller, G.H., 1978: Former ice
    shelves in the Canadian High Arctic. Journal of Glaciology, 20:
    393-404.

Feyling-Hanssen, R.W., 1976: A mid-Wisconsin interstadial on
    Broughton Island, Arctic Canada, and its foraminifera. Arctic and
    Alpine Research, 8:161-182.
Flint, R.F., 1971: Glacial and Quaternary Geology, John Wiley and Sons,
    New York. 892 pp.

Gilbert, R., 1982: Contemporary sedimentary environments on Baffin
    Island, N.W.T., Canada: glaciomarine processes in fiords of eastern
    Cumberland Peninsula. Arctic and Alpine Research, 14:1-12.
Gilbert, R., 1984: The fiordlands of northeastern Canada. In: Memorial
    Volume for J.L. Robinson. Tantalus Press, Vancouver, 27-37
Gilbert, R. and MacLean, B., 1983: Geophysical studies based on
    conventional shallow and Huntec high resolution seismic surveys of
    fiords on Baffin Island. J.P.M. Syvitski and C.P. Blakeney
    (compilers). Sedimentology of Arctic Fiords Experiment: HU82-031 Data
    Report, volume 1. Canadian Date Report of Hydrography Ocean Sciences
    No. 12. 935 pp.
Grant, A.C. 1975: Geophysical results from the continental margin off
    southern Baffin island. In: Yorath, C.J., Parker, E.R., and Glass,
    D.J. (eds.), Canada's Continental Margin and Offshore Petroleum
    Exploration. Canadian Society of Petroleum Geologists, Memoir
    4:411-431.
Grant, A.C. and Manchester, K.S., 1970: Geophysical investigations in
    the Ungava Bay-Hudson Strait region of northern Canada. Canadian
    Journal of Earth Sciences, 7:1062-1076.

Gregory, J.W., 1913: The Nature of Origin of Fiords. John Murray, London. 542 pp.

Haaggett, P., 1965: Locational Analysis in Human Geography. Edward Arnold, London. 339 pp.

Haggett, P., Cliff, A.D., and Frey, A., 1977: Locational Analysis in Human Geography, 2nd Edition. Edward Arnold, London. 605 pp.

Hay, W.W., 1981: Sedimentological and geochemical trends resulting from the breakup of Pangea. Oceanologica Acta Supplement, 4:135-147.

Haynes, V.M., 1972: The relationship between the drainage areas and sized of outlet troughs of the Sukkertoppen Ice Cap, west Greenland. Geografiska Annaler, 54A, 66-75.

Holtedahl, H., 1967: Notes on the formation of fjords and fjord valleys. Geografiska Annaler, 49A, 188-203.

Hughes, T.J., 1983: On the disintegration of ice shelves; the role of fracture. Journal of Glaciology, 29, 98-117.

Johnson, D.W., 1915: The Nature and Origin of Fiords. Science, 41:537-543.

Johnson, G.L., McMillan, N.J., Rasmussen, M., Campsie, J., and Dittmer, F., 1975: Sedimentary rocks dredged from the southwest Greenland continental margin. In: Yorath, D.J., Parker, E.R., and Glass, D.J. (eds.), Canada's Continental Margins and Offshore Petroleum Exploration. Canadian Society of Petroleum Geologists, Memoir 4:391-432.

Johnston, R.L., 1978: Multivariate Statistical Analysis in Geography. Longman, London. 280 pp.

Keen, C.E., Barrett, D.L., Manchester, K.S., and Ross, D.I., 1972: Geophysical studies in Baffin Bay and some tectonic implications. Canadian Journal of Earth Sciences, 9:239-256.

Kerr, J. Wm., 1980: Structural framework of Lancaster Aulacogen, Arctic Canada. Geological Survey of Canada, 319, 24 pp.

King, L.J., 1969: Statistical Analysis in Geography. Prentice-Hall, Englewood Cliffs, New Jersey. 288 pp.

Krumbein, W.C. and Graybill, F.A., 1965: An Introduction to Statistical Models in Geology. McGraw-Hill, New York. 475 pp.

Laine, E.P., 1980: New evidence from beneath the western North Atlantic for the depth of glacial erosionin Greenland and North America. Quaternary Research, 14:188-198.

Lastochkin, A.N., 1978: Submarine valleys on the northern continental shelf of Eurasia. Polar Geography, 240-250.

Le Pichon, X., Sibuet, J.C., and Francheteau, J., 1977: The fit of the continents around the North Atlantic Ocean. Tectonophysics, 38:169-209.

Linton, D.L., 1963: The forms of glacial erosion. Transactions of the Institute of British Geographers, 33:1-28.

Locke, W.W., 1979: Etching of hornblende grains in Arctic soils: an indicator of relative age and paleoclimate. Quaternary Research, 11:197-212.

Locke, W.W., 1980: The Quaternary geology of the Cape Dyer area, southernmost Baffin Island, Canada. Unpublished Ph.D. thesis, University of Colorado, Boulder. 331 pp.

Løken, O.H. and Hudgson, D.A., 1971: On the submarine geomorphology along the east coast of Baffin Island. Canadian Journal of Earth Sciences, 8:185-195.

MacLean, B. and Falconer, R.K.H., 1979: Geological/geophysical studies in Baffin Bay and Scott Inlet-Buchan Gulf and Cape Dyer- Cumberland Sound areas of the Baffin Island Shelf. Geological Survey of Canada, Paper 79-1B, 231-244.

MacLean, B., Jansa, L.F., Falconer, R.K.H., and Srivastava, S.P., 1977:
    Ordovician strata on the southestern Baffin Island shelf revealed by
    shallow drilling. Canadian Journal of Earth Sciences, 14:1925-1939.
Manchester, K.S., and Clarke, D.B., 1973: Geological structure of
    Baffin Bay and Davis Strait as determined by geophysical techniques.
    In: Pitcher, M.G. (ed.), Arctic Geology. American Association of
    Petroleum Geologists, Memoir 19, 536-541.
Martin, R., 1973: Crataceous-Tertiary rift basin of Baffin Bay-
    continental drift without sea-floor spreading. In: Pitcher, M.G.,
    (ed.), Arctic Geology. American Association of Petroleum Geologists,
    Memoir 19:500-505.
McGill, G.E., and Stromquist, A.W., 1974: A model for graben formation
    of subsurface flow; Canyonlands National Park, Utah. Department of
    Geology and Geography, Contribution No. 15, University of
    Massachusettes, Amherst. 79 pp.
Mears, A.K., 1972: Glacial geology of crustal properties of the
    Nedlukseak Fiord region, east Baffin Island, Canada. Unpublished
    M.S. thesis, University of Colorado, Boulder. 86 pp.
Menard, H.W., 1964: Marine Geology of the Pacific. McGraw-Hill, New
    York. 271 pp.
Miller, G.H., Andrews, J.T., and Short, S.K., 1977: The last
    interglacial-glacial cycle, Clyde Foreland, Baffin Island, N.W.T.:
    stratigraphy, biostratigraphy and chronology. Canadian Journal of
    Earth Sciences, 14:2824-2857.
Mode, W.N., 1980: Quaternary stratigraphy and palynology of the Clyde
    Foreland, Baffin Island, N.W.T., Canada. Unpublished Ph.D. thesis,
    University of Colorado, Boulder. 219 pp.

Nelson, A.R., 1981: Quaternary glacial and marine stratigraphy of the
    Quivitu Peninsula, northern Cumberland Peninsula, Baffin Island.
    Geological Society of American Bulletin, 92, Part II, 1143-1261.
Nelson, A.R., 1982: Aminostratigraphy of Quaternary marine and
    glaciomarine sediments, Quivitu Peninsula, Baffin Island. Canadian
    Journal of Earth Sciences, 19:945-961.
Nie, N.H., Hull, C. H., Jenkins, J.G., Steinbrenner, K., and
    Bent D.H., 1975: SPSS Statistical Package for the Social Sciences,
    2nd Edition. McGraw-Hill, New York, 675 pp.
Nielsen, T.F.D. and Brooks, D.K., 1981: The East Greenland rifted
    continental margin: an examination of the coastal flexure. Journal
    Geological Society London, 138:559-568.
Nilson, T.H., 1973: The relation of joint patterns to the formation of
    fjords in western Norway. Norsk Geologisk Tidsskrift, 53:183-194.

Pheasant, D.R., 1971: The glacial chronology and glacioisostasy of the
    Narpaing-Quajon Fiord area, Cumberland Peninsula, Baffin Island.
    Unpublished Ph.D. thesis, University of Colorado, Boulder. 232 pp.
Pickard, G.L. and Stanton, B.R., 1980: Pacific fjords--a review of
    their water characteristics. In: Freeland, H.J., Farmer, D.M., and
    Levings, C.D., (eds.), Fjord Oceanography. Plenum Press, New York,
    1-52.
Reid, I., and Falconer, R.K.H., 1982: A seismicity study in northern
    Baffin Bay. Canadian Journal of Earth Sciences, 19:1581-1531.

Sanderson, T.J.O., 1979: Equilibrium profile of ice shelves. Journal
    of Glaciology, 22:435-461.
Sanford, B.V., Grant, A.C. Wade, J.A., and Barss, M.S., 1979: Geology
    of eastern Canada and adjacent areas, 1:2,000,000. Geological Survey
    of Canada, Map 1401A.
Sokal, R.R. and Sneath, P.H.A., 1963: Principles of Numerical
    Taxonomy. W.H. Freeman and Co., San Francisco. 359 pp.

Steckler, M.S. and Watts, A.B., 1981. Subsidence history and tectonic evolution of Atlantic-type continental margins. In: Scrutton, R.A., (ed.), Dynamics of Passive Margins. American Geophysical Union, Geodynamics Series, 6, 184-196.

Stone, R., 1960: A comparison of the economic structure of regions based on the concept of distance. Journal of Regional Science, 20:1-20.

Stromquist, A.W., 1976: Geometry and growth of grabens, lower Red Lake canyon area, Canyonlands National Park, Utah. Department of Geology and Geography, University of Massachusettes, Amherst. 119 pp.

Sugden, D.E., 1974: Landscapes of glacial erosion in Greenland and their relationship to ice, topographic and bedrock conditions. Institute of British Geographers, Special Publication No. 7:177-195.

Sugden, D.E., 1978: Glacial erosion by the Laurentide Ice Sheet. Journal of Glaciology, 20:367-391.

Sugden, D.E., and John, B.S., 1976: Glaciers and Landscape. John Wiley and Sons, New York. 376 pp.

Szabo, B.J., Miller, G.H., Andrews, J.T., and Stuiver, M., 1981: Comparison of uranium-series, radiocarbon and amino acid data from marine molluscs, Baffin Island, arctic Canada. Geology, 9:451-457.

Vogt, P.R., 1980: Post-rift continental accretion: Arctc, Northest Atlantic, and Labrador Seas: largely the work of glacial age ice streams? Berliner Geowissenschafliche Abhandlungen, Reihe A/Band 19. Internationales Alfred Wegener Symposium, Berlin, 243-244.

Vogt, P.R., Schneider, E.D., and Johnson, G.L., 1969: The crust and upper mantle beneath the sea. Geophysical Monograph, No. 13:556-617.

Wilson, J.T., 1963: Hypothesis of earth's behavior. Nature, 198: 925-929.

APPENDIX

FIORD PARAMETER DEFINITIONS

Item 1--Fiord Length--The sum of the lengths of the straight line
segments drawn through the center of the fiord. Lines are constructed
minimizing the number of bends by starting new line segments only when
a line intersects a fiord wall. The new segment also begins in the
center of the fiord. Fiords shorter than 5 km and longer than 125 km
were excluded from the analyses.
Item 2--Fiord Mouth Breadth--The length of a line crossing the fiord at
its mouth. The points of contact from which this line is derived are
defined by the first change in direction (plan view) along the coast at
the fiord mouth, pinpointed by the intersection of the tangents to this
break in slope.
Item 3--Maximum Elevation Along the Fiord--The maximum elevation of
land within one fiord breadth on either side of the fiord. Where no
spot height is given, the elevation of the last contour line plus one
half the contour interval is used.
Item 4--Number of Bends--The number of bends along the length of the
fiord. The joining of two straight line segments defines one bend.
Item 5--General Fiord Orientation--The general orientation of the
fiord, measured in the direction of a vector extending from the
midpoint of the fiord to the fiord mouth. A 360° scale is used. When
the fiord is composed of only two segments, the general orientation is
equal to the resultant vector.
Item 6 and 7--Subsidiary Segment Orientations--The orientations of the
separate line segments used to measure fiord length, provided these
differ from the general orientation by 20° or more. The longest line
segment is considered first, then the segment which makes the highest
angle with the general orientation. Measured similarly to Item 5.
Item 8--Outer Coast Orientation--The orientation of the fiord mouth as
defined by the straight line segment used to measure fiord mouth
breadth. The orientation is measured on a 180° scale, clockwise from
North. When used in conjunction with the general orienation, the
direction of the fiord with respect to the coastline results.
Item 9--Presence or Absence of Cross-Valleys--Cross-valleys are defined
as side-valleys linking two fiords. No distinction is made between
submerged and emerged cross-valleys, although by far the majority are
submerged. If the floor of an emerged valley exceeded 152 m (500
feet), it was not considered a cross-valley.
Item 10--Cross-Valley Orientation--The orientation of the cross-valley
measured clockwise from North on a 180° scale.
Item 11--Islands at or near the Fiord Mouth--This variable indicates
the presence or absence of islands (and by inference, shallow sills)
near fiord mouths. When a land mass is greater than or equal to 13 km,
it was considered to separate two fiords, rather than represent an
island in the main fiord.
Item 12--Maximum Island Elevation--The maximum elevation of islands at
or near the fiord mouth. Where spot heights are not given, the
elevation of the last contour line plus one half the contour interval
is used.
Item 13--Access to Interior Plateau--Indicates whether or not a fiord
has access to the interior plateau, and the type of access. Five
categories are defined: 1) No access; 2) Direct access--the fiord is
connected to the interior plateau by a valley continuing from the fiord
head; 3) Local access only--the fiord has access to a local
accumulation area, but no direct access to the interior plateau;
Indirect access--this involved cases where tributaries could have

supplied ice--otherwise there would be no access (e.g. where the fiord ends in a steep headwall, but tributary valleys may have supplied ice); and 5) Access with others--this applies to fiords that originate in other fiords and have access by way of these other fiords.

Item 14--Fiord Mid-Breadth--The breadth of the fiord measured at half its length. The width is measured by connecting two closely spaced points on both sides of the half length mark and measuring the orthogonal to this line.

Item 15--Fiord Head Breadth--The width of the fiord head is measured at a point representing one-quarter of the length of the last straight line segment from the fiord head to its mouth.

Item 16--Density Down Fiord--The number of fiords within the first 50 km down-coast from a particular fiord.

Items 17 to 20--Bend Angles--The magnitude of the angles created by the intersection of the line segments used to measure length were recorded by frequency.

    Item 17--the number of bends with angles from 0 - 45°.
    Item 18--the number of bends with angles from 46 - 90°.
    Item 19--the number of bends with angles from 91 - 135°.
    Item 20--the number of bends with angles from 136 - 180°.

Item 21--Fiord Area--The area of the fiord as outlined by the present water level and the line segment used to measure the mouth width. Areas were taken directly from topographic maps using a digitizing planimeter.

Item 22--Trough--The presence or absence of an offshore trough is determined from hydrographic charts.

Item 23--Trough Applicable--Indicates the suitability of fiords in determining the presence or absence of a coastal trough (Item 22). Fiords opening along the coast were considered applicable, whereas fiords opening into other fiords were not.

Item 24--Maximum Depth Within the Fiord--The maximum depth sounding within each fiord was measured using hydrographic charts.

Item 25--Sills--The presence or absence of sills at the fiord mouth. Sills were identified by areas of shallow depth soundings, surrounded by deeper soundings.

Item 26--Minimum Depth within Fiord Mouth--The minimum depth sounding inside the fiord mouth.

Item 27--Depth 5 km from Fiord Head--The depth sounding at a distance of 5 km from the fiord head.

Item 28--Shape Index--The overall shape or elongation of a fiord defined as: $1.27 \ (Area)/(length)2$

Item 29--Fiord Volume--An estimate of the volume of a fiord defined as:

$$\frac{(Maximum \ elevation \ along \ the \ fiord + maximum \ depth \ in \ fiord)}{2} \times Area$$

5 Grain-size characteristics of Quaternary sediments, Baffin Island region

J. T. Andrews

In Chapter 1 we outlined the sedimentary environments of Baffin Island and surrounding shelves and seas and suggested the importance of these different records to our understanding of Quaternary events. In this chapter I pull together some two decades of data acquistion on the grain-size characteristics of sediments that have been collected by myself and other colleagues. Grain-size data are in themselves not totally diagnostic of the depositional and post-depositional history of sedimentation. Modern sedimentological research quite rightly stresses the concept of facies or sediment associations and in one recent publication entitled Facies Models (ed. Walker, 1979) it was noticeable that relatively little attention was paid to grain-size. This is most probably because, in the words of Walker (1979, p. 1) "Unfortunately, statistical methods are unsuited to clastic rocks, where most of the information (sedimentary and biological structures) cannot readily be quantified." There is little doubt that textural data are but one facet of sedimentological research; nevertheless an important one.

Research on Baffin Island has certainly not ignored the emphasis on lithofacies; indeed several workers have made substantial contributions to our understanding of glacial marine shallow water facies by detailed descriptions and interpretations of tens of kilometers of raised glacial marine/glacial/beach sediments exposed in wave-cut cliffs along the outer east coast of Baffin Island (Nelson, 1978, 1981; Mode, 1980; Mode et al., 1983; Miller et al., 1977; Brigham, 1983). In addition, specific research has been carried out on the Baffin Island sandur deposits (Church, 1972), on glacial marine sediments in fiords and major bays (Gilbert, 1982; Osterman, 1982; Osterman and Andrews 1983), on glacial sediments and processes (Andrews, 1963; Dowdeswell, 1982) and to a lesser extent on coastal sediments (Miller et al., 1980). In addition, Church et al. (1979) have discussed in detail aspects of debris slope accumulations - a facet of the Baffin Island environment which I do not consider in this chapter.

The volume of glacigenic sediments on Baffin Island is not large. LANDSAT mapping indicates that in many places surficial sediments are either thin or absent (Müller, 1980; Colvill, 1982; Chapter 3 this volume). In the area of light aerial scour (Sugden, 1978; Fig. 3.2 this volume) the surface sediment is a reworked felsenmeer or blockfield and is again of limited thickness. Although glacial fluvial and alluvial fills can be substantial in many of the valleys and fiords (cf Church, 1972) sediment thicknesses of more than 100 m are rare. Probably the greatest volume of sediment is contained in the raised glacial marine deltas and associated outwash plains that mark the transition from glacial to marine deposition (Church, 1972, 1978; Gilbert, 1982), and in glacial marine sediments deposited in glacierized fiord basins (Gilbert and MacLean, 1983; Lewis et al., 1977). The total volume of Quaternary sediment stored on land or in nearshore basins is unknown but my estimate would be of the order of 4300 km^3.

BAFFIN ISLAND SEDIMENTARY ENVIRONMENTS

Figure 5.1 is a schematic diagram of the environments discussed in this chapter. Some explanation of the diagram is required. The sketch illustrates a glacierized fiord. The major bedrock lithologies are a variety of Precambrian granites, granite gneisses, and metasediments (e.g. Jackson and Taylor 1972). However, at least in the major sounds and bays, Paleozoic limestones (Silurian and Ordovician) are downfaulted and constitute substantial portions of the floors of Frobisher Bay, Cumberland Sound, and Hudson Strait (Grant, 1975; McLean and Falconer, 1979). Finally, away from the east coast of Baffin Island, Mesozoic and Tertiary marine sediments crop-out on the shelf and constitute a large sediment wedge primarily of pre-Miocene rocks.

Figure 5.1 shows only limited glaciation. Large areas of till of different ages are exposed seaward of the present glacial margin and the effects of glacial isostatic rebound are clearly evident along most shorelines by the presence of raised marine beaches, raised deltas, and in places extensive exposures of glacial marine muds (cf. Andrews et al., 1970). Major meltwater streams drain from the glacier and outwash trains characterize many valley bottoms at present (Church, 1972) whereas substantial terrace deposits indicate that paraglaciation

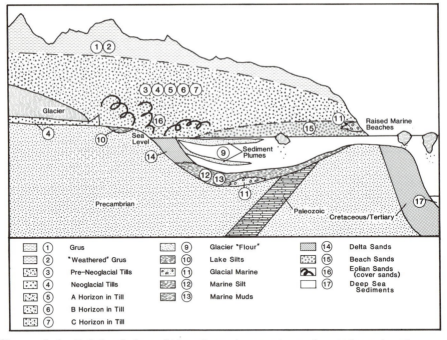

①	Grus	⑨	Glacier "Flour"	⑭	Delta Sands
②	"Weathered" Grus	⑩	Lake Silts	⑮	Beach Sands
③	Pre–Neoglacial Tills	⑪	Glacial Marine	⑯	Eolian Sands (cover sands)
④	Neoglacial Tills	⑫	Marine Silt	⑰	Deep Sea Sediments
⑤	A Horizon in Till	⑬	Marine Muds		
⑥	B Horizon in Till				
⑦	C Horizon in Till				

Figure 5.1 Model of depositional environments and settings in the
eastern Canadian Arctic. The numbers for each environment/sediment are
used in other figures and within the text.

(Church and Ryder, 1972) was a major source of sediment transfer during
deglaciation. Large raised outwash terraces characterize many of the
major valleys on both the east and west coasts of the island (Andrews,
1965; Church, 1972, 1978). Coarse clastic sediments (sand size and
coarser) are usually stored on land, but silts and clays are
transported in suspension and enter lakes or fiords where they are
deposited as lacustrine or marine muds (Gilbert and Church, 1983;
Davis, 1980). Some of these sediments may be transported through the
nearshore basins and out into the deep-sea (#17, Fig. 5.1) where they
will be deposited in conjunction with a rain of ice-rafted detritus
from icebergs and from sediment transported in and on sea ice floes
(Fillon et al., 1981, 1982).

    In recent years there has been a realization that there is a
significant eolian component in the Baffin Island surface sediments.
In 1954 Thompson described the layered sands of Pangnirtung Pass which
can attain thicknesses of 8 m and which certainly seem to have a strong
niveo-eolian origin (e.g. Andrews et al., 1979). In addition along the
outer coast of Baffin Island Miller et al. (1977) described the Scott

Inlet sands as a wedge of eolian sand that mantles the surface of the outer coast. Finer particles in the silt/clay range are transported to higher elevations, well above the floodplains, and this influx may be the origin for silt accumulations that occur on top of pebbles and cobbles in late Pleistocene and older soils (Locke, 1980; see Chapter 12 this volume).

GRAIN-SIZE MEASUREMENTS AND DATA ANALYSIS

Grain-size measurements at INSTAAR are usually undertaken by R. Kihl, senior technician in the Sedimentological Laboratory. During the years there have been some changes in operating procedures and in the procedures that have been followed. Table 5.1 briefly outlines the main procedures that we follow. In the last two years analysis of the silt and clay fractions (63 μm to finer than 0.49 μm; 4φ to 11φ) has been greatly speeded up by the use of a Sedigraph 5000D Particle Size Analyzer. Repeated testing of the Sedigraph results against the accepted laboratory norm, that is the pipette method, indicates that the Sedigraph closely duplicates the pipette analysis (Fig. 5.2). The terminology in this paper follows that of Wentworth with the sand fraction being defined as grains between 2000 and 62.5 μm. Other boundaries and terms are illustrated as Table 5.2. For many purposes grain-size data are reported as percentages of the <2000 μm fraction although in certain sediments, such as tills or other diamictons, the distribution of clasts coarser than 2000 um has also been determined.

Statistical parameters for each sample have most frequently been expressed in terms of the Folk and Ward (1957) graphic moment measures although of late we have opted for a calculation of means, standard deviation, skewness, kurtosis, and delta variable based on moment measures (Swan et al., 1979; Leroy, 1981). However, in our study statistical parameters are mostly based on the graphical approach noted above. Textural descriptions are based on Shepard's (1954) ternary diagram for sand, silt, and clay.

Each sample was characterized by 23 variables. Clay percents were not included as they can be obtained by a simple differences calculation given the percentages of sand and silt. A total of 622 samples were included in the data bank. For each sample a genetic origin for the sediment was coded based on the written laboratory card, theses, or publication description. From this list of about 25 discrete environments of deposition there were sufficient samples to present data on 17 major sediment types. These not only include "original" sediments but also include information on textural changes

TABLE 5.1. Proceedures used in the INSTAAR Sedimentological Laboratory (1969-present).

Determination		Method	Reference
Sand %	1969-1979	Sand analysis by sieving methods. Silt and clay analysis by sedimentation procedure using the pipet alequnt method.	Manual of Sedimentary Petrography-Krumbein and Pettijohn 1938
Silt %	1980-Present	Sand analysis by sieving methods. Silt and clay analysis by sedimentation procedure using the Sedigraph 5000D Particle Size Analyzer. Particle concentration measured by a finely collineated x-ray beam.	INSTAAR Sedimentology Laboratory Report - June 1980
Organic carbon %			
Carbonate %			
Hygroscopic moisture %	1969-Present	15 hr. 1105°C. [Oven-dry Wt. Basis]	
Organic Carbon %	1969-1980	Loss on ignition. Wt. loss 105°C 450°C	
Organic Matter %	1981-Present	Walkley-Black Method [Postassium Derhements Method] Reported as readily oredegable organic matter and organic carbon. Heat by dilution only. Total recovery factor 1.30 not used.	Jackson, M.L., 1958 Soil Chemical analyzer. Prentice-Hall, Inc. Sixth printing by the author, Department of Soil Science, University of Wisconsin, Madison, Wisconsin 53076
Carbonate %	1969-Present	Gasometric determinations of carbon dioxide using the Chittick apparatus.	Journal Sed. Pet. Vol 32, No. 3, pp. 520-529. Sept. 1962.

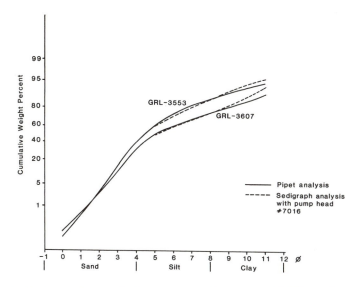

Figure 5.2   Comparison of pipette and Sedigraph 5000D analyses (from R. Kihl).

TABLE 5.2.   Grain-size diameters and phi intervals
             for textural terms

Diameter (μm)	Phi (φ)	Description
2000	-1	very coarse sand
1000	0	coarse sand
500	1	medium sand 250
	2	fine sand
125	3	very fine sand
625	4	coarse silt
313	5	medium silt
156	6	fine silt
78	7	very fine silt
39	8	coarse clay
19.5	9	medium clay
9.8	10	fine clay
4.9	11	

in the A, B and C horizons of soil developed on tills (e.g. Birkeland, 1978; Bockheim, 1980; Isherwood, 1975). A data file of all samples was then established on the University of Colorado's CDC Computer System. Analysis was performed using the SPSS (Statistical Package for the Social Sciences (Nie et al., 1975). Specific programs that were used for this contribution included Condescriptive and Discriminant Analysis. Descriptive statistics for each sediment were generated from the program Condescriptive and these were used to construct a series of "average" histograms of grain-size distributions for the various environments (Fig. 5.1). These histograms are determined on the percentages by weight in the >2000 μm fraction. The percentages of the total sample >2000 μm is also shown.

These data form an initial background for understanding the various sedimentary environments that have existed, and in many cases still exist today across the eastern Canadian Arctic. The model for the environments of deposition (Fig. 5.1) also suggest the possibility of examining the data against the model of McLaren (1981, p. 611) who states: "In a system of related environments, these trends (in mean grain-size, sorting, and skewness) can be used to identify both the probable source and the probable deposit and, by inference, the net sediment transport paths among sedimentary deposits."

CHARACTERIZATION OF THE DIFFERENT SEDIMENTS

1. & 2. Gruss:

Gruss is defined as the in situ weathering products which result from the physical and chemical breakdown of rock. In Arctic areas surface weathering of boulders is apparent after 6 to 8 ka* and indeed the rate of surface weathering is commonly used as one criterion for the establishment of a relative age sequence (Boyer and Pheasant, 1975; Dyke, 1977; Locke this volume). Gruss is a characteristic deposit of the higher mountain summits which have experienced a long history of subaerial weathering and have been little affected by recent glacial erosion. Many mountain summits lie within weathering zone I (Pheasant and Andrews, 1973; Boyer and Pheasant, 1975; Locke this volume) - the oldest recognized landscape in the outer coastal mountains of Baffin Island.

----

*ka is equal to "thousands of years"

The samples were collected at several locations, in one case, excavation indicated that the gruss was quite deep and consisted of an upper layer of "unweathered" minerals, whereas at depth the gruss was heavily oxidized and possessed the characteristics of a B horizon. The grain-size distribution of the sediment (#1 and #2, Fig. 5.3) indicates the gruss, on average, has a modal class in the very coarse sand and texturally is classified as a clayey sand (Figs 5.4 & 5.5). Note in particular the relatively large amount of clay-sized material that is finer than 10ϕ (Fig. 5.3). The one sample of weathered gruss has a more symmetric log-normal distribution with a modal class of medium sand. The percentage of grains larger than -1ϕ (2000 μm) is 14 to 7% of the total sample. The weathered gruss has a distinct build-up in the silt and clay with a minor mode in the coarse silt. Both types of gruss are very poorly sorted (Figs 5.6 & 5.7) (for abbreviations used in the figures see Table 5.3). The origin of silt and clay-sized particles in arctic soils is a matter of interest and debate (Isherwood, 1975; Locke, 1980; Tedrow, 1977). In places the silt forms caps on the upper surfaces of pebbles and cobbles and it thus appears to be at least partially the result of mechanical infiltration of fine-grained sediments, probably associated with eolian processes.

3, 5, 6, & 7 Pleistocene till:

"Pleistocene" till is used in this chapter as a general descriptor for tills that pre-date the neoglaciation, that is for deposits that usallly date from 5 ka or earlier. These sediments are commmonly associated with former fiord glaciers, or from areas close to the margin of the former Laurentide Ice Sheet. The tills have been collected from sections and from soil pits. In contrast, neoglacial tills have been collected from the termini of existing valley and cirque glaciers which are invariably ice-cored. Pleistocene tills have been classified in our data bank as "till" (undifferentiated) and from the A, B, and C horizons of soils developed on lateral and terminal moraines of Pleistocene age. Pleistocene till (Fig. 5.3) consists on average of 24% by weight of clasts larger than 2000 μm. The sediment has a broad modal peak in the medium to fine sand category and are texturally a clayey sand (Figs 5.4 & 5.5). In comparing the average textures of sediments from the A, B, and C horizons (Fig. 5.3) one of the most noticeable features of the A and B horizon textures is the increased percentage of very fine clay. This may represent either an increase in this grade due to weathering (Moss et al., 1981) or be associated with the infiltration of eolian dust. Locke (1980) has suggested that in eastern Baffin Island there is a significant eolian component in the soils.

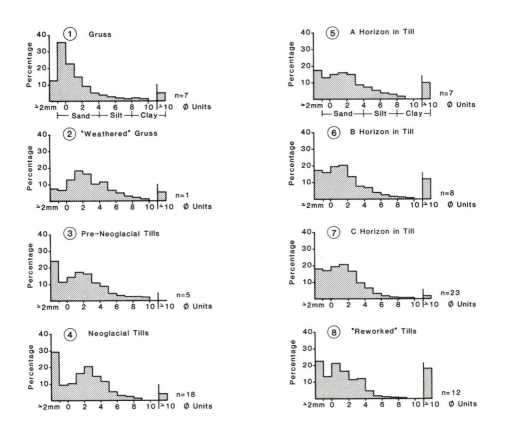

Fig. 5.3. Histograms of the average sediment percents by weight in phi
(φ) class intervals from -1 to finer than 10 or 11φ. To the left is
shown the percentage of the total sediment coarser than -1φ (coarser
than 2000 μm).

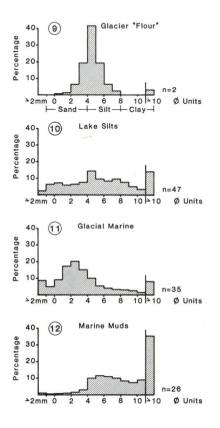

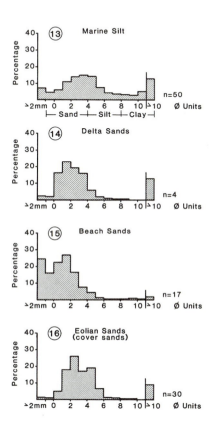

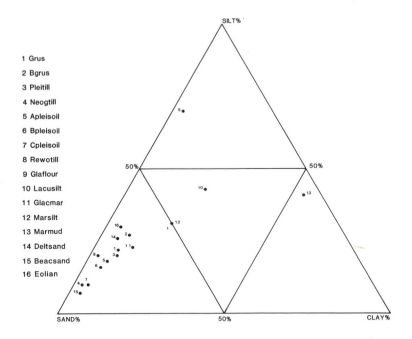

1 Grus
2 Bgrus
3 Pleitill
4 Neogtill
5 Apleisoil
6 Bpleisoil
7 Cpleisoil
8 Rewotill
9 Glaflour
10 Lacusilt
11 Glacmar
12 Marsilt
13 Marmud
14 Deltsand
15 Beacsand
16 Eolian

Figure 5.4 Triangular graph showing the sand/silt/clay percentages for the average sediment distributions shown as Figure 5.3.

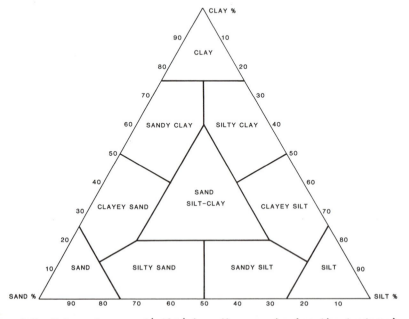

Figure 5.5 Triangular sand/silt/clay diagram showing the textural classes used in this study (from Shepard, 1954).

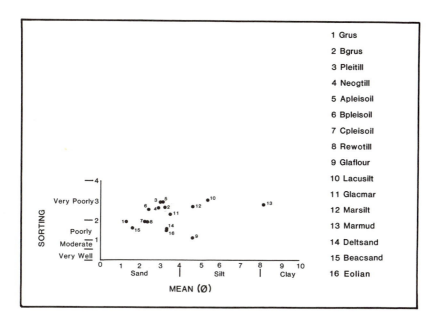

Figure 5.6 Graphic mean versus graphic sorting of the sediments.
Numbers refer to the environments/sediments noted as Figure 5.1.

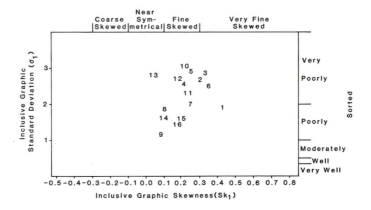

Figure 5.7 Graphic sorting versus graphic skewness for the histograms
shown as Figure 5.3.

The textures for Pleistocene tills were largely determined from samples collected from pits dug in the crest of moraines. It might be argued, with some justification, that such tills may have been strongly affected by subaerial processes during and after sedimentation. They could represent, for example, surface melt-out or flow tills. There are few tills exposed in the coastal cliff sequences along eastern Baffin Island and most diamictons that crop-out in these sections are glacial marine in origin (Nelson, 1978, 1981; Brigham, 1983; Mode, 1980; Mode et al., in press). Dowdeswell (1982) has, however, collected samples of sediment from the basal layers of the Watts Glacier, southern shore of Frobisher Bay. A composite diagram from 6 basal ice samples is shown as Figure 5.8a. These samples were obtained in the field by melting ice and then forcing the water through a millipore filter with a small air pump. Dowdeswell (personal communication) suggests that the results might underrepresent the amount of silt and clay-sized particles because of clogging of the filter paper. This possible effect has to be born in mind in comparing Figure 5.8 with Figure 5.3 (3, 5, 6, & 7). Even in this basal transport position 5/6 samples are classified as a sand (Shepard, 1954), with the other sample classified as a silty sand. The modal classes in these basal tills are the coarse and medium sands with 20.4 and 20.8% respectively. The total silt plus clay are slightly less than 10% on average. This composite diagram of the sediment within the basal transport zone is remarkably similar to the other grain-size histograms for tills (Fig. 5.3 (3 & 7). A simple nonparametric test for the degree of similarity between these histograms utilizes Sorenson's similarity coefficient (Andrews, 1972). The comparison between Figure 5.8a and the other Pleistocene tills gives similarity values between 97 and 77%, with a value of 97% for the comparison between Dowdeswell's basal tills and the tills from the C (unaltered) horizon.

These data therefore suggest that tills from the Precambrian shield terrain of Baffin Island are texturally sands with strong modes in the coarse to medium sand fraction. They are deficient in silt and clay, and yet.... Church (1972) and Dowdeswell (1982) have both measured significant levels of silts and clays being carried in suspension by meltwater streams emerging from glaciers on Baffin Island. Recent seismic surveys in Baffin Island fiords during the 1982 Hudson cruise further indicates that the floors of the fiords are covered by thick piles of silts and clays (Gilbert and MacLean, 1983). The interesting question is: why is the silt and clay produced by glacial abrasion not being retained within the till matrix more effectively (Lind, 1983)? This question will be returned to later.

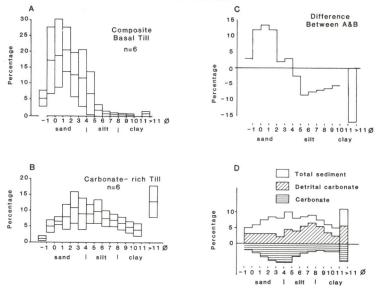

Figure 5.8   Histograms of various sediments from eastern Baffin
Island. a. Composite till from zone of basal traction, Watts Glacier
(Dowdeswell, 1982).   The bars give the range ±1 standard. deviation.
b. Grain-size diagram for 6 samples of carbonate-rich (+20% till from
southern Baffin Island (Miller, unpublished data). c. Differences in
percentages between the till and A and B. d. Histogram of the
grain-size distribution for a single till sample for the total sediment
and for the detrital carbonate (above line). Below the line is shown
the grain-size distribution of the non-carbonate fraction. Note the
shift in the modal values.

There is a major contrast in the textures of tills derived from
the Precambrian Shield and those derived by the erosion of the
Paleozoic limestones that surround Baffin Island. This difference is
illustrated on Figure 5.8b where a composite till from southeastern
Baffin Island is shown (Miller, unpublished data). In this area (see
Osterman et al., this volume) the tip of Meta Incognita was overridden
by northerly moving ice that had crossed Paleozoic outcrops in Ungava
Bay and within Hudson Strait. Six samples of this carbonate-rich till
(Fig. 5.8b) indicate that the sediment is greatly enriched in silt,
clay, and there is even a large percentage of very fine clay (finer
than 11ϕ). To investigate the grain-size characteristics of the
carbonate-rich tills further, some samples were studied both before and
after an acid leach. In this way we can see what is the grain-size
distribution of the glacially eroded carbonate and non-carbonate
components (Figs 5.8c, d). In addition this figure also shows the
difference between the Precambrian and carbonate-rich composite tills.
Notice how the Precambrian till is proportionally enriched in the sand
fraction, whereas the carbonate-rich tills are strongly enriched in

silt, clay, and very fine clay. However, despite the presence of these
fines the modal class is still in the sand fraction, although it has
shifted to the fine sand grade.

4. Neoglacial till:

Neoglacial tills have been sampled for soil studies and weathering
studies (Isherwood, 1975; Boyer, 1972; Birkeland, 1978) and for
textural information (Anderson, 1976; Dowdeswell, 1982). Much of these
data have been presented in the form of sand, silt, and clay
percentages - detailed textural data are limited. Comparison of the
histograms indicates that there is not significant difference between
Neoglacial tills (which invariably reflect short distances of travel)
and Pleistocene tills that were the product of much larger-scale
glaciation (see Chapter 3). The neoglacial tills are somewhat enriched
in the >2000 μm fraction and are slightly more fine-grained. A and C
horizons from neoglacial moraines all show (n = 2 and 3 respectively)
an increase in the percentage of silt that is most probably a product
of eolian processes (Table 5.4).

8. Reworked till:
This category of till is not very satisfactory and yet it does
represent a field judgement that cannot be ignored. "Reworked till" is
used to describe a sediment that has, in the judgement of the field
worker, been largely derived from existing sediments. A case in point
is the Platform till on Broughton Island (Andrews et al., 1976), which
is considered to have been largely derived from the reworking of
glacial marine sediments. The multi-sediment origin of reworked till
is evident from Figure 5.3. These tills, on average, are both coarser
and finer grained than either the Neoglacial or Pleistocene tills.

Sandur

Once till is released either underneath the glacier or at the front of
the glacier, the major transportation agent is initially that of
running water, largely in the form of meltwater. In this process the
coarse fractions frequently become an important component of the sandur
(Church, 1972) and may be contributed directly or indirectly into the
littoral zone or into deeper water (Fig. 5.1). Church (1972, 1978), in
particular, has paid specific attention to the sedimentology and
hydrology of the sandurs on Baffin Island, and thus this particular
facet of the sediment cycle is only treated briefly.

Church (1972, p. 121) presented a histogram of the average surface sediment from the sandur that he had studied. That portion of the figure less than 2000 μm is recalculated and reproduced in Figure 3.9. The sandur surface sediment is bimodal with a dominant peak in the -6 to -5ϕ range and a subsidiary peak between 1 and 1.5ϕ (Church, 1972). In the matrix <2000 μm (Fig. 3.9) the sand is largely coarse to medium in size (Table 3.2). Church (1972, p. 124) noted that in only one sample was the fraction finer than 63 μm a significant proportion of the total sample and that was caused by the washing-in of marine silts. (Church, 1972, Fig. 83), discussed cumulative grain-size diagrams in terms of depositional environments.

Sediments that are removed from the glaciers by meltwater are usually stored in two different settings (Fig. 5-1). Sands and gravels are deposited in the sandurs, in deltas, along beaches, and some travel seaward as debris in glacial and sea ice. Fine-grained sediments, silts, and clays travel further largely in suspension, and finally settle out as lake muds, glacial marine or marine muds in the nearshore zone, or they can reach the ocean basins. Settling of these fine-grained particles in the sea is enhanced by flocculation (Krank, 1980; Gilbert, 1982). Logically we will consider the grain-size characteristics of the sand facies and then discuss the various silt and clay deposits.

14. Delta sands; 15. beach sands:

Sands from raised marine delta foresets are poorly sorted sands (Figs 5.3, 5.4, & 5.6), with a modal class in the 1-2ϕ category. They contain relatively little silt but they have a distinctive tail in the very fine clay fraction that, on average, approaches 14% by weight. This very fine clay equals or exceeds the total amount of silt-sized particles.

The origin of the very fine clay is presumably from glacial abrasion but its presence in these "clean" sands requires explanation. I suggest that the most likely explanation is that the fine clay enters the sand during winter months when the fiord head would be frozen. The silts and coarse clays which are the main grain-sizes associated with glacial abrasion are rapidly flocculated but the very fine clay, or some fraction of it, remains in suspension and slowly settles during the course of the winter. Because of its small diameter (less than 0.1 μm) these particles could infiltrate into the sands.

TABLE 5.4. Means and Standard Deviations (% by Weight) for Different Phi Intervals and Other Properties of Baffin Island Sediments

Sediment	φ-1		0		1		2		3	
1. Grus	35.4	21.2	22.6	6.2	14.7	6.3	9.0	4.7	4.8	3.5
2. Bgrus	6.8	–	13.0	–	18.4	–	16.3	–	10.4	–
3. Pleitill	11.8	5.0	14.4	3.6	17.7	3.3	16.5	10.4	11.1	2.3
4. Neogtill	9.7	1.1	10.2	2.1	16.3	2.9	20.2	11.1	14.1	0.4
5. Apleisoil	13.0	5.2	14.9	3.7	16.1	3.3	14.6	14.1	8.8	2.4
6. Bpleisoil	16.4	7.4	19.6	7.2	20.1	5.4	13.6	8.8	7.8	4.1
7. Cpleisoil	17.2	6.3	19.4	5.8	20.8	4.2	16.6	7.8	9.9	3.8
8. Rewotill	13.7	10.0	15.2	7.2	21.4	12.3	16.4	9.9	11.5	7.2
9. Glaflour	0.1	0.07	0.2	0.3	1.1	1.3	6.8	11.5	19.1	4.4
10. Lacusilt	6.2	7.5	7.5	8.4	5.9	6.1	6.2	19.1	7.8	4.2
11. Glamar	5.0	4.7	9.3	7.6	17.1	13.1	20.0	7.8	15.4	12.1
12. Marsilt	3.7	3.4	6.4	4.9	11.0	8.5	14.2	15.4	15.0	8.4
13. Marmud	0.5	0.8	0.7	1.3	1.2	2.0	1.9	15.0	3.2	3.9
14. Deltsand	2.4	2.6	6.9	6.3	16.6	14.8	23.6	3.2	19.8	7.6
15. Beacsand	15.9	8.6	22.7	7.9	26.9	8.2	16.5	19.8	7.8	7.6
16. Eolian	1.2	1.8	5.2	6.1	18.1	13.8	26.2	7.8	17.2	10.2

φ	10	9		8		7		6		5		4	
		–	1.3	0.7	1.9	1.0	1.4	1.8	1.8	2.5	2.6	3.7	3.8
		–	1.4	–	2.3	–	3.1	–	5.0	–	7.4	–	11.9
		1.3	2.3	1.1	2.3	1.3	2.6	1.4	3.2	1.6	4.8	2.8	9.1
		–	–	0.2	1.5	1.0	2.4	1.3	3.3	1.6	5.4	0.9	11.8
		–	–	1.1	1.9	1.8	3.2	2.2	3.8	2.9	5.3	3.4	7.4
		0.4	0.8	0.8	1.3	1.6	1.8	2.7	2.7	4.6	4.3	7.4	7.2
		–	0.9	0.1	0.9	1.1	1.1	1.6	1.8	2.5	3.0	3.9	6.6
		0.5	0.6	0.7	0.7	1.1	1.3	1.8	2.2	2.4	4.7	9.0	11.8
		–	–	–	–	2.5	2.7	4.1	6.3	3.5	19.5	13.0	41.9
		2.2	4.9	3.7	7.2	4.7	9.7	4.6	9.0	4.6	9.6	7.2	14.5
		1.5	2.4	1.9	2.8	3.1	3.5	3.3	3.8	3.7	5.3	7.9	10.4
		2.7	3.5	2.6	3.8	2.7	4.0	3.1	5.0	4.7	7.8	9.5	14.5
		1.6	7.6	1.8	8.4	2.5	10.1	2.8	10.9	4.2	11.7	6.3	10.0
		1.1	1.0	1.2	1.9	1.9	0.7	2.9	2.3	6.1	6.2	11.7	16.0
		0.8	0.7	0.8	0.7	1.2	1.1	0.8	0.8	1.4	1.4	3.6	4.5
		0.8	0.7	0.9	1.0	1.4		2.6	2.0	3.9	4.6	13.7	19.9

Raised beach sands have a distinct gravel component that appproaches 25% by weight on average (Fig. 5.3). The modal class is similar to that identified in the deltaic sands although the sand fraction is coarser and there are less silt and clay-sized particles. Most of the samples included in this analysis come from the area of Broughton Island, Baffin Island where the tidal fetch is less than 1.2 m, the fast-ice extends across the littoral zone for between 8-9 months of the year, and where under present conditions, heavy pack ice inhibits wave-action during all but 1 to 2 months of the year. The beach sands also have a small, very fine, clay component that might be part of the original sediment, or it might be attributable to eolian processes. Most of the beaches included in this survey would date between 8 to 10 ka and have thus been out of the influence of wave and tidal forces for much of the Holocene.

9. Glacial flour; 10. Lake silts; 12. Marine silt; 13. Marine muds; & 11. Glacial marine

Two samples of suspended sediment are hardly representative, but they do provide initial insights into the textural characteristics of sediment transported away from the glacier via meltwater streams. The histogram shows a very dominant mode in the coarse silt (4-5ϕ) class. The sample is the "best" sorted of any included in this study (Fig. 5.6).

Rivers and streams from glacial and nonglacial sources enter lakes which constitute efficient sediment traps (e.g. Davis, 1980; Mode, 1980; Gilbert and Church, 1983). Average lake sediment textures (Fig. 5.3) are amongst the most poorly sorted of all the sediments collected on Baffin Island (Fig. 5.6). This is because lakes receive a variety of sediment inputs, including silts and clays carried in suspension, sands and gavels carried into the lake during high spring/summer floods, and a mixture of sediments carried down onto the ice or into the lake by such processes as wet snow slides, avalanches, mud flows, and other debris flows (e.g. Davis, 1980). This variability is well expressed by the figures for the standard deviations of each size class (Table 5.4).

In an environment such as the inshore waters of Baffin Island the designation of samples as glacial marine, marine silt, or marine muds represents a combination of genetic/textural descriptions that in reality probably represent a continuum of sediment characteristics from the near proximal to the extreme distal position (Osterman, 1982; Osterman and Andrews, 1983; Nelson, 1981; Mode et al., 1983; Gilbert,

1982, 1983). In the context of this paper, "glacial marine" sediments largely represent shallow water, ice proximal sediments whereas marine silts and muds represent deeper water facies.

When compared to Pleistocene tills (Fig. 5.3) glacial marine sediments are finer, have less gravel, and have 10% of the sediment by weight in the finer than 10ϕ class. The mode is medium sand. Marine silts (remember that these names refer to field/laboratory descriptions) are texturally finer than glacial marine sediments with a mode in the fine sand/coarse silt size range, and with 19% of the sediment finer than 10ϕ. However, even in marine silts there is a substantial percentage of the sediment in the sand fraction (Fig. 5.4). The marine mud facies is noteworthy because of the very large peak in the finer than 11ϕ class (Fig. 5.3). The sediment is also deplete in sand (Figs 5.3 & 5.4). The histograms of Figure 5.3 represent disaggregated sediments and it should be born in mind that Gilbert (1982) has shown that the actual grain-size distribution during settling is strongly influenced by flocculation; hence these histograms should not be taken to represent the grain-size of these sediments during actual deposition, as information about the source of the sediment.

The marine silt facies largely comes from collections of intermediate to proximal bottomset silts, frequently deposited in 50 to 150 m of water. In contrast the marine muds have been sampled from piston cores taken in at least 400 to 600 m of water in sounds and deep troughs on the Baffin Island shelf. Thus the marine mud facies represents deeper water sedimentation but not necessarily ice-distal sedimentation (cf. Osterman, 1982; Osterman and Andrews, 1983). The sand and gravel in the marine muds, at least in Baffin Island, most probably comes from ice-rafting of sediment either in the form of glacier ice, or possibly as significant sediment included within, and on, the surface of drifting ice during the annual break-up.

The carbonate content of marine silts and marine muds is variable off Baffin Island but it has been recognized as an important diagnostic element of the sediment for several years (Marlowe, 1966; Piper, 1973; Aksu, 1981; Osterman and Andrews, 1983). Generally speaking, the presence of peaks of carbonate in marine cores within the Baffin Island nearshore zone or in the deeper waters of Baffin Bay represents a glacial element associated with major ice advances across Paleozoic limestone terrains (see Klasen, Chapter 15, this volume). Aksu (1981) reported that in his ice-rafted facies, up to 50% of the sediment was detrital carbonate, and similarly high values have been reported from cores in the area of Frobisher Bay/Resolution Island (Osterman, 1982;

Osterman and Andrews, 1983; McMillan, 1971). The average carbonate content in the raised glacial marine sediments on Baffin Island is 3.5% which is similar to a mean of 3.2% for marine silts. However, the deeper water marine muds average 20 ± 13% by weight of carbonate. Osterman and Andrews (1983) examined the grain-size of a series of samples from core HU77-159 from 570 m of water in the middle reaches of Frobisher Bay. Carbonate content varied between about 6 and 50%, and the input of carbonate (mg cm-2 yr-1) varied between 150 and 5 over the last 12 ka. Analysis showed that there was little difference in the grain-size of the carbonate and noncarbonate clasts (Osterman and Andrews, 1983, Fig. 6).

Deep-sea sediments:

Aksu (1981) and Fillon et al. (1981, 1982) have described the sediments collected in piston cores from Baffin Bay and the Labrador Sea (see Chapters 7 and 8). Some of these sediments are very sandy and represent a significant input of ice-rafted debris (IRD) into the deep marine basins. Such sandy sediments appear to be notably absent from marine cores collected in deeper troughs on the Baffin Island shelf. Aksu (1981, personal communication, 1982) provided 4 samples from the ice-rafted facies of Baffin Bay (Fig. 5.9, C1). Although these 4 samples constitute a very small sample, Aksu believes they are broadly representative of the ice-rafted facies within Baffin Bay. In texture they also compare favorably with the sediments discussed by Fillon (Chapter 8, this volume; also Fillon et al., 1981). The samples average 16.9% by weight of gravel; in terms of the matrix the sand:silt:clay percentages are 35.9, 23.5, and 29.7 and are classified as sand silt-clay; the sediments are thus texturally similar to the carbonate tills of southern Baffin Island (Fig. 5.8b). Within Baffin Bay one of the features of the ice-rafted facies is the high percentage of detrital carbonate in the sediment (Aksu, 1981; Aksu, this volume). The grain-size of these, and other deep-sea sediments from the northern northwest North Atlantic, thus stand in contrast with the marine muds that occur in thick stratigraphic sequences in the fiords and within the troughs along the eastern coast of Baffin Island. Part of the apparent contrast might be related to the time element, i.e. comparing Holocene sediments in the inshore area with Pleistocene sediments offshore. Although this might be part of the answer, it is certainly not complete because the extent of glaciation along the northeastern margin of the Laurentide Ice Sheet at 8-10 ka was not radically different from that during the early Wisconsin, although the ice did extend to the outer coast during this latter interval (Nelson, 1982; Klassen, this volume; Aksu, this volume). In the near future we

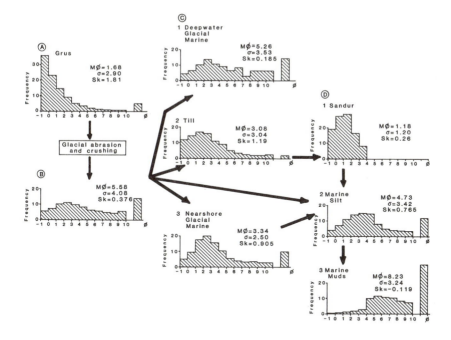

Figure 5.9 Transport paths and associated changes in the grain-size distribution of sediment. Values for mean, standard deviation etc., are based on moment measures.

anticipate that there will be sufficient stratigraphic and chronological control on inshore and offshore (deep water) marine cores that we will be able to reconstruct the nature of sedimentation at specific intervals of time. At that time we will be in a better position to examine the significance of some of the differences noted above.

16. Eolian sediments (cover sands):

In the eastern Canadian Arctic there have been few reports on eolian sediments, although the increased import of the process is gradually becoming clear. In his Ph.D. thesis on Pangnirtung Pass, Thompson (1954) discussed "the riddle of the layered sands". One explanation that he advanced was that they were eolian in origin, but alternatives were considered. Further study of these exposures and others were carried out in 1973 by Miller and Andrews as part of a survey of the new Canadian National Park (Stuckenrath et al., 1979; Andrews et al., 1979). Many sections contained buried soils and peats and these were extensively 14C dated (Struckenrath et al., 1979). All 14C dates

turned out to be middle to late Holocene in age. Dyke (1977) and Dyke
et al. (1982) suggested that the radiocarbon dates clustered into
discrete intervals that delimited "warm/wet" periods. Thus the
intervening sands represented eolian activity during "cold/dry"
periods. Along the outer eastern coast of Baffin Island Miller et al.
(1977) reported that a wedge of eolian sands lies on the surface of the
glacial forelands. The unit thins rapidly away from the coast.
Thicknesses are usually 1 m or less along the outer coast whereas in
Pangnirtung and Kingnait Passes total sediment thickness can reach 4-6
m.

Gilbert (1982, 1983) noted the presence of a large eolian sand
body on the south side of the Maktak sandur and he further suggested
that sand-sized particles in short gravity cores near the fiord head
might originate as wind-blown sand.

The data bank includes two descriptions of eolian sediments with a
total of 51 samples. Where the samples are massive clean sands the
term "eolian" was used. However, in many areas the sands are layered.
Small beds thin and thicken irregularly and, in addition, slight
increases in organic content (and some macrofossils) indicate that the
sediment has been built up on a partly vegetated surface. Because of
the poor sorting in sediments from a site near Windy Lake, Pangnirtung
Pass, Andrews et al. (1979) suggested that these sediments represented
niveo-eolian deposits (cf. Cailleux, 1976). The processes responsible
for niveo-eolian sedimentation include the transport of sand, silt, and
clay particles from exposed dry surfaces in winter and the lodging of
these grains in the snowcover. Soper, for example, in his report of a
sledge trip through Panginrtung Pass in winter, noted that the sledges
had difficulty running because of the high sand content in the snow!
During the spring melt the wind-blown sediment gradually accumulates as
the snow melts and may even be transported a short distance during the
final stages by surface waters running across frozen ground.
Histograms of eolian sediment and niveo-eolian sediment (Fig. 5.3)
indicates that the sediment is texturally a clayey sand. It is poorly
sorted (Fig. 5.7) and is slightly coarser than the sand described by
Gilbert (1982, Fig. 4) from cores in Maktak Fiord. The modal class is
fine sand but there is a small secondary mode of coarse silt and a
distinct tail of grains finer than 0.908 μm (10ϕ). Evidence for
sediments of this size-range is not shown by Gilbert (1982) in his
cumulative diagrams of fiord and eolian sediments from Baffin Island.

DISCUSSION

It must be born in mind that the samples upon which this study is based were not collected as part of a particular, focused enquiry. Therefore the statistical basis for drawing sweeping conclusions is not altogether sound. Nevertheless, I feel that the information in this chapter has value and can be used as a basis for future studies on the nature of Quaternary sediments on Baffin Island and the surrounding sea floors. Two aspects deserve further comment--the <u>first</u> is concerned with the problem of differentiating between till and glacial marine sediment which is a substantial problem in many areas, and in many parts of the geological record; the <u>second</u> is to discuss the change in grain-size through a sequence of erosional, transportational, and depositional events (McLaren, 1981).

Till versus Glacial marine sediments:

Ladim and Frakes (1968) used discriminant analysis to differentiate betwen till, alluvial-fan, and outwash deposits. In their study they used derived parameters of the sample distributions namely graphic means, standard deviation, skewness, and kurtosis. In this study I have used the weight percents in the different phi classes (e.g. Fig. 5.3). Two group stepwise discriminant analysis successfully classified 85% of the sediments. The two most important variables in discriminating between till and glacial marine deposits were the percentages of material larger than 2000 μm, followed by the percentage of sediment in the fine sand class, and thirdly by the percentages in the coarse silt. Eight of the 12 sediment class intervals contributed to the discriminating equation which in unstandardized form appears as:

$D = 1.86 + .05$ (% coarser than 2000 μm) $+ 0.07$ (% very coarse sand) $- 0.09$ (% medium sand) $- 0.6$ (% very fine sand) $- .19$ (% coarse silt) $+ .41$ (% very fine clay).

The group mean D value for Pleistocene till is 1.12 and for the glacial marine sediments is -0.64. Although this equation gives a reasonable separation of the two sediments, only further testing will demonstrate whether it has general applicability within the eastern Canadian Arctic.

Trends in mean grain-size, sorting, and skewness:

McLaren (1981) has proposed an interesting scheme for changes in
grain-size distributions from a "source" sediment. The processes
involve the selected erosion, transportation and deposition of sediment
from a common source. Figure 5.1 suggests that on Baffin Island we
have the potential for discussing sediment changes from an original
glacially abraded and crushed sediment to another end member which is
the fine-grained silts and clays deposited in deep water in the inshore
zone. Ice-rafted deep-sea sediments (Aksu, 1981) do not appear to be
part of this continuum. An initial assumption that must be made is
that the source sediment can be reconstituted by averaging the various
constituent grain-size assemblages. In terms of a direct comparison
with McLaren's work our samples extend out to finer than 11ϕ whereas
McLaren (1981, p. 613) limited his example to between -1 and 7ϕ. The
hypothetical source sediment (Fig. 5.9) is not rectangular. However,
in our case I suggest that the initial grain-size might be represented
by grus, that is the initial available grain-size is determined by the
basic nature of the bedrock lithology.

Figure 5.9 sketches a flow path for sediments. In stage B the
local bedrock is crushed and abraded beneath a warm-based or
warm-freezing ice sheet/glacier (see Chapter 3). I contend, or
hypothesize, that this process produces a sediment as graphed in Figure
5.9b in which a considerable porportion of silt and clay-sized
particles is produced. Much of these finer-grained components are not
retained within the glacier system (Lind, 1983; Vivian, 1980) but they
are expelled along various paths to sediment sinks (i.e. route B --- C1
--- D2, --- B --- D2, or B --- C2 --- D3). In this progression the
till is a residual deposit essentially winnowed of fine sediment by
meltwater operating at the micro- and macro-level or removing silt and
clay-sized clasts (Vivian, 1980; Lind, 1983;). With transport along
the system, the sediments get finer-grained and progress from being
positively skewed (on the phi scale) to being negatively scaled (e.g.
D3) that is toward a lower energy regime.* (see Sly et al., 1983.)
The progressive changes in sediment between C1 and C2 to D1, D2 and/or
D3 are reasonable, however, the ice-rafted facies in the deep basins of
Baffin Bay is significantly coarser than nearshore marine muds but has
less of a tail of fine particles.

---

*The measures on Figure 5.9 are moment measures, not graphically
derived.

This study suggests that grain-size data still have an important role to play in the interpretation of sedimentary environments. The samples were not collected for the specific purpose of this study and no corrections have been attempted to thwart any actual or perceived biases, in the sample collections. It is hoped that the study adds further information on the characteristics of Baffin Island Quaternary sediments and thus extends existing work, such as Church's study of sandur (Church, 1972) or the various studies of glacial marine sediments (Nelson, 1981; Gilbert, 1982; Mode et al., 1983; Osterman, 1982). I suggest that future studies should aim at providing data, such as are illustrated in Figure 5.9, but where specific time-planes are taken, such as the present, the start of neoglaciation, the Cockburn substage, and other important times of change in the paleo-environment of the eastern Canadian Arctic. By examining sediments deposited within specific intervals we can work toward the development of appropriate facies models for this high latitude, glacial and marine environment.

ACKNOWLEDGEMENTS

I wish to thank the many INSTAAR personnel who contributed, albeit unwittingly to the data bank! Julian Dowdeswell has my sincere thanks for compiling the sediment records and pursuing recalcitrants who did not wish to cooperate! This chapter is a contribution to NSF grant EAR-81-21296 and DPP-81-16048 and much of the data used in this study has been collected over the years with NSF support of the "Baffin Island Program".

REFERENCES

Aksu, A., 1981: Late Quaternary stratigraphy, paleoenvironmentology and sedimentation history of Baffin Bay and Davis Strait. PhD thesis, Dalhousie University, Halifax. 771 pp.
Anderson, L.W., 1976: Rates of cirque glacier erosion and sources of glacial debris, Pangnirtung Fiord area, Baffin Island, N.W.T., Canada. MSc thesis, University of Colorado, Boulder. 78 pp.
Andrews, J.T., 1963: The cross-valley moraines of the Rimrock and Isortoq river valleys, Baffin Island, N.W.T.: A descriptive analysis. Geographical Bulletin, 19:49-77.
Andrews, J.T., 1965: Glacial geomorphological studies on north-central Baffin Island, Northwest Territories, Canada. PhD thesis, University of Nottingham, England. 476 pp.
Andrews, J.T., 1972: Multivariate analysis of Quaternary deposits using nomical scale data: ordination and information and graph theoretical methods. In: Yatsu, E. and Falconer, A. (eds.), Research Methods in Pleistocene Geomorphology. GeoAbstracts, Norwich, UK, 186-214.
Andrews, J.T., Buckley, J.T., and England, J.H., 1970: Late-glacial chronology and glacio-isostatic recovery, Home Bay, east Baffin Island. Geological Society of America Bulletin, 81:1123-1148.

Andrews, J.T. et al., 1976: Alternative models of early and middle-Wisconsin events, Broughton Island, Northwest Territories, Canada: toward a Quaternary chronology. In: International Geological Correlation Program #24, I.U.G.S./U.E.S.C.O., Bellingham/Prague, 12-61.

Andrews, J.T., Webber, P.J., and Nichols, H., 1979: A late Holocene pollen diagram from Pangnirtung Pass, Baffin Island, N.W.T., Canada. Review of Palaeobotany and Palynology, 27:1-28.

Birkeland, P.W., 1978: Soil development as an indication of relative age of Quaternary deposits, Baffin Island, N.W.T., Canada. Arctic and Alpine Research, 10:733-747.

Bockheim, J.G., 1980: Properties and relative age of soils of southwestern Cumberland Peninsula, Baffin Island, N.W.T., Canada. Arctic and Alpine Research, 11:289-306.

Boyer, S.J., 1972: Pre-Wisconsin, Wisconsin and Neoglacial ice limits in Maktak Fiord, Baffin Island: a statistical analysis. MSc thesis, University of Colorado, Boulder. 117 pp.

Boyer, S.J. and Pheasant, D.R., 1974: Delimitation of weathering zones in the fiord area of eastern Baffin Island. Geological Society of America Bulletin, 85:805-810.

Brigham, J.K., 1983: Stratigraphy, amino acid geochronology, and correlation of Quaternary sea-level and glacial events, Broughton Island, arctic Canada. Canadian Journal of Earth Sciences, 20:577-598.

Cailleux, A., 1976: Formes et depots niveo-eoliens sur le pied de glace, a Poste-de-la-Baleine, Quebec Subarctique. Revue Geographie Montrial, 213-219.

Church, M., 1972: Baffin Island sandurs: a study of arctic fluvial processes. Geological Survey of Canada, Bulletin 216, 208 pp.

Church, M., 1978: Palaeohydrological reconstructions from a Holocene valley fill. Canadian Society Petroleum Geologists, Memoir 5, Fluvial Sedimentology, 743-772.

Church, M. and Ryder, J.M., 1972: Paraglacial sedimentation: a consideration of fluvial processes conditioned by glaciation. Geological Society of America Bulletin, 83:3059-3072.

Church, M., Stock, R.F., and Ryder, J.M., 1979: Contemporary sedimentary environments on Baffin Island, N.W.T., Canada: debris slope accumulation. Arctic and Alpine Research, 11:371-402.

Colvill, A.J., 1982: Glacial landforms at the head of Frobisher Bay, Baffin Island, Canada. M.A. thesis, University of Colorado, Boulder. 202 pp.

Davis, P.T., 1980: Late Holocene glacial, vegetational, and climatic history of Pangnirtung and Kingnait fiord area, Baffin Island, N.W.T., Canada. PhD thesis, University of Colorado, Boulder. 399 pp.

Dowdeswell, J.A., 1982: Debris transport paths and sediment flux through the Grinnell Ice Cap, Frobisher Bay, Baffin Island, N.W.T., Canada. MA thesis, Univeristy of Colorado, Boulder, 176 pp.

Dyke, A.S., 1977: Quaternary geomorphlogy, glacial chronology, and climatic and sea-level history of southwestern Cumberland Peninsula, Baffin Island, Northwest Territories, Canada. Unpublished Ph.D. thesis, University of Colorado, Boulder. 185 pp.

Dyke, A.S., Andrews, J.T., and Miller, G.H., 1982: Quaternary geology of Cumberland Peninsula, Baffin Island, District of Franklin. Geological Survey of Canada, Memoir 403, 32 pp.

Fillon, R.H., Fjull, W.E., and Ehrlich, R., 1982: High resolution
    granulometric studies of deep-sea sediments and the nature of
    ice-rafted debris. EOS, Transactions American Geophysical Union, 63,
    Abstracts, p. 984.
Fillon, R.H., Miller, G.H., and Andrews, J.T., 1981: Terrigenous sand
    in Labrador Sea hemipelagic sediments and paleoglacial events on
    Baffin Island over the last 100,000 years. Boreas, 10:107-124.
Folk, R.L. and Ward, W., 1957: Brazos River bar: a study in the
    significance of gain size parameters. Journal Sedimentary Petrology,
    27:3-26.

Gilbert, R., 1982: Contemporary sedimentary environments on Baffin
    Island, N.W.T., Canada: Glaciomarine processes in fiords of eastern
    Cumberland Peninsula. Arctic and Alpine Research, 14:1-12.
Gilbert, R., 1983: Sedimentary processes of Canadian arctic fiords.
    Sedimentary Geology, 36:147-175.
Gilbert, R. and Church, M., 1983: Contemporary Sedimentary environments
    of Baffin Island, N.W.T., Canada: reconnaissance of lakes on
    Cumberland Peninsula. Arctic and Alpine Research, 15:321-332.
Gilbert, R. and McLean, B., 1983: Geophysical studies based on
    conventional shallow and Huntec high resolution seismic surveys of
    fiords on Baffin Island. In: Syvitski, J.P.M. and Blakeney, C.P.
    (compilers), Sedimentology of Arctic Fiords: HU 82-031 Data Reports
    Volume 1. Canadian Date Report of Hydrography and Ocean Sciences No.
    12 Minister of Supply and Service, Canada, 51-1 to 15-19.
Grant, A.C., 1975: Geophysical results from the continental margin off
    southern Baffin Island. In: Yorath, C.J., Parker, E.R. and Glass,
    D.J. (eds.), Canada's continental margins and offshore petroleum
    exploration. Canadian Society of Petroleum Geologists, Memoir 6,
    411-431.

Isherwood, D.J., 1975: Soil geochemistry and rock weathering in an
    arctic environment. PhD thesis, University of Colorado, Boulder, 173
    pp.

Kranck, K., 1980: Experiments on the significance of flocculation in
    the settling of fine grained sediments in still water. Canadian
    Journal of Earth Sciences, 17:1517-1526.
Krumbein, W.C. and Pettijohn, F.J., 1938: Manual of sedimentary
    petrography. Appleton-Century Crofts, New York. 549 pp.

Landim, P.M.B. and Frakes, L.A., 1968: Distinction between tills and
    other diamictons based on textural characteristics. Journal
    Sedimentary Petrology, 38:1213-1223.
Leroy, S.D., 1981: Grain-size and moments measures: a new look at Karl
    Pearson's ideas of distributions. Journal of Sedimentary Petrology,
    51:625-630.
Lewis, C.F.M. et al., 1977: Marine geological and geophysical
    activities in Lancaster Sound and adjacent fiords. Geological Survey
    of Canada, Paper 77-1A, 495-506.
Lind, E.K., 1983: Paleoecology and deglacial history of Cape
    Rammelsberg area, southern Baffin Island, N.W.T., Canada. M.Sc.
    thesis, University of Colorado, Boulder. 217 pp.
Locke, W.W. III, 1980: The Quaternary geology of the Cape Dyer area,
    southeasternmost Baffin Island, Canada. PhD thesis, University of
    Colorado, Boulder. 331 pp.

MacLean, B. and Falconer, R.K.H., 1979: Geological/geophysical studies
    in Baffin Bay, Scott Inlet - Buchan Gulf and Cape Dyer - Cumberland
    Sound areas of the Baffin Island shelf. In: Current Research, Part B,
    Geological Survey of Canada, Paper 79-1B, 231-244.

Marlowe, J.I., 1966: Mineralogy as in indicator of long-term current fluctuations in Baffin Bay. Canadian Journal of Earth Sciences, 3:191-201.

McLaren, P., 1981: An interpretation of trends in grain size measures. Journal Sedimentary Petrology, 51:611-624.

McMilan, N.J., 1973: Surficial geology of Labrador and Baffin Island shelves. Geological Survey of Canada, Paper 71-23, 451-468.

Miller, G.H., Andrews, J.T., and Short, S.K., 1977: The 1st interglacial/glacial cycle, Clyde Foreland, Baffin Island, N.W.T.: stratigraphy, biostratigraphy, and chronology. Canadian Journal of Earth Sciences, 14:2824-2857.

Miller, G.H., Locke, W.W. III, and Locke, C., 1980: Physical characteristics of the southeastern Baffin Island coastal zone. Geological Survey of Canada, Paper 80-10, 251-265.

Mode, W.N., 1980: Quaternary stratigraphy and palynology of the Clyde Foreland, Baffin Island, N.W.T., Canada. PhD dissertation, University of Colorado, Boulder. 219 pp.

Mode, W.N., Nelson, A.R., and Brigham, J.K., 1983: Sedimentologic evidence for Quaternary glaciomarine cyclic sedimentation along eastern Baffin Island, Canada. In: Molnia, B.F. (ed.), Glacial-marine sedimentation. Plenum Press, New York, 495-534.

Moss, A.J., Green, P. and Hutka, J., 1981: Static breakage of granitic detritus by ice and water in comparison with breakage by flowing water. Sedimentology, 28:261-272.

Müller, D.S., 1980: Glacial geology and Quaternary history of southeast Meta Incognita Peninsula, Baffin Island, Canada. M.Sc. thesis, University of Colorado, Boulder. 211 pp.

Nelson, A.R., 1978: Quaternary glacial and marine stratigraphy of the Qivitu Peninsula, northern Cumberland Peninsula, Baffin Island, Canada. PhD dissertation, University of Colorado, Boulder. 215 pp.

Nelson, A.R., 1978: Quaternary glacial and marine stratigraphy of the Qivitu Peninsula, northern Cumberland Peninsula, Baffin Island: Geological Society of America Bulletin, 92 Part I, 512-518, Part II, 1143-1281.

Nelson, A.R., 1982: Aminostratigraphy of Quaternary marine and glaciomarine sediments, Qivitu Peninsula, Baffin Island. Canadian Journal Earth Sciences, 19:945-961.

Nie, N.H. et al., 1975: Statistical package for the Social Sciences. McGraw-Hill, New York. 675 pp.

Osterman, L.E., 1982: Late Quaternary history of southern Baffin Island, Canada: a study of foraminifera and sediments from Frobisher Bay. PhD dissertation, University of Colorado, Boulder. 380 pp.

Osterman, L.E., and Andrews, J.T., 1983: Changes in glacial-marine sedimentation in core HU77-159, Frobisher Bay, Baffin Island, N.W.T.: a record of proximal, distal and ice-rafting glacial-marine environments. In: Molnia, B.J. (ed.), Glacial-marine sedimentation. Plenum Press, New York, 451-494.

Pheasant, D.R. and Andrews, J.T., 1973: Wisconsin glacial chronology and relative sea level movements, Narpaing Fiord Broughton Island area, eastern Baffin Island N.W.T. Canadian Journal Earth Science, 10:1621-1641.

Piper, D.J.W., 1973: A late Quaternary stratigraphic marker in the central basin of Baffin Bay. Maritime Sediments, 9:62-63.

Shepard, F.P., 1954: Nomenclature based on sand-silt-clay ratios. Journal Sedimentary Petrology, 24:151-158.

Sly, P.G., Thomas, R.L., and Pelletier, B.R., 1983: Interpretation of moment measures derived from water-lain sediments. Sedimentology, 30:219-233.

Stuckenrath, R., Miller, G.H., and Andrews, J.T., 1979: Problems of radiocarbon dating Holocene organic-poor sediments, Cumberland Peninsula. Arctic and Alpine Research, 11:109-120.

Sugden, D.E., 1978: Glacial erosion by the Laurentide ice sheet. Journal of Glaciology, 20:367-392.

Swan, D., Clague, J.J., and Luternauer, J.L., 1979: Grain-size statistics II: evaluation of grouped moment measures. Journal Sedimentary Petrology, 49:487-500.

Tedrow, J.D.F., 1977: Soils of the Polar Landscape. Rutgers University Press, Rutgers, N.J. 638 pp.

Thompson, H.R., 1954: Pangnirtung Pass: an exploratory geomorphology. PhD thesis, McGill University, Montreal, P.Q., 227 pp.

Vivian, R., 1980: The nature of the ice-rock interface: the results of investigation on 20,000 m^2 of the rock bed of temperate glaciers. Journal Glaciology, 25, 267-277.

Walker, R.G. (ed.), 1979: Facies Models. Geoscience Canada, Reprint Series Number 1, Geological Association of Canada, Toronto, 211 p.

6 Geology of the Baffin Island shelf

Brian MacLean

The continental shelf of eastern Baffin Island forms the western margin of the northern part of the Labrador Sea, of Davis Strait, and of most of Baffin Bay (Fig. 6.1).

The southeastern part of the shelf between Hudson Strait and Cape Dyer includes an inner shelf area with water depths to 200 m seaward from which depths gradually increase to about 500 m at the shelf edge (Canadian Hydrographic Service Chart 5.04). The shelf reaches a maximum width of 230 km east and northeast of the entrance to Cumberland Sound and then quickly narrows toward Cape Dyer. Seaward from the shelf edge in the southern part of the area, depths increase rapidly into the Northern Labrador Sea, but northward between Cumberland Sound and Cape Dyer in the area of the Davis Strait sill, the shelf edge becomes less well defined. The Baffin shelf south of Cape Dyer is morphologically different from that to the north; relief is lower, and water depths are generally greater with bank areas <200 m in depth mainly confined to the western or nearshore part of the shelf. An elongate local depression with depths to 385 m occurs between Frobisher Bay and Cumberland Sound within the inner shelf zone which locally widens and contains shoal areas surrounding two offshore islands (Monumental and Lady Franklin). There is, however, no indication of an extensive marginal channel system such as that which borders much of the inner zone of the Labrador and Norwegian shelves. Water depths to 1289 m in Cumberland Sound and to 615 m in Frobisher Bay are greater than on the adjacent shelf. These presumably result from a combination of tectonics and glacial overdeepening (see Chapters 3 and 4). A narrow sill of variable geological composition at a depth of about 500 m separates an area of deep water (to 980 m) in the outer part of Hudson Strait from deeper water farther seaward.

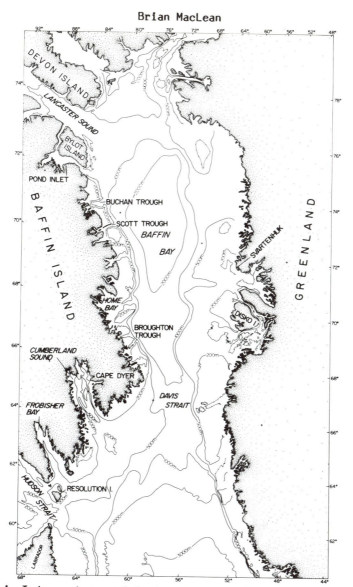

Figure 6.1 Index and general bathymetric map of Baffin Bay-Davis
Strait region.

North of Cape Dyer in Baffin Bay, the Baffin Island shelf in
general is narrower, averaging about 50 km in width, and the shelf edge
for the most part is shallower and represented by the 300 m bathymetric
contour. This section of shelf is characterized by (a) the occurrence
of numerous shallow banks <200 m in depth that commonly extend
virtually the full width of the shelf; and (b) by numerous submarine
troughs with depths to 800 m or more that extend across the shelf from

fiord mouths and separate the various banks. The composition of the
banks is known to include metamorphic and sedimentary bedrock (e.g.
MacLean et al., 1981) and younger overlying sediments presumably mainly
of Quaternary age. The fiords and submarine troughs are considered to
have been carved in part by glacial erosion along pre-existing
(probably Tertiary) drainage systems as suggested by Fortier and Morley
(1956); Ives and Andrews (1963); and Pelletier (1966) (see Chapter 4).
Moraines have been deposited along the margins of some of these troughs
(Løken and Hodgson, 1971; MacLean, 1978; Gilbert, 1982). For the most
part there is no conspicuous outbuilding of the shelf or slope at the
seaward end of the submarine troughs. The slopes of the trough walls
are variable, for example at Scott Trough where the overall slope of
the walls is in the order of 8° -10°, vertical or near vertical
sections, apparently associated with more resistant strata, occur
locally.

Seaward from the shelf edge in Baffin Bay, as far north as Buchan
Trough, depths increase rapidly down the adjacent continental slope,
but in northwestern Baffin Bay the gradient of the continental slope
becomes much more gradual.

SUMMARY OF INVESTIGATIONS OF BAFFIN ISLAND SHELF

The geology of the Baffin Island shelf has been investigated by various
workers employing a variety of geophysical and sampling techniques.
Many of the studies were carried out during the last decade by the
Atlantic Geoscience Centre, Bedford Institute of Oceanography. These
include a reconnaissance shallow seismic and magnetometer survey of the
southeastern Baffin Island shelf (Grant, 1974, 1975a, b), refraction
and reflection seismic investigations in Baffin Bay and Davis Strait
(Keen and Barrett, 1972; Keen et al., 1972, 1974; Falconer, 1977;
Jackson et al., 1979) and in the northern Labrador Sea and Davis Strait
(Srivastava, 1978, Srivastava et al., 1981, 1982). The geology of the
Baffin shelf subsequently has been further delineated by means of
single channel seismic reflection (655 cm^3 compressed air source),
Huntec high resolution seismic, gravity, and magnetometer surveys in
conjunction with shallow corehole drilling using the BIO underwater
electric drill (e.g. MacLean et al., 1977, 1981). Similar methods were
employed in 1980 during a more detailed study of the geology of the
shelf off Cumberland Sound in cooperation with the Canadian
Hydrographic Service which, in 1980 and 1981, undertook more detailed
bathymetric charting of the southeastern Baffin Island shelf. The
region is also under exploration by industry, and investigators
associated with petroleum companies as well as other government

agencies have reported on the area, e.g. McMillan (1973), shelf
morphology and geology; Beh (1975), geology and evolution of Davis
Strait and western Baffin Bay; Hood and Bower (1973), aeromagnetic
reconnaissance of Davis Strait and adjacent areas; and Klose et al.
(1981), data from two exploratory wells drilled offshore southeastern
Baffin Island.

The work by Atlantic Geoscience Centre indicated above was
mainly related to investigation of the bedrock geology, but at the same
time acoustic data and some samples were acquired to provide
reconnaissance level information on the surficial sediments. These
data were augmented by textural data from samples collected from
various parts of the southeastern Baffin offshore area in 1978 by the
Department of Fisheries and Oceans in connection with studies of
shrimp. These studies were followed in 1981 by an Atlantic Geoscience
Centre program to obtain additional regional samples to define further
the distribution of surficial sedimentary units on the southeastern
Baffin shelf. Some facets of the geology of the Baffin shelf seabed
were locally examined from the research submersible Pisces IV (MacLean,
1982).

Studies of the surficial geology of the Labrador Shelf (e.g.
Josenhans, 1981), and of cores from the Labrador Sea (e.g. Fillon et
al., 1981) and from outer Frobisher Bay (Osterman, 1982, Osterman et
al., 1980) have contributed to the knowledge of conditions in the
region during the Quaternary.

REGIONAL BEDROCK GEOLOGY

On Shore

A brief review of the geology of the adjacent land masses is a
necessary background for discussion of the geology offshore. Where not
specifically referenced the general outline of the onshore geology in
this section has been drawn from Douglas (1969) and Escher et al.
(1970).

Archean and Proterozoic metamorphic rocks compose most of the
Baffin Island landmass (Fig. 6.2). Precambrian rocks also make up most
of the landmass of northern Labrador, West Greenland, and parts of
eastern Devon and Ellesmere Islands.

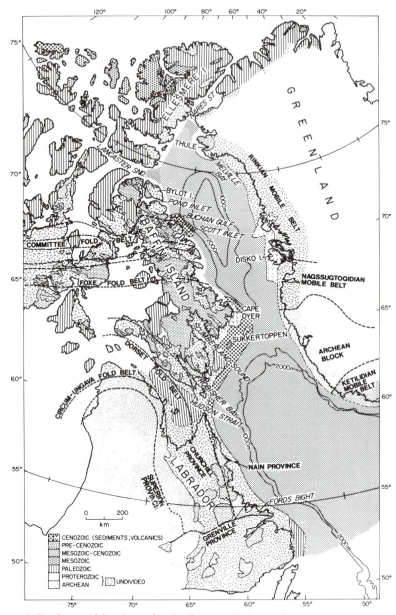

Figure 6.2 Generalized geological map of Labrador Sea-Davis
Strait-Baffin Bay region (from Srivastava et al., 1981) (with
permission).

        Lower Paleozoic strata of Ordovician-Silurian age occur on parts
of western and northwestern Baffin Island. Cambrian strata occur
locally in the latter area. Middle-late Ordovician carbonate rocks
also occur as outliers near the head of Frobisher Bay (Miller et al.,

1954) and at Atpatok Island in Ungava Bay (Workum et al., 1976). Paleozoic sedimentary rocks are widely distributed throughout the Canadian Arctic Islands, where they range in age from Cambrian to Permian, and also occur in northern Greenland where they range in age from Eocambrian to Devonian (Dawes, 1976). A small outcrop of breccia containing lower Paleozoic marine rocks occurs near Sukkertoppen on the west coast of Greenland.

Cretaceous and Tertiary (Albian to possibly Eocene) shallow marine and paralic sediments occur in the Pond Inlet - Bylot Island area at the northern end of Baffin Island (Jackson and Davidson, 1975, Jackson et al., 1975, Clarke and Daae, 1979, Miall et al., 1980). A small outcrop of lacustrine to marginal marine sediments of Paleogene age lies on the Precambrian crystalline rocks of north-central Baffin Island (Andrews et al., 1972). Farther to the south, Paleocene subaqueously erupted volcanic breccia and subaerial volcanic flows rest in part on Precambrian metamorphic rocks and in part on Paleocene terrestrial sediments in a narrow belt along eastern Baffin Island from Cape Dyer northwestward for 85 km (Clarke and Upton, 1971). Paleocene volcanic rocks occur in the Disko Island-Svartenhuk area of West Greenland where they have been extensively studied (e.g. Rosenkrantz and Pulvertaft, 1969; Clarke and Pederson, 1976), and on the adjacent West Greenland shelf (Denham, 1974; Clark and Pederson, 1976; Brett and Zarudzki, 1979). Mesozoic and Cenozoic strata of Barremian to Danian age also occur in the Disko-Svartenhuk area of West Greenland (Henderson et al., 1976).

Strata ranging in age from Triassic to Upper Tertiary also occur extensively in the Sverdrup Basin of the Canadian Arctic Islands (Douglas, 1969; Trettin and Balkwill, 1969).

Geology Offshore Baffin Island

The discussion that follows is based on a regional subdivision of the shelf.

Southeastern Baffin Shelf. The bedrock geology of the southeastern Baffin Island shelf as interpreted on the basis of geophysical and sample data (MacLean et al., 1982) is indicated in Figure 6.3. Precambrian metamorphic rocks of the eastern Baffin Island land mass form the bedrock offshore adjacent to the coast in most areas, and also form a basement high that extends northeastward from the entrance to Frobisher Bay.

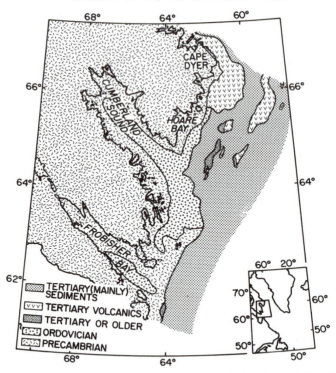

Figure 6.3 Geological map of southeastern Baffin Island Shelf (from MacLean et al., 1982) (with permission).

Middle-late Ordovician (Caradoc) limestone strata deposited in marine near shore to outer shelf-upper bathyal environments underlie much of the shelf between Frobisher Bay and Cumberland Sound where they have been sampled at five localities (MacLean et al., 1977; MacLean, 1978). Ordovician strata are also interpreted on the basis of seismic reflection, magnetic, and gravity data to be present in the floors of Frobisher Bay (also supported by submersible observations, MacLean, 1982) and Cumberland Sound, and to extend northward along the shelf toward Cape Dyer (Grant, 1975b; MacLean and Falconer, 1979; MacLean et al., 1982). Lower Paleozoic rocks including strata possibly of Silurian age also appear to underlie the outer part of Hudson Strait (Grant, 1975b; Grant and Manchester, 1970). To the north in Baffin Bay, Keen et al. (1974) and Jackson et al. (1977) considered that high-velocity material beneath the western margin of the Bay may be indicative of the presence of Paleozoic rocks. The Precambrian and Lower Paleozoic rocks that underlie the shelf are presumed to extend seaward in the subsurface to the continent-ocean transition zone.

Strata thought to be mainly of Tertiary age on the basis of palynological data from shallow drill core samples off Cumberland Sound (MacLean et al., 1982) and data from the Aquitaine et al. Hekja O-71 and Esso H.B. Gjoa G-37 wells (Klose et al., 1981) form the bedrock along much of the central and outer parts of the southeastern Baffin shelf. Mesozoic or older strata may be present in the subsurface, but are thought to outcrop only locally, e.g. associated with structures off Cumberland Sound.

The geology of the southeastern Baffin shelf from Cumberland Sound northward is complex. There, the stratigraphic section has been disturbed by the emplacement of a series of four northeast-southwest trending subsurface ridge-like structures. Data from shallow core samples indicate that sedimentary rocks of late Paleocene-early Eocene age and volcanic rocks are present in these structures, and that strata at least as old as middle Cretaceous appear also to be represented. Seismic profiles indicate that some of these structures resemble piercement features. Although sedimentary diapirism may have been involved, the various data suggest that emplacement of the structures may have been mainly due to massive upward intrusion of volcanic dyke material along northeast-southwest fractures developed as a consequence of translational motion between Greenland and North America (MacLean et al., 1982). Emplacement of the structures apparently was not complete until after early Eocene time. The region subsequently has been erosionally bevelled. The outer two ridges off Cumberland Sound are on trend and apparently continuous with a structural high in mid-Davis Strait considered by Srivastava et al. (1981, 1982) to be composed of attenuated continental crust, modified oceanic crust, or elements of both.

A horizontally layered sedimentary unit that reaches 100 m or more in thickness unconformably overlies Tertiary and older rocks off Cumberland Sound and part of the outer shelf to the southeast. This unit in part is overlain by material interpreted to be glacial till and is presumed to be of late Tertiary or early mid-Pleistocene age. The unit appears to have been subjected to moderately extensive erosion that resulted in removal of large amounts of material and the development of terraces, channels, and other erosional features. Age(s) of these events has not been defined (D. Praeg, personal communication, 1982).

Offshore from Cape Dyer volcanic rocks on the shelf (MacLean et al., 1978) are more extensive and continuous in comparison to the occurrences of Paleocene volcanics onshore eastern Baffin Island studied by Clarke and Upton (1971).

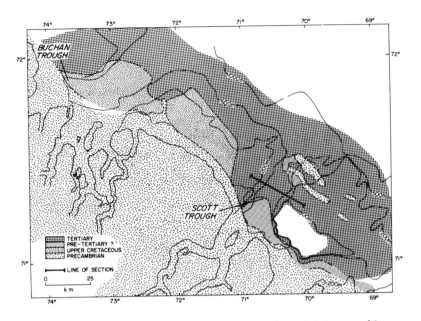

Figure 6.4 Geological map of Scott Inlet-Buchan Gulf area (from
MacLean et al., 1981) (with permission).

Northeastern Baffin Shelf. North of Cape Dyer, the geology of the
northeastern Baffin shelf is best known in the Scott Inlet - Buchan
Gulf area (Fig. 6.4). The Mesozoic-Cenozoic succession there includes
shallow marine Upper Cretaceous (Campanian) strata sampled at Buchan
Trough and nearshore marine Tertiary (late Eocene-early Oligocene)
strata sampled at Scott Trough (MacLean and Falconer, 1979; MacLean et
al., 1981). Older strata also may be present. Jackson et al. (1979),
from seismic refraction data, inferred a sedimentary section at least 4
km thick in a structural depression offshore Scott Inlet. A gravity
low associated with this structure is consistent with the
interpretation of a thick sedimentary section in this area.

     The geology of the shelf between Scott Trough and Cape Dyer is
less well defined than the areas to the north or south, but existing
data suggest that the sedimentary section there includes Tertiary
strata as well as probable Cretaceous strata. Older sedimentary rocks
possibly are represented, but if so, their occurrence would seem to be
mainly confined to the subsurface.

     At the northern end of the Baffin shelf, seismic data east of
Lancaster Sound indicate a gently seaward dipping sedimentary sequence
that contrasts with highly disturbed strata farther west considered by
Keen et al., (1974) to be possibly of Paleozoic or Mesozoic age.

The landmass of eastern Baffin Island in general stands structurally high relative to the shelf. In the east and northeast elevations of Paleocene volcanic rocks and late Eocene sediments on shore and offshore suggest that the coastal landmass of Baffin Island in that area has undergone some 600 m of vertical movement since late Eocene time (MacLean and Falconer, 1979). On the southeastern shelf between Frobisher Bay and Cumberland Sound Ordovician strata appear, at least locally, to be in fault contact close to the coast with the now structurally higher Precambrian rocks of eastern Baffin Island. Cumberland Sound and Frobisher Bay are fault controlled graben and half-graben features. Seismic reflection and sample data from the southeastern shelf (MacLean et al., 1982) indicate that the shelf in that region has been extensively bevelled by erosion in post early Eocene time, possibly during the Oligocene as suggested from Labrador shelf well data (F. Gradstein and G.L. Williams, personal communication, 1981). Initial development of the submarine troughs on the northeastern Baffin shelf may be related to this erosional episode. The shelf and parts of the coastal land areas have also been affected by subsequent isostatic and eustatic changes.

PLATE TECTONIC HISTORY

Formation of the Frobisher Bay half graben and Cumberland Sound graben may relate to crustal stretching during initial rifting in the northern Labrador Sea-Davis Strait region during the Cretaceous or possibly earlier. Opening is believed to have commenced in the northern Labrador Sea during the late Cretaceous (Maastrichtian); to have begun in Baffin Bay during the Paleocene accompanied by massive outpouring of basalt on- and off-shore in the Davis Strait region; and to have continued in Labrador Sea-Davis Strait-Baffin Bay until early Oligocene time (Srivastava 1978; Srivastava et al., 1981; McWhae 1980).

The actual mechanism of separation and the nature of the crust, as well as problems in reconstruction of the plate positions, however, have been the subject of controversy in the geological literature. Some workers, e.g. Keen et al. (1972, 1974), Srivastava (1978), Srivastava et al. (1981), and Jackson et al. (1979), believe that much of the area should be underlain by oceanic crust generated by sea-floor spreading; other workers, e.g. Kerr (1967, 1980), Grant (1975a, 1980), and Umpleby (1979), have suggested that foundered or attenuated continental crust predominates.

Movement between Greenland and North America as interpreted by Srivastava (1978) and Srivastava et al. (1981), included an initial

rotational motion, superceded during the late Paleocene by a large
translational motion as Greenland commenced to move in a northeasterly
direction in response to spreading between Eurasia and Greenland.
Emplacement of volcanic rocks and formation of the ridge structures off
Cumberland Sound on the Baffin shelf are believed to relate to
northeast-southwest fractures developed as a consequence of the
translational motion (MacLean et al., 1982). Klose et al. (1981)
indicate that a transform fault zone dominates tectonic aspects of the
shelf in this area.

SURFICAL SEDIMENTS

Southeastern Baffin Shelf

Six surficial units on the seabed of the southeastern Baffin Island
shelf have been tentatively recognized of the basis of acoustic
information and textural data derived from widely spaced surface grab
samples. These informal units include:
Unit 1 - mainly exposed bedrock, but probably locally includes gravel.
This unit mainly is associated with areas of Precambrian metamorphic
rocks on the inner shelf and areas of volcanic rocks offshore from Cape
Dyer;
Unit 2 - sand and gravel with little or no fine sediment. This unit
occurs locally as a thin cover over the underlying bedrock or till and
is found mainly on parts of the inner shelf in water depths less than
approximately 200 m;
Unit 3 - glacial till or diamictons. Interpretation of this unit as a
till is based mainly on acoustic expression, principally the lack of
coherent acoustic events characteristic of till deposits on the Scotian
Shelf and other Canadian east coast offshore areas (e.g. King, 1970,
King et al.; 1972). Deposits on the southeastern Baffin shelf (Fig.
6.5) range in thickness from a few meters to accumulations of 100 m or
more in apparent moraines (Fig. 6.6) and multiple till sections (Figs
6.7 and 6.8). Till occurrences on the southern part of the shelf seem
to be mainly on the outer part of the inner shelf. They also are in
the outer part of Frobisher Bay and Cumberland Sound (see Fig. 6.5).
Deposits that acoustically resemble till locally extend farther seaward
on the shelf off Cumberland Sound. Northward from there toward Cape
Dyer till appears to become progressively more confined to the inner
shelf.
Unit 4 - A unit consisting mainly of muddy sand and gravel is prevalent
across much of the outer part of the southeastern shelf. (Preliminary
data from samples indicate that silt and clay content occasionally may
exceed 50%.) The unit in large part apparently lies directly on

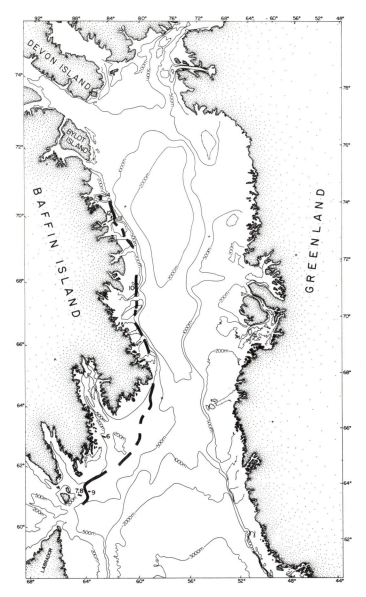

Figure 6.5 Map showing approximate seaward limit of glacial till
(indicated by heavy discontinuous line) on the Baffin Island shelf as
inferred from shallow single channel and Huntec high-resolution
seismic- reflection data. Numbers and accompanying small lines indicate
locations and orientations of profile figures in this chapter.

Tertiary bedrock (Fig. 6.9) and on the inferred late Tertiary-early
Pleistocene material. In places, however, it also appears to mantle
glacial till. Thickness of the unit varies from less than a meter to

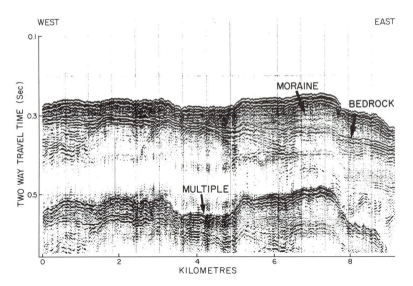

Figure 6.6 Seismic reflection profile showing inferred moraine on the shelf near the entrance to Cumberland Sound (see Fig. 6.5 for location).

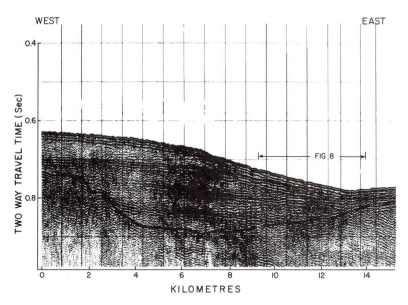

Figure 6.7 Seismic reflection profile showing what are interpreted as four successive tills overlying bedrock on the shelf northeast of Resolution Island (see Fig. 6.5 for location).

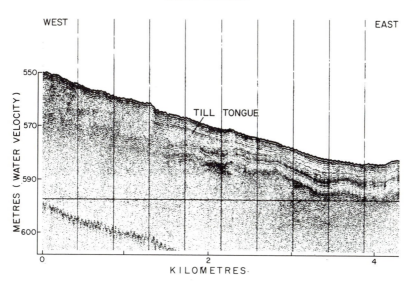

Figure 6.8 Huntec high resolution profile of part of the area
illustrated in Figure 6.5 showing a till tongue (term employed by King,
1980, to describe similar feature on Section Shelf) which near the
centre of the diagram apparently is underlain as well as overlain by
parts of the stratified sedimentary sequence to the right. The till
tongue indicates ice lift off position.

several tens of meters. The surface of this unit is very irregular on
a scale of 1-7 m due to extensive scouring by grounding icebergs (Fig.
6.9). Such scouring is a prevalent feature of the Baffin shelf, but is
particularly evident on this unit.
Unit 5, 6 - Basinal sediments composed primarily of silt (5) or clay
(6) are confined mainly to parts of the floors of Frobisher Bay and
Cumberland Sound, except for an area on the shelf east and southeast of
the entrance to Frobisher Bay (see Chapter 20).   Acoustic data suggest
that clay bottoms in these areas are less extensive than those of
silt. Stratified sediments, whose composition and relationships have
not yet been fully defined by samples, occur locally off Frobisher Bay
and Cumberland Sound and may be correlatives of the silt unit or unit
4.

Northeastern Baffin Shelf

The distribution of surficial sediments on the northeastern Baffin
Island shelf is less well known than on the shelf to the south.
However, the different morphology of the northeastern Baffin Island
region which is characterized by shallow banks and transverse submarine

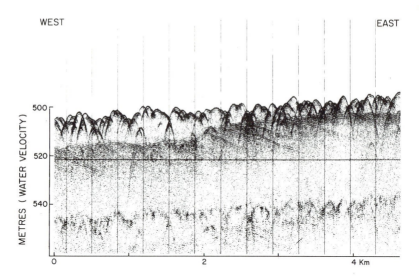

Figure 6.9 Huntec high resolution profile showing intensely iceberg scoured sediments (muddy sand and gravel unit) overlying Tertiary bedrock (see Fig. 6.5 for location).

troughs offshore, deep coastal fiords, and the presence onshore of extensive coastal foreland deposits composed of Quaternary marine and glacial sediments, suggests that variations in the sediment distribution pattern on this part of the shelf compared to that of the southeastern Baffin shelf are likely. Available data on the sediments are mainly acoustical, from conventional single channel and high-resolution seismic systems. There is as yet little textural data from samples in most areas. However, the data do permit some general observations and inferences.

Bedrock commonly is exposed at the seafloor on the inner shelf except where locally mantled by soft sediments in areas of deep water or where Quaternary foreland deposits apparently extend offshore.

Sediments that acoustically resemble glacial till have been observed on profiles across various parts of the northestern shelf (Fig. 6.5). Accumulations of this material in places attain 60 + m in thickness and resemble morainal developments (Fig. 6.10). Seismic reflection data (Fig. 6.11) (MacLean and Falconer, 1979) support Løken and Hodgson's (1971) interpretation from echo sounder profiles of moraines along the margins of Scott Trough. Gilbert (1982) has also noted probable moraines on the flank of Broughton Trough. In some localities, e.g. the outer part of Home Bay, acoustic data suggest the

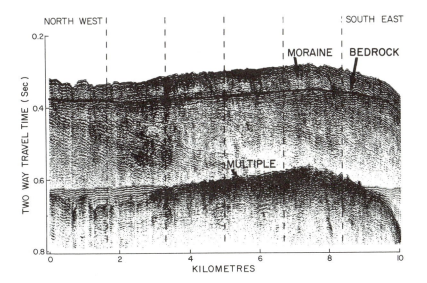

Figure 6.10 Seismic reflection profile showing sediment accumulation believed to be a moraine on the northeastern Baffin shelf a short distance north of Home Bay (see Fig. 6.5 for location).

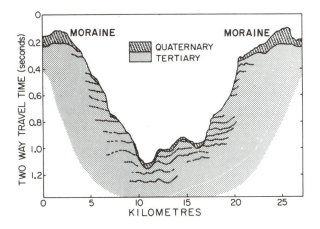

Figure 6.11 Line drawing of seismic reflection profile across the central part of Scott submarine trough illustrating apparent moraines that border the trough (see Fig. 6.5 for location) (with permission).

possible presence of more than one till separated by stratified sediment. Seismic data from the shelf off Cape Adair, between Scott and Buchan Troughs, indicate a section comprising approximately 80 m of apparently irregularly stratified and unstratified material which thins

seaward. Observations from the submersible <u>Pisces IV</u> in 1981 in this
general area indicated a seafloor comprising mainly fine sediment, but
gravel fragments (to boulder size) along the margins of iceberg scours
apparently displaced from underlying sediments during formation of the
scours (Fig. 6.12) suggest the presence of underlying coarser
material. Gravelly sand, and silty sediments including some fine
sediment resulting from burrowing by organisms have been observed at
some other localities, such as on the shelf adjacent to the south side
of outer Scott Trough. The banks appear to be variously mantled by
sediments observed to range up to some 80 m in thickness.

    The floors of the submarine troughs in general consist of fine
sediment. In Scott Trough, for example, these sediments locally attain
60 m in thickness and generally appear to lie directly on the bedrock.
This unit locally includes some weakly stratified or unstratified
material. Till possibly underlies the stratified material at a very
few localities at Scott Trough in contrast to an apparently wider
distribution beneath the stratified sediments in the floor Buchan
Trough (Fig. 6.13).

DISCUSSION OF OFFSHORE UNITS

<u>Till</u>: The presence of sediments that have been interpreted as tills
(unit 3) on the basis of their acoustic character, and apparent
morainal developments indicates that grounded glacial ice extended onto
the Baffin Island shelf during one or more intervals during the
Quaternary. Acoustic data suggest that repeated advances and retreats
occurred in some localities; a prime example is the area northeast of
Resolution Island where four tills totalling some 160 m in thickness
appear to be present (Fig. 6.7). The source of the tills in this area
may be complex. G. Miller (personal communication, 1982 and Chapter
20) suggests that ice moving out of Hudson Strait over-rode part of
southern Baffin Island. One or more of the multiple tills northeast of
Resolution Island may be associated with this ice movement. Evidence
from several eastern Baffin Island localities indicates that ice had
extended at least to outer coastal areas several times during the
Pleistocene (e.g. Miller et al., 1977). Land data, e.g. Miller et al.
(1977), Miller (1980), Andrews and Ives (1978) indicate that although
glacial ice presumably moved seaward through the fiords, late Foxe
(Wisconsin) ice did not over-ride the eastern part of northeastern
Baffin Island, but apparently did extend closer to the outer coast
farther south (see Chapter 20). Foraminifera in cored sediments from
outer Frobisher Bay dated at 11 ka indicate that ice conditions may
have been similar to those at present (Osterman et al., 1980). A

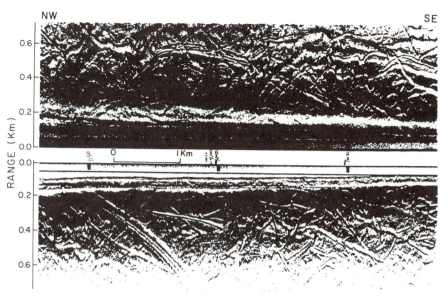

Figure 6.12 Sidescan sonar record across a section of the seafloor in 135 m of water a short distance north of Scott Trough (see Fig. 6.5 for location) illustrating the intense scouring   of the seabed by grounding icebergs that is evident on much of the Baffin Island shelf.

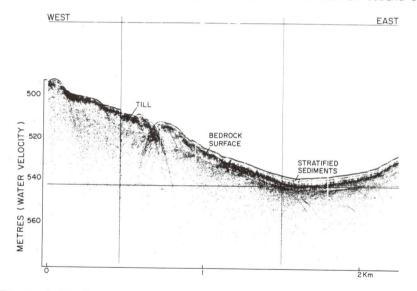

Figure 6.13 Huntec high resolution profile showing relations of inferred till with underlying Upper Cretaceous rocks and overlying younger sediments in the floor of Buchan Trough (see Fig. 6.5 for location).

similar age for the Hall moraine on Hall Peninsula (Miller, 1980),
however, indicates a readvance of late Foxe ice relatively nearby. A
major moraine system farther inland on Hall Peninsula between the heads
of Frobisher Bay and Cumberland Sound dated at 8.2 ka (Blake, 1966,
Dyke, 1979) apparently marks a later ice limit. Although dates to
define the actual age of the glacial material on the shelf are
presently lacking, data from sediments onshore suggest that the till on
the shelf is of pre-late Wisconsin age.

Muddy sand: The origin and mode of deposition of the muddy sand unit
(4) containing gravel that lies mainly seaward of the till, but in
places apparently also lies on till on the southeastern shelf is
uncertain. Possibly it may in part have been deposited from an ice
shelf or in part derived from reworking of material in shallower areas
during periods of lowered sea level. The unit presumably also includes
more recent ice-rafted components. This unit like others on the Baffin
shelf has been intensely scoured to a depth of several meters by
grounding icebergs. The unit may correlate with unscoured stratified
sediments in deeper water off Frobisher Bay and Cumberland Sound, but
it is not yet clear whether the absence of acoustically visible
layering within the muddy sand unit results entirely from the intense
reworking by grounding icebergs, or from other factors.

Sand and gravel. Bare bedrock and gravel and sand seabottom areas
devoid of finer sediment components (units 1, 2) on the inner shelf are
thought to have formed due to exposure, reworking, and removal of
sediments during a lower than present stand of sea level, such as that
which King (1970) found recorded in the sediment distribution pattern
of the Scotian Shelf. Andrews (1980) has noted evidence for Quaternary
sea levels $\geq$ 100 m lower than at present near Broughton Island and
believes further evidence exists near the mouth of Cumberland Sound
that suggests sea level was lower than at present during a large part
of the Foxe Glaciation. D. Praeg (personal communication 1981, 1982)
has noted apparent removal of till above 180 m in the Lady
Franklin-Monumental Islands area of the shelf between Frobisher Bay and
Cumberland Sound, as well as the presence of numerous terrace-like
features on the acoustic profiles across the shelf. While the latter
range widely in depth and may relate to other events, some occur near
the $\geq$ 100 m range suggested by Andrews from the Broughton I. area, and
some in the 180 m range marking apparent removal of till in the Lady
Franklin area.

<u>Silt, clay</u>. The floors of deep basins such as Frobisher Bay and
Cumberland Sound are blanketed with post-glacial muds that are included
in units 5 and 6.

Iceberg scour marks are observable on sidescan sonar records and
high resolution seismic profiles across the sediment covered areas of
the Baffin shelf down to depths of some 500 m or greater (e.g. Lewis et
al., 1980). Some of these features are believed
to be relict, others are the product of present day iceberg
groundings. Although depth of water, iceberg size, seabed gradient,
current regime, etc. are factors in the frequency of grounding and
iceberg ploughing capability, various acoustic data and follow-up
observations from the submersible <u>Pisces IV</u> indicate that depth of
scouring into the seabed is also very much related to the hardness and
resistance of the various seabed sediment types. In addition to
intense reworking of sediments by grounded icebergs, and to a lesser
extent by organisms, the character of the sediments on the Baffin
Island shelf also has been affected in places by currents, and by
deposition of ice-rafted material.

ACKNOWLEDGEMENTS

The assistance and cooperation of the Captains, officers, crews, and
scientific staffs of CSS <u>Hudson</u> and M.V. <u>Pandora II</u> in the collection
of the various data are gratefully acknowledged. I am also indebted to
J.T. Andrews and D.B. Praeg for helpful comments and discussion and to
H. Josenhans and A.C. Grant for review of the manuscript.

REFERENCES

Andrews, J.T., 1980: Progress in relative sea level and ice sheet
    reconstructions, Baffin Island, N.W.T. for the last 125,000 years.
    In: <u>Earth rheology, isostasy and eustasy</u>, Edited by N.A. Morner. John
    Wiley & Sons, N.Y., 175-200.
Andrews, J.T., Guennel, G.K., Wray, J.L. and Ives, J.D., 1972: An early
    Tertiary outcrop in north central Baffin Island, Northwest
    Territories, Canada: environment and significance. <u>Canadian Journal
    of Earth Sciences</u>, 9:233-238.
Andrews, J.T. and Ives, J.D., 1978: "Cockburn" nomenclature and the
    late Quaternary history of the eastern Canadian Arctic. <u>Arctic and
    Alpine Research</u>, 10:617-633.

Beh, R.L., 1975: Evolution and geology of western Baffin Bay and Davis
    Strait, Canada. In: Canada's continental margins and offshore
    petroleum exploration, Edited by C.J. Yorath, E.R. Parker and D.J.
    Glass. <u>Canadian Society of Petroleum Geologists, Memoir</u> 4:453-476.

Blake, W., Jr., 1966: End moraines and deglaciation chronology in
    northern Canada, with special reference to southern Baffin Island.
    Geology Survey of Canada Paper 66-26. 31 pp.
Brett, C.P. and Zarduzki, E.F.K., 1979: Project Westmar - a shallow
    marine geophysical survey on the west Greenland continental shelf in
    the region 64° to 69°30'N and a preliminary interpretation of the
    data. Grønlands Geologiste Undersøgelse, Rapport, 87. 29 pp.

Clarke, B.W. and Daae, H.D. 1979: Cretaceous and Tertiary geology of
    Bylot Island, N.T.T. Canadian Society of Petroleum
    Geologists-Canadian Society of Exploration Geophysists, Joint
    convention, Calgary, Alberta, Program and Abstracts.
Clarke, D.B. and Upton, B.G., 1971: Tertiary basalts of Baffin Island:
    field relations and tectonic setting. Canadian Journal of Earth
    Sciences, 8:248-258.
Clarke, D.B. and Pedersen, A.K., 1976: Tertiary volcanic province of
    West Greenland. In: Geology of Greenland. Edited by A. Escher and
    W.S. Watt. Grønlands Geologiste Undersøgelse, Copehagen, 364-385.

Dawes, P.R., 1976: Precambrian to Tertiary of Northern Greenland. In:
    Geology of Greenland. Edited by A. Escher and W.S. Watt. Grønlands
    Geologiste Undersøgelse, Copenhagen, 248-303.
Denham, L.R., 1974: Offshore geology of northern west Greenland (69° to
    75°N) Grønlands Geologiste Undersøgelse, Rapport 63. 24   pp.
Douglas, R.J.W., 1969: Geological map of Canada. Geological Survey of
    Canada, Map 1250A.
Dyke, A.S., 1977: Quaternary geomorphology, glacial chronology, and
    climatic and sea-level history of southwestern Cumberland Peninsula,
    Baffin Island, Northwest Territories, Canada. Ph.D. thesis,
    University of Colorado, Boulder. 185 p.
Dyke, A.S., 1979: Glacial and sea-level history of southwestern
    Cumberland Peninsula, Baffin Island, N.W.T., Canada. Arctic and
    Alpine Research, 11:179-202.

Escher, A., Henriksen, N., Dawes, P.R. and Weidick, A., 1970:
    Tectonic/Geological map of Greenland. Geological Survey of Greenland.

Falconer, R.K.H., 1977: Marine geophysical and geological research in
    Baffin Bay and the Labrador Sea, CSS Hudson 1976. In: Report of
    Activities, Part B, Geological Survey of Canada, Paper 77-1B,
    255-260.
Fillon, R.H., Miller, G.H., and Andrews, J.T., 1981: Terrigenous sand
    in Labrador Sea hemipelagic sediments and paleoglacial events on
    Baffin Island over the last 100,000 years. Boreas, 10:107-124.
Fortier, Y.O., and Morley, L.W., 1956: Geological unity of the Arctic
    Islands. Royal Society of Canada Transactions, 50 (sec. III); 3-12.

Gilbert, R., 1982: The Broughton Trough of eastern Baffin Island.
    Canadian Journal of Earth Sciences, 19:1599-1607.
Grant, A.C., 1975a: Structural modes of the western margin of the
    Labrador Sea. In: Offshore Geology of Eastern Canada, Edited
    byW.J.M. van der Linden and J.A. Wade. Geological Survey of Canada,
    Paper 74-30, 2:217-231.
Grant, A.C., 1975b: Geophysical results from the results from the
    continental margin off southern Baffin Island. In: Canada's
    continental margins and offshore petroleum exploration, Edited by
    C.J. Yorath, E.R. Parker, and D.J. Glass. Canadian Society of
    Petroleum Geologists, Memoir 6:411-431.
Grant, A.C., 1980: Problems with plate tectonics: Labrador Sea.
    Bulletin of Canadian Petroleum Geology, 28:252-278.

Grant, A.C., and Manchester, K.S., 1970: Geophysical investigation in the Ungava Bay - Hudson Strait region of northern Canada. Canadian Journal of Earth Sciences, 7:1062-1076.

Henderson, G., Rosenkrantz, A., and Schiener, E.J., 1976: Cretaceous and Tertiary sedimentary rocks of West Greenland. In: Geology of Greenland, Edited by A. Escher and W.S. Watt. Grönlands Geologiste Undersögelse, 488-505.

Hood, P.J., and Bower, M.E., 1973: Low-level aeromagnetic surveys of the continental shelves bordering Baffin Bay and the Labrador Sea. In: Earth science symposium on offshore Eastern Canada, Edited by P.J. Hood. Geological Survey of Canada, Paper 71-23, 573-598.

Ives, J.D., and Andrews, J.T., 1963: Studies in the physical geography of north-central Baffin Island, N.W.T. Geographical Bulletin, 19:5-48.

Jackson, G.D., and Davidson, A., 1975: Bylot Island map-area, District of Franklin. Geological Survey of Canada, Paper 74-29, 12 pp.

Jackson, G.D., Davidson, A., and Morgan, W.C., 1975: Geology of the Pond Inlet map-area, Baffin Island, District of Franklin. Geological Survey of Canada, Paper 74-25, 33 pp.

Jackson, H.R., Keen, C.E., and Barrett, D.L., 1977: Geophysical studies on the eastern continental margin of Baffin Bay and in Lancaster Sound. Canadian Journal of Earth Sciences, 14:1991-2001.

Jackson, H.R., Keen, C.E., Falconer, R.K.H., and Appleton, K.P., 1979: New geophysical evidence for sea floor spreading in central Baffin Bay. Canadian Journal of Earth Sciences, 11:2122-2135.

Josenhans, H., 1981: Evidence for pre-late Wisconsinan glaciation in Labrador Shelf - Cartwright Saddle Region. University of Colorado Institute of Arctic and Alpine Research, Arctic Workshop, March 1981, Program and Abstracts, 24-25.

Keen, C.E., and Barrett, D.L., 1972: Seismic refraction studies in Baffin Bay: an example of a developing ocean basin. Royal Astronomical Society, Geophysical Journal, 30:253-271.

Keen, C.E., Barrett, D.L., Manchester, K.S., and Ross, D.I., 1972: Geophysical studies in Baffin Bay and some tectonic implications. Canadian Journal of Earth Sciences, 9:239-256.

Keen, C.E., Keen, M.J., Ross, D.I., and Lack M., 1974: Baffin Bay: small ocean basin formed by sea floor spreading. American Association of Petroleum Geologists, Bulletin 58:1089-1108.

Kerr, J.W., 1967: A submerged continental remnant beneath the Labrador Sea. Earth and Planetary Science Letters, 2:283-289.

Kerr, J.W., 1980: Did Greenland drift along Nares Strait? Bulletin of Canadian Petroleum Geology, 28:279-289.

King, L.H., 1970: Surficial geology of the Halifax-Sable Island map-area. Canada, Dept. of Environment, Marine Science Paper 1. 16 pp.

King, L.H., MacLean, B., and Drapeau, G., 1972: The Scotian Shelf submarine end moraine complex. In: Marine geology and geophysics. XXIV International Geological Congress, Section 8, 237-249.

Klose, G.W., Zinkan, G.G. Malterre, E., and McMillan, N.J., 1981: Petroleum exploration offshore southern Baffin Island, northern Labrador Sea, Canada. In: Third International Symposium on Arctic Geology, Calgary, Program and Abstracts, p. 76.

Lewis, C.F.W., MacLean, B., Falconer, R.K.H., 1980: Iceberg scouring in Labrador Sea and Baffin Bay; a reconnaissance of regional variability. In: Proceedings First Canadian Conference on Marine Geotechnical Engineering, Calgary, Edited by W.J. Eden, 79-94.

Løken, O.H., and Hodgson, D.A., 1971: On the submarine geomorphology along the east coast of Baffin Island. Canadian Journal of Earth Science, 8:185-195.

MacLean, B. 1978: Marine geological-geophysical investigations in 1977 of the Scott Inlet and Cape Dyer - Frobisher Bay areas of the Baffin Island continental shelf. In: Current Research, Part B, Geological Survey of Canada, Paper 78-1B, 13-20.

MacLean, B. 1982: Investigations of Baffin Island shelf from surface ship and research submersible in 1981. In: Current Research, Part A, Geological Survey of Canada, Paper 82-1A, 445-447.

MacLean, B., and Falconer, R.K.H., 1979: Geological/geophysical studies in Baffin Bay, Scott Inlet - Buchan Gulf and Cape Dyer - Cumberland Sound areas of the Baffin Island shelf. In: Current Research, Part B, Geological Survey of Canada, Paper 79-1B, 231-244.

MacLean, B., Jansa, L.F., Falkconer, R.K.H., and Srivastava, S.P., 1977: Ordovician strata on the southeastern Baffin Island shelf revealed by shallow drilling. Canadian Journal of Earth Sciences, 15:1925-1939.

MacLean, B., Falconer, R.K.H. and Clarke, D.B., 1978: Tertiary basalts of western Davis Strait; bedrock core samples and geophysical data. Canadian Journal of Earth Sciences, 15:773-780.

MacLean, B., Falconer, R.K.H., and Levy, E.M., 1981: Geological, geophysical and chemical evidence for natural seepage of petroleum off the northeast coast of Baffin Island. Bulletin of Canadian Petroleum Geology, 29:75-95.

MacLean, B., Srivastava, S.P., and Haworth, R.T., 1982: Bedrock structures off Cumberland Sound, Baffin Island shelf: core sample and geophysical data. In: Arctic geology and geophysics. Editors A.F. Embry and H.R. Balkwill. Canadian Society of Petroleum Geologists, Memoir 8:279-295.

McMillan, N.J., 1973: Shelves of Labrador Sea and Baffin Bay, Canada. In: Future petroleum provinces of Canada. Edited by R.G. McCrossan. Canadian Society of Petroleum Geologists, Memoir 1:473-515.

McWhae, J.R.H., 1981: Structure and spreading history of the northwestern Atlantic region from the Scotian Shelf to Baffin Bay. In: Geology of the Atlantic borderlands and basins. Edited by J. Wm. Kerr and A.J. Fergusson. Canadian Society of Petroleum Geologists, Memoir 7:299-332.

Miall, A.D., Balkwill, H.R., and Hopkins, W.R. Jr., 1980: Cretaceous and Tertiary sediments of Eclipse Trough, Bylot Island area, Arctic Canada, and their regional setting. Geological Survey of Canada, Paper 79-23, 20 pp.

Miller, G.H., 1980: Late Foxe glaciation of southern Baffin Island, N.W.T., Canada. Bulletin Geological Society of America, 91:399-405.

Miller, A.K., Youngquist, W. and Collinson, C., 1954: Ordovician cephalopod fauna of Baffin Island. Geological Society of America, Memoir 62, 166 pp.

Miller, G.H., Andrews, J.T. and Short, S.K., 1977: The last interglacial-glacial cycle, Clyde foreland, Baffin Island, N.W.T.: stratigraphy, biostratigraphy, and chronology. Canadian Journal of Earth Sciences, 14:2824-2857.

Osterman, L.E., 1982: Late Quaternary history of southern Baffin Island, Canada: a study of foraminifera and sediments from Frobisher Bay. University of Colorado, Ph.D. thesis, 380 p.

Osterman, L.E., Andrews, J.T. and Hart, T., 1980: Biostratigraphy, litho-stratigraphy and paleomagnetism of a late Quaternary core (Hu 77-159) from outer Frobisher Bay, Baffin Island. Geological Association of Canada, Annual Meeting, Halifax, N.S. Program and Abstracts, 73 pp.

Pelletier, B.R., 1966: Development of submarine physiography in the Canadian Arctic and its relation to crustal movements. In: Continental drift, Edited by G. D. Garland. Royal Society of Canada, Special Publication, 9:77-101.

Rosenkrantz, A., and Pulvertaft, T.C.R., 1969: Cretaceous-Tertiary stratigraphy and tectonics in northern West Greenland. American Association of Petroleum Geologists, Memoir 12:883-898.

Srivastava, S.P., 1978: Evolution of the Labrador Sea and its bearing on the early evolution of the North Atlantic. Royal Astronomical Society, Geophysical Journal, 52:313-357.

Srivastava, S.P., Falconer, R.K.H., and MacLean, B., 1981: Labrador Sea, Davis Strait, Baffin Bay: geology and geophysics-a review. In: Geology of the Atlantic borderlands and basins, Editors J.W. Kerr and A.J. Fergusson. Canadian Society of Petroleum Geologists, Memoir 7:333-398.

Srivastava, S.P., MacLean, B., Macnab, R.F., and Jackson, H.R., 1982: Davis Strait: structure and evolution as obtained from a systematicgeophysical survey. In: Arctic geology and geophysics, Editors A.F. Embry and H.R. Balkwill. Canadian Society of Petroleum Geologists, Memoir 8:267-278.

Trettin, H.P. and Balkwill, H.R., 1979: Contribution to the tectonic history of the Innuitian Province, Arctic Canada. Canadian Journal of Earth Sciences, 16:748-769.

Umpleby, D.C., 1979. Geology of the Labrador Shelf. Geological Survey of Canada, Paper 79-13, 39 p.

Workum, R.H., Bolton, T.E., and Barnes, C.R., 1976: Ordovician geology of Akpatok Island, Ungava Bay, District of Franklin. Canadian Journal of Earth Sciences, 13:157-178.

Survey — Part II

In Part II of this volume, the four chapters discuss the patterns of glacial erosion, landform development, and glacial and marine sediments. Together these chapters constitute an important source of information on the nature of glacial erosion and the rate or amount of glacial erosion during the Cenozoic. Mapping the patterns of glacial erosion across Baffin Island and surrounding areas (Chapter 3) indicates that glacial scouring is confined to the terrain below c. 350 m asl. The high upland surfaces are primarily unscoured although there is limited evidence for glaciation in the form of striations and ice-molded landforms. Within these high uplands, fiords and valleys penetrate inland--these have been sites of active glacial erosion, although how much of the typical fiord form is associated with glacial erosion and how much is a result of structural style? The LANDSAT mapping of glacial landforms indicates that the major areas of erosion in the eastern Canadian Arctic occur in the lowlands where the ice would be thickest and basal temperatures probably reached the pressure melting point. Bedrock erosion from these areas would have led to the production of glacial sediment, however, the amount of glacial sediment within the areas of scour is sparse. The lightly scoured uplands intervene along the glacial flow paths that link the areas of heavy scouring with the fiords and waters of the east coast. Interpretation of the landforms, and glaciological modelling suggest that the ice was cold-based at elevations greater than 350 m, hence glacial sediment from the lowlands to the west would have to be transported englacially prior to their release by bottom or surface melting in the fiords. The isochrones of reconstructed ice sheets indicate that the rate of sediment transport would be slow.

These comments are intended to establish the framework for a brief discussion on the rate and amount of glacial erosion in the eastern Canadian Arctic and the rate of transfer of these materials to the shelf and deep sea. As noted in Chapters 3, 4, and 6 these are important questions that pertain to the question of the effeciency of glacial erosion and the rate of development. Some authors (Pisias and Moore, 1981; Prell, 1982) have suggested that the changes in the amplitude and frequency of O-18 changes in the deep sea are associated with a change in the physiography of the area that underlaid the Laurentide Ice Sheet. Pisias and Moore (1981) and Prell (1982) both noted such a change in the isotopic signal in deep-sea records, and suggested that this change might have been associated with "deep erosion" by the Laurentide Ice Sheet and the development of a marine-based ice sheet complex over the Laurentide Region. These ideas also need to be considered in light of Laine's argument (1980) that glacial erosion by the Atlantic flowing sectors of the Laurentide Ice Sheet had resulted in an overall lowering of the region by 100 m. In the restricted case of the eastern Canadian Arctic (Chapter 3) approximately one-half of the region shows virtually no evidence for glacial erosion. In addition, if the Precambrian shield is traced westward to its contact with the Paleozoic limestones this surface declines in elevation gently toward Foxe Basin and there is no evidence for a major change in the intensity of erosion across the contact. Relative relief in the area west of the no scour/scour contact is moderate and probably averages 200-300 m, whereas within the area of no scour it averages 50-150 m as an estimate. The source for the 100 m or so of the bedrock erosion is thus far from clear.

There are some very important points to be made about the glacial geology of Baffin Island as viewed from the perspective of the Continental Shelf (Chapter 6), especially the contrast between the bathymetry and morphology of the Baffin Island Shelf and that to the south along the coast of Labrador. The absence of a "coast-parallel" marginal trough along eastern Baffin Island suggests that the style and extent of glaciation has been different, that is, probably les extensive glaciation of the shelf by grounded ice. Nevertheless, there are major troughs that extend, although not in a simple manner, from the fiords across the shelf. These troughs are eroded into the shelf sediments and may be (must be?) caused by glacial erosion. It remains to be resolved how this can be reconciled with the presence of a sill at the fiord mouth or with the evidence for limited Quaternary glaciation of the shelf (see Chapters 13, 14, and 17). In a similar vein, the presence of diamictons on the shelf of eastern Baffin Island, interpreted as tills (Chapter 6), potentially represents a departure of the view espoused by the land-based glacial geologists that glaciation was largely restricted to the fiords throughout the Quaternary. However, it must be noted that in the discussion of glacigenic sediments (Chapter 5) terrestrial till on Baffin Island were texturally similar to ice-rafted facies in Baffin Bay. Thus the critical questions are: How easy is it to discriminate between tills and glacial marine diamictons on the basis of texture or seismic signal? The answer must surely lie in the combining of terrestrial and marine mapping of surficial sediments, in the dating of events in both environments, and in the search for "marker beds" that will allow the physical tracing of one into the other.

REFERENCES

Laine, E.P., 1980: New evidence from beneath the western North Atlantic for the depth of glacial erosioin in Greenland and North America. Quaternary Research, 14:188-198.

Pisias, N.F. and Moore T.C., Jr., 1981: The evolution of Pleistocene climate: a time series approach. Earth and Planetary Science Letters, 52:450-458.

Prell, W.L., 1982: Oxygen and carbon isotope stratigraphy for the Quaternary of Hole 502B: evidence for two modes of isotopic variability. Initial Reports of the Deep Sea Drilling Project, Vol. LXVIII:455-464.

PART III
QUATERNARY PALEOCEANOGRAPHIC AND ICE CORE RECORDS

7   Climatic and oceanographic changes over the past 400,000 years: evidence from deep-sea cores on Baffin Bay and Davis Strait

A. E. Aksu

Baffin Bay is situated in the high Arctic, between 65°N and 77°N
latitudes. It is connected to the Arctic Ocean through Smith, Jones,
and Lancaster Sounds, and to the Atlantic Ocean through Davis Strait.
The Bay is about 1300 km long and 450 km wide with a maximum water
depth slightly exceeding 2300 m. The surface water circulation in the
Baffin Bay region results from the outflow of water from the Arctic
Ocean, and it is subject to seasonal variations (Timofeyev, 1960). In
summer in the Davis Strait the warmer West Greenland Current flows
northward along the west coast of Greenland. The Baffin Land Current
originating from polar water flows southward along the east coast of
Baffin Island. These two opposite flowing currents form a large
counter-clockwise gyre in central Baffin Bay. In the winter, however,
a large portion of the West Greenland Current is deflected westward
around Davis Strait, and the Baffin Land Current dominates the surface
circulation.

The distribution pattern of major oceanic currents in the North
Atlantic Ocean experienced major modifications during the Upper
Quaternary. Prior to the opening of Hudson Strait at about 8000 yr BP
(Prest, 1968), Baffin Bay was the only route of the excess Arctic water
outflow. Today, this discharge manifests itself as the cold Labrador
Current and is partially responsible for the colder climate along the
eastern Canadian continental margin. Similarly, the warmer West
Greenland Current is responsible for the maritime climates around
southwest Greenland. Therefore, Baffin Bay provides a unique
opportunity to study the development of major oceanic current systems
in the North Atlantic, and their significance during glacial inception
and decay.

This paper examines the Upper Quaternary climatic and
oceanographic changes in the Baffin Bay region. It is based on

analyses of 26 piston and gravity sediment cores collected from water depths of about 700 m to 2300 m. The paper is divided into three parts: (I) Stratigraphy; (II) Paleoceanography and III) Paleoclimatology.

MATERIALS AND TECHNIQUES

Piston and gravity cores were collected for this study by Bedford Institute of Oceanography research vessel CSS Hudson on cruises 69-050, 74-026, 76-029, 77-024, 77-027 and 78-029 (Fig. 7.1). Cores were stored on horizontal racks at 4°C onboard ship and at Dalhousie University until they were split. Visual description of cores was made; x-radiographs were then taken to determine textural variations, primary and secondary sedimentary structures, stratification and coring disturbances. Cores were routinely sampled for: (1) sedimentological analyses (all cores); (2) paleomagnetic properties of sediments (10 cores); (3) volcanic ash studies (22 cores); (4) micropaleontological studies (16 cores); 5) oxygen and carbon isotope determinations (2 cores); and 6) C-14 dating (4 samples from two cores). Details of the analytical procedures are given in Aksu (1981).

I. STRATIGRAPHY

Sediments

The Upper Quaternary deposits in central Baffin Bay are predominantly ice-rafted and hemipelagic sediments. Thin graded sands and silt to mud couplets, interpreted as distal turbidites, also occur in this area. On the slopes, around the margin of the bay, the dominant sediments are debris flow deposits and proximal turbidites derived from ice-rafted sediments on the shelf (Aksu, 1981). On the basis of color, grain size, and mineralogical analyses, the sediments observed in the cores are divided into six lithofacies. Figure 7.2 illustrates the distribution of these facies in time and space. The sources of sediments and the processes involved in the deposition of two of these six lithofacies (facies A and B) are pertinent to the understanding of water masses and current directions; therefore they are briefly discussed below.

Facies A. This facies consists of dusky yellowish-brown (10YR 2/2) to pale yellowish-brown (10YR 6/2) gravelly sandy muds and a few sand/silt

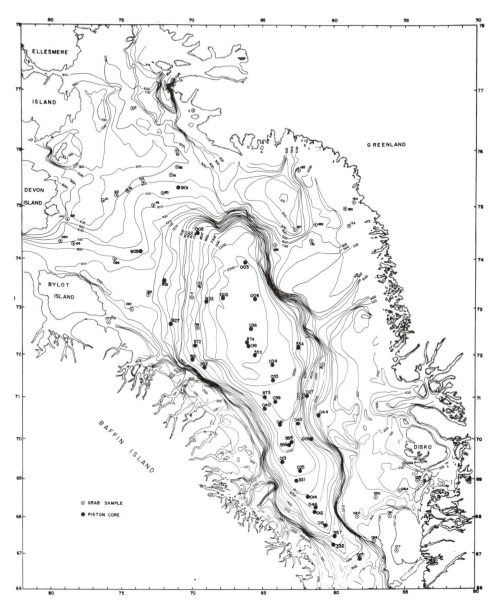

Figure 7.1 Bathymetry of the Baffin Bay area. Depth contours are in
meters. Solid circles represent piston cores, open circles represent
grab samples and gravity cores.

to mud couplets. It occurs predominantly in the northern and western
portion of the bay and become less abundant and thinner toward the the
south and east. Facies A is generally absent in the cores from the

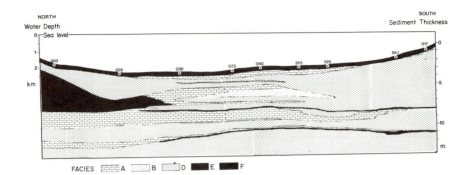

FACIES ☐A ☐B ☐D ■E ■F

Figure 7.2 Schematic north-south section through Baffin Bay and Davis
Strait showing distribution of lithofacies. Section length is
approximately 750 km. Facies A consists of yellowish brown gravelly
sandy muds and sand/silt to mud couplets. Facies B is composed of red
massive gravelly sandy muds and mottled muds. Facies C consists of
brownish black and olive black gravelly sandy muds. Facies D is
composed of yellow green and olive gray sorted sand, silt and sand/silt
to mud couplets. Facies E consists of brown fine muds. Facies F is
composed of reddish brown-brown gravelly sandy mud and/or mud couplets,
sand/silt to mud couplets and fine muds.

Greenland slope and the Davis Strait. The petrology of the gravel
fraction and the identified sand and clay mineral assemblages indicated
that the predominant source of this facies is carbonate rocks. The
minor occurrence of mineral assemblages other than carbonates indicated
varying amounts of mixing with different sources. Due to the general
absence of carbonate outcrops along the coastal area of Baffin Island
and Greenland, the source of the carbonate-rich debris in the
unconsolidated sediments of Baffin Bay has been attributed to long
distance ice transport from the Paleozoic carbonates of the Arctic
Islands (Marlowe, 1966; Grant, 1971; Aksu, 1981).

Facies B. This facies consists of a grayish-red (10R 4/2), generally
massive gravelly sandy mud. Facies B also occurs predominantly in the
northern Baffin Bay cores, becoming less abundant to the south and
east. Only one core from the lower slope off Greenland contained one
thin bed of facies B, and this facies is absent in Davis Strait. The
identified mineral assemblages and the surface texture of quartz grains
indicated that the prominant source of facies B is fluvial-deltaic
sediment particularly rich in montmorillonite, mixed layer clays and
kaolinite, and a polycyclic heavy mineral suite (Aksu, 1981).
Palynomorphs from bulk samples suggest that much of the source rock is
of Mesozoic and early Cenozoic age (G. L. Williams, 1979; pers. comm.)

Facies B represents an admixture of several sources: the hematite stained, rounded quartz grains are predominantly derived from the Proterozoic and Meso-Cenozoic siliclastic rocks that encircle Lancaster Sound. Expandable minerals and kaolinite have originated from the siliclastic rocks of the Sverdrup Basin and Bylot Island areas and some sediments may be derived from continental shelf sediments. Carbonate debris is mainly derived from Ordovician-Silurian limestone and dolomites. Many textural, structural, fabric, mineralogical and micropaleontological characteristics suggest an ice-rafted origin for the deposition of much of facies A and B (Aksu, 1981).

Correlation

Volcanic Ash. Sand and silt sized volcanic ash, disseminated in fine mud, is identified in the SE Baffin Shelf, Davis Strait and Baffin Bay cores. It consists primarily of colorless, and light brown, platy, bubble-wall shards with lesser amounts of black pumice. Similar tephra materials were first noted in the North Atlantic sediments by Boeggild (1900). The geographical and down core distribution of the tephra in the North Atlantic has been studied by Bramlette and Bradley (1941), Black (1969), Binns (1972) and Ruddiman and Glover (1972, 1975). Recent discoveries of ash in Labrador Sea and Baffin Bay sediments have been reported by Chough (1978), Aksu and Piper (1979), and Fillon and Duplessy (1980). Three stratigraphically distinct ash horizons are attributed to relatively brief periods of volcanic activity in the North Atlantic (probably in Iceland or Jan Mayen Island). The ages of these episodes of ash deposition have been estimated to be 9300 yr BP, 64,700 yr BP and 340,000 yr BP (Ruddiman and Glover, 1975; Rudduman, 1977).

The down core variation of ash concentration in 22 cores from SE Baffin Island Shelf, Davis Strait and Baffin Bay was examined (Fig. 7.3). Because rare ash grains appear in most of the samples throughout the cores, background concentration levels were estimated to be 5-10 shards/g and 3-4 shards/g for SE Baffin Island Shelf/Davis Strait and Baffin Bay cores respectively.

Three levels of significant ash abundance were identified. Level I occurs near the core tops and it is generally characterized by a broad single peak in Baffin Bay but up to three less prominent peaks (Ia, Ib and Ic) around Davis Strait and the SE Baffin Island Shelf (cores 012, 557, 017, 01-2 and 02-1). Ash abundance in this level varies between 10-150 shards/g sediment, except in core 044 which

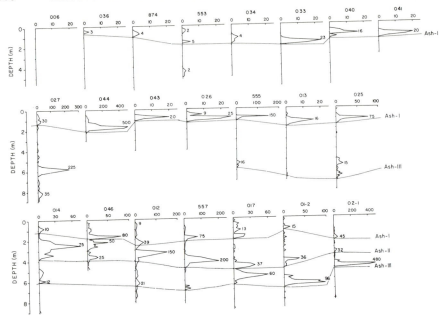

Figure 7.3 Graph of ash abundance (shards per gram sediment) versus depth in 22 cores from Baffin Bay, Davis Strait and SE Baffin Island shelf. Actual shard counts are indicated next to the major ash peaks.

contained approximately 500 shards/g sediment. Level II. In eight cores from the SE Baffin Island Shelf and Davis Strait a second peak was identified. The maximum thickness of the ash-bearing zone is generally less than 100 cm. The ash abundance in this level is about 30 shards/g sediment for SE Baffin Island Shelf, and increases to about 75-200 shards/g sediment in Davis Strait cores. However, the background level also increases towards Davis Strait and level II becomes less distinct in this area. Level III occurs in Davis Strait, Baffin Bay and SE Baffin Island Shelf cores, and is absent in central Baffin Bay cores; except in core 553 where a few shards occur around 450 cm depth in the core and may correlate with level III. This level is characterized by a single peak up to 100 cm thick at around 400 cm to 700 cm depth, in the cores, and the ash abundance varies between 75 and 480 shards/g sediment. In two cores (02-1 and 01-2) a double peak was identified with less abundant ash in the younger peak (Fig. 7.3).

Areal distribution of ash. The maximum abundance of ash level III is found in cores from the SE Baffin Island Shelf, and decreases gradually towards Davis Strait. This level also decreases abruptly northward

toward Baffin Bay and it is generally absent north of 71°N latitude.
The distribution of ash level II is very similar to that of level III
except that the maximum occurrence is found around Davis Strait. A
more complex distribution pattern is observed in ash level I as it
contains up to three separate subzones (Ia, Ib, Ic), and the
distinction between the three ash subzones is not always possible.
However, the available data indicate that Ib and Ic extend northwards
to about 72°N latitude (core 553) and they are absent in northern
Baffin Bay. Their abundances vary significantly from core to core and,
and unlike ash levels II and III, the maximum occurrence is found
between 69°N and 71°N latitudes. An order of magnitude decrease is
found in ash abundances in levels Ib and Ic going from cores 025, 555
and 044 of the southern bay to cores 041, 040 and 033 of the central
bay. This decrease and eventual disappearance is most pronounced
around 71°N latitude. The distribution pattern of Ia is very similar
to that of Ib. The abundance peaks in this subzone are often obscured
by the background noise; however, the overall northward decrease is
detectable.

Marine microfossils

The size fraction larger than 63 um was examined for benthic and
planktonic foraminifera according to standard micropaleontological
counting procedures. Results were converted to "individuals per gram
dry weight sediment." The down core abundance values of planktonic and
benthic foraminifera were plotted on semilogarithmic graphs.

Biogenic Zones

Cores are divided into benthic (B) and planktonic (P) subzones based on
the occurrence and abundance of microfossils. Striking correlations,
based solely on abundance, can be seen in cores from central Baffin
Bay, i.e. cores 006, 036, 553, 033, 040, 027, and 043. The correlation
of Davis Strait and SE Baffin Island shelf cores is less clear since
there is generally a higher background level of material of biogenic
origin.

Zone 1. In all cores the interval between 5 - 40 cm contains high
amounts of skeletal tests and is subdivided as planktonic Subzone P1
and benthic Subzone B1. In Subzone P1, diatoms predominate, typically
Coscinodiscus spp. and Nitzschia spp.; few radiolarians are present,
and planktonic foraminifera are absent. Subzone B1 is composed

entirely of arenaceous foraminifera; typically Trochamina lobata, Texturlaia earlandi, T. torquata, Tritaxis atlantics, Trochamina nana, T. rotaliformis, Eggerella advena, Reophax arctica, R. curtus, Cribrostomoides crassimargo, C. jeffreysi, and Spiroplectammina biformis. Subzones P1 and B1 can be clearly correlated from the northernmost core (006) in Baffin Bay to Davis Strait (017), (Figs 7.4 and 7.5). P1 can possibly be extended to SE Baffin Island shelf core 02-1.

Zone 2. Below Zone 1, a zone of high abundance of biogenic tests, characterized by two or three abundance peaks, is identified. The arenaceous foraminifera, diatoms and radiolarians disappear abruptly in Zone 2. The fauna in benthic Subzone B2 is composed mainly of calcareous species, typically Triloculina trihedra, Quinqueloculina stalkeri, Q. seminulum, Pyrgo williamsoni, Islandiella islandica, I. norcrossi, I. helenae, and lesser amounts of Elphidium sp. (E. excavatum forma clavata). Haynesina orbiculare, Melonis zaandamae. T. triheldra, Q. stalkeri, Q. seminulum and P. Williamsoni occur in Subzone B2 and together with I. helenae and I. islandica dominate over the rest of the benthic fauna. Subzone B2 clearly correlates from core 006 of Northern Baffin Bay to core 025 of Southern Baffin Bay (Fig. 7.4). Towards the south, the correlation becomes less obvious due to the higher background level of foraminifera; however, the above-mentioned benthic assemblage is used to extend the correlation to Davis Strait (core 017).

    Planktonic Subzone P2, of similar extent to B2, extends from 40 - 100 cm and the fauna is composed predominantly of Neogloboquadrina pachyderma (sinistral) with less abundances of N. pachyderma (dextral), Globigerina quinqueloba, G. bulloides, and N. pachyderma/N. dutertrei. In southern Baffin Bay and Davis Strait the lower levels of subzone P2 also contain statistically significant quantities of diatoms (mainly Coscinodiscus sp.) and radiolarians (mainly Diplocyclas davisiana). Subzone P2 can also be correlated based on abundance and species assemblages from the northernmost core 006 to Davis Strait (core 017).

Zone 3. There is a sharp decrease and eventual disappearance of biogenic tests below Zone 2. In northern Baffin Bay (from cores 006 to 040) an almost barren zone is identified from 100 cm - 300 cm (Fig. 7.4 and 7.5). In southern Baffin Bay cores (from 555 to 017, and possibly to 02-1) there is a decline in the total number of biogenic tests from the extreme top of the cores, but below 50 cm there is a high fluctuation in test abundance and the background level is higher than

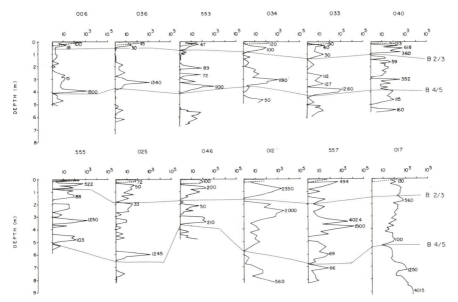

Figure 7.4 Benthic foraminiferal abundance (foraminifera per gram sediment) versus depth in 12 cores from Baffin Bay and Davis Strait. Foraminiferal abundance scale is semi-logarithmic. B2/3 and B4/5 are benthic foraminiferal zone transitions corresponding to F1/2 etc. on Figure 7.8. Shaded areas represent 100% arenaceous foraminiferal faunas.

in northern Baffin Bay. Thus the distinction between Zones 2 and 3 is not clear. However, the ash level I occurs at the lower part of Zone 2 in central bay cores; thus it is used to define the boundary more clearly between these zones.

The upper limit of benthic Subzone B3 is characterized by the complete disappearance of I. trihedra, Q. seminulum, Q. stalkeri and P. williamsoni of Subzone B2. The benthic fauna in subzone B3 is composed mainly of Elphidum sp., H. orbiculare, I. islandica, I. helenae, Fursenkoina fusiformis, Cassidella complanata and Epistominella exigua. The first occurrence of Bulmina aculeata and B. marginata is found near the bottom of this zone. The lower boundary of this subzone is identified by a sudden increase in total benthic abundance in almost all cores. The planktonic fauna of Subzone P3 is entirely composed of N. pachyderma (sinistral), where the total numbers are an order of magnitude less than those in Subzone P2. The correlation of P3 extends from core 006 to core 025, and it is much less clear in Davis Strait.

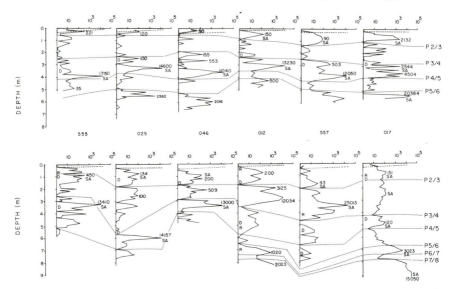

Figure 7.5 Planktonic foraminiferal abundance (foraminifera per gram sediment) versus depth in 12 Baffin Bay and Davis Strait cores. Foraminiferal abundance scale is semi-logarithmic. P2/3, P3/4, P4/5, P5/6, P6/7 and P7/8 are planktonic foraminiferal zone transitions corresponding to F1/2 etc. on Figure 7.8. Shaded areas denote intervals with abundant radiolarians and diatoms. SA = subarctic planktonic foraminiferal assemblage.

Zone 4. A high abundance of biogenic tests characterized by two or four abundance peaks is identified as Zone 4. The boundary between Zones 3 and 4 is marked by the high occurrence of diatoms and the radiolarians in cores from southern Baffin Bay and Davis Strait. The lower boundary of this zone is defined by a sharp decrease in total abundance and apparent disappearance of benthic and plantonic foraminifera around 400 - 500 cm depth in the northern and central bay cores. Furthermore, in southern Baffin Bay and Davis Strait this boundary is also defined by the lower boundary of ash level III.

The fauna in benthic Subzone B4 is composed of I. islandica, I. helenae, Oolina spp,. Lagena spp., E. exigua, B. aculeata, B. marginata, Virgulina spp., and Elphidum spp. B4 clearly correlates from core 006 to core 557. The fauna of the planktonic Subzone P4 is very similar to that of Subzone P2. This subzone also correlates from core 006 to core 557.

Zone 5. A zone similar to Zone 3 is identified immediately below Zone 4. The majority of Baffin Bay cores only penetrate to the top of this zone, however, it is fully represented in cores 036, 034, 040 and possibly in cores 012, 557 and 017. The benthic fauna is composed of Virgulina spp., E. exigua, B. aculeata, B. marginata, Oolina spp., Lagena spp., Elphidium spp., and Islandiella sp. The planktonic fauna (P5) is entirely formed of N. pacyderma (sinistral).

Zone 6. Following Zone 5, a zone similar to 2 and 4 (both in benthic and planktonic faunas) is identified in at least six cores. It is marked by a prominent increase in numbers of biogenic skeletal tests. This zone is probably completely penetrated in cores 557 and 017.

Zone 7. This zone is identified below Zone 6 in cores 557 and 017 and contains abundant diatoms and calcareous foraminifera with very few radiolarians. The benthic fauna is similar to that of Zones 4 and 6, except that Zone 7 also contains a few arenaceous foraminifera. The calcareous planktonic fauna is similar to that of Subzone P3; and the siliceous fauna is similar to that of P1.

Zone 8. Following Zone 7 down core, two very broad abundance peaks are identified only in core 017. The benthic and planktonic faunas are very similar to those of Zones 2, 4 and 6.

Oxygen and carbon isotopes

The oxygen and carbon isotope variations of planktonic foraminifera were examined down cores 040 and 017. These cores represent typical lith-, bio- and tephra-stratigraphic sequences for Baffin Bay and Davis Strait. They do not include major resedimented beds, and bioturbational mixing is insignificant. The results of the analyses are listed in Tables 7.1 and 7.2. Figures 7.6 and 7.7 illustrate the O-18 PDB and C-13 PDB values versus core depth. A three-point moving average is used to smooth O-18 PDB values so as to eliminate the noise and to simplify the identification of isotope stage boundaries.

The isotope curve for cores 017 and 040 can be compared with those previously established for the Atlantic, Caribbean and Mediterranean regions. However, the fluctuations in the O-18/O-16 ratios of N. pachyderma in my cores exhibit cycles with amplitudes much greater (about $4^\circ/_{\circ\circ}$) than the observed global isotope changes. Approximately 10 major intervals of light or heavy isotope composition are recognized in the cores. These should be correlatable with similar

TABLE 7.1.   Oxygen and carbon isotope analyses of N. pachyderma in core
76-029-040.

DEPTH (cm)	0-18	C-13	DEPTH (cm)	0-18	C-13
15-20	3.35	0.90	295-300	1.13	0.55
25-30	1.90	0.29	305-310	3.60	0.13
30-35	0.78	-0.36	315-330	4.99	0.40
35-40	3.11	0.30	325-330	2.45	-0.13
40-45	3.90	0.53	335-340	0.72	-1.17
45-50	3.98	0.43	345-350	3.33	0.18
55-60	3.81	0.71	355-360	3.45	0.07
65-70	3.27	0.36	365-370	3.11	0.08
70-75	4.32	0.31	375-380	4.38	0.78
75-80	2.02	0.60	385-390	2.67	-0.44
80-85	3.96	0.41	395-400	2.38	0.24
85-90	2.44	-0.79	405-410	-2.39	-3.12
95-100	1.52	-0.61	415-420	-0.39	-1.19
105-110	0.56	-1.45	425-430	0.95	-1.25
115-125	1.69	-2.61	435-440	-2.05	-0.34
125-135	0.19	-1.93	445-450	1.79	-0.53
135-145	1.34	-2.44	455-460	3.63	0.17
150-155	2.13	-3.73	465-470	3.27	-0.27
160-165	4.28	-2.44	475-480	4.67	2.56
170-175	1.41	-0.57	485-490	4.16	0.34
175-180	-0.67	-0.15	495-500	2.75	-0.33
185-195	-0.79	-2.27	505-515	1.94	-0.50
195-200	0.81	-1.26	515-525	2.66	0.08
205-210	0.52	-1.06	525-530	3.43	0.54
215-225	-1.77	-2.10	535-540	3.36	0.91
225-235	-2.02	-4.48	545-550	0.94	-0.34
235-245	-1.53	-1.75	555-560	2.60	0.01
245-255	-2.59	-3.01	565-570	1.04	-0.35
255-265	-2.63	-2.41	575-580	2.70	0.16
265-270	2.18	-0.36	585-590	3.02	0.73
275-280	0.74	-0.95	595-600	3.42	0.28
285-290	-3.26	-3.52			

intervals in standard open ocean isotope curves. Some indication of
the age of the Baffin Bay cores is needed to establish the approximate
duration of each major isotopic event before such a correlation can be
attempted.

Absolute Dating

Four intervals in two Baffin Bay cores have been C-14 dated (Table
7.3). A radiocarbon age was obtained from the base of the upper
hemipelagic brown mud. This analysis was carried out on several large
gastropod shells where the C-13 analysis indicated unaltered normal
marine carbonates. Three other radiocarbon ages based on total organic
carbon were obtained from two cores to delineate the oxygen isotope

TABLE 7.2.   Oxygen and carbon isotope analyses of N. pachyderma in core 77-027-017.

DEPTH (cm)	0-18	C-13	DEPTH (cm)	0-18	C-13
5-9	3.12	0.29	485-490	4.46	0.65
10-15			495-500	3.82	1.04
30-35	3.12	0.29	505-510	4.71	0.13
43-47	0.90	0.44	515-520	1.06	-1.24
55-60	2.31	0.85	525-530	2.21	-0.41
65-70	1.99	0.05	535-540	0.90	-1.41
75-80	3.30	0.18	545-550	1.41	-1.52
85-90	3.53	0.84	555-560	1.71	-1.21
95-100	3.52	0.73	565-570	0.81	-1.55
105-110	3.95	0.40	575-580	2.22	0.13
115-120	3.01	-0.01	585-590	0.85	-1.18
125-130	4.02	-0.75	595-600	1.58	-0.96
135-140	2.55	-0.19	605-610	0.93	-1.01
145-150	3.90	0.65	615-620	1.56	-0.97
155-160	2.43	-0.10	625-630	-0.24	-2.07
165-170	3.62	0.38	635-640	3.86	0.22
175-180	3.65	0.17	645-650	3.67	0.09
185-190	4.14	-0.17	655-660	3.66	0.26
195-200	2.98	-0.21	665-670	3.76	-0.14
205-210	3.92	0.63	675-680	2.16	-0.63
215-220	4.08	0.66	685-690	3.80	0.18
225-230	3.76	0.68	695-700	3.85	0.34
235-240	3.70	-0.67	705-710	2.22	0.35
245-250	4.17	0.46	715-720	3.63	-0.21
255-260	4.23	-0.13	725-730	3.93	0.42
265-270	2.78	0.19	735-740	3.09	-0.32
275-280	3.03	0.10	745-750	1.29	-1.23
285-290	3.74	0.34	755-760	2.96	-0.26
295-300	3.57	-0.27	765-770	0.67	-2.51
305-310	2.35	-0.75	775-780	0.90	-0.17
315-320	1.44	-1.62	785-790	1.87	1.36
325-330	2.43	-0.13	795-800	3.07	1.05
335-340	2.43	-0.19	805-810	2.81	0.05
345-350	1.34	-1.26	815-820	3.37	0.41
355-360	2.67	-1.10	825-830	3.02	-0.33
365-370	2.15	-1.49	835-840	3.59	0.13
375-380	1.61	-1.18	845-850	4.00	-0.07
385-390	1.68	-0.85	855-860	3.46	-0.26
395-400	2.82	1.59	865-870	3.85	-0.34
405-410	1.40	0.59	875-880	3.60	0.16
415-420	0.00	1.59	885-890	4.50	2.00
425-430	0.70	1.78	895-900	4.28	0.78
435-440	2.23	0.10	905-910	3.54	-0.21
445-450	2.23	2.95	915-920	3.95	-0.38
455-460	1.18	-0.68	925-930	4.18	-0.32
465-470	3.49	-0.13	935-940	4.29	0.52
475-480	3.49	-0.11			

boundaries. The C-13 in these analyses suggested that carbon analyzed was also of marine origin. The sedimentation rates calculated on the

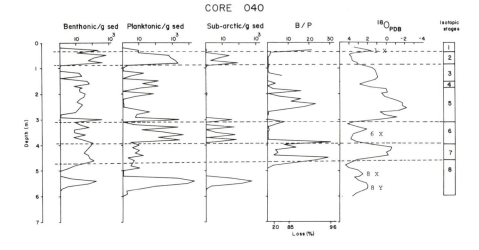

Figure 7.6 Benthic foraminifera per gram sediment (column 1);
planktonic foraminifera per gram sediment (column 2); sub-arctic
planktonic foraminifera per gram sediment including N. pachyderma
dextral, G. quinqueloba, N. pachyderma/N. dutertrei (column 3); ratio
of benthic to planktonic foraminifera (column 4) and 0-18 record are
plotted versus depth in core 040. Note the low biogenic carbonate
stages coincide with high benthic/planktonic ratios and interglacial
stages.

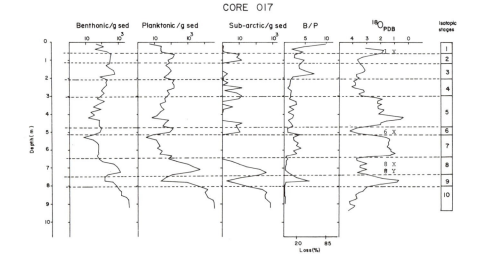

Figure 7.7 Benthic foraminifera per gram sediment (column 1);
planktonic foraminifera per gram sediment (column 2); subarctic
planktonic foraminifera per gram sediment including N. pachyderma
dextral, G. bulloides, G. quinqueloba, N. pachyderma/N. dutertrei
(column 3); ratio of benthic to planktonic foraminifera (column 4) and
^{18}O record are plotted versus depth in core 017. Note the low biogenic
carbonate stages coincide with high benthic/planktonic ratios and
interglacial stages.

TABLE 7.3. Radiocarbon dates from marine cores.

Core	Depth (in cm)	Lab	Lab #	Age (Yr BP)
77-027-017	50-55	Geochron Laboratories	GX6399	10,530 ± 280
76-029-040	60-75	Geochron Laboratories	GX7203	$31,200^{+1800}_{-1200}$
77-027-017	130-150	Geochron Laboratories	GX7204	20,490 ± 820
77-027-017	120-140	Geochron Laboratories	GX7254	21,030 ± 710

TABLE 7.4. Chronology of the oxygen isotope stages.

Stage Boundaries	Depth In Core (cm)		Yr. BP (Shackleton) Opdyke 1973
	040	017	
1-2	35	75	13,000
2-3	95	180	32,000
3-4	150	225	64,000
4-5	180	300	75,000
5-6	305	470	128,000
6-7	400	515	195,000
7-8	460	640	251,000
8-9	-	740	297,000
9-10	-	810	347,000

basis of four dates are 1.92-2.40 cm/1000 years and 4.76-5.20 cm/1000 years for cores 040 and 017, respectively.

Because of the absence of calcareous foraminifera in core top samples, the complete record of isotope stage 1 is not present. Cores 040 and 017 exhibit isotope records extending into isotope stages 8 and 10 respectively. The ages of isotope stage boundaries listed in Table 7.4 are adopted from Shackleton and Opdyke (1973).

Melt Water Spikes

Perhaps one of the most striking features of these records are the major isotopically light peaks superimposed on the characteristic ocean isotope curves. These are interpreted as melt water spikes and are clearly identified near the 1/2 transition, within isotope stage 6, and

at least two major peaks in isotope stage 8 in both cores. These peaks
are labelled as 1x, 6x, 8x and 8y on Figures 7.6 and 7.7. The common
characteristic of these peaks in core 040 is that they all have maximum
light values of about +0.5°/₀₀. These figures are comparable to
the isotope values of interstadial stage 3. In core 017, melt water
spike 1x exhibits similar 0-18 values of about 1°/₀₀, whereas peak
8x and 8y are around 2.5°/₀₀ and peak 6x is much heavier
(4°/₀₀). In this core, only the 0-18 values of peaks 8x and 8y are
comparable to that of stage 3. On the basis of a constant rate of
deposition, the duration of these melting periods are estimated as
follows:

Meltwater Peak	1x	6x	8x	8y
Core 040	4,100 yr	8,300 yr	10,400 yr	10,700 yr
Core 017	5,700 yr	8,800 yr	7,600 yr	7,600 yr

Correlation of Ash and Isotope Stages

The chronology of the ash levels in Baffin Bay is primarily based on
the 0-18 time framework, and the available C-14 dates (Fig. 7.8). Ash
level Ia occurs in isotope stage 1, immediately above the 1/2
transition. This level is C-14 dated as 10,530±710 yr BP. Ash level
Ib is found in the middle of isotope stage 2, the base of which is C-14
dated as 21,030±710 yr BP. Ash level 1c is identified in isotope stage
3, occurring near the 3-4 boundary. This boundary is dated at 64,000
yr BP by Shackleton and Opdyke (1973) and, if this date is correct, ash
level Ic is probably slightly younger than 64,000 yr BP. Ash levels
Ia, Ib and Ic are correlated (although poorly in some cores) over the
Davis Strait area, whereas in Baffin Bay a rather large broad peak is
identified occurring near the base of isotope stage 3. Thus the base
of ash level I in the Baffin Bay area is based on the first significant
occurrence of level Ic in Davis Strait. Placing the ash levels within
the framework of the isotope stratigraphy suggests that ash level Ib
correlates with ash zone A of the Labrador Sea cores (Fillon and
Duplessy, 1980), and that levels Ia and Ic may correlate with the two
most prominent ash beds 1 and 2 in the North Atlantic Ocean (Ruddiman
and Glover, 1972, 1975).

Ash level II occurs near the top of isotope stage 6 in the 5/6
transition and is estimated to be younger than 128,000 yr BP. This
level is not recognized in Labrador Sea cores, presumably because it
was not penetrated (Fillon and Duplessy, 1980). On the other hand, a
few scattered shards are identified in core K708-8 of the North

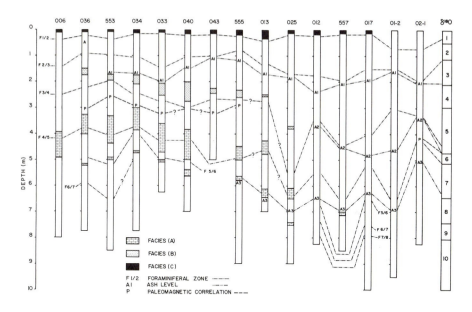

Figure 7.8 Summary stratigraphic correlation. A1 = ash level 1, A2 =
ash level 2, A3 = ash level 3; P = paleomagnetic reversal or
anomalously low inclination zones: F1/2, F2/3, F3/4, F4/5, F5/6, F6/7
and F7/8 are biogenic zone boundaries. Isotopic stages are from core
040 and 017.

Atlantic around termination II (Ruddiman and Glover 1972 and Ruddiman,
pers. commun., 1980) and may correlate with it. The first occurrence
of ash level III is near the 7-8 isotope transition (partly in stage
8), but the maximum abundance is found early in isotope stage 7. This
ash zone is estimated to be younger than 251,000 yr BP. There is no
definite time equivalent of this ash level in the North Atlantic
cores. A minor ash zone is noted in cores V27-20, V23-42 and K708-8 in
the North Atlantic cores in isotope stage 8 (Ruddiman, pers. commun.,
1980) which might correlate with ash level III in the Davis Strait
region. However, if the dating of level III is correct it may be a
completely different ash zone, restricted to the Davis Strait region.

    Biogenic zones 1, 5 and 7 correspond to interglacial isotope
stages 1, 7 and 9 and zones 2, 4, 6 and 8 represent the glacial isotope
stages 2, 6, 8 and 10. Biogenic zone 3 includes isotope stages 3, 4
and 5.

II. PALEOCEANOGRAPHY

Paleocirculation Patterns

Planktonic foraminiferal assemblages, deposition of volcanic ash
levels, and ice rafting are used as evidence for the presence of water
masses and paleocurrent direction during the last 400,000 years in the
Baffin Bay region.

Subarctic planktonic foraminifera occur as a significant
proportion of the total planktonic foraminiferal assemblage in glacial
isotope stages 2, 6, 8 and 10. The percentages of the subarctic fauna
in these intervals range from a maximum of 15% in Davis Strait to about
7% in Baffin Bay (Aksu, 1981). A similar situation is also found in
Labrador Sea cores in stages 2 and 4 (Fillon and Duplessy, 1980).
Plankton tow data from Baffin Bay and Labrador Sea indicated that the
assemblages found in these stages are very similar to the present day
West Greenland Current fauna. Northward advection of North Atlantic
subarctic water through the eastern Labrador Sea to high latitudes
(72°N) in Baffin Bay during glacial stages 2, 6, 8 and 10 is therefore
implied. These data also suggest that Baffin Bay was partially pack
ice free during the summers of these isotopic stages.

The geographical pattern of ash distribution in North Atlantic
sediments is very similar to that of ice-rafted detritus (Ruddiman and
Glover, 1972, 1975; Ruddiman, 1977). These authors concluded that the
ash is predominantly transported as debris in glacial ice, and later
injected into the paleocurrent systems. Many of the shards examined in
Baffin Bay and Davis Straight cores were between 63 and 149 um; also
favoring a similar mechanism of transportation. There is also a strong
positive correlation between the distribution patterns of volcanic ash
and the inferred flow pattern of the proto-West Greenland Current
(Fillon and Duplessy, 1980). Comparison of the ash data with oxygen
isotope curves indicated that ash level 1b occurs in isotope stage 2,
implying open water in Baffin Bay and the presence of the proto-West
Greenland current; and independently confirming the northward advection
of North Atlantic subarctic water suggested by the faunal data.
Furthermore, during isotope stage 2, summer surface water temperatures
must be above 0° C in order to melt the icebergs. In general, the
abundance of ash in Baffin Bay sediments decreases northward and it is
absent north of 71-72°N latitude; as well as in the western bay. This
trend suggests that either: (a) substantial melting occurred in Baffin
Bay during the time of ash deposition and possibly all bergs crossing
Davis Strait northward melted in Baffin Bay, or (b) some melting took
place prior to the deflection of the West Greenland Current to the west

around 71-72°N latitude, and the surface water was sufficiently cooled
to delay melting until bergs reached warmer waters. However, the
findings of Fillon and Duplessy (1980) indicated similar penetration of
North Atlantic subarctic water into the Labrador Sea during isotope
stage 2 with summer temperatures greater or equal than about 5°C.
Therefore, the first solution appears to be more likely for Davis
Strait and Baffin Bay, although temperatures may be slightly cooler.

Separate ash peaks (a, b and c) in level 1 are clearly identified
in Davis Strait core 017 but they occur as a single (occasionally
double) peak in the Baffin Bay cores. The following paleoceanographic
interpretations are conceivable: (a) all three ash peaks are present
in Baffin Bay cores, but due to the difference in sedimentation rates
the resolution of individual peaks in Baffin Bay is less clear; (b)
Baffin Bay peak 1 is a sum of ash peaks 1a, 1b and 1c delayed by
oceanographic factors; (c) ash peak 1c is absent in Baffin Bay; and the
ash level 1 correlates with Davis Strait peak 1a plus 1b. The third
possibility can easily be rejected on the basis of the C-14 date in
core 040 which dates the peak of ash level 1 at 31,200 yr BP. In
addition there is no known ash deposition in either the Labrador Sea or
the North Atlantic Ocean during this time. The first interpretation
implies that the bay was partially ice free during the deposition of
these three peaks. The combination of the first and second
alternatives is believed to be the most plausible explanation for the
deposition of ash level 1. During isotope stage 4, the O-18 data
indicate that Baffin Bay had extensive all-year sea-ice cover (Aksu,
1981) and the opening of the bay did not occur until the early to
middle of stage 3 in central Baffin Bay, delaying the ash deposition in
this area. On the other hand, the early advection of the proto-West
Greenland Current probably accelerated the clearing of the ice around
Davis Strait as early as late isotope stage 4, permitting the
deposition of ash level 1c.

With the clearing of the ice in stage 3, and with the consequent
re-establishment of the current regime, the shard remnants of this
level were then deposited in Baffin Bay. During isotope stage 3, the
majority of this age was deposited in the southern Labrador Sea (Fillon
and Duplessy, 1980). The ages of the principal depositions of ash in
North Atlantic sediments are estimated to be 9300 yr BP and 64,700 yr
BP (Ruddiman and Glover, 1975), of which the second date is very close
to the age of isotope 3/4 boundary (64,000 yr BP). The first
occurrence of ash in level 1 in Baffin Bay core 040 is about 10 cm
above the 3/4 transition. Approximately 40,000 to 45,000 years after
the first deposition of ash level 1c in central Baffin Bay, another ash
pulse occurred, dated at around 21,030 yr BP in Davis Strait core 017

(level 1b) and at about 20,000 yr BP in the Labrador Sea cores (Fillon and Duplessy, 1980). This was followed by the last North Atlantic ash pulse around 9300 yr BP in North Atlantic sediments (Ruddiman and Glover, 1972), and Labrador sea sediments (Fillon and Duplessy, 1980); and deposited around 10,530 yr BP in Davis Strait sediments. The last two ash pulses occurring approximately 10,000 years apart, together with the tail of level 1c, produced a large and broad peak in Baffin Bay primarily due to the very slow rate of deposition and bioturbation. The time lag between North Atlantic and Baffin Bay ash deposition is estimated to be about 1000 to 2000 years.

The ash data discussed above suggest that the proto-West Greenland Current was one of the major components of the current system since early isotope stage 3 in the Davis Strait and Baffin Bay region. Westward deflection of this current in Baffin Bay was around 72°N latitude during this time period, approximately 2-3° southward of that of present day, and the current was presumably considerably weaker. Higher ash abundances towards the east imply that the current was to a great extent confined to eastern Baffin Bay.

There are two other episodes of ash deposition which provide direct evidence for the presence of the proto-West Greenland Current in Baffin Bay. Ash level 2 abruptly disappears north of 68.5°N latitude, suggesting much less penetration of the proto-West Greenland Current in the bay during the isotope 5/6 transition and early in interglacial period 5. Ash level 3 decreases significantly in abundance around 65°N latitude; although the ash-bearing zone can be traced as far as 70°N latitude. This pattern of ash distribution suggests that a large proportion of the proto-West Greenland Current was deflected southward around 66°N latitude, although a weaker branch of the same current penetrated into southern Baffin Bay to about 70°N latitude during early interglacial stage 7b. The interesting question is, then, if the West Greenland current could penetrate into Baffin Bay to about 72°N latitude during several glacial periods, what prevented further northward flow of this current during the interglacial periods?

The sediments in isotope stages 5 and 7 are predominantly ice-rafted in origin, except for those deposited in early stage 5. The distribution pattern and mineralogy indicate that the sediments of facies B and A are mainly derived from outcrops along the northwestern and western margin of Baffin Bay (Aksu, 1981). The southernmost occurrence of facies B in stage 5 is about 69.5°N latitude (core 025) and that of facies A in stage 7 is about 67.5°N latitude (core 557). The presence of a southerly flowing current occupying most of the western and central Baffin Bay and the western part of southern Baffin

Bay is therefore implied. There is a remarkable correlation between the suggested northernmost penetration of the West Greenland Current during the deposition of ash levels 2 and 3 and the southernmost occurrence of facies B in interglacial stage 5 and facies A in interglacial stage 7. It is suggested that during the early interglacial periods of stage 5 and 7 and possibly 9, a polar front developed around northwestern Baffin Bay. With the increasing melt water input from the surrounding ice caps, together with the partial or complete clearing of the Arctic channels, the excess water of the Arctic started to flow into Baffin Bay, establishing the equivalent of the Baffin Land Current. Delta 0-18 and C-13 values of planktonic foraminifera which lived during these periods and the presence of substantial ice-rafted debris, suggests that these water masses probably had low salinities and cooler temperatures (above 0°C).

Similar ice-rafted debris is also found in the early stage 1, 1/2 transition, interstadial stage 3 and interstadial periods within glacial stages 6 and 8, suggesting similar circulation patterns. The ash levels discussed above also occur in NW Labrador Sea (SE Baffin Island shelf) indicating that a branch of the proto-West Greenland Current was deflected westward before crossing Davis Strait northward.

Synthesis of Oceanic currents (Fig. 7.9)

Synthesis of the major oceanic currents in the Baffin Bay region during the last 400,000 years shows a cyclic pattern. During glacial periods, one major current dominated the circulation pattern in the Baffin Bay region. This current, which was initiated by the northward advection of the North Atlantic subarctic water, penetrated into Baffin Bay to about 72°N latitude. A polar front probably marked by strong temperature and salinity gradients occupied the northern portion of the bay, causing the abrupt westward deflection of the proto-West Greenland Current. The equivalent of Baffin Island current was absent during the full glacial periods, simply due to freezing up of the Arctic channels and the substantial reduction of excess water in the Arctic Ocean during these periods; consequently the proto-West Greenland Current encroached against the eastern shore of Baffin Island (Miller et al., 1977). A certain amount of cooling must have occurred before this current returned to the Atlantic Ocean; however, the bottom water masses inferred by the preservation of biogenic carbonate (Aksu, 1983, in press) suggests that substantial cooling and consequently dense bottom water formation did not occur during these periods. This implies that the current was considerably warmer after crossing Davis Strait.

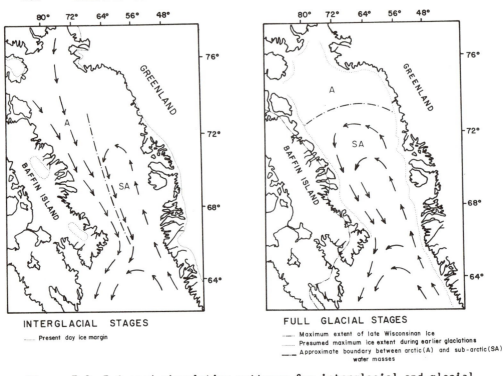

INTERGLACIAL   STAGES
.... Present day ice margin

FULL   GLACIAL STAGES
.... Maximum extent of late Wisconsinan Ice
.... Presumed maximum ice extent during earlier glaciations
_.._ Approximate boundary between arctic(A) and sub-arctic(SA)
water masses

Figure 7.9 Inferred circulation patterns for interglacial and glacial
periods (details explained in text). A = arctic water mass, SA =
subarctic mater mass. Maximum extent of Late Wisconsinan Ice is after
Andrews and Ives (1978); presumed maximum ice extent during earlier
glaciations is after Weidick (1976); present ice margin is after the
Quaternary map of Greenland (Geological Survey of Greenland) and the
Glacial map of Canada (Geological Survey of Canada, Map 1253A).

A branch of the proto-West Greenland Current was deflected
westward around the northern Labrador Sea, where it was joined with the
southward flowing branch of the same current. The paleoceanographic
model presented above implies the absence of the proto-Labrador Current
and possibly the presence of slightly warmer current along the eastern
margin of Canada during the full glacial stages.

During interglacial periods, however, two major currents dominated
the surface circulation in the Baffin Bay region. The proto-Baffin
Land Current, which was initiated by the opening of the Arctic channels
and was enhanced by melt water discharge, flowed southward along the
eastern margin of Baffin Island. The warmer proto-West Greenland
Current flowed northward along the west coast of southern Greenland and
one branch of it was deflected westward around the northern Labrador
Sea. However, a weaker branch penetrated into Baffin Bay to about 70°N

latitude. After crossing Davis Strait southward, the proto-Baffin Land
Current was joined with the warmer proto-West Greenland Current and
flowed southward along the eastern margin of Canada similar to the
present day Labrador Current.

III. PALEOCLIMATOLOGY

Glacial History of the Baffin Bay Area

The investigation of the glacial history of the Baffin Bay area is
primarily based on the interpretation of the oxygen isotope curves.
Additional information is obtained from micropaleontological, volcanic
ash and sedimentological data. Approximately 80 to 95% of the
variation in the oxygen isotope composition of N. pachyderma is
estimated to be the result of variation in the isotope composition of
Baffin Bay water. However, the magnitude of the isotope changes from
peak glacial to peak interglacial periods is found to be larger than
those of open ocean records, and thus it is concluded that core 040 and
017 provide further information on temperature and salinity effects.

The results indicated that four long-term climatic cycles varying
in length from 67,000 to 115,000 years occurred in the past 400,000
years in Baffin Bay and Davis Strait; each leading to interglacial
conditions that started at about 13,000; 128,000; 251,000 and 347,000
yr BP. Cold climatic pulses lasting less than 15,000 years are
superimposed on these cycles. The most prominent of these are the
brief cold periods that reached their peaks at about 72,000, 118,000
and 206,000 yr BP; the last two of these periods are located within the
interglacial stages 5 and 7 respectively. Similarly, short duration
(20,000 years) warm climatic episodes representing significant
deglaciations are also identified, again superimposed on the major
cycles. These occur particularly in early stage 1 and within glacial
stages 6 and 8 respectively.

Comparison with other oceanic data

The inferred climatic record of Baffin Bay is compared with that of the
Labrador Sea (Fillon and Duplessy, 1980), and the Norwegian and
Greenland Seas (Kellog, 1977; Kellog et al., 1978). In Figure 7.10,
the oxygen isotope records of cores 040 and 017 are plotted together
with those of cores HU75-58, HU75-51, V28-56 and V28-14. Oxygen
isotope records are preferentially used as a climatic correlative tool,

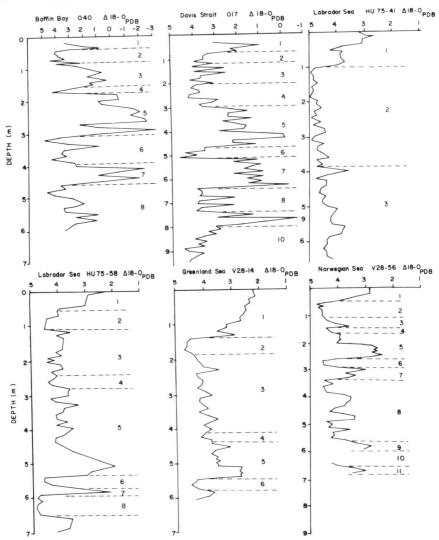

Figure 7.10 Comparison of oxygen isotope record from Baffin Bay to open ocean cores adjacent to Baffin Bay. Labrador Sea cores HU75-41 and HU75-58 from Fillon and Duplessy (1980) and Fillon (unpublished data) Greenland Sea and Norwegian Sea cores from Kellogg (1977) and Kellogg et al. (1978).

because of 0-18 fluctuations in marine microfossils primarily represent the isotope composition of the ocean water which is largely a function of the waxing and waning of the ice caps (Broecker and von Donk, 1970). Because oceanic mixing is relatively fast (1500 years, Broecker et al., 1960) the variations in 0-18 must occur almost synchronously throughout the oceans.

The stratigraphy of the Labrador, Norwegian and Greenland Sea
cores is based on micropaleontological and volcanic ash data. The
chronology of the isotope stage boundaries is adopted from the
Equatorial Pacific core V28-238 of Shackleton and Opdyke (1973). In
general, there is good agreement between Labrador Sea, Norwegian Sea,
Greenland Sea and Baffin bay cores in the timing of the isotopically
determined global glacial and interglacial stages (Fig. 7.10). The
most impressive correlation among the isotope curves is the
isotopically light peak that occurred near the 1/2 boundary. The other
interesting correlation is the near glacial period recorded within
isotope stage 7 in cores V28-56 and 040.

There is also an excellent correlation in the 0-18 values of the
glacial periods, all occurring around 4-5°/₀₀. This suggests that
the isotope composition of the water masses that occupied Baffin Bay
during stages 2, 4, 6, 8 and 10 were very similar to those in the
Labrador Sea during stages 2, 4, and 6, and the North Atlantic Ocean
during stages 2, 4, 6, 8 and 10. These glacial ^{18}O values are at least
2 to 4°/₀₀ heavier than in Caribbean Sea cores (Broecker and von
Donk, 1970; Emiliani, 1966), but very similar to those of the central
North Atlantic (Tiede, 1977) and equatorial Pacific cores (Ninkovitch
and Shackleton, 1975). The enrichment of oceanic waters during glacial
stages may be attributed to substantial mass transfer from the North
Atlantic Ocean and the Baffin Bay area to the adjacent ice caps. The
intensity of the interglacial 0-18 values, however, shows significant
geographical variations. For example, 0-18 values in stage 5 show a
clear enrichment of about 2.5°/₀₀ in 0-18 from core 040 to core 017
and about 2.1°/₀₀ from core 017 to core HU75-58. This enrichment
in 0-18 towards the south is attributed to the efficiency of oceanic
mixing, as well as continuous or long duration melt water outflow from
the northern oceans.

The oxygen isotope curves of cores 040 and 017 suggest that three
major and several minor periods of rapid ice accumulation occurred on
the land masses encircling Baffin Bay. This is supported by two major
independent lines of evidence. (1) The abrupt changes in source of
sediment and sedimentary processes: the mineralogical data (Aksu,
1981) indicate that sediments deposited during the glacial stages are
predominantly polycyclic. These sediments are primarily derived from
the surrounding shelves and are deposited via turbidity current and
debris flows which suggest higher activities on the adjacent land. In
contrast, the interglacial deposits are characterized by ice-rafting
and hemipelagic sedimentation. In these sediments, mixing of different
assemblages is less common and facies deposited during these periods
possess mineralogically distinct sources (Aksu, 1981). (2) Major

changes in the surface circulation pattern: during glacial periods,
one major current, initiated by the northward advection of subarctic
North Atlantic water, dominated the circulation pattern in the bay.
The cold Baffin Land Current originating from the Arctic Ocean during
interglacial periods was absent during glacial stages.

The major rapid ice growth periods occurred at the transitions
from stages 9 to 8; 7 to 6; and 5 to 4. Comparison of the isotope
curves with the summer insolation curves of Berger (1978) indicated
that the solar radiation values which were about 20 to 25 Langleys/yr
above present radiation during stages 5, 7 and 9 dropped to values of
about 20 to 30 Langleys/yr below modern values during the transition to
and within stages 4, 6 and 8. Because the existence of glaciers mainly
depends on the amount of solar radiation received during the summer
(Berger, 1978), the glaciers developed during periods of reduced
seasonal contrast with cooler summers and warmer winters. They decayed
during periods of enhanced contrast, suggesting some correlation
between the changes in the earth's orbital parameters and the evolution
of continental glaciers. Similar but minor phases of ice accumulation
also occurred at the substage 5e/5d and stage 3/2 boundaries. The
extrapolated ages of the sediment thicknesses, which record the major
isotope transitions from interglacial to full glacial, is about 8000 to
13,000 years, suggesting substantial mass transfer in a very short time
period.

SUMMARY

There are two major factors which are particularly necessary for rapid
ice growth: low summer temperatures in the high latitudes and
substantial amounts of available moisture. The presence of sub-polar
water masses in Baffin Bay and Davis Strait during most of glacial
stages has been discussed here. It has also been suggested that
changes in the paleocirculation patterns are largely controlled by the
variations in climate and that these changes are rapid. Decreasing
summer insolation values and the positive feedback from the albedo
caused by expanding year round snow cover, favor enhanced atmospheric
cooling during the transition to glacial periods in the Baffin Bay
region. This cold air in contact with a comparatively warmer surface
water mass probably initiated significant evaporation from the Baffin
Bay and Davis Strait area.

The data discussed here suggest a warm Baffin Bay surrounded by an
ice-covered land mass. It is also suggested that summers during the
transition to, and probably during, full glacial periods had an

anomalous surface temperature contrast with an extremely strong thermal
gradient developed between the warm ocean and the ice and snow on the
land. This paleocirculation pattern implies the presence of warmer
water masses along the eastern Canadian margin during the full glacial
periods, probably forming strong thermal fronts between the ocean and
the Laurentide ice sheet. In the present day North Atlantic Ocean
cyclonic low atmospheric pressure systems develop along thermal fronts
in lower latitudes and they are the major source of atmospheric
precipitation (Blackmon et al., 1977).

It is possible that middle to high latitude low pressure systems
also developed and were channeled northward along these strong thermal
fronts. Such oceanic and atmospheric configurations are very favorable
for additional moisture transport to the developing ice sheets in the
high Arctic. Supporting evidence for a warm ocean during glacial
periods is also present in Labrador Sea cores for at least parts of
isotope stages 2 and 4 (Fillon and Duplessy, 1980). Similarly, the
presence of warm water masses along the Labrador coast and Grand Banks
during isotope stage 2 and stage 4/5 transition is demonstrated by Alam
(1979) and Ruddiman et al. (1980). Ruddiman et al. (1980) also showed
that 75 to 90% of the late Wisconsinan global ice volume was
accumulated during the stage 4/5 transition within about 10,000 years.
They suggested that the local oceanic warmth during this period
provided a major source of moisture for the ice sheet growth on eastern
North America. However, their data differ from those of Baffin Bay in
that Baffin Bay surface water stayed relatively warm during the
majority of the glacial cycles (2, 6, 8 and 10), providing a local
source of moisture to the ice sheets; whereas data from North Atlantic
indicate a prominent cooling trend towards full glacial stage 4. In
contrast, the Greenland Sea data do not contain evidence for sub-polar
foraminiferal fauna during the glacial stages spanning the last 450,000
years (Kellogg et al., 1978).

ACKNOWLEDGEMENTS

I thank D.J.W. Piper of the Bedford Institute of Oceanography, and Ann
A.L. Miller of Dalhousie University, for their useful discussions and
critical reading of the manuscript. Special thanks are due to the
Master, Officers and crew and scientific personnel of the Bedford
Institute of Oceanography and Dalhousie University cruises over the
last five years. The laboratory work was supported by NSERC grants to
D.J.W. Piper.

REFERENCES

Aksu, A.E., 1981: Late Quaternary stratigraphy, paleoenvironmentology and sedimentation history of Baffin Bay and Davis Strait, Ph.D. thesis, Dalhousie University, Department of Geology. 771 pp.

Aksu, A.E., in press: Holocene and Pleistocene dissolution cycles in deep-sea cores of Baffin Bay and Davis Strait: paleoceanographic implications. Marine Geology.

Aksu, A.E. and Piper, D.J.W., 1969: Baffin Bay in the last 100,000 years. Geology, 7:245-248.

Alam, M., 1979: The effect of Pleistocene climatic changes on the sediments around the Grand Banks. Ph.D. thesis, Dalhousie University, Department of Geology. 295 pp.

Berger, A.L., 1978: Long-term variations of caloric insolation resulting from the earth's orbital elements. Quaternary Research, 9:139-167.

Binns, R.E., 1972: Composition and derivation of pumice on post-glacial standlines in northern Europe and western Arctic. Bulletin Geological Society of America, 83:2303-2324.

Blake, W. Jr., 1969: Studies of glacial history in Arctic Canada. I. Pumice, radiocarbon dates and differential postglacial uplift in the eastern Queen Elizabeth Islands. Canadian Journal of Earth Science, 7:634-664.

Blackmon, M.L., Wallace, J.M., Lau, N.C. and Mullen, S.L., 1977: An observational study of the northern hemisphere winter circulation. Journal of the Atmospheric Sciences, 34:2040-2053.

Boeggild, O.B., 1900: The deposits of the sea-bottom. Reports of the Danish Ingolf-Expedition, 1(3):1-89.

Bramlette, M.N. and Bradley, W.H., 1941: Geology and biology of the north Atlantic deep-sea cores between Newfoundland and Ireland. 1. Lithology and geological interpretation. U.S. Geological Survey Professional Paper 196-A, 1-34.

Broecker, W.S., Gerard, R., Ewing, M. and Heezen, B.C., 1960:Natural radiocarbon in the Atlantic Ocean. Journal of Geophysical Resources, 65:2903-2931.

Broecker, W.S. and von Donk, J., 1970: Insolation changes, ice volumes, and the ^{18}O record in deep sea cores. Review Geophysics and Space Physics, 8:169-198.

Chough, S., 1978; Morphology, sedimentary facies and processes of the Northwest Atlantic Mid-Ocean channel between 61° and 52° N, Labrador Sea. Ph.D. Thesis, McGill University, Department of Geological Sciences. 167 pp.

Emiliani, C., 1966: Paleotemperature analysis of Carribbean cores P6304-8 and P6304-9 and a generalized temperature curve for the past 425,000 years. Journal of Geology, 74:109.

Fillon, R.H. and Duplessy, J.C., 1980: Labrador Sea Bio-, Tephra-, and Oxygen isotope stratigraphy and Late Quaternary paleoceanographic trends. Canadian Journal of Earth Sciences, 17:831-854.

Grant, A.C., 1971: Distributional trends in the recent marine sediments of northern Baffin Bay. Maritime Sediments, 7(2):41-63.

Kellogg, T.B., 1977: Paleoclimatology and paleo-oceanography of the
   Norwegian and Greenland Seas. The last 450,000 years. Marine
   Micropaleontology, 2:235-249.
Kellogg, T.B., Duplessy, J.C. and Shackleton, N.J., 1978: Planktonic
   foraminiferal and oxygen isotope stratigraphy and paleoclimatology of
   Norwegian sea deep-sea cores. Boreas, 7:61-73.

Marlowe, J.I., 1966: Mineralogy as an indicator of long-term current
   fluctuations in Baffin Bay. Canadian Journal of Earth Science,
   3:191-201.
Miller, G.H., Andrews, J.T. and Short, S.K., 1977: The last
   interglacial-glacial cycle, Clyde foreland, Baffin Island, N.W.T.:
   Stratigraphy, biostratigraphy and chronology. Canadian Journal of
   Earth Science, 14(12):2824-2857.

Ninkovitch, C.D. and Shackleton, N.J., 1975: Distribution,
   stratigraphic position and age of ash layer "L" in the Panama
   Basin region. Earth and Planetary Science Letters, 27:20-34.

Prest, V.K., 1968: Quaternary geology. In: Geology and Economic
   Minerals of Canada. R.J.W. Douglas (ed.), Dept. of Energy, Mines and
   Resources, Canada, p. 675-765 (838 pp).

Ruddiman, W.F., 1977: Late Quaternary deposition of ice-rafted sand in
   the sub-polar North Atlantic (Lat. 40° to 65°N). Bulletin Geological
   Society of America, 88:1813-1827.
Ruddiman, W.F. and Glover, L.K., 1972: Vertical mixing of ice rafted
   volcanic ash in North Atlantic sediments. Bulletin Geological Society
   of America, 83:2817-2836.
Ruddiman, W.F. and Glover, L.K., 1975. Subpolar north Atlantic
   circulation at 9,300 yr. B.P. Faunal evidence. Quaternary Research,
   5:361-389.
Ruddiman, W.F., McIntyre, A., Niebler-Hunt, V. and Durazzi, J.T., 1980:
   Oceanic evidence for the mechanism of rapid Northern Hemisphere
   glaciation. Quaternary Research, 13:33-64.

Shackleton, N.J. and Opdyke, N.D., 1973: Oxygen isotope and
   paleomagnetic stratigraphy of Equatorial Pacific core V 28-238:
   oxygen isotope temperatures and ice volumes on a 10^5 year and 10^6
   year scale. Quaternary Research, 3:39-55.

Tiede, J., 1977: Aspects of the variability of the glacial and
   interglacial North Atlantic eastern boundary current (last 150,000
   years). Meteorology Forsch-Ergebnisse C.P., 1-36.
Timofeyev, V.T., 1960: Water masses of the Arctic Basin.
   Leningrad:Hydrometeor, Izd., 190 pp. Transl. L.K. Coachman,
   University of Washington, M61-17-1961.

8 Northwest Labrador Sea stratigraphy, sand input and Paleoceanography during the last 160,000 years

R. H. Fillon (with oxygen isotope stratigraphy by Jean Claude Duplessy)

Previously published studies of Late Pleistocene paleoceanography in the Labrador Sea (Fillon and Duplessy, 1980; Fillon et al., 1981) have focused on the eastern portion which presently underlies the relatively warm West Greenland Current. In this paper, data from a northwestern Labrador Sea piston core, HU75-58 (62°46'N; 59°22'W; 1057 meters water depth - Fig. 8.1) will be presented and compared with reconstructions of paleoceanographic events inferred from eastern Labrador Sea cores. Core HU75-58 was recovered approximately 200 km due east of Lok's Land, southeastern Baffin Island. The core site is on a portion of the continental margin that bulges seaward, separating depressions that run obliquely down the slope toward the southeast. The core site therefore minimizes the chances of significant downslope transport having influenced the depositional record.

Modern surface circulation in the area (summarized in Fig. 8.1) is influenced by the cold, low salinity (polar) Baffinland Current which flows along the shelf and upper slope and by warmer, higher salinity (subpolar) Labrador Sea water farther offshore and at depth. The latter is derived in part from the warm, saline West Greenland Current (Dunbar, 1951; Lazier, 1973). Most of the icebergs in transit over the core site (Fig. 8.1) are probably from Disko and Melville Bays in northwest Greenland (U.S. Naval Oceanographic Office, 1968; Murray, 1968). Because the Baffinland Current and its southern continuation, the Labrador Current are so cold (c. 2° to 4°C maximum in summer; M.E.D.S.*, written communication, 1980), iceberg disintegration is minimal north of about 54°N (Fillon, 1977). Holocene sediments in basins on Sagelek Bank, for example, are muds which contain <0.5% gravel by weight. This suggests that the influx of ice-rafted debris

---

*Marine Environmental Data Service, Environment Canada, Ottawa

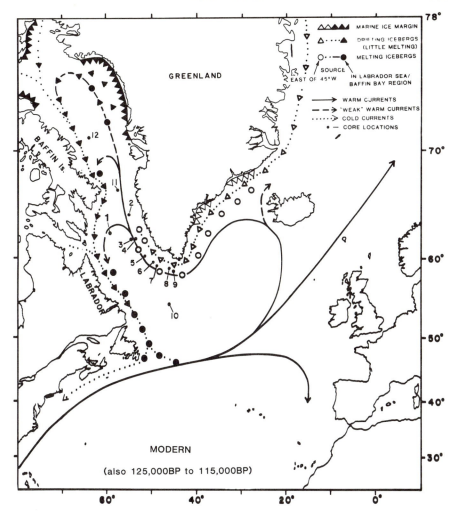

Figure 8.1 Map of the Labrador Sea, Baffin Bay and the northwest
Atlantic Ocean showing locations of important cores (dots) as follows:
(1) HU75-58; (2) HU75-48; (3) HU75-42; (4) HU75-41; (5) V16-227; (6)
V27-26; (7) HU75-37; (8) HU75-31; (9) HU75-29; (10) V27-20; (11)
HU77-17; and (12) HU76-40. Also shown schematicaly are the modern
surface circulation and principal iceberg calving areas within the
Labrador Sea/Baffin Bay region and elsewhere in the northwest Atlantic
(Dunbar, 1951; Murray, 1968; U.S. Naval Oceanographic Office, 1968;
Lazier, 1973).

is relatively unimportant compared to other sources of sedimentation on
the northern Labrador shelf. Icebergs further offshore in the vicinity
of the core site, especially those with deep draft, however, may come
into contact with the warmer, denser (because of higher salinities)
Labrador Sea water (c. 8° to 10°C maximum surface temperature in

summer, M.E.D.S., written communication, 1980) and hence may undergo considerable melting with consequent deposition of ice-rafted debris at the core site.

The objectives of this paper are: to provide a detailed correlation between core HU75-58 and eastern Labrador Sea cores; to determine the sand input rates and paleoceanographic conditions that were maintained in the northwestern Labrador Sea during the last interglacial-glacial-interglacial cycle and to compare those findings with records from the eastern Labrador Sea and North Atlantic Ocean. Finally, we shall also suggest a stratigraphic framework for comparing the results of Labrador Sea studies with work carried out in Baffin Bay by Aksu (1981; this volume).

TERRIGENOUS SAND

Analytical Procedure

Terrigenous sand is an important component of hemipelagic sediments in the Labrador Sea composing up to 40 percent by weight of samples. In contrast, the most abundant biogenic sand-size constituents, foraminifera and radiolaria, are relatively scarce, being one or two orders of magnitude less abundant. Thus, little error (uniformly less than 10 percent and typically less than 1 percent) is involved in estimating terrigenous sand content simply by taking the total weight of material retained on a 63 μm sieve. This procedure was followed with all samples. Additional weight fractions were determined, also by sieving, for 125 μm to 1 mm and 1 mm to 8 mm. The coarsest material (> 1 mm) was eliminated from calculations of absolute sand input because of its low abundance in cores and because it is often dominated by single, large, erratic grains (Watkins et al., 1974; Fillon, 1977; Ruddiman, 1977). Absolute sand input rates (mg/cm^2/10^3 yr) were calculated from percent sand data based on dry bulk sample weight using the ratio of dry weight to wet volume.

Potential Sources of Sand

In core HU75-58, and eastern Labrador Sea cores, terrigenous sand and minor amounts of gravel occur for the most part as dispersed particles in a mottled, apparently hemipelagic, mud matrix. The coarseness of much of the dispersed sand and gravel suggests that ice-rafting has been an important mechanism of sediment transport in the Labrador Sea.

A large and variable very fine sand fraction (63 to 125 um) could have
been included in ice-rafted till (e.g. Mills, 1977), most tills from
the Precambrian on Baffin Island are essentially sands (Andrews, this
volume), but also could have been transported offshore, along with
silt-size and finer particles suspended in plumes of glacial meltwater
(Rees, 1966) or as wind blown-dust (Windom, 1969; Folger, 1970). Sieve
and pipette grain-size analyses of core HU75-42 indicate that high
percentages of fine sand (63 to 125 μm) are accompanied by coarse silt
(31 to 63 μm) maxima (Fillon et al., 1981).

Because many studies have employed specific sand-size fractions as
optimum indicators of ice-rafting in deep-sea sediments (Kent et al.,
1971; Watkins et al., 1974; Keany et al., 1976) and others have
considered that all sand sizes are potentially related to terrestrial
glacial events (Fillon, 1977; Ruddiman, 1977), it is important to
consider just how non ice-rafted fine terrigenous sand might be related
to the terrestrial glacial-interglacial cycle in the Labrador Sea
region.

If the fine sand component was deposited from glacial meltwater,
fine sand maxima should correspond to times of maximum glacial extent
when the ice margin was close to or at the shelf edge. Subglacial or
supraglacial meltwater charged with clay, silt and fine sand-size
particles, injected close to the shelf edge could have reached the
adjacent slope by spreading either on the sea surface or along the
seabed as turbidity currents depending on the effective density of the
sediment-freshwater mixture, current velocities and the turbulence of
the surface layer (Figs 8.2a & b).

During periods when the continental ice margin was well back from
the coast, glacial meltwater would have reached the sea only after
passing through numerous sediment traps, such as lakes, fiords, and
deep innershelf depressions (Fig. 8.2d). Hence, the likelihood of
significant amounts of fine sand-size glacial sediments reaching the
deep sea would have been minimal. Episodes of maximum suspended
sediment input to the deep sea and resultant turbidites or debris flows
should therefore correspond in time to periods of maximum ice extent
and so to ice-rafted debris maxima.

Wind-blown dust either blown directly over the ocean or onto sea
ice which later drifts seaward and melts might also provide a
significant input of fine sand to the deep sea, if a sufficient
unglaciated, unvegetated, seasonally snow-free expanse of fine-grained
sediment is available on land to provide an upwind source of
particles. Rapid terrestrial ablation of an ice sheet with the

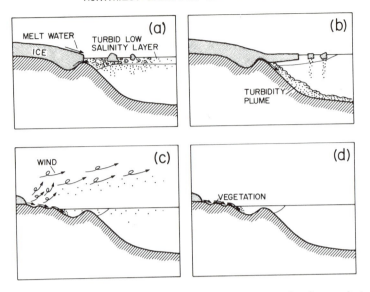

Figure 8.2 Schematic representations of possible glacier-related
mechanisms of fine sand transport from the continent to the ocean; (a)
turbid glacial meltwater/low-salinity surface layer transport and
ice-rafting; (b) turbid hyperpycnal inflow and ice-rafting; (c) wind
deflation and eolian transport, and (d) turbid glacial meltwater
discharge passing through lakes in a vegetated landscape. Hypothetical
boundary between normal salinity sea water and meltwater-diluted sea
water is indicated in each case by a light line. (from ¯illon et al.,
1981).

resultant development of large outwash plains would best meet this
criterion (Fig. 8.2c).

Total sand input to the deep sea at subpolar latitudes, regardless
of transport mechanism, can therefore be considered to reach maximum
levels during regional glacial maximum and/or rapid growth or retreat
phases as long as melting remains constant and to drop to a minimum
when ice sheets are small or absent. This suggests that a great deal
of information on continental glaciers and relative rates of iceberg
disintegration can be extracted from deep-sea core data by examining
the entire fine to medium sand-size fraction (Fillon, 1977; Ruddiman,
1977; Fillon et al., 1981), even though some of the material may not,
strictly speaking, be ice-rafted.

Bio-Tephro-Litho-Stratigraphy

Eastern Labrador Sea cores (Fig. 8.1) have been correlated using
variations in volcanic ash, benthic foraminiferal abundance, the

relative abundance of the radiolarian <u>Diplocyclas</u> (<u>Cycladophora</u>)
<u>davisiana</u> and weight percent of sand (Fillon and Duplessy, 1980; Fillon
et al., 1981). Analytical techniques are detailed elsewhere (Fillon
and Duplessy, 1980; and Fillon et al., 1981). The correlation of all
of these parameters except for <u>D</u>. <u>davisiana</u> can be extended westward
across the Labrador Sea to HU75-58. Very few radiolaria were found in
HU75-58, but this was expected because the two closest eastern Labrador
Sea cores (HU75-48 and HU75-54; Fig. 8.1) also contain poor radiolarian
faunas (Fillon and Duplessy, 1980).

<u>Ash</u>. Clear, platey, bubble-wall volcanic glass shards in a peak
concentration of 20 shards/gram were found in core HU75-58, at a
depth of 190 cm (Fig. 8.3). This compares with the range of maximum
concentrations in eastern Labrador Sea ash zones A and B of from about
10,000 shards/gram (zone B) in the extreme southeast to about 20
shards/gram (zone B) in the most northwesterly core and from about 300
shards/gram (zone A) in the southeast to about 2 shards/gram (zone A)

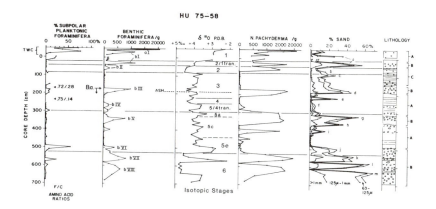

Figure 8.3 Percent abundance of subpolar planktonic foraminifera (>63
μm, numbers of benthic foraminifera per gram of total sediment,
planktonic (<u>N</u>. <u>pachyderma</u>, sinistral) foraminiferal 0-18 values,
numbers of <u>N</u>. <u>pachyderma</u> per gram of total sediment, percent of sand
and granules (63-125 μm, 125 μm to 1.0 mm, and >1.0 mm) and oxygen
isotopic stages for piston core HU75-58 and its accompanying
trigger-weight core (TWC). Also shown are the location of the peak
abundance level of volcanic ash zone B, the upper limit of occurrence
of the benthic foraminifera <u>Bulimina aculeata</u> (B.a.) and schematic
description of the lithology. Facies A, B and C are described in the
text. Amino acid ratios, free and combined (F/C) were made available
by G.H. Miller (written communication, 1980) and were based on <u>Astarte</u>
(220 cm level AAL-1251 and 176 cm level AAL-1021). Benthic
foraminiferal and sand percentage abundance peaks are labeled to
correspond with apparently synchronous events in eastern Labrador Sea
cores (Fillon and Duplessy, 1980; Fillon et al., 1981).

in the most northwesterly core. Because, as it will be shown from
other data, core HU75-58 clearly penetrates into sediments that are
stratigraphically situated below the level of ash zone B (c. 59,000 BP)
it is considered that the ash concentration at 190 cm in HU75-58
correlates with eastern Labrador Sea ash zone B. Ash zone A in HU75-58
is represented by only a few ash grains in the trigger-weight core and
the piston core-top (concentration $\leq$ 2/gram). A third ash zone not
previously seen in the Labrador Sea is indicated at 580 cm
(concentration $\leq$ 2/gram).

Benthic Foraminifera. Five calcareous, benthic foraminiferal maximum
abundance zones were recognized in the eastern Labrador Sea. A single
arenaceous, benthic foraminiferal abundance zone (aI) was also
recognized coincident with the upper part of calcareous abundance zone
bI. In the eastern Labrador Sea, zone bI typically contains the
highest abundances of benthic foraminifera. It is consistently
recognized very close to the top of the cores. In HU75-58,
counterparts to zones aI and bI were recognized in the trigger core and
near the top of the piston core (Fig. 8.3). Foraminiferal abundance
zone bIII characteristically coincides with, or occurs slightly below,
the peak abundance level of ash zone B. There is a prominent benthic
abundance maximum in this position in HU75-58 and accordingly, it is
correlated with bIII (Fig. 8.3).

Benthic foraminiferal abundance values below 250 cm increase to
about 200 tests/gram, maintain this level down to 300 cm and
subsequently increase to a peak of about 800 tests/gram at 340 cm. A
very similar trend is seen in the closest of the northeastern Labrador
Sea cores, HU75-54 (Fillon and Duplessy, 1980). In that core, the
lower, highest abundance part of this pre-ash zone B foraminiferal
abundance maximum was designated zone bV and the upper part as zone
bIV. In Figure 8.3, this correlation is extended to HU75-58, but it
will be demonstrated later that only the upper limit of zone bV appears
to be an approximately synchronous datum in the Labrador Sea.

There are three additional abundance peaks below zone bV in
HU75-58 which apparently lack counterparts or were not penetrated in
the eastern Labrador Sea. These zones are designated bVI, bVII, and
bVIII.

The benthic foraminiferal fauna in HU75-58 is principally composed
of the following taxa: Cibicides lobatulus*, C. wuellerstorfi*,
Pullenia bulloides*, Epistominella exigua*, Buccella frigida*,
Virgulina fusiformis, Cassidulina carinata, C. teretis, Islandiella

helanae*, Cassidulina reniforme (Globocassidulina crassa var.
reniforme), Bulimina aculeata, Elphidium excavatum*, Nonionella
atlantica*, and N. labradoricum. C. carinata, C. teretis, I. helenae,
C. reniforme, and G. crassa are described and synomized by Rodrigues et
al. (1980) and Sejrup and Guibault (1980). Their usage is followed
here. It is considered that this assemblage is consistent with the
1057 meter water depth at the core site. All those species marked with
an * have actually been found living in water depths of from 700 meters
to 2,800 meters on the northeast Newfoundland slope and rise (Schafer
and Cole, 1982). It seems likely, therefore, that the downslope
migration of tests has not played a significant role in modulating
foraminiferal abundance through the cored section. Later in the paper,
oxygen isotopic evidence will be presented which confirms that core
HU75-58 contains essentially in situ mid-slope assemblages of benthic
tests.

   Bulimina aculeata is common in the lower parts of cored sections
in the eastern Labrador Sea, but is absent from the upper parts. The
last appearance of this species ranges from mid-isotopic stage 3 in
core HU75-41 (c. 45,000 BP) to the upper part of benthic zone bV in
core HU75-54 (c. 80,000 BP). It consistently reaches its most frequent
occurrence within zone bV (Fillon and Duplessy, 1980). In core
HU75-58, the last appearance of B. aculeata is just above the ash zone
at 190 cm (Fig. 8.3). This is strong support for identification of the
ash zone as ash zone B. B. aculeata also reaches high levels of
abundance in core HU75-58 within benthic foraminiferal abundance zone
bV. Zones bVI and bVII contain abundant B. aculeata but were not found
in the eastern labrador Sea.

Lithostratigraphy. Sand content (weight-percent of sand) variations
were studied in eight eastern Labrador Sea cores (Fillon et al.,
1981). Seven zones of increased sand content were assigned letter
designations from "a" (youngest) to "g" (oldest), peak "g" being
coincident, in part, with the lower portion of benthic foraminiferal
abundance zone bV. These sand content maxima were shown to be
correlative within the framework of oxygen isotope, volcanic ash and
benthic foraminiferal stratigraphic markers. Core HU75-58 contains
four prominent sand percentage peaks, (coincident with foraminiferal
zone bI, just above and below ash zone B, and partly within
foraminiferal zone bV) which correspond stratigraphically to sand peaks
"a", "d", "e", and "g" in the eastern Labrador Sea. Additional
intervening sand peaks in HU75-58 have been labeled accordingly (Fig.
8.3) using the eastern Labrador Sea records as a guide (Fillon et al.,
1981). Six additional sand peaks appear below peak "g" in HU75-58.
These have been designated "h" through "m".

A schematic lithologic record is displayed adjacent to the HU75-58 sand-content curve in Figure 8.3. It is based on both visual description of the split core and on x-radiographs. Facies "B" is characterized by a high sand and gravel content, for the most part dispersed randomly within the "mud" matrix. Facies "C" has a sand and gravel content similar to facies "B" but in facies "C" current modification of the seabed is indicated by frequent fine sand laminae. Facies "A" contains significantly less sand and gravel than facies "B" and "C". There are occasional laminae of sponge spicules in facies "A". Eastern Labrador Sea cores all show lower sand percentages than those observed in HU75-58 and tend to resemble facies "A" throughout most of their length although they lack sponge spicule laminae.

PALEOCEANOGRAPHIC IMPLICATIONS

Inferences about Labrador Sea paleoceanography have been drawn from the input of ice-rafted sand and ice-rafted volcanic ash (Fillon et al., 1981). The clear, platey, bubble-wall volcanic glass shards which make up ash zones A and B in the Labrador Sea and ash zones 1 and 2 in the North Atlantic range in size from fine silt to granule size. The coarser fractions certainly could not have reached the Labrador Sea in transport through the atmosphere from likely source volcanoes in Iceland or Jan Mayen Island (Binns, 1972; Ruddiman and Glover, 1972). Neither is it likely that rafting by sea ice could have moved the ash for the distances required. This leaves iceberg transport as the most probable means of long distance transport for the ash (Ruddiman, 1977). It has been inferred, therefore, that the presence of large quantities of zone B ash in the eastern Labrador Sea is indicative of a surface circulation pattern which rounded the southern tip of Greenland, much like that of the modern West Greenland Current (Fillon et al., 1981).

In contrast to sand content (percent sand) which is strongly dependent on the capacity of bottom currents to remove finer particles and on the overall rate of input of finer sediment (Ledbetter and Watkins, 1978), the input ($mg/cm^2/10^3$ yr) of terrigenous sand is probably a primary function of changes in sand transport which, as discussed earlier, are closely related to glaciation on the surrounding continents. Certainly, the fraction coarser than 1 mm strongly suggests rafting by icebergs as a major transport mechanism. Significant transport of very coarse sand and granules from intertidal areas by freezing onto the base of sea ice (Amos, 1978) or from the shelf-edge by downslope transport processes (Chough, 1978) is not indicated. These forms of transport would also have picked up and

redeposited inshore micro- and macro- fossils; and, because we find no
evidence of important downslope reworking of microfossils, especially
of intertidal taxa, we conclude that the bulk of the coarser than 1 mm
fraction was transported by icebergs. Further, because down-core
variation in the greater than 1 mm fraction is consistently
proportional to the variation in the 125 μm to 1 mm fraction (Fig.
8.3), it is probable that much of that size fraction was also
iceberg-rafted.

OXYGEN ISOTOPE STRATIGRAPHY

Planktonic foraminiferal 0-18 variations were determined for the deep
living planktonic foraminifera Neogloboquadrina pachyderma
(sinistral). Two benthic species, Cibicides wuellerstorfi and Bulimina
aculeata, were also analyzed over a portion of the section between 450
cm and 600 cm. Samples contained about 0.3-0.5 mg of carbonate (25
tests of N. pachyderma). The foraminiferal tests were cleaned in an
ultrasonic bath, then roasted for 45 minutes at 400°C and finally
reacted with 100% H_3PO_4 at 50°C (Duplessy, 1978). The evolved CO_2 was
purified by passing through a trap at -100°C and analyzed in a V.G.
Micromass 602C mass spectrometer. Replicate analyses of a standard
carbonate sample show that the reproducibility of this treatment is
subject to a standard deviation of 0.07°/oo.

The isotopic records are reported in Table 8.1 and Figures 8.3 and
8.4. Core HU75-58 shows a typical isotopic record extending into stage
6 (Emiliani, 1955). Isotopic stage 1 is only represented by a single
core-top sample in the piston core, but, stage 1 is well represented in
the trigger weight core (TWC in Fig. 8.3). Isotopic stage 2 is
recorded between 50 cm and 85 cm. As in the eastern Labrador Sea cores
HU75-41 (Fig. 8.5) and HU75-42, the stage 3/2 transition is taken just
above the first light isotopic peak (Fillon and Duplessy, 1980).

The isotopic record of HU75-58, from just above the level of ash
zone B at c. 170 cm to c. 350 cm is correlated with the early isotopic
stage 3 to late isotopic stage 5 portion of the 0-18 record of HU75-42
in Figure 8.6. Isotopic stage 4 is recognized in HU75-58 (from 220 cm
to 255 cm) by its heavy 0-18 value and by the presence, just above, of
Labrador Sea volcanic ash zone B (coincident with benthic foraminiferal
abundance zone bIII) which has been correlated with eastern Labrador
Sea cores and through them with Ruddiman and Glover's (1972) North
Atlantic ash zone 2 (Fillon and Duplessy, 1980). Also, in core
HU75-58, as in eastern Labrador Sea core HU75-42, isotopic stage 4 is
found to lie just below sand percentage peak "e".

TABLE 8.1. Results of oxygen isotopic analyses (0-18 vs. P.D.B.) for
the planktonic foraminifera Neogloboquadrina pachyderma (sinistral) and
the benthic foraminifera Cibicides wuellerstorfi corrected to Uvigerina
(+0.64°/oo) and Bulimina aculeata. TWC refers to trigger-weight
core.
* = insufficient foraminifera in sample.
===

Sample Depth (cm)	N. pachyderma (°/oo)	C. wuellerstorfi	B. aculeata
0-2 TWC	+2.36		
2-3 TWC	+2.45		
3-4 TWC	+2.80		
5-6 TWC	+2.72		
10-12 TWC	+2.87		
20-22 TWC	+2.93		
30-32 TWC	+2.93		
35-37 TWC	+3.98		
10-12	+2.99		
20-22	+3.91		
30-32	+4.01		
40-42	+3.97		
50-52	+4.40		
60-62	+4.45		
70-72	*		
80-82	+4.52		
90-92	+3.54		
100-102	+4.04		
110-112	+3.78		
120-122	*		
130-132	*		
140-142	+3.85		
150-152	+4.08		
160-162	+3.85		
170-172	+4.38		
180-182	+4.41		
190-192	+3.88		
200-202	+3.81		
210-212	+4.25		
220-222	+4.03		
230-232	+4.14		
240-242	+4.38		
250-252	+4.21		
260-262	+3.56		
270-272	+3.63		
280-282	+3.66		
290-292	+4.26		
300-302	+4.06		
310-312	+3.27		
320-322	+3.78		
330-332	+3.82		
340-342	+4.09		
350-352	+4.04		
360-362	+3.64		
370-372	+4.07		
380-382	+3.46		
390-392	*		
400-402	+3.79		
410-412	+ .73		

TABLE 8.1. - continued

Sample Depth (cm)	N. pachyderma (°/₀₀)	C. wuellerstorfi (°/₀₀)	B. aculeata (°/₀₀)
420-422			
430-432	+4.10		
440-442	*		
447-450	+3.61		+3.89
450-452	+3.48		
457-460	+3.85		+3.77
460-462	*		
467-470	+3.87		+4.01
470-472	*		
477-480	+3.81		+3.14
480-482	*		
485-488	+3.86		+3.94
487-488	+4.00		
488-489	+3.50		
489-490	+3.68		
490-491	+3.64		
490-492	+2.97		
492-493	+2.95		
500-502	+2.77	+2.98	+3.13
505-507	+2.97		
510-512	+2.84	+3.08	+3.04
515-517	*		
520-522	+2.97	+3.17	+3.12
530-532	+4.54	+3.40	+3.46
540-542	+4.55	+3.91	+4.13
550-552	+4.26	*	+4.43
560-563	+3.16	+3.83	+3.96
570-572	+2.11	+3.73	*
580-582	+4.71	*	
590-592	+4.81	+4.16	+4.23
600-602	+4.63		+4.52
610-612	+4.81		
620-622	+4.64		
630-632	*		
640-642	+4.70		
650-652	+3.50		
660-662	*		
670-672	*		
680-682	+3.62		
690-692	+4.10		

In the eastern Labrador Sea, the substage 5a/4 transition and substage 5a are each associated with widespread benthic foraminiferal abundance maxima (zones bIV and bV in Fillon and Duplessy, 1980). Benthic foraminiferal zone bV occupies the upper part of a well-defined isotopic substage 5a peak in core HU75-42. In HU75-58, isotopic substage 5a is probably centered on the light isotopic peak 310 cm which reaches a value 1.25°/₀₀ lighter than stage 2. This places substage 5a in HU75-58 above the highest benthic foraminiferal

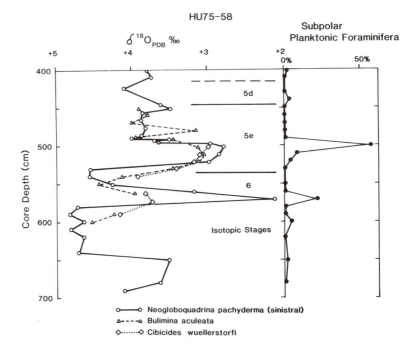

Figure 8.4 Planktonic (N. pachyderma, sinistral) and benthic (Bulimina aculeata and Cibicides wuellerstorfi) foraminiferal 0-18 values and isotopic stratigraphy for the late isotopic stage 6 to early isotopic stage 5 interval in core HU75-58 compared with the percentage of subpolar planktonic foraminifera (>63 μm). The benthic isotopic record defines the stage 6/5e boundary at c. 535 cm thereby placing the planktonic 0-18 minimum at 570 cm within isotopic stage 6. C. wuellerstorfi is corrected to Uvigerina (+0.64°/oo).

abundance of zone bV. Foraminiferal abundance zone bV in HU75-42 can then be equivalent to only the upper, lower-abundance portion of zone bV in HU75-58. The occurrence of sand peak "g" in the upper part of benthic foraminiferal abundance zone bV in core HU75-58 and in the lower part of zone bV in core HU75-42 is also compatible with an inferred chronological discontinuity in the lower limit of zone bV. This suggests an enhanced lower bV zone in HU75-58 as is probably also the case for the long and prominent bV zones in shallow northeastern Labrador Sea cores HU75-48 and HU75-54 (Fillon and Duplessy, 1980, Fig. 7, p. 842).

The isotopic stage 5a/4 transition in HU75-58 (between 255 cm and 300 cm) is not uniform. Just as in HU75-42 (Fillon and Duplessy, 1980), it is marked by a peak roughly 0.5 to 0.8°/oo lighter than the heaviest values in stage 4 (Fig. 8.6). A similar shape for substage 5a and the 5a/4 transition was reported for eastern North

HU 75-41

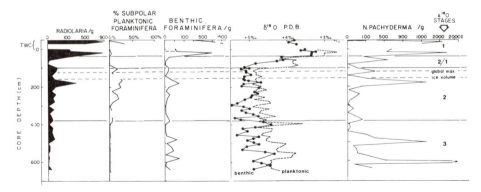

Figure 8.5   Numbers of radiolaria per gram of total sediment (shaded), percent abundance of subpolar planktonic foraminifera (>63 μm), numbers of benthic foraminifera per gram of total sediment, benthic (<u>Melonis pompiloides</u> and <u>Pullenia</u> spp.)and planktonic (<u>Neogloboquadrina pachyderma</u>, sinistral) foraminiferal 0-18 values, numbers of <u>Neogloboquadrina pachyderma</u> per gram of total sediment and oxygen isotopic stage boundaries (solid lines) for piston core HU75-41 (Fillon and Duplessy, 1980) and its accompanying trigger-weight core (TWC). The dashed horizontal lines probably include the isotopic stage 2 global ice-volume maximum (see discussion in Fillon and Aksu, this volume).

Atlantic core RC9-225 which also has a high sedimentation rate comparable to that in HU75-42 (Ruddiman et al., 1980). The heavy isotopic interval just below substage 5a in HU75-58 (330 cm to 355 cm) and the heavy interval labeled 5b in HU75-42 (Fig. 8.6) have similar magnitudes suggesting that substage-level events recorded by <u>N.</u> <u>pachyderma</u> in mid-isotopic stage 5 are stratigraphically useful within the Labrador Sea.

     Heavy, "glacial" values characterize the <u>N. pachyderma</u> 0-18 record in HU75-58 from 580 cm to 640 cm (c. 0.3°/₀₀ heavier than stage 2). We refer this portion of the sequence to isotopic stage 6 (Figs 8.3 & 8.4). Just above 580 cm, however, 0-18 values jump dramatically to a very light +2.1°/₀₀ (lighter than the stage 1 minimum in the trigger-weight core) and then increase briefly to stage 2 levels between 530 cm and 540 cm before shifting again to "interglacial" levels in the interval from 490 cm to 520 cm. Either one of these isotopically light peaks could be reasonably assigned to "interglacial" substage 5e on the basis of planktonic 0-18 amplitudes alone, but clearly they are not both likely to represent substage 5e, given the benthic 0-18 values across the interval (Fig. 8.4). There are several possible explanations if we consider only the planktonic data: (1) the

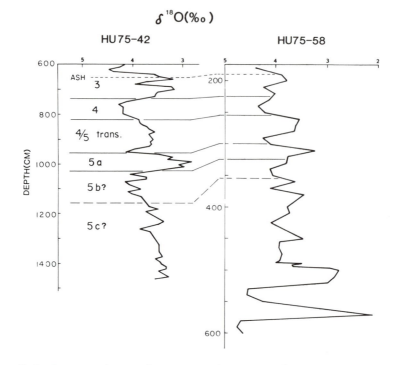

Figure 8.6 Suggested correlation of planktonic (N. pachyderma, sinistral) foraminiferal 0-18 records from the northeastern Labrador Sea (core HU75-42) and northwestern Labrador Sea (core HU75-58) for the interval from late isotopic stage 5 up to and including the volcanic ash datum (Labrador Sea ash zone B). The isotopic stratigraphy of HU75-42 is discussed in detail elsewhere (Fillon and Duplessy, 1980). Refer to Figures 8.4 and 8.5 for the complete isotopic record of HU75-58.

lower peak is substage 5e and the upper peak is substage 5c; (2) the lower peak is substage 5e while the upper peak was created by foraminiferal reworking from substage 5e sediments deposited somewhere else on the continental slope; (3) the upper peak is substage 5e, the heavy values immediately below it are late stage 6 and the lower "peak" represents the downward displacement of foraminiferal tests from the substage 5e peak by burrowing organisms (this explanation implies that "burrow-fill" material was sampled between 560 and 570 cm; but as no burrows were detected in the core itself or in the x-radiographs, this seems unlikely); (4) both peaks represent substage 5e while the interval of heavy values between them records foraminiferal reworking of tests from stage 6 sediments deposited elsewhere on the slope; (5) the upper peak represents substage 5e while the lower peak represents a previously undocumented isotopic event in the late stage 6.

The benthic foraminifera $\underline{B}$. $\underline{aculeata}$ and $\underline{C}$. $\underline{weullerstorfi}$ were analyzed to provide additional insight into the stage 6--substage 5e portion of the isotopic record (Fig. 8.4). The benthic 0-18 record exhibits a total change of about 1.5°/₀₀ from 500 cm to 600 cm. The lightest benthic values occur in the upper planktonic 0-18 minimum while benthic 0-18 values corresponding to the lower planktonic 0-18 minimum are at least 0.7°/₀₀ heavier. In contrast, to the planktonic data, the benthic data from both species examined appear to reflect a rather normal isotopic stage 6/5e transition. Accordingly, it may be inferred that $\underline{B}$. $\underline{aculeata}$ and $\underline{C}$. $\underline{wuellerstorfi}$ have not been transported downslope differentially. This is additional support for a largely $\underline{in}$ $\underline{situ}$ benthic fauna.

Using the benthic 0-18 data (Fig. 8.4`, we are able to place the isotopic stage 6/5e transition at about 535 cm in the core. This implies that of the above listed "explanations" for the two isotopically light peaks in the planktonic 0-18 curve, numbers (1) and (2) can be immediately ruled out. Explanation (4) is rejected because the lower peak contains benthics with 0-18 values that cannot have been related to substage 5e. Explanation (3) can also be ruled unlikely on the basis of the isotopic data because it requires that planktonic microfossils were reworked into a "burrow-filling" while the benthics were somehow left behind. It seems more likely that only the upper planktonic 0-18 minimum represents substage 5e while the lower planktonic 0-18 minimum is the result of a local perturbation of temperature and/or salinity in the upper part of the water column which predated the substage 5e global ice volume minimum.

The isotopic substage 5e/5d transition in HU75-58 can be placed between about 440 and 490 cm in both the planktonic and benthic records (Fig. 8.4). The substage 5b/5a transition (in the planktonic record, Fig. 8.3) is probably at about 340 cm. Between 340 cm and 440 cm, there is an interval in which planktonic 0-18 values vary over approximately 0.6°/₀₀. Substage 5c appears to be centered within this zone at about 380 cm (Fig. 8.3).

ABSOLUTE SAND INPUT RATES

To separate "sand content" which is intimately related to current velocity, from "sand input" which, as discussed above, tends to be affected more by glacial or proglacial processes, Ruddiman (1977) averaged North Atlantic absolute sand input rates ($mg/cm^2/10^3$ yr) down-core for seven "climatically important" time blocks between

TABLE 8.2. Average sand input rates in core HU75-58 calculated for 9 time blocks.

Time Blocks	Upper Limit (Yr BP)	Core Depth (cm)	Number of Samples in Time Block	Sand Input (mg/cm^2/ 1000 yr)
8*	0	0 (TWC)	9	1110
7	13,000	40	4	426
6	25,000	70	5	846
5	40,000	120	6	1285
4	57,000	180	7	748
3	70,000	235	5	329
2	80,000	290	15	777
1	115,000	450	8	978
0†	125,000	520	15	2291
	(c. 160,000)**			

Time blocks after Ruddiman (1977)
* Added by Fillon et al. (1981)
† Added in this paper
** The lower limit of time block 0 (mid-isotopic stage 6)

125,000 BP (the isotopic stage 6/5e transition) and 13,000 BP (the isotopic stage 2/1 transition). To produce comparable data for the Labrador Sea, Ruddiman's chronological subdivisions for down-core averaging have been used in this study and two additional time blocks added, one at the top of the section to cover the postglacial interval, 13,000 BP to the present (Fillon et al., 1981), and another at the base of the section to cover late stage 6, c. 160,000 BP to 125,000 BP. Ruddiman's (1977) time block boundaries (Table 8.2) were determined in HU75-58 by extrapolating sedimentation rates from the nearest isotopic stage boundaries which were assigned ages according to Shackleton and Opdyke's (1973) time scale. Because the base of HU75-58 has been determined to lie within isotopic stage 6, a median age for stage 6 c. 160,000 BP (within a range of 190,000 BP to 130,000 BP) was employed in the computations of sand input rates.

Sand input rates shown for the eastern Labrador Sea from 80,000 BP to the present in Figure 8.7 (Fillon et al., 1981) are regional averages for a sequence of cores from sites along the West Greenland margin (Fig. 8.1). The sand input rate shown for the period preceding 80,000 BP, however, is based solely on core HU75-42. It was originally

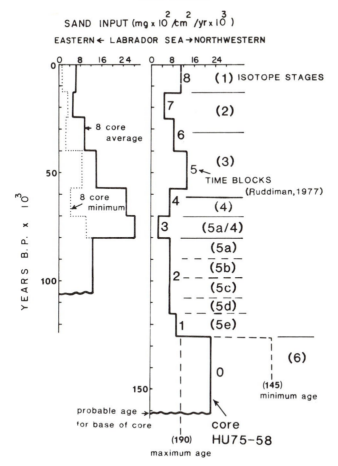

Figure 8.7 A comparison of eastern and northwestern Labrador Sea time
block averaged sand input rates. Boundaries of time blocks (labeled 0
to 8) are listed in Table 8.2. Also shown are marine isotopic stage
boundaries. The northwestern Labrador Sea input rate curve is for core
HU75-58. The eastern Labrador Sea input rates are mean values for 8
cores (2, 3, 4, 5, 6, 7, 8, and in Fig. 8.1). The dotted curve
portrays the lowest single-core input rate for each time block.

assumed that the core did not reach isotopic substage 5c (Fillon and
Duplessy, 1980), but from a correlation of the 0-18 record of HU75-42
with that of HU75-58 (Fig. 8.6), it seems likely that the lowermost two
meters of HU75-42 lie somewhere within isotopic substage 5c. This
increases our age estimate for the base of HU75-42 from c. 97,000 BP
(Fillon and Duplessy, 1980) to c. 106,000 BP and therefore decreases
the average sand input rate for that section of the record from an
original estimate of 1893 mg/cm^2/10^3 yr (Fillon et al., 1981' to the
c. 1231 m/cm^2/10^3 yr (Fig. 8.7).

The largest differences between average eastern and northwestern (HU75-58) Labrador Sea sand input rates over the last 100,000 years occurred during the interval from 80,000 BP to 57,000 BP (Fig. 8.7). Within that time span, from 80,000 BP to 70,000 BP, core HU75-58 records a much lower sand input rate than the lowest for any individual eastern Labrador Sea core (dashed line in Fig. 8.7). Also, from 70,000 BP to 57,000 BP, the input rate of sand to the HU75-58 core site was lower than at five out of seven eastern Labrador Sea core sites (Fillon et al., 1981). This is in strong contrast with the interval from 57,000 BP to the present, during which HU75-58 sand input rates fairly closely matched or exceeded average eastern Labrador Sea rates in each time interval. It can be inferred that the low sand input rates at the HU75-58 core site from 80,000 BP to 57,000 BP represent a real decrease in northwestern Labrador Sea sand input because of the generally low sand content (percent sand) observed in that inverval (Fig. 8.3). If strong, local bottom currents had been responsible for a net removal of sand from the HU75-58 core site during the period 80,000 BP to 57,000 BP, they would also have preferentially removed finer particles, leaving behind a coarse sand and gravel lag. The presence of facies A, a generally muddy sediment (typically less than 10 percent sand) with a few sponge spicule laminae is not at all compatible with net sand removal by bottom currents.

In core HU75-58 the average input rate of 978 $mg/cm^2/10^3$ yr recorded in interglacial isotopic substage 5e (125,000 BP to 115,000 BP) is close to the value of 1110 $mg/cm^2/10^3$ yr registered for the period 13,000 BP to 0 BP. The highest sand input rates in HU75-58 are found in late isotopic stage 6 when a mean stage 6 age of c. 160,000 BP is assumed for the base of the core. Even the lowest estimated input rates for stage 6, however, those computed assuming that the base of the core nearly reaches isotopic stage 7 (c. 190,000 BP), are higher than those recorded for the "glacial" periods 25,000 BP to 13,000 BP and 70,000 BP to 57,000 BP (isotopic stages 2 and 4).

Planktonic Foraminifera

Information on input rates of ice-rafted sand can be combined with paleoceanographic inferences based on volcanic ash distribution and on the paleoecology of planktonic foraminifera to reconstruct a relatively detailed history of paleoceanographic change in the Labrador Sea. The percentage of subpolar planktonic foraminiferal tests preserved in cores compared to the polar species N. pachyderma has formed the basis of previous reconstructions of Labrador Sea paleoceanography (Fillon and Duplessy, 1980). The subpolar fauna (>63 μm) typically consists

of: Globigerina bulloides, Globigerina falconemsis, Globigerina
quinqueloba, and Globigerinita bradyii (uvula) with occasional
specimens of Globigerinita glutinata. N. pachyderma (right coiling) is
omitted from this list of subpolar taxa because dextral forms of N.
pachyderma in the Labrador Sea are mostly concentrated among the small,
thin-walled "normal form" population described by Vilks (1974) as
living in the upper 200 meters of the water column. It is the larger,
deeper-living "quadrate" and "kummerform" phenotypes of dextral N.
pachyderma (>150 um) that have been traditionally used as subpolar
indicators (e.g., Kipp, 1976; McIntyre et al., 1976). In view of the
strong possibility that the environmental significance of dextral N.
pachyderma may vary with size and test morphology in the Labrador Sea,
dextral N. pachyderma has not been included with the other subpolar
taxa in determining percent subpolar planktonic foraminifera in studies
of Labrador Sea sediments (Fillon and Duplessy, 1980). In any case,
right-coiling N. pachyderma constitutes less than 5% of total N.
pachyderma in eastern and northwestern Labrador Sea slope and rise
core-tops, compared with up to 50% (of total planktonic foraminifera)
for subpolar planktonic foraminifera exclusive of right-coiling N.
pachyderma. Core-tops from basins on the Labrador Shelf which underlie
the present course of the cold, polar Labrador Current (Vilks, 1980)
are dominated virtually completely by N. pachyderma (c. 0% subpolar).
The dramatic contrast between planktonic foraminiferal faunas in slope
and rise cores which underlie subpolar Labrador Sea and West Greenland
Current water and Labrador Shelf cores from beneath the polar Labrador
Current points to the value of subpolar planktonic foraminifera as
recorders of paleoceanographic conditions. Plankton-tow data on living
planktonic foraminifera from the central Labrador Sea (Be and
Tolderlund, 1971; Tolderlund and B , 1971) and the Labrador Shelf
(Vilks, 1980) mirror the core-top data cited above.

    Differential dissolution of subpolar planktonics in the eastern
Labrador Sea has been demonstrated in the down-core record, based on
variations in the ratios of benthic to planktonic foraminifera. Fillon
and Duplessy (1980) concluded that a dramatic decrease in subpolar
planktonic foraminifera within benthic zone bIV from c. 100% in the
extreme southeastern Labrador Sea to less than 5% in the northeastern
Labrador Sea was probably due to differential dissolution (Berger,
1970) associated with a shallower than modern lysocline. The absence
of subpolar planktonic foraminifera in sediments is therefore not
necessarily indicative of an absence of subpolar water, but, the
presence of subpolar planktonic foraminifera, in our opinion,
unambiguously indicates the presence of subpolar water.

PALEOCEANOGRAPHY

<u>25,000 BP to 9,800 BP</u> (isotopic stage 2 and the stage 2/1
transition/time block 7 and part of 8)

High percentages of subpolar planktonic foraminifera in the eastern
Labrador Sea and Baffin Bay during isotopic stage 2 up until c. 16,000
BP (Fig. 8.5 and Fillon and Aksu, this volume) suggest that summer
surface water temperatures there were probably greater than or equal to
5°C (Fig. 8.8a)--the modern limit of the subarctic zone of abundant

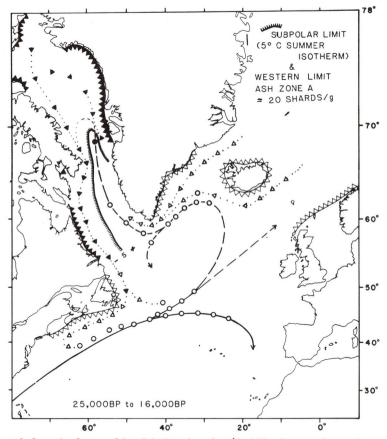

Figures 8.8 a-d Generalized Labrador Sea/Baffin Bay and northwest
Atlantic Ocean paleoenvironmental reconstructions representing "modal"
conditions of surface circulation and iceberg drift. Symbols used are
the same as in Figure 8.1 which can also be taken to represent the
interval c. 125,000 BP to c. 115,000 BP. The reconstructions were
drawn to accommodate microfaunal data (Fig. 8.3) and sand input rates
(Fig. 8.7) as well as published interpretations of North Atlantic sand
input, the distribution of watermass sensitive micro and macrofauna,
the concentrations of ice-rafted volcanic ash in North Atlantic, and
Labrador Sea cores and Wisconsinan (and equivalent) ice limits (see
text and Fillon et al., 1981 for references and a full discussion of
the strategy employed in producing these maps).

subpolar planktonics (Bé and Tolderlund, 1971). Fillon and Duplessy (1980) concluded that this warmth was related to the northward advection of warm water rather than to the *in situ* summer warming of a stable low salinity surface layer because there was evidence in the benthic isotopic record of a simultaneous bottom-water cooling. The bottom-water cooling implies an unstable surface water mass that cooled and sank in winter and therefore indicates open water conditions thoughout most of the year. The absence of subpolar planktonic foraminifera from the northwestern Labrador Sea during isotopic stage 2 is apparently not associated with strong dissolution (low benthic/ planktonic ratios of c. 0.1). We infer that advection of warm water into the Labrador Sea was limited in effect to the West Greenland sector. The very long abundance of iceberg-rafted volcanic ash in the upper half of isotopic stage 2 (ash zone A) in core HU75-58 further suggests that few Icelandic icebergs in transit up the west coast of Greenland reached the northwestern Labrador Sea in the interval 25,000 BP to 16,000 BP.

The period from about 16,000 BP to 9,800 BP includes an isotopically light interval in the planktonic 0-18 record which probably represents a low salinity episode which affected much of the North Atlantic (cf. Ruddiman and McIntyre, 1981b). It should be noted that the major pulse of ash zone 1 ash deposition in the North Atlantic (Ruddiman and Glover, 1972) post-dates the ash zone A maximum, dating from about 9,800 BP (Fillon et al., 1981) and is not well represented in the northeastern Labrador Sea (Fillon and Duplessy, 1980). This implies the absence of a northerly flowing current off West Greenland. In the North Atlantic the earlier, less well-defined eastern Labrador Sea ash zone A event is barely discernible against the lower, mixing-induced limb of the North Atlantic ash zone 1 peak (see Fillon and Duplessy, 1980).

Sand input into the northwestern Labrador Sea during time block 7 was low, about the same as off West Greenland (Fig. 8.7). The ice-rafted debris component was evidently supplied chiefly by ash-free icebergs originating in Baffin Bay. The generalized Labrador Sea/Baffin Bay and northwest Atlantic Ocean paleoenvironmental reconstruction for the period c. 25,000 BP to c. 16,000 BP (cf. Fillon et al., 1981) has now been updated and is presented in Figure 8.8ᐤ. The illustrated northward advection of warm surface water along the West Greenland margin is also supported by warm, sea-surface temperatures in the central subpolar North Atlantic during middle isotopic stage 2 (Ruddiman and McIntyre, 1981a: Fillon and Aksu, this volume). The period c. 16,000 BP to c. 9,800 BP is depicted in Figure 8.8b.

57,000 BP to 25,000 BP   (isotopic stage 3 / time blocks 5 and 6)

This interval in the eastern Labrador Sea and in the northwestern North
Atlantic was marked by large increases in the relative abundance of D.
davisiana (Fillon and Duplessy, 1980; Morley and Hays, 1979). In the
Labrador Sea, isotopic stage 3 sediments are also characterized by a
relative scarcity of benthic foraminifera, N. pachyderma and subpolar
planktonic foraminifera. Fine sand laminae in HU75-58 indicate the
presence of strong bottom currents or turbidity currents on the
continental slope during stage 3. Sand input rate averages for time
blocks 5 and 6, however, were similar in the eastern and the
northwestern Labrador Sea. In both regions, sand input was
substantially higher than during isotopic stage 2 (time block 7).

    In the central subpolar North Atlantic summer sea-surface
temperatures were at minimum levels throughout most of the period from
about 45,000 BP to 30,000 BP (Ruddiman and McIntyre, 1981a). Ruddiman
and McIntyre have interpreted cyclic temperature minima there to
reflect the impact of cold melt products from disintegrating
continental glaciers. The higher levels of sand input in the Labrador
Sea may thus be due to increased iceberg and meltwater fluxes from
circum-Labrador Sea glaciers whereas the apparently lower foraminiferal
productivity and peculiar radiolarian fauna may be related in some way
to cold, low-salinity surface-water (Robertson, 1975). If a portion of
the sand input associated with facies C (Fig. 8.3) is related to
turbidity flows, these too could be tied to a heavy discharge of turbid
meltwater onto the southeast Baffin Island Shelf. The history of
glacial activity in southeastern Baffin Island is unfortunately
incomplete through this time interval as it is represented by a major
unconformity in raised marine sediments on the outer coast (Miller et
al., 1977; Fillon et al., 1981; Nelson, 1981; 1982). Isolated raised
marine sediments from about 40,000 BP have been found in northern
Labrador and Baffin Island (Clark and Miller, person. commun., 1983).

    The picture that emerges for most of the interval 57,000 BP to
25,000 BP in the Labrador Sea is one dominated by conditions decidedly
unlike those of today. Copious injections of melt-products with a
resultant stable, low-salinity surface layer and winter sea-ice cover
probably dominated the circulation regime, effectively reducing the
northward convective flow of surface water into the Labrador Sea by
inhibiting the winter chilling and sinking of surface waters in the
northeastern Labrador Sea. This would be similar in principle to the
drift ice stagnation stage (c. 1810 to 1860 AD) described by Vibe
(1967) for the eastern Labrador Sea in which the Canadian Current
dominates circulation through Davis Strait. It should be emphasized,

however, that in trying to represent a 33,000 year long period of time by a single reconstruction, it can only be hoped to capture an "average" environment for the period (Fig. 8.8b).

A basically similar paleoceanographic pattern is thought to have dominated the deglacial, isotopic stage 2/1 transition interval from c. 16,000 BP to c. 9,800 BP (Fillon et al., 1981; Fillon and Aksu, this volume). The early part of that transition period was marked in the eastern Labrador Sea by generally reduced numbers of microfossils and a peak in the relative abundance of D. davisiana (Fillon and Duplessy, 1980; Fillon et al., 1981) (Fig. 8.5).

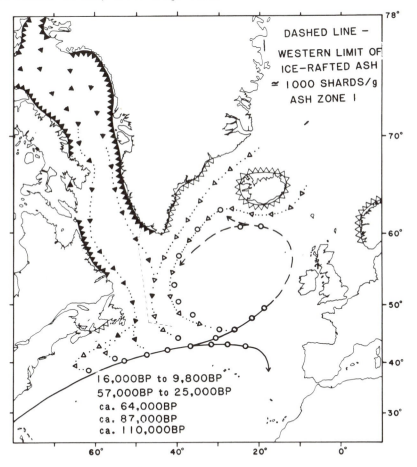

DASHED LINE –

WESTERN LIMIT OF ICE-RAFTED ASH ≈ 1000 SHARDS/g ASH ZONE I

16,000BP to 9,800BP
57,000BP to 25,000BP
ca. 64,000BP
ca. 87,000BP
ca. 110,000BP

Figure 8.8b   (See Fig. 8.1 for key)

The period of maximum global ice-volume and generally low North
Atlantic temperatures c. 18,000 BP (McIntyre et al., 1976) contains
another episode of low microfossil abundance (bracketed by dashed lines
in Fig. 8.5). Although this interval does not include a D. davisiana
maximum in the eastern Labrador Sea the low radiolarian abundance, as
it is unaffected by $CaCO_3$ dissolution, may be suggestive of a
relatively rigorous environment (see Fillon and Aksu, this volume).

80,000 BP to 57,000 BP (the isotopic stage 5a/4 transition, isotopic
stage 4 and earliest stage 3 / time blocks 3 and 4)

This interval in the eastern Labrador Sea contains a subpolar
planktonic foraminiferal maximum during the isotopic stage 5a/4
transition (c. 75,000 BP) and a high concentration of volcanic glass
shards (ash zone B - c. 59,000 BP) in earliest stage 3 (Fillon and
Duplessy, 1980). Both of these events constitute evidence of a
proto-West Greenland Current. The comparatively low concentration of
ash in HU75-58 and in the most northwesterly of the eastern Labrador
Sea cores HU75-54 (Fig. 8.1), however, clearly points to a limited
transfer of surface-water from the eastern to the western Labrador Sea.

The small subpolar foraminiferal maximum during the earliest part
of the isotopic stage 5a/4 transition in core HU75-58 is associated
with a low benthic/planktonic ratio (0.1) indicating minimal
dissolution. Later in the 5a/4 transition period, however, dissolution
apparently increased sharply producing benthic/planktonic ratios of
about 1.0 and subpolar planktonic foraminifera are absent. Although
calcareous microfossils are generally not very abundant in northwestern
Labrador Sea core HU75-58 during the stage 5a/4 transition, benthic
productivity is indicated by abundant siliceous sponge spicules (from
240 cm to 280 cm in the core).

There are fewer subpolar planktonic foraminifera within the
isotopic stage 5a/4 transition in HU75-58 than are associated with ash
zone B. It does not necessarily follow, however, that more warm
surface-water reached the northwestern Labrador Sea in earliest stage 3
than during the isotopic stage 5a/4 transition, because severe
dissolution of subpolar tests in the northern Labrador Sea during the
latter part of the 5a/4 transition is indicated for core HU75-58 and
has also been documented for the eastern Labrador Sea (Fillon and
Duplessy, 1980).

Sand input during time blocks 3 and 4 in the northwestern Labrador
Sea was much lower than the eastern Labrador Sea average (Fig. 8.7).

Considering the large east-west difference in ice-rafted volcanic ash
concentrations within ash zone B, this is not surprising. It provides
additional evidence of a strong east-west thermal gradient in the
Labrador Sea during both the stage 5a/4 transition and ash zone B
events. Apparently only a few of the icebergs in transit northward up
the West Greenland margin were transported across the Labrador Sea and
into the southerly drift of a proto-Baffinland Current. Those that did
make the crossing and those which entered the southerly flow after
calving from glaciers terminating in Baffin Bay evidently supplied very
little debris. The large amount of sand deposited in the eastern
Labrador Sea during this interval compared with relatively low input in
the North Atlantic (Ruddiman, 1977) was explained by a "closed gyre"
circulation pattern which would tend to trap icebergs within the
Labrador Sea (Fillon et al., 1981). This "closed gyre" reconstruction,
updated to account for the low northwestern Labrador Sea sand input and
low subpolar percentages, is illustrated in Figure 8.8c. It is
probable that the "closed gyre" circulation pattern would have
maintained somewhat warmer sea-surface temperatures in the eastern and
central Labrador Sea than those which prevailed during stage 2 (Fig.
8.8a), principally because of the direct input of Gulf Stream water.

In the central subpolar North Atlantic both the stage 5a/4
transition (c. 75,000 BP) and the c. 59,000 BP ash zone 2 level were
marked by warm sea-surface temperatures, while a period of cold
temperatures (indicative of abundant glacial melt products) is centered
just on the stage 4/3 transition (Ruddiman and McIntyre, 1981a) at
about 64,000 BP. There is a D. davisiana maximum associated with late
stage 4 in the Labrador Sea (Fillon and Duplessy, 1980) and it may be
that conditions in the Labrador Sea during the latter part of isotopic
stage 4 (c. 64,000 BP) were as polar as those at the stage 2/1
transition. Thus a portion of isotopic stage 4 may also have resembled
Figure 8.8b. (In citing ages here, we have applied the chronology of
Shackleton and Opdyke, 1973, to Ruddiman and McIntyre's, 1981a, 0-18
stratigraphy so as to be consistent with our Labrador Sea chronology.)

125,000 BP to 80,000 BP (isotopic stage 5 / time blocks 1 and 2)

This period encompasses isotopic substages 5a to 5e, but because
substages 5c to 5e are not adequately represented in eastern Labrador
Sea cores, reconstructions of paleoceanographic events in the eastern
Labrador Sea will have to be based on extrapolations from the
northwestern North Atlantic. Core V27-20 (Fig. 8.1) is particularly
useful in this regard (Ruddiman, 1977) since it has already been
compared with eastern Labrador Sea cores and shown to correspond

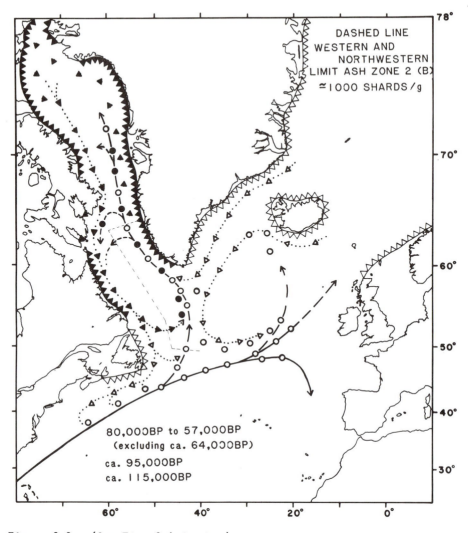

Figure 8.8c   (See Fig. 8.1 for key)

closely in a number of parameters through substage 5a including:
percent subpolar planktonic foraminifera; volcanic ash abundance;
relative abundance of D. davisiana; and percent sand (Fillon and
Duplessy, 1980; Fillon et al., 1981). The V27-20 record of percent
abundance of subpolar planktonic foraminifera (Ruddiman, 1977) also
corresponds with the central subpolar North Atlantic sea-surface
paleotemperature record published by Ruddiman and McIntyre (1981a). In
comparing the two records, it should be noted that the latter record is
based on a transfer function which utilizes important variations in the
taxonomic composition of the whole greater than 150 um size fraction of

planktonic foraminifera, but is keyed most heavily to dissolution-resistant forms (Ruddiman and Glover, 1975; Kipp, 1976). This technique has been shown to predict accurate paleotemperatures even in partially dissolved assemblages whereas the percent abundance of subpolar planktonics is known to be strongly dissolution dependent.

The central subpolar North Atlantic paleotemperature record (Ruddiman and McIntyre, 1981a) indicates the presence of warm conditions centered late in isotopic substage 5a (i.e. on the 5a/4 transition, c. 75,000 BP), during isotopic substage 5b (c. 95,000 BP) and through the interval spanning isotopic substage 5e and the substage 5e/5d transition (c. 125,000 BP to c. 113,000 BP). Cold conditions are indicated for isotopic substages 5d (c. 110,000 BP) and early 5a (c. 87,000 BP). Core V27-20 records large subpolar planktonic foraminiferal percentage peaks at the isotopic substage 5a/4 transition (Ruddiman et al., 1980) and within substage 5e (Ruddiman, 1977), possibly extending into the substage 5e/5d transition (Ruddiman and McIntyre, 1979). The subpolar peak at the 5a/4 transition in core V27-20 has been correlated to a similar event in southeastern Labrador Sea core HU75-31 (Fillon and Duplessy, 1980 and Fig. 8.1).

As mentioned in the last section, northwestern Labrador Sea core HU75-58 exhibits a small but significant (c. 10%) subpolar planktonic foraminiferal maximum just after the isotopic substage 5a 0-18 minimum (i.e. early in the 5a/4 transition). There is a much larger (c. 60%) subpolar maximum lower in the core at a point marking the upper limit of the prominent substage 5e 0-18 minimum (Fig. 8.3 and 8.4). Of these two subpolar events, the one late in substage 5a is associated with a low benthic/planktonic ratio (c. 0.1) suggesting minor dissolution whereas the event in substage 5e, although larger, is associated with a high benthic/planktonic ratio of about 4.0, implying strong dissolution. We infer from this that the pre-dissolution composition of the isotopic substage 5e planktonic fauna in the northwestern Labrador Sea contained an even greater subpolar component than is indicated in Figure 8.4. At least part of substage 5e, in the northwestern Labrador Sea was subpolar in character. During the 0-18 minimum of substage 5e it may well have been as warm as the present interglacial which contains a subpolar planktonic maximum of c. 79% accompanied by a benthic/planktonic ratio of over 2.0. Similarly, substage 5e planktonic foraminiferal faunas in the northwestern North Atlantic (V27-20) and central subpolar North Atlantic appear to have been just as warm as stage 1 faunas (Ruddiman, 1977; Ruddiman and McIntyre, 1981a). This evidence from the North Atlantic and northwestern Labrador Sea implies that the eastern Labrador Sea was also in an interglacial mode during substage 5e. Figure 8.1 therefore probably adequately portrays ocean circulation during the substage 5e climatic optimum (c. 120,000 BP).

The 5e/5d transition (c. 115,000 BP) and substage 5d (c. 95,000 BP), although still relatively warm, were somewhat cooler than 120,000 BP throughout the northwestern/central-subpolar North Atlantic (Ruddiman and McIntyre, 1979; 1981a) and therefore presumably throughout the Labrador Sea also, with conditions perhaps resembling Figure 8.8c. For example, substages 5b and early 5d in core HU75-58 are "cool", exhibiting low percentages of subpolar planktonic foraminifera (<5%) and low benthic/planktonic ratios (c. 0.3). The substage 5b interval in eastern Labrador Sea core HU75-42, however, appears "warm" because several samples contain c. 10 percent subpolar planktonic foraminifera in spite of evidently strong dissolution (Fillon and Duplessy, 1980).

Early isotopic substage 5a was "cold" in the central subpolar North Atlantic and it is marked by a D. davisiana maximum in the eastern Labrador Sea (Fillon and Duplessy, 1980), probably indicating that conditions like those outlined in Figure 8.8b persisted briefly c. 87,000 BP. A similar reconstruction may also be appropriate for "glacial" isotopic substage 5d (c. 110,000 BP) which appears cold in the central subpolar North Atlantic and which is also marked by a D. davisiana maximum there (Morely and Hays, 1979).

Average sand input rates in the northwestern Labrador Sea during time blocks 1 and 2 (Fig. 8.7) were similar to time block 8 and time block 4 rates respectively. In the North Atlantic, time block 1 and 2 sand input rates were highest off southeastern Greenland and in a zone stretching from the southwestern Labrador Sea to c. 50°N, 25°W in the Central North Atlantic (Ruddiman, 1977). North Atlantic and Labrador Sea sand input data for time blocks 1 and 2 then are compatible with the "modern analog" and "closed gyre" paleoceanographic reconstructions suggested for these time intervals (Figs 8.1 & 8.8c).

C. 160,000 BP to c. 125,000 BP (late isotopic stage 6/time block 0)

Summer sea-surface temperatures in the central subpolar North Atlantic were cool from c. 160,000 BP to c. 150,000 BP and from c. 135,000 BP to c. 125,000 BP, but were relatively warm in the interval c. 150,000 BP to c. 135,000 BP (Ruddiman and McIntyre, 1981a). In the northwestern Labrador Sea, core HU75-58 records a zone of significantly increased subpolar planktonic foraminiferal percentages within stage 6 (between 560 cm and 620 cm). The latest portion of this subpolar interval is

coincident with a "step" in the transition of benthic O-18 values from
stage 6 levels to substage 5e levels and with a very strong planktonic
O-18 minimum (Fig. 8.4). The peak subpolar percentage (c. 20%) at 570
cm corresponds to a 6.6°C transfer-function estimate of summer
sea-surface temperature (calculated from the sample by W.F. Ruddiman,
pers. commun., 1980). This compares with the modern average summer
sea-surface temperature at the core site of 6.8°C (M.E.D.S., written
communication, 1980). The value of 6.6°C could be a minimum
temperature estimate because the transfer-function utilizes only the
greater than 150 μm fraction of planktonic foraminifera (Ruddiman and
Glover, 1975) whereas subpolar planktonic foraminifera in the Labrador
Sea tend to be small. The benthic/planktonic ratio at the 570 cm level
is close to 1.0 (Fig. 8.3), further implying significant preferential
dissolution of subpolar tests. Actual sea-surface paleotemperatures in
the northwestern Labrador Sea during this part of late isotopic stage 6
then clearly approximated interglacial levels.

We consider that the late stage 6 subpolar interval in the
northwestern Labrador Sea is probably correlative with the 150,000 BP
to 135,000 BP warming the central subpolar North Atlantic. The strong
planktonic O-18 minimum in the northwestern Labrador Sea which occurred
during the latter part of the subpolar episode must therefore date from
about 140,000 BP. Its magnitude, c. 1.6°/₀₀ lighter than the
benthic record can be explained either by a 5°C surface-water warming,
a 2.0 p.p.t local decrease in surface salinity due to meltwater
discharge, or some combination of the two effects (neither is
impossibly large).

The interval from c. 135,000 BP to c. 125,000 BP which was cool in
the central subpolar North Atlantic is marked by very low (<.01)
benthic/planktonic ratios and virtually 0% subpolar foraminifera in
core HU75-58 (Fig. 8.3). Conditions were therefore probably similar to
those which obtained during late isotopic stage 2 in the northwestern
Labrador Sea (Fig. 8.8a and/or Fig 8.8b). Additional work is now
underway to try to resolve the true paleoceanographic significance of
the late stage 6 O-18 minimum and a similar, c. 140,000 BP, planktonic
O-18 event in the Gulf of Mexico (Fall, 1980) which is also possibly
meltwater related.

Assignment of a c. 140,000 BP age for the negative planktonic O-18
spike in late isotopic stage 6 produces an extrapolated age for the
base of the core (assuming a constant accumulation rate) of c. 170,000
BP. Comparing this age estimate with Figure 8.7 confirms a relatively
high average sand input rate for the interval from the base of the core
to c. 125,000 BP. This suggests the episodic co-occurrence during late

stage 6 of a high iceberg flux and, up until about 135,000 BP, surface
water warm enough to significantly melt icebergs. A similar set of
conditions in the eastern Labrador Sea stimulated sand input to very
high levels between c. 80,000 BP and c. 57,000 BP (Fig. 8.8c). The
record in HU75-58, however, suggests that warm surface water and
melting icebergs penetrated farther to the west during part of late
stage 6 than during the 5a/4 transition or the period bracketing the
deposition of ash zone B. A paleoceanographic reconstruction
explaining warmth in the northwestern Labrador Sea during late stage 6
(c. 135,000 BP to c. 150,000 BP) is reproduced in Figure 8.8d).

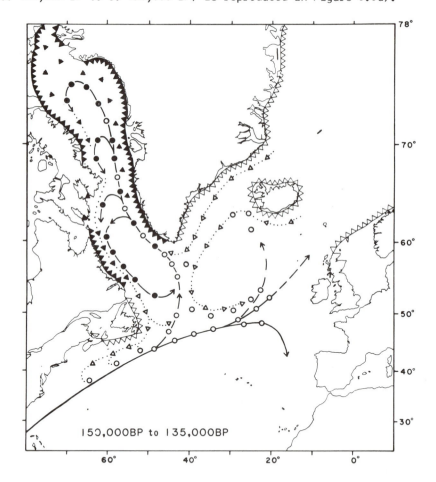

Figure 8.8d (See Fig. 8.1 for key)

A Note on Correlation with Baffin Bay Stratigraphy

The strong O-18 minimum in the late isotopic stage 6 planktonic O-18
record of northwestern Labrador Sea core HU75-58 dictates a
reassessment of published planktonic O-18 records from two Baffin Bay
cores (Fig. 8.1; Aksu, 1981). The O-18 records of these two cores
exhibit down-core variations of up to 6.0°/₀₀ compared to the
1.65°/₀₀ estimated to account for ice sheet growth (Duplessy,
1978). The bulk of the planktonic O-18 signal in Baffin Bay is
therefore undoubtedly due to the local input of meltwater (Aksu,
1981). It would obviously be easy in such records to confuse global
ice-volume minima with meltwater maxima. A correlation of Baffin Bay
and Labrador Sea O-18 records, which suggests an alternative placement
of the isotopic stage 6/5e boundary in the Baffin Bay cores, is
presented in Figure 8.9. In making the correlation and in constructing
the time-scale, only the interglacial-to-glacial isotopic stage
transitions, which are not as likely as glacial-to-interglacial stage

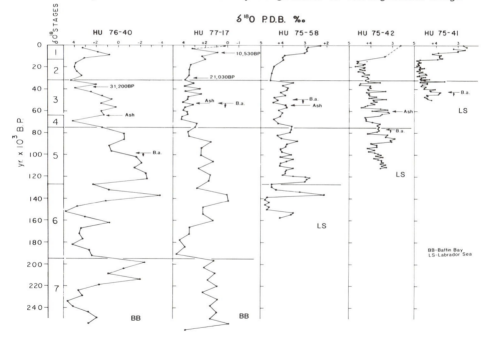

Figure 8.9 Suggested correlation of planktonic (N. pachyderma,
sinistral) foraminiferal 0-18 records from Baffin Bay (BB) and the
Labrador Sea (LS). The Baffin Bay 0-18 records and associated C-14
dates are from Aksu (1981). Also shown is the level of peak
concentration of Labrador Sea ash zone B (North Atlantic ash zone 2)
and the upper limit of occurrence of the benthic foraminifera B.
aculeata (B.a.). The limits of B. aculeata in the Baffin Bay cores
were provided by A.E. Aksu (pers. commun., 1981).

transitions, to be disrupted by meltwater events were assumed to be synchronous. The Baffin Bay cores were aligned using the 7/6, 5/4 and 3/2 isotopic stage boundaries and linear extrapolation was applied between those boundaries. The stage 6/5e transition in Labrador Sea core HU75-58 was located independently using the benthic O-18 record (Fig. 8.4). The alignment of O-18 records in Figure 8.9 places the largest negative peaks in the two Baffin Bay cores in late stage 6 at just about 140,000 BP and suggests that the late stage 6 planktonic negative O-18 spike is a regional phenomenon. Substage 5e in cores HU76-40 and HU77-17 would then be centered at about 260 cm and 375 cm respectively with corresponding peak O-18 values of c. -2.6°/₀₀ and c. +1.6°/₀₀ (cf. Aksu, 1981; this volume) which are still lighter than substage 5e in core HU75-58.

Adjustment of the isotopic stage 6/5e boundary in Baffin Bay does not significantly affect Aksu's (1981) identification of warm (subpolar) planktonic foraminiferal faunas within glacial isotopic stages 6 and 2 and across the isotopic substage 5a/4 transition. His findings have been accommodated in the reconstructions of ocean circulation presented in Figures 8.8a, 8.8c, and 8.8d. The nature of the isotopic stage 2 subpolar interval in the Labrador Sea and Baffin Bay is more fully examined in an accompanying paper (Fillon and Aksu, this volume).

SUMMARY

Core HU75-58 from the northwestern Labrador Sea consists of alternating coarse, probably ice-rafted sediments rich in planktonic foraminifera, and fine-grained sediments containing abundant siliceous sponge spicules and lesser numbers of foraminifera. A curve of Neogloboquadrina pachyderma oxygen isotopic composition encompasses a range of O-18 values from +2.0°/₀₀ to +4.5°/₀₀. Stratigraphic interpretation of the planktonic O-18 curve and supplemental O-18 analyses of benthic foraminifera provide a detailed chronology and place the base of the core in mid-isotopic stage 6 (c. 160,000 BP). This chronology is supported by the correlation of benthic foraminiferal abundance maxima, peaks in terrigenous sand content and a volcanic ash horizon in core HU75-58 with isotopically calibrated eastern Labrador Sea cores.

Time block averaged sand input rates (mg/cm^2/10^{3}yr) were calculated for core HU75-58 using the O-18 time scale and compared to eastern Labrador Sea rates (Fillon et al., 1981). Sand input rates in the Labrador Sea and North Atlantic Ocean (Ruddiman, 1977) provide a powerful paleoceanographic tool when integrated with variations in the percent abundance of subpolar planktonic foraminifera which favor subarctic water masses (warmer than 5°C in summer). Paleoceanographic reconstructions based on the resulting integrated data base suggest that the northwestern Labrador Sea has generally been substantially cooler than the northeastern Labrador Sea because of a persistent counterclockwise surface circulation pattern which advected warm North Atlantic water northward along the West Greenland margin and delivered a southward flow of cooler water to the western Labrador Sea.

During the period c. 25,000 BP to c. 16,000 BP the surface circulation was similar in some respects to the modern West Greenland Current - Baffinland/Labrador Current system but differed in that less warm subpolar water was transferred from the eastern to the western Labrador Sea (Fig. 8.8a). This was most likely the result of cooler-than-modern temperatures in subpolar waters of the eastern Labrador Sea and northwestern North Atlantic.

Surface circulation during the period c. 57,000 BP to c. 25,000 BP is considered to have been predominantly polar with a low salinity meltwater cap depressing temperatures, enhancing sea-ice formation and retarding vertical mixing (Fig. 8.8b). The Labrador Sea then would have resembled in many respects the northern reaches of Baffin Bay today (north of 75°N). Similar polar circulation patterns are also suggested for: c. 16,000 BP to 9,800 BP, c. 64,000 BP, c. 87,000 BP, and c. 110,000 BP.

From c. 80,000 BP to 57,000 BP (except for a brief polar interval c. 64,000 BP), c. 95,000 BP, and c. 115,000 BP a counterclockwise "closed-gyre" surface circulation pattern probably existed in the Labrador Sea. It would have been characterized by the large-scale entrainment of southeasterly drifting western Labrador Sea water (and icebergs) into the northerly flow off southwestern Greenland (Fig. 8.8c). In order to maintain a closed-gyre system, surface salinities in the cold, western Labrador Sea would have had to have been relatively high to balance the density of Labrador Sea surface outflow with the warm but saline Atlantic surface inflow. Indeed, at such times, meltwater discharge into the North Atlantic appears to have been relatively low (Ruddiman and McIntyre, 1981a).

Surface circulation c. 120,000 BP is inferred to have closely resembled the modern (Fig. 8.1) whereas the period immediately preceding the c. 120,000 BP global ice volume minimum (c. 135,000 BP to 125,000 BP) probably resembled the period 25,000 BP to 9,800 BP (Figs 8.8a & 8.8b). Between c. 150,000 BP and c. 135,000 BP surface waters

in the northwestern Labrador Sea and North Atlantic were warm.
Terrigenous sand input, apparently mostly from melting icebergs, also
reached a maximum in the northwestern Labrador Sea during that period.
A closed-gyre surface circulation pattern is indicated but more warm
water must have crossed into the northwestern Labrador Sea (Fig. 8.8d)
than is indicated for the closed-gyre model depicted in Figure 8.8c,
otherwise, density-balance constraints on surface inflow and outflow in
the two models are similar. The c. 150,000 BP to c. 135,000 BP
reconstruction calls for a relatively warm, i.e. subarctic, nearshore
environment along much of eastern Baffin Island combined with
significant glacier growth and iceberg calving. An attempt has been
made to correlate core HU75-58 to cores in Baffin Bay (Aksu, 1981)
using isotopic records. The results suggest a redefinition of the
substage 6/5 transition in Baffin Bay.

ACKNOWLEDGEMENTS

This work was funded by U.S. National Science Foundation grant
AFM-8200424, by support to R.H.F. while at the Geological Survey of
Canada and by support to J.C.D. through Centre National de la Recherche
Scientifique (France). R. Harmes and R. Kral provided valuable
technical assistance and the officers and crew of CSS <u>Hudson</u> made
collection of the cores possible.

REFERENCES

Aksu, A.E., 1981: Late Quaternary stratigraphy paleoenvironmentology
    and sedimentation history of Baffin Bay and Davis Strait. Ph.D.
    dissertation, Dalhousie University, Halifax, Nova Scotia, Canada. 585
    pp.
Amos, C.L., 1978: The post glacial evolution of the Minas Basin,
    N.S., a sedimentological interpretation. <u>Journal of Sedimentary
    Petrology</u>, 48:965-982.

Bé, A.W.H. and Tolderlund, D.S., 1971: Distribution and ecology of
    living planktonic foraminifera in surface waters of the Atlantic and
    Indian Oceans. In: Funnell, B.M., and Riede, W.R. (eds.),
    <u>Micropaleontology of the Oceans</u>. Cambridge University Press, London,
    England, 105-148.
Berger, W.H., 1970: Biogenous deep-sea sediments: fractionation by
    deep-sea circulation. <u>Geological Society of America Bulletin</u>, 81:
    1385-1402.
Binns, R.E., 1972: Composition and derivation of pumice on
    post-glacial strandlines in northern Europe and the western Arctic.
    <u>Geological Society of America Bulletin</u>, 83:2303-2324.

Chough, S.K., 1978: Morphology, sedimentary facies and processes of the
    northwest Atlantic Mid-Ocean Channel between 61° and 52° N, Labrador
    Sea. Ph.D. dissertation, McGill University, Montreal, 167 pp.

Dunbar, M.J., 1951: Eastern Arctic waters. Fisheries Research Board of
   Canada, Bulletin No. 88: 1-131.
Duplessy, J.C., 1978: Isotope studies. In: Gribbin, J. (ed.), Climatic
   Change. Cambridge University Press, Cambridge, England, 46-67.

Emiliani, C., 1955: Pleistocene Temperatures. Journal of Geology,
   63:538-578.

Fall, W.F., 1980: Glacial meltwater in-flow into the Gulf of Mexico
   during the last 150,000 years, implications for isotope stratigraphy
   and sea level studies. M.S. thesis, University of South Carolina,
   Columbia, South Carolina, 38 pp.
Feyling-Hanssen, R.W., 1977: The Clyde Foreland Formation: a
   micropaleontological study of Quaternary stratigraphy. Maritime
   Sediments, Special Publication 1, 315-377.
Feyling-Hanssen, R.W., 1980: Microbiostratigraphy of young Cenozoic
   marine deposits of the Qivituq Peninsula, Baffin Island. Marine
   Micropaleontology, 5:153-184.
Fillon, R.H., 1977: Ice-rafted detritus and paleotemperature: Late
   Cenozoic relationships in the Ross Sea region. Marine Geology,
   25:73-93.
Fillon, R.H. and Duplessy, J.C., 1980: Labrador Sea bio-, tephro-,
   oxygen isotopic stratigraphy and Late Quaternary paleoceanographic
   trends. Canadian Journal of Earth Science, 17:831-854.
Fillon, R.H., Miller, G.H. and Andrews, J.T., 1981: Terrigenous sandin
   Labrador Sea hemipelagic sediments and paleoglacial events on Baffin
   Island over the last 100,000 years. Boreas, 10:107-124.
Folger, D.W., 1970: Wind transport of land-derived mineral, biogenic
   and industrial matter over the North Atlantic. Deep-Sea Research,
   17:337-352.

Keany, J. Ledbetter, M., Watkins, N.D. and Huang, T.C., 1976.
   Diachronous deposition of ice-rafted debris in sub-Antarctic deep-sea
   sediments. Geological Society Society of America Bulletin,
   87:873-882.
Kent, D., Opdyke, N.D. and Ewing, M., 1971: Climate change in the North
   Pacific using ice-rafted detritus as a climatic indicator. Geological
   Society of America Bulletin, 82:2741-2745.
Kipp, N.G., 1976: New transfer function for estimating post sea-surface
   conditions from sea-bed distribution of planktonic foraminiferal
   assemblages in the North Atlantic. In: Cline, R.M. and Hayes, J.D.,
   (eds.), Investigation of Late Quaternary Paleoceanography and
   Paleoclimatology, Geological Society of America, Memoir 145:3-41.

Lazier, J.R., 1973: The renewal of Labrador Sea water. Deep Sea
   Research, 20:341-353.
Ledbetter, M. and Watkins, N.D., 1978: Separation of primary
   ice-rafted ebris from lag delposits, utilizing manganese micronodule
   accumulation rates in abyssal sediments of the Southern Ocean.
   Geological Society of America Bulletin, 89:1619-1629.

McIntyre, A., Kipp, N.G., Be, A.W.H., Crowley, T., Kellog, T.,
   Gardner,J.V., Prell, W. and Ruddiman, W.F., 1976: Glacial North
   Atlantic 18,000 years ago: a CLIMAP reconstruction. In: Cline, R.M.
   and Hays, J.D. (eds.), Investigation of Late Quaternary
   Paleoceanography and Paleoclimatology, Geological Society of America,
   Memoir 145:43-46.
Miller, G.H., Andrews, J.T. and Short, S.K., 1977: The last
   interglacial-glacial cycle. Clyde Foreland, Baffin Island   N.W.T.;
   stratigraphy, biostratigraphy and chronology. Canadian Journal of
   Earth Sciences, 14:2842-2857.

Mills, H.H., 1977: Textral characteristics of drift from some representative Cordilleran glaciers. Geological Society of America Bulletin, 88:1135-1143.

Morley, J.J. and Hays, J.D., 1979: Cycladophora davisiana: a stratigraphic tool for Pleistocene North Atlantic and interhemispheric. Earth and Planetary Science Letters, 44:383-389.

Murray, J.E., 1968: The drift, deterioration and distribution of icebergs in the North Atlantic Ocean. In Ice Seminar: A conference sponsored by the Petroleum Society of the Canadian Institute of Mining and Metalurgy (CIM). Calgary, Alberta, Canada. CIM Special Volume 10, 3-18.

Nelson, A.R., 1981: Quaternary glacial and marine stratigraphy of the Qivitu Peninsula, northern Cumberland Peninsula, Baffin Island, Canada. Geological Society of America Bulletin, Part II, 92:1143-1261.

Nelson, A.R., 1982: Aminostratigraphy of Quaternary marine and glaciomarine sediments, Qivitu Peninsula, Baffin Island. Canadian Journal of Earth Sciences, 19:945-961.

Robertson, J.H., 1975: Glacial to interglacial oceanographic changes in the northwest Pacific, including a continuous record of the last 400,000 years. Ph.D. thesis, Columbia University, New York, N.Y., 355 pp.

Rodrigues, C.G., Hooper, K. and Jones, P.C., 1980: The apertural structures of Islandiella and Cassidulina. Journal of Foraminiferal Research, 10:48-60.

Ruddiman, W.F., 1977: Late Quaternary deposition of ice-rafted sand in the subpolar North Atlantic (latitude 40° to 65°N). Geological Society of America Bulletin, 88:1813-1827.

Ruddiman, W.F. and Glover, L.K., 1972: Vertical mixing of ice-rafted volcanic ash in North Atlantic sediments. Geological Society of America Bulletin, 83:2817-2836.

Ruddiman, W.F. and Glover, L.K., 1975: Subpolar North Atlantic circulation at 9300 yr B.P.: faunal evidence. Quaternary Research, 5:361-389.

Ruddiman, W.F. and McIntyre, A., 1979: Warmth of the subpolar North Atlantic Ocean during Northern Hemisphere ice-sheet growth. Science, 204:173-175.

Ruddiman, W.F., McIntyre, A., Niebler-Hunt, V. and Durazzi, J.T., 1980: Oceanic evidence for the mechanism of rapid northern hemisphere glaciation. Quaternary Research, 13:33-64.

Ruddiman, W.F. and McIntyre, A., 1981a: Oceanic mechanisms for amplification of the 23,000-year ice-volume cycle. Science, 212:617-627.

Ruddiman, W.F. and McIntyre, A., 1981b: The North Atlantic Ocean during the last deglaciation. Palaeogeography, Palaeoclimatology, Palaeoecology, 35:145-214.

Schafer, C.T. and Cole, F.E., 1982: Living benthic foraminifera distributions on the continental slope and rise east of Newfoundland, Canada. Geological Society of America Bulletin, 93:207-217.

Sejrup, H.P. and Guibault, J.P., 1980: Cassidulina reniforme and C. obtusa (Foraminifera), taxonomy, distribution and ecology. Sarsia, 65:79-85.

Shackleton, N.J. and Opdyke, N.D., 1973: Oxygen isotope and paleomagnetic stratigraphy of equatorial Pacific core V28-238: oxygen isotope temperatures and ice volumes on a 10^5 year and 10^6 year scale. Quaternary Research, 3:39-55.

Tolderlund, D.S. and Be, A.W.H., 1971: Seasonal distribution of
    planktonic foraminifera in the western North Atlantic.
    Micropaleontology, 17:297-329.
U.S. Naval Oceanographic Office, 1968: Oceanographic Atlas of the North
    Atlantic Ocean, Section 3, Ice. Washington, D.C., publication number
    700.

Vilks, G., 1974: The distribution of planktonic foraminifera in the
    sediments and water of the Northwest Passage and northern Baffin Bay:
    a tool for paleoceanographic synthesis. In: Offshore Geology of
    Eastern Canada, Geological Survey of Canada, Paper 74-30, pp.
    109-121.
Vilks, G., 1980: Postglacial basin sedimentation on Labrador Shelf.
    Geological Survey of Canada Paper 78-28, pp. 1-28.

Watkins, N.D., Keany, J., Ledbetter, M.T. and Huang, T.C., 1974:
    Antarctic glacial history from analyses of ice-rafted deposits in
    marine sediments: new model and initial tests. Science, 186: 533-536.
Windom, H.L., 1969: Atmospheric dust records in permanent snowfields:
    implications to marine sedimentation. Geological Society of America
    Bulletin, 80:761-782.

9 Evidence for a subpolar influence in the Labrador Sea and Baffin Bay during marine isotopic stage 2

Richard H. Fillon and Ali E. Aksu

Marine isotopic stage 2 (Emiliani, 1955) records a global enrichment of 0-18 with respect to 0-16 in seawater. Most of this enrichment has been attributed to the glacial-period accumulation of 0-16-enriched high latitude precipitation in Northern Hemisphere ice sheets (Duplessy, 1978). The lower limit of isotopic stage 2 in deep-sea cores has been dated from about 32,000 BP (Shackleton and Opdyke, 1973) to about 23,000 BP (Pastouret et al., 1978). The upper limit of stage 2, the isotopic stage 2/1 transition, exhibits a range of ages from about 11,000 BP to about 13,000 BP (Broecker, 1965; Pastouret et al., 1978). In North America, Europe and Asia the period of time represented by stage 2 was marked by the growth of large continental ice sheets in middle latitudes (Prest, 1970; Andersen, 1981) whereas the amount of ice at higher latitudes is a topic of much debate (Miller and Dyke, 1974; Boulton, 1979; Denton and Hughes, 1981; Grosswald, 1980).

The greatest volume of continental ice as indicated by the lowest sea level and the maximum 0-18 enrichment recorded in deep-sea cores occurred approximately simultaneously in late isotopic stage 2 between about 15,000 BP and 18,000 BP (Bloom, 1971; Dreimanis and Karrow, 1972; Pastouret et al., 1978). An 18,000 BP "glacial maximum" datum has been used in deep-sea cores to reconstruct a synoptic picture of ocean temperatures and circulation patterns (McIntyre et al., 1976; CLIMAP, 1976). These reconstructions clearly depict paleoceanographic conditions but there is a question of how representative the 18,000 BP glacial maximum interval was when compared to the rest of isotopic stage 2, especially the period from 23,000 BP or 32,000 BP to 18,000 BP which lead up to the glacial maximum.

There is evidence of two earlier 10,000± year long periods
of marked glacial growth corresponding to interglacial to glacial
changes in O-18 at the isotopic substage 5e/5d and 5a/4 transitions
(centered approximately at 115,000 BP and 75,000 BP respectively).
These transitions were accompanied by western North Atlantic
sea-surface temperatures which, in the region from 40°N to 60°N, were
as warm as today's (Ruddiman and McIntyre, 1979; Ruddiman et al.,
1980). Warm (subpolar) surface waters also penetrated into the
Labrador Sea and Baffin Bay during these periods of ice growth (Fillon
and Duplessy, 1980; this volume; Aksu, this volume; Andrews et al.,
1981).

Warm surface waters were thus clearly present in close proximity
to the centers of glaciation in North America and Europe concurrent
with rapidly increasing ice volume. A persistent relationship is
evident between warmth in the North Atlantic and glacial growth back at
least to 250,000 BP (Ruddiman and McIntyre, 1981a). Evidence of warm
water in the North Atlantic during periods of glacial growth supports
the basic conditions of ice sheet growth postulated by Andrews et al.
(1972), Barry et al. (1975), Lamb and Woodroffe (1970), Adam (1975),
and Johnson and McClure (1976) which are all compatible with strong
meridional atmospheric circulation controlled and directed by large
land-sea thermal gradients.

In this paper, we will address the question of whether the
isotopic stage 2 glacial maximum was also closely preceded by an
interval of relative warmth in the high latitude northwestern North
Atlantic (Labrador Sea and Baffin Bay).

SUBPOLAR PLANKTONIC FORAMINIFERA

The taxonomy and paleoceanographic significance of subpolar planktonic
foraminifera (>63 um) in the Labrador Sea and Baffin Bay have been
discussed elsewhere (Fillon and Duplessy, 1980; this volume; Aksu,
1981; this volume). Basically, high percentages of living subpolar
planktonic foraminifera occur in the relatively warm (5° to 10° C in
summer) West Greenland Current and central Labrador Sea watermasses (Bé
and Tolderlund, 1971); whereas the cold (<5° C in summer), polar waters
of the Labrador and Baffin land currents typically contain very few
subpolar planktonic foraminifera (Vilks, 1980). It should be noted
here that the spines and protoplasm of living planktonic foraminifera
ensure the representation of even small shelled species such as
Globigerinita bradyi (uvula) and Globigerina quinqueloba in plankton
tow samples. After reproduction, however, planktonic foraminifera lose

their spines and, once on the seafloor, carbonate dissolution often modifies the foraminiferal assemblage by selectively destroying the delicate smaller subpolar planktonic tests and concentrating the more resistant and generally larger tests of the polar species Neogloboquadrina pachyderma (Berger, 1970). When carried to the extreme, differential dissolution can remove virtually all subpolar tests thereby removing all evidence of the presence of subpolar surface waters. It is therefore unwise to examine subpolar planktonic foraminiferal populations only in the greater than 150 μm size fractions as is commonly done for temperate and tropical assemblages (e.g. Kipp, 1976) because this further emphasizes the polar aspect of subpolar assemblages.

During the isotopic substage 5a/4 interglacial-to-glacial transition, differential dissolution, indicated by high benthic/planktonic ratios, was active in the northeastern Labrador Sea creating a dramatic decrease in subpolar planktonic foraminifera (e.g. 5% in core HU75-42) compared to southeastern Labrador Sea/northwest North Atlantic cores such as HU75-31 which contain nearly 80% subpolar planktonic foraminifera (Fillon and Duplessy, 1980). To the north of the Davis Strait sill in southern Baffin Bay, differential dissolution was evidently not as vigorous as in the northestern Labrador Sea. In southern Baffin Bay core HU77-17, the isotopic substage 5a/4 transition is marked by subpolar planktonic foraminiferal percentages as high as 20%(Aksu, 1981).

In general, there was less dissolution in the Labrador Sea during isotopic stage 2 than during the isotopic sub-stage 5a/4 transition and a stage 2 subpolar planktonic foraminiferal maximum is a consistent feature of cores from the eastern Labrador Sea (Fillon and Duplessy, 1980). Fillon and Duplessy interpreted this as evidence for the northward advection of warm, subpolar surface water (5° to 10°C in summer) into the eastern Labrador Sea during isotopic stage 2. We will now re-examine the broader paleoceanographic implications of that conclusion in the light of additional data from the northwestern Labrador Sea and Baffin Bay.

STRATIGRAPHIC CONTROL

Planktonic foraminiferal O-18 records across isotopic stage 2 are available for three Labrador Sea cores and two Baffin Bay cores (Fillon and Duplessy, 1980; Aksu, 1981; this volume). Core numbers and locations are given in Figure 9.1, sampling densities in Figure 9.2. In addition to the planktonic (Neogloboquadrina pachyderma) O-18

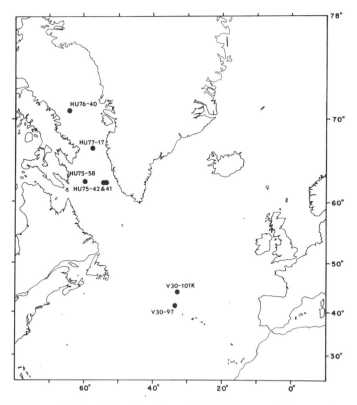

Figure 9.1 Map of the Labrador Sea, Baffin Bay and the northwest
Atlantic Ocean showing core locations. The modern surface circulation
pattern is depicted in Chapter 8 (Fig. 8.1).

records, 0-18 records based on benthic foraminifera (Melonis
pompilioides and Pullenia spp.) are also available for Labrador Sea
cores HU75-41 and HU75-42. The detailed benthic record of HU75-41
shows a regular stage 2/1 transition in contrast to the planktonic
record (Fig. 8.5) which contains an isotopically light peak similar to
those seen in many high-resolution planktonic records from other parts
of the world (Pastouret et al., 1978). An isotopically light
planktonic 0-18 peak on the stage 2/1 transition is seen in all
Labrador Sea/Baffin Bay cores except HU75-42 which is missing a part of
the core-top (see Fig. 8.9).

The lower limit of isotopic stage 2 in cores HU75-41 and HU75-42,
the stage 3/2 transition, is picked on the upper limb of the first
large negative planktonic 0-18 peak below stage 1. The stage 3/2
boundary can be recognized in this way in the planktonic records of all
of the Labrador Sea/Baffin Bay cores used in this study (see Fig.

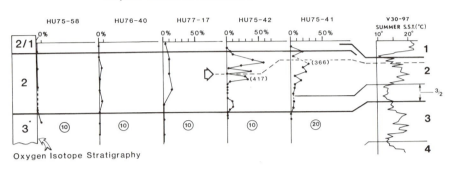

Figure 9.2 Comparison of isotopic stage 2 subpolar planktonic
foraminiferal percentages (>63 μm) in five Baffin Bay/Labrador Sea
cores (Fig. 9.1) with summer sea-surface temperatures (S.S.T.) for the
central subpolar North Atlantic recorded in core V30-97 (re-drawn from
Ruddiman and McIntyre, 1981a). The cores are plotted from right to
left in order of decreasing subpolar percentages, i.e. the inferred
direction of surface water flow (see text). Subpolar percentages
greater than 5 to 10% imply the presence of subpolar surface-water (5°
to 10°C in summer). The oxygen isotope stratigraphy for the Baffin
Bay/ Labrador Sea cores is based on planktonic 0-18 records while that
for V30-97 (Ruddiman and McIntyre, 1981a) is based on a record of
benthic 0-18. The two are related in Figure 9.3 via a detailed
comparison of the benthic 0-18 record for core V30-97 with the benthic
record for core HU75-41 (Fillon and Duplessy, 1980). The benthic stage
3/1 boundary picked in core V30-97 (light solid line) is projected into
core HU75-41 while the planktonic 3/2 boundary picked in HU75-41 (heavy
solid line) is projected into core V301-97 as indicated in Figure 9.3.
The large arrow and dashed line marks the approximate location of the
c. 18,000 BP level of maximum global ice volume in three of the cores.
The total number of foraminifera counted (in brackets) and total
planktonic forminiferal abundance (see Fig. 8.5) are highest just below
the ice volume maximum. The numbers in circles indicate the sample
interval (in cm) for each core.

8.9). The planktonic stage 3/2 transition corresponds to a point in
the benthic records at which 0-18 values are lighter than at any other
level within the record of stage 2 to that point (Fillon and Duplessy,
1980; Fig. 8.5). Although benthic 0-18 values below the planktonic
isotopic stage 3/2 transition (Fig. 9.3) are intermittently as heavy as
in stage 2, the down-core trend in stage 3 is clearly to lighter
benthic values also.

In Figure 9.2, Labrador Sea and Baffin Bay subpolar planktonic
foraminiferal data are aligned using planktonic oxygen isotope
stratigraphy. Isotopic stage boundaries have been selected in all

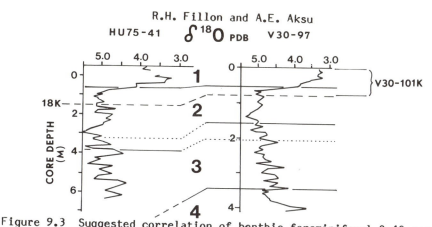

Figure 9.3 Suggested correlation of benthic foraminiferal 0-18 records
from the northeastern Labrador Sea (core HU75-41, 2381 m water
depth)and the central subpolar North Atlantic (a composite of cores
V30-101K, 3519 m water depth, an V30-97, 3371 m water depth). The
former is from Fillon and Duplessy (1980), the latter from Ruddiman and
McIntyre (1981a). The c. 18,000 BP (18K) level of maximum global ice
volume is indicated (see also Fig. 9.2). The isotopic stage 3/2
boundary in HU75-41 (solid line) was picked primarily in reference to
the planktonic 0-18 record (not shown); the 3/2 boundary in V30-97
however is based solely on the benthic record (solid line). The dotted
lines show the approximate differences in positions of the 3/2
boundaries in the two cores (see text for a fuller discussion). The
portion of the HU75-41 0-18 record above 0 cm is from the
trigger-weight core.

cores based on the planktonic records because benthic 0-18 records are
only available for two of the cores. To stress that we are dealing
with a planktonic 0-18 stratigraphy, we show the width of the
isotopically light 0-18 peak which marks the planktonic stage 2/1
transition (the interval labeled 2/1 in Fig. 9.2). In HU75-41, the
benthic 0-18 stage 2/1 boundary lies in the upper one-third of the
planktonic 2/1 transition zone.

To verify that the 0-18 stratigraphic boundaries used in Figure
9.2 can be related to those in the North Atlantic, we have drawn a
comparison of the benthic 0-18 record of Labrador Sea core HU75-41 with
the combined benthic 0-18 records of cores V30-97 and V30-101K
(Ruddiman and McIntyre, 1981a) from the central subpolar North Atlantic
(Fig. 9.1). The two curves (Fig. 9.3) resemble each other closely in
form and magnitude. Both records exhibit about a 1.9 °/₀₀ isotopic
shift across the stage 2/1 boundary. The glacial maximum 18,000 BP
level, picked by C-14 dating in V30-97/V30-101K (Fig. 9.3) occupies a
position similar to that estimated from an assessment of the planktonic
and benthic 0-18 curves in core HU75-41 (Fillon and Duplessy, 1980).
The stage 3/2 boundary in V30-97, however, is not at an equivalent

position to that in HU75-41. Application of the criteria we used in
identifying the stage 3/2 boundary in the benthic records of HU75-41
requires a downward shift of the boundary in V30-97 (Fig. 9.3). By the
same token, if the stage 3/2 boundary is placed just below the heaviest
benthic 0-18 values in HU75-41, as in the case in V30-97, an upward
adjustment of the boundary in HU75-41 is required (Fig. 9.3). In spite
of this uncertainty over the precise placement of the stage 3/2
boundary in the benthic 0-18 records of the two cores it is, however,
clear that they contain similar histories of short-period isotopic
variation within stage 2. It is probable, therefore, that the
"planktonic" stage 3/2 boundary in HU75-41 (at c. 4.0 meters) and in
other Labrador Sea/Baffin Bay cores (Fig. 9.2) is chronologically
equivalent with the 2.0 meter level in V30-97 (Fig. 9.3).

     The c. 18,000 BP glacial maximum levels in cores HU75-41 and
HU75-42 (Fillon and Duplessy, 1980) differ in position relative to the
stage 2/1 and stage 3/2 boundaries (Fig. 9.2), thus there is no reason
to expect that stage 2 accumulation rates have been constant within
every core. For that reason, we have not attempted to superimpose a
detailed chronology on events recorded within stage 2 nor even to
interpolate the 18,000 BP level in cores HU75-58, and HU77-17 (Fig.
9.2).

ISOTOPIC STAGE 2 PALEOCEANOGRAPHY

Records of the abundance of subpolar planktonic foraminifera (as
percent of total planktonics) within isotopic stage 2 are shown for the
three Labrador Sea cores and two Baffin Bay cores in Figure 9.2. Also
shown in this figure is the record of estimated summer sea-surface
temperatures for central subpolar North Atlantic core V30-97 (Ruddiman
and McIntyre, 1981a). It is apparent from Figure 9.2 that subpolar
foraminifera were most abundant in the eastern Labrador Sea when the
central subpolar North Atlantic was at its warmest. The subpolar North
Atlantic was coolest during the isotopic stage 2/1 transition while, by
contrast, temperatures were 2° to 3°C warmer at 18,000 BP and as much
as 10°C warmer during mid-isotopic stage 2 as compared to the minimum
stage 2/1 transition temperatures. The portion of V30-97 which we
consider to be correlative with early stage 2 in Labrador Sea core
HU75-41 is characterized by cool temperatures nearly equal to those
which marked the stage 2/1 transition (Fig. 9.2).

     In general the patterns of subpolar planktonic foraminiferal
abundance within stage 2 in cores HU75-41, HU75-42 and HU77-17 are
similar. Each record is characterized by low percentages of subpolar

planktonic foraminifera in early stage 2 and higher percentages in late
stage 2. The high resolution records obtained from closely adjacent
cores HU75-41 and HU74-42 show an additional similarity in that both
record the largest total number of planktonic tests counted for a
sample (a measure of total abundance, see Fig. 8.5) just below the
glacial maximum. This level of maximum abundance corresponds to
near-peak subpolar percentages of 30% in each case. Both cores also
show an increase in subpolar planktonic percentages just above the
glacial maximum reaching 20% in HU75-41 at the base of the isotopic
stage 2/1 transition zone and 58% in HU75-42 about 30 cm below the
stage 2/1 transition zone. Specific differences in the late stage 2
portions of the curves are probably mostly attributable to the
different number of samples counted in that interval in the two cores.
More detailed studies of the stage 2 record would benefit from an
increased density of faunal counts in the 60 cm to 200 cm low
accumulation rate section of HU75-41. The present sample density is
sufficient, however, to meeet our principal objective of assessing
general trends in isotopic stage 2 paleoceanography.

Isotopic stage 2 subpolar planktonic foraminiferal percentages are
low in cores HU76-40 and HU75-58 (Fig. 9.2) while benthic/planktonic
ratios are also low (cf. Fig. 8.3). This implies only moderate
carbonate dissolution and perhaps a real decrease in subpolar influence
toward those areas in northern Baffin Bay and the western Labrador Sea
that are presently dominated by cold polar waters (Dunbar, 1951). The
distribution of subpolar planktonic foraminifera therefore suggests the
presence of a counterclockwise circulation pattern much like the modern
one, in which warm, subpolar surface water (5° to 10°C in summer) was
advected northward from the North Atlantic through the eastern Labrador
Sea and into southern Baffin Bay, while cold, polar surface water (<5°C
in summer) flowed southward out of northern Baffin Bay and into the
western Labrador Sea (cf. Fig. 8.8a).

The presence of a well-defined subpolar planktonic foraminiferal
peak in core HU75-42 which postdates the glacial maximum implies the
presence of a residual subpolar surface-water source in the North
Atlantic Ocean after 18,000 BP. Because the paleotemperature record of
V30-97 (Fig. 9.2) indicates that central subpolar North Atlantic summer
suface-water temperatures were still as warm as 15°C immediately after
the glacial maximum, i.e. until c. 15,000 BP according to Ruddiman and
McIntyre's (1981a) C-14 date chronology, the central subpolar North
Atlantic may therefore have been warm enough up until that time to
serve as a source of subpolar Labrador Sea water. Ruddiman and
McIntyre (1981a) discovered a systematic phase relationship between the
sea-surface temperature record in core V30-97 and Berger's (1978)

orbital precision index. This provides an independent means of dating
the stage 2 warm interval in the North Atlantic. Examination of the
precessional index curve places the initiation of the isotope stage 2
warm interval at about 26,000 BP ± 1500 calendar years and its
termination at about 16,000 BP ± 1500 calendar years, which is
compatible with the C-14 dates. Correspondence within the 0-18
stratigraphic framework of: (1) the initiation of a warm North
Atlantic with the initial incrase of subpolar planktonic foraminifera
in the Labrador Sea (Fig. 9.2) and (2) the end of the North Atlantic
warm interval with the final decrease in subpolar planktonic
foraminifera in the Labrador Sea (Fig. 9.2) strongly implies that the
inferred warming in the northeastern Labrador Sea and southern Baffin
Bay also encompasses the period from c. 26,000 BP to c. 16,000 BP.

Long 0-18 records from ice cores drilled in ice caps around Baffin
Bay can be compared with the inferred stage 2 oceanic warming in the
eastern Labrador Sea and southern Baffin Bay. Both the Camp Century,
Greenland record (Dansgaard et al., 1971) and the Devon Island, Canada
record (Paterson et al., 1977) span isotopic stage 2. Devon Island ice
core "72" contains a long Holocene section in which 0-18 values average
approximately -26°/₀₀ (SMOW). Below this, 0-18 values rapidly
become lighter, reaching c. -35°/₀₀ but then recover in the
interval from 4.0 m to 3.5 m above bedrock to around -29°/₀₀, only
3°/₀₀ lighter than the Holocene, before again decreasing
down-core. Paterson and colleague's correlation of the Devon Island
record, with the 0-18 record and chronology determined for the Camp
Century ice core (Dansgaard et al., 1971) places the pre-Holocene (4.0
m to 3.5 m) section of maximum 0-18 enriched ice, in the period from
c. 23,000 BP to 18,000 BP. This 0-18 enriched interval may well be
related to warmer air temperatures over Baffin Bay and over the thin
Devon Island Ice Cap. A corresponding 0-18 enrichment in the Camp
Century record is well defined but remains 9°/₀₀ lighter than the
Holocene, apparently due to the accumulation of the pre-Holocene ice
recovered at Camp Century on a higher portion of the Greenland Ice
Sheet (see discussion in Paterson et al., 1977).

The minimum summer surface temperature of subpolar water in
today's ocean is about 5°C - the modern limit of the subarctic zone of
abundant subpolar planktonics (Bé and Tolderlund, 1971). This implies
that from c. 26,000 BP to c. 16,000 BP, the 5°C sea-surface summer
isotherm extended into southern Baffin Bay. The virtual absence of
subpolar planktonic foraminifera within isotopic stage 2 in northern
Baffin Bay core HU76-40 and northwestern Labrador Sea core HU75-58
where dissolution appears to have been only moderate places northern

and western limits on the 5°C summer surface water isotherm (see Fig. 8.8a).

Subpolar planktonic foraminiferal percentages in cores HU75-41 and HU75-42 fluctuate markedly within late isotopic stage 2 reaching near zero values close to the glacial maximum level as well as within or close to the isotopic stage 2/1 transition (Fig. 9.2). The fluctuations are sufficiently rapid that an accurate estimate of faunal composition at a level which is exactly the global glacial maximum is impossible without more precise isotopic and faunal curves and a study of the effects of bioturbational mixing. However, it is clear that at least during portions of stage 2 within ± 20 cm (less than about ± 2000 years) of the glacial maximum there was an important subpolar planktonic faunal component in the Labrador Sea. It can also be inferred from Figure 9.2 that any episodic replacement of subpolar surface water by polar surface water in the eastern Labrador Sea during late isotopic stage 2 (including the glacial maximum) but prior to the isotopic stage 2/1 transition peak in the planktonic record was relatively short-lived.

DISCUSSION AND CONCLUSIONS

The distribution of subpolar planktonic foraminifera within isotopic stage 2 in the eastern Labrador Sea and southern Baffin Bay strongly implies that significant advection of warm surface water into the region preceded the last global glacial maximum. This warm period in the eastern Labrador Sea and southern Baffin Bay probably coincided with an episode of relative warmth in the central subpolar North Atlantic lasting from c. 26,000 BP to c. 16,000 BP which has been documented by Ruddiman and McIntyre (1981a). Isotopic stage 2 is therefore similar to glacial isotopic stages 4 and 5d, in which the global glacial maxima were also preceded by significant warmth in the high latitude northwestern Atlantic/Labrador Sea/Baffin Bay region (Ruddiman and McIntyre, 1979; Ruddiman et al., 1980; Fillon and Duplessy, 1980; Aksu, 1981). The warm northern waters in each case would have enhanced the delivery of moisture to the continental ice sheets as envisaged by Lamb and Woodroffe (1970), Andrews et al. (1972), Barry et al. (1975), Johnson and McClure (1975), and Ruddiman et al. (1980).

Labrador Sea temperatures as indicated by subpolar planktonic foraminifera were also relatively warm (at least episodically) for some time after the c. 18,000 BP isotopic stage 2 glacial maximum, i.e. until c. 16,000 BP which marks the approximate beginning of the light

0-18 peak seen in planktonic records of the stage 2/1 transition (Fig. 9.2). A similar history of isotopic stage 2 warmth just after the glacial maximum is indicated for the central subpolar North Atlantic where relatively warm sea-surface temperatures are recorded as late 16,000 BP to 15,000 BP (Ruddiman and McIntyre, 1981a). Ruddiman and McIntyre (1981b) applied an iterative mixing-model to the faunal data in core V30-97 (and other North Atlantic cores) which suggests that the abundance and composition of the planktonic foraminiferal assemblage after 16,000 BP can be consistently explained by a period of very low planktonic productivity lasting until approximately 13,000 BP. This they interpret to have been the direct result of a maximum deglacial influx of meltwater and icebergs into the North Atlantic. It is quite possible that the light 0-18 peak in the planktonic record of core HU75-41 (Fig. 8.5, which is inferred to span the interval c. 16,000 BP to c. 13,000 BP) was created by a decrease in sea-surface salinity associated with the meltproduct discharge maximum. About 1.0°/oo of the observed excess decrease in planktonic foraminiferal 0-18 could be accounted for by a salinity reduction of 1.0°/oo (assuming an approximate -35°/oo 0-18 composition for the meltwater). The reconstruction of polar surface circulation in the northwestern North Atlantic, orginally attributed to the period c. 13,000 BP to 9,800 BP (Fillon et al., 1981) is therefore updated to encompass the somewhat longer period from 16,000 BP to 9800 BP (Fig. 8.8b).

While warm water was entering the eastern Labrador Sea and southern Baffin Bay during isotopic stage 2, it is apparent that the Foxe Basin/Baffin Island dome (Ives and Andrews, 1963; Dyke et al., 1982) did not reach its maximum extent until sometime after 11,000 BP. In fact, ice extent on Baffin Island appears to have been relatively constant between 18,000 BP and 8,000 BP (Quinlan, this volume). Accelerated ice growth on Baffin Island in response to moisture derived from warm Baffin Bay water may only have occurred in the period after 9,800 BP, when the first "warm" (subarctic) water molluscs appear in raised marine sediments on eastern Baffin Island (Miller, 1980), and before about 5,000 BP, when the initiation of the modern cold Canadian Current (Baffinland Current) is recorded in cores from Frobisher Bay and the southeast Baffin Island shelf (Osterman, 1982). Warm water restricted to the eastern Labrador Sea/Baffin Bay region is evidently insufficient to greatly stimulate glacier growth on Baffin Island.

One critical difference between oceanographic conditions during stage 2 and during other Wisconsin (and pre-Wisconsin?) glacial stages which saw vigorous glacial expansion on Baffin Island, may be that during the earlier glacials warm subpolar water was able to penetrate into the western portions of the Labrador Sea and Baffin Bay. This is

indicated by the common occurrence of subarctic molluscan faunas in
close proximity to older deposits on Baffin Island (Andrews et al.,
1981; Miller et al., 1977) which suggests the loss or reduction of the
Canadian current. Cores HU76-40 and HU75-58 indicate that a cold
current was present off Baffin Island from c. 30,000 BP to c. 10,000 BP
and this may have reduced precipitation over eastern Baffin Island as
it does today (see Chapter 2, this volume, Climatology).

ACKNOWLEDGEMENTS

This work was funded partly by U.S. National Science Foundation grant
AFM-82000242. We are indebted to the officers and crew of CSS Hudson
for successful coring cruises in 1975, 1976 and 1977.

REFERENCES

Adam, D.P., 1975: Ice ages and the thermal equilibrium of the Earth
    II. Quaternary Research, 5:161-171.
Aksu, A.E., 1981: Late Quaternary stratigraphy paleoenvironmentology
    and sedimentation history of Baffin Bay and Davis Strait. Ph.D.
    dissertation, Dalhousie University, Halifax, Nova Scotia, Canada,
    pp. 1-585.
Andersen, B.G., 1981: Late Weichselian ice sheets in Eurasia and
    Greenland. In: Denton, G.H. and Hughes T.(eds.), The Last Great Ice
    Sheets, John Wiley and Sons, New York, pp. 3-65.
Andrews, J.T., Barry, R.G., Bradley, R.S., Miller, G.H. and Williams,
    L.D., 1972: Past and present glaciological responses to climate in
    eastern Baffin Island. Quaternary Research, 2:303-314.
Andrews, J.T., Miller, G.H., Nelson, A.R., Mode, W.N. and Locke, W.W.,
    III, 1981: Quaternary near-shore environments on eastern Baffin
    Island, N.W.T. In: Mahanney, W.C. (ed.), Quaternary Paleoclimates.
    Geo Books, University of East Anglia, Norwich, U.K., pp. 13-44.

Barry, R.G., Andrews, J.T. and Mahaffy, M.A., 1975: Continental
    icesheets: Conditions for growth. Science, 190:979-981.
Bé, A.W.H. and Tolderlund, D.S., 1971: Distribution and ecology of
    living planktonic foraminifera in surface waters of the Atlantic and
    Indian Oceans. In: Funnell, B.M., and Riede, W.R. (eds.),
    Micropaleontology of the Oceans. Cambridge University Press, London,
    England, pp. 105-148.
Berger, A.L., 1978: Long-term variations of caloric insolation
    resulting from the earth's orbital elements. Quaternary Research,
    9:139-167.
Berger, W.H., 1970: Biogenous deep-sea sediments: fractionation by
    deep-sea circulation. Geological Society of America Bulletin,
    81:1385-1402.
Bloom, A.L., 1971: Glacial-eustatic and isostatic controls at sea level
    since the last glaciation. In: K.K. Turedian, (ed.), The Late
    Cenozoic Glacial Ages. Yale University Press, New Haven, Connecticut,
    355-379.
Boulton, G.S., 1979: A model Weichselian glacier variation in the North
    Atlantic region. Boreas, 8:373-395.

Broecker, W.S., 1965: Isotope geochemistry and the Pleistocene climatic record. In: Wright, H.E., Jr. and Frey, D.G., eds., The Quaternary of the United States, Princeton University Press, Princeton, New Jersey, 737-753.

CLIMAP Project Members, 1976: The surface of the ice-age earth. Science, 191:1131-1137.

Dansgaard, W., Johnsen, S.J., Clausen, H.B. and Langway, C.C., Jr., 1971: Climatic record revealed by the Camp Century ice core. In: Turekian, K.K. (ed.), The Late Cenozoic Glacial Ages, Yale University New Haven, Connecticut, 37-56.

Denton, G.H. and Hughes, T.J., 1981: The Last Great Ice Sheets. John Wiley and Sons, New York. 484 pp.

Dreimanis, A. and Karrow, P.F., 1972: Glacial history of the Great Lakes-St. Lawrence region, the classification of the Wisconsin(an) Stage, and its correlatives. International Geological Congress, 24th, Montreal 1972, Section 12, 5-15.

Dunbar, M.J., 1951: Eastern Arctic waters. Fisheries Research Board of Canada, Bulletin, 88:1-131.

Duplessy, J.C., 1978: Isotope studies. In: Gribbin, J. (ed.), Climatic Change. Cambridge University Press, Cambridge, England, 46-67.

Dyke, A.S., Andrews, J.T., and Miller, G.H., 1982: Quaternary geology of Cumberland Peninsula, Baffin Island, District of Franklin. Geological Survey of Canada Memoir, 403. 32 pp.

Emiliani, C., 1955: Pleistocene Temperatures. Journal of Geology, 63:538-578.

Fall, W.F., 1980: Glacial meltwater in-flow into the Gulf of Mexico during the last 150,000 years, implications for isotope stratigraphy and sea level studies. M.S. thesis, University of South Carolina, Columbia, South Carolina. 38 pp.

Fillon, R.H. and Duplessy, J.C., 1980: Labrador Sea bio-, tephro-, oxygen isotopic stratigraphy and Late Quaternary paleoceanographic trends. Canadian Journal of Earth Science, 17:831-854.

Fillon, R.H., Miller, G.H. and Andrews, J.T., 1981: Terrigenous sand in Labrador Sea hemipelagic sediments and paleoglacial events on Baffin Island over the last 100,000 years. Boreas, 10:107-124.

Grosswald, M.G., 1980: Late Weichselian ice sheet of northern Eurasia. Quaternary Research, 13:1-32.

Ives, J.D. and Andrews, J.T., 1963: Studies in the physical geography of north central Baffin Island, N.W.T. Geographical Bulletin, 19:5-48.

Johnson, R.G. and McClure, B.T., 1976: A model for Northern Hemisphere continental ice sheet variation. Quaternary Research, 6:325-353.

Kipp, N.G., 1976. New transfer function for estimating post sea-surface conditions from sea-bed distribution of planktonic foraminiferal assemblages in the North Atlantic. In: Cline, R.M. and Hayes, J.D., (eds.), Investigation of Late Quaternary paleoceanography, paleoclimatology. Geological Society of America, Memoir 145, 3-41.

Lamb, H.H. and Woodroffe, A., 1970: Atmospheric circulation during the last Ice Age. Quaternary Research, 1:29-58.

McIntyre, A., Kipp, N.G., Bé, A.W.H., Crowley, T., Kellog, T., Gardner, J.V., Prell, W. and Ruddiman, W.F., 1976: Glacial North Atlantic 18,000 years ago: a CLIMAP reconstruction. In: Cline, R.M. and Hays,J.D. (eds.), Investigation of Late Quaternary Paleoceanography and Paleoclimatology. Geological Society of America, Memoir 145, 43-46.

Miller, G.H., 1980: Late Foxe glaciation of southern Baffin Island, N.W.T., Canada. Geological Society of America Bulletin, 91:399-405.

Miller, G.H. and Dyke, A.S., 1974: Proposed extent of Late Wisconsin Laurentide ice on eastern Baffin Island. Geology, 2: 125-130.

Miller, G.H., Andrews, J.T. and Short, S.K., 1977: The last interglacial-glacial cycle. Clyde Foreland, Baffin Island N.W.T.:stratigraphy, biostratigraphy and chronology. Canadian Journal of Earth Sciences, 14:2842-2857.

Mills, H.H., 1977: Textural characteristics of drift from some representative Cordilleran glaciers. Geological Society of America Bulletin, 88:1135-1143.

Morley, J.J. and Hays, J.D., 1979: Cycladophora davisiana: a stratigraphic tool for Pleistocene North Atlantic and interhemispheric. Earth and Planetary Science Letters, 44:383-389.

Murray, J.E., 1968: The drift, deterioration and distribution of icebergs in the North Atlantic Ocean. In Ice Seminar: A conference sponsored by the Petroleum Society of the Canadian Institute of Mining and Metalurgy (CIM). Calgary, Alberta, Canada. CIM Special Volume 10:3-18.

Nelson, A.R., 1981: Quaternary glacial and marine stratigraphy of the Qivitu Peninsula, northern Cumberland Peninsula, Baffin Island, Canada. Geological Society of America Bulletin, Part II, 92: 1143-1261.

Nelson, A.R., 1982: Aminostratigraphy of Quaternary marine and glaciomarine sediments, Qivitu Peninsula, Baffin Island. Canadian Journal of Earth Sciences, 19:945-961.

Osterman, L.E., 1982: Late Quaternary history of southern Baffin Island,Canada: a study of foraminifera and sediments from Frobisher Bay. Ph.D. dissertation, University of Colorado, Boulder, Colorado. 380 pp.

Pastouret, L., Chamley, H., Delibrias, G., Duplessy, J.C. and Thiede, J., 1978: Late Quaternary climatic changes in western tropical Africa deduced from deep-sea sedimentation off the Niger delta. Oceanologica Acta, 1:217-231.

Paterson, W.S.B., Koerner, R.M., Fisher, D., Johnsen, S.J., Clausen, H.B., Dansgaard, W., Bucher, P. and Oeschger, H., 1977: An oxygen isotope climatic record from the Devon Island ice cap, Arctic Canada. Nature, 166:508-511.

Prest, V.K., 1970: Quaternary geology. In: Douglas, R.J.W. (ed.), Geology and Economic Minerals of Canada, Geological Survey of Canada, Economic Geology Report No. 1, 675-764.

Ruddiman, W.F. and McIntyre, A., 1979: Warmth of the subpolar North Atlantic Ocean during Northern Hemisphere ice-sheet growth. Science, 204:173-175.

Ruddiman, W.F., McIntyre, A., Niebler-Hunt, V. and Durazzi, J.T., 1980: Oceanic evidence for the mechanism of rapid northern hemisphere glaciation. Quaternary Research, 13:33-64.

Ruddiman, W.F. and McIntyre, A., 1981a: Oceanic mechanisms for amplification of the 23,000-year ice-volume cycle. Science, 212: 617-627.

Ruddiman, W.F. and McIntyre, A., 1981b: The North Atlantic Ocean during the last deglaciation. Palaeogeography, Palaeoclimatology, Palaeoecology, 35:145-214.

Shackleton, N.J. and Opdyke, N.D., 1973: Oxygen isotope and paleomagnetic stratigraphy of equatorial Pacific core V28-238: oxygen isotope temperatures and ice volumes on a 10^5 year and 10^6 year scale. Quaternary Research, 3:39-55.

Vilks, G., 1980: Postglacial basin sedimentation on Labrador Shelf. Geological Survey of Canada Paper 78-28, 1-28.

10 Marine palynology of Baffin Bay

P. J. Mudie and S. K. Short

Marine palynology is the study of the plant microfossils (palynomorphs) found in marine sediments, including their transport, distribution and biostratigraphic significance (Heusser, 1978; Williams, 1978). The most common palynomorphs in the recent marine sediments of Baffin Bay are pollen, moss spores and dinoflagellate cysts (Mudie, 1980, 1981, 1982). The pollen and spores are of terrestrial origin and their presence in marine sediments provides a link between terrestrial and oceanic paleoenvironments. Dinoflagellate cysts are spores of yellow-brown algae (Division Pyrrophyta). The chemical composition of the cyst walls is similar to that of pollen; therefore, the cysts and pollen may be extracted from sediments using the same processing techniques. Some dinoflagellate species live in freshwater environments but most dinoflagellates are primary producers near the ocean surface, where they may be almost as abundant as diatoms (Bujak and Williams, 1980). Thus, dinoflagellates in marine sediments are proxy-indicators of paleoceanographic conditions at the sea surface. The acid-resistant nature of the cyst walls makes dinoflagellates important biostratigraphic markers for sedimentary environments where carbonate dissolution limits the preservation of better-known planktonic microfossils, e.g. foraminifera and coccoliths. The deep-water sediments of Baffin Bay are a good example of such a limiting environment (Aksu, 1983b).

Marine palynological studies of polar regions are still in their infancy: there are only about 10 publications on Quaternary palynomorphs in arctic marine sediments and few studies have been made in Baffin Bay (Mudie, 1980, 1981). Any attempt to infer climate from fossil biota in the eastern Canadian Arctic should be based on detailed knowledge of the present distributions and ecology of arctic faunas and floras (Andrews et al., 1981). This paper therefore provides the

following information: (1) a detailed account of the distribution and
ecology of pollen and dinoflagellate cysts in recent sediments of
Baffin Bay; (2) initial interpretations of the paleoenvironments
represented by palynomorphs in Late Quaternary sediment cores from
Baffin Bay; and (3) an outline of major areas requiring more data in
order to establish marine palynology as a reliable biostratigraphic
tool for Arctic research.

SAMPLE DISTRIBUTION AND ENVIRONMENT

    Figure 10.1 shows the distribution of palynology samples that will
be described in this study of Baffin Bay. The samples are from gravity
core tops or grabs taken on various cruises from 1970 - 1981 (Table
10.1); sediment volumes of 5 - 20 cm^3 were made available for study
through the co-operation of scientists at Bedford Institute of
Oceanography (B. MacLean, C.T. Schafer) and at Dalhousie University
(A.E. Aksu). With a few exceptions (samples 13, 14 and 24), the
sediments are muddy, with textures ranging from clayey silt to sandy
mud; the other samples are organic-rich fine sands. Muddy sands and
gravels from the West Greenland Shelf above 200 m were excluded because
they are almost barren of palynomorphs. Samples from the large troughs
on the northern West Greenland Shelf were also excluded because these
may be areas of massive debris flows (Aksu, pers. comm.). Sample
density over the remaining area averages about 1 site per 100 km^2.

    Although most of the Baffin Bay study region lies north of the
Arctic Circle, the climate and oceanography (Fig. 10.2) shows a strong
gradient from west to east. This gradient is related to the inflow of
the warm West Greenland Current from the North Atlantic, and to the
outflow of the arctic Baffin Current along the eastern margin of Baffin
Island. The climatic influence of the W. Greenland Current is clearly
seen in the northeasterly trend of the -9°C mean annual isotherm and in
the extent and duration of sea ice which covers most of the region in
winter (Fig. 10.3). The -9° isotherm also has special significance
with respect to marine pollen distributions in the Baffin Bay region:
the isotherm corresponds closely to the boundary between the southern
area which is dominated by >50% North Atlantic airmasses in July (Barry
and Chorley, 1971) and the northern area which is characterized by
about 50% N. Atlantic and 50% Arctic Ocean air masses (Arctic
Archipelago Air).

    Onshore, the mean annual air temperature on Baffin Island shows a
latitudinal gradient from -9°C at Frobisher Bay and -11°C at Broughton
Island, to -12° at Clyde Inlet and -15° at Pond Inlet (Petterson et
al., 1956). On the West Greenland Coast, the temperature gradient is

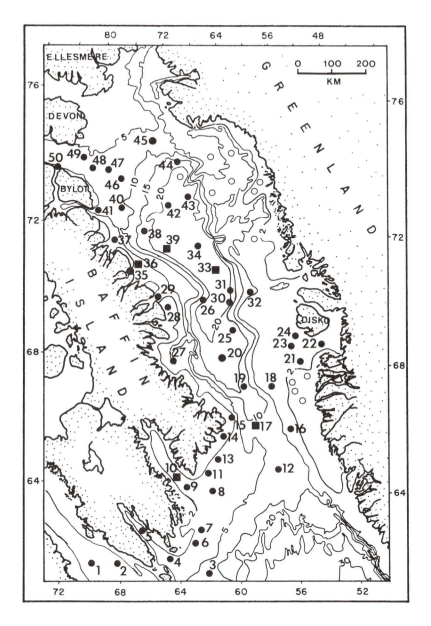

Figure 10.1 Location map showing the sample sites (dots and squares) and bathymetry of the Baffin Bay region. Isobath numbers indicate metres x 100. Squares mark sites of cores described in text. Circles denote samples with few or no palynomorphs.

Table **10-1.** Location and depth of sample sites in Fig.**10-1**

MAP #	SAMPLE	LATITUDE	LONGITUDE	DEPTH(m)
1	C78-08-85	61 30.20N	70 5.40W	240.0
2	C78-08-88	61 43.50	67 59.70	262.0
3	C78-08-97	61 59.90	62 55.30	435.0
4	HU77-021-156	61 51.50	64 12.30	487.0
5	HU77-021-159	62 50.50	67 2.40	570.0
6	81-045-44	62 11.90	63 0.70	365.0
7	C78-08-71	63 29.20	62 3.20	241.0
8	C78-08-63	63 59.50	61 52.40	222.0
9	C78-08-59	64 31.10	64 8.90	267.0
10	C78-08-60	64 45.50	64 50.30	460.0
11	81-055-88	64 45.10	60 46.10	365.0
12	70-028-12	65 00.00	58 00.00	500.0
13	C78-08-43	65 29.10	61 6.00	291.0
14	81-045-38	62 18.50	64 10.80	143.0
15	C78-08-34	67 4.20	61 17.30	175.0
16	C78-08-30	66 27.60	56 19.20	177.0
17	77-027-017	66 54.09	58 17.71	935.0
18	C78-08-24	67 56.30	59 19.80	475.0
19	77-027-15	67 48.67	60 31.92	1691.0
20	76-029-25	69 12.30	62 25.50	1910.0
21	70-028-62	68 15.50	55 58.00	333.0
22	70-028-27	68 53.00	53 44.20	750.0
23	70-028-48	68 25.00	56 30.00	250.0
24	70-028-35	69 19.00	54 53.20	181.0
25	76-029-26	69 59.30	61 38.00	1710.0
26	77-027-13	69 26.90	63 31.73	1889.0
27	HU78-029-37	68 15.50	65 12.90	457.0
28	HU78-029-36	70 8.80	66 48.70	99.0
29	81-045-31	70 52.20	68 6.20	197.0
30	76-029-43	70 23.50	62 30.00	2002.0
31	77-027-27	71 2.50	62 4.40	1933.0
32	76-029-44	70 35.40	61 4.90	920.0
33	76-029-034	70 42.40	64 58.70	2275.0
34	76-029-36	72 33.80	65 52.40	2336.0
35	HU78-029-023	71 2.20	71 29.80	603.0
36	HU78-029-024	71 13.20	70 45.60	832.0
37	81-045-25	71 42.80	72 14.50	146.0
38	70-028-74	72 30.00	69 30.00	1900.0
39	80-028-3A	71 50.00	67 00.00	2250.0
40	70-028-83	73 00.00	69 25.00	900.0
41	70-028-73	72 40.00	76 00.00	500.0
42	77-027-05	73 12.10	67 49.42	2200.0
43	80-028-2A	72 48.20	66 28.20	2410.0
44	77-027-02	74 30.40	69 45.80	1646.0
45	78-029-901	75 23.40	71 5.60	650.0
46	70-028-86	74 5.00	75 30.00	805.0
47	70-028-88	74 14.00	76 30.00	750.0
48	70-028-89	74 15.00	78 30.00	720.0
49	70-028-90	74 18.00	79 00.00	680.0
50	81-045-21	73 45.90	81 9.80	459.0

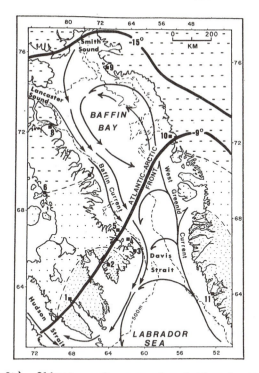

Figure 10.2 (left) Climate and ocean circulation in the Baffin Bay
region. Heavy solid lines are mean annual air temperatures (°C); thin
solid lines indicate flow direction of the surface currents. Shaded
areas indicate mean annual precipitation: close stipple >51 cm yr^{-1};
open stipple 25 - 50 cm yr^{-1}; dashes <25 cm yr^{-1}. Squares correspond
to the following weather stations: 1 Frobisher Bay; 2 Pangnirtung; 3
Cape Dyer; 4 Padloping Island; 5 Broughton Island; 6 Longstaff Bluff; 7
Clyde River; 8 Pond Inlet; 9 Thule; 10 Upernavik; 11 Godthaab. (right)
Isotherms (°C) for mean summer surface water during August and
September. Data from Dunbar 1951 and Marine Environmental Data
Service, Canada. Bold dashed line indicates the variable extent of the
3° isotherm.

less steep: it ranges from -1° at Godthaab to -9° at Uvernalik and -11°
at Thule. Mean annual precipitation is more evenly distributed on
either side of Baffin Bay, however, and it shows a fairly close
correspondence to the tundra vegetation subzones of Elliot-Fisk et al.,
(1982) and Fredskild (1973). The 51 cm isohyet roughly separates the
Low Arctic Tundra from the Transitional (Middle) Arctic Tundra, and the
25 cm isohyet marks the border between the Transitional Arctic and High
Arctic Tundra vegetation zones.

The physical oceanography of Baffin Bay is described in detail by
Dunbar (1951), Muench (1971) and by Coachman and Aagaard (1974). The
major features of the nearshore environments in summer are summarized

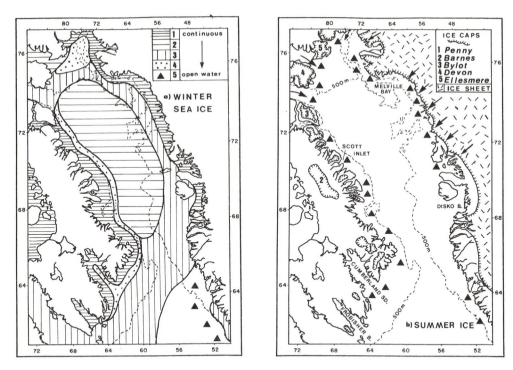

Figure 10.3(a) Distribution of sea ice in Baffin Bay during winter
(after Osterman, 1980). (1) continuous ice cover; (2) consolidated
pack ice; (3) unconsolidated pack ice; (4) semi-permanent open water;
and (5) open water with icebergs. (b) Major iceberg concentrations in
Baffin Bay during summer. Triangles indicate large numbers of ice
bergs; arrows show main sources of ice bergs. Ice berg data are from
Pelletier et al. (1975).

by Andrews et al. (1981) for Baffin Island and by Andersen (1981) for
the Disko Bay area of West Greenland. Over most of Baffin Bay, the
vertical and horizontal distribution of temperature, salinity and
nutrients is quite stable and three water layers can be defined by T-S
(temperature-salinity) and nutrient characteristics (Jones and Levy,
1981; Coote and Jones, 1982). However, the stable deep water is
overlain by a thin surface layer with very variable T-S characteristics
(Platt et al., 1982).

    The main factors affecting phytoplankton production and
microplankton assemblages in surface sediments of Baffin Bay are (i)
ice-cover, summer meltwater and iceberg distribution (Fig. 10.3); (ii)
surface water temperature (Fig. 10.2); and (iii) water mass structure
and chemistry. A strong pycnocline from 5 - 10 m below the water
surface effectively limits primary production to the surface layer. In

addition, the light compensation level for these latitudes lies at or above about 20 m depth; therefore, the surface layer has special importance with regard to phytoplankton productivity. In this study, then, four water layers are recognised, as defined by the following characteristics.

1. Surface Layer (0 - 20 m). In winter, the T-S of this layer is generally <-1°C and 32% throughout Baffin Bay; in summer, it ranges from about 3-7°C and 32-33°/₀₀ over the West Greenland Current to <2° and 30-31°/₀₀ over the Baffin Current. In spring and summer, salinity often fluctuates temporarily with the local distribution of ice meltwater and rain(Tan and Strain, 1980); salinity may decrease to 29°/₀₀ offshore and to 20°/₀₀ in fjords (Gilbert, 1982). The passage of icebergs offshore may cause a temporary lowering of temperature to -1°C.

2. Upper Water Layer (10 - c. 150 m). This is a mixture of surface water from the Arctic and Atlantic Ocean. The T-S ranges from about -1.8-4° and 31-34°/₀₀. Dissolved O_2 in this layer is high (c. 9ml/l); major nutrients (PO_4, NO_3 and SiO_2) are low in summer due to biological productivity (Platt et al., 1982).

3. Intermediate Water Layer (c. 150 - 1200 m). This layer is formed by Atlantic Water which cools and mixes with Arctic water north of Davis Strait. The T-S ranges from about 0-2°C and 34-34.5°/₀₀. Oxygen levels (c. 6%o) are lower than at the surface; nutrients show a 50% increase from the low levels at the top of this layer to higher values at the bottom.

4. Baffin Bay Bottom Water Layer (c. 1200 - 2400 m). This layer has very uniform T-S characteristics (c. 0.4°, 34.5 - 34.7%o), low oxygen (c. 4.7ml/l), and high nutrient levels that are almost twice the surface values. The origin of the Bottom Water is uncertain but it may represent Arctic Ocean Water of Atlantic origin. Low temperature, and high dissolved CO_2 make the bottom water corrosive to calcareous organisms.

In contrast to the relatively stable water structure in the main part of Baffin Bay, oceanographic conditions are very dynamic over the sills south and north of Baffin Bay. Northern Davis Strait is a mixing zone of the Baffin and West Greenland Currents. Here, both surface and deep flows are variable; although there is a net southwest surface flow, there is often an eastward drift of the deep water (Esso Resources Canada Ltd., 1979) and complex internal tides have been recorded in western Davis Strait. The area from Lancaster to Smith

Sound is a mixing zone between water from the Arctic Channels, Arctic
Ocean and Baffin Bay. Large spatial variations in currents have been
found between Smith Sound and Bylot Island (Collin, 1965; Fissel et
al., 1982). In 1979, strong intrusive currents in eastern Lancaster
Sound were associated with transient large-scale meanders off Bylot and
Baffin Island and anticyclonic eddies were found on the western edge of
the Baffin Current (Fissel et al., 1982). Another anomalous feature of
the northern Baffin Bay region is the persistence in winter of an
extensive polynya off Ellesmere Island (Dunbar and Dunbar, 1972).

MODERN POLLEN DISTRIBUTION

     Quaternary pollen, fern and moss spores in the samples from Baffin
Bay (Fig. 10.1) were identified and counted following the methods
described by Mudie (1982). All slides were scanned completely for
pollen-spore content. Only three samples (no. 39, 41 and 42) contained
no Quaternary pollen and spores, even though their dinoflagellate
content indicates the sediments at these sites are potentially suitable
for pollen deposition. Almost half of the slides, however, contained
less than 50 pollen and spores; therefore, many of the pollen data
(Table 10.2) are unsuitable for detailed statistical analysis. The
isopolls in Figures 10.5 to 10.6, however, are believed to indicate the
main trends shown by pollen-spore distributions in the Baffin Bay
region.

     Figure 10.4 shows that there is generally a strong negative
gradient in the nearshore-offshore distribution of pollen and spores,
similar to that found other regions of the Eastern Canadian continental
margin (Mudie, 1982). In Baffin Bay, the offshore decrease in
pollen-spore concentration is most rapid where the continental shelf
(500 m isobath) is narrowest. However, this trend is noticeably
reversed at the entrance to Lancaster Sound and in Davis Strait where
pollen concentrations show isolated peaks offshore. These are both
areas of intense mixing and oscillating internal water movements which
may concentrate particulate organic matter and account for the high
offshore pollen-spore concentrations. These anomalous areas are also
marginal to waters that become ice-free earlier than most parts of
Baffin Bay and it is possible that early removal of the ice cover
increases the chance of pollen influx from aerial transport.

     Figures 10.5 and 10.6 show isopolls for 8 of the most common
pollen and spore taxa in the surface sediments of Baffin Bay. In order
to compare these with data previously described for Baffin Bay
sediments (Mudie, 1982), the marine pollen percentages are based on a

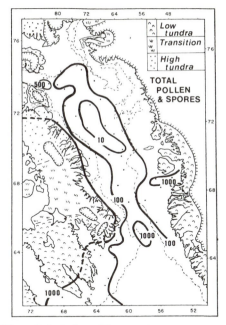

Figure 10.4 Isopolls of total pollen and spore concentrations per cubic centimeter of wet sediment, in relation to Tundra vegetation zones in the Baffin Bay region. Hachured lines indicate ice caps.

sum of total Quaternary pollen and spores, excluding mosses other than Sphagnum. For comparison of the marine pollen and onshore pollen distributions, pollen percentages are also shown for the Devon Island Ice-cap (Lichti-Federovich, 1974), and for Baffin Island and Ungava (Elliot-Fisk et al., 1982). The pollen sum used by Lichti-Federovich is the same as for the marine samples. The pollen sum used by Elliot-Fisk, however, excludes spores and Cyperaceae, which are locally over-represented in moss polsters; therefore, the onshore isopolls and marine data can only be related in a general way. It should also be noted that where total counts for the marine samples are less than 25, the sites were not included in the isopercentage lines; for the other sites, at least 75 grains were counted.

Figure 10.5 illustrates isopolls for four taxa which produce abundant wind-dispersed pollen but have restricted distributions within the Tundra vegetation (Alnus and Betula) or grow only south of the Tundra border (Picea and Pinus). All these isopolls show a decrease in pollen representation from south to north, with the main peaks lying south of the summer position of the Atlantic - Arctic airmass boundary. Alnus (Fig. 10.5a) is mostly represented by small A. crispa-type pollen. This taxon has the most limited distribution in

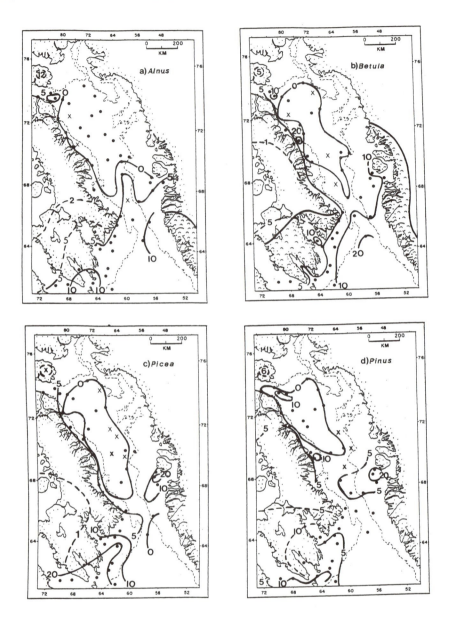

Figure 10.5 Isopolls for selected wind-pollinated Low Tundra shrubs (a and b) and for Boreal Forest conifers (c and d). Numbers indicate pollen percentages; x indicates the presence of a taxon in very low numbers. Onshore, solid lines mark the present growth limits of Alnus and Betula. Data for onshore isopolls (dashed lines) are from Elliot-Fisk et al. (1982) and Lichti-Federovich (1974).

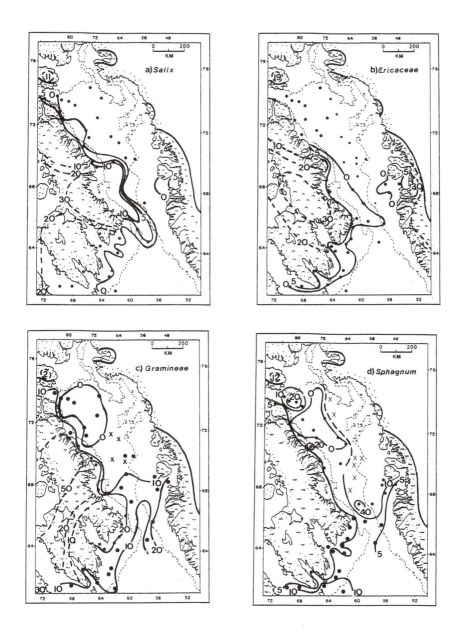

Figure 10.6 Isopolls for selected insect-pollinated Tundra shrubs (a and b), wind-pollinated herbs (c) and Sphagnum moss (d). Numbers indicate pollen or spore percentages; x indicates the presence of a taxon in very low numbers. Onshore, solid lines mark the present growth limits of the plant taxa. Data for onshore isopolls (dashed lines) are from Elliot-Fisk et al. (1982); Lichti-Federovich (1974); Fredskild (1973).

the marine sediments, with peaks occurring close to its growth limit in northern Labrador and c. 200 km south of Disko Island in Greenland. In general, Alnus pollen is confined to marine sediments within about 200 km of the coastline. It is rare north of latitude 68°, except for a small peak off Lancaster Sound which may correspond to its strong representation in the youngest ice (past 1400 yrs) of the Devon Ice-cap (Lichti-Federovich, 1975; McAndrews, 1982). Isopolls of Betula (mostly B. nana-type grains of <21 μm diameter) show a distribution roughly similar to that of Alnus, but Betula is better represented offshore, especially in the Davis Strait area (Fig. 10.5b). The tendancy for Betula pollen to be more widely dispersed in marine sediments of the Arctic than in temperate regions (Mudie, 1982) may reflect the small size of tundra birch pollen and drier air at the time of flowering in the Arctic.

Most Picea and Pinus pollen grains in the sediments of Baffin Bay are small (35 - 60 um) for these tree pollen genera and they are therefore referred to as Picea mariana-type and Pinus banksiana-type, respectively (see Table 10.2). The species identity of these small grains is uncertain because of the wide size ranges of North American Picea and Pinus pollen (Elliot-Fisk et al., 1982). When considering the long distance transport of tree pollen, however, it is useful to retain a distinction between these small grains and the larger grains (>60 um) that are typical of Picea glauca and Pinus strobus in eastern Canada (Mudie, 1982).

Isopolls of the combined Picea taxa show distinct peaks off southeast Baffin Island and southwest of Disko Island (Fig. 10.5c). In the first area, most of the grains are well preserved, and the ratio of P. mariana- to P. glauca-type grains is >1.0. Off Disko Island, however, there is a higher proportion of poorly preserved (broken) P. glauca-type grains and the ratio is often <1.0. The difference between the two areas may be important since the only likely source of Picea pollen off W. Greenland is North Americ. The well-preserved pollen off Baffin Island suggests wind transport over a relatively short distance (c. 500 km) from Labrador-Ungava, whereas the poor preservation off Disko Island may reflect the circuitous aerial pathway (>2000 km) that is required for Picea to travel to West Greenland (Barry et al., 1981).

Almost all the Pinus pollen in Baffin Bay (Fig. 10.5d) is of the small P. banksiana type, in contrast to the high frequency of larger P. strobus-type grains found on the continental shelves south of latitude 52°N (Mudie, 1982). This feature may reflect size sorting over the long distances (>1500 km) from the nearest Pinus sources south and west of Hudson Bay. The high ratio of Picea to Pinus pollen in the

Table 10-2. Relative abundance of Quaternary pollen and spores in surface samples from the Baffin Bay region.
+ = <0.5% present.

SAMPLE NUMBER	1	2	3	4	5	6	7	8	9	10	11	12	13	14	15	16	17	18	19	20	21	22	23	24	25
Number counted	70	57	71	64	205	50	85	76	74	75	37	20	70	19	60	43	41	22	32	44	36	72	97	14	39
TREE																									
Picea glauca-type	7	4	3	6	3	-	9	7	3	4	4	-	-	-	-	-	2	-	-	-	-	-	-	-	-
Picea mariana-type	16	19	14	6	17	18	6	16	3	10	7	-	-	-	-	-	3	9	-	-	6	3	10	11	-
Pinus strobus-type	1	2	-	-	-	-	-	-	-	-	-	-	-	-	2	-	-	-	-	6	2	-	-	-	-
Pinus banksiana-type	6	9	6	12	4	9	7	8	5	2	7	-	2	-	2	2	2	5	8	6	17	3	22	22	31
TREE/SHRUB																									
Alnus	6	9	-	19	9	-	1	1	3	4	-	10	-	-	2	12	3	5	15	-	5	6	2	-	6
Betula - 21 um	1	+	-	5	3	-	1	1	-	-	-	-	-	-	2	2	5	-	-	-	-	-	-	11	-
Betula - 21 um	3	4	9	7	5	-	6	7	5	12	7	20	4	-	5	7	10	5	-	17	6	11	4	-	-
Juniperus	-	2	-	-	-	-	-	-	-	-	-	-	-	-	-	-	-	-	-	-	-	-	-	-	-
Salix	1	2	-	-	3	-	1	3	2	2	-	5	2	-	5	-	11	-	8	-	-	6	-	-	-
Ericaceae	1	5	-	-	3	-	3	1	3	2	-	-	-	-	2	2	2	-	-	-	-	3	2	-	-
HERB																									
Ambrosia	1	-	-	-	-	-	-	-	-	-	3	-	-	-	-	2	-	-	-	-	-	-	-	-	-
Artemisia	-	-	3	-	1	-	1	3	-	-	-	-	-	-	2	-	-	4	-	6	-	2	-	-	-
Compositae	1	2	-	-	-	-	-	1	5	-	-	-	-	-	-	-	-	-	-	-	-	-	-	-	6
Dryas	1	-	-	-	-	-	1	-	3	-	4	-	4	-	7	-	-	5	-	-	3	3	-	-	-
Chenopodium	-	-	-	-	-	-	1	-	-	-	3	-	-	-	-	-	-	-	-	-	-	-	-	-	-
Cyperaceae	7	2	14	-	10	-	8	9	11	4	7	10	10	8	13	12	-	14	-	-	17	-	10	-	12
Gramineae	10	7	14	6	11	-	14	14	22	18	18	15	18	33	20	23	12	9	8	17	17	30	10	-	12
Plantago	-	-	-	2	-	-	-	-	-	-	-	-	-	-	-	2	2	-	-	-	-	-	-	-	-
Rosaceae	7	11	6	6	5	20	10	8	19	14	4	15	18	-	15	12	2	18	-	5	-	11	-	11	6
Rumex/Oxyria	-	2	3	5	3	-	3	-	-	-	4	-	4	-	2	2	13	-	8	5	-	3	4	11	1
Saxifragaceae	1	2	-	-	-	-	3	-	3	4	-	-	-	-	2	2	-	-	-	-	-	-	-	-	-
Various	14	7	17	2	7	9	19	14	16	12	10	5	30	25	20	7	2	23	-	-	11	8	4	-	-
FERN/MOSS																									
Lycopodium	1	4	3	-	3	1	3	3	-	4	4	10	-	-	2	2	13	5	8	11	16	3	12	11	12
Polypodiaceae	6	4	-	-	5	8	1	3	-	4	4	10	2	8	-	12	5	4	-	5	-	6	2	-	-
Selaginella	-	-	-	7	-	-	1	-	2	-	-	-	-	-	-	-	-	-	-	-	-	-	-	-	-
Sphagnum	6	9	14	18	5	9	7	3	8	18	5	5	3	-	3	7	7	-	46	33	7	3	18	22	25
Other mosses (% Total)	69	18	13	-	41	29	24	40	58	-	-	-	-	46	8	-	59	33	-	-	47	36	60		
Pre-Quaternary (% Total)	1	0	0	54	11	35	7	1	3	2	14	0	0	0	0	46	0	28	10	4	0	5			

marine sediments south of Baffin Island may also reflect the greater
travel distance of the Pinus grains. It is notable that this ratio is
reversed north of Davis Strait and west of the Atlantic-Arctic airmass
boundary, both in the marine and onshore sites. The significance of
this feature is presently unclear, but it is possible that better
understanding of the importance of southerly wind transport to Baffin
Bay (Nichols et al., 1978; Andrews et al., 1979; Barry et al., 1981)
may be gained by analysis of the relation between exotic pollen size,
preservation, and the wind transport pathways to Baffin Island.

Isopolls of plant taxa which have widespread distributions in the
Arctic are shown in Figure 10.6. Arctic willows (Salix) are
low-growing shrubs which are primarily insect-pollinated, although the
small exposed catkins allow for some wind transport. Some of the
Ericaceae-type pollen may be from Ericales herbs but most of the pollen
in the marine sediments appears to be of the Vaccinium type (Comtois
and Larouche, 1981) which are produced by low-growing,
insect-pollinated shrubs. The marine isopolls of Salix (Fig. 10.6a)
and Ericaceae pollen (Fig. 10.6b) show similar patterns in the Baffin
Bay region, with percentages of >5% being generally confined to
nearshore areas. Salix, however, shows a wide peak corresponding to
its onshore pollen maximum in central Baffin Island. It is notable
that unlike the wind-pollinated trees and shrubs, these insect-
pollinated taxa do not show peaks off southeast Baffin Island although
their pollen is abundant onshore in Labrador-Ungava (Elliot-Fisk et
al., 1982). Cyperaceae and Rosaceae pollen, which is also poorly
adapted for wind transport, shows a similar restricted offshore
distribution (Mudie, 1982). Other insect-pollinated herbs with
widespread distributions in the Arctic, e.g. Dryas and Saxifraga, also
have restricted distributions offshore. In general, these
predominantly insect-pollinated taxa appear to be the best pollen
markers of Arctic nearshore marine environments (Vilks and Mudie,
1978).

The Gramineae isopolls (Fig. 10.6c) represents the marine
distribution pattern typical of wind-pollinated Arctic herb taxa.
Oxyria and Rumex pollen (Table 10.2) shows similar distribution
patterns in Baffin Bay. Gramineae pollen is widely distributed in the
marine sediments from Davis Strait southwards. The pollen is also well
represented in nearshore sediments north of Davis Strait, but here
distribution further offshore is very patchy, suggesting sporadic
deposition from Arctic Archipelago airmasses.

The distribution of Sphagnum (Fig. 10.6d) is representative of the
very widespread but variable occurrence of moss spores in sediments of

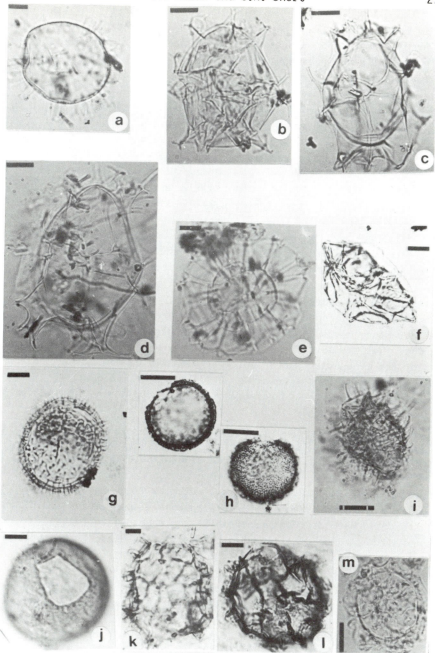

Figure 10.6 Light microphotographs of indicator dinoflagellates and other cysts in surface sediments of Baffin Bay. Scale bar = 10 um. a) Operculodinium centrocarpum; b) Spiniferites membranaceus; c) S. frigidus; d) S. elongatus; e) Nematosphaeropsis labyrinthea; f) ?Pterosperma species 1; g) Multispinula minuta; h) Leiospaera species A: pylome in focus on left; surface ornamentation on right; ?Protoperidinium species A; j) Brigantedinium simples; k & l) Polykrikos schwartzii m) Polykrikos species A. Reference slide numbers are listed in Table 10.3.

Baffin Bay. Onshore, Sphagnum moss vegetation is most abundant in the
Low Arctic Tundra and Subarctic regions, and this is reflected in the
marine isopolls off southeast Baffin Island. However, erratic Sphagnum
peaks are also present north of Davis Strait which suggests that many
of these spores are transported long distances by Arctic air and/or
iceberg meltwater. In the Transitional and High Arctic Tundra, other
mosses, e.g. Polytrichum and Dicranum, are more widespread than
Sphagnum. These small moss spores (Table 10.2) are almost ubiquitous
in the Baffin Bay region and there is no clearly discernable pattern to
their percentage frequencies. In contrast to these widespread moss
spore distributions, Lycopodium spores have a relatively restricted
distribution in Baffin Bay (Table 10.2) and they are rarely found in
marine sediments north of the -9°C isotherm. Onshore, Lycopodium and
other ferns are rare or absent north of the Low Arctic Tundra, and
Lycopodium spores are most abundant both onshore and in marine
sediments of the Labrador-Ungava region (Mudie, 1982). The spores are
not common off southeast Baffin Island which suggests that wind is not
an important transport agent for Lycopodium. However, relatively high
percentages of Lycopodium spores (5 - 13%) are found off Disko Island
and in eastern Davis Strait. This unusual distribution pattern
suggests that the West Greenland Current may be an important transport
pathway for Lycopodium spores which are locally common in the Godthaab
area of West Greenland (Fredskild, 1973) but are infrequent in air
pollen samples from the Canadian arctic-subarctic region (Ritchie and
Lichti-Federovich, 1967).

As explained earlier, the variable and often low pollen-spore
concentrations make the Baffin Bay data unsuitable for rigorous
statistical analysis. However, because Q-mode factor analysis (Klovan
and Imbrie, 1971) operates mainly on variables with high percentage
values and systematic geographical variations, factor analysis of the
data reported here produces essentially the same pollen factor
assemblages as those described for the arctic to subarctic vegetation
regions of eastern Canada by Mudie (1982). The Arctic factor
assemblage (Factor 2 of Mudie, 1982) has high scores for tundra herbs
(mainly Oxyria), Gramineae, Rosaceae and Cyperaceae, and maximum
loadings on the Baffin Shelf from Bylot Island to Cumberland Sound.
From Frobisher Bay south to about latitude 56°N, the marine sediments
are characterized by a Subarctic factor assemblage (Factor 4 of Mudie,
1982), in which Sphagnum, Alnus, Betula, Lycopodium and P. mariana-type
pollen has high factor scores. This factor assemblage is also found
off West Greenland. Addition of new data from the Baffin Shelf and
Davis Strait further distinguishes a Transitional Arctic factor
assemblage with high factor scores for Gramineae and maximum loadings
on the inner shelf from Clyde Inlet to Frobisher Bay.

In summary, then, two widespread marine pollen provinces can be distinguished in Baffin Bay and these correspond to the regions north and south of the summer boundary between Arctic Archipelago and Atlantic-Arctic airmasses. Off Baffin Island, the area between the relatively dry High Arctic and moister Low Arctic sub-regions of the Tundra is further characterised by the Transitional Arctic factor assemblage. Air transport appears to be the main pathway by which pollen enters the arctic marine environment, but ocean currents and meltwater distribution probably play an important role in restricting the distribution of insect-pollinated Arctic pollen types and Lycopodium spores to the continental shelves.

MODERN DINOFLAGELLATE DISTRIBUTION

Quaternary dinoflagellate cysts are common to abundant in all the surface samples from the Baffin Bay region (Figs 10.6 and 10.7). In general, cyst concentrations decrease offshore, with peaks occurring in deep-water channels at the mouths of large fjords or bays. The lowest concentrations are found in the fine-grained muds of Baffin Basin below 1500 m, which suggests that cyst densities are strongly related to water depth. However, there is also a tendancy for higher cyst concentrations to occur in zones of current convergence on the southern margin of Davis Strait and eastern Lancaster Sound. This implies that cyst density is also positively related to the presence of frontal zones between ocean currents, and the low density of cysts in Baffin Basin may reflect weaker mixing in the central gyre of northern Baffin Bay.

There are insufficient phytoplankton data for Baffin Bay to establish a definite relationship between motile and cyst-stage dinoflagellate concentrations. It seems significant, however, that the peaks in Figure 10.8 lie close to plankton survey areas in which blooms of peridinioid dinoflagellates have been reported, including Disko Bay (Grøntved and Seidenfaden, 1938), southern Hudson Strait to eastern Davis Strait (Esso Resources, 1979) and off Ellesmere Island (Grøntved and Seidenfaden, 1938). This relationship between motile and cyst-stage dinoflagellates suggests that encystment occurs in response to conditions associated with high phytoplankton productivity, and that most of the cysts are not transported more than about 200 km from their areas of production.

The apparent association between high cyst concentrations and ocean fronts in Baffin Bay is supported by studies around the British Isles which show that summer blooms of Gyrodinium aureolum occur in

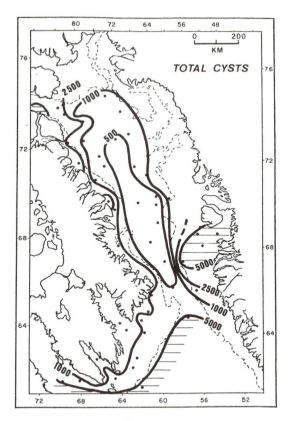

Figure 10.7 Isopleths of total dinoflagellate cyst concentrations per
cubic centimeter of wet sediment. Shading indicates areas with
concentrations of 5000 to 7500 cysts per cubic centimeter.

areas of high mixing along thermal fronts near the shelf break (Pingree
and Mardell, 1981) and around large islands (Simpson et al., 1982).
Dinoflagellates, including <u>Peridinium</u> (probably <u>Protoperidinium</u>),
<u>Gonyaulax</u> and <u>Prorocentrum</u> species, are common in the late winter and
spring sea ice of Frobisher Bay (Hsiao, 1979), but they have not been
reported as generating algal blooms. In nearshore areas of the
northern Baltic Sea, however, spring blooms of <u>Gonyaulax</u> <u>catenata</u> have
been found in water at the ice base (Müller-Haeckel, 1981). These
blooms are normally inhibited by nanoplankton that are abundant in the
ice, but it is now clear that in the absence of these inhibitors,
dinoflagellates such as <u>G</u>. <u>catenata</u> and <u>Gymnodinium</u> are capable of high
productivity under low temperatures and light intensities.
<u>Protoperidinium</u> <u>depressum</u> (<u>P</u>. <u>antarcticum</u>) is also known to live at
temperatures of -2°C in aphotic conditions under the Ross Ice Shelf
(Azam et al., 1979), but is is not certain whether this ice-shelf

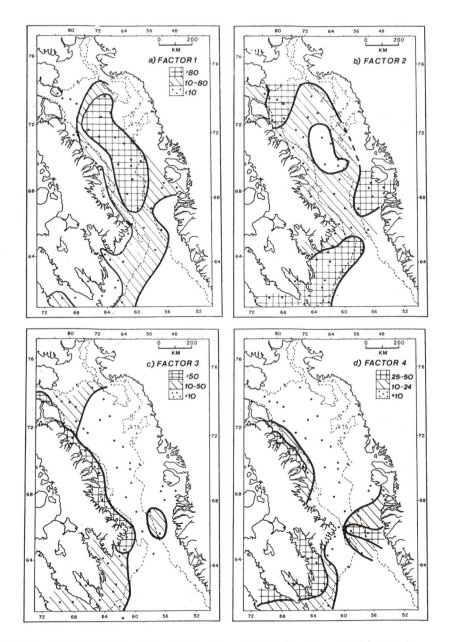

Figure 10.8   Distribution of dinoflagellate factor assemblages in
Baffin Bay.   Shading indicates the varimax factor loading values x 100
as shown in the legends.   Note that the loading values for Factor 2 are
the same as those in the legend for Factor 1.

population is indigenous or advected from the Ross Sea. Studies of
longer duration than those of Hsiao (1979) are needed to determine
whether sub-ice dinoflagellate blooms occur in Baffin Bay.

Twenty-six species of Quaternary dinoflagellate cysts have
been found in the surface sediments of Baffin Bay; these species and
their probable motile-stage taxa are listed in Table 10.3 and
illustrated in Figure 10.6. Four other cyst taxa are also common to
abundant: Leiosphaera species A, ?Protoperidinium species A, ?Dinocyst
species C and ?Pterosperma species 1. It is presently uncertain
whether these four taxa are cysts of dinoflagellates, but their
acid-resistant walls and their preservation in Holocene or older marine
sediments (Mudie, 1980; Miller et al., 1982) make them useful
palynostratigraphic markers. Hence, these cysts are included here and
they are described in detail by Mudie (1980, Appendix 4).

Q-mode factor analysis was used to analyse the geographical
variation among the dinoflagellate cyst assemblages in the Baffin Bay
samples. Raw data for this analysis are the percentage abundances of
the 30 taxa listed in the factor score matrix (Table 10.4). The input
data matrix also contains 58 samples from neritic and deep-water sites
south of Hudson Strait, which were included in order to confirm the
apparent phytogeographical affinities and species relationships
suggested by the Arctic samples.

The factor analysis, with varimax rotation, produces four
dinoflagellate factor assemblages (Fig. 10.8) which account for 90% of
the between-sample variation. Factor 1 (Fig. 10.8a) accounts for 43%
of the variance and is dominated by Operculodinium centrocarpum,
Spiniferites ramosus, Tectatodinium pellitum and Spiniferites
frigidus. The first 3 species are known to be cysts of Gonyaulax
species (Wall and Dale, 1968). Spiniferites frigidus appears to
intergrade with S. elongatus (Harland, 1982a); therefore, it is
probably a cyst of G. scrippsae. Factor 2 (Fig. 10.8b) accounts for
34% of the variance and is dominated by Multispinula minuta which is
probably a cyst of a Protoperidinium species (Harland et al., 1981).

Factors 1 and 2 together account for 77% of the covariance among
the recent dinoflagellate cyst assemblages off eastern Canada. This is
due to the widespread distribution of the main species contributing to
these factors, which are O. centrocarpum and M. minuta (Table 10.4).
O. centrocarpum has an almost circum-global distribution in shelf and
offshore marine sediments, but abundances of more than 50% are usually
found only in sediments below or on the northern margin of the North
Atlantic Drift Current, including the warm slope water off southeastern

Table 10-3. Taxonomic correlations of motile-stage (theca name) and cyst-stage (cyst name) dinoflagellates, and selected references to illustrations of the cysts.
* indicates correlation confirmed by laboratory cultures; ? indicates uncertain correlation and/or taxonomic identity. Reference slides for Fig.10-10 are at AGC.

THECA NAME	CYST NAME	REFERENCE
*Gonyaulax digitale (Pouchet)Kofoid	Spiniferites bentori (Rossignol) Wall & Dale	Harland 1977
*G. grindleyi (Reinecke)Von Stosch	Operculodinium centrocarpum Deflandre & Cookson)Wall	Fig.10-10a; PJM76-029-42,D36/3
*G. spinifera (Claparede et Lachmann)Diesing	Spiniferites mirabilis (Rossignol) Sargeant	Miller et al.1982
	S. membranaceus(Rossignol) Sargeant	Fig.10-10b; PJM76-029-42,3a/8
	S. ramosus (Ehrenberg)Loeblich & Loeblich	Reid & Harland 1977
	Nematosphaeropsis labyrinthea (Ostenfeld) Reid	Fig.10-10e; PJM75-009-118G,4/1
	Planinosphaeridium membranaceum Eisenack	Dale 1976
*Gonyaulax scrippsae Kofoid	Spiniferites bulloideus(Deflandre & Cookson)Sargeant	Dale 1976
	S. elongatus Reid	Fig.10-10d; PJM-M10,S39/2
?Gonyaulax scrippsae	S. frigidus Harland & Reid	Fig.10-10c; PJM76-029-42,3a/1
*Gonyaulax tamarensis Lebour	G. tamarensis hypnocyst	Miller et al. 1982
*Protoperidinium conicum (Gran) Balech	Multispinula quanta Bradford ex Harland & Reid	Reid & Harland 1977
?Protoperidinium species indet.	M. minuta Harland & Reid	Fig.10-10g; PJM-SC97,3a/12
?Protoperidinium divaricatum (Meunier)Parke & Dodge	?Protoperidinium species A	Fig.10-10i; PJM70-90,P44/3
*Protoperidinium conicoides (Paulsen)Balech	Brigantedinium simplex(Wall)Reid ex Harland & Reid	Fig.10-10j; PJM-SC97,3a/15
*Protoperidinium avellana(Meunier) Balech	Brigantedinium cariacoense(Wall) Reid	Dale 1976
?P.punctulatum (Paulsen)Balech		
?P.denticulatum(Gran & Braarud) Balech		
*Protoperidinium oblongum (Aurivillius)Balech	Votadinium calvum Reid	Reid 1977
*Peridinium faeroense Paulsen	Peridinium faeroense cyst form	Dale 1976
Scrippsiella faeroense (Paulsen) Balech & Soares	?Scrippsiella cysts	
*Glenodinium trochoideum Stein	?Scrippsiella faeroense cyst form	Dale 1976
Scrippsiella trochoidea(Stein) Loeblich III	Scrippsiella trochoidea cyst form	
*Peridinium cinctum(Mueller) Ehrenberg	Peridinium cinctum cyst form	Miller et al. 1982
*P. limbatum (Stokes)Lemmermann	P. limbatum cyst form	Miller et al. 1982
Amphidinium lacustre Stein	? Amphidinium cf. lacustre	Miller et al. 1982
*Diplopsalis lenticulata f. minor Paulsen	Dubridinium species	Reid 1977
Zygabikodinium lenticulatum (Paulsen)Loeblich & Loeblich		
?Gymnodinium species	Leiosphaera species A	Fig.10-10h; PJM78-029-23,2a/1,2
?Polykrikos schwartzii Bütschli	Polykrikos schwartzii cyst form	Fig.10-10k; PJM70-90,3a/27
	?Polykrikos schwartzii cyst form	Fig.10-10 1; PJM70-90;3a/26
?Polykrikos species	Polykrikos species A	Fig. 10-10m; PJM80-030-5,D46/3
?Pterosperma reticulatum Ostenfeld	?Pterosperma species 1	Fig.10-10f; PJM78-005-95,1/4

Table 10-4. Scaled varimax factor scores for dinoflagellate facto
assemblages shown in Fig. 10-9.

TAXON	FACTOR			
	1	2	3	4
Bitectatodinium tepikiense	.15	-.03	.02	.07
Brigantedinium cariacoense	.11	-.06	-.07	-.62
Brigantedinium simplex	.01	.23	.52	-4.48
Brigantedinium species	.16	-.07	1.61	-2.54
?Dinocyst species C	.01	-.001	.07	-.40
Amphidinium species A	.04	.01	.47	.23
Dubridinium species	.11	.02	.12	-.58
Gonyaulax tamarensis	.05	.06	-.05	-.21
Hemicystodinium zoharyi	.22	-.03	.03	.20
Leiosphaera species A	-.01	-.22	5.13	1.24
Leptodinium species	.20	.01	.001	.04
Multispinula quanta	-.002	.137	-.02	-.14
Multispinula minuta	-.201	5.45	.19	.22
Nematosphaeropsis labyrinth	.27	.08	-.07	.09
Operculodinium centrocarpum	5.40	.20	-.05	.08
Peridinium faeroense	.05	.14	.31	-.11
Peridinium species	.06	.14	.17	-.37
Planinosphaeridium membran-aceum	.02	.002	.12	-.04
Polykrikos species	.004	.04	.55	-.07
?Protoperidinium species A	-.01	.16	.12	.18
Spiniferites bentori	.12	-.003	.12	.10
Spiniferites bulloideus	.14	.001	-.019	.08
Spiniferites elongatus	.107	.002	-.006	.10
Spiniferites frigidus	.33	.03	.15	.11
Spiniferites membranaceus	.09	.03	-.03	-.01
Spiniferites mirabilis	.11	-.01	.001	.08
Spiniferites ramosus	.47	-.05	-.03	.26
Tectatodinium pellitum	.33	-.07	.04	.14
Votadinium calvum	.09	-.01	.03	-.67
?Pterosperma species 1	.15	.21	-.21	-.20

Canada (Reid and Harland, 1977; Mudie, 1981). The dominance of Factor
1 in central Baffin Bay and southeastern Davis Strait is therefore
consistent with the North Atlantic influence of the water masses in
these areas. Spiniferites ramosus and Tectatodinium pellitum are also
common in boreal regions of the North Atlantic (Reid and Harland,
1977), but S. frigidus is more or less confined to Arctic regions where
there is a strong influence of warm currents, e.g. the southern Barents
Sea (Harland, 1982a) and southwestern Beaufort Sea (Harland et al.,
1981). Multispinula minuta is almost ubiquitous in nearshore and
continental margin sediments from Beaufort Sea to central Nova Scotia
(Mudie, 1981) and it is also found in the Barents Sea (Harland,
1982a). It is most abundant (>50%), however, in shelf sediments
overlain by water of Canadian Arctic origin. In Baffin Bay, this is
clearly reflected in the dominance of Factor 2 off Lancaster Sound and
southern Baffin Island. The dominance of Factor 2 off Disko Island
implies that M. minuta may also be abundant in areas of strong mixing
between Arctic and North Atlantic water.

Factor 3 (Fig. 10.8c) accounts for 7% of the variance and it is
dominated by Leiosphaera species A, Brigantedinium and Polykrikos
species. The motile stage of Leiosphaera sp. A has not yet been
determined by laboratory study, but its small size, simple morphology
and excystment aperture suggest that it is a Glenodinioid cyst. The
Brigantedinium cysts are resting spores of the genus Protoperidinium
(Harland, 1982b). The Polykrikos species include cysts of P.
schwartzii (Harland, 1981) and a smaller cyst with thinner walls and
more delicate reticulate ornamentation. Leiosphaera sp. A is a
dominant species of nearshore marine sediments in the Canadian Arctic,
from Beaufort Sea (Mudie, unpubl.) to northern Labrador Shelf, and it
is very rare further south and east in the North Atlantic.
Brigantedinium species are widespread in nearshore and shelf sediments
of arctic and boreal regions (Harland, 1977; Harland et al., 1981;
Harland, 1982a). Polykrikos schwartzii is widely distributed in
sediments of cool temperate North Atlantic waters (Harland, 1981) but
it is most abundant in sediments near the entrances to fjords from
Newfoundland to northern Baffin Island (Mudie, unpubl.). The dominance
of Factor 3 on the northeast Baffin Island Shelf seems to reflect the
High Arctic affinity of Leiosphaera sp. A and the heavily fjorded
coastline of northeast Baffin Island.

Factor 4 (Fig. 10.8d) accounts for 9% of the variance and is
dominated by cysts of Brigantedinium, Gonyaulax tamarensis and
Peridinium species. The motile stage affinity and distribution of
Brigantedinium cysts has been described above. G. tamarensis cysts are
the resting spores of Gonyaulax tamarensis, including G. tamarensis

var. excavatum. Gonyaulax tamarensis cysts are widespread in bays and
inlets from Massachusetts to Hudson Bay (Prakash et al., 1971) and in
British Columbia fjords (Turpin et al., 1978), and blooms of this
species are often associated with intervals of heavy runoff (Prakash
and Rashid, 1968; Bursa, 1961). The Peridinium species is a diverse
group of cysts which have a thin-walled periphragm enveloping a thicker
walled inner body, and which have transapical excystment apertures.
These cysts include Peridinium cinctum, which occurs in streams of
northern Baffin Island (Polunin, 1948), and other unspecified
freshwater Peridinium species. High percentages of these cysts are
typically found in estuarine or fluvial-dominated fjord environments
from Nova Scotia (Miller et al., 1982) to northern Baffin Island. The
dominance of Factor 4 in Baffin Island fjords and adjacent shelf
sediments therefore appears to reflect the high runoff in these areas.
The dominance of Factor 4 in sediments of eastern Davis Strait may
reflect the influence of large rivers in West Greenland or transport in
icebergs calved from Greenland or Baffin Island fjord glaciers.

In summary, the dinoflagellate cyst assemblages in the recent
sediments of Baffin Bay constitute 4 phytogeographical groups that
correspond to different surface water T-S conditions summarized below.

Assemblage 1, dominated by Operculodinium centrocarpum and other cysts
of Gonyaulax species, characterizes sediments overlain by water that is
warmer than 2°C in summer and has a salinity of >32°/₀₀. This
subarctic assemblage corresponds to areas dominated by the warm West
Greenland Current.

Assemblage 2, dominated by the arctic cyst Multispinula minuta, is
prevalent under a wider range of summer T-S conditions (<1°-9°C and
31-33°/₀₀) and is most abundant in offshore areas that have an
unconsolidated pack-ice cover in winter (Fig. 10.3a). This assemblage
corresponds to areas characterized by high mixing of Canadian Arctic
water and warm West Greenland water.

Assemblage 3, dominated by Leiosphaera species A, characterizes
nearshore sediments overlain by summer water that is very cold (<1°C)
and of relatively low salinity (30-31.5°/₀₀) and it is found only
in areas covered by continuous ice in winter. This assemblage
corresponds to nearshore areas dominated by the Baffin Current.

Assemblage 4 is dominated by Brigantedinium cysts, Gonyaulax tamarensis
and Peridinium cysts, all of which are most abundant where surface
water salinity is periodically very low (<30°/₀₀ in summer) due to
high freshwater runoff. This assemblage mainly characterises Baffin

Island fjords and adjacent shelf sediments but it also occurs further offshore in eastern Davis Strait.

Because of the present sparseness of accurate taxonomic data for motile-stage dinoflagellates in Arctic plankton samples, it is difficult to assess the extent to which the dinoflagellate cysts in Baffin Bay are indigenous or transported into the Bay by the major ocean currents. It is notable, however, that the recorded occurrences of Gonyaulax grindleyi (= Protoceratium reticulatum) and G. spinifera are for warm water areas off Disko Bay and southwest of Melville Bay (Grøntved and Seidenfaden, 1938) which are largely free of ice in summer (Fig. 10.3b) and that Gonyaulax scrippsae, which probably produces the indigenous cyst species Spiniferites frigidus, has been found only in areas north of Davis Strait. It is also notable that Protoperidinium species which produce the brown-walled cysts of Brigantedinium and Multispinula are reported as rare beneath the West Greenland Current but common in cold nearshore waters off northwest Greenland, eastern Lancaster Sound and eastern Davis Strait. Hence, the geographical distribution of the two major cyst assemblages, Factors 1 and 2, at least corresponds broadly to the distribution of Gonyaulax species and Protoperidinium species in the surface water phytoplankton from offshore areas of Baffin Bay.

The possibility of reworking from pre-Pleistocene shelf sediments can also be eliminated as a source for several of the main indicator cyst species. The youngest pre-Pleistocene shelf sediments in the Baffin Bay region are probably of early Oligocene age (MacLean et al., 1981). Operculodinium centrocarpum has a stratigraphic range of Paleocene to Recent in North Atlantic sediments (Williams and Bujak, 1977) but the Tertiary cysts are usually larger and have fewer spines than the Pleistocene - Recent cysts (Mudie, umpubl. data) and O. centrocarpum is absent or very rare in Paleogene sediments from exploration wells off Greenland (Croxton, 1979). Multispinula minuta, Spiniferites frigidus, Leiosphaera species A and Gonyaulax tamarensis cysts have not been found in sediments older than mid-Pleistocene (Bujak, in press; Mudie, unpubl.). Therefore, although many of the Baffin Bay sediment samples contain reworked dinoflagellates of Senonian to Eocene age (Aksu, 1981), and reworked Tertiary pollen is common, the apparent absence of Neogene to mid-Pleistocene marine sediments makes it very unlikely that the recent dinoflagellate cyst assemblages contain a significant number of reworked specimens.

PALEOECOLOGICAL STUDIES

Five cores were selected for initial studies of Quaternary pollen and dinoflagellate assemblages in the Quaternary sediments of Baffin Bay (Fig. 10.1). Two of the cores are from the Davis Strait area: Core HU 77-159 (site 5) from Frobisher Bay was studied for pollen by S.K. Short; Core 17 (site 17) from northern Davis Strait was studied for both pollen and dinoflagellates by P.J. Mudie. Three short cores were investigated (by P.J. Mudie) from areas north of Davis Strait: Core 73 (site 36) is from a submarine trough off Scott Inlet; Core 3A (site 39) is from western Baffin Basin; Core 34G (site 33) is from central Baffin Basin. Many problems were encountered in the preparation of palynological data from these cores, including (1) the difficulty of extracting palynomorphs from sediments of variable texture and mineralogy; (2) large between-sample variations in palynomorph concentrations which partly reflect variable sedimentation rates due to ice-rafting, gravity flows, etc.; and (3) frequent well-preserved pre-Quaternary bisaccate and triporate pollen which is not easily distinguished from imperfect grains of Quaternary pollen. The problem of recognising reworked Tertiary pollen has not yet been satisfactorily resolved, but the initial results of our studies are considered interesting enough to present here as a framework for more detailed work in the future.

Core HU 77-159:

This piston core (HU 77-021-159) was recovered from a water depth of 570 m in Frobisher Bay (Site 5, Fig. 10.1). The lithology, foraminiferal stratigraphy, and methods of dating the core have been described in detail by Osterman (1982). The core consists of 9.7 m of olive gray to light gray silty or clayey mud, with rare sandy-pebbly layers and scattered larger clasts.

Fifteen samples were analysed for pollen at 50 to 100 cm intervals in a preliminary study. The full pollen diagram shows the relative abundances of all the Quaternary pollen and spore taxa found in core HU77-159. Figure 10.9 is a summary diagram showing the percentages of the main pollen taxa and the ratio of pre-Quaternary to Quaternary pollen and spores in relation to the core lithology and radiocarbon dates. Two pollen assemblage zones and three subzones are recognised, as described below.

Zone I (0 - 350 cm) is generally characterised by relatively high percentages of Betula, Alnus and Picea pollen, and by very low ratios

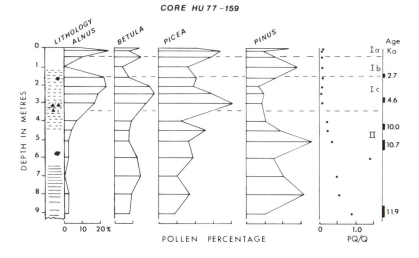

CORE HU 77-159

POLLEN PERCENTAGE

Figure 10.9 Summary diagram for the main Quaternary tree and shrub taxa in Core HU 77-159, and PQ/Q, the ratio of total pre-Quaternary to Quaternary pollen and spores. Pollen sum is total Quaternary pollen and spores, excluding unidentifiable taxa. Lithology and radiocarbon dates (in Ka BP) are from Osterman (1982). Roman numerals indicate informal pollen assemblage zones (acme zones) described in the text. Lithology symbols: unshaded = bioturbated mud; broken lines = silty mud; dots = sandy mud; thin lines = laminated clayey mud; triangles = gravel; black polygons = pebble or cobble.

of pre-Quaternary to Quaternary pollen (= PQ/Q values). The radiocarbon dates suggest that the sediments in Zone I were continuously deposited during the past c. 4600 years. The absence of samples between 300 and 400 cm depth unfortunately makes it impossible to determine the exact position and age of the zone 1/2 boundary. Zone I may be subdivided on the basis of a major change in pollen percentages from 50 - 150 cm core-depth and minor differences in the pollen composition above and below this interval. Subzone Ia (0 - 50 cm) has moderately high percentages of Alnus, Betula and Picea, suggesting a brief recent interval of relatively mild climatic conditions. Subzone Ib (50 - 150 cm) shows large decreases in these pollen taxa and a corresponding increase in Pinus. This subzone, which has an age of c. 1000 to 2750 yrs BP, may correspond to the cold climatic interval recorded in the Burton Bay Cliff peat, upper Frobisher Bay (Short and Jacobs, 1982). Subzone Ic (150 - 350 cm) contains peaks in Alnus, Betula and Picea percentages, suggesting that it corresponds to a warmer, wetter climatic interval. It is unfortunate that the 360-cm level sample was lost during preparation;

that level was interpreted by Osterman (1982) as marking the change
from ice-distal and ice-rafted deposits at c. 7.5 KaBP.

Zone II (350 - 910 cm) is clearly marked by a low percentage of
Alnus, relatively high representation of Pinus, larger percentages of
Gramineae, Cyperaceae and Filicales, and by PQ/Q values of 0.25 or
more. The radiocarbon dates suggest that the sediment was deposited
very rapidly between the end of the Hall glacial advance (c. 10.8 Ka)
and the Gold Cove readvance (c. 10.1 Ka). The large number of
pre-Quaternary grains indicates major reworking of the sediments,
especially before and after the major ice advance (700-800 cm
interval), with the peak at 600 cm corresponding to the start of ice
draw-back. During the Zone II interval, the nearest large sources of
contemporary pollen were probably the Mackenzie Delta and periglacial
areas south of latitude 51°N. Glacial meltwater may also have
contributed Quaternary pollen, since large quantities of Betula and
Pinus occur in early Holocene samples of the Devon Island Ice Cap
(McAndrews, 1982). More frequent sampling of the 300-400 cm and
600-800 cm intervals in core HU77-159 may clarify the importance of
meltwater as a pollen source.

Core 17:

This piston core (77-027-017) is from a water depth of 935 m on the
north slope of Davis Strait (Site 17, Fig. 10.1). The lithology and
micropaleontology of this core have been described in detail by Aksu
(1981; this volume) whose radiocarbon dates, foraminifera and oxygen
isotope data are shown in Figures 10.10 and 10.11. The core contains
about 10 m of graded olive-gray sand, silt and mud, with a thin surface
layer of brown diatomaceous mud overlying a carbonate lithofacies (Aksu
and Piper, 1979).

A. Pollen Stratigraphy

Figure 10.10 shows the pollen and spore concentrations, percent
trace pollen and relative abundances of selected Quaternary pollen in
Core 17. Pollen assemblage zones have been tentatively delimited,
mainly to simplify the description of major features in the pollen
profile and their correlation with other cores. More closely spaced
sample intervals are required before a formal palynozonation can be
defined.

Figure 10.10 shows that there are 3 major peaks in "absolute"
pollen concentrations (APC) which occur in zones I, III and VI and

CORE 77-027-017: POLLEN

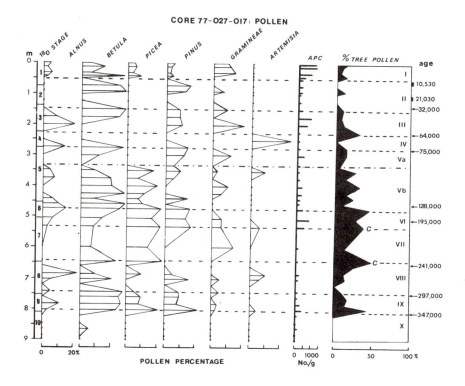

Figure 10.10 Pollen percentage diagram for the main tree, shrub and herb taxa in Core 17, total concentrations of Quaternary pollen and spores (APC) and percentage of tree pollen (= Picea + Pinus + Carya). C = Carya. Solid rectangles show C-14 dated intervals; arrows mark isotopic stage boundaries. All dates are in years BP, as determined by Aksu (1983a).

roughly correspond to peaks in % tree pollen (including Pinus, Picea, and Canya). Zone I (0 - 50 cm) is characterized by shrub and tree pollen percentages similar to those in the modern marine sediments south of the present summer position of the Atlantic- Arctic air front. Gramineae dominate the herb pollen, and the absence of Picea (other taxa not shown include Salix, Rosaceae, Rumex, Lycopodium, Cruciferae, Compositae and Cyperaceae). Zone III lies within isotopic Stage 3 which has been assigned an age range of 32 Ka to 64 Ka (Aksu, this volume). The high pollen concentrations and dominance of Betula in zone III of Core 17 suggest a climatic interval warmer than present, and may correspond to the Ellesmere Island interstadial with dates of c. 30-40 Ka (Blake, 1982) and the Loks Land interval (Miller; this volume).

CORE 77-027-017

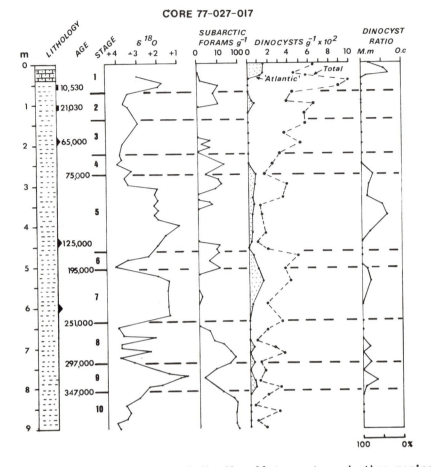

Figure 10.11 Summary diagram of dinoflagellate cysts and other marine microfossil data for Core 17. Total = total cysts; "Atlantic" = North Atlantic dinoflagellates. Dinocyst Ratio is the percentage ratio of Multispinula minuta (M.m.) to Operculodinium centrocarpum (O.c.). Stage 1, 2, 3, etc. are the isotopic stages of Aksu (1980). Ages in years BP and volcanic ash horizons (triangles) are from Aksu (1983a). Lithology symbols: unshaded = diatomaceous mud; rectangles = carbonate mud; broken lines = silty mud; dots = sand.

Zone IV (200-275 cm) is another palynofacies which contains very low numbers of Quaternary pollen, with Salix, Rumex, and moss spores being the most common taxa. This zone, which also marks the highest occurrence of Artemisia, corresponds to marine isotopic Stage 4.

Zone V (275-470) contains relatively high Quaternary pollen concentrations and high tree pollen percentages. The boundary between Zone V and VI has been semiarbitrarily set to coincide with the maximum peaks in <u>Betula, Picea</u>, and tree pollen percentages, which lie just below the isotopic State 5/6 boundary. Unfortunately, Stage 6 is not very clearly demarcated in Core 17 (see Fig. 10.11) and there is some evidence of an erosional surface between 450 and 500 cm core-depth (Aksu, pers. comm.). It is therefore possible that Zone V should include the tree pollen peaks at 525 and 650 cm (Zone VII in Fig. 10.10). In the absence of independent dating evidence, however, Zone V is presently taken as corresponding to isotopic Stage 5.

    Zone V has been divided into 2 subzones, based on major differences in pollen percentages of the upper and lower sediment interval. Subzone Va (300 - 330 cm) is marked by a relatively small peak in % tree pollen and the absence of <u>Picea</u>. <u>Pinus</u> and Gramineae are dominant, but other taxa in zone Va include <u>Salix</u>, Cyperaceae, <u>Rumex</u>, <u>Saxifraga</u>, Compositae, <u>Lycopodium</u>, and <u>Sphagnum</u>, suggesting a relatively cold climate similar to that found in the present Transitional Arctic Tundra vegetation zone on Baffin Island. Subzone Vb (330 - 470 cm) is marked by the co-occurrence of high percentages of <u>Picea</u> and <u>Betula</u> pollen, with <u>Alnus</u> and <u>Pinus</u> also being well represented. These pollen assemblages are similar to those in zone I of Cores 17 and HU 77-159, and they suggest a relatively warm climate similar to the present Low Arctic. Higher percentages of <u>Betula</u> in subzone Vb, however, may indicate a wider distribution of <u>Betula</u> shrubs than presently found on Baffin Island (see Fig. 10.6b), hence suggesting that the climate was warmer and wetter than the present. If zone V in Core 17 corresponds to marine isotopic Stage 5, it has an age range of c. 75 to 131 Ka and is probably correlative with the Kogalu amino zone (Andrews et al., 1981). Pollen in soil horizons below the Kogalu marine sediments also suggest a warmer, wetter climate for Baffin Island about 130,000 yrs BP (Mode, this volume).

    Zone VI (470-510 cm) in Core 17 is poorly defined, partly because of insufficient sediment for fine-scale sampling, and perhaps because of erosion in this interval. It is marked by a temporary decline in % tree pollen and by a reduction in <u>Alnus</u> and <u>Betula</u>, although total pollen concentrations remain high. This suggests a cooler climate during this interval. Zone VII (510-650 cm), which corresponds to isotopic Stage 7, is marked by two peaks in % tree pollen, both of which include a few grains of <u>Carya</u> pollen in addition to strong representation of <u>Picea</u>. Closer sampling intervals are required to determine whether the large variations in <u>Betula</u> and Gramineae have any real significance.

    Zones VIII (650-750 cm) and X (800-900 cm) correspond to isotopic Stages 8 and 10, respectively, both of which represent cold climatic

intervals in the North Atlantic. In Core 17, these zones are marked by
low pollen concentrations (including barren intervals in Zone X), and
tree pollen is sparse or absent. Artenisia shows a conspicuous peak in
Zone VIII, which is similar to that found in Zone IV; both of the
Artemisia peaks correspond to N. Atlantic glacial stages. In contrast,
zone IX (750-800 cm) shows an increase in pollen concentrations and %
tree pollen, with conspicuous peaks of Betula and Picea pollen like
those at the base of zone 5. Zone IX corresponds to isotopic stage 9,
which represents a warm climate interval in the North Atlantic, and has
an age range of 297-347 Ka according to Aksu (this volume).

B. Dinoflagellate Stratigraphy

     Figure 10.11 shows the concentrations of Quaternary dinoflagellate
cysts in Core 17 and the relative abundances of the two most common
species, Multispinula minuta and Operculodinium centrocarpum. The
diagram also shows the oxygen isotope data obtained from sinistrally
coiled Neogloboquadrina pachyderma by Aksu (1980) and the
concentrations of subarctic planktonic foraminifera (mainly dextrally
coiled N. pachyderma, Globigerina quinqueloba and G. bulloides) which
are the basis of Aksu's paleoceanographic interpretation of Core 17.

     Quaternary dinoflagellate cysts are common in all the samples from
Core 17. The most conspicuous features in Figure 10.11 are (1)
fluctuations in the abundance of "Atlantic" dinoflagellates; (2)
changes in the ratio of the warm water indicator O. centrocarpum to the
arctic indicator, M. minuta; and (3) large peaks in dinoflagellate
abundances which occur near the boundaries of the interglacials, i.e.
just above the boundary between isotopic Stages 1 and 2, and below the
Stages 5/6 boundary.

     The "Atlantic" dinoflagellates in this study refer to the group of
species which are presently common in Baffin Bay beneath the outer West
Greenland Current (Factor 1, Fig. 10.8a) but have peak percent
abundances in temperate regions of the North Atlantic Ocean. These
species include Operculodinium centrocarpum, Nematosphaeropsis
labyrinthea, Spiniferites ramosus, S. elongatus, S. bulloideus and S.
membranaceus. In Core 17, these species occur mainly in sediments
deposited during isotopic Stages 1, 5, 7 and 9. Multispinula minuta,
the main species in the Canadian Arctic assemblage (Factor 2, Fig.
10.8b) is present in all the samples from Core 17 but it becomes
dominant only in sediments of isotopic Stages 2 to 4, 6, upper Stage 8
and Stage 10, where it is associated with minor amounts of
Brigantedinium species and occasional occurrences of Peridinium
faeroense and Leiosphaera species A.

The "Dinocyst Ratio" (Fig. 10.11) is the ratio between the relative abundances of the dominant species in factor assemblages 1 and 2, i.e. O. centrocarpum and M. minuta. Changes in this ratio clearly correspond to fluctuations in the "Atlantic" cyst concentrations, with the proportion of M. minuta being inversely related to the "Atlantic" cyst peaks. The high total dinoflagellate concentrations at depths of about 50 cm and 470 cm mainly reflect very high numbers of M. minuta, which precede the peaks of "Atlantic" dinoflagellates and correspond closely to the stage 1/2 and 5/6 boundaries. On the basis of data shown in Figure 10.8, it is believed that these dinoflagellate peaks reflect intervals of strong mixing along fronts between water masses of Arctic and Atlantic origin. Other possible explanations for the high dinoflagellate concentrations, however, may include the influx of humic compounds (Prakash and Rashid, 1968) during glacial meltwater events or reduced competition from diatoms in the photic zone during the outflow of turbid meltwater. The occurrence of "meltwater" peaks near the stage 1/2 and 5/6 boundaries, does not agree with the data of Fillon et al. (1981), who interpret sand layers in Labrador Sea cores as indicating a major meltwater event at the stage 4/5a boundary. Smaller peaks coinciding with the presence of "Atlantic" dinoflagellates in Stages 2, upper 5 and 8, however, may be associated with influxes of Atlantic water during or preceding North Atlantic glacial phases.

In general, the "Atlantic" dinoflagellates and subarctic foraminifera in Core 17 show an inverse relationship. There are two alternative explanations for this feature: (i) the acid-resistant dinoflagellates truly reflect glacial-interglacial changes in surface water temperature, whereas the subarctic foraminifera merely reflect intervals of optimum calcite preservation and are not reliable indicators of surface temperature in interglacial stages; or (2) the "Atlantic" dinoflagellates in Baffin Bay are not indigenous markers of surface water temperature but merely reflect influxes of "exotic" populations during intervals of increased Atlantic water inflow. Data from Core 3A (see below) and carbonate dissolution data (Aksu 1983) tend to support the first explanation, but many more cores from Baffin Bay and the Labrador Sea must be studied before this important difference between dinoflagellate and foraminiferal data can be resolved.

Core 3A:

This short core (36 cm) is from a large box core (80-028-3) taken from a water depth of 2288 m on the west side of Baffin Basin (Site 39, Fig. 10.1). The box core recovered an undisturbed sample of very soft

sediment which undoubtedly includes the modern surface layer. The
micropaleontology of Core 3A is being studied by I.A. Hardy at the
Atlantic Geoscience Centre, whose preliminary data are cited here.
Foraminifera and siliceous microfossils were obtained from 1 cm-thick
samples (35 cm^3 volume) that were washed using a sieve of 63 um mesh
size.

The sediments in Core 3A (Fig. 10.12) consist of an upper (0 - 14
cm) siliceous mud unit and a lower (14 - 36 cm) carbonate-rich unit
with abundant sand and granules. The upper mud unit is reddish brown
(10 YR 3/3) with rare granules above and a grayish clay layer at the
base, with a very sharp contact between the clay and the lower sandy
unit. Benthic foraminifera, radiolaria and sponge spicules are common
to abundant in the upper unit which contains no planktonic
foraminifera. The top of the sandy unit (14 - 18 cm) is also
siliceous, with rare benthic foraminifera. At 18 cm, however, abundant
sinistral N. pachyderma appear and they persist in variable amounts to
the base of the core. Figure 10.12 shows the concentrations of
subarctic foraminifera that are found in the core.

Quaternary pollen is very rare throughout Core 3A, which is
consistent with its deep-water, offshore location to the north of the
Atlantic-Arctic air front. The rare palynomorphs include Salix, Rumex,
Cyperaceae, Gramineae and Polytrichum spores above 20 cm core-depth,
and very rare Picea in the lower sediment unit. The pollen-spore
counts are too low (<20/ sample) to attempt a pollen analysis of Core
3A, but the occurrences described above are consistent with the High
Arctic location of the core site and they suggest that Core 3A contains
little redeposited terrigenous or shelf sediment.

Quaternary dinoflagellate cysts occur throughout Core 3A (Fig.
10.12) although the concentrations are very low below 15 cm
core-depth. Despite the fact that the cyst concentrations are much
lower than in Core 17, the inverse relationship between "Atlantic"
dinoflagellates and subarctic planktonic foraminifera is almost
identical to that found at the top of Core 17. In Core 3A, a large
peak in dinoflagellate abundance also occurs in about the same
stratigraphic position as in Core 17, i.e. just above the highest
occurrence of subarctic foraminifera. Initial results of oxygen
isotope studies of Core 3A, the lithology and foraminiferal
stratigraphy strongly indicate that the base of the clay layer marks
the last glacial-interglacial boundary, and that the carbonate
sediments are correlative with the marine isotopic Stage 1/2 transition
(Aksu, pers. comm.)

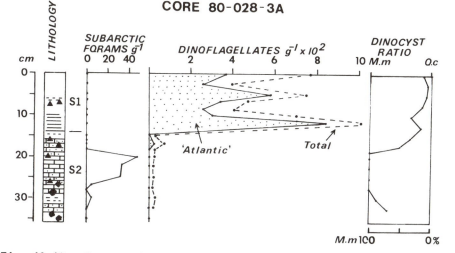

Fig. 10.12. Summary diagram of dinoflagellate cysts and selected
foraminifera data for Core 3A. Dinoflagellate cyst and foraminifera
data have the same meaning as in Figure 10.11. S1 and S2 probably
correspond to isotopic Stages 1 and 1/2 transition of Aksu (1980).
Lithology symbols: unshaded = siliceous mud; dashed lines = silt
laminae; fine lines = clay; triangles = granules; polygons = gravel;
rectangles = carbonate mud; dots = sand.

peak in dinoflagellate abundance also occurs in about the same
stratigraphic position as in Core 17, i.e. just above the highest
occurrence of subarctic foraminifera. Initial results of oxygen
isotope studies of Core 3A, the lithology and foraminiferal
stratigraphy strongly indicate that the base of the clay layer marks
the last glacial-interglacial boundary, and that the carbonate
sediments are correlative with the marine isotopic Stage 1/2 transition
(Aksu, pers. comm.)

     In general, the dinoflagellate stratigraphy for Core 3A is almost
identical to that found at the top of Core 17. However, two
differences exist which may have very important paleoenvironmental
implications. First, the dinoflagellate abundance peak at 12 cm depth
in Core 3A is due to the Atlantic species O. centrocarpum, not the
Canadian Arctic species M. minuta, which forms the peak in Core 17.
Since O. centrocarpum is rare in surface waters of <30%o (Mudie,
unpub. data; Miller et al., 1982), the peak of O. centrocarpum in Core
3A suggests that surface water was more saline in western Baffin Basin
than in Davis Strait during the early Holocene. Secondly, the peak in
Core 3A and the underlying sediments to a core depth of 18 cm contain
high percentages of the cyst species Spiniferites frigidus, which is

indigenous to the Arctic and is presently associated with the warmer
water areas of Baffin Bay (Factor 1, Fig. 10.8A). The postglacial
increase in this indigenous dinoflagellate species indicates that the
peaks of "Atlantic" dinoflagellates in Core 3A and 17 correspond to
local warming of surface water in Baffin Bay, rather than greater
influx of "exotic" North Atlantic cysts. These detailed palynological
data for the Holocene and isotopic Stage 1/2 transition in central
Baffin Bay thus provide further evidence that the subarctic
foraminifera peaks may be better indices of carbonate chemistry changes
in the waters of Baffin Bay than indicators of surface water
temperature. In Core 3A, a large peak in dinoflagellate abundance also
occurs in about the same stratigraphic position as in Core 17, i.e.
just above the highest occurrence of subarctic foraminifera. Initial
results of oxygen isotope studies of Core 3A, the lithology and
foraminiferal stratigraphy strongly indicate that the base of the clay
layer marks the last glacial-interglacial boundary, and that the
carbonate sediments are correlative with the marine isotopic Stage 1/2
transition (Aksu, pers. comm.)

Core 73:

This piston core (80-028-73) was recovered from a water depth of 432 m
on the south flank of Scott Trough (MacLean et al., 1981), where thin
unconsolidated Quaternary sediments overlie bedrock of probable late
Eocene age. The location of Core 73 is close to site 36 shown on
Figure 10.1. The core comprises 2.1 m of soft olive gray sandy mud
with scattered gravel in the top 100 cm, and alternating beds of stiff
gravelly gray mud and soft black mud below. Nearshore benthic
foraminifera (mainly Elphidium excavatum and Cassidulina reniforme) are
present in variable amounts throughout the core (I. Hardy, unpubl.
data). Therefore, it was hoped that Core 73 would provide a continuous
pollen record for the northern Baffin Shelf which could be compared to
that for Frobisher Bay.

The data in Table 10.5, however, show that this hope was not
fulfilled, and they illustrate some of the complexities of
biostratigraphic studies in Baffin Bay. The surface sediment (0 - 5
cm) in Core 73 contains Quaternary pollen and dinoflagellate
assemblages which are quite typical for modern High Arctic neritic
sediments: Betula, Picea, Ericaceae and Sphagnum are the dominant
pollen and spores, while Multispinula minuta, Brigantedinium and
Leiosphaera dominate the dinoflagellate assemblage. The PQ/Q ratio is
1.2. At 16-25 cm, the Quaternary pollen contains more Ericaceae (32%)
and less Betula, but the pollen assemblage is otherwise similar to the

Table 10-5. Summary of biostratigraphic data from Core 80-028-73.
            r=rare(1-5/slide); f=frequent (6-15); c=common (16-30);
            a=abundant (>30).

		CORE DEPTH (cm)								
MICROFOSSILS		0	25	50	75	100	125	150	175	200
		POLLEN AND SPORES								
QUATERN -ARY	Alnus	r	r	f	c	f	c	-	-	-
	Betula	a	f	c	f	c	a	-	-	-
	Picea	a	a	c	f	f	r	-	-	-
	Ericaceae	f	a	f	a	c	f	-	-	-
PQ/Q		1.2	1.2	1.1	0.8	2.3	1.3	a	c	c
		DINOFLAGELLATES								
QUATERN -ARY	M.minuta	a	-	r	r	-	r	-	-	-
	Brigantedin -ium species	a	-	r	r	r	r	-	-	-
	Leiosphaera	c	-	-	-	-	-	-	-	-
PQ/Q		0.6	6.0	3.3	4.5	8.0	15.5	c	a	a
FORAMINIFERA		Elphidium excavatum; Cassidulina reniforme								

surface. In contrast, Quaternary dinoflagellates are very rare, even
though Quaternary foraminifera are common. Abundant well-preserved
Cretaceous dinoflagellates are present in this sample. At 50 - 55 cm,
Quaternary pollen concentrations are relatively high: Betula and Picea
are dominant; Ericaceae and Alnus are subdominant. The PQ/Q pollen
ratio is 1.06, with abundant well-preserved Cretaceous taxa being
present, including Metasequoia and Sciadopitys. Quaternary
dinoflagellates are very rare but Cretaceous cysts are common. Similar
palynomorph assemblages are found in samples down to 125 cm; however,
samples from 150 cm to the base or the core are barren of Quaternary
palynomorphs, although well-preserved pre-Quaternary palynomorphs are
common to abundant.

At present, no convincing interpretation can be offered to explain
the palynological assemblages below 16 cm in Core 73. The Quaternary
pollen percentages do not match those of the Clyde foreland - Scott

Inlet area (Miller et al., 1977) except for the dominance of Ericaceae
in Core 73 and in various interglacial or interstadial soil horizons
onshore (Mode, this volume). Ice-furrowed lateral moraines appear to
flank the south wall of Scott Trough (MacLean, 1978) and these may be a
source of redeposited interstadial pollen and foraminifera, as well as
pre-Quaternary palynomorphs. However, the sparsity of Quaternary
dinoflagellates is puzzling. A better understanding of the
ecostratigraphy of Core 73 might be gained from study of the pollen and
dinoflagellates in the Clyde foreland raised marine sediments and till
deposits. It may also be important to study the onshore and marine
pollen deposits with epifluorescence microscopy (Bujak and Davies,
1981) in order to assess whether homeomorphic Quaternary and Tertiary
Betula, Alnus, Ericaceae and bisaccate pollen are present.

Core 34G:

     This gravity core (76-029-034G), from a water depth of 2275 m on
the east side of central Baffin Basin (site 33, Fig. 10.1), contains
1.2 m of yellowish brown mud, with scattered gravel and sandy laminae
above 40 cm, and alternating gray or brownish gray sandy and clay
layers below 40 cm (Fig. 10.13). A carbonate lithofacies occurs at 40
- 55 cm; therefore, it was hoped that this long gravity core would
provide a detailed palynostratigraphy of the Holocene and late Foxe
glacial - interglacial transition.

     Figure 10.13 summarizes the dinoflagellate, pollen and planktonic
foraminiferal data obtained from Core 34G. From 0 to 20 cm, the total
dinoflagellate concentrations, abundance of "Atlantic" species, and the
"Dinocyst Ratio" are similar to those found at the top of Core 17.
These features and the absence of planktonic foraminifera indicate
Holocene interglacial sediments. The fine sandy mud at 25 cm, however,
is barren of foraminifera and palynomorphs and, although the underlying
carbonate facies (37-55 cm core depth) contains common planktonic and
benthic foraminifera, subarctic foraminifera are absent.

     The sediments below 60 cm in Core 34G contain brown sandy layers
which are barren of foraminifera and palynomorphs. These barren zones
alternate with brownish gray or gray muds containing thin, silty
laminae, sinistral Neogloboquadrina pachyderma and Quaternary
dinoflagellates. The dinoflagellates comprise a low diversity
assemblage of M. minuta and Brigantedinium simplex, with very rare
occurrences of Spiniferites frigidus and Operculodinium centrocarpum.
These predominantly arctic marine microfossil assemblages suggest that
the sediments were deposited during the late Foxe glacial. However,

CORE 76-029-034

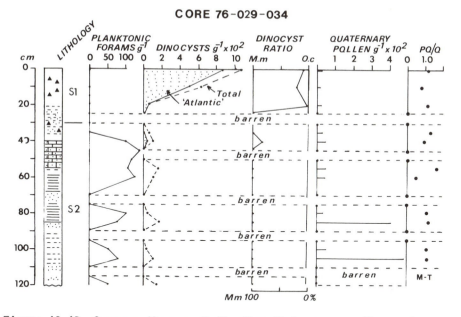

Figure 10.13 Summary diagram of dinoflagellate cyst, pollen and
planktonic foraminifera data for Core 34G. Total = total cysts;
Atlantic = North Atlantic taxa. Dinocyst Ratio is the same as in
Figure 10.11. Planktonic forams are the number of N. pachyderma
sinistral per gram. PQ/Q is the ratio of Prequarternary to Quaternary
pollen and spores. S1 and S2 are isostopic stages. Lithology symbols
are the same as in Figure 10.12, with triangles indicating fine gravel.

Quaternary pollen concentrations are relatively high in the mud layers
below 60 cm, with Betula and Alnus being the dominant taxa. This
suggests that the laminated sediments in the lower half of Core 34G
consist of graded distal turbidites, in which the sandy bases are
barren of microfossils and the upward fining mud layers contain
redeposited interstadial pollen. Thus Core 34G illustrates that the
complexities involved in palynological studies of Baffin Bay marine
sediments are not confined to the till-covered shelf and upper slope
environments but are also manifest in some parts of the deep-water
basin.

CONCLUSIONS

Study of Quaternary pollen and spores in marine sediments from Baffin
Bay shows that steep gradients in pollen concentrations limit the
resolution with which paleoclimatic interpretations can be made for the
offshore area. However, pollen concentrations should be sufficiently

high in muddy nearshore and shelf sediments to obtain counts of >100
per sample for both glacial and interglacial climatic phases. South of
the present summer position of the Atlantic-Arctic air front, pollen
concentrations are usually high enough to obtain meaningful pollen
diagrams that record large-scale climatic changes within interglacial
and interstadial episodes. If this air front moves southward during
the glacial episodes, however, deep-water sediments are likely to
contain insufficient pollen for detailed paleoecological
interpretation.

Isopolls for the main tree and wind-pollinated shrubs show
north-south gradients which appear to be strongly related to the summer
position of the Atlantic-Arctic air front; the ratio of Picea:Pinus
pollen also changes from >1.0 south of the air front to <1.0 in the
north. Isopolls for insect-pollinated arctic shrubs and herbs show
steep east-west gradients which suggest that their distribution is more
strongly related to surface runoff and ocean current transport than air
transport. Pollen of wind- pollinated arctic herbs is most abundant in
nearshore areas, but it also show sporadic abundances north of the
Atlantic- Arctic air front. Sphagnum and Polytrichum spores are almost
ubiquitous in the marine sediments but their abundance varies in a
non-systematic manner which reduces their value for paleoecological
studies.

Initial studies of pollen profiles for marine sediment cores from
southern Baffin Bay can be broadly correlated with pollen profiles from
onshore areas. Pollen in late glacial to Holocene sections of Cores HU
77-159 and Core 17 shows similar changes in dominant pollen taxa, with
a mid-Holocene peak in Betula suggesting a short warm climatic episode,
and with a dominance of Betula Pinus and high PQ/Q ratios marking the
late glacial interval from c. 20 - 10 Ka. In Core 17, most of the
glacial stades contain very low numbers of Quaternary pollen and spores
and low tree pollen percentages. The abundance of Betula and Alnus in
early isotopic Stage 5 and Stage 9 sediments strongly resembles the
pollen assemblages of soils below the Kogalu sediments and suggests a
warmer, wetter climate than presently found in Baffin Bay. Picea
pollen, however, appears to be the best single indicator of N. Atlantic
interglacial stages: its cylical occurrence in Core 17 from southern
Baffin Bay corresponds closely to the major Late Quaternary forest
development phases recorded in northwest Central Europe (Frenzel 1973)
and Pacific North America (Hensser and Shackleton 1979).

Overall, the pollen assemblages in Core 17 suggest that the
onshore climate was relatively warm (Low Arctic or Subarctic) during
the interglacials and that the surface of Baffin Bay was relatively ice

free during these times. However, it is not clear if the reduction of pollen during glacial stages reflects (1) more extensive sea ice cover in Baffin Bay; (2) a change in air circulation resulting in much lower pollen influx from southern ice-free regions; (3) increased export of pollen from Baffin Bay in low salinity surface meltwater; or (4) more extensive glacial ice cover on the land around Baffin Bay. Dinoflagellate data tend to support the first two explanations; in contrast, the subarctic foraminifera and lithological data for Labrador Sea cores (Fillon et al., 1981) suggest that the latter may pertain. The pollen data for Core 34G, however, illustrate the need for more work on the relationship between pollen concentrations and sediment texture in graded mud beds before the low concentration on absence of Quaternary pollen can be safely interpreted as indicating the termination of pollen influx during glacial intervals. Finally, the need to find better means for recognizing reworked homeomorphic pre-Quaternary pollen types is well illustrated by the confusing data obtained from Core 73.

Study of dinoflagellate cysts from the surface sediments of Baffin Bay shows that concentrations of these marine palynomorphs are everywhere much higher than the terrigenous pollen and spores. Maximum total cyst concentrations appear to be associated with areas of strong mixing between ocean fronts and around large islands. However, more studies are needed on sub-ice dinoflagellate populations, and on the response of arctic dinoflagellates to the availability of humic compounds before a definite causal relationship can be established.

Q-mode factor analysis shows that 4 assemblages of dinoflagellate cysts distinguish different environments within the Baffin Bay region. Factor 1, dominated by O. centrocarpum and other species which are common in temperate to boreal regions of the North Atlantic, is confined to sediments below surface water that is warmer than 2°C in summer and has a salinity of >32°/₀₀. Factor 2 is dominated by Multispinula minuta which has its primary distribution in Canadian Arctic shelf sediments and appears to be tolerant of a wide range of summer surface temperatures (<1°-9°C) and salinity (31-33°/₀₀). Factor 3 is dominated by Leiosphaera species A, for which a theca-stage origin remains to be determined by laboratory culture. However, the association of this cyst with areas covered by continuous ice in winter and cold (<1°C), low salinity (30 - 31.5°/₀₀) water in summer suggests that it may be a sea ice alga. Factor 4 is dominated by cysts of Brigantedinium, Gonyaulax tamarensis and freshwater Peridinium species, all of which are associated with nearshore areas of high surface runoff.

Study of dinoflagellates in Cores 17 and 3A shows very similar
sequences of total cyst concentrations, abundance of "Atlantic" cysts
and changes in the ratio of % $\underline{0}$. centrocarpum to % $\underline{M}$. minuta.   In both
cores, the Holocene sediments are dominated by $\underline{0}$. centrocarpum and
other Factor 1 cysts, thus indicating relatively warm, saline surface
water, while the late glacial sediments are dominated by $\underline{M}$. minuta, the
Canadian arctic species.   In Core 17, $\underline{M}$. minuta and Brigantedinium
species are dominant throughout the Foxe glacial stage, and in glacial
stage sediments older than isotopic Stage 5; this suggests that cold,
low salinity water prevailed during the glacial intervals.   During
Stages 5, 7, and 9, however, $\underline{0}$. centrocarpum and the "Atlantic" species
reappear in large numbers, indicating the return of warm surface water
conditions.

The abundance of "Atlantic" cysts thus show an inverse relation to
the abundances of subarctic foraminifera in the Baffin Bay cores.   This
raises critical questions regarding the paleoceanographic
interpretation of these different planktonic microfossils: (1) are the
organic-walled "Atlantic" dinoflagellates non-indigenous and only
transported into Baffin Bay during interglacial stages when they are
abundant in the North Atlantic; or (2) do the calcareous subarctic
foraminifera only seem to indicate warmer water during the glacial
stages because a change in sediment/water chemistry allows for their
preservation at these times whereas they are not preserved during the
interglacials? The first question needs many more dinoflagellate
studies to be carried out for a series of cores from Baffin Bay to the
Northwest Atlantic before it can be answered.   Regardless of the
correct paleoecological interpretation, however, it seems clear that
Quaternary dinoflagellates are useful biostratigraphic markers for
Eastern Canadian Arctic sediments, and their value should increase with
more laboratory and field research on the taxonomy and ecology of
modern dinoflagellates in Arctic waters.

REFERENCES

Aksu, A.E. and Piper, D.J.W., 1979: Baffin Bay in the past 100,000 yr.
    Geology, 7:245-248.
Aksu, A.E., 1980: Late Quaternary stratigraphy, paleoenvironments and
    sedimentation history of Baffin Bay and Davis Strait. Doctoral
    Thesis, Dalhousie University, Nova Scotia. 771 pp.
Aksu, A.E., 1983: Holocene and Pleistocene dissolution cycles in
    deep-seacores of Baffin Bay and Davis Strait: paleoceanographic
    implications. Marine Geology, 53:331-348.
Andersen, O.G. Norden., 1981: The annual cycle of temperature,
    salinity, currents and water masses in Disko Bay and adjacent waters,
    West Greenland. Meddelelser om Grönland, Bioscience 5:3-33.

Andrews, J.T., Webber, P.J. and Nichols, H., 1979: A late Holocene
    pollen diagram from Pangnirting Pass, Baffin Island, N.W.T., Canada.
    Review of Paleobotany and Palynology, 27:1-28.
Andrews, J.T., Miller, G.H., Nelson, A.R., Mode, W.N. and Locke,
    W.W.III., 1981: Quaternary near-shore environments on eastern Baffin
    Island, N.W.T. In: Mahaney, W.C. (ed.), Quaternary Paleoclimate,
    13-44. Geoabstracts, Norwick, England.
Azam, F., Beers, J.R., Campbell, L., Carlucci, A., Holm-Hansen, O. and
    Reid, F.M.H., 1979: Occurrence and metabolic activity of organisms
    under the Ross Ice Shelf, Antarctica, at Station J9. Science, 203:
    451-453.

Baker, S.R. and Friedman, G.M., 1973: Sedimentation in an arctic marine
    environment: Baffin Bay between Greenland and the Canadian Arctic
    Archipelago. Geological Survey of Canada Paper 71-23, 471-498.
Barry, R.G. and Chorley, R.J., 1971: Atmosphere, Weather and Climate.
    2nd Edn. Methuen and Co. Ltd., London. 375 pp.
Barry, R.G., Elliott, D.L. and Crane, R.G., 1981: The paleo-climatic
    interpretation of exotic pollen peaks in Holocene records from the
    eastern Canadian Arctic: a discussion. Review of Palaeobotany and
    Palynology, 33:153-167.
Blake, W.Jr., 1982: Terrestrial interstadial deposits, Ellesmere
    Island, N.W.T., Canada. American Quaternary Association, 7th Biennial
    Conference, Program and Abstracts, 73.
Bujak, J.P. and Williams, G.L., 1980: Dinoflagellates, the grass of the
    sea. Geos, Fall issue, 2-5.
Bujak, J.P. and Davies, E.H., 1981: Neogene dinoflagellate cysts from
    the Hunt Dome Kopanoar M-13 well, Beaufort Sea, Canada. Bulletin of
    Canadian Petroleum Geology, 29:420-425.
Bujak, J.P. (in press). Cenozoic dinoflagellate cysts and acritaarchs
    from Bering Sea and northern North Pacific. Marine Micropaleontology.
Bursa, A., 1961: Phytoplankton of the Calanus Expeditions in Hudson Bay
    1953 and 1954. Journal of the Fisheries Research Board of Canada
    18:51-83.

Coachman, L.K. and Aagaard, K., 1974: Physical oceanography of arctic
    and subarctic seas. In: Y. Herman (ed.), Marine Geology and
    Oceanography of the Arctic Seas. Springer-Verlag, New York. 397 pp.
Collin, A.E., 1965: Oceanographic observations of Nares Strait,
    northern Baffin Bay, 1963, 1964. Bedford Institute of Oceanography
    Report BIO 65-5, 9 pp.
Comtois, P. and Larouche, A., 1981: Morphologie pollinique des Ericales
    du Quebec. Le Naturaliste Canadien, 108:245 - 262.
Coote, A.R. and Jones, E.P., 1982: Nutrient distributions and their
    relationships to water masses in Baffin Bay. Canadian Journal of
    Fisheries and Aquatic Science, 39:1210-1214.
Croxton, C.A., 1979: Palynological studies offshore West Greenland with
    preliminary results from the Kangamuit 1 well. Geological Survey of
    Greenland Report No. 95, Report of Activities 1978, 45-49.

Dale, B., 1976: Cyst formation, sedimentation, and preservation:
    factors affecting dinoflagellate assemblages in recent sediments from
    Trondheimsfjord, Norway. Review of Palaeobotany and Palynology,
    22:39-60.
Dunbar, M.J., 1951: Eastern Arctic waters. Fisheries Research Board of
    Canada, Bulletin No. 88, 131 pp.
Dunbar, M. and Dunbar, M.J., 1972: The history of the North Water.
    Proceedings of the Royal Society of Edinburgh (B), 72:232-240.

Elliot-Fisk, D.L., Andrews, J.T., Short, S.K., and Mode, W.N., 1982: Isopoll maps and an analysis of the distribution of the modern pollen rain, eastern and central northern Canada. Geographic physique et Quaternaire, 36:91-108.

Esso Resources Canada Ltd. 1979: Supplement environmental impact statement for exploratory drilling in Davis Strait region. Report to Government of Canada. 78 pp.

Fillon, R.H., Miller, G.H. and Andrews, J.T., 1981: Terrigenous sand in Labrador Sea hemipelagic sediments and paleoglacial events in Baffin Island over the last 100,000 years. Boreas, 10:107-124.

Fissel, D.B., Lemon, D.D. and Birch, J.R., 1982: Major features of the summer near-surface western Baffin Bay, 1978 and 1979. Arctic, 35:180-200.

Fredskild, B., 1973: Studies in the vegetational history of Greenland Meddelelser om Grønland, Bd. 198, Nr. 4. 245 pp.

Frenzel, B., 1973: Climatic fluctuations of the Ice Age. Case Western Reserve University Press, Cleveland and London. 306 pp.

Gilbert, R., 1982: Contemporary sedimentary environments on Baffin Island, N.W.T., Canada: glaciomarine processes in fjords of eastern Cumberland Peninsula. Arctic and Alpine Research, 14:1-12.

Grøntved, J. and G. Seidenfaden., 1938: The Godthaab Expedition 1928: the phytoplankton of the waters west of Greenland. Meddelelser om Grønland, 82(5). 380 pp.

Harland, R., 1979: Recent and late Quaternary (Flandrian and Devension) dinoflagellate cysts from marine continental shelf sediments around the British Isles. Paleontographic. Abt. B, 164, 87-126.

Harland, R., Reid, P.C., Dobell, P. and Norris, G., 1980. Recent and sub-recent dinoflagellate cysts from the Beaufort Sea, Canadian Arctic. Grana, 9:211-225.

Harland, R., 1981: Cysts of the colonial dinoflagellate Polykrikos schwartzii Butschli 1873, (Gymnodiniales), from Recent sediments, Firth of Forth, Scotland. Palynology, 5:65-79.

Harland, R., 1982a: Recent dinoflagellate cyst assemblages from the southern Barents Sea. Palynology, 6:9-18.

Harland, R., 1982b: A review of recent and Quaternary organic-walled dinoflagellate cysts of the genus Protoperidinium. Palaeontology, 25:369-397.

Heusser, L., 1978: Spores and pollen in the marine realm. In: Haq, B.U. and Boersma, A. (eds.), Introduction to Marine Micropaleontology. Elsevier, New York, 327-339.

Heusser, L.E. and Shackleton, N.J., 1979: Direct marine-continental correlation: 150,000-year oxygen isotope-pollen record from the North Pacific. Science, 204:837-839.

Hsiao, S.I.C., 1979: Phytoplankton and sea ice microalgal data from Frobisher Bay, 1971-1978. Fisheries and Environment Canada, Fisheries and Marine Service Data Report No. 155, 82 pp.

Jones, E.P. and Levy, E.M., 1981: Oceanic CO_2 increase in Baffin Bay. Journal of Marine Research, 39:405-416.

Klovan, J.E. and Imbrie, J., 1971: An algorithm and FORTRAN-IV Program for large-scale Q-mode factor analysis and calculation of factor scores. Mathematical Geology, 3:61-77.

Lichti-Federovich, S., 1974: Pollen analysis of surface snow from the Devon Island ice cap. Geological Survey of Canada, Paper 74-1, A, 197-199.

MacLean, B., 1978: Marine geological investigations in 1977 of the Scott Inlet and Cape Dyer - Frobisher Bay areas of the Baffin Island continental shelf. Geological Survey of Canada, Paper 78-1B, 13-20.

MacLean, B., Falconer, R.K.H. and Levy, E.M., 1981: Geological, geophysical and chemical evidence for natural seepage of petroleum off the northeast coast of Baffin Island. Canadian Petroleum Geology, 29:75-95.

McAndrews, J.H., 1982: Pollen analysis of the 1973 ice core from Devon Island Ice Cap, Canada. American Quaternary Association, 7th Biennial Conference, Program and Abstracts, 133.

Miller, A.A.L., Mudie, P.J. and Scott, D.B., 1982: Holocene history of Bedford Basin, Nova Scotia: Foraminifera, dinoflagellate and pollen records. Canadian Journal of Earth Sciences, 19:2342-2367.

Miller, G.H., Andrews, J.T. and Short, S.K., 1977: The last interglacial cycle, Clyde foreland, Baffin Island, N.W.T.: stratigraphy, biostratigraphy and chronology. Canadian Journal of Earth Sciences, 14:2824-2857.

Mudie, P.J., 1980: Palynology of later Quaternary marine sediments, Eastern Canada. Doctoral Dissertation, Dalhousie University, Halifax, N.S. 638 pp.

Mudie, P.J., 1981: Dinoflagellate cysts in Holocene sediments, Eastern Canadian Arctic. Program and Abstracts, 14th Annual Meeting of American Association of Stratigraphic Palynologists, New Orleans, 36.

Mudie, P.J., 1982: Pollen distribution in recent marine sediments, eastern Canada. Canadian Journal of Earth Sciences, 19:729-747.

Muench, R.D., 1971: The physical oceanography of the northern Baffin Bay region. Arctic Institute of North America, The Baffin Bay-North Water Project, Scientific Report 1, 150 pp.

Muller-Haeckel, A., 1981: The low-light adapted dinoflagellate Gonyaulax catenata. Sarsia, 66:267-272.

Nichols, H., Kelly, P.M. and Andrews, J.T., 1978: Holocene palaeowind evidence from palynology in Baffin Island. Nature, 273:140-142.

Osterman, L.E., 1982: Late Quaternary history of southern Baffin Island, Canada: a study of foraminifera and sediments from Frobisher Bay. PhD thesis, University of Colorado, Boulder. 380 pp.

Pelletier, B.R., Ross, D.I., Keen, C.E. and Keen, M.J., 1975: Geology and Geophysics of Baffin Bay. Geological Survey of Canada, Paper 74-30, 2:247-258.

Pettersen, S., Jacobs, W.C. and Haynes, B.C., 1956: Meteorology of the Arctic. United States Naval Operations for Polar Projects, Washington, D.C. 207 pp.

Pingree, R.D. and Mardell, G.T., 1981: Slope turbulence, internal waves and phytoplankton growth of the Celtic Shelf-break. Philosophical Transactions of the Royal Society of London, A:302, 663-682.

Polunin, N., 1948: Botany of the Canadian Eastern Arctic. Part III. Vegetation and Ecology. National Museum of Canada Bulletin 104, 304 pp.

Prakash, A. and Rashid, M.A., 1968: Influence of humic substances on the growth of marine phytoplankton: dinoflagellates. Limnology and Oceanography, 13:598-606.

Prakash, A., Medcof, J.C. and Tennant, A.D., 1971: Paralytic shellfish poisoning in eastern Canada. Fisheries Research Board of Canada Bulletin 177, 87 pp.

Reid, P.R. and Harland, R., 1977: Studies of Quaternary dinoflagellatecysts from the North Atlantic. In: Elsik, W.C. (ed.), Contributions of stratigraphic palynology. Vol. 1, Cenozoic palynology. American Association of Stratigraphic Palynologists, Contributions Series 5A, 147-169.

Ritchie, J.C. and Lichti-Federovich, S., 1967: Pollen dispersal phenomena in arctic-subarctic Canada. Review of Paleobotany and Palynology, 3:255-266.

Short, S.K. and Jacobs, J.D., 1982: A 1100 year paleoclimatic record from Burton Bay-Tarr Inlet, Baffin Island. Canadian Journal of Earth Sciences, 19:398-409.

Simpson, J.H., Tett, P.B., Argote-Espinosa, M.L., Edwards, A., Jones, K.J. and Savidge, G., 1982: Mixing and phytoplankton growth around an island in a stratified sea. Continental Shelf Research, 1:15-31.

Tan, F.C. and Strain, P.M., 1980: The distribution of sea ice meltwater in the eastern Canadian Arctic. Journal of Geophysical Research, 85 (C4):1925-1932.

Turpin, D.H., Dobell, P.E.R. and Taylor, F.J.R., 1978: Sexuality and cyst formation in Pacific strains of the toxic dinoflagellate Gonyaulax tamarensis. Journal of Phycology, 14:235-238.

Wall, D. and B. Dale., 1968: Modern dinoflagellae cysts and evolution of the Peridiniales. Micropaleontology, 14:265-304.

Williams, G.L. and Bujak, J., 1977: Cenozoic palynostratigraphy of offshore Eastern Canada. In: W. Elsik, (ed.), Contributions of Stratigraphic Palynology, v. 1. Cenozoic Palynology. American Association of Stratigraphic Palynologists, Contribution Series No. 5A, 14-47.

Williams, G.L., 1978: Dinoflagellates, acritarchs and tasmanitids. In: Haq, B.U. and Boersma, A. (eds.), Introduction to Marine Micropaleontology. Elsevier, New York, 293-326.

11   The Devon Island ice core and the glacial record

R. M. Koerner and D. A. Fisher

Ice cores are repositories of past atmospheric conditions at each ice cap drill site. Ice core parameters are an integration of atmospheric conditions obtained along the entire air mass trajectory of the parameter studied, be it water vapour (0-18), dust, chemical elements, or pollen. Thus ice cores provide valuable proxy data for studies of climate change.

The Polar Continental Shelf Project began its ice-coring programme in 1964 under the direction of S. Paterson. The first core was through Meighen Ice Cap in 1965 (M-65). Subsequently, cores have been taken on the Devon Island Ice Cap in 1971, 1972, and 1973 (D-71, D-72, D-73) and on Agassiz Ice Cap on Northern Ellesmere in 1977 and 1979 (A-77 and A-79). The details of each core are listed in Table 11.1.

The Meighen Ice Cap provided a discontinuous climate record of some 4000 years (4 ka) through analyses of dirt concentrations, oxygen isotopes, bubble and ice texture. The results are summarized in Table 11.2.

In this paper we will concentrate on the analyses of the Devon Island Ice Cap cores as these are the most relevant to the climatic history of the Baffin Bay region. The Devon Island ice cores cover a time span of about 100,000 years from the present back to, questionably, the last interglacial. The time scale for the cores has been discussed at length elsewhere (Paterson et al., 1977; Koerner and Fisher, 1981). It is based on radioactive methods (Si^{32}, C-14), measurement of the vertical strain rate of the ice at the drill site (Paterson, 1976), and analysis at several depths of seasonal particulate cycles (Koerner, 1977a). We believe the time scale is accurate to ±5% to 5 ka BP. Beyond that the cores have been cross-correlated with those from Camp Century in Greenland (Fig. 11.1)

TABLE 11.1. Ice cores, Polar Continental Shelf Project

LOCATION	YEAR	SURFACE ACCUMULATION RATE g cm⁻²y⁻¹	MEAN ANNUAL AIR TEMP °C	BED TEMPERATURE °C	SURFACE ELEVATION m. asl.	DISTANCE FROM ICE DIVIDE KM.	CORE LENGTH M.
Meighen Ice Cap	1965	20	-17	-16	270	0.0	121
Devon Island Ice Cap	1971	23	-23	-18	1800	0.3	212
	1972	23	-23	-18	1800	0.9	299
	1973	23	-23	-18	1800	0.9	299
N. Ellesmere Agassiz Ice Cap	1977	17	-24	-17	1670	1.2	338
	1979	11.5	-22	-19	1670	0.2	139

All cores are from surface to bedrock except for the 1971 core.

TABLE 11.2. Climate history of hte Meighen Ice Cap, Northwest
Territories

Depth Interval (m)	Estimated Time Interval (years BP)	Climate and Balance at the Core site
0	Present - 80	Ablation surface formed during period of negative balance which removed a maximum of 13 m of ice at the core site
0-24	80 - 390	Positive balance.
24-44	390 - 560	Relatively little summer melt, coldest period in ice cap's history, positive balance.
44-54	560 - 660	Core site on a well-drained slope, positive balance.
54	660 - 2500/2000	Long period of negative balance which reduced the ice cap in size.
54-116	2500/2000 - ?	Overall positive balance with several short periods of negative balance.
116-121	? - 3000/4500	Positive balance, with the ice cap covering most of Meighen Island.

where higher accumulation rates and a thicker ice sheet have allowed
dating of the core to beyond 10 ka BP (Hammer et al., 1978). For ice
older than 20 ka the time scale was developed by Fourier spectral
analysis of the 0-18 values (Dansgaard et al., 1971) and, more recently
by comparison with a deep sea record (Dansgaard et al., 1983).

Because of the increasing inaccuracy of the time scale beyond 5 ka
BP (and especially beyond 10 ka BP) and possible flow discontinuities
in the core, we will refer only to the major parameters in the older
section of the core.

PREVIOUS WORK

We will now outline published results to serve as background to the
main context of the paper.

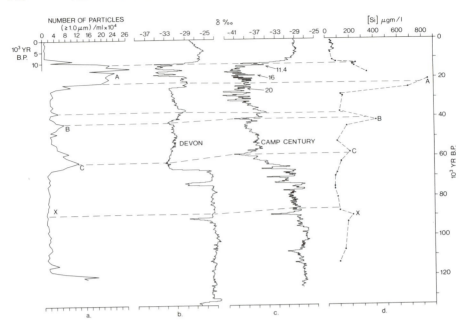

Figure 11.1 Microparticle, Si and oxygen isotope concentrations in Devon Island Ice Cap and Camp Century cores. The peaks marked at A, B, C, and D are referred to in the text. The arrowed numbers refer to more accurate dates based on annual layer thicknesses (Hammer et al., 1978).

Holocene (0-10 ka BP)

The Devon Island Ice Cap, at least throughout the Holocene and the last ice age, is believed to have remained independent from both the Laurentide ice sheet and the other Queen Elizabeth Islands' ice masses (Paterson, 1977; Fisher, 1979). As an independent ice cap it is unlikely that the ice cap at the drill site location was more than 100-200 m thicker than it is at present because the dimensions of the ice cap are constrained by deep channels at its margins.

The Devon Island drill sites are within 1-2 km of the top of the flowline. In contrast the Camp Century, and more recently the Dye 3, ice cores in Greenland were drilled from sites several kilometers down the flowline. Various substantial thickness changes have been suggested for the Camp Century flowline over the Holocene/Wisconsin transition (e.g. Raynaud and Lorius, 1973; Fisher, 1979).

Consequently, we feel that of all the Northern hemisphere cores the Devon Island ice core record for the Holocene and the

Holocene/Wisconsin transition periods best represents a mainly climatic effect, only mildly affected by thickness changes during the ice cap's history.

The Devon 0-18 record shows the familiar Holocene/Wisconsin (Ocean Isotope Stages 1/2 = OIS 1/2) 0-18 transition. The magnitudes of this 0-18 change vary from place to place with Devon Island showing the smallest change as it represents a dominantly climatic one. The most striking feature of the Holocene 0-18 record is the steadily decreasing 0-18 values from 4.6 ka BP to the 'Little Ice Age' some 200 BP. The warming phase of this century terminated in the early 1960s. Our mass balance records for the Devon Island Ice Cap show no significant trend since the small cooling step 20 years ago. The overall Holocene cooling over the past 4.5 K years is 2.7-3.5°C.

A study of ice layering in the Devon ice cores has been interpreted in terms of summer climate (Koerner, 1977c). Warm summers effect more ice layering and cold summers fewer. This record shows a relatively warm period 0.4 to 0.6 ka BP and a cold period 0.1 to 0.4 ka BP. There is a 20 to 50 year lag of summer climate change behind the annual average climate change as given by the 0-18s.

Although we have detected a solar effect in our 0-18 records (Fisher and Koerner, 1980; Fisher, 1983) we have not determined the driving force behind the cooling between 5 ka to the present. There are no significant trends in the concentration of dust and acid in the cores, thus ruling out a volcanic cause. However, there is a decreasing level in the concentration of certain ions and metals over the 5 ka period (e.g., Mg and Na are presently half the 5 ka BP concentration). In this case the nearby 'North Water' (a large polynya between Greenland, Devon and Southern Ellesmere Island) appears to be exerting a diminishing effect on atmospheric circulation. A decrease in formation of marine aerosols from the North Water, indicated by the falling ion and cation concentration in the Devon Island cores, suggests that Baffin Bay may be now contributing less precipitation to the ice cap. As precipitation from local sources has less negative 0-18s than that from further south a decreasing supply of local moisture may play a part in the 5 ka record of decreasing 0-18s at the drill site.

Pre-Holocene (>10 ka BP)

Figure 11.1 shows the 0-18 profile of the lower parts of the Devon Island cores. There are large 0-18 steps at (c.) 10 ka and 65 ka BP, very negative 0-18 values at 12.8 ka BP and isotopically warmer 0-18 values than present in the lowermost >, 65 ka BP, section.

The microparticle concentrations (particles > 1.0 um in diameter) show peaks at 12-22, 40, and 60 ka. Similar peaks in Si concentrations have been found in the Camp Century core (Cragin et al., 1977), and have been attributed to periods of high turbidity caused either by exposure of continental shelves with lower sea levels and/or the presence of poorly vegetated areas around the margin of the Laurentide and Fenno-Scandian ice sheets (Cragin et al., 1977; Fisher, 1979). The peaks rise above generally high particulate/Si concentrations in what we identify as OIS 2, 3, and 4. The same high dust levels have been found in the glacial stages of all High Polar ice cores (Thompson, 1977; Hammer et al., 1978; Lorius et al., 1979).

RELATIONSHIP BETWEEN THE DEVON ISLAND ICE CORES AND BAFFIN ISLAND GLACIAL GEOLOGY

Baffin Island was dominated by the northeast part of the Laurentide ice sheet (Foxe Dome) during the last ice age (Dyke et al., 1982) whereas Devon Island maintained an independent ice cap. We need not, therefore, expect a priori a direct correlation between the two areas. It has been suggested that early buildup of the ice on Baffin Island, possibly of a proto-Barnes ice cap, was dependent on high snow accumulation rates (Barry et al., 1975; Andrews et al., 1974). The high snow accumulation rates were themselves dependent on the presence of a warm ocean offshore. There is ocean core evidence to support this argument (Ruddiman et al., 1980; Aksu, this volume). The Laurentide ice sheet itself, once established, is seen as forming an effective moisture barrier to the High Arctic. Thus the main ice sheet's development could have caused a retreat of the early expanded Barnes Ice Cap (Foxe Dome?) and the Queen Elizabeth Island ice masses from 70 ka BP into the full glacial (OIS 2, 3, and 4) period. However, while there may have been substantially different responses of the southern and north to northeasterly margins of the Laurentide ice sheet there may have been mainly a lag difference between those northerly margins and the Queen Elizabeth Island ice masses.

The Holocene Period

Throughout the following discussion we will refer to time series dated by C-14 techniques in terms of C-14 years. Ice cores and the large part of ocean cores are dated in terms of calendar years. Because of the problems involved in all forms of dating it should be borne in mind that imprecise dating precludes the possibility of accurate cross-correlations unless they are made between elements of the same core (e.g., Ruddiman et al., 1980; Heusser and Shackleton, 1979).

During the Holocene period (OIS-1) the ice masses on Baffin Island retreated slowly from a 8 ka BP maximum. Since then there have been small but significant advances at 5, 1.7, 0.7, and 0.25 ka BP; fjord heads were not deglaciated until 6.3 - 5.0 ka BP (Andrews et al., 1970; Andrews and Ives, 1978). In Southern Ellesmere Island the outer east coast of Ellesmere Island north of 78°36'N and much of Makinson Inlet were glacier-free 9 ka BP. Glacier ice retreated from the northern end of the inlet 7 ka BP. There was less ice than present 2.6 to 5.2 ka BP in an area north of Makinson Inlet. Advances occurred 1.0 ka and 0.1 ka BP (Blake, 1981).

The deglaciation of Devon Island began 9.5 ka BP (Barr, 1971) and the first evidence of deglaciation of Jones Sound dates at 9.0 ka BP (Blake, 1975). All of the dates in this discussion so far are C-14 determined.

The initial retreats from the glacial period and into the Holocene (OIS 2 into 1) are not dissimilar between the Southern Queen Elizabeth Islands and eastern Baffin Island but there is no evidence for a 5.0 ka BP advance in Southern Ellesmere. The 5.0 ka BP advance on eastern Baffin Island occurs during a warm period on Southeast Ellesmere when there was a high influx of driftwood through the, by then, relatively ice-free channels of the Southern Queen Elizabeth Islands (Blake, 1981).

The glaciers in Southern Ellesmere, Western Axel Heiberg and Northern Ellesmere Islands are now at their maximum positions for the past 0 to 5.0 ka (Blake, 1981: Muller, 1966; Hattersley-Smith, 1972) and there is evidence from southern Bylot that at least one glacier there is at its maximum position for the past 8.0 ka (Dilabio & Shilts, 1979).

How does this compare to the ice core 0-18 record? If we interpret the 0-18 record as reflecting temperature and, through that the mass balance of the Devon Island Ice Cap, we would date the main phase of deglaciation at 10 ka BP with minimum ice volume at 4.6 ka BP. The warmest part of the ice core record based on the 0-18 values and also on the high concentration of marine aerosols (indicating a relatively open Baffin Bay) is centered on 5 ka BP. Since that time, the ice cores suggest glacier readvance to a 0.2 ka BP maximum. Taking into consideration the errors inherent in comparing calendar and C-14 dates, the agreement between the glacial geology and ice core record for the southern Queen Elizabeth Islands is good.

However, as we have discussed elsewhere (Koerner and Fisher, 1981) our ice core records do not indicate the separate Holocene advance and retreat phases of the past few thousand years. Furthermore, the mass balance of the northwest part of the Devon Island Ice Cap and of other Queen Elizabeth Island ice caps is close to a steady state now (Koerner, unpublished data). Yet the present 0-18 value is isotopically colder than for most of the past 5 ka. There is simply not a direct linear relationship between 0-18 (and maybe temperature) and mass balance. We will develop this point more fully in the next section.

THE LATE WISCONSIN/HOLOCENE OXYGEN ISOTOPE STEP (OIS - 2 to 1)

The large shift in 0-18 values from isotopically cold to isotopically warm ice ending about 10 ka BP has been seen in all ice cores from both the Arctic and Antarctic. There is a strong contrast in the ion concentration between ice from above and below the 0-18 step. The ion concentration in OIS-2 ice is the lowest in the Devon cores and contrasts to the highest levels just above it in 5 ka old ice. We attribute the high ion concentration in ice 5 ka old to the openness of the nearby Baffin Bay (Blake, 1981). The open water there probably had the effect of bringing higher precipitation rates to the slopes facing Baffin Bay. It has been suggested that the same effect allowed the growth of relatively thick ice caps to form on these same east-facing slopes of Eastern Devon and Southeast Ellesmere Islands so that the ice there may be thicker now than in the late glacial periods (OIS-2) (Koerner, 1977b). In contrast the pre-Holocene ice (OIS-2, 3, and 4) may have been formed under much lower precipitation rates with no nearby moisture sources in Baffin Bay.

Previously it was calculated that 8% of the present precipitation at the Devon Island drill sites is of local origin (Koerner and Russell, 1979); local in terms of Baffin Bay or, more specifically, the North Water. If, as seems likely, Baffin Bay contributed no moisture during OIS-2, 3, and 4 then $2^\circ/_{oo}$ of the $7^\circ/_{oo}$ climatic 0-18 step at 10 K a BP (Paterson, 1979) may be attributable to a change in the local moisture availability. We then have a non-Baffin Bay climatic effect of $5^\circ/_{oo}$.

It has been common practice to divide the O-18 'step' at OIS-1/2 into several components. The two basic ones are climatic and topographic (e.g. Hooke and Clausen, 1982). The calculations are essentially imprecise and potentially misleading. Dansgaard et al. (1973) have demonstrated an empirical relationship between ice cap elevation and O-18. This relationship has been used to estimate the elevation (adiabatic) effect on the OIS-1/2 O-18 step (see Paterson, 1977); less negative O-18 implies, in part, a lower drill site elevation. However, the argument presupposes that the O-18/elevation relationship hold true for both the Holocene and glacial stages. Although the relationship may obtain today on the slopes of the Greenland ice sheet, and also on the eastern side of the Devon Island Ice Cap (Koerner and Russell, 1979), it does not obtain for most of the present ice masses of the Queen Elizabeth Islands (Koerner, 1979). In the latter case depletion is from isobaric processes; O-18 values bear no significant relationship to elevation.

Consider, as an example, the OIS-1/2 O-18 step in the ice at the edge of the Barnes Ice Cap (Hoke and Clausen, 1982). The step is from $-38°/_{oo}$ in glacial stage ice to $-23°/_{oo}$ in Holocene ice. $7°/_{oo}$ of the step is attributed to a climatic change effect (after Paterson et al, 1977), $3°/_{oo}$ to a superimposed ice fractionation effect (Koerner et al., 1973) and the remaining $5°/_{oo}$ to an elevation effect. A $5°/_{oo}$ O-18 change is equivalent to an 800 m change in elevation. As the Barnes Ice Cap is a remnant of the much larger Foxe Dome we do not know the flowpath of the glacial stage ice and can at best only assume that 800 m is a minimum elevation effect. However, whereas we have argued that it is impossible to make an accurate determination of the elevation effect on O-18 over the OIS-1/2 transition, we would point out the O-18 change from glacial stage to Holocene ice is greatest at those stations that have most likely accurate determination of the elevation effect on O-18 over the OIS-1/2 transition, we would point out the O-18 change from glacial stage to Holocene ice is greatest at those stations that have most likely undergone a change in elevation, i.e. Barnes Ice Cap and Camp Century.

Fisher (1976), from cross-correlation of the Devon Island cores and the Camp Century core O-18 profile, considered the OIS-1/2 O-18 steps were synchronous. Since then, cross-correlation of particle peaks between the cores (Fisher, 1979) has substantiated the synchroneity of the two O-18 steps. We have also found the step occurred close to 10 ka BP on northern Ellesmere Island (Fisher et al., 1983). Yet the ice masses in each of the 3 cases are different in size and location and the climate/glacier response lags must be quite different. The synchroneity of the three O-18 steps on northern

Ellesmere, Devon Island, Northeast Greenland (Camp Century) (and also at Dye-3 in southern Greenland), despite different glacier response times, suggests very strongly that oxygen isotopes are related to the climate of a very large area and not to the mass balance of the drill site's ice cap. A similar argument further weakens the existence of an elevation effect on the 0-18 step as the different response time of each ice mass would effect a non-synchronous elevation-determined 0-18 change.

The question remains as to what sort of climatic change would cause such a rapid and large change in the 0-18 value of the ice. We have essentially a 7°/₀₀ 0-18 change (this incorporates the 'local' change wich is climatically caused too) taking place over about 100 years in the Devon cores. Most of this change occurred over about 500 ka years. In Antarctica (Epstein et al., 1970; Lorius et al., 1979; Barkov et al., 1975) the OIS-1/2 0-18 change is not as rapid. A rapid climatic change occurring 12-15 ka BP can be seen in the European record (Woillard, 1979 and 1982) and in the ocean record (e.g. Sancetta et al., 1973; Ruddiman et al., 1980).

What could cause such a large step in the various climatic records? A catastrophic collapse of the main Laurentide ice sheet occurred 8 ka BP (C-14 years). Deglaciation was rapid occurring over a few hundreds of years (Andrews and Ives, 1972). However, the deglaciation must have been in the form of rapid melting of the ice sheet for some distance in from the margins rather than calving along its northern margins and subsequent export of floating ice through Hudson Strait, since there is a lack of evidence for increasing 'ice-rafted' sediments in the Labrador Sea (Fillon et al., 1981), and also because increased influx of melting ice along the path of the northern 'ice masses' moisture supply would cause a dramatic 'cold' spike in the ice-core 0-18 records; this is not the case. There also is good evidence of an increased influx of Laurentide ice sheet meltwater into the Gulf of Mexico which peaked 11,500 C-14 years ago (Emiliani et al., 1975). The 8 ka BP (C-14 years) deglaciation event is largely related to the influx of marine water into what is now Hudson's Bay. It is associated with the formation of an extensive moraine system (Cochrane) indicative of ice sheet readvance. Such a readvance appears unlikely to be associated with a volume increase but rather with an area increase. Possibly at this point in time atmospheric circulation patterns changed dramatically; it is this change we see in the 10 ka BP 0-18 step. The dramatic nature of the 0-18 step and correlative changes in vegetation and the ocean on a global scale further demonstrate that the various ice-core 0-18

profiles are integrations of climatic change over areas of continental
dimensions as well as of local changes in ice cap size and climate.

PRE-HOLOCENE TIMES

Time scale

We equate the very negative 0-18 section of our Devon Island ice cores
with OIS-2, 3, and 4. The less negative section extending to 2.7 m
above the bed has been considered to represent OIS-5 (Paterson et al.,
1977). We will illustrate later that pollen concentrations in the ice
suggest that the lowermost 0.5 m may represent interglacial OIS-5e.

Dansgaard et al. (1983) have changed the Camp Century time scale
to fit the ocean-core record. Thus they change the 0-18 change
previously at 60-70 ka BP to 115-125 ka BP. Geometrically the fit
between the ice core and certain ocean cores is thereby improved. The
isotopically warmest 0-18s are then identified as interglacial (OIS-5e)
rather than glacial (OIS-5a-d) ice.

We disagree with the new Camp Century time scale. If we compare
the concentration of microparticles (>0.65 um) in the Devon Island
cores with the Si concentration in Camp Century core and then with the
Si concentration rates in the ocean (Fillon et al., 1981) we find 3
distinct 'dirt' peaks. In the ocean record where dating assumes
constant sedimentation to the Brunhes/Matuyama reversal, the peaks are
dated at 23, 42, and 52 ka BP. Fisher (1976) originally dated the
Devon Island cores beyond 5 ka BP by cross-correlation of the 0-18
profile to that of Camp Century. Subsequently, he found (Fisher, 1979)
that microparticle peaks in the Devon core at 12-22, 42, and 60 ka BP
substantiated the 0-18 correlation.

Pollen counts in meltwater generated by the drilling process
(Lichti- Federovich, 1975; McAndrews, in press) have shown a
significantly low concentration of pollen between Devon Island ice
dated from the Camp Century time scale as 5-6 ka BP and that at c. 120
ka BP. High pollen concentrations are found in ice deposited 0-5 ka
BP. The pollen is of exotic origin from south of the northern limit
for Alnus. The early Holocene Pollen drought we attribute to the
remaining presence of the Laurentide Ice Sheet in the pollen source
area and then a subsequent revegetation lag. The pollen drought
throughout OIS-2, 3, and 4 may be explained again by the continuing
presence of ice, of varying areal dimensions, in the pollen source

region to the south. The absence of pollen in Devon ice which
Dansgaard et al. (1983) identify as interglacial (OIS-5e) seriously
questions the new Camp Century time scale.

The microparticle Si sand-sediment and pollen records strongly
suggest the earlier Camp Century and hence the Devon Island ice core
time scale, is reasonable. In fact, recent evaluation of the pollen
data shows that part of the lowermost core (dated at > 120 ka BP by
cross-reference to the spectrally determined Camp Century core) has
pollen concentrations similar to those of ice formed in the last 500
years. The lowermost ice in our Devon cores therefore may represent
the last interglacial (OIS-5e) although pollen from a single core
increment is weak evidence.

Comparisons with the glacial geology record

We will now compare the Devon Island 0-18 profile in pre-Holocene ice
with the glacial record for the same period. The 0-18 record shows a
full glacial period between 70 and 10 ka BP but warmer than present
0-18s in ice which our preferred time scale and the pollen record
identify as the early part of the last glacial period (i.e. OIS-5).
The isotopically warm 0-18s in the pre-70 ka BP ice indicate once more
that there is not a simple and direct relationship between 0-18 and ice
sheet mass balance. Paterson et al. (1977) considered the occurrence
of isotopically warm 0-18 in ice deposited at a time when sea level was
lower than present (i.e. OIS-5) may be attributable to a different
moisture source. A greater influx of local moisture (i.e. from Baffin
Bay) would present this effect. Alternatively, a higher preponderance
of summer snow would effectively 'warm' the 0-18s. Barry et al. (1975)
suggest that early ice sheet growth is related to higher snow
accumulation rates in the eastern Arctic associated with a warm ocean
moisture source. Ruddiman et al. (1980) have provided evidence for
such an hypothesis. The pollen drought in what we identify as ice
formed in OIS-5 suggests the ice coverage at this time (or at least a
seasonally very persistent snow cover) extended to more than the
eastern Arctic. Snow accumulation, dominated by increased summer
and/or locally derived snowfall, would account for the otherwise
anomalously isotopically warm 0-18s in the OIS-5 ice.

The Baffin Island glacial record (e.g. Andrews, 1979; Miller, this
volume) and that for the nearby Bylot Island (Klassen, 1981, this
volume) is substantially different from that of the southern margins of
the Laurentide ice sheet and hence to the various records of climate
change indicated in Figure 11.2. Because of the limitations of the ice

core record we cannot interpret it in terms of ice extent in the southern part of the Queen Elizabeth Islands. The obvious correlation between the various climatic records when we consider the major events, i.e. at c. 12, 15-30, 70 and > 110 ka BP show that they were of global extent. Respectively the dates represent a major shift in air mass circulation dependent (?) on the demise of the Laurentide and Fenno-Scandian ice sheets, the coldest period of the last glacial period, the onset of full glacial conditions over North America and Fenno-Scandia, and the last Interglacial.

We do not 'see' the pre-Holocene events on Eastern Baffin Island or Bylot Island in our ice cores. This is because, in terms of sensitivity to the ice core record, the extent of the ice masses on Eastern Baffin, large as they may seem at first sight, are too local to override the more powerful effect on the main Laurentide ice sheet with its dominating feedback effects on the atmospheric circulation.

CONCLUSION

We adhere to the spectrally determined Camp Century time scale, and pollen evidence in the Devon Island core and cross-correlation between ice core particulate Si concentrations and ocean core sand sedimentation rates suggests this is reasonable. Consequently, we find no major discrepancies between the ice core and glacial geology records for the past 100 k years. We see inception of the last glacial phase at c. 100,000 BP, onset of the main glaciation c. 70 ka BP with the coldest phase 20 ka BP. Transition from the last glacial to the present interglacial ended in terms of ice core oxygen isotopes, particulates and cation concentrations at 10 ka BP. The warmest period of the present interglacial peaked at around 5 ka BP. There has been a period of unusual warmth this century which ended in the early 1960s. Since then our mass balance records have been unable to detect any trend in High Arctic climate.

It appears that ice core records do not simply represent local conditions but provide a much more valuable integration of atmospheric and water vapour characteristics over areas of possibly continental, and in some cases, hemispheric proportions.

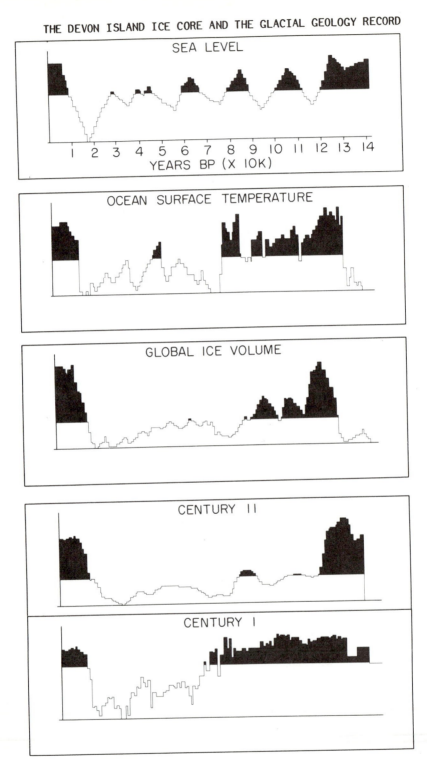

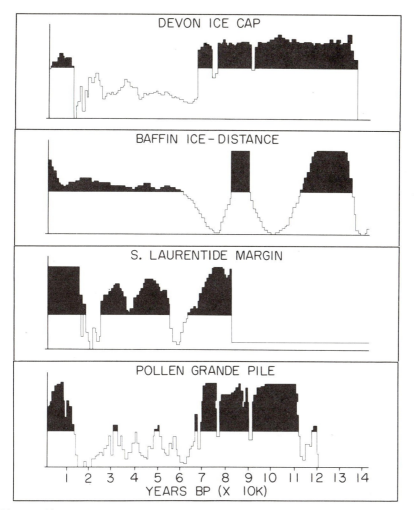

Figure 11.2 Proxy climate data for the period present to 140 ka
BP. The vertical axes are dimensionless but drawn to express
climate in terms of increasing warmth upwards. Thus global ice
volume is shown in inverse with increasing volume downwards; the
same argument is used for Baffin ice-distance and the Southern
Laurentide margin. The time scale for the southern Laurentide
margin is according to the original reference between present and
60 ka BP but beyond that we used Stuiver et al.'s (1978) date for
the St. Pierre interstade. The Pollen, Grande Pile time scale
beyond 70 ka BP is a linear extension of that for the 50-70 K a BP
section. The references are as follows: Sea level (Bloom et al.,
1974), ocean surface temperature (Sancetta et al., 1973), global
ice volume (Ruddiman et al., 1980), Century II is the Camp Century
time scale based on the ocean record (Dansgaard et al, 1983),
Century I is based on the spectrally determined time scale
(Dansgaard et al., 1971), Baffin ice distance (Andrews, 1979),
South Laurentide margin (Dreimanis and Goldthwaite, 1973), Pollen
Grande Pile (Woillard, 1982).

REFERENCES

Aharon, P., J. Chappell, and W. Compston, 1980: Stable isotopes and sea
level data from New Guinea supports Antarctica ice-surge theory of
ice ages. Nature, 283:649-651.
Andrews, J.T., T.G., Barry, R.S. Bradley, G.H. Miller, and L.D.
Williams, 1972: Past and Present Glaciological Responses to Climate
in Eastern Baffin Island. Quaternary Research, 2:303-314.
Andrews, J.T. and J.D. Ives, 1972: Late and post-glacial events (<
10,000 BP) in the eastern Canadian Arctic with particular reference
of the Cockburn moraines and breakup of the Laurentide Ice Sheet.
In: Vasoir, Y. and colleagues (eds.), 'Climatic changes in Arctic
areas during the last ten thousand years'. Acta Universitat
Oulensis, (Series A), No. 3, Geology, 1:149-174.
Andrews, J.T., S. Funder, C. Hjort, J. Imbrie, 1974: Comparison of the
Glacial chronology of Eastern Baffin Island, East Greenland, and the
Camp Century Accumulation Record. Geology, 2:355-358.
Andrews, J.T. and J.D. Ives, 1978: 'Cockburn' nomenclature and the late
Quaternary history of the Eastern Canadian Arctic. Arctic and
Alpine Research, 10:617-633.
Andrews, J.T., 1970: Progress in Relative Sea Level and Ice Sheet
Reconstructions Baffin Island, N.W.T. for the last 125,000 Years. In:
Mörner, N.A. (ed.), Earth Rheology, Isostascy and Eustasy. John Wiley
& Sons, New York.

Barkov, N.I., G.F. Gordienko, Ye.S. Korotkevich, and V.M. Kotlayakov,
1975: Oxygen-isotope studies of the 500-meter ice core from the
Vostok borehole: Information Bulletin Soviet Antarctic Expedition,
90:34-49.
Barr, W., 1971: Postglacial Isostatic Movement in Northeastern Devon
Island: A Reappraisal. Arctic, 24:249-268.
Barry, R.G., J.T. Andrews, and M.A. Mahaffy, 1975: Contintental
Ice sheets: Conditions for Growth. Science, 190:979-981.
Blake, W., Jr., 1970: Studies of glacial history in Arctic Canada. I.
Pumice, radiocarbon dates, and differential postglacial uplift in the
eastern Queen Elizabeth Islands. Canadian Journal Earth Sciences,
7:634-664.
Blake, W., Jr., 1975: Radiocarbon Age Determinations and Postglacial
Emergence at Cap Storm, Southern Ellesmere Island, Arctic Canad.
Geografiska Annaler, 57(A), 1-71.
Blake, W., Jr., 1981: Neoglacial Fluctuations of Glaciers, Southeastern
Ellesmere Island, Canadian Arctic Archipelago. Geografiska
Annaler,63(A), 201-218.

Cragin, J.H., M.M. Herron, C.C. Langway Jr., and G. Klouda, 1977:
Interhemispheric comparison of changes in the composition of
atmospheric precipitation during the late Cenozoic era, in M.
Dunbar., Polar Oceans, p. 682. Montreal, McGill University.

Dansgaard, W., S.J. Johnsen, H.B. Clausen, and C.C. Langway, 1977:
Climatic record revealed by Camp Century ice core, In: Turekian,
K.E., ed., Late Cenozoic Glacial Ages. New Haven, Conn., Yale Univ.
Press. p. 37-56.
Dansgaard, W., S.J. Johnsen, H.B. Clausen, and N. Gunderstrup, 1973:
Stable isotope glaciology: Meddelelser om Grønland, 197(2):1-53.
Dansgaard, W., H.B. Clausen, N. Gundestrup, and N. Reech, 1983: A new
Greenland Deep Ice Core (Record Reconciled with the Camp Century and
Deep Sea Records). Science 218:1273-1277

DiLabio, R.N.W. and W.W. Shilts, 1979: Composition and dispersal of
    debris by modern glaciers, Bylot Island, Canada. In: Schluchter, Ch.,
    ed., Proceedings of an INQUA Symposium on Genesis and Lithology of
    Quaternary Deposits (Zurich), A.A. Balkema, Rotterdam, 145-155.
Dyke, A.S., L.A. Dredge, and J.-S. Vincent, 1982: Configuration and
    Dynamics of the Laurentide Ice Sheet during the Late Wisconsin
    Maximum. Geographie physique et Quaternaire, 36:5-14.

Emiliani, C., S. Gartner, B. Lidz, K. Eldridge, D.K. Elvey, T.C. Haung,
    J.J. Stipp, and M.F. Swanson, 1975: Paleoclimatological Analysis of
    Late Quaternary Cores from the Northeastern Gulf of Mexico. Science,
    189:1083-1088.
Epstein, S., R.P. Sharp, and A.J. Gow, 1970: Antarctic ice sheet:
    stable isotope analyses of Byrd Station cores and interhemispheric
    climatic implications. Science, 168:1570-1572.
Fillon, R.H., G.H. Miller, and J.T. Andrews, 1981: Terrigenous sand in
    Labrador Sea hemipelagic sediments and paleoglacial events on Baffin
    Island over the last 100,000 years. Boreas, 10:107-124.
Fisher, D.A., 1976: A study of two 0-18 records from Devon Ice Cap,
    Canada, and comparison of them to Camp Century 0-18 record,
    Greenland, Ph.D. thesis, Univ. Copenhagen. 287 pp.
Fisher, D.A. and R.M. Koerner, 1980: Some aspects of climatic change in
    the High Arctic during the Holocene as deduced from ice cores.
    Abstracts. Quaternary Climatic Change Symposium. York Univ., Toronto,
    1979. p. 33. Quaternary Paleoclimate, W.C. Mahaney, ed.,
    GeoAbstracts, 349-371.
Fisher, D.A., 1979: Comparison of 10^5 years of Oxygen Isotope and
    Insoluble Impurity Profiles from the Devon Island and Camp Century
    Ice Cores. Quaternary Research 11:299-304.
Fisher, D.A., R.M. Koerner, W.S.B. Paterson, W. Dansgaard, N.
    Gundestrup, and N. Reeh, 1983a: Effect of wind scouring on climatic
    records from icecore oxygen isotope profiles. Nature, 301:205-209.
Fisher, D.A., 1983: Carbon-14 composed to oxygen isotope records from
    Camp Century, Greenland and Devon Island, Canada. Climate Change,
    4:419-426.

Hammer, C.V., 1977: Dating of Greenland ice cores by microparticle
    concentration analyses, in Isotopes and Impurities in Snow and Ice,
    Proc. of the Grenoble Symposium, 1975: IAHS Pub. 118, 297-301.
Hammer, C.V., H.B. Clausen, W. Dansgaard, N. Gundestrup, S.J. Johnsen,
    and N. Reech, 1978: Dating of Greenland ice cores by fow models,
    isotopes, volcanic debris and continental dust. Journal of
    Glaciology, 20:3-26.
Hattersley-Smith, G., 1972: Climatic change and related problems in N.
    Ellesmere Island, N.W.T., Canada, in Vasari, Y., H. Hyvarinen, and
    S. Hicks, eds., Climatic Changes in Arctic Areas during the last Ten
    Thousand Years, Proc. of a symposium held at Oulanka and Keva,
    Finland, Oct. 1971: Acta Universitat, Oulu University, Oulu, Finland,
    no. 3, 137-148.
Heusser, L.E., and N.J. Shackleton, 1979: Direct Marine-Continental
    Correlation: 150,000-year Oxygen Isotope-Pollen Record from the North
    Pacific. Science, 204:837-839.
Hooke, R. LeB., and H.B. Clausen, 1982: Wisconsin and Holocene 0-18
    variations Barnes Ice Cap, Canada. Geological Society of America
    Bulletin, 93:784-789.

Klassen, R.A., 1981: Aspects of the Glacial History of Bylot Island,
    District of Franklin. In: Current Research, Part A, Geological Survey
    of Canada, Paper 81-1A, 317-326.

Koerner, R.M., 1968: Fabric analysis of a core from the Meighen Ice Cap, Northwest Territories, Canada. Journal of Glaciology, 7:421-430.

Koerner, R.M., W.S.B. Paterson, and J.R. Krouse, 1973: Profile in ice formed between the equilibrium and firn lines. Nature Physical Science, 245:137-140.

Koerner, R.M., and W.S.B. Paterson, 1974: Analysis of a core through the Meighen Ice Cap, Arctic Canada, and its paleoclimatic implications. Quaternary Research, 4:253-263.

Koerner, R.M., 1977a: Distribution of microparticles in a 299 m core through the Devon Island Ice Cap, N.W.T., Canada. In: Isotopes and impurities in snow and ice. Proc. of the Grenoble Symposium 1975. IASH Pub. 118:371-376.

Koerner, R.M., 1977b: Ice thickness measurements and their implications with respect to past and present ice volumes in the Canadian High Arctic ice caps. Canadian Journal of Earch Sciences 14:2697-2705.

Koerner, R.M., 1977c: Devon Island Ice Cap; core stratigraphy and paleoclimate. Science, 15-18.

Koerner, R.M. and R.P. Russell, 1979: 0-18 variations in snow on the Devon Island Ice Cap, Northwest Territories, Canada. Canadian Journal of Earth Sciences, 16:1419-1427.

Koerner, R.M., 1979: Accumulation, Ablation and Oxygen Isotope Variations on the Queen Elizabeth Island Ice Caps, Canada. Journal of Glaciology, 22:25-41.

Koerner, R.M. and Fisher, D.A., 1981: Studying climatic change from Canadian High Arctic ice cores, Syllogeus. 33. Climate Change in Canada. ed. C.R. Harington, National Museum of Man, Ottawa, Canada. 195-218.

Lichti-Federovich, S., 1975: Pollen analysis of ice core samples from the Devon Island Ice Cap. Geological Survey of Canada, Part A, Paper 75-1, 441-444.

Lorius, C., L. Merlivat, J. Jouzel, and M. Pourchet, 1979: A 30,000-year isotope climatic record from Antarctic ice. Nature, 280:644-648.

McAndrews, J.H., in press: Pollen analysis of the 1973 Ice Core from Devon Island Ice Cap, Canada.

Müller, F., 1966: Evidence of climatic fluctuation on Axel Heiberg Island, Canadian Arctic Archipelago, in Fletcher, J.O., ed., Proc. of the Symposium on the Arctic heat budget and atmospheric circulation: Santa Monica, Calif., Rand Corp., 136-156.

Paterson, W.S.B., 1976: Vertical strain-rate measurements in an arctic ice cap and deductions from them: Journal of Glaciology, 17:3-12.

Paterson, W.S.B., 1977: Extent of the Late-Wisconsin Glaciation in Northwest Greenland and Northern Ellesmere Island. Quaternary Research 8:180-190.

Paterson, W.S.B., R.M. Koerner, D. Fisher, S.I. Johnsen, H.B. Clausen, W. Dansgaard, P. Butcher, and M. Oeschger, 1977: An oxygen-isotope climatic record from the Devon Island Ice Cap, Arctic Canada, Nature, 266:508-511.

Paterson, W.S.B., 1981: The Physics of Glaciers. Pergamon Press, Oxford, 380 pp.

Raynaud, D. and C. Lorius, 1973: Climatic implications of total gas content in ice at Camp Century. Nature, 243:283-284.

Raynaud, D. and B. Lebel, 1979: Total gas content and surface elevation of polar ice sheets. Nature, 281:289-291.

Ruddiman, W.F., A. McIntyre, V. Niebler-Hunt, and J.T. Durazzi, 1980: Oceanic evidence for the mechanism of rapid Northern Hemisphere glaciation. Quaternary Research, 13:33-64.

Sancetta, D.J., J. Imbrie, and N.G. Kipp, 1973: Climatic record of the
    past 130,000 years in North Atlantic deep-sea core V23-82:
    correlation with the terrestrial record. Quaternary Research,
    3:110-116.
Stuiver, M., C.J. Heussel, and I.C. Yang, 1978: North American Glacial
    History Extended to 75,000 Years Ago. Science, 200:16-21.

Thompson, L.G., 1977: Variations in microparticle concentration, size
    distribution and elemental composition found in Camp century,
    Greenland, and Byrd Station, Antarctica, deep ice cores. In: Isotopes
    and impurities in snow and ice. Proc. of the Grenoble, Symposium,
    1975. IASH Pub. 118:351-364.

Woillard, G., 1979: Abrupt end of the last interglacial in northeast
    France. Nature, 281:558-562.
Woillard, G., 1982: Carbon-14 dates at Grande Pile Correlation of Land
    and Sea Chronologies. Science, 215:159-161.

Survey — Part III

The five chapters in this section have contributed new and important evidence bearing on the oceanographic, atmospheric, and glacial conditions of the Baffin Island, Baffin Bay, West Greenland region. Together with some of the contributions in Part IV (i.e. Feyling-Hanssen, Miller, Klassen, Kelly, Mode, Osterman et al.) these five chapter represent a "window" into conditions during the last interglacial/glacial cycle. Of particular interest is the use of selected cores to study the co-varying changes in a variety of physical and biological parameters. A careful reading of these chapters indicates that although there are areas of consensus there are also vital areas where the records are apparently leading to contradictory conclusions. We are not talking here about "right" or "wrong" but about different elements of a complex oceaographic/atmospheric/ glaciological response. Thus, as Mudie and Short observe, there is an antiphase relationship between the numbers of subpolar planktonic foraminifera and "Atlantic" dinogflagellates. They proposed several mechanisms to explain this relationship but were in favor of a hypothesis that suggested that the numbers of foraminifera in the cores was controlled by dissolution, not by surface oceanographic conditions. Under the present oceanographic regime in Baffin Bay dissolution is occurring and thus it is tempting to equate episodes of dissolution with conditions similar to those that have prevailed during the last 5000 years or so of the Holocene (e.g. Jennings et al., 1984). With such an analog in mind then periods of flow of a "Canadian Current" can be inferred. During the Holocene the development of this cold "acid" current occurred sometime in the last 5000 years (Osterman, 1982) and may be associated with the outflow of surface water from the Arctic Ocean. The stable oxygen isotope stratigraphy of core HU77-027-017 (Aksu, this volume) provides a chronology into which ocean and atmospheric changes can be placed. Thus the biostratigraphy of this core (Aksu, this volume; Mudie an Short, this volume) must also be associated with the statements of Koerner and Fisher (Chapter 11) on the character of the local precipitation source, and most importantly, the presence/absence of far-travelled pollen in the ice core.

Koerner and Fisher noted the presence of pollen in the lower section of the Devon Island ice core followed by a pollen drought. Mudie and Short show that Picea pollen is present in core #017 during early isotope stage 5 but disappears from the sediment at the 5b/5a boundary. This is coincident with a rise in the number of subpolar foraminifera and "Atlantic" dinocysts continue throught stage 5 and only die out during marine isotope stage 4. "Atlantic" dinocysts again appear in core 017 during stage 2 nd are partly coincident with a rise in the number of subarctic foraminifera (Aksu and Fillon; Mudie and Short). The suggestion of "warm" subarctic conditions off Davis Strait during marine isotope stage 5 and into stage 4 strengthens the arguments made by Koerner and Fisher (Chapter 11) that the chronology of the Devon Islnad ice core is approximately correct and should not be shifted to agree with the new stretched chronology proposed by Dansgaard et al. As noted by Koerner and Fisher, the high isotopic ratios in the ice core suggest the importance of a local moisture source in the precipitation reaching the summit of the ice cap.

An important question at this point, and one that will be implicit during the discussions in certain chapters in Part IV is: Can the chronologies from the cores in Baffin Bay, Davis Strait, and northern Labrador Sea (Aksu & Fillon) be related to glacial events? Aksu has mapped the facies in a series of cores from Baffin Bay and in particular noted the occurrence of Facies A--a facies dominated by detrital carbonates probably associated with the glacial erosion of Paleozoic limestones in various areas of Arctic Canada north and west of Baffin Bay (see Chapter 2 or 6 for geology map). Aksu (1981) reported that facies A in the cores occurs during the latter part of stage 5 and into stage 4. In the Labrador Sea, Fillon's discussion of core HU75-58 adds further information to our knowledge of events in the northern Labrador Sea, and thus provides an important link between the studies in Baffin Bay with those in the middle sector of the Labrador Sea between South Greenland and Newfoundland. The accumulation of sand-sized particles in core HU75-58 is linked (although not in a simple fashion) to glaciation around the margin of Baffin Bay/Davis Strait. In terms of the timing of the onset of glaciation the highest rate of sand accumulation occurred during stage 5a and the 5/4 transition. This implies that icebergs were free to drift and melt during this interval of ice accumulation. This episode is followed by one of the low sand accumulation.

The information present in Chapters 7 through 11 has to be used in any evaluation of conditions that pertained during the development of the Wisconsin Glaciation (and local equivalents). Denton and Hughes (1983) proposed that the ice sheet developed under a "cold" scenario when Baffin Bay would have been frozen over. This hypothesis is contrasted with the "warm" scenario (Andrews et al., 1981) which suggested that the development of glaciation over the eastern Canadian Arctic was coincident with subarctic water offshore eastern Baffin Island. If Baffin Bay and Davis Strait were covered by permanent ice during the initial development of terrestrial glaciation we might expect one or more of the following:

1. A cessation in ice rafting as icebergs would no longer be free to drift and melt;

2. The appearance in cores of substantially barren zones where biological productivity would fall close to zero apart from the rare influx of organisms under the ice cover through ocean currents;

3. An absence of pollen in marine sediments as Mudie and Short have inferred that most pollen is due to wind transport. Pollen falling on the ice might eventually be melted-out and released but the presence of an ice shelf in itself would tend to restrict the passage of southerly air masses into the middle and high arctic.

In examining the information presented in Part III it appears that the above conditions were not met during marine isotope stage 5c, b, a and into stage 4. There is, however, a marked change in conditions during the latter half of isotope stage 4. The evidence suggests that if glaciation of Baffin Island and Arctic Canada commenced during stages 4 or 5, than the initial development was associated with an unimpeded circulation within Baffin Bay and conditions "subarctic" in nature. Within stage 4 it is possible that there was substantial ice cover over the ocean. Aksu's scenario for glacial/nonglacial conditions thus appear reasonable despite the concerns of Mudie and Short on the exact meaning of subarctic foraminifera occurrences.

REFERENCES

Andrews, J.T., Miller, G.H., Nelson, A.R., Mode, W.N., and Locke, W.W., III, 1981: Quaternary near-shore environments on eastern Baffin Island, N.W.T. In: Mahaney, W.C. (ed.), Quaternary Paleoclimate. Geoabstracts, Norwich, U.K., 13-44.

Denton, G.H. and Hughes, T.J., 1983: Milaknovitch theory of Ice Ages: Hypothesis of Ice-Sheet linkage between regional insolation and global climate. Quaternary Research, 20:125-144.

Jennings, A. et al., 1984: A Late Holocene dissolution event and changes in sediment sources of Late Pleistocene and Holocene shelf sediments, Baffin Island, N.W.T., Canada. GSA Annual Meeting, 1984, Reno Nevada. GSA Abstracts with Programs. In prep.

Osterman, L.E., 1982: Late Quaternary history of southern Baffin Island, Canada: A study of foraminifera and sediments from Frobisher Bay. Ph.D. dissertation, University of Colorado, Boulder. 380 pp.

Part IV

PLEISTOCENE GLACIAL AND NONGLACIAL STRATIGRAPHY

12 Weathering and soil development on Baffin Island

W. W. Locke, III

The Quaternary history of Baffin Islnd has been established using every means currently possible for stratigraphic reconstruction. Where circumstances have permitted, absolute dating methods such a C-14 (Løken, 1966; Andrews and Ives, 1978) and uranium-series (Andrews et al., 1975) dating have been used to define and correlate stratigraphic events. More recently, amino-acid analysis has provided a relatively inexpensive and efficient method for the correlation of glaciomarine sedimentary sequences (Miller, this volume). All of these methods are recent additions to the arsenal of stratigraphic techniques, and are applicable only where fossiliferous marine units are present. For terrestrial glacial stratigraphy, such marine units must be directly traceable to their subaerial conterparts. Because of the difficulty of such correlation, much of the terrestrial record must be established as it was prior to the widespread use of such techniques--by studies of surface weathering and soil development.

Studies of surface weathering and soil development have a long history in the eastern Canadian Arctic. Prior to the advent of radiometric dating, they provided the only method for relative dating. Until the 1970s, however, weathering was viewed only in a qualitative sense (e.g., Daly, 1902), thus limiting the resolution of the "stratigraphy" developed thereby. Present techniques of assessment of weathering and soil development have the potential to serve as accurate indicators of relative age, with an absolute resolution of about ± 25%. The major advantage of such methods is that they may be applied on any deposit, by any person, at little further expense than the cost of fieldwork. Their major drawback is that much additional effort must be made to improve the resolution of the techniques from their present ± 100%, and that they will probably never achieve the precision of the present absolute dating techniques.

SURFACE WEATHERING

Introduction

Surface weathering studies may be defined as the observation of
characteristics of boulder or bedrock surfaces, or of the morphology of
the deposit itself. Initially (e.g. Daly, 1902), such studies were
purely descriptive, and served to differentiate "weathered" from
"fresh" or "unweathered" rock surfaces. Later (e.g. Dyke, 1979)
observations were applied to a scale which is presumed to be
progressive with time, thus the age of a surface was determined
relative to other nearby surfaces. It is important to note the two
assumptions inherent in surface weathering studies: 1) The difference
in degree of surface weathering between any two deposits ("weathering
break") is primarily a function of age, 2) A progressive scale of
weathering can be defined along which, over time, a surface will pass.

Each of these primary assumptions carries with it at least one
secondary assumption. It is understood that, in many cases, the
additional variables of lithology, sedimentology, morphology, climate,
vegetation (in non-Arctic areas), and observer bias may obscure the
effects of time in the weathering progression. It is assumed that the
design of any study will minimize the effects of these variables, and
optimize that of time. Recent work (e.g. Sugden and Watts, 1977)
implies that some weathering breaks may result from differences in
glaciological properties (i.e. warm-based vs. cold-based ice) which
would, of course, violate the primary assumption. The assumption of a
progressive scale includes with it two secondary assumptions: first,
that the starting point in the progression is unique and is common to
all studies, and second, that there is a single weathering progression
which again is common to all studies. None of these assumptions is
completely justified. The utility of weathering studies depends on how
much they are violated, and to what degree any violations are
understood.

History

In order to understand the present state of weathering studies, it is
important to review past work. Weathering studies in the eastern
Canadian Arctic began with Daly (1902), and consisted of descriptive
differentiation between glaciated and unglaciated terrain. That even
such a simple yes/no decision may be difficult to accomplish is
attested to by the conflict between Daly (1902) and Coleman (1920), who

favored a limited glaciation of Labrador on the basis of extreme weathering forms on the high coastal interfluves, and Odell (1933) and Tanner (1944), who favored complete glaciation of the same area because of the presence of erratics on similar interfluves. On Baffin Island, Mercer (1956) also saw evidence of zones of weathering. He discussed the theories of Daly (1902), Coleman (1920), Odell (1933), and Tanner (1944), and concluded that on southern Baffin Island there was evidence for both recent and past glacierization and for a zone which showed no evidence of active glaciation. The admission that the highest zone may have been covered, but by stagnant ice, was an attempt to reconcile the two points of view. However, the debate still exists, with many authors (e.g. Ives and Borns, 1971) describing an "unglaciated" weathering zone, but Sugden and Watts (1977) suggesting that such zones may well have been glaciated.

Detailed fieldwork in Labrador (Ives, 1957, 1958a and b, 1963; Löken, 1962; Andrews, 1963) established a consistent trend; in most areas there is an altitudinally controlled sequence of (1) mature mountain-top detritus, with few (if any) erratic boulders above, (2) frost-shattered glaciated bedrock with scattered till and numerous erratics above, and (3) a series of prominent ice-marginal features (Ives, 1976). Initially, this sequence was interpreted only directly, as geomorphology. Gradually, however, the indirect, stratigraphic, implication of the differences in weathering became obvious. The differentiation of weathering zones was still largely descriptive, and based on the presence or absence of specific features, such as moraines and erratics.

The work of Pheasant (1971), Boyer (1972), Pheasant and Andrews (1972), and Boyer and Pheasant (1974) added several new dimensions to weathering studies in the eastern Canadian Arctic. First, they (Boyer, 1972) defined the first progressive surface weathering scheme (Table 12.1). Second, they used multiparametric methods in an attempt to provide semi-independent lines of evidence as to the degree of weathering. Third, they used statistical methods to analyze their data. Finally, the data supported the three-fold subdivision of topography into unglaciated (Zone I), formerly glaciated (Zone II), and recently glaciated (Zone III) areas. This conclusion supported that of Loken (1962), in defining the uppermost zone as unglaciated, rather than glaciated, as Ives (1957, 1958a and b) had concluded in Labrador.

The expansion of the two-part weathering record to three parts implies the ability to differentiate between glacial events. It was natural to assume that only events as great as a glaciation (ACSN, 1961) could be differentiated (Burke and Birkeland, 1979), thus the

TABLE 12.1.   WEATHERING SCALES

Surface Characteristics	Boyer, 1972*	Dugdale, 1972	Dyke, 1977	Birkeland, 1978	Locke, 1980	Mode, 1980	Müller, 1980	Brigham, 1981
Fresh	1	1	1		0	1	1	1
Oxidized	-	-	2		1	2	2	2
Grain Relief	2	2	3	weathered/unweathered	2	3	3	3
Grains Removable	3	3	4		3	3	3	3
Grains Easily Removable	3	3	5		3	3	3	-
Grains Aggregates Removable	3	3	-		4	4	4	4
Relief > 1 mm	-	-	-		-	5	5	5
Incipient or Micropits	-	-	6	pitted/no pits	-	6	5	5
Macropits	-	-	7		5	7	6	6
Surface Grus	4	4	8		6	-	-	6
Complete Grus	5	5	9		6	-	-	6
Disintegration	6	-	9		-	-	-	-

*also Nelson, 1978
Note: Dashes indicate criteria not mentioned the classification;
        numbers indicate weathering classes.

youngest zone was presumed to represent the "last glaciation" of the area, and the intermediate zone all prior glaciations. By analogy to the southern margin of the Laurentide Ice Sheet, the lowest weathering zones thus represented the limit of Wisconsin ice in the eastern Canadian Arctic. This analogy is not necessarily correct (below).

Although Boyer and Pheasant (1974) were not able to subdivide Zone III (corresponding to the last, or Foxe, glaciation), the presence within that zone of glacial features at varying distances from ice sources suggested the need for such subdivision. Other workers placed less emphasis on the definition of Zones I and II, and more on the characterization of Zone III.

The first such attempt was that of Dugdale (1972). Using a variant of the progressive scale (Table 12.1), and following Nelson's (1954) assumption of linearity in weathering rates, he interpreted the absolute age of minor cirque glacier fluctuations. Although no events of corresponding ages have been reported by other workers in the area, and Dugdale himself believes that lithologic variation may have caused the observed weathering differences (pers. comm., 1978), his interpretations have not been conclusively disproved.

Dyke (1977, 1979) subdivided the deposits of the last glaciation in the Pangnirtung Fiord/Penny Ice cap area into three groups, not including Neoglaciation, on the basis of a 9-point weathering scheme (Table 12.1). Reevaluation of Dyke's (1977) data and the conclusions of Locke (1979) suggest that the Duval moraines can be subdivided (Mid- and Early-Foxe of Locke, 1979), but that only the mid-Foxe moraines were deposited during the last 125,000 years. (See Szabo et al., 1982; Nelson, 1980; and Brigham, 1981 for discussions of the age of Mid- and Early Foxe events.) Birkeland (1978) had some success in differentiating events within Zone III on the basis of yes/no weathering descriptions (Table 12.1), as did Locke (1980) using a 7-point weathering scale.

Further attempts to subdivide Zone III in a relative manner have not always been successful. Although Nelson (1978) was not able to subdivide his field area on the central Baffin Bay coast of Baffin Island on the basis of weathering, he felt "the lack of major differences in boulder weathering suggest all landforms below the 85 m marine limit do not differ greatly in age" (p. 32), thus there was no failure in the technique.

In contrast, Mode (1980) distinguished three weathering zones in the Clyde area. The youngest/lowest zone is Holocene in age (by C-14), the middle zone includes the deposits of several glacial events, and the oldest/highest zone may not have been glaciated. Although the zones were qualitatively different, individual sites within the zones could not be reliably discriminated.

Müller (1980), following Mercer (1956) on southern Baffin Island, found three distinct weathering zones plus a fourth zone, differentiated on the basis of a lack of erratics. His youngest zone is C-14 dated at less than c. 9,000 yr. BP--the older zones are equated to the last one or more glaciations(s), and unglaciated terrain.

In summary, weathering studies by many single workers on Baffin Island have been successful at differentiating deposits of different ages on a scale of 10^4 to 10^6 years--stadial to glacial events. There are, however, both significant improvements to be made, particularly in the realm of technique and analysis, and problems to be resolved, most notably in the understanding of the variables and mechanisms involved in the weathering process, in the improvement in the resolution of weathering studies, and in the correlation between authors and study areas.

The State of the Art

Weathering studies in the Canadian Arctic have been undertaken purely as local indicators of relative ages of glacial events. To that end, they have been successful. To weathering studies should go major credit for the concepts of glacial limits (e.g. Daly, 1902), nunatak theory (e.g. Ives and Borns, 1971), multiple glaciations (e.g. Boyer and Pheasant, 1974), and multiple advances within a glaciation (e.g., Dyke, 1979). In only few cases (e.g. Nelson, 1978) have weathering studies been ineffective for local relative stratigraphic control.

However, weathering studies have not been effective for regional correlation of glacial events. The only major published attempt at such correlation (Denton and Hughes, 1981) provides clear evidence that no obvious correlations exist. There are two reasons why correlations between weathering studies are not possible now. Firstly, each researcher has designed a weathering scheme to fit his or her own needs (Table 12.1). In most cases such schemes are not directly comparable. Secondly, most of the criteria in the quantitative (numerical) weathering scales tend to be qualitative (e.g. "grains easily removable", "incipient pitting", "grussified"). In the words of Boyer (1972, p. 27); "estimates of weathering are always relative and the references at any given time had to be the material just previously studied. For this reason it was impossible to make all criteria constant." The lack of consistent criteria both within and among weathering studies may explain much of the variability between two studies using similar methods on surfaces of similar age (Fig. 12.1). The contrast in weathering results evident in Figure 12.1 may have resulted from differences in parent material, and weathering processes (marine terraces vs. moraines) and age differences as well as differences in technique. A summary of the results of Dyke (1979), Brigham (1980), Locke (1980), Mode (1980), and Müller (1980) (Fig. 12.2) indicates that, within broad limits, weathering occurs at approximately synchronous rates across southern Baffin Island. However, the lack of precision in these studies makes local correlation of individual deposits difficult, and regional correlation unacceptable.

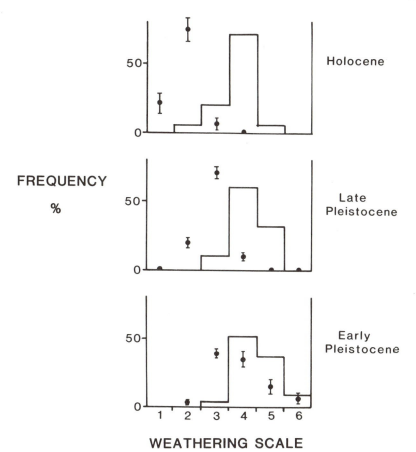

Figure 12.1 Weathering results of Müller (1980) (closed circles, bar indicates one standard error on the mean) and Mode (1980) (histogram)

There are, of course, several other factors which make correlation difficult. The work of Dyke (1977, 1979) and Locke (1980), among others, indicates that a three-fold division of weathered terrain is artificial. The existence of more than three different degrees of weathering implies that the ages of the landscapes in a tripartite system may vary from place to place, depending on the placement of boundaries. Boundary placements should reflect major weathering "breaks". Given the potential variability in actual ice chronologies, lithology of bedrock and surface boulders, and environment of weathering, even major weathering breaks may vary in age from place to place.

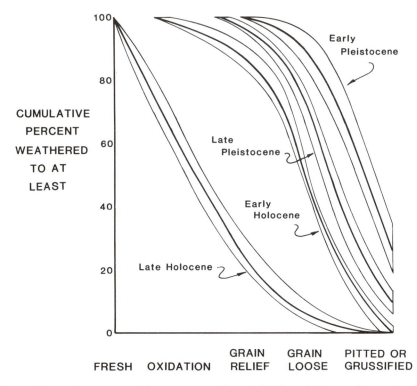

Figure 12.2 Representative curves of cumulative degree of weathering.
Error envelopes include one standard error on the mean; standard
deviations of all older deposits overlap.

     The inability to correlate weathering zones on a regional scale
does not render such studies worthless. Their strength in local
studies has been proven many times, and the universal applicability of
weathering analyses ensures their continued use. However, until more
directly-comparable studies are undertaken, no regional correlation of
weathering data is justified.

The Future of Surface Weathering Studies

If weathering studies are to retain an honored position in surficial
geology, they must keep pace with other methods, both relative and
absolute, of Quaternary stratigraphy. For weathering studies, this
means three major improvements: in understanding, in resolution, and
in correlation.

     The processes of rock weathering, although partially understood
individually (e.g. Ollier, 1969; Birkeland, 1974, Chap. 3), are not at

all understood in combination and in a natural setting. An example of
this lack of understanding can be found in the two variants of the
progressive weathering scheme which end with grus (e.g. Boyer, 1972)
and pitting (Mode, 1980) respectively. Why the bedrock in one area
grussifies and in another does not is not known, but it may affect the
propensity for the rock surface to show loose grains or spalling
fragments--other criteria in the progressive scale. Without a clear
understanding of the variables which affect rock weathering fundamental
conflicts between workers will always exist.

The resolution of weathering studies is surprisingly good--with
units differing by only as little as a factor or two in age being
discernable in some studies (Dugdale, 1972; Dyke, 1977, 1979).
Although this achievement is laudable, it can almost certainly be
improved upon. The first requirement for improvements in the
resolution of weathering studies is that directly measurable
quantitative criteria be used wherever possible. Rather than
describing a boulder surface as oxidized or unoxidized, its color
should be estimated using a color chart. Such a procedure would
minimize discrepancies due to observer variability or bias. Other
quantitative variables which could be used include the length of the
long/short axis of prominent boulders (measured with a caliper), the
minimum/mean/maximum radius of corner curvature (measured with a
template), the minimum/mean/maximum thickness of rinds (measured with
an ocular), and the mean/maximum depth/diameter/volume of weathering
pits (measured with calipers). A multiparametric approach (Burke and
Birkeland, 1979) is more likely to generate useable data than an
approach which trusts to a single variable. Analysis of strictly
quantitative data is easier than that of qualitative data, because it
allows for meaningful analysis of variance and significance limits. In
this way, the work of Boyer and Pheasant (1974) has not been surpassed
in the eastern Canadian Arctic. Given a quantitative, multiparametric
approach, capable of rigorous statistical analysis, all that is needed
is large sample size. Exhaustive sampling techniques are perhaps the
least glamorous, yet most productive of field tools, because they offer
a guaranteed improvement of resolution for the time invested.
Adherence to a strictly quantitative scale or series of scales should
allow a single worker, in a restricted area, to generate a relative
weathering chronology of high resolution. In addition, such techniques
shoud permit direct correlation between workers, again within a
restricted area. Correlation across thousands of kilometers on the
basis of surface weathering is far more difficult because of the
obvious, but unknown, effects of variations in climate and lithology.
It is likely that correlation of weathering zones in the immediate
future will be on the basis of associated C-14, U-series, and

amino-acid ratios. If correlations so derived will permit the
quantification of climatic and lithologic effects, then the reasoning
can be reversed and the surfaces will be dated on the basis of their
weathering characteristics. This should not be expected in the near
future, however, because of the massive amount of field and laboratory
work which will be necessary.

Summary

Qualitative descriptions of weathering and surface materials formed the
basis for the first subdivisions of glacial/nonglacial and recently
glaciated/formerly glaciated/unglaciated terrain in the eastern
Canadian Arctic. No agreement exists as to the presence, in some
areas, of such weathering zones. More detailed semiquantitative and
quantitative studies permitted the subdivision of the most recent zone
by several workers. It appears, however, that increased resolution and
correlation are nearly impossible using present techniques. Future
weathering studies will require a totally quantitative, multiparametric
approach if they are to improve on past results.

SOIL DEVELOPMENT

Introduction

In contrast to surface weathering, where the effects of time (and other
variables) are obvious, the development of soils with time is not
immediately evident to the untrained observer. The need for
excavation, the ubiquity of frost stirring, and the subtle nature of
soil development even in stable deposits combine to limit its
application to the problem of dating. In part offsetting the
difficulty of acquiring soil data is the amount of data available--
physical, chemical, and mineralogical changes can each be assessed in
numerous ways. Soil characterization is inherently multiparametric,
thus is ideal for studies of relative age. Some characteristics (e.g.
organic matter percent) reach steady-state conditions rapidly; others
(e.g. clay mineralogy) have not been shown to reach equilibrium. Such
variety should be ideal for studies of relative or absolute age.

In common with surface weathering, the use of soil development as
a dating criterion assumes; (1) that time is the dominant variable, and
(2) that a progression can be defined through which, with time, any
soil must pass. Implicit in both types of study is the assumption that

the deposit has remained stable (non-cryoturbated) since deposition. This latter assumption is of particular importance with respect to soil development because many criteria are strongly affected by depth in the profile, which minor cryoturbation can greatly alter.

History of Use

Arctic pedology is an unsung science, primarily due to the inaccessibility of the area and the lack of practical applications for the results of such study. Many texts either ignore arctic soils (Steila, 1976), or characterize them as bog or tundra soils (e.g. Eyre, 1969). With the advent of World War II and the development of the arctic oil fields, North American work has stressed the engineering properties of arctic soils, but again, the mode of development of stable arctic soils was not studied.

Most of the work on stable arctic soils, their classification and description, has been accomplished by J.C.F. Tedrow and co-workers (Tedrow, 1973, 1977; Tedrow et al., 1958). Tedrow, however, has done little in the way of evaluation of the time variable in soil formation. Soils develop in response to five major variables: parent material, climate, topography, biota, and time. In order to quantify the effects of time, the other variables must be kept constant. This is the first fundamental assumption with respect to soil development, and the one which is probably least valid.

Despite the drawbacks, soil development has been used as a measure of relative age of comparable deposits--coarse-grained, well-drained, moraine crests. Initial work, as in surface weathering, was focused on semi-qualitative criteria, such as the depth of oxidation (Andrews and Miller, 1972). Further investigations have examined, as in studies of soils in temperate areas, changes in soil morphology (Birkeland, 1978), chemistry (Bockheim, 1979), and mineralogy (Isherwood, 1975; Mabee, 1978; Locke, 1979). The motives for such studies have been twofold: firstly, to understand the important processes in the development of arctic soils, and secondly, to use soil development as a relative dating tool. Because each study has focused on certain, specific characteristics of soils, it is not possible to define the evolution of soil development studies, but only to state where work has been accomplished and where it needs to be accomplished.

The State of the Art

In contrast to the boulder weathering studies, which have languished since the statistical studies of Boyer and Pheasant (1974), soil development studies are increasing in number, complexity, and resolution. The major reason for this popularity lies in the ability to sample the soils and subject them to rigorous laboratory analysis. Most of the current work on the time variable in soil formation uses lab results rather than field data. The techniques being applied to analyses of arctic soils (Table 12.2) are a fair cross-section of techniques presently applied in studies of soil development world-wide. Non-utilized techniques are generally a function of the small number of researchers in arctic soils and their research preferences. However, the very slow development of arctic soils limits the usefulness of some techniques.

Field criteria. Field criteria of soil development which have been applied in the Arctic include the depth of oxidation (Miller and Andrews, 1972; Birkeland, 1978) and soil texture (Birkeland, 1978; Bockheim, 1979). Although both criteria are in part a function of time, other of the soil-forming variables combine to limit their usefulness. The depth of oxidation is affected by the present and past precipitation, by the type and amount of weathering minerals present, by the particle size distribution in the soil, and by exotic inputs of salts and iron. Soil textures become finer with time, but the accumulation of fines may be due to a complex combination of aeolian inputs, weathering, parent material, and translocation by water and/or frost. A linear or other simple relationship between these criteria and time cannot be assumed.

Past studies show that, in well-developed soils, the depth from the surface to 5Y (2.5Y) dry hues increases from 0 (0) cm in recent deposits to 20 (10) cm in early Neoglacial deposits to 90+ (25) cm in early Holocene or older deposits. Nearby soils with slightly better drainage or coarser texture, however, may consist of 5Y hues (dry) throughout (Birkeland, 1978; Locke, 1980.

Neither horizon order nor structure has been considered important in the characterization of stable arctic soils, except as it recognizes changes in the other field criteria (above). The reason for this omission lies in the necessary selection of only excessively well-drained, lightly vegetated, coarse-grained soils for study. Such parent material restrictions eliminate significant variability in horizon order and structure.

TABLE 12.2. Criteria of soil development
==

Field Criteria		Lab Criteria	
Horizons present	X	ph	+
Depth of oxydation	+	% O.M.	+
Structure	?	Clay Mineralogy	?
Texture	+	Particle Size	?
		Unstable Minerals	
		Abundance	X
		Weathering	+
		Soil Chemistry	
		Iron	+
		Phosporous	+

Note: + indicates criteria which have been used or have the potential
         for use in arctic soils
       ? indicates criteria which have not yet shown potential
       X indicates criteria which have proven unreliable

Laboratory criteria. Laboratory techniques which have been applied to
stable arctic soils include most of those which have been applied at
lower latitudes. Not suprisingly, conclusions reached in studies of
temperate soils appear to be equally valid in the Arctic, although
rates are different. For example, values of soil pH, which may reach a
steady state in as little as 150 years in a cool, maritime climate
(Glacier Bay, Crocker and Major, 1955), may require as much as 75,000
years to reach a steady state in the eastern Arctic (Locke, 1980). The
major conclusions are that most criteria of soil development show
decreasing development with increasing depth, and that the rate of
increase in soil development decreases with time (Bockheim, 1980). A
summary of the most notable results follows.

The reported values of pH for unweathered till fall generally
between 4.5 and 5.5 (Birkeland, 1978; Bockheim, 1979; Locke, 1980).
With time and the accumulation of organic acids, surface pH values
decrease to 4 to 5; this appears to represent an equilibrium value. At
depth, however, pH values between 6 and 6.5 are common in soils in
Sunneshine (Locke, 1980) and Duval (Birkeland, 1978) drift, which may
date from c. 75,000 BP. Bockheim (1979) has suggested marine aerosols
as a probable source for soluble salts which may reduce soil acidity.
Although 75,000 years may be necessary to achieve equilibrium,
quantification of the effects of parent material, continentality, and
biologic agents is required.

Organic matter percentages in Baffin Island soils are uniformly
low, being normally less than 1% at depths of over 5 cm in the soil.
Microclimatic variables, which control both the growth of organisms and
eolian inputs, are believed to be responsible for most variabilty. It

is most likely that steady state conditions are achieved in hundreds, rather than thousands of years.

In temperate climates, the presence of pedogenic clays is a hallmark of well-developed, therefore old, soils. In the Arctic, however, no persuasive evidence of pedogenic clays has been found. Variations among identified clay species (Table 12.3) appear more likely the result of variations in parent material and/or eolian inputs than pedogenesis. Nevertheless, some evidence of temporal changes in clay mineralogy has been found. Isherwood (1975) found relatively high percentages of vermiculite in her oldest (pre-Wisconsinan) soils, and invoked pedogenesis in explanation. Locke (1980) found that the percentages of chlorite and an intergrade of mica and vermiculite of hydrated mica increased with age (Table 12.4). He suggested that the chlorite may represent an eolian addition as it is the most common clay mineral in North Atlantic waters (Biscaye, 1965). Because of the difficulties in the identification of clay minerals and in their quantitative determination, it is unlikely that clay mineral analyses will soon be an accepted geochronologic tool.

Particle size distribution within a soil, with time, should show an increasing percentage of fines as a result of weathering processes. Because of the slow rates of such processes and the generally unstable nature of arctic soils, no such trends have been identified. One trend which has been identified, however, is that of changing distribution of particles within the soil (Locke and Mabee, 1979). Initial silt: clay percentages in Neoglacial tills fall in the range 10-20%: 0-7%. With time, surface horizons are depleted in fines and subsurface horizons are enriched. The range of silt (clay) percentages changes to 7-25% (0-9%) in early Holocene soils and 4-35%: 0-13% in early Holocene and older soils, with depletion of fines at the surface and accumulation at depth. Detailed particle-size analyses (Locke and Mabee, 1979; Locke, in prog.) provide evidence that redistribution of particles finer than 32 ^{e}m is taking place. The most commonly cited mechanism of such redistribution, freeze-thaw (Fitzpatrick, 1956; Tedrow, 1977; Bockhkeim, 1979), was rejected in favor of illuviation by rainwater. Whatever the mechanism, fine-particle translocation does occur in stable arctic soils. It is, however, sensitive to climate, parent material, and eolian inputs, and as such cannot be expected to hold much potential as a geochronologic indicator.

Although most of the constituents of arctic soils appear to be stable over periods of 10^5 years, some minor constituents are unstable. Two such constituents are plagioclase and hornblende. Plagioclase weathering was examined by Isherwood (1975) as the ratio of

Table 12.3. Comparative x-ray data from Baffin Island

Author	Location	Depth	Clay Mineralogy[1]
Isherwood (1975)	NE Cumberland Peninsula	Surface	Mica > verm > kaol
		Depth	Mica > kaol > verm
Moore (1978)	East-Central Baffin	Surface	Mica > m/ve > chlo > verm
		Depth	Mica > m/va > chlo > verm
Bockheim (1979)	NW Cumberland Peninsula	Surface	Kaol > mica > verm > chlo > m/ve
		Depth	Mica > kaol > verm > chlo y m/ve
Locke (1980)	SE Cumberland Peninsula	Surface	Mica > m/ve > chlo > kaol > verm y smec
		Depth	Mica > m/ve > chlo

[1]Verm = vermiculite, koal = kaolinite, m/ve = interstratified mica/vermiculate or hydromica, chlo = chlorite, smec = smectite

NOTE: All results are from early Holocene or older soils, most are an Arctic Brown/Polar Desert intergrade (Pergelic Cryochrepts).

TABLE 12.4. Clay mineralogy of selected soils (after Locke, 1980)

Profile	Depth	Clay mineral Percentages[1]						Age
		Mica	M/ve	Chlo	Kaol	Verm	Smec	
76-27	0-25 cm	72	16	5	3	2	3	Shannagh (early Neoglacial)
	25-66+	87	6	4	2	0	1	
75-44	1-12 cm	71	16	5	5	1	2	Gilbert (early Holocene)
	12-20	54	26	10	2	4	4	
	20-30	73	13	10	2	0	1	
	30-40+	80	8	13	0	0	0	
	0-5 cm	54	24	10	6	2	2	Sunneshine (upper Pleistocene)
	5-24	65	13	15	3	3	1	
	24-30	71	13	14	0	2	1	
	50-54+	77	11	8	4	0	0	

[1]Mica = illite/muscovite for 75-42 and 44, biote for 76-27. M/ve indicates interstratified mica/vermiculite or hydromica, chlo = chlorite, koal - kaolinite, verm = vermiculite, smec = smectite. All values are approximate, but trends should be valid.

x-ray peak heights of silt-sized particles of plagioclase and quartz (Fig. 12.3). The results are consistent with weathering increasing with time and decreasing with depth. The lack of agreement with a logarithmic curve may result from a combination of eolian additions at the surface and particle translocation (above) at depth. Inasmuch as this technique is related to parent material at depth, it is relatively insensitive to parent material variations.

Locke (1976, 1979) examined the depth of etching of hornblende grains (Fig. 12.4). Barring major differences in the sensitivity of different hornblende species to dissolution, this technique too is relatively insensitive to parent material variability. Because only sand-sized particles are examined, it should also be insensitive to particle translocation (note logarithmic curves in Fig. 12.4). This technique may be the best technique yet developed for determining the relative degree of soil development, and thus age. More recent studies (Locke, 1980), however, indicate that the effect of climatic changes across as small an area as Cumbrland Peninsula ($\sim$ 200 km diameter) may cause significant differences in etching of correlative deposits.

WEATHERING INDEX

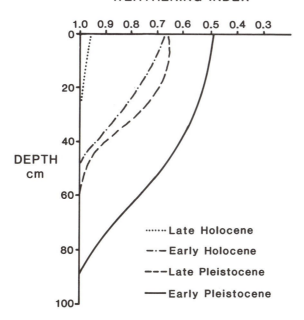

Figure 12.3 Weathering Index as a function of depth and age in arctic soils (after Isherwood, 1975). The weathering index is defined as the ratio of quartz (002) x - ray peak height to the plagioclase (001) peak height, for a sample, divided by the similar ratio for parent material.

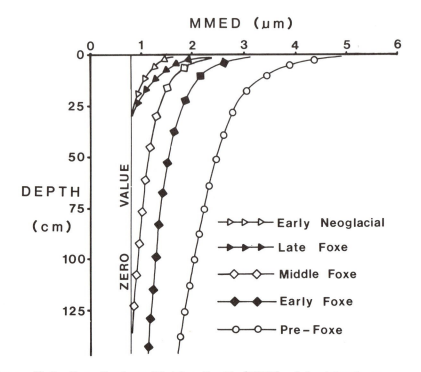

Figure 12.4 Mean Maximum Etching Depth (MMED) of homblende as a
function of depth and age in arctic soils (from Locke, 1979). Late
Foxe = early Holocene, middle, early, and pre-Foxe = Pleistocene
deposits of increasing age (with permission).

Almost any element or compound may be analyzed for in soils; many
have potential as indicators of soil development. Among those which
have been used in the past are iron (Isherwood, 1975; Moore, 1978;
Bockheim, 1979), aluminum (Isherwood, 1975; Moore, 1978), manganese
(Moore, 1978), and phosporous and exchangeable cations (Ca, Mg, Na, K;
Bockheim, 1979). Each of these elements, and probably others, shows
subtle trends with time, however, the difference in parent material
(mineralogy, particle size) and environment combine to minimize the
usefulness of such trends in geochronology.

Summary

Studies of soil development have generally used quantitative criteria
which augur well for future studies. However, no one technique has
been applied by enough researchers, or in enough localities, to warrant
acceptance as the primary soil development criterion. Of techniques

now applied, the study of the weathering of unstable minerals appears least affected by variables other than time, although climatic change affects all such techniques. Because of the effects of variables such as parent material size and mineralogy, eolian input, microclimate, and climatic change, no schematic profiles can be drawn to illustrate model soil development in the eastern Canadian Arctic. The general trends discussed above may be present and clear in some soils and obscured in others. At this point, correlation within a study solely on the basis of degree of soil development is barely possible, and correlation between studies on that basis is not possible.

The Future of Soil Development Studies

Studies of soil development have great potential in the field of relative dating. Like surface weathering, they may be multiparametric, and offer potential for rigorous statistical analysis. Soil development studies are also subject to variability in factors other than time (Table 12.5). Notice that all soil studies are subject to contamination by wind-blown material (with the possible exception of the etching of sand-sized grains) and to changes in climate (particularly the effective precipitation, or rainfall available for leaching). The next step is to continue detailed studies of soil variability to quantify these effects. Sampling at depth increments of 10 cm might serve to define the variability in parameters, particularly such variability as may have been caused by translocation of fines or by freeze-thaw sorting.

CONCLUSIONS

Studies of surface weathering and soil development have been undertaken in the eastern Canadian Arctic primarily to serve as tools for relative dating. This use implies two assumptions, which are usually left unstated. These assumptions are:

1. That the difference in any weathering criterion between two sites is primarily a function of the time elapsed between deposition or stability.
2. That the difference in any weathering criterion between two sites can be assessed on a scale, calibration of which will yield a relative age.

Both of these assumptions are certainly false, in some cases. For example, the first assumption is violated by the postulated effects of

TABLE 12.5. Variability in weathering studies

CRITERIAN	Cause of Variability				
	PM	EOLIAN	MICRO-ENVIRON-MENT	TRANSLO-CATION	CLIMATE
ph	+	++	+	-	+
Organic Matter		+	++	+	+
Clay Mineralogy	++	++	-	+	+
Particle Size	++	++	+	++	+
	-	+	+	+	+
Mineral Weathering					
Soil Chemistry	++	++	+	+	+

NOTE: ++ means high potential variability due to the given cause
+ means moderate potential variability
- means little effect

warm-based ice (eroding) and cold-based ice (protecting) coexisting (Sugden and Watts, 1977). It is violated as well by evidence that, in the long term, climatic effects may dominate time as a control on rates of weathering (Locke, 1980 and in progress). The second assumption is commonly violated in soil studies, where parent material variations render the use of absolute scales impossible. Yet, empirically, the techniques work. This suggests that, with further refinement of techniques, particularly those of soil-development assessment, and with a continued awareness of the operative assumptions, weathering studies will continue to be a useful tool for terrestrial Quaternary stratigraphy in the eastern Canadian Arctic.

REFERENCES

American Commission on Stratigraphic Nomenclature, 1961: Code of stratigraphic nomenclature. American Association of Petroleum Geologists, 45:645-655.
Andrews, J.T., 1963: End moraines and late-glacial chronology of the northern Nain-Okak section of Labrador. Geografiska Annaler, 45:158-171.

Andrews, J.T. and G.H. Miller, 1972: Chemical weathering of tills and
    surficial deposits in east Baffin Island, N.W.T., Canada. Internat.
    Geog. 1972, paper submitted to the 22nd International Geographical
    Congress I: Sections I: Geomorphology, 5-7.
Andrews, J.T. and J.D. Ives, 1978: "Cockburn" nomenclature and the late
    Quaternary history of the eastern Canadian Aarctic. Arctic and Alpine
    Research, 10:617-633.
Andrews, J.T., B.J. Szabo, and W. Isherwood, 1975: Multiple tills,
    radiometric ages, and assessment of the Wisconsin glaciation in
    eastern Baffin Island, N.W.T., Canada: a progress report. Arctic and
    Alpine Research, 7:39-59.

Birkeland, P.W., 1974: Pedology, Weathering, and Geomorphological
    Research. Oxford University Press, NY, 285 pp.
Birkeland, P.W., 1978: Soil development as an indication of relative
    age of Quaternary deposits, Baffin Island, N.W.T., Canada. Arctic and
    Alpine Research, 10:733-747.
Biscaye, P.E., 1965: Mineralogy and sedimentation of recent deep-sea
    clay in the Atlantic Ocean and adjacent seas and oceans. Geological
    Society of America Bulletin, 76:803-832.
Bockheim, J.G., 1979: Properties and relative age of soils of
    southwestern Cumberland Peninsula, Baffin Island, N.W.T., Canada.
    Arctic and Alpine Research, 11:289-306.
Bockheim, J.G., 1980: Solution and use of chronofunctions in studying
    soil development. Geoderma, 24:71-85.
Boyer, S.J., 1972: Pre-Wisconsin, Wisconsin, and Neoglacial ice limits
    in Maktak Fiord, Baffin Island: a statistical analysis. Unpublished
    M.S. thesis, University of Colorado. 117 pp.
Boyer, S.J. and D.R. Pheasant, 1974: Delimitation of weathering zones
    in the fiord area of eastern Baffin Island, Canada. Geological
    Society of America Bulletin, 85:805-810.
Brigham, J.K., 1981: Stratigraphy, amino acid geochronology, and
    genesis of Quaternary sediments, Broughton Island, E. Baffin Island,
    Canada. Unpublished M.S. thesis, University of Colorado, Boulder. 199
    pp.
Burke, R.M. and P.W. Birkeland, 1979: Reevaluation of multiparameter
    relative dating techniques and their application to the glacial
    sequences along the eastern escarpment of the Sierra Nevada,
    California. Quaternary Research, 11:21-51.

Coleman, A.P., 1920: Extent and thickness of the Labrador Ice Sheet.
    Geological Society of America Bulletin, 31:319-328.
Crocker, R.L. and J. Mayor, 1955: Soil development in relation to
    vegetation and surface age at Glacier Bay, Alaska. Journal of
    Ecology, 43:427-448.

Daly, R.A., 1902: The geology of the northeast coast of Labrador.
    Harvard College Museum, Comparitive Zoology Bulletin, 38:205-270.
Denton, G.H. and T.J. Hughes (eds.), 1981: The Last Great Ice Sheets.
    John Wiley, NY, 484 pp.
Dugdale, R.E., 1972: The Quaternary history of the northern Cumberland
    Peninsula, Baffin Island, N.W.T. Part III: The late glacial deposits
    of Sulung and Itidlirn valleys and adjacent parts of the
    Maktak/Narpaing trough. Canadian Journal of Earth Sciences,
    9:366-374.
Dyke, A.S., 1977: Quaternary geomorphology, glacial chronology, and
    climate and sea-level history of southwestern Cumberland Peninsula,
    Baffin Island, Northwest Territories, Canada. Unpublished Ph.D.
    thesis, University of Colorado, Boulder. 184 pp.

Dyke, A.S., 1979: Glacial and sea-level history of southwestern Cumberland Peninsula, Baffin Island, N.W.T., Canada. Arctic and Alpine Research, 11:179-202.

Eyre, S.R., 1968, Vegetation and Soils: A World Picture (2nd edition), Aldine Publishing Company, Chicago, 328 pp.

Fitzpatrick, E.A., 1956: An indurated soil horizon formed by permafrost. Journal of Soil Science, 7:248-254.

Isherwood, D.J., 1975: Soil geochemistry and rock weathering in an arctic environment. Unpublished Ph.D. thesis, University of Colorado, 173 p.
Ives, J.D., 1957: Glaciation of the Torngat Mountains, Northern Labrador, Arctic, 10:66-87.
Ives, J.D., 1958a: Glacial geomorphology of the Torngat Mountains, Northern Labrador. Geographical Bulletin, 12:47-75.
Ives, J.D., 1958b: Mountain-top detritus and the extent of the last glaciation in northeastern Labrador-Ungava. Canadian Geographical Journal, 12:25-31.
Ives, J.D., 1963: Field problems in determining the maximum extent of Pleistocene glaciation along the eastern Canadian seaboard--a geographer's point of view. In: Love, A. and D. Love (eds.), North Atlantic Biota and their History, Pergamon Press, New York, 337-354.
Ives, J.D., 1976: The Saglek moraines of northern Labrdor: A commentary. Arctic and Alpine Research, 8:403-408.
Ives, J.D. and H.W. Borns, Jr., 1971: Thickness of the Wisconsin ice sheet in southeast Baffin Island, Arctic Canada. Zeitshrift für Gletscherkunde und Glazialgeologie, 7:167-74.

Locke, W.W., 1979: Etching of hornblende grains in Arctic soils: an indicator of relative age and paleoclimate. Quaternary Research, 11:197-212.
Locke, W.W., 1980: The Quaternary geology of the Cape Dyer area, southernmost Baffin Island, Canada. Unpublished Ph.D. thesis, University of Colorado, Boulder. 33 pp.
Locke, W.W. and S.B. Mabee, 1979: Fine particle migration in coarse-grained Arctic soils. Abstracts 71st annual meeting of ASA, CSSA, and SSSA, Fort Collins, Colorado, 160 pp.
Løken, O., 1962: On the vertical extent of glaciation in northeastern Labrador-Ungava. Canadian Geographer, 6:106-119.
Løken, O., 1966: Baffin Island refugia older than 54,000 years. Science, 153:1378-1380.

Mabee, S.B., 1978: The use of magnetite alteration as a relative age dating technique: preliminary results. Unpublished M.S. thesis, University of Colorado, Boulder. 185 p.
Mercer, J.H., 1956: Geomorphology and glacial history of southernmost Baffin Island. Geological Society of America Bulletin, 67:553-570.
Miller, G.H., 1980: Late Foxe glaciation of southern Baffin island, N.W.T., Canada. Geological Society of America Bulletin, 91:399-405.
Miller, G.H., J.T. Andrews, and S.K. Short, 1977: The last interglacial-glacial cycle, Clyde Foreland, Baffin Island, N.W.T.: Stratigraphy, biostratigraphy, and chronology. Canadian Journal of Earth Sciences, 14:2824-2857.
Mode, W.N., 1980: Quaternary stratigraphy and palynology of the Clyde foreland, Baffin Island, N.W.T., Canada. Unpublished Ph.D. thesis, University of Colorado, Boulder, 219 pp.
Moore, T.R., 1978: Soil formation in northeastern Canada. Annals Association of American Geographers, 68:518-534

Müller, D., 1980: Glacial geology and quaternary history of southeast Meta Incognita Peninsula, Baffin Island, Canada. Unpublished M.S. thesis, University of Colorado, Boulder. 211 pp.

Nelson, A.R., 1978: Quaternary glacial and marine stratigraphy of the Qivitu Peninsula, northern Cumberland Peninsula, Baffin Island, Canada. Unpublished Ph.D. thesis, University of Colorado, Boulder. 299 pp.

Nelson, A.R., 1980: Chronology of Quaternary landforms, Qivitu Peninsula, northern Cumberland Peninsula, Baffin Island, N.W.T., Canada, Arctic and Alpine Research, 12:265-286.

Nelson, R.L., 1954: Glacial geology of the Frying Pan River drainage, Colorado. Journal of Geology, 62:325-342.

Odell, N.E., 1933: The mountains of northern Labrador. Geographical Journal 82:193-211, and 315-326.

Ollier, C.D., 1969: Weathering. Oliver and Boyd, Edinburgh. 304 pp.

Pheasant, D.R., 1971: The glacial chronology and glacio-isostasy of the Narpaing/Quajon Fiord area, Cumberland Peninsula, Baffin Island. Unpublished Ph.D. thesis, University of Colorado, Boulder. 232 pp.

Pheasant, D.R. and J.T. Andrews, 1972: The Quaternary history of northern Cumberland Peninsula, Baffin Island, N.W.T., Part III, Chronology of Narpaing and Quajon Fiords during the last 120,000 years. 24th International Geological Congress, 1972, Section 12, 81-88.

Steila, D., 1976: The Geography of Soils. Prentice Hall, Inc., N.J. 222 pp.

Sugden, D.E. and S.H. Watts, 1977: Tors, felsenmeer, and glaciation in northern Cumberland Peninsula, Baffin Island. Canadian Journal of Earth Sciences, 14:2817-2823.

Szabo, B.J., G.H. Miller, and J.T. Andrews, 1982: Dating Quaternary raised marine deposits of Baffin Island, Arctic Canada: Uranium-series and amino acid results on marine molluscs. Geology, 9:451-457.

Tedrow, J.C.F., 1973: Soils of the Polar region of North America. Biuletyn Peryglacjalny 23:57-65.

Tedrow, J.C.F., 1977: Soils of the Polar Landscapes. Rutgers University Press, New Brunswick, NJ. 638 pp.

Tedrow, J.C.F., J.V. Drew, D.E. Hill, and L.A. Douglas, 1958: Major genetic soils of the Arctic slope of Alaska. Soil Science, 9:33-45.

Tanner, V., 1944: Outlines of the geography, life and customs of Newfoundland-Labrador. Acta Geographica, 8:1-907.

13 Late Cenozoic marine deposits of East Baffin Island and East Greenland: microbiostratigraphy, correlation, and age

Rolf W. Feyling-Hanssen

On the northeast coast of Baffin Island there is a low foreland between the outermost part of Clyde Inlet and the mountains between Kogalu-River and Eglington Fiord (Figs 13.1 and 13.2). This foreland, called Clyde Foreland (Feyling-Hanssen, 1967, 1976a, b), rises gently inland to a height of approximately 100 m at the foot of the nearest mountains, and is composed of unconsolidated and slightly consolidated deposits varying in grain size from very coarse bouldery gravel to clay. These deposits have been termed the Clyde Foreland Formation, the type locality being the 20-40 m high coastal cliff facing Baffin Bay (Feyling-Hanssen, 1976a).

Ever since Goldtwait (1964) recorded unconsolidated deposits with shells and fragments of marine molluscs and interlayered till beds in the conspicuous cliffs of Clyde Foreland, these and other marine deposits of eastern Baffin Island have been considered to be exclusively Quaternary in age (e.g. Löken, 1966a, b; Andrews and Miller, 1972; Pheasant and Andrews, 1972; Andrews et al., 1975; Nelson, 1978, 1980, 1981). Tills indicate glaciations and glaciations usually belong in the Quaternary. The species of marine molluscs which occur in the deposits are also such as occur in the Quaternary. However, repeated visits to the coastal area of northeastern Baffin Island and thorough examination of the microfossil, in this case foraminiferal, content of collected sediment samples seem to prove that only the upper part of the sediments in the cliffs is of Quaternary age. The lower part is of Upper Tertiary age (Feyling-Hanssen, 1980a, b).

Three major localities in northeastern Baffin Island have been micropaleontologically investigated: Clyde Foreland, Qivituq Peninsula (320 km south-east of Clyde Foreland), and Broughton Island (70 km to the south-east of Qivituq Peninsula). A single sample from Pigojoat

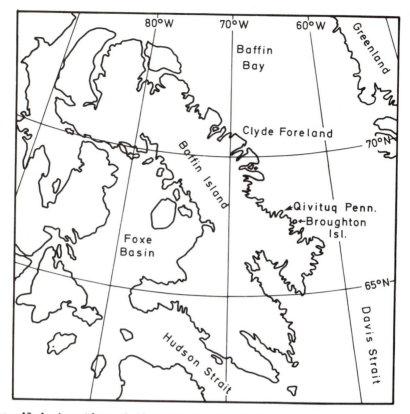

Figure 13.1 Location of Clyde Foreland, Qivituq Peninsula and
Broughton Island. Pigojoat is situated between Qivituq Peninsula and
Broughton Island.

(17.5 km north-west of Cape Broughton on Broughton Island) was also
analyzed, and finally a comparison was made with an interglacial
locality in East Greenland.

CLYDE FORELAND

The deposits in the cliff along Clyde Foreland are more or less
horizontally stratified. Locally the layers may be tilted, minor
faults also occur, and folds may be present. Many of these distortions
were probably caused by glacial tectonism. Other disturbances may have
been caused by wave or river erosion undermining the cliff. Thus in a
valley cliff close to the mouth of Coast Cliff River (Fig. 13.3) a
large block of sediment, from the river to the plain on top of the
cliff, has attained a seaward tilt. Such dislocations are in general

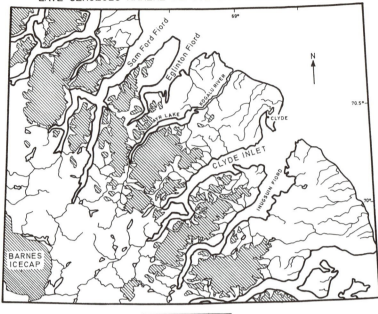

Figure 13.2 Clyde Foreland between the outer part of Clyde Inlet and
the coastal mountains between Kogalu River and Eglinton Fiord.
Present-day glaciation hatched.

small. Still, visual correlation of layers from one locality to
another along the cliff is not simple. Slumped material and mud flows
obliterate the continuation of the strata in many places, and lateral
change in the characteristics of certain layers may render it
impossible to recognize these layers only a short distance away.
Therefore, in order to obtain a better understanding of the
stratigraphy of the deposits, sections were cleared in convenient
places where there was little or no slumped material on the cliff
face. Thirty such profiles were measured, described and sampled along
the coastal cliff of Clyde Foreland. They are numbered from I, at the
mouth of Kogalu River, to XXX, at the erosional cliff of the Hook River
and are indicated in Figure 13.3. Heights are given in meters above
sea level.

Sediment samples for foraminiferal analysis were collected, mostly
from fine-grained deposits, as coarser sediments will ususally be
barren. In each profile the samples were numbered from the top of the
section downwards. Many of the samples, even those which were quite
rich in molluscan shells, mostly of Hiatella arctica (Linne) were
barren of foraminifera or contained only specimens with corroded
tests. This was probably caused by dissolution of calcium carbonate by

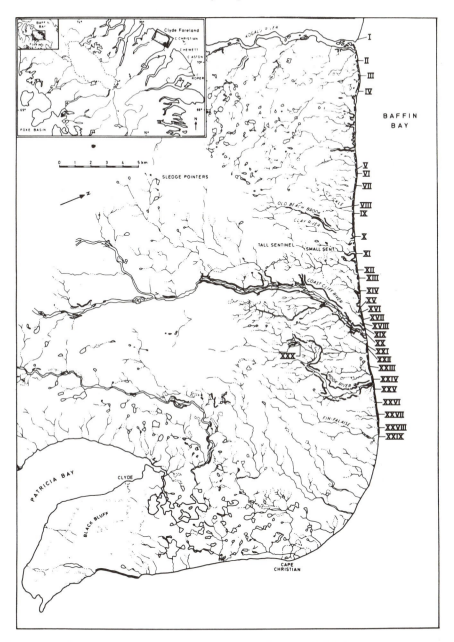

Figure 13.3 Location of investigated profiles (I-XXX) at the coastal
cliff and river bluffs of Clyde Foreland (from Feyling-Hanssen, 1976a).

acidic ground water which in some sediments completely dissolved the tiny tests of foraminifera, and in others dissolved the more delicate ones, leaving those with larger and thicker walls with badly etched surfaces. But some samples contained remarkably rich assemblages of well-preserved benthic foraminifera. Through the identification of certain index species, and by analyzing the faunal composition of each assemblage it was possible to subdivide the deposits into four units, viz. a Nonion tallahattensis zone, a Cibicides grossa zone, a Cassidulina terestis zone, and a Islandiella islandica zone, youngest. A complete succession of these zones did not occur in any single profile, only parts of it. Examples of the foraminiferal zonations of four of the profiles are given in Figures 13.4 - 13.5. A general sediment signature for different parts of the section is placed to the left (Fig. 13.6). To the right of the sediment column the position of samples is indicated, and in the broad middle part of the percentage distribution of a few species of foraminifera in these samples (Fig. 13.6).

By arranging fossiliferous samples from different profiles of the cliff into a chronological series a key chart of the foraminiferal zonation of the Clyde Foreland Formation was compiled (Fig. 13.7). A total of 95 different species of foraminifers were recognized in the deposits, but only a few frequent or stratigraphically significant species were entered in the chart. To the right of this range chart there are three curves indicating the number of different species in each sample (including those which are not listed in the range chart of the profiles); faunal diversity (which, according to Walton, 1964, is "the number of ranked species in a counted assemblage, whose cumulative percentage accounts for 95% of the total assemblage"--high diversity would, in general, indicate stable and favourable environmental conditions, whereas low diversity would reflect unstable and severe conditions in one way or another); and the number of foraminiferal specimens in 100 g of a sediment sample. The number of benthonic specimens per 100 g sediment seldom exceeded 5000. A high number of well-preserved specimens may reflect low sedimentation rate. As this number was generally low in the investigated samples, the sedimentation rate was probably high. A description of the foraminiferal zones is given below. Representative microphotographs of some species are illustrated as Plate 13.1 (end of chapter).

The Nonion tallahattensis zone

This zone seems to be poorly represented in the Clyde Foreland cliff. It is characterized by the presence of a small Nonion species which was referred to Nonion tallahattensis Bandy, 1949 (Feyling-Hanssen, 1976a),

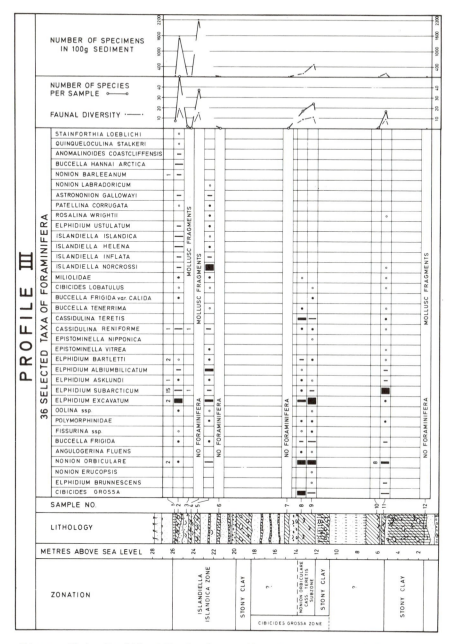

Figure 13.4 Profile III showing sediment sequence and frequency distribution of foraminifera (legend Fig. 13.6). Foraminiferal zonation to the left (redrawn from Feyling-Hanssen, 1976a).

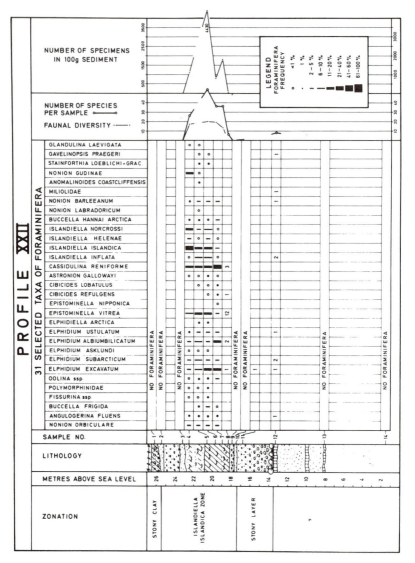

Figure 13.5 Profile XXII. The <u>Islandiella islandica</u> zone well represented (legend Fig. 13.6).

but which has affinity also to <u>Nonion akitaensis</u> Asano, 1950. Large <u>Rotalia columbiensis</u> (Cushman, 1925), <u>Rosalina wrightii</u> (Brady, 1881) and <u>Trichohyalus bartletti</u> (Cushman, 1933) are present as well as quite numerous <u>Miliolinella chukchiensis</u> Loeblich & Tappan, 1953. <u>Elphidium albiumbilicatum</u> (Weiss, 1954) and <u>E. subarcticum</u> Cushman, 1944, are frequent. <u>Nonion orbiculare</u> (Brady, 1881) normally occurs in a compressed form, known from Upper Pliocene deposits in the Netherlands

Foraminifera
frequency   Lithology

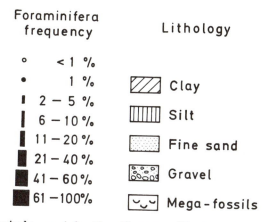

Figure 13.6   Symbols used in the diagrams (Feyling-Hanssen, 1980a)
(with permission).

(cf. Doppert, 1980, Pl. IV, Fig. 2). Elphidium excavatum (Terquem,
1876, cf. Feyling-Hanssen, 1972; Miller, 1982)--frequent in most
samples from the other zones of Clyde Foreland--is practically absent
in the Nonion tallahattensis zone. Cibicides grossa Ten Dam and
Reinhold, 1941, does occur in the zone. The type locality and type
section of the Nonion tallahattensis zone is profile X of the Clyde
Foreland Cliff, the layer from 6.8 m to 6.5 m above sea level is the
type stratum and the foraminiferal assemblage of sample no. 4 its type
assemblage (Feyling-Hanssen, 1976a).

Samples of this zone occur below the Cibicides grossa zone in the
lowest part of the cliff. The Nonion tallahattensis zone is thus, the
oldest unit recognized in the Clyde Foreland cliff. The faunal
diversity of the type assemblage is 31, which is the highest found
among foraminiferal assemblages in that region.

The Cibicides grossa zone

This zone is characterized by the presence of the large, conspicuous
Cibicides grossa Ten Dam and Reinhold, 1941 (referred to as Cibicides
rotundatus by Feyling-Hanssen, 1976a). Cassidulina teretis Tappan,
1951, usually occurs together with it and so do Nonion orbiculare and
Elphidium subarcticum. Elphidium excavatum is often frequent, and
Cassidulina reniforme Nørvang, 1945 (= C. subacuta (Gudina, 1966), cf.
Sejrup & Guilbault, 1980) occurs throughout the zone. Elphidium
bartletti Cushman, 1933, is found mostly in the upper part of the zone,
whereas Nonion erucopsis Todd, 1957, is present almost throughout. The

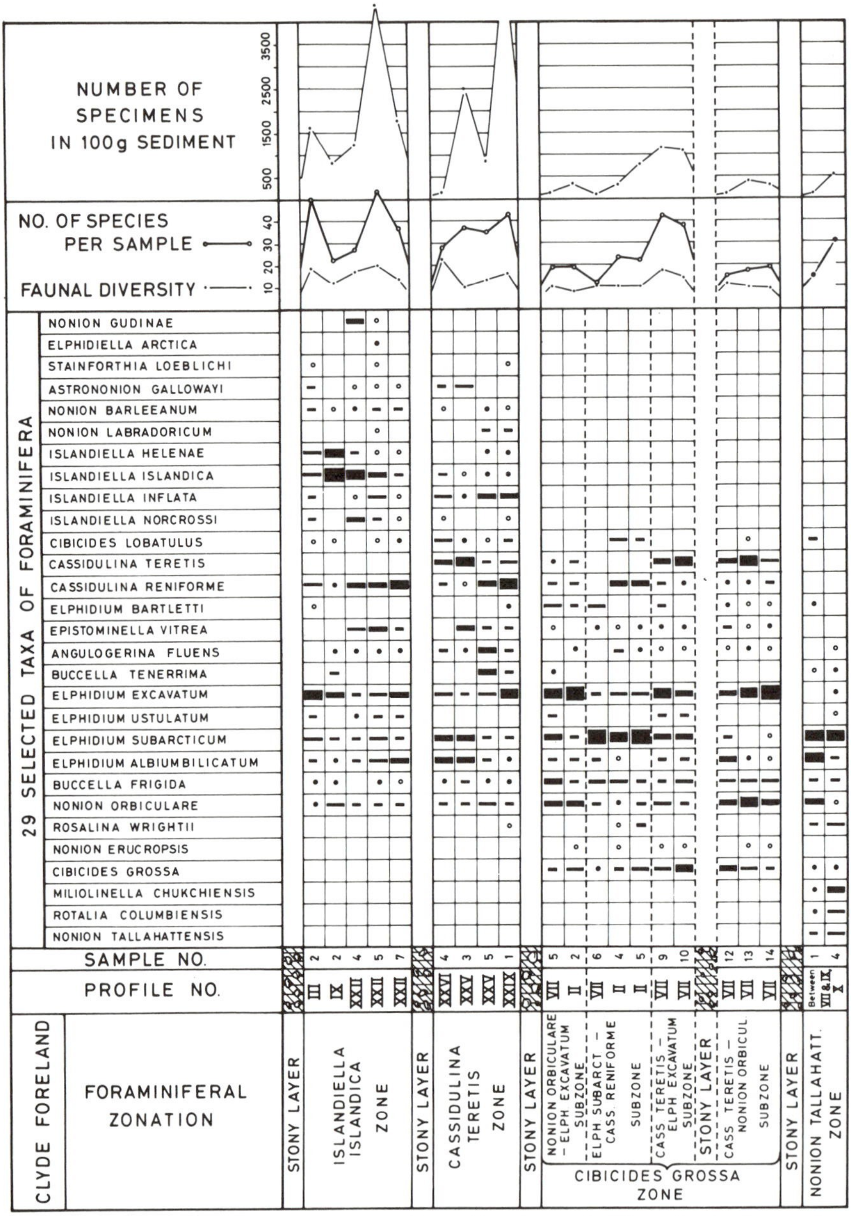

Figure 13.7 Range and frequency distribution of 29 selected species of foraminifera in Clyde Foreland samples. Zonation to the left. See text (from Feyling-Hanssen, 1976a).

layers from 23.6 m to 10.2 m in profile VII were chosen as the type stratum of the Cibicides grossa zone, its type locality and type section being profile VII of the Clyde Foreland Cliff.

The Cibicides grossa zone was subdivided into four subzones, the uppermost being the Nonion orbiculare - Elphidium excavcatum subzone, characterized by high frequency of these two species. Below this occurred the Elphidium subarcticum - Cassidulina reniforme subzone. This was underlain by the Elphidium excavatum - Cassidulina teretis subzone, which overlay the Nonion orbiculare - Cassidulina teretis subzone. These subzones may reflect changing water depths during deposition of the Cibicides grossa zone.

Specimens of Islandiella inflata (Gudina, 1966) and Islandiella norcrossi (Cushman, 1933) are absent or very rare in the Cibicides grossa zone. Nonion orbiculare is usually much more compressed in this zone than in Pleistocene and Recent deposits. The upper boundary of this zone is placed at the disappearance of Cibicides grossa in the Clyde Foreland Cliff.

The Cassidulina teretis zone

This zone is characterized by the frequent occurrence of Cassidulina teretis Tappan (cf. Feyling-Hanssen & Buzas, 1976). Islandiella inflata (Gudina), Angulogerina fluens Todd, 1947 and Epistominella vitrea Parker, 1953, are common as, occasionally, are Islandiella islandica (Nørvang, 1945) and Astrononion gallowayi Loeblich and Tappan. Nonion orbiculare again occurs in a more compressed form than in younger and Recent deposits. The layers from 16.5 m to 10.1 m above sea level in profile XXV were chosen as the type stratum of the Cassidulina teretis zone, its type locality and type section being profile XXV of the Clyde Foreland Cliff. The upper boundary of this zone is placed at the disappearance of Cassidulina teretis in the Clyde Foreland Cliff. The Cassidulina teretis zone was called the Cassidulina teretis subzone of the Islandiella islandica zone by Feyling-Hanssen (1976a, b) and was raised to zone rank in 1980 (Feyling-Hanssen, 1980a, b).

The Islandiella islandica zone

This zone is characterized by assemblages with high frequencies of the large, conspicuous species Islandiella islandica (Nørvang, 1945). At the same time other species of the genus Islandiella are also frequent, particularly Islandiella helenae Feyling-Hanssen and Buzas, 1976, but also Islandiella norcrossi Cushman, 1933, and Islandiella inflata (Gudina, 1966). Other characteristic species of the zone are Nonion barleeanum (Williamson, 1858) and Elphidium ustulatum Todd, 1957.

Epistominella vitrea is common in many samples of the zone as is
Astrononion gallowayi. Elphidium excavatum and Cassidulina reniforme
are usually very frequent, and Nonion orbiculare occurs mostly in the
broad typical form described by Brady (1881, 1884) and which is
commonly found in Pleistocene and Recent deposits of the Arctic. There
is a consistent occurrence of Elphidium subarcticum and also of E.
albiumbilicatum.

The type locality and type section of the Islandiella islandica
zone is profile IV of the Clyde Foreland Cliff, the layers from 28.7 m
to 27.7 m above sea level, its type stratum, and the assemblage of
sample no. 2 the type assemblage. In 1976 this zone was described as a
subzone of an Islandiella zone, but was later raised to zone rank
(Feyling-Hanssen, 1976a, b, 1980a, b). All the species occurring in
the Islandiella islandica zone of Clyde Foreland are known from Recent
assemblages of Arctic environment (cf. Phleger, 1952; Leslie, 1965;
Vilks, 1969; Osterman, 1982; Østby and Nagy, 1982) as well as from
Pleistocene deposits in the Arctic and in many other regions of the
northern hemisphere, e.g. Canada, Scandinavia, Kola Peninsula, and
nearshore areas of Siberia, Taymir and Chukotka.

Age of the Clyde Foreland Formation

The deposits of Clyde Foreland have been considered to be wholly
Quaternary in age (Löken, 1966a, b; Feyling-Hanssen, 1967, 1976a, b;
Andrews and Miller, 1976). Some still consider this to be the case
(i.e., Miller et al, 1977; Mode, 1980). This conclusion has been
backed up by both radiometric and amino-acid datings of mollusc shells
and by the occurrence or apparent occurrence of till layers in the
coastal cliff.

Absolute dating methods. A shell sample collected by O.H. Löken near
profile XXVI from a layer which would be the Cassidulina teretis zone
of the present investigation (Löken, 1966b; Feyling-Hanssen, 1976a, p.
330) had a C-14 age of > 50,000 years before present. As a conspicuous
till layer, indicative of a prominent glaciation, occurred below the
Cassidulina teretis zone; the layers of this zone were thought to
represent an interglacial period comparable to the Sangamon
Interglacial of North America, the Kazantzevo Interglacial of arctic
Soviet Union, and the Eem Interglacial of northern Europe. A [230]Th
data of the same sample gave > 115,000 BP, and a later check on the
same sample gave 133,000±10,000 BP (Szabo, 1978 p. 17). The oldest
uranium-series dating of this unit is 185,000 BP (Mode, 1980). These

dates may not contradict a Sangamon age for the zone, and amino-acid ratios from the same unit seem to support a Sangamon age (Miller et al., 1977, pp. 2830-2834). Another shell sample, collected by the present author from the Cassidulina teretis zone of profile XXVI, had a C-14 age of 48,000 BP and a ^{230}Th age of 61,000 BP (Feyling- Hanssen, 1980a).

A shell sample from the Islandiella islandica zone of profile IX in the middle part of the Clyde Foreland cliff, collected by the present author, gave a C-14 age of 40,000±1740 BP, and a sample collected by G.H. Miller from the same unit gave an age of 41,400±500 BP. This seemed to place the Islandiella islandica zone in the middle part of the Wisconsin glaciation. Mode (1980, p. 154) suggests that his Kogalu amino zone, which he correlates with the middle part of the Islandiella islandica zone may date the marine oxygen isotope stages 4-5 d (cf. Shackleton and Opdyke, 1973).

It should be remembered, however, that radiocarbon ages greater than 25,000 to 30,000 years, particularly on shell material, may not be reliable and may be mere minimum estimates (Stuiver, 1978, p. 13). This may also apply to the quite widely dispersed uranium series dates of the Cassidulina teretis zone (Feyling-Hanssen, 1980a, pp. 166-169). This zone may be considerably older. The Islandiella islandica zone, or parts of it, may also be older than indicated by the radiocarbon dates. As to amino-acid ratios of the Clyde Foreland samples, it is emphasized that Miller et al. (1977) did not attach definite ages to them. It must be admitted that there are still uncertainties to overcome with respect to the temperature dependency and thermal history of the shells used in the analyses. Miller (this volume) suggests a possible maximum age for the Cape Christian amino zone, correlated with the Cassidulina teretis zone, of 500,000 years BP. The lower-lying units--the Cibicides grossa zone and the Nonion tallahattensis zone--were considered old, but still of Quaternary age. A shell sample collected by O.H. Løken from the Nonion tallahattensis zone of profile X had an infinite C-14 date of > 39,000. Mode (1980) correlated four of his older aminozones with the Cibicides grossa zone and the Nonion tallahattensis zone, but did not suggest ages.

Till? Glacigenic deposits are to be expected in the Clyde Foreland cliffs, at least towards both ends of the cliff, less in its middle part. It is seen (Fig. 13.2 and 13.3) that the course of the numerous streams which flow across the lowland outline stages in the extent of two former glacier lobes; (1) a large Clyde glacier, or probably a combined Clyde-Inugsuin glacier which fanned out over the eastern part

of the lowland and into Baffin Bay off the mouth of Clyde Inlet, and
(2) a smaller Ayr Valley glacier (Feyling-Hanssen, 1967, 1976a; Miller
et al., 1977; Mode, 1980). The streams originated as marginal
melt-water channels. Therefore most till-like layers with cobbles and
boulders in the Clyde Foreland cliff were interpreted as tills
representing glaciations between units of marine deposits. They were
termed "Drifts" and numbered, e.g. "Drift 3", a conspicuous
coarse-grained layer which occurs below the Cassidulina teretis zone of
profile XXV and XXVI. Where heavy distortion occurs in and below a
boulder-bearing layers and if also a till fabric is present (cf.
Froese, 1967), its interpretation as till seems reasonable, but with
some of the cobble-bearing layers, particularly in the older part of
the Clyde Foreland cliff, this is not the case and the layers have
erroneously been registered as till. Some units may be glacial marine
deposits (e.g. Feyling-Hanssen, 1976a, p. 350), but most of the
semistratified bouldery gravels may represent high-energy beach
deposits. In the coastal cliff on the Qivituq Peninsula, southeast of
Clyde Foreland, stony layers with pebbles, cobbles and boulders in
sandy or silty clay matrix could be traced for tens of meters. They
had a till-like appearance and had been described as such (Andrews et
al., 1975), but in many instances they contained excellently preserved
fossils and overlied stratified sediments without any sign of
distortion. In gullies at right angles to the coast such stony layers
were seen to dip seawards, and in one place a stony layer of this kind
could be traced backwards in a gully cliff up to the foreland surface,
where it ended in a beach ridge (Feyling-Hanssen, 1980a).

Many of these cobble- and boulder-bearing strata might therefore
be considered as beach deposits. Some of them may be glacial marine
deposits and scattered cobbles and boulder in a fine-grained sediment
may be ice-rafted, but most of these stony layers have no connection
with glaciers or glacial ages (cf. Nelson, 1978, 1980, 1981). On the
other hand, glaciations may well have started in the Upper Tertiary in
these arctic regions.

Microbiostratigraphical correlation and dating

The Clyde Foreland foraminiferal assemblages do not correlate with
those from Miocene (Oligocene?) deposits of Carter Creek in
northeastern Alaska (Todd, 1957). There is some similarity between the
Islandiella islandica zone and Upper Quaternary assemblages from the
northern Soviet Union (Gudina, 1976) and from Scandinavia
(Feyling-Hanssen, 1964; Feyling-Hanssen et al., 1971). The other three
zones do not, however, compare with the Upper Quaternary of Northern

Europe. Positive correlation is found, however, with foraminiferal successions in some North Sea borings, and on this basis Feyling-Hanssen (1980a, b) suggested that the Cibicides grossa zone and the Nonion tallahattensis zone are of Pliocene age. The lower part of the Cassidulina teretis zone may also belong to the Pliocene. A boundary between the Pliocene and the Pleistocene might therefore be found within this zone (Fig. 13.8).

A commercial boring in the central North Sea revealed assemblages of fossil foraminifera with frequent large Cibicides grossa together with i.a., Cassidulina teretis, Cassidulina reniforme and Elphidium excavatum at about 900 m below sea floor. At about 700 m Cassidulina teretis and C. cf. teretis dominated the assemblages. Cassidulina reniforme and Elphidium excavatum, among others, occurred and a few Islandiella helenae appeared. In the above-lying unit of the boring Islandiella helenae became dominant, and in layers immediately above, Elphidium ustulatum was characteristic.

In a boring in the Fladen Ground area farther north in the North Sea a similar succession occurred: A lower part, about 200 m below sea floor, had large Cibicides grossa and frequent Cassidulina teretis. Higher up in the borehole Cassidulina teretis predominated, and above this unit, at about 170 m Islandiella helenae was frequent, and occurred together with Islandiella islandica in this borehole, whereas Cassidulina reniforme was rare. Similar sequences have been observed also in other North Sea borings.

King (1980) has published a range chart "Provisional microfaunal zonation--North Sea Cainozoic" in which Cibicides lobatulus grossa (= Cibicides grossa) characterized his zone TB 17. This zone represents the Late Pliocene and the uppermost part of the Early Pliocene (King, 1980; King et al., 1981). The overlying zone TB 18 has as its index species Elphidiella hannai (Cushman and Grant, 1927), and its lower part is characterized also by the occurrence of Elphidiella oregonensis (Cushman and Grant, 1927). This lower part of the zone is distinguished as TB 18 a. The Pliocene/Pleistocene boundary is placed in the Elphidiella oregnensis subzone of zone TB 18 (cf. also van Voorthuysen et al., 1972).

Doppert (written communication, 1980) correlates his subzone FA2, the Buccella-Cassidulina subzone of the biozonation of the Dutch Neogene, with King's zone TB 17 of the North Sea, i.e., the Cibicides lobatulus grossa zone, and, in accordance with King, considers it Upper Pliocene in age. It equals the Reuverian and the Oosterhout Formation and was (Doppert et al., 1979) correlated with the Belgian biozone BF

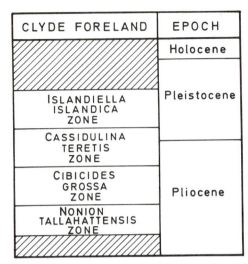

Figure 13.8   Ages of the Clyde Foreland biostratigraphic zones.

N6, the Elphidiella hannai-Cribrononion excavatum zone of De Meuter and
Laga (1977), equaling the Oorderen Sands, the Kruisschans Sands and the
Merksem Sands. At the transition between the Dutch subzone FA_2 and the
overlying subzone FA_1, the Ammonia-Quinqueloculina subzone, a zonule
F_1 with Elphidiella oregonensis is distinguished, and the boundary
between the Upper Pliocene and the Lower Pleistocene is placed within
this zonule (Doppert, 1980).

The shallow-water species Elphidiella orgonensis was not observed
in the Baffin Island sequences. A boundary comparison using this
marker is therefore not possible. However, the present author feels
that Pleistocene conditions were not established either in the North
Sea or in Baffin Island waters until the arctic species Islandiella
helenae had replaced Cassidulina teretis in the foraminiferal
assemblages or had at least entered the assemblages. This happens in
the upper part of the Cassidulina teretis zone of the Clyde Foreland
and in the upper part of the unit with frequent Cassidulina teretis in
the North Sea borings. Further investigation is needed to solve the
boundary problem in shelf deposits. A subdivision of the Cassidulina
teretis zone seems necessary in order to determine which part, or
parts, of it belong to the Pliocene and which to the Pleistocene.

However, I suggest there is little doubt about assigning a
Pliocene age for the Cibicides grossa zone of Clyde Foreland. It
correlates with Cibicides grossa units of North Sea successions which I
had the opportunity to examine, and it correlates with the Upper

Pliocene Cibicides lobatulus grossa zone (TB 17) established by King
(1980; King et al., 1981). If these faunal changes should turn out to
be metachronous rather than synchronous, on a larger scale, this would
render the Cibicides grossa zone of Clyde Foreland even older rather
than younger than their North Sea correlatives because severe
conditions capable of exiling the Tertiary Cibicides grossa would
probably occur earlier at Clyde Foreland than in the North Sea.

Thus the Nonion tallahattensis zone and the Cibicides grossa zone
of the Clyde Foreland Formation, which occur usually in the lower part
of the cliffs, belong to the Pliocene. Also the Cassidulina teretis
zone is most probably of Pliocene age, but an upper part of this zone
may belong to the Pleistocene. The Islandiella islandica zone may
represent different parts of the Pleistocene, both stadials and
interstadials seem to be present (Fig. 13.8, cf. also Fig. 13.18).

QIVITUQ PENINSULA

A microbiostratigraphical investigation similar to that at Clyde
Foreland was undertaken at the Qivituq Peninsula (Fig. 13.1) 320 km to
the southeast, at 68°N latitude and 65°W longitude (Feyling-Hanssen,
1978, 1980). Wave-eroded cliffs here face northeast, north, and west
exposing unconsolidated or weakly consolidated deposits of different
grain-size, in places with shells and fragments of marine molluscs,
echinoids and cirripedes. The surface above the cliff is flat and
gently rising towards the southwest and west and carries distinct,
blockstrewn and more or less closely set raised beach ridges in the
near coastal area. Remnants of raised marine terraces and of deltas
are seen inland (Fig. 13.9).

The height of the cliff varies from 33 m above sea level in the
west to 4 m asl in the east, again rising toward Kangaajuk Point in the
extreme east. Thirty-three sections (marked with numbered circles in
Fig. 13.9) were cleared and measured, and 70 sediment samples
collected. The location of samples from each section is indicated with
a short horizontal mark and a number in Figure 13.10. Even though the
samples were taken from the most promising parts of the profiles only
33 of them contained fossil foraminifera. The numbers assigned to the
fossiliferous samples are placed within squares in Figure 13.10.

Nelson (1978, 1980, 1981, 1982) studied in detail the sediments
outcropping in the coastal cliff of the Qivituq Peninsula. In the
present study, therefore, only some general field observations on the
sediments in the 23 sampled sections are illustrated in Figure 13.11.

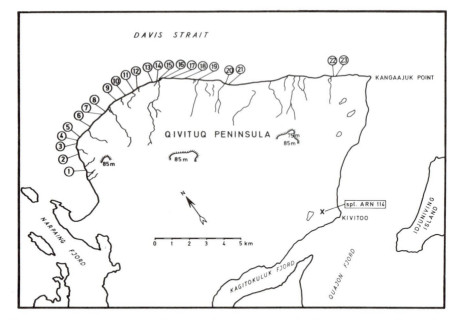

Figure 13.9 Qivituq Peninsula. Micropaleontologically investigated sections of the coastal cliff indicated by rings with numbers (from Feyling-Hanssen, 1980a) (with permission).

It is seen that sand dominates. It is more or less horizontally stratified with occasional slight dipping seaward or eastwards. In some parts of the sand there are interlayered wavy silt beds, 5-10 cm thick. Thicker layers of silt, 0.5-2 m thick, occur in most of the sections. Stony layers with scattered or concentrated pebbles, cobbles and boulders occur in many places. They are considered to be beach deposits or glacio-marine deposits. No till layer was found in the cliff of the peninsula (Feyling-Hanssen, 1978, 1980).

Zonation

The foraminiferal analyses of the fossiliferous samples of the Qivituq Peninsula are presented in the range chart of Figure 13.12 (symbols explained in Fig. 13.6). The samples from the different sections of the cliff have been arranged in stratigraphical order with the oldest samples lowest. Sample numbers and corresponding section number are listed to the left of the chart. Of the 123 species of foraminifera which in total were observed in the samples from the Qivituq Peninsula, only 43 taxa are listed in the chart of Figure 13.12. As with the Clyde Foreland chart (Fig. 13.7) the selection was based on frequent

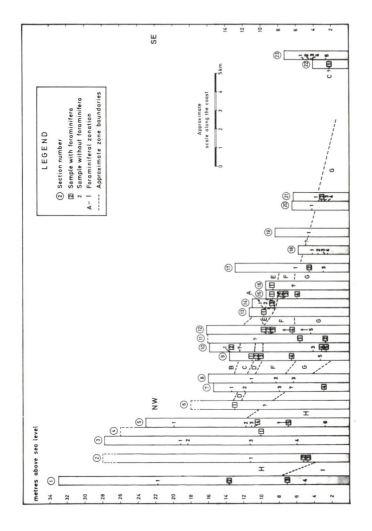

Figure 13.10 Diagram of the Qivituq Peninsula cliff sections. Samples
indicated by short horizontal mark and number; numbers of fossiliferous
samples in squares. Foraminiferal zonation indicated by capital
letters (from Feyling-Hanssen, 1980a) (with permission).

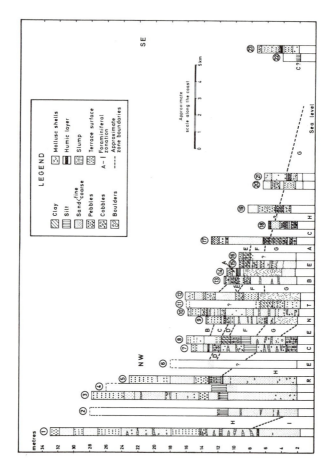

Figure 13.11 The sediment of the sections (from Feyling-Hanssen, 1980a) (with permission).

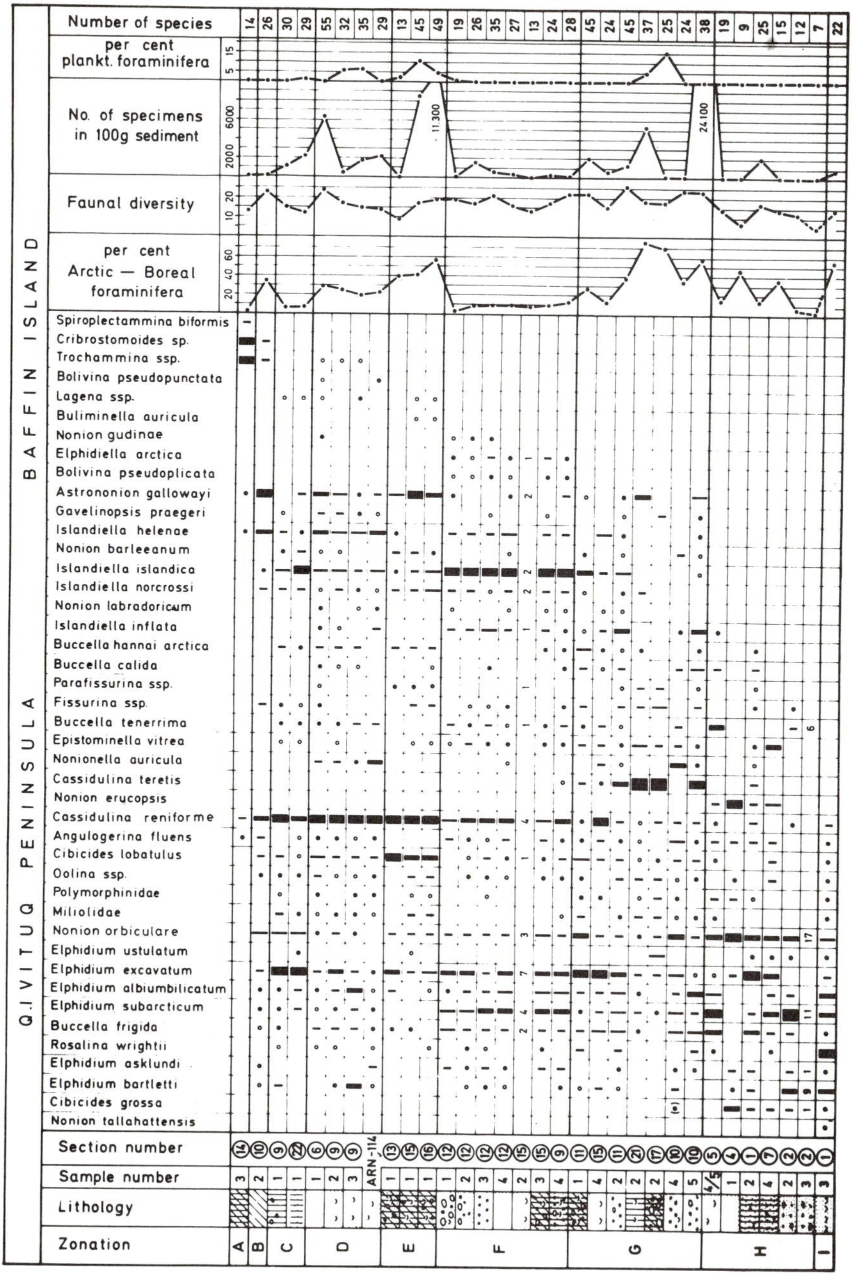

Number of species

per cent plankt. foraminifera

No. of specimens in 100g sediment

Faunal diversity

per cent Arctic — Boreal foraminifera

BAFFIN — ISLAND

Spiroplectammina biformis
Cribrostomoides sp.
Trochammina ssp.
Bolivina pseudopunctata
Lagena ssp.
Buliminella auricula
Nonion gudinae
Elphidiella arctica
Bolivina pseudoplicata
Astrononion gallowayi
Gavelinopsis praegeri
Islandiella helenae
Nonion barleeanum
Islandiella islandica
Islandiella norcrossi
Nonion labradoricum
Islandiella inflata
Buccella hannai arctica
Buccella calida
Parafissurina ssp.
Fissurina ssp.
Buccella tenerrima
Epistominella vitrea
Nonionella auricula
Cassidulina teretis
Nonion erucopsis
Cassidulina reniforme
Angulogerina fluens
Cibicides lobatulus
Oolina ssp.
Polymorphinidae
Miliolidae
Nonion orbiculare
Elphidium ustulatum
Elphidium excavatum
Elphidium albiumbilicatum
Elphidium subarcticum
Buccella frigida
Rosalina wrightii
Elphidium asklundi
Elphidium bartletti
Cibicides grossa
Nonion tallahattensis

QIVITUQ PENINSULA

Section number

Sample number

Lithology

Zonation

Figure 13.12 Range and frequency distribution of 43 selected taxa of foraminifera in Qivituq Peninsula samples. Foraminiferal zonation indicated by capital letters at left. Legend Fig. 13.6. Curves at right explained in text (redrawn from Feyling- Hanssen, 1980a).

occurrence and/or stratigraphic significance; some species are grouped under genera or families (Feyling-Hanssen, 1980, pp. 157-158).

There are four curves, to the right of the range chart in Figure 13.12. They help in establishing a zonation and contribute to a paleoecological explanation of the zones. The second and third diagrams, faunal diversity and number of specimens per 100 g sediment, were explained in connection with the key chart for Clyde Foreland, and the number of different species per sample is given on the extreme right. The first curve, however, needs some explanation. It illustrates the percentages of arctic-boreal benthonic species (Gudina and Evserov, 1973) and indicates ameliorated conditions. In the present investigation the following species are used: Guttulina yamazakii Cushman and Ozawa, 1930, Sigmoidella pacifica Cushman and Ozawa, 1928, Bolivina pseudopunctata Höglund, 1947, Buccella calida (Cushman and Cole, 1930), Rosalina wrightii (Brady, 1881). Rosalina vilardeboana d'Orbigny, 1839, Epistominella nipponica Kuwano, 1967, Epistominella vitrea Parker, 1953, Eoeponidella pulchella (Parker, 1952), Gavelinopsis praegeri (Heron- Allen and Earland, 1913), Cibicides lobatulus (Walker and Jacob, 1798), Cibicides pseudoungerianus (Cushman, 1922), Cibicides refulgens Montfort, 1808, Nonionella auricula Heron-Allen and Earland, 1930, Nonion barleeanum (Williamson, 1858), Astrononion gallowayi Loeblich and Tappan, 1953, Elphidium albiumilicatum (Weiss, 1954), and Elphidium incertum (Williamson, 1858). To this list are added Cassidulina teretis Tappan, 1951, Cibicides grossa Ten Dam and Reinhold, 1941, and Nonion erucopsis Todd, 1957. These species have not been found living in shelf waters of the present day (Feyling-Hanssen, 1980).

Some of the samples contained planktonic foraminifera, mostly Globigerina pachyderma (Ehrenberg, 1861). Their percentage in the total assemblage, benthonic + planktonic, is also illustrated in Figure 13.12. Based on the frequency distribution of the foraminiferal species of the different samples and on the trend of the curves just explained, the following microbiostratigraphic subdivision was undertaken (Feyling-Hanssen, 1980a).

Zone I. This unit was represented only by one sample, taken from a fossil shell bed containing numerous fragments of Hiatella arctica. The foraminiferal assemblage was dominated by Rosalina wrightii, which occurred with a high proportion of Elphidium albiumbilicatum. Other frequent species were Elphidium subarcticum, E. bartletti, and also Nonion orbiculare. Nonion tallahattensis accounted for 1% of the assemblage. The assemblage is correlatable with the Nonion

tallahattensis zone of Clyde Foreland. Both zones are poorly
represented and there are differences in faunal composition, but some
of the characterizing species are the same in both zones.

Zone H. Characteristic species of this unit are the large _Cibicides_
grossa and also _Nonion erucopsis_. _Elphidium_ bartletti, _E._ subarcticum
and _Nonion orbiculare_ occurred frequently and so did _Buccella_ frigida.
Elphidium asklundi should also be mentioned. Most of the Qivituq
samples from this zone were poor in foraminifera. There is a clear
correlation between zone H and the _Cibicides grossa_ zone of Clyde
Foreland, but a closer correlation with the subzones of the _Cibicides_
grossa zone is uncertain.

Zone G. An abundance of _Cassidulina teretis_ is characteristic of this
zone, but some Qivituq samples with a lower frequence of this species
are also referred to zone G. Characteristic species include the
appearance of _Islandiella inflata_, _I._ helenae, _I._ islandica and _I._
norcrossi. _Astrononion galIowayi_, _Angulogerina fluens_, _Cibicides_
lobatulus, and _Epistominella vitrea_ are all well represented. The
genera _Oolina_ and _Fissurina_ are common, and so are species of
Miliolidae and _Polymorphinidae_. The percentage curve of arctic-boreal
foraminifera indicates a decreasing water temperature in the upper part
of zone G. Arctic forms such as _Elphidium_ excavatum, _Cassidulina_
reniforme, _Islandiella islandica_, and _I._ helenae become frequent here.
If the boundary between the Pliocene and the Pleistocene epochs occurs
within zone G, it should probably be placed in this upper part of the
zone, e.g. between sample no. 4 of section 15 and sample no. 2 of
section 11. Zone G of the Qivituq Peninsula correlates with the
Cassidulina teretis zone of Clyde Foreland.

Zone F. This unit is a cold-water zone characterized by large
specimens of _Islandiella helenae_ together with frequent _Islandiella_
islandica and common occurrence of _I._ inflata and _I._ norcrossi.
Elphidium subarcticum, _E._ excavatum, and _Cassidulina reniforme_ are
commmon, and _Elphidium asklundi_ and _Elphidiella arctica_ (Parker and
Jones, 1864) occur in nearly all the samples. The percentage of
arctic-boreal specimens stays low throughout zone F, most probably an
indication of cold water--high-arctic conditions. Sedimentation rates
seem to have been high, as the number of specimens per 100 g sediment
is low, and the water has been shallow or quite shallow, as
shallow-water specimens are frequent and planktonic specimens absent.

The absence of planktonic specimens could also indicate a year-round ice cover.

Zone F probably represents an ice age or a cold interval of the Quaternary period, most probably of the Lower Quaternary. It correlates in part with the Islandiella islandica zone of Clyde Foreland.

Zone E. This unit reflects distinctly ameliorated conditions compared with those of the underlying zone F. Islandiella species are still present, but to a considerably lesser extent. Islandiella helenae is almost absent. Characteristic of the zone are high frequencies of Cibicides lobatulus and Astrononion gallowayi and the presence of Buccella hannai arctica Voloshinova, 1960, and Nonion barleeanum. There is a firm representation of specimens of the genera Oolina, Fissurina, and Parafissurina. Miliolids are common. The percentage of arctic-boreal specimens is from 40% to 57% indicating an amelioration which most probably could only have occurred during an interglacial. Planktonic foraminifera, mostly sinistrally coiling Globigerina pachyderma also occur in this zone, probably an indication of increased water depths as well as of an extended period of ice-free water.

There is a general similarity between zone E of the Qivituq Peninsula and the Islandiella islandica zone of Clyde Foreland, particularly as Islandiella species and Nonion barleeanum occur in both zones. But there are also distinct differences; in particular the high frequency of Cibicides lobatulus and Astrononion gallowayi, and rareness of Islandiella helenae, in zone E, was not observed in the Islandiella islandica assemblages of Clyde Foreland. There are also differences in the percentage of arctic-boreal specimens. In zone E it varies from 40-57% whereas in the Islandiella islandica zone of Clyde Foreland the range is from 6-31%. The lowermost samples of this zone (Fig. 13.7) reflect more pronounced amelioration (22-31%) than the upper two samples (6-15%). It may turn out that zone E represents a later stage in foraminiferal assemblage composition than those represented in the Islandiella islandica zone of Clyde Foreland and that zone E represents a Quaternary interglacial of unknown age. This interglacial is here named the Qivituq Interglacial. Its type locality is the middle part of the coastal cliff of Qivituq Peninsula, its type section is section no. 16, and its type stratum and type assemblage is represented by sample no. 1 at 8.7 m above sea level.

The abrupt change in paleoecological conditions between zone F and zone E may indicate that a hiatus, probably of considerable length, occurs between the two zones.

Zone D. This unit is also ameliorated, but less distinctly so than the underlying zone E. Nonionella auricula Heron-Allen and Earland, 1930, is characteristic of these assemblages. Cassidulina reniforme is frequent, as it was in zone E, but Cibicides lobatulus and Astrononion gallowayi have decreased in numbers, and Islandiella helenae is again firmly represented. The changes infer decreased water temperatures, as also demonstrated by the proportions of arctic-boreal specimens (Fig. 13.12), and reduced water depths suggested by re-established Elphidium representation.

Zone D may represent a transitional unit at the culmination of interglacial conditions before the onset of glacial conditions. The presence of four Islandiella species gives these assemblages a general similarity to the Islandiella islandica assemblages of Clyde Foreland, but as with zone E, there are differences. Assemblages from the Islandiella islandica zone of profile XXII, Clyde Foreland, show some similarity to those of zone D.

Zone C. This unit, though poorly sampled, indicates the reestablishment of high-arctic conditions. Elphidium excavatum forma clavata, Cassidulina reniforme, and Islandiella islandica are frequent. A C-14 dating of a shell sample collected close to and apparently just above one of the zone C samples gave 9280±120 BP (Nelson, 1978, p. 179). This may indicate that zone C or parts of it is of very late Wisconsinian age. Similar layers have not been observed in the Clyde Foreland material.

Zone B. This unit is represented only by one fossiliferous sample. Its assemblage is dominated by Astrononion gallowayi, Islandiella helenae, and Cassidulina reniforme. Nonion orbiculare is also firmly represented and arenaceous specimens of the genera Trochammina and Cribrostomoides are present. The age of this unit is probably Holocene or close to the transition between the Holocene and Pleistocene (Feyling-Hanssen, 1980a). A correlative of zone B was not observed at Clyde Foreland.

Zone A. This unit is also represented only by one sample, sample no. 3 of section 14, which was collected immediately below a humic layer rich in plant debris. The sample is poor in foraminifera, and arenaceous forms (Cribrostomoides, Trochammina, Recurvoides, Spiroplectammina) outnumber the calcareous ones. This may have been caused by dissolution of calcareous tests by acidic ground water percolating through the overlying peaty layer. Zone A probably represents a Holocene regressive facies of the marine series of the Qivituq Peninsula.

Thus the marine deposits outcropping in the coastal cliff of the Qivituq Peninsula are of Upper Cainozoic age: Zone A belongs to the Holocene, zone C through F to different parts of the Pleistocene zone B to the transition Holocene/Pleistocene, zone G to the Pliocene/ Pleistocene transition, and zones H and I to the Pliocene. Nelson (1978b, 1981) established an amino acid stratigraphy for the lithostratigraphical units of the Qivituq Peninsula consisting of eight Aminozones, Aminozone 1, youngest, and Aminozone 8, oldest. He correlated these zones with the biostratigraphical zones of the present author: Aminozone 1 with zone A, B, and C; Aminozone 2 with zone D; Aminozone 3 with zone E and F; Aminozone 4 and 5 with zone G; Aminozone 6 with zone H; Aminozone 7 with zone I; Aminozone 8 was considered older than the biostratigraphical zone I.

Aminozone 3, and part of Aminozone 2, were also correlated with the Islandiella islandica zone of Clyde Foreland; aminozone 4 and 5 with the Cassidulina teretis zone; aminozone 6 with the Cibicides grossa zone, and aminozone 7 with the Nonion tallahattensis zone.

Aminozone 4 and the upper part of aminozone 5 were also correlated with the Cape Christian Member of Miller et al. (1977) which was considered to represent the last world-wide interglacial, which Andrews and Barry (1978) correlated with the marine oxygen isotope stage 5e (Schackleton and Opdyke, 1973). Nelson (1981) found this reasonable, but expressed some doubt. Referring to the reinterpretation of the uranium-series and amino-acid data by B.J. Szabo and others (1981), he expressed the possibility that these units (= the Cassidulina teretis zone and zone G) "are at least as old as marine-isotope stage 7." In 1982 Nelson concludes that the oldest aminozones of the Qivituq Peninsula (7 and 8) may well predate the Quaternary (cf. also Brigham, 1980; Miller, this volume).

PIGOJOAT

This locality is found 17.5 km northwest of Cape Broughton on Broughton Island (Figs. 13.1 and 13.13). A sandy sediment sample with molluscan shells of Hiatella arctica and Mya truncata was collected by P. Thompson Davis of the University of Colorado, Boulder, from the lower half of a 25 m high cliff. The shells were dated by C-14 at 44,800±500 BP and the foraminiferal content of the sediment was analyzed

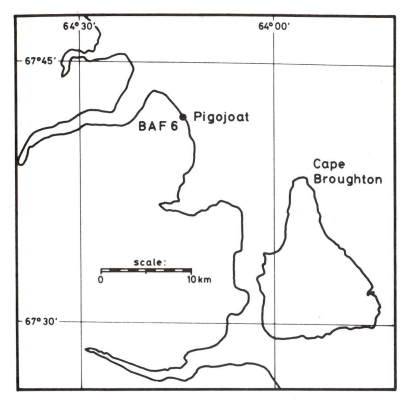

Figure 13.13 Location of Pigojoat.

(Feyling-Hanssen, 1980b). 866 specimens of foraminifera were contained in 43 g (dry weight) sediment. By extrapolation a 100 g sample would contain 2000 specimens. The number of different species was 34, the faunal diversity 23, and the content of arctic-boreal specimens 23%. The analysis is produced in Table 13.1.

The strong representation of the genus Islandiella, in total 20% of the assemblage, places this sample in the Pleistocene of Baffin Island. Its C-14 age indicates Mid-Wisconsinian age, but again it should be remembered that radiocarbon ages greater than 25,000 to 30,000 years may not be reliable and most probably should be considered minimum ages, rather than definite dates. Comparing the present assemblage with those from the Qivituq Peninsula, the best correlation is found with zone D. The sample from Pigojoat thus originates from an ameliorated interval in the Pleistocene between a distinct interglacial, below, and a glacial stage, above. Its age is unknown, but if the Qivituq Interglacial should turn out to be contemporaneous with the pronounced Sangamon Interglacial, the Pigojoat sample could be of Early Wisconsinian age.

TABLE 13.1. Foraminiferal assemblage, sample GRL-139 S, Pigojoat
==

Species	Percentage
Elphidium subarcticum Cushman	21
Cassidulina reniforme Nörvang	13
Elphidium albiumbilicatum (Weiss)	11
Islandiella islandica (Nörvang)	8
Elphidium excavatum (Terquem)	5
Islandiella helenae Feyling-Hanssen and Buzas	4
Islandiella norcross (Cushman)	4
Rosalina wrightii (Brady)	4
Buccella frigida (Cushman)	4
Islandiella inflata (Gudina)	3
Epistominella vitrea Parker	2
Elphidiella arctica (Parker and Jones)	2
Buccella calida (Cushman and Cole)	2
Nonion orbicula (Brady)	2
Bolvinia pseudoplicata Heron-Allen and Earland	2
Cibicides lobatulus (Walker and Jacob)	1
Elphidium bartletti Cushman	1
Miliolinella subrotunda (Montagu)	1
Gavelinopsis praegeri (Heron-Allen and Earland)	1
Quinqueloculina seminulum (Linne)	1
Astrononion gallowayi Loeblich and Tappan	1
Elphidium asklundi Brotzen	1
Fissurina ventricosa (Wiesner)	1
Patellina corrugata Williamson	1
10 other combined species	2
Indeterminate specimens	2

866 specimens were counted in 43 g sediment

BROUGHTON ISLAND

Sediment samples were collected 1974 at 10 sites on Broughton Island (Figs 13.1, 13.13, 13.14) from layers, most of which had been previously dated by radiocarbon assay at ages ranging from 28,200 BP to 46,950 BP (Andrews and Miller, 1972; England and Andrews, 1973; Pheasant and Andrews, 1973; Andrews, 1975; Feyling-Hanssen, 1976c). A later compilation of radiometric dates and age estimates can be found in Brigham (1980). The height of the sampling sites ranged from 5 m to 32 m above present-day sea level.

Those samples contained assemblages of fossil benthonic foraminifera. Planktonic foraminifera occurred only occasionally except in the sample from site 9 on the east side of Broughton Island where 80 specimens of Globigerina pachyderma occurred in a counted assemblage of 668 benthonic specimens. Quite deep water is indicated by this assemblage, though most probably less than 200 m. The other assemblages suggest shallower water conditions.

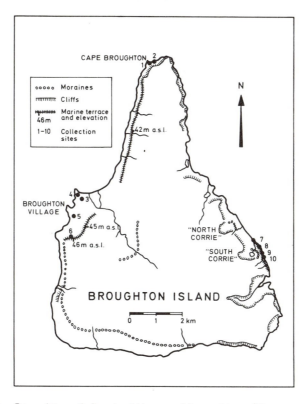

Figure 13.14 Broughton Island with sampling sites (from
Feyling-Hanssen, 1976c).

All the samples, except one from site 10, are characterized by a
strong representation of the genus Islandiella, and are thus probably
of Pleistocene age. They all indicate ameliorated marine-climatic
conditions. This and the radiocarbon ages lead to the supposition that
they originated from the middle part of the last ice age, the
Wisconsinian. A Cape Broughton Interstadial was erected. This
interstadial was thought to represent a long-lasting Mid-Wisconsinian
ameliorated interval correlatable with similar findings in North
America, Northern Europe and northern Soviet Union (Feyling-Hanssen,
1976c; Andrews et al., 1976, 1st alternative model). If, however, the
radiocarbon dates are considered minimum dates, the Broughton Island
samples and the Cape Broughton Interstadial may turn out to be
older--Early Wisconsin or Pre Wisconsin or some of the samples Early
Wisconsin and some Pre Wisconsin, some of Sangamon age (cf. Andrews et
al., 1976, alternative 2 and 3 and Brigham, 1980). The closest
biostratigraphic correlation is found with the assemblages of zone D
and zone E of the Qivituq Peninsula (Feyling-Hanssen, 1980a, b).

The foraminifera assemblages of the Broughton Island samples are characterized by rich representation of the genus Islandiella and in general by many shallow-water specimens. But otherwise quite marked differences in composition occur. Results of the analyses are presented in Feyling-Hanssen (1976c). In Table 13.2 of the present paper, results are presented of the analysis of a 100 g sample (0807601) collected in 1976 from a bulldozer-scraped area 12 m above sea level near the Hudson Bay Company establishement in Broughton Village (close to Site 3 in Feyling-Hanssen, 1976c, site C of Andrews et al., 1976). The assemblage is characterized by extremely high frequency of Islandiella inflata. There were 554 specimens of benthonic foraminifera in 100 g sediment, 30 different species were present, the faunal diversity was 14-15, the faunal dominance 21. Boreo-arctic specimens accounted for 14%, and shallow-water specimens for 25% of the assemblage. Molluscan shell fragments of Mya truncata and Hiatella arctica occurred with the sample.

The foraminiferal assemblage of site 10, sample no. 26087402, from the east side of Broughton Island differs from those of the other samples of Broughton Island by its high frequency of occurrence of Cassidulina teretis. The analysis of this sample is reproduced here in Table 13.3. The total number of species and forms was 20, the number of specimens per 100 g sediment 730, the faunal diversity 11 and the percentage of arctic-boreal specimens 37. The deposit represented by this sample is thus referable to the Cassidulina teretis zone of Clyde Foreland and zone G of the Qivituq Peninsula.

LODIN ELV, EAST GREENLAND

S. Funder and K. Strand Petersen collected sediment samples from two sections at the mouth of the stream Lodin Elv on the north side of Scoresby Sund in East Greenland (Fig. 13.15). Subsequent analyses of the foraminiferal content of the samples by the present author indicated a Pliocene-Pleistocene age for the sediment. A Lodin Elv Formation was established with section 1 on the north side of the stream as type locality (Feyling-Hanssen et al., 1983). The thickness of the sequence is 38 m measured from the present stream bed at 24 m above sea level to its top at 62 m above sea level.

The foraminiferal composition of the fossil assemblages of the samples from the two sections at Lodin Elv made a grouping of the samples into three main units possible: Zone III with no foraminifera, Zone II with high frequency of Cassidulina cf. teretis, and Zone I with

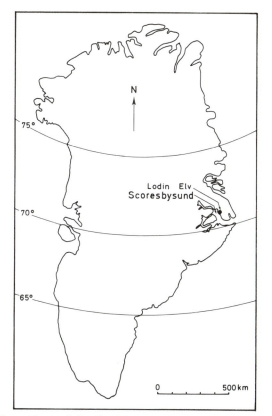

Figure 13.15 Location of Lodin Elv, East Greenland.

firm representation of <u>Cibicides</u> <u>grossa</u>. These units conform with the
lithologies of the two investigated sections (Fig. 13.16).

    Zone III, represented by one sample from the uppermost sand-gravel
of section 2, is barren of foraminifera.   This may be a primary
feature, or it may be caused by postdepositional leaching and
dissolution of foraminiferal tests. The sediment of zone III
constitutes a fluvial terrace, and a radiocarbon date indicates that it
is of Holocene age (Feyling-Hanssen et al., 1983).

    Zone II is characterized by a high frequency of <u>Cassidulina</u> cf.
<u>teretis</u> (25 to 52%). Under this taxon is grouped distinct <u>Cassidulina</u>
<u>teretis</u> Tappan as found at Clyde Foreland and at the Qivituq Peninsula
on Baffin Island, but also forms which in edge view show a wavy
periphery as found in <u>Cassidulina</u> <u>laevigata</u> d'Orbigny. In contrast to
<u>C</u>. <u>laevigata</u>, however, the central area of the Lodin Elv specimens
remains uncovered by the umbilical part of the chamber walls--as in <u>C</u>.

TABLE 13.2. Foraminiferal assemblage, sample 08087601, Broughton Village

Species	Percentage
Islandiella inflata (Gudina)	21
Cassidulina reniforme Nørvang	20
Elphidium subarcticum Cushman	11
Islandiella islandica (Nørvang)	9
Elphidium excavatum (Terquem)	6
Elphidium albiumbilicatum (Weiss)	6
Elphidium asklundi Brotzen	5
Rosalina wrightii (Brady)	4
Islandiella helenae Feyling-Hanssen and Buzas	3
Cibicides lobatulus (Walker and Jacob)	3
Quinqueloculina seminulum (Linne)	2
Nonion orbiculare (Brady)	2
Buccella frigida (Cushman)	2
Elphidiella arctica (Parker and Jones)	1
Islandiella norcrossi (Cushman)	1
Angulogerina fluens Todd	1
Miliolinella subrotunda (Montagu)	1
13 other combined species	2

554 specimens were counted in 100 g of the sample

TABLE 13.3. Foraminiferal assemblage, of site 10, sample no. 26087402, east Broughton Island

Species	Percentage
Cassidulina reniforme Nørvang	24
Cassidulina teretis Tappan	20
Elphidium subarcticum Cushman	17
Nonion orbiculare (Brady)	9
Elphidium excavatum forma boreale Nuzhdina	7
Astrononion gallowayi Loeblich and Tappan	5
Islandiella norcrossi (Cushman)	4
Elphidium excavatum forma clavata Cushman	4
Angulogerian fluens Todd	2
Islandiella islandica (Norvang)	2
Cibicides lobatulus (Walker and Jacob)	2
Elphidium albiumbilicatum (Weiss)	1
Buccella calida (Chushman and Cole)	1
Epistominella vitrea Parke	1
6 other combined species	1

363 specimens were counted in 50 g of sample

teretis. Such deviating forms also occur together with typical C. teretis in deep borings from the North Sea above Cibicides grossa assemblages there. In addition to Cassidulina cf. teretis Epistominella vitrea is frequent (7 to 31%) and so is Elphidium excavatum (8-33%). Stainforthia loeblichi occurs in all the samples of the zone, whereas Cassidulina reniforme and Nonion orbiculare are

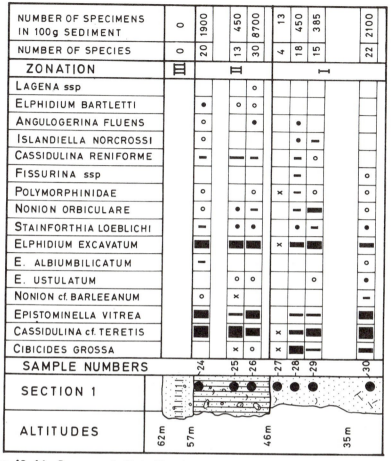

	III	-24	-25	-26	-27	-28	-29	-30
NUMBER OF SPECIMENS IN 100g SEDIMENT	0	1900	450	8700	13	450	385	2100
NUMBER OF SPECIES	0	20	13	30	4	18	15	22
ZONATION	III	II	II	II	I	I	I	I
LAGENA ssp				o				
ELPHIDIUM BARTLETTI		•	o	o				
ANGULOGERINA FLUENS		o		•		•		
ISLANDIELLA NORCROSSI		o				•	—	
CASSIDULINA RENIFORME		—	—	—		—	o	
FISSURINA ssp						—		o
POLYMORPHINIDAE		o		o	×	—	o	o
NONION ORBICULARE		o	•	—		—	▉	o
STAINFORTHIA LOEBLICHI		—	•	•		•	—	•
ELPHIDIUM EXCAVATUM		▉	▉	▉	×	▉	▉	▉
E. ALBIUMBILICATUM		—						o
E. USTULATUM			o	o			o	•
NONION cf. BARLEEANUM		o	×					—
EPISTOMINELLA VITREA		▉	—	▉		—	—	▉
CASSIDULINA cf. TERETIS		▉	▉	▉	×	▉	▉	▉
CIBICIDES GROSSA			×	o	×	▉		—
SAMPLE NUMBERS		-24	-25	-26	-27	-28	-29	-30

Figure 13.16   Range and distribution of 16 selected taxa of foraminifera in samples from section 1 of the Lodin Elv Formation, East Greenland.

rare.   There is a total of 53 species of foraminifera in zone II (13 to 30 per sample).   The number of specimens per 100 g sediment varies from 350 to 8700.

Zone II of the Lodin Elv Formation seems to correlate with the Cassidulina teretis zone of the Clyde Foreland Formation and with zone G of the Qivituq Peninsula.   Cassidulina teretis is common to abundant in their assemblages, and in the field these zones occur next above the Cibicides grossa bearing units.   This is also so at Lodin Elv in East Greenland.

Zone I of the Lodin Elv Formation is characterized by a strong representation of the species Cibicides grossa (9-36%). This large and prominent species is absent or occurs only accidentally in the overlying zone II. Cassidulina cf. teretis also is frequent in zone I (20-44%), and so are Elphidium excavatum (11-25%), Epistominella vitrea (8-21%), and Nonion orbiculare (12% in one sample). Nonion cf. barleeanum (which is close to N. erucopsis Todd) makes 2% in one and is present in another sample in the zone. Zone I of the Lodin Elv Formation, with its large Cibicides grossa, seems almost identical with assemblages from the lower part of the Cibicides grossa zone of the Clyde Foreland Formation, where also large specimens of C. grossa are frequent and at the same time Cassidulina teretis is abundant. In other assemblages from Clyde Foreland Cassidulina teretis may be less frequent or even lacking in the Cibicides grossa zone (cf. Fig 13.7). Zone I also correlates with zone H of the Qivituq Peninsula (cf. Fig. 13.12). Zone H carries large Cibicides grossa, but Epistominella vitrea is less frequent than in zone I, and Cassidulina teretis is absent in most zone H samples.

As described above (cf. also Feyling-Hanssen, 1980a, b) the Cibicides grossa zone and zone H were placed in the Upper Tertiary--more precisely in the Pliocene. By comparison, therefore, zone I of the Lodin Elv Formation is considered to be of Pliocene age (Feyling-Hanssen et al., 1983). The Cassidulina teretis zone of the Clyde Foreland and zone G of the Qivituq Peninsula were considered to represent a transitional unit between the Tertiary and Quaternary (cf. Feyling-Hanssen, 1980a, b). The present foraminiferal zone II of the Lodin Elv Formation is also referred to this unit. The correlation thus arrived at by comparison of foraminiferal assemblages from Clyde Foreland, Qivituq Peninsula and Lodin Elv with such from borings in the North Sea (Fig. 13.17) is tentatively illustrated in Figure 13.18.

Fragments of the mollusc species Hiatella arctica from zone II of the Lodin Elv Formation were analyzed for their amino acid diagenesis by G.H. Miller, University of Colorado, Boulder. They show consistently high ratios, and indicate that zone II of Lodin Elv is older than any other deposit in East Greenland from which amino-acid data are available. Comparing this with results from Alaska, Miller (personal communication with Funder) suggested 1 million years as a minimum age for the sample.

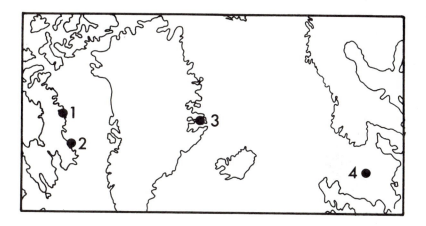

Figure 13.17 Localities correlated: 1, Clyde Foreland; 2, Qivituq
Peninsula; 3, Lodin Elv; 4, Central North Sea.

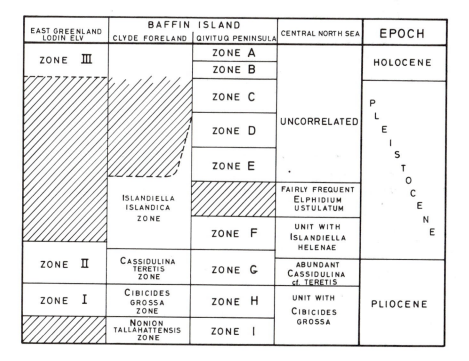

| EAST GREENLAND LODIN ELV | BAFFIN ISLAND | | CENTRAL NORTH SEA | EPOCH |
	CLYDE FORELAND	QIVITUQ PENINSULA		
ZONE III		ZONE A	UNCORRELATED	HOLOCENE
		ZONE B		
		ZONE C		P L E I S T O C E N E
		ZONE D		
		ZONE E		
	ISLANDIELLA ISLANDICA ZONE		FAIRLY FREQUENT ELPHIDIUM USTULATUM	
		ZONE F	UNIT WITH ISLANDIELLA HELENAE	
ZONE II	CASSIDULINA TERETIS ZONE	ZONE G	ABUNDANT CASSIDULINA cf. TERETIS	PLIOCENE
ZONE I	CIBICIDES GROSSA ZONE	ZONE H	UNIT WITH CIBICIDES GROSSA	
	NONION TALLAHATTENSIS ZONE	ZONE I		

Figure 13.18 Correlation and dating of microbiostratigraphical
units from the Arctic and the North Sea.

REFERENCES

Andrews, J.T., 1976: Radiocarbon date list III, Baffin Island, N.W.T., Canada. INSTAAR Occasional Paper No. 21, University of Colorado, Boulder, Colorado, 1-49.

Andrews, J.T. and Barry, R.G., 1978: Glacial inception and distintegration during the last glaciation. Earth and Planetary Science Letters Annual Review, 6:205-228.

Andrews, J.T., Feyling-Hanssen, R.W., Miller, G.H., Schlüchter, C., Stuiver, M., and Szabo, B.J., 1976: Alternative models of early and middle Wisconsin events, Broughton Island, Northwest Territories, Canada: Toward a Quaternay chronology. In Easterbrook and Sibrava, eds., Quaternary glaciation in the northern hemisphere: Project 73-1-24, IGCP, Report 3, Bellingham, Washington and Prague, 28-61.

Andrews, J.T., Funder, S., Hjort, C., and Imbrie, J., 1974: Comparison of the glacial chronology of eastern Baffin Island, East Greenland, and the Camp Century accumulation record. Geology, 2:355-358.

Andrews, J.T. and Miller, G.H., 1972: The Quaternary history of northern Cumberland Peninsula, east Baffin Island, N.W.T. Part X: Radiocarbon date list. Arctic and Alpine Research, 4:261-277.

Andrews, J.T. and Miller, G.H., 1976: Quaternary glacial chronology of the eastern Canadian Arctic: A review and a contribution on amino acid dating of Quaternary molluscs from the Clyde Cliffs. In: Mahanney, W.C. (ed.), Quaternary stratigraphy of North America. Dowden, Hutchison, & Ross. Stroudsburg, PA, 1-32.

Andrews, J.T., Szabo, B.J., and Isherwood, W., 1975: Multiple tills, radiometric ages, and assessment of the Wisconsin glaciation in Eastern Baffin Island, N.W.T, Canada: A progress report. Arctic and Alpine Research, 7:39-59.

Brigham, J.K., 1980: Stratigraphy, amino acid geochronology, and genesis of Quaternary sediments, Broughton Island, east Baffin Island, Canada. M.S. thesis, University of Colorado, Boulder, CO. 199 pp.

Dam, A. Ten, and Reinhold, T., 1941: Die stratigraphische Gliederrung des niederländischen Plio-Pleistozäns nach Foraminiferen. Mededelingen van de Geologische Stichting, Serie C-V, 1:1-66.

Doppert, J.W. Chr., 1980: Lithostratigraphy and biostratigraphy of marine neogene deposits in the Netherlands. Mededelingen Rijks Geologische Dienst, 32-16:255-311.

Doppert, J.W. Chr., Laga, P.G., Meuter, F.J. de, 1979: Correlation of the biostratigraphy of marine neogene deposits, based on benthonic foraminifera, established in Belgium and the Netherlands. Mededelingen Rijks Geologische Dienst, 31-1:1-8.

England, J.H. and Andrews, J.T., 1973: Broughton Island - a reference area for Wisconsin and Holocene chronology and sea level changes on eastern Baffin Island. Boreas, 17-32.

Feyling-Hanssen, R.W., 1964: Foraminifera in Late Quaternary deposits from the Oslofjord area. Norges geologiske Undersökelse 225:1-377.

Feyling-Hanssen, R.W., 1967: The Clyde Foreland. Field Report North-Central Baffin Island 1966, ed. O.H. Löken, Department of Energy, Mines and Resources, Geographical Branch, Ottawa, 35-55.

Feyling-Hanssen, R.W., 1972: The foraminifer Elphidium excavatum and its variant forms. Micropaleontology 18, 3:337-354.

Feyling-Hanssen, R.W., 1976a: The Clyde Foreland Formation, a
  micro-paleontological study of Quaternary stratigraphy. 1st.
  International Symposium on Benthonic Foraminifera of Continental
  Margins, Pt.B: Paleoecology and Biostratigraphy. Maritime Sediments,
  Special Publication 1:315-377.
Feyling-Hanssen, R.W., 1976b: The stratigraphy of the Quaternary Clyde
  Foreland Formation, Baffin Island, illustrated by the distribution of
  benthic foraminifera. Boreas, 5:77-94.
Feyling-Hanssen, R.W., 1976c: A Mid-Wisconsinan interstadial on
  Broughton Island, Arctic Canada, and its foraminifera. Arctic and
  Alpine Research, 8, 2:161-182.
Feyling-Hanssen, R.W., 1978: Microbiostratigraphy of marine deposits of
  Qivituq Peninsula. In: Andrews, J.T. (ed.), Final Report to the
  National Science Foundation, 37-54.
Feyling-Hanssen, R.W., 1980a: Microbiostratigraphy of young Cenozoic
  marine deposits of the Qivituq Peninsula, Baffin Island. Marine
  Micropaleontology, 5:153-184.
Feyling-Hanssen, R.W., 1980b: An assemblage of Pleistocene foraminifera
  from Pigojoat, Baffin Island. Journal of Foraminiferal Research,
  10:266-285.
Feyling-Hanssen, R.W. and Buzas, M.A., 1976: Emendation of Cassidulina
  and Islandiella helene new species. Journal of Foraminiferal
  Research, 6(2):154-158.
Feyling-Hanssen, R.W., Funder, S., and Petersen, K.S., 1983: The
  Lodin Elv Formation; a Plio/Pleistocene occurrence in Greenland.
  Bulletin of the Geological Society of Denmark, 31:81-106.
Feyling-Hanssen, R.W., Jørgensen, J.A., Knudsen, K.L., and Andersen,
  A.-L.L., 1971: Late Quaternary Foraminifera from Vendsyssel, Denmark
  and Sandnes, Norway. Bulletin of the Geological Society of Denmark,
  21, 2-3:67-317.
Froese, A., 1967: 3-dimensional analysis and interpretation of till
  fabric from the tills of the Clyde Foreland cliffs, Baffin Island.
  Unpublished manuscript, Department of Geography, Middlesex College,
  University of Western Ontario, London, Canada, i-vii and 1-58.

Goldthwait, R.P. 1964: Deglaciation of the Clyde River area, Baffin
  Island. Unpublished manuscript, Institute of Polar Studies, Ohio
  State University.
Gudina, V.I., 1976: Foraminifera, stratigraphy, and paleozoogeography
  of the marine Pleistocene of the northern U.S.S.R. Akad. Nauk SSSR,
  Siberian Branch, Trud. Inst. Geol. Geoph., 314, 126 pp.
Gudina, V.I. and Evserov, U.Y., 1973: Stratigraphy and foraminifera of
  Late Pleistocene of Kola Peninsula. Nauka Siberian Branch, Trans.
  Inst. Geol. Geoph., 175, 148 pp.

King, C., 1980: Provisional microfaunal zonation - North Sea
  Cainozoic. In: Vinken, R. and Meyer, K.-J (eds.), Report no. 6 of
  "The N.W. European Tertiary Basin," IGCP project 124. 420 pp.
King, C., Bailey, H.W., King, A.D., Meyrick, R.W., and Roveda, V.L.,
  1981: North Sea Cainozoic. In: Jenkins, D.G. and Murray, J.W. (eds.),
  Stratigraphical Atlas of fossil foraminifera. Ellis Horwood Limited,
  Chichester, 294-298.

Leslie, R.J., 1965: Ecology and paleoecology of Hudson Bay
  Foraminifera. Bedford Institute of Oceanography, Dartmouth, N.S.
  Rep. B.J.O. 65-6, 1-192.
Loeblich, Jr., A.R. and Tappan, H., 1953: Studies of Arctic
  Foraminifera. Smithsonian Miscellaneous Collection, 121, 7:1-150.
Løken, O.H., 1966a: Geomorphological observations along the east
  coast. Field report North Central Baffin Island, 905-18-8. Department
  of Mines and Technical Survey, Geographical Branch, Ottawa, 11-20.

Löken, O.H., 1966b: Baffin Island refugia older than 54,000 years. Science, 153:1378-1380.

Meuter, F.J. de and Laga, P.G., 1977: Lithostratigraphy and biostratigraphy based on benthonic foraminifera of the Neogene deposits of northern Belgium. Bull. Soc. Belge G ol. 85(4):133-152.

Miller, A.A.L., Scott, D.B., and Medioli, F.S., 1982: Elphidium excavatum (Terquem): Ecophenotypic versus subspecific variation. Journal of Foraminiferal Research, 12,2:116-144.

Miller, G.H., Andrews, J.T., and Short, S.K., 1977: The last interglacial-glacial cycle, Clyde Foreland, Baffin Island, N.W.T.: stratigraphy, biostratigraphy, and chronology. Canadian Journal of Earth Science, 14, 12:2824-2857.

Mode, W.M., 1980: Quaternary stratigraphy and palynology, Clyde Foreland, Baffin Island. Ph.D. thesis, University of Colorado, Boulder, CO. 217 pp.

Nelson, A.R., 1978: Stratigraphy of the Qivituq Peninsula, Northern Cumberland Peninsula, Baffin Island. In: Andrews, 1978: Final report to the National Science Foundation, 95-114.

Nelson, A.R., 1978: Quaternary glacial and marine stratigraphy of the Qivitu Peninsula, northern Cumberland Peninsula, Baffin Island, Canada. Ph.D. thesis, University of Colorado, Boulder, CO. 297 pp.

Nelson, A.R., 1980: Chronology of Quaternary landforms, Qivitu Peninsula, northern Cumberland Peninsula, Baffin Island, N.W.T., Canada. Arctic and Alpine Research, 12:265-286.

Nelson, A.R., 1981: Quaternary glacial and marine stratigraphy of the Qivitu Peninsula, northern Cumberland Peninsula, Baffin Island, Canada. Geological Society of America Bulletin, Pt. II, 92, 8: 1143-1261, Pt. I:512-518.

Nelson, A.R., 1982: Aminostratigraphy of Quaternary marine and glaciomarine sediments, Qivitu Peninsula, Baffin Island. Canadian Journal of Earth Science, 19:945-961.

Østby, K.L. and Nagy, J., 1982: Foraminiferal distribution in the western Barents Sea, Recent and Quaternary. Polar Research, 1:53-95.

Osterman, L.E., 1982: Late Quaternary history of southern Baffin Island, Canada: A study of foraminifera and sediments from Frobisher Bay. Ph.D. thesis, University of Colorado, Boulder, CO. 380 p.

Pheasant, D.R. and Andrews, J.T., 1972: The Quaternary history of the northern Cumberland Peninsula, Baffin Island, N.W.T.: Part VIII, Chronology of the Narpaing and Quajon Fiords, during the past 120,000 years. 24th International Geological Congress, Section 12:81-88.

Phleger, F.B., 1952: Foraminifera distribution in some sediment samples from the Canadian and Greenland Arctic. Contributions from the Cushman Foundation for Foraminiferal Research, 3(2):80-89.

Sejrup, H.-P. and Guilbault, J.-P., 1980: Cassidulina reniforme and C. obtusa (foraminifera), taxonomy, distribution and ecology. Sarsia, 65: 79-85.

Shackleton, N.J. and Opdyke, N.D., 1973: Oxygen Isotope and palaeomagnetic stratigraphy of Equatorial Pacific Core V28-238: Oxygen Isotope temperatures and Ice volumes on a 10^5 year and 10^6 year scale. Quaternary Research, 3:39-55.

Smith, R.K., 1978: Systematics of the North American high northern latitude very shallow cold water foraminiferal fauna. Archives de Science, 31(2):133-161.

Stuiver, M., 1978: Radiocarbon dating of old material. In: Andrews, 1978: Final report to the National Science Foundation, Grant No. Ear-74-01857, 12-13.

Szabo, B.J., 1978: Uranium-series dates. In: Andrews, 1978: Final report to the National Science Foundation, Grant No. Ear-74-01857, 17.

Szabo, B.J., Miller, G.H., Andrews, J.T., and Stuiver, M., 1981: Comparison of uranium-series, radiocarbon and amino acid data from marine molluscs, Baffin Island, Arctic Canada. Geology, 9:451-457.

Todd, R., 1957: Foraminifera from Carter Creek, northeastern Alaska. United States Geological Survey, Professional Paper 294-F, 221-235.

Vilks, G., 1969: Recent foraminifera in the Canadian Arctic. Micropaleontology, 15(1)35-60.

Voorthuysen, J.H. Van, Toering, K. and Zagwign, W.H., 1972: The Plio-Pleistocene boundary in the North Sea basin, revision of its position in the marine beds. Geologie en Mijnbouw, 51(6):627-639.

Walton, W.R., 1964: Recent foraminiferal ecology and paleo-ecology. In: Imbrie, J. and Newell, N.D. (eds.), Approaches to Paleoecology, 151-237.

PLATE 13.1. (facing page) Figures 13.1-13.2: <u>Islandiella islandica</u> (Nørvang), side and edge view of a specimen from zone E of the Qivituq Peninsula. x 75. Figure 13.3: <u>Islandiella helenae</u> Feyling-Hanssen and Buzas, side view of a specimen from Pigojoat. x 75. Figures 13.4-13.5: <u>Islandiella inflata</u> (Gudina), opposite sides of a specimen from the <u>Cassidulina teretis</u> zone of Clyde Foreland. x 65. Figures 13.6 13.7: <u>Cassidulina teretis</u> Tappan, side view of two specimens from the <u>Cibicides grossa</u> zone of Clyde Foreland. x 75. Figures 13.8-13.10: <u>Cibicides grossa</u> Ten Dam and Reinhold, edge, umbilical side and spiral side of a specimen from the <u>Cibicides grossa</u> zone of Clyde Foreland. x 50. Figures 13.11-13.12: <u>Nonion cf. tallahattensis</u> Bandy, edge and side view of a specimen from the <u>Nonion tallahattensis</u> zone of Clyde Foreland. x 80. Figure 13.13: <u>Miliolinella chuckchiensis</u> Loeblich and Tappan, side view of a specimen from the <u>Nonion tallahattensis</u> zone of Clyde Foreland. x 75.

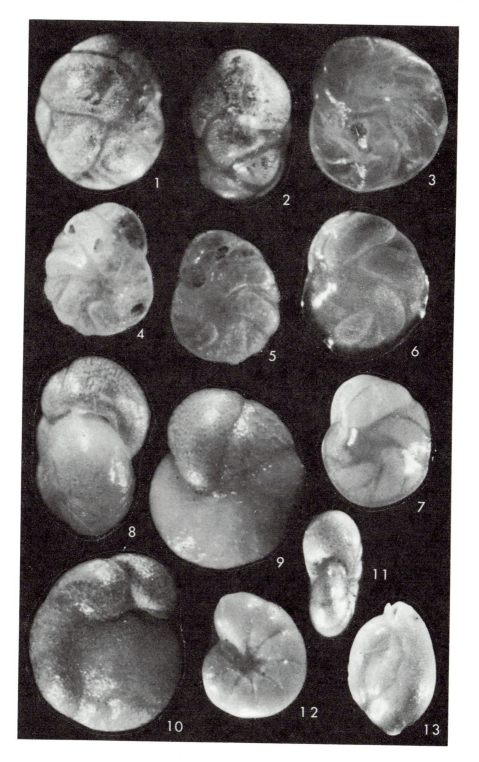

14 Aminostratigraphy of Baffin Island shell-bearing deposits

Gifford H. Miller

The preservation on eastern Baffin Island of a long and complex interbedded Quaternary glacial and marine record was first noted during the 1950 Arctic Institute of North America's expedition. However, it was not until the mid 1960s that Løken (1966) was able to provide limiting C-14 dates suggesting that the last deglaciation of at least a portion of the east coast occurred prior to 54 ka ago, and that evidence of several earlier glaciations lay preserved in a succession of Quaternary sediments exposed in wave-eroded cliffs fronting Baffin Bay on the Clyde Foreland. Feyling-Hanssen (1967, 1976a) studied the micropaleontology of the Clyde Foreland sediments and compiled a preliminary biostratigraphic zonation for what he termed the Clyde Foreland Formation. He obtained a C-14 date on the uppermost marine unit of 40,000±1740 (GSC-796), indicating that essentially the entire sedimentary stack was at or beyond the limits of C-14 capabilities. Feyling-Hanssen later studied the micropaleontology of similar sediments exposed on Broughton Island (Feyling-Hanssen, 1976b) and the Qivituq Foreland (Feyling-Hanssen, 1980). The lack of an independent absolute dating method inhibited establishing an absolute time scale, although Feyling-Hanssen's (this volume) most recent opinion based on diagnostic foraminiferal taxa is that some of the units may be as old as late Pliocene.

In 1974 a four-year multidisciplinary dating program was initiated to test a number of methodologies for dating the "old" fossiliferous units exposed along eastern Baffin Island. The conclusions of this campaign are summarized below:
(1) C-14 dating. Radiocarbon dating of carbonate fossils provides reliable absolute age estimates for shells of late Foxe/Holocene age (< 13 ka), but all older units are apparently beyond the limits of the method (> 40 ka). Finite apparent ages obtained for these units (35 to

50 ka) should be considered minimum ages. Stuiver (unpublished) performed a series of radiocarbon-dating experiments, determining the C-14 age on five successive 20% leaches of several large (1 kg) shell collections. He found that the apparent ages increased with each successive leach, and that this progression could be explained by a simple diffusion model of carbonate exchange. Stuiver concluded that even the C-14 age of the innermost 20% fraction should be considered as a minimum date.

(2) Uranium - series. A series of mollusc collections dated by the uranium-thorium and -protactinium methods yielded greater ages than did the C-14 datings of the same samples. However these dates occasionally conflicted with the litho-, bio- and/or amino-stratigraphy. Our current concensus (Szabo et al., 1981) is that the dates represent reliable minimum age estimates.

(3) Amino acid geochronology. Isoleucine epimerization ratios (aIle/Ile--also formerly abbreviated allo/iso) were determined for the bivalve molluscs Hiatella arctica (L) and Mya truncata (L) to test whether the extent of epimerization could be used as an index of relative age. Our initial results demonstrated that: (1) aIle/Ile ratios were consistent within a single stratum; (2) that lower (younger) aIle/Ile ratios characterized shells from overlying strata whereas higher (older) ratios were found in shells from underlying strata (Andrews and Miller, 1976); and (3) that the ratios were consistent with the geological and paleontological observations. A compilation of all ratios from the Clyde Foreland area yielded a composite relative chronostratigraphic framework (Miller et al., 1977) that served as a model for the glacial history of Baffin Island.

The conversion of aIle/Ile ratios from relative to absolute ages requires an independent age determination. Our initial chronology was calibrated to a preliminary U-series mollusc date from the Cape Christian marine sands of c. 130 ka (Miller, et al., 1977). Subsequently, the U-series age of this unit was reevaluated to be $\geq$ 180 ka (Szabo et al., 1981). Additional work on the temperature dependency of epimerization kinetics has shown that the thermal history would have to have been unrealistically high (about 0°C) to explain the observed aIle/Ile ratio in M. truncata from this unit with an age of 130 ka. Using a realistic thermal assumption, an age of c. 500 ka for this unit has been proposed (Andrews and Miller, in press).

We conclude that aIle/Ile ratios provide a reliable relative chronostratigraphic base for a limited geographic region, and have termed this usage aminostratigraphy (Miller and Hare, 1980). Recent work on the temperature dependency of the isoleucine epimerization rate constant (see below) permits the calculation of absolute age envelopes for observed aIle/Ile ratios based on limiting thermal assumptions.

Objectives

The purpose of this paper is to present new data from several
inter-related projects that have been major foci of the INSTAAR Amino
Acid Geochronology Laboratory over the last eight years.

(1) Alteration of aIle/Ile ratios by relatively subtle changes in
the method of preparing shell samples for analysis and a compilation of
aIle/Ile ratios for samples prepared under different procedures.

(2) A discussion of the geographic range along the eastern
Canadian Arctic over which amino acid ratios can be used directly as
correlation indices.

(3) An update of the original aminostratigraphic subdivision of
the Clyde Foreland Formation, with standard aIle/Ile ratios for each
member.

(4) Derivation of Arrhenius parameters for isoleucine
epimerization in the genus Mya.

(5) Amino-acid-derived age envelopes for the Baffin Island
aminozones.

ALTERATION OF OBSERVED AILE/ILE RATIOS BY FRACTIONATION
DURING SAMPLE PREPARATION

Preparation of carbonate fossils for amino acid analysis is a
relatively straightforward procedure involving the isolation of a
particular structural layer and/or portion of the shell for analysis,
removal of surface contaminants from the sample(generally a 33% acid
leach or brief sonication), dissolution of the cleaned shell in
purified HCl with separate preparations for the Free (naturally
hydrolyzed) and Total (Free plus peptide-bound amino acids) amino acid
assemblages. The Total fraction is pyrolyzed in 6N HCl under a
nitrogen atmosphere at 110°C for 22 hr, then taken to dryness in a
vacuum desiccator. The Free fraction is simply dried to remove excess
acid before analysis. There are, however, a variety of specific
procedures to accomplish these ends, each with subtle differences. The
effect of these differences on Total aIle/Ile ratios could not have
been predicted a priori.

Recent work in our laboratory has demonstrated that minor changes
in the sample preparation procedure can produce major alteration of the
observed Total aIle/Ile ratio. This effect became apparent as type
sites were re-analyzed under different preparation procedures over a
ten-year period; some of those analyses yielded Total aIle/Ile ratios
that were not reproducible within statistically acceptable limits with
ratios from previous preparations. This effect was not observed in the

Free ratios, hence instrumental variation can be excluded.

An in-depth investigation of the cause of the observed discrepancies (detailed in Miller et al., 1982) revealed that fractionation of the amino acid population could occur during sample preparation. When a cleaned shell sample is dissolved in acid, the resultant solution contains naturally released Free amino acids and polypeptide residues of a wide molecular weight range. The aIle/Ile ratio is highest within the Free amino acid population, decreasing in polypeptides of increasing molecular weight. The lowest aIle/Ile ratios occur in the highest molecular weight polypeptide residues which contain the greatest proportion of Ile in internal sites (cf. Kriausakul and Mitterer, 1978). The high molecular weight fraction evidently interacts with the walls of the glass vial in which it is contained, adhering even after vigorous sonication. Our sample preparation prior to 1982 included dissolving a single cleaned fragment, then withdrawing aliquots for both Free and Total fractions. In so doing, the sample inadvertently fractionated by leaving the high molecular weight residues adhering to the original vial wall, thereby enriching the transferred solutions in aIle. Because the Free fraction is independent of peptide-bound amino acids, the Free aIle/Ile ratio is not affected by this transfer and fractionation.

The following example (Table 14.1) illustrates the effect of fractionation. A single valve of M. truncata from the type Kuvinilk marine sands at Clyde was analyzed without any transfers. The observed Total aIle/Ile was 0.070 (AAL-2796D). Another fragment of the same valve weighing 100 mg was dissolved in 1.0 ml 6N HCl and five successive 0.2 ml aliquots were withdrawn. The first aliquot (AAL-2817 A1) was used to determine the Free ratio, the following four aliquots (-2817 A1 to A4) were prepared for the Total aIle/Ile ratios. Finally, the empty vial in which the shell was dissolved received 0.5 ml 6N HCl and was prepared as if it were a normal sample to test whether any amino acid residues remained on the glass walls (-2817 A5). The Total aIle/Ile ratio in these five fractions decreases with each successive aliquot; the empty test tube had slightly higher amino acid concentrations than did the 0.2 ml aliquots, but the aIle/Ile ratio was 0.026, only slightly above a modern shell value. The mean ratio of all five fractions is 0.069, indistinguishable from that of the untransferred preparation (0.070; AAL-2796 D). An additional 60 mg fragment of the same valve was dissolved in 1.0 ml 6N HCl and two 0.5 ml aliquots withdrawn, one each for Free and Total fractions (AAL-2817 B, Table 14.1). The Free ratio is similar to others from this unit, but the Total ratio is higher than in AAL-2796 D, again suggesting fractionation during transfer. A final 30 mg fragment of the same

TABLE 14.1. AIle/Ile ratios in Free and Total fractions in a
single valve of Mya truncata from the Kuvinilk member of the Clyde
Foreland Formation showing the effects of fractionation during sample
preparation (see text for discussion).

Lab ID	AIle/Ile	
	Total	Free
AAL-2796D	0.070	0.53
AAL-2817A1	0.085	0.54
A2	0.082	
A3	0.081	
A4	0.068	
A5	0.026	
B	0.075	0.5
C1	0.075	
C2	0.066	

shell was dissolved in 0.5 ml 12N HCL and 0.2 ml transferred to a
separate vial for a Total preparation (AAL-2817 C1). The remaining 0.3
ml was hydrolyzed in the same vial in which the shell was dissolved
(AAL-2817 C2). The observed aIle/Ile ratios (Table 14.1) also reflect
a relatively high ratio in the transferred fraction (C1) and a lower
ratio in the original vial (C2). The mean ratio, however, is again
0.070.

These data confirm that a significant portion of high molecular
weight polypeptide residues adhere to the walls of a glass vessel;
additional experiments demonstrated that even with sonication, these
residues will not be removed by transfer of the surrounding solution,
regardless of the strength of the acid. The aIle/Ile ratio, and
presumable D/L ratios of other amino acids, are significantly lower in
the adsorbed residues relative to the transferred solution. Depending
on the number of transfers, a wide range of aIle/Ile ratios may be
obtained from a given shell. Consequently, it would appear essential
that the preparation for the Total fraction be completed through the
pyrolysis step in the same vessel in which the shell is initially
dissolved.

Conversion tables for aIle/Ile ratios obtained under various sample preparation procedures

The INSTAAR laboratory has determined aIle/Ile ratios on shells from
over 3500 collections, not only for in-house projects, but also under
various arrangements for a wide range of scientists from other

institutions. Because the changes in preparation can alter the reported Total aIle/Ile ratios, determinations made at different times may not be directly comparable. In an effort to quantify the effects of these changes, 92 samples have been analyzed by two or more preparation procedures. These comparisons include a range of aIle/Ile ratios for the three most commonly utilized taxa in this laboratory: Mya truncata (50 collections), Hiatella arctica (28 collections), and Arctica islandica (14 collections). The aIle/Ile ratios for these collections are tabulated in Appendix Tables 14.A, 14.B, and 14.C.

Three categories of preparation are listed in the conversion tables. Preparation A is the procedure currently in use at the INSTAAR laboratory. It has been the standard preparation since ID AAL-2790. Preparation B covers most of the rest of the samples processed by the INSTAAR lab except for lab IDs AAL-1300 to AAL-1550. Samples analyzed under Preparation B had various specific differences in certain steps, but all contained at least one post-dissolution, pre-hydrolysis transfer. The ratios appear to be generally comparable for all Preparation B samples. Preparation C (lab IDs AAL-1300 to 1550) all yield ratios below expected values, although the cause for this difference has not yet been resolved. In preparations A and B, the Free ratios are statistically indistinguishable, whereas Free ratios under Preparation C tend to be about 10% lower.

The alteration of aIle/Ile ratios by fractionation requires that the sample contain a substantial portion of high molecular weight polypeptide residues. This state will occur throughout the early portion of protein diagenesis, but should become less pronounced in samples with aIle/Ile Total ratios above c. 0.4 to 0.5. Although few samples of this range are reported herein, analyses of Glycymeris with Total aIle/Ile ratios between 0.4 and 0.7 yielded similar results when prepared by both procedures A and B.

In general, there does not appear to be a clear dependency of the conversion factor on site temperature or aIle/Ile ratio for samples below 0.3 Total. For M. truncata the ratio of aIle/Ile observed under Preparation A relative to Preparation B is 0.61±0.16 (n = 28) and in Preparation A relative to C it is 1.31±0.21 (n = 16). For H. arctica the conversion factors are 0.78±0.19 (n = 9) for Preparation A/Preparation B and 1.35±0.20 (n = 22) for Preparation A/Preparation C. In A. islandica it is 0.62±0.09 (n = 14) for Preparation A/Preparation B. These factors can be used, with appropriate error terms, to convert most reported Total aIle/Ile values to ratios equivalent to the same sample processed under Preparation A.

GEOGRAPHIC LIMITS ON DESIGNATED AMINOZONES

A fundamental question in any application of aminostratigraphy is over what geographic range can aIle/Ile ratios be used directly as correlation indices? Kennedy et al. (1982) have addressed this question for the US west coast and Wehmiller and Belknap (1982) discussed the latitudinal temperature controls on amino acid ratios along the US east coast.

In this section, I will argue that aIle/Ile ratios can be used directly along the entire 1500 km stretch of coastal eastern Baffin Island between Bylot Island in the north and Resolution Island in the south and possibly into northeasternmost Labrador (Fig. 1). A rigorous test of this hypothesis would require an evaluation of the long-term thermal regime over the range of available sites. Lacking such data, a first approximation can be reached by examining the temperature data over the period of instrumental record. Although mean annual temperatures could be compared, Andrews et al. (1981) have argued that the mean July temperature may be a better index as the epimerization reaction is more sensitive to high than low temperatures.

In Table 14.2 the mean annual air temperatures and mean July temperatures are tabulated for weather stations in the eastern Canadian Arctic. July temperatures are between 4 and 5 °C for all east coast sites between Pond Inlet and NE Labrador, whereas the sites away from the coast (Pangnirtung and Frobisher) are significantly warmer. Mean annual temperatures range from -14 to -6°C along the same transect. Arguing that the summer temperatures most strongly controls the rate of epimerization, it follows that all east coast regions of Baffin Island experience similar epimerization rates at present. Consequently, aIle/Ile ratios can be used directly to correlate disjunct sites across this region. The similarity in July temperatures in northern Labrador suggests that the correlation may be extended directly to that region. Mean July temperatures increase rapidly southward along the Labrador coast and caution must be used for sites along the Labrador coast.

Additional confirmation that this hypothesis is broadly correct comes from aIle/Ile ratios in C-14-dated early Holocene sites across the region (see subsequent discussion of Eglinton member). The ratios in shells from NE Labrador are not significantly different from the ratios in shells from the Broughton Island region; mean July temperatures at the two sites are similar, whereas mean annual temperatures are 5°C higher in Labrador.

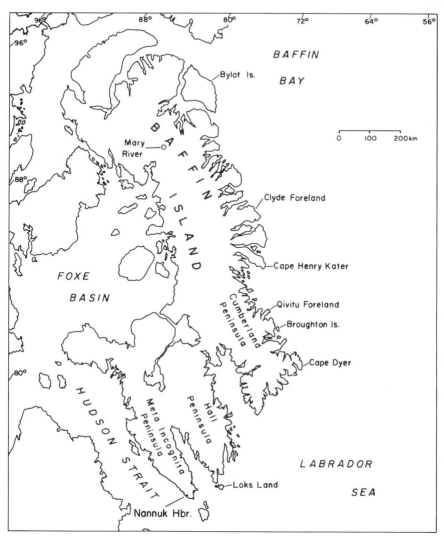

Figure 14.1 Location map of the eastern Canadian Arctic showing
sites referred to in the text.

REVISED AMINOSTRATIGRAPHY OF THE CLYDE FORELAND FORMATION

In this paper, I propose that the Clyde Foreland Formation, originally
defined as the unconsolidated deposits of variable grainsize and
lithology that mantle much of the lowlying foreland between Clyde and
Eglinton fiords (Feyling-Hanssen, 1966; 1967) be extended and the
definition modified to include all of the unconsolidated sediment below
the local marine limit along eastern Baffin Island. The Clyde Foreland

TABLE 14.2. Temperatures for coastal stations along eastern Baffin Island and NE Labrador over the period of instrumental record. Site locations are on Figure 14.1.

Site	Temperature (°C)	
	Mean Annual	Mean July
Pond Inlet	-14.6	5.3
Clyde River	-12.3	4.5
Cape Hooper	-11.9	4.2
Broughton Is.	-11.4	4.5
Cape Dyer	-10.3	5.2
Pangnirtung	-9.0	7.8
Brevoort Is.	-8.9	4.2
Frobisher	-9.2	7.8
Resolution Is.	-6.4	3.2
NE Labrador	c. -6	c. 5

Formation contains marine, glacial-marine, fluvial and organic units, including buried soils. Subdivision locally can be based on standard stratigraphic criteria and recognizable unconformities, but because of the discontinuous nature of the deposits, these techniques are not widely applicable. The overall subdivision and correlation of members of the Clyde Foreland Formation is based on characteristic aIle/Ile ratios measured from indigenous proteinaceous residues preserved in carbonate fossils, primarily in the marine sediments.

Recent studies (Mode 1980; Nelson, 1982; Brigham, 1983; Miller unpubl.) demonstrating the occurrence of previously unrecognized members (aminozones) within the Clyde Foreland Formation, and alteration of the observed Total aIle/Ile ratios by changes in the sample preparation procedures, have prompted a reevaluation of the regional stratigraphic framework for Quaternary depositional events in the Eastern Canadian Arctic. A program to reanalyze shells from the type and para-type deposits of the original members of the Clyde Foreland Formation and from newly defined units by the Preparation A procedure has been recently completed. These samples are exclusively M. truncata and H. arctica. The difference in racemization rate between the two genera is not statistically discernable.

Pre-Cape Christian Units

Abundant deposits pre-dating the Cape Christian marine sands are exposed in the sections along the Clyde and Qivituq forelands (Miller

et al., 1977; Nelson, 1978; 1981; Mode, 1980). However, no concerted
effort has been made to correlate the older units between these areas
or to reevaluate the amino acid data. Two collections of the youngest
pre-Cape Christian member at Clyde (Qakijaanga member, Mode, 1980) have
been reanalyzed. The results (Table 14.3) show a 42% decrease in Total
aIle/Ile ratios from the values reported by Miller et al. (1977).

TABLE 14.3. AIle/Ile determinations in the youngest pre-Cape
Christian member at Clyde (Qukijaanga member).

Locality	AAL # or ref	Species	aIle/Ile	
			Total	Free
Clyde	2844	M.t.	0.111±0.005 (3)	0.81±0.08 (2)
Clyde	3010	M.t.	0.121 (1)	0.64 (1)
	Mean values		0.116±0.005 (4)	0.72±0.08 (3)

Number of individuals analyzed given in parentheses.

Cape Christian Member: 0.092±0.012 (Total); 0.64±0.02 (Free)

The primary shell-bearing unit within this member is the Cape Christian
marine sands. Numerous analyses have been made on shells collected by
Miller (in 1974) and Mode (in 1977 and 1978). Eight valves of M.
truncata were analyzed under preparation A. The mean aIle/Ile (Total)
ratio of 0.092±0.012 is 38% lower than the original value (0.16). The
mean Free ratio in these shells is 0.63±0.03, indistinguishable from
the original value of 0.66±0.05 (Miller et al., 1977). Mode (1980,
Table 3.1) reported aIle/Ile ratios on 24 valves in nine preparations
of Hiatella and Mya from deposits of the Cape Christian marine sands.
An additional series of sixteen valves of Hiatella analyzed in 1977
(Miller, unpub) gave a mean Free ratio of 0.64±0.02. The mean Free
aIle/Ile ratio from the five series of analyses of 57 individual shells
is 0.64 ± 0.02 (Table 14.4).

Kuvinilk Member: 0.060±0.009 (Total); 0.54±0.03 (Free)

The marine sediments of the Kuvinilk member contain abundant in situ
valves of H. arctica and scattered M. truncata. Four different
preparations (all of Preparation A) of 19 individual valves of M.
truncata from the type beds at Clyde gave mean aIle/Ile ratios of
0.060±0.009 (Total) and 0.52±0.045 (Free). Additional Free ratios are
available from earlier preparations by Miller et al. (1977), Brigham
(1980, 1983, and unpubl.), and Mode (1980). The available analyses for
the Free aIle/Ile ratio total 165 individual preparations of either Mya
or Hiatella with a mean ratio of 0.54±0.03 (Table 14.5).

TABLE 14.4. AIle/Ile determinations in the Cape Christian member.

Locality	AAL # or ref.	Species	aIle/Ile	
			Total	Free
Clyde	3332	M.t.	0.096±0.015 (4)	0.63±0.02 (4)
Clyde	2845	M.t.	0.088±0.008 (4)	0.64±0.02 (2)
Clyde	Mode (1980)	H.a. & M.t.		0.63±0.02 (24)
Clyde	Miller et al. (1977)	H.a.		0.66±0.05 (11)
Clyde	147	H.a.		0.62±0.05 (16)
Mean values			0.092±0.012 (8)	0.64±0.02 (57)

Number of individuals analyzed given in parentheses.

TABLE 14.5. AIle/Ile determinations in the Kuvinilk member.

Locality	AAL # or ref	Species	aIle/Ile	
			Total	Free
Clyde	3331	M.t.	0.056±0.005 (4)	0.56±0.03 (4)
Clyde	2796	M.t.	0.058±0.008 (5)	0.52±0.04 (5)
Clyde	2847	M.t.	0.064±0.009 (4)	0.49±0.04 (4)
Clyde	2773	M.t.	0.061±0.009 (6)	0.50±0.05 (5)
Clyde	2817	M.t.		0.53±0.01 (3)
Clyde	2816	M.t.		0.55±0.02 (3)
Clyde	Mode (1980)	H.a. & M.t.		0.56±0.03 (9)
Clyde	Brigham (1983)	H.a.		0.50±0.05 (92)
Clyde	Brigham (unpubl.)	H.a.		0.56±0.05 (6)
Clyde				0.53±0.02 (11)
Clyde	Miller et al., 1977	H.a.		0.56±0.03 (23)
Mean values			0.060±0.009 (19)	0.54±0.03 (165)

Number of individuals analyzed given in parentheses.

Cape Broughton Member: 0.045± 0.002 (Total) ; 0.43± 0.04 (Free)

The marine sediments at Cape Broughton, northern tip of Broughton
Island, were originally described by Feyling-Hanssen (1976) and defined
as the Cape Broughton Interstadial of middle Wisconsinan age. Brigham
(1983) reinterpreted the sedimentary environment as being proximal to
an ice tongue in Broughton Harbour. She defined the Quaternary
deposits on Broughton Island as the Broughton Island Formation, with
marine sediments at the cape defined as the Cape Broughton member.

TABLE 14.6.   AIle/Ile determinations in the Cape Broughton member.
===

| Locality | AAL # or ref | Species | aIle/Ile | |
			Total	Free
Broughton Is.	3329	M.t.	0.045±0.001 (4)	0.49±0.04 (4)
Clyde*		M.t.	0.057±0.10 (3)	0.39±0.03 (3)
Clyde*	2843	M.t.	0.063±0.02 (5)	0.42±0.07 (5)
Clyde	Mode (1980)			0.42±0.023 (8)
Broughton Is.	Brigham (1983)	H.a.		0.42±0.03 (24)
	Mean values		0.045±0.001 (4)	0.43±0.04 (39)

*Total values not used in computing mean ratios.
Number of individuals analyzed given in parentheses.

Here the Cape Broughton sediments are defined as a subdivision of the
Clyde Foreland Formation.   U-series dates on shells suggest an age in
excess of 140 ka (Szabo, et al., 1981).   AIle/Ile ratios in shells
collected from the type site and an adjacent correlative were reported
by Andrews et al. (1976) and Brigham (1983).   Free and Total aIle/Ile
ratios from a new preparation of four M. truncata valves from the Cape
Broughton sediments are listed in Table 14.6.   Brigham (1983) lists a
mean Free ratio based on 24 individual shells of H. arctica of
0.42±0.03.   Mode (1980) identified a correlative unit in the exposures
at Clyde with a Free ratio of 0.42±.02 that he named the Sajjugiaq
member.   A new preparation of three M. truncata from Mode's collections
of his Sajjugiaq member gave rather divergent ratios (Table 14.6), but
which agree within the statistical uncertainties.   With the intent of
keeping the member names pronounceable, I have elected to call this
aminozone the Cape Broughton member rather than use Mode's (1980)
appelation of Sajjugiaq, with the type site designated as the sediment
between 14 and 32 m asl at Cape Broughton (as defined by Brigham, 1983,
p. 586).

Kogalu Member: 0.029±0.005 (Total); 0.30±0.02 (Free)

Sediment considered to have been deposited during the Kogalu aminozone
represents the most ubiquitous of the "old" deposits on eastern Baffin
Island.   In most regions these deposits are related to the last major
continental glaciation to reach the continental shelf.   Kogalu
correlatives have been identified on SE Meta Incognita and Hall
peninsulas and intermittently along most of the low-lying headlands
between Cape Dyer and Bylot Island (Fig. 14.1).   AIle/Ile ratios of the
Kogalu aminozone are found in both ice-proximal sediment related to the
last deglaciation of the outer coast as well as in clearly younger,

non-glacial deposits that record normal marine sedimentation.   Both
Brigham (1980, Fig. 4.7) and Mode (1980, Fig. 3.2) suggested that the
Kogalu aminozone is bimodal, with clusters of Free ratios at 0.30 and
0.24.   In general, the lower ratios come from deposits postdating ice
recession; higher ratios are associated with ice-proximal deposits.

TABLE 14.7. AIle/Ile determinations in the Kogalu member.

Locality	AAL # or ref	Species	aIle/Ile	
			Total	Free
Broughton Is.	3330	M.t.	0.034±0.01 (3)	0.29±0.03 (3)
Cape Raper	3333	M.t.	0.034±0.03 (4)	0.30±0.01 (4)
McBeth Fd	3334	M.t.	0.025±0.02 (4)	0.28±0.02 (4)
Clyde	2846	M.t.	0.025±0.03 (4)	0.33±0.02 (4)
McBeth Fd	2009	M.t.	0.029±0.02 (3)	0.29±0.01 (3)
Henry Kater* (late Kogalu)	3011	H.a.	0.027 ±0.02 (3)	0.21±0.02 (3)
Clyde Foreland	Mode (1980)		-	0.30±0.04 (28)
Clyde Foreland	Miller et al. (1977)	H.a. & M.t.	-	0.29±0.03 (24)
Quajon Fd.	Andrews et al. (1976)	H.a.	-	0.30 (3)
Cape Dyer	Locke (1980)	H.a. & M.t.		0.29±0.04 (7)
Pangnirtung	Dyke (1977)	H.a.	-	0.28 (3)
Qivitu	Nelson (1982)	M.t. & H.a.	-	0.30±0.03 (12)
Hall Peninsula	Miller (1980)	H.a.	-	0.30±0.02 (3)
Broughton Is.	Brigham (1983)	H.a.	-	0.33±0.04 (27)
Bylot Is.	Klassen, this vol.	M.t.	-	0.31±0.04 (6)
	Mean values		0.029±0.004 (18)	0.30±0.02 (134)

*not used in calculating Total or Free mean values.
Number of individuals analyzed given in parentheses.

    New preparations were made of _in situ_ _M._ _truncata_ from six sites
analyzed previously and correlated with the Kogalu member based on the
physical stratigraphy and amino acid ratios. Five of these sites are
typical ice-proximal deglacial deposits related to the last episode of

continental-shelf glaciation. The mean Free aIle/Ile ratio in 18 valves of M. truncata from these sites is 0.30±0.02 (Table 14.7), indistinguishable from previous studies (e.g. Miller et al., 1977), whereas the Total ratios are 0.029±0.004, 64 percent of the earlier determinations (0.045±0.007).

The sixth site in this series (AAL-3011, Table 14.7) is from Cape Henry Kater where a sandy, normal-marine delta, 20 m asl, with a warm, subarctic fauna, stratigraphically overlies the ice-proximal glacial-marine sediment related to recession of ice from the Ayr Lake advance. Previous analyses had indicated the ice-proximal unit correlated with the Kogalu aminozone, but the overlying delta had slightly lower aIle/Ile ratios. This relationship is confirmed by the new analyses, particularly the Free ratios that are statistically lower than in the ice-proximal Kogalu deposits. Neither the Free nor Total ratios, however, are as as low as those of the next younger Loks Land member.

Determinations of Free aIle/Ile ratios are available from earlier analyses at nine regions along eastern Baffin Island (Table 14.7). The mean ratios are indistinguishable from the five new preparations, resulting in a Free ratio for the Kogalu aminozone of 0.30±0.02 based on 134 analyses of M. truncata or H. arctica.

Loks Land Member: 0.020±0.004 (Total), 0.14±0.02 (Free)

Although no sites have been found on the Clyde Foreland that yield aIle/Ile ratios intermediate between the Kogalu and the early Holocene Eglinton members, such deposits have been reported elsewhere in the eastern Canadian Arctic. Klassen (this volume) reports intermediate Free (0.17) and Total (0.025, Prep B) ratios for deposits C-14 dated > 43 ka (GSC-3410) that were related to a limited glaciation in the Pond Inlet area, and Blake (1980) reported finite radiocarbon ages of about 40 ka on algae associated with shells with similarly intermediate aIle/Ile ratios (0.15 Free; 0.023 Total, Prep B) at Cape Storm, southeast Ellesmere Island. Ives (1977) reported a radiocarbon date in excess of 40 ka on shells from the northern Labrador coast; aIle/Ile ratios on the same collection were only slightly higher than in Holocene shells from that area.

A similar deposit has been recently sampled on Loks Land, southeastern Baffin Island (Fig. 14.1). Shells were collected from a sandy marine delta on eastern Loks Land that overlies a low-lying plain mantled by glacial-marine sediment containing abundant limestone

erratics and fragmented mollusc shells. Paired <u>Macoma</u> <u>calcarea</u>
collected from the delta gave a C-14 date of 41,900+7100/-3700
(QC-446), which at the two sigma level is infinite. The entire
collection was used by the dating laboratory, hence no amino acid
ratios are available. A second collection of whole valves and large
fragments of <u>M</u>. <u>truncata</u> and <u>H</u>. <u>arctica</u> was submitted for C-14 dating
with a subsample of <u>M</u>. <u>truncata</u> prepared for amino acid analysis. The
C-14 date is 32,500+2400-1800 (GX-8591) which is again infinite at the
two sigma level. AIle/Ile ratios (AAL-3521, 3522; Table 14.8) are
intermediate between ratios in Kogalu and Holocene shells.

    A correlative site is located near the head of Nannuk Harbour,
southeastern Meta Incognita Peninsula. Large fragmented <u>H</u>. <u>arctica</u>
collected from shelly till 180 m asl yielded a C-14 age > 37 ka
(GX-8942) and aIle/Ile ratios similar to those in the Loks Land samples
(Table 14.8).

TABLE 14.8. AIle/Ile determinations in the Loks Land member.

Locality	AAL # or ref	Species	aIle/Ile Total	aIle/Ile Free
Hall Penn (in till)	3000	H.a. & M.t.	0.029±0.04 (3)	0.14±0.01 (3)
Loks Land	2462	H.a.	-	0.14 (1)
Loks Land	2461	M.t.	-	0.14±0.01 (2)
Loks Land	2642*	H.a.	0.022±0.004 (3)	0.17±0.02 (3)
Loks Land	3522	M.t.	0.024 (1)	0.12 (1)
Loks Land	3521	H.a.	0.021 (2)	0.12 (1)
Bylot Is.	932*	H.a.	0.014±0.002 (3)	0.14±0.03 (3)
Bylot Is.	1639+	M.t.	0.016±0.002 (3)	0.11±0.02 (3)
Navy Board Inlet	906	H.a.	0.015±0.003 (3)	0.14±0.01 (3)
Mean values			0.020±.04 (18)	0.14±0.02 (18)

*Total ratio obtained by multiplying recorded value by 0.78 (Table
  14.B).
+Total ratio obtained by multiplying recorded value by 0.62 (Table
  14.A).
Number of individuals analyzed given in parentheses.

The glacial-marine sediment and overlying deglacial delta on Loks Land are here designated as the type locality of the Loks Land member of the Clyde Foreland Formation. They represent a glacial advance and retreat, in general less extensive than the preceding Ayr Lake till of the Kogalu member, but more extensive than the late Foxe advance. The age of the Loks Land member is at or beyond the limits of C-14 dating ($\geq$ 40 ka), but is characterized by aIle/Ile Free and Total ratios (0.14, 0.020) well below those of the Kogalu member. The molluscan fauna is composed of undiagnostic panarctic species, in marked contrast to the numerous thermopholous taxa in the upper marine sands of the Kogalu member.

Eglinton Member: 0.013±0.001 (Total), ND (Free)

The Eglinton member represents marine, terrestrial organic (peat) and eolian sedimentation that occurred on eastern Baffin Island in late Foxe and Holocene time. A new series of aIle/Ile ratios (Table 14.9) has been recently obtained on a representative suite of C-14 dated earliest Holocene collections of M. truncata. Four collections of three valves each were analyzed; all have been C-14 dated. No detectable (ND) aIle was found in any of the Free fractions and the aIle/Ile Total ratio averaged 0.013±0.002, 65 percent of the figure used previously (0.020).

Modern: 0.011±0.001 (Total), ND (Free)

A sample of recent M. truncata collected from northern Cumberland Peninsula was analyzed to quantify the degree of isoleucine epimerization that is induced during the preparation procedure. As in other tests of this kind, no aIle was observed in the Free fraction (in fact > 99.9% of all amino acids are peptide bound). The mean aIle/Ile Total ratio in three different valves is 0.011±0.001 (Table 14.10). The same ratio in a series of modern M. truncata from Greenland (supplied by S. Funder and K-S Peterson) averaged slightly higher, whereas the ratio in six H. arctica collected live in Arctic Alaska are similar to the Baffin Island Mya (Table 14.10).

DERIVATION OF ARRHENIUS PARAMETERS FOR ISOLEUCINE EPIMERIZATION IN THE GENUS MYA

The racemization or epimerization of a protein L-amino acid to the D-amino acid configuration is a reversible first order reaction that

TABLE 14.9. AIle/Ile determinations in the Eglinton member.

Locality and C-14 age	AAL # or ref	Species	aIle/Ile Total	Free
Kipisa, Cumberland Sound 8750±100 (GSC-2508)	3407	M.t.	0.0131±0.0013 (2)	N.D. (3)
Kingnait Fd., Cumberland Sound 8680±140 (GSC-2478)	3408	M.t.	0.0129±0.0002 (3)	N.D. (3)
Qivitu, Cumberland Peninsula 9280±120 (GSC-2479)	3409	M.t.	0.0118±0.0003 (3)	N.D. (3)
Warwick Sd, Frobisher Bay 10760±30 (Miller, 1980)	3410	M.t.	0.0138±0.0004 (3)	N.D. (3)
NE Labrador 9110±470 (GX-9293; Clark, 1983)			0.0126±0.0010 (3)	N.D. (3)
		Mean value	0.0128±0.0007 (15)	N.D. (15)

N.D. = no detectable aIle in the sample.
Number of individuals analyzed given in parentheses.

TABLE 14.10. AIle/Ile determinations in modern specimans of *Mya truncata* and *Hiatella arctica*.

Locality	AAL #	Species	aIle/Ile Total	Free
Baffin Island	3411	M.t.	0.0108±0.0009 (3)	N.D. (3)
Arctic Alaska	2871	H.a.	0.0106±0.0012 (3)	N.D. (3)
Arctic Alaska	3448	H.a.	0.0111±0.0010 (3)	N.D. (3)
	Mean value		0.0108±0.0002 (9)	N.D. (9)

N.D. = no detectable aIle in the sample.
Number of individuals analyzed given in parentheses.

DERIVATION OF ARRHENIUS PARAMETERS FOR ISOLEUCINE EPIMERIZATION IN THE
GENUS MYA

The racemization or epimerization of a protein L-amino acid to the
D-amino acid configuration is a reversible first order reaction that
can be expressed in the general format as:

$$L\text{- amino acid} \underset{k_2}{\overset{k_1}{\rightleftharpoons}} D\text{- amino acid} \tag{14.1}$$

where k_1 and k_2 are the forward and reverse rate constants,
respectively. Integrating this equation (assuming no net gain or loss
of reactants) yields the equation (from Schroder and Bada, 1976):

$$\ln \left(\frac{1 + D/L}{1 - K' \; D/L} \right) = (1 + K') \; k_1 t + C \tag{14.2}$$

where:

   D/L = the ratio of D- to L- amino acids measured in the
        laboratory

   $K' = \dfrac{k_2}{k_1}$

   C = left side of the equation evaluated at time zero (death of
        organism)

   k_1 = forward rate constant

   t = time elapsed since death of the organism

   We can evaluate K' (the reciprocal of the equilibrium constant) by
empirically determining the D/L ratio in racemic shells (shells for
which equation 14.1 has reached equilibrium). The values reported in
the literature and determined in our laboratory range between 1.25 and
1.35; a mean value of 1.30 is used.

$$K' = \frac{1}{K} = \frac{1}{1.30} = 0.77$$

   The constant C can be evaluated by measuring the D/L ratio in
modern shells. Several such determinations for isoleucine
epimerization in modern Mya averaged 0.011 (Table 14.13) which,when
substituted into the left side of equation 14.2, yield C = 0.0194.

Substituting these constants back into equation 14.2 gives:

$$\ln\left(\frac{1 + D/L}{1 - 0.77\ D/L}\right) - 0.0194 = 1.77\ k_1 t \tag{14.3}$$

To reduce this expresssion further it is necessary to determine the temperature dependency of the forward rate constant, k_1. As in most chemical reactions, the reaction rate (k_1) is strongly affected by temperature. This relationship can be expressed by the Arrhenius equation:

$$k_1 = Ae^{(-Ea/RT)} \tag{14.4}$$

where A is a constant, often referred to as the frequency factor
    Ea is the activation energy
    R is the gas constant (1.987)
    T is absolute temperature (K).

The linear relationship between log k_1 and T^{-1} allows the deriviation of A and Ea if k_1 can be determined for a sufficient range of temperatures. This relationship only holds over those temperatures for which Ea is constant (<160°C) and furthermore implies that the reaction mechanisms are the same for all temperatures of interest. Because of the complex reactions involved in the degradation of protein in a carbonate matrix (e.g. Kriausakul and Mitterer, 1978), we cannot be assured that this assumption is fully satisfied, but until strong evidence is presented to the contrary I will assume that high-temperature pyrolysis experiments (100 to 160°C) experience the same reaction mechanisms as do shells in the natural environment (-15 to +25°C), and that the Arrhenius equation is the best first approximation of the temperature sensitivity for the isoleucine epimerization reaction rate in molluscan fossils.

The Arrhenius equation can be solved by obtaining data on the reaction rate for isothermal experiments over a wide temperature range, then fitting the log k_1 :1/T data set by a least-squares linear regression. We undertook a series of high-temperature pyrolysis experiments in which cleaned fragments of modern Mya truncata were heated for various times in a controlled-temperature oven at isothermal settings of 80, 100, 120, 140, and 160°C. At 80°C the reaction proceeds so slowly that nearly a year was required to obtain significant changes in the aIle/Ile ratios. At temperatures above 160°C, decomposition reactions complicate the reaction kinetics. Values for k_1 at these temperatures are listed in appendix Table 14.D.

To provide control for the lower temperatures we determined the
alle/Ile ratios in radiocarbon-dated samples from sites with a
wide range of modern temperatures. We assumed that the mean air
temperature over the period of instrumental record at each site is a
close first approximation of the integrated thermal history since shell
depostion. However, for the arctic sites we have used a temperature
2°C higher because of the tendency for permafrost temperatures to
average about 2°C above the mean air temperature (e.g. Washburn, 1980,
Table 5). These data points are listed in appendix Table 14.E.

Fitting log k_1 and the reciprocal of the absolute temperature
(Fig. 14.2) by at least-squares linear regression yields the
expression:

$$\log k_1 = 16.45 - \frac{6141}{T} \quad \text{(coefficient of correlation} = 0.9994) \quad (14.5)$$

Thus the values for Ea and A in equation 14.4 are:

Ea = 28100 cal mole^{-1}
A = 16.45

Equation 14.5 describes the temperature dependency of the
isoleucine epimerization reaction rate constant. It can be substituted
into equation 14.3 to yield an expression for which the unknowns are
D/L, t and T. Solving for T yields the temperature-prediction
equation:

$$T = \frac{6141}{16.45 - \log \left(\dfrac{\ln \left(\dfrac{1 + D/L}{1 - 0.77\,D/L} \right) - 0.0194}{1.77\,t} \right)} \quad (14.6)$$

whereas solving for t yields the age-prediction equation:

$$t = \frac{\ln \left(\dfrac{1 + D/L}{1 - 0.77\,D/L} \right) - 0.0194}{1.77 \; 10^{\left(16.45 - \frac{6141}{T} \right)}} \quad (14.7)$$

These equations are dependent on the following assumptions: (a)
That isoleucine epimerization kinetics can be approximated as a first
order reversible reaction. This is probably valid for samples with

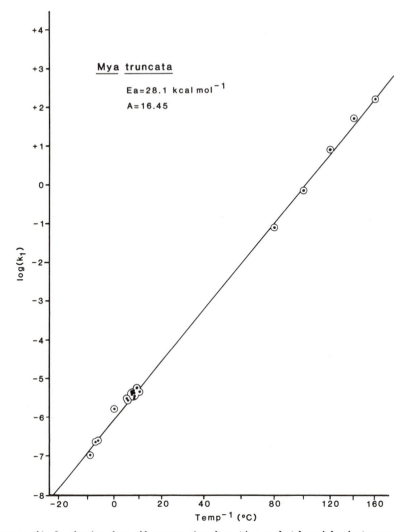

Figure 14.2 Arrhenius diagram showing the relationship between the isoleucine epimerization rate and temperature. The relationship is defined by the epimerization rate observed in pyrolyzed shells at temperatures of 80, 100, 120, 130, 140 and 160°C (Table 14.D) and that measured in C-14 dated Late glacial shells from a wide range of present temperature regimes (Table 14.E). Temperatures were plotted as the reciprocal of absolute temperature (°K) but are converted back to °C for ease of comprehension. The least-squares linear regression through the data points defines the Arrhenius parameters Ea (28.1 kcal mole^{-1}) and A (16.45).

aIle/Ile values between 0 and c. 0.35. At later stages of diagenesis
the reaction becomes distinctly non-linear. (b) That there is no net
loss of reactants with time. Quantitative analyses of amino acid
concentrations with increasing sample age do show a slow systematic
depletion in amino acid levels in increasingly older samples, but this
does not appear to significantly affect the fit of the equations within
the specified D/L range (0 to 0.35). (c) That the reactions occurring
during high-temperature pyrolysis experiments are governed by the same
mechanisms as at ambient environmental temperatures. It is not yet
possible to validate this assumption fully, but no data have been
presented to the contrary, and the linear relationship over an 80°C
temperature range suggests it is a reasonable assumption.

Consequently, I propose that equations 14.6 and 14.7 provide a
reasonable first approximation of the relationship between D/L, time
and temperature during the early stages of protein diagenesis in the
genus Mya.

In their review of amino acid geochronology and geothermometry,
Williams and Smith (1976) emphasized the dependency of the temperature-
and age-equations on the precision of the activation energy (Ea)
calculation, claiming, for example, that a 1% error in Ea corresponds
to a 3°C change in the predicted temperature. Although this is
strictly correct, we have found that for Mya truncata any activation
energy between 27.9 and 28.4 kcal mole^{-1} can be combined with an
appropriate value for A to yield essentially identical age/temperature
predictions. Thus a 2% change in activation energy that theoretically
could cause a 6°C difference in predicted temperature in fact causes
only 0.1°C difference. Consequently, it is not essential that the
precise activation energy be determined, but that an appropriate
combination of values for both Ea and A be determined that will best
satisfy all of the control points.

AGE CALCULATIONS FOR BAFFIN ISLAND AMINOZONES

The conversion from D/L (aIle/Ile) ratios to absolute age is arguably
the most tenuous aspect of amino acid geochronology. Having derived
the equations relating D/L, time and temperature (equations 14.6 and
14.7) for Mya truncata, it is now possible to calculate first
approximations for the ages of aminozones defined earlier. Such
calculations first require an estimate of the thermal term (T).

A proper value for T cannot be easily derived. Instrumental air
temperature records have been available only since the late 1950s and

no mean annual sediment temperatures have been reported. The temperature that governs amino acid racemization will be an integration of equation 14.6 through all of the temperature fluctuations experienced by the fossil since death of the organism. This integrated temperature is generally referred to as the effective diagenetic temperature (EDT). It is not related to the temperature at which the shell lived.

The nearest comparable stations that report both mean annual air and sediment temperature are at (1) Mary River, north-central Baffin Island, (2) Resolute, Cornwallis Island, and (3) Alert, Ellesmere Island (Washurn, 1980, Table 3.5). The mean difference between sediment temperature at 6 to 30 m and adjacent mean annual air temperature is +3.4°C (range +1.9 to +6.3 °C). The EDT will thus deviate from the annual air temperature by at least +2°C.

The EDT is also influenced by the fluctuations in regional temperature on both an annual and longer-term basis. The nonlinear relationship between the rate constant, k_1, and temperature (described by equation 14.5) indicates that high temperatures cause an exponential increase in racemization rate. In all cases, a fluctuating thermal history will increase the EDT over the mean temperature, with the difference between them related to the amplitude of the fluctuations. It is unlikely that this effect will result in a departure of more than 1 or 2°C at the most. The net result of this effect and the observation that mean permafrost temperatures are warmer than mean air temperatures is that the temperatures predicted by equation 14.6 or used in equation 14.7 will probably be 3 to 7°C above the corresponding air temperatures.

In Table 14.11, ages for the Baffin Island aminozones are calculated for a range of EDTs from -5 to -13°C. This range can be further constrained by limiting radiometric age determinations on the younger units. Temperatures above -7°C can be excluded because they yield ages that directly conflict with the radiometric dates on the Loks Land and/or Kogalu units (i.e. they are too young). A temperature of -13°C results in ages that are much older than would be generally accepted based on the excellent geomorphic preservation and limited soil development of moraines from the advance associated with the Kogalu aminozone. Thus the EDT range that results in ages that cannot be dismissed as obviously implausible is -7 to -11°C. Implicit in this analysis, is the assumption that an appropriate EDT for the last glacial cycle (last 125 ka) is a close first approximation of an EDT for the last 700 ka or more. Thus an acceptable temperature envelope

TABLE 14.11. Amino acid ages for various effective diagenetic
temperatures for Baffin Island aminozones.

Aminozone	Mean aIle/Ile	Effective Diagenetic Temperature (EDT) (°C)						
		-5	-7	-9	-9.5	-11	-13	
Modern	0.011	-	-	-	-	-	-	
Eglinton	0.012 0.013 0.014	3 6 9	4 8 13	6 13 19	7 4 21	9 19 28	14 29 43	
Loks Land	0.020	20	29	44	49	66	100	
(late Kogalu)	0.027	-	-	69	-	-	-	
Kogalu	0.029	51	75	110	125	170	260	Calculated Ages (Ka)
Cape Broughton	0.045	96	140	210	240	320	480	
Kuivinilk	0.060	140	200	305	340	460	690	
Cape Christian	0.092	230	340	500	560	750	1150	
youngest pre-Cape Christian	0.116	290	430	650	720	970	1500	
oldest pre-Cape Christian	0.55	1600	2300	3500	3800	5200	7900	

for the Kogalu aminozone can be applied to the older aminozones as
well.

We can speculate on the best temperature approximation for the
Kogalu aminozone by considering the geologic relationships.
Ice-proximal deposits with Kogalu aIle/Ile ratios are associated with
the last major glaciation to reach the continental shelf along eastern
Baffin Island (e.g. Miller et al., 1977; Mode, 1980; Locke, 1980; Dyke
et al., 1982; Brigham, 1983; Klassen, this volume). The associated
fauna is generally pan-arctic. In at least one site, these deposits
overlie an interglacial soil (pollen indicating terrestrial
temperatures above the Holocene maximum: Brigham, 1983; Mode, this
volume) that formed when relative sea level was close to present. The
ice-proximal Kogalu deposits are stratigraphically overlain by normal
open-marine deposits at lower elevations that contain subarctic
molluscan and foraminiferal assemblages (e.g. Feyling-Hanssen, 1976;
Miller et al., 1977) indicative of inshore marine conditions warmer
than present.

There are two primary alternatives for the timing of the glaciation within the Kogalu aminozone. It could have occurred either late in deep-sea isotope stage 6 or within stage 5. In the former case, the overlying warm-fauna deposits would be the local stage 5e interglacial unit, whereas the underlying soil might be isotope stage 7. Arguing most strongly against this hypothesis is the limited glacial event at the analogous isotope stage 2/1 transition and the warmer-than-present pollen assemblage in the underlying soil.

The alternative hypothesis would place the interglacial soil as the local stage 5e deposit, with terrestrial temperatures above the Holocene optimum as has been inferred elsewhere in the world, and the subsequent glacial event would have occurred within a younger substage of stage 5. The overlying warm-marine deposits reflect a time when inshore marine waters were similar to those along west Greenland today, but terrestrial temperatures need not have been higher than present. Such conditions would occur at present if it were not for the cold Canadian Current that flows out of the Arctic Ocean, through the High Arctic channels southward along the Baffin Island coast, forcing the returning warm West Greenland Current off-shore. If the build-up of glacial ice in the High Arctic (Innuitian Ice Sheet) filled the inter-island channels and did not recede when the NE sector of the Laurentide Ice Sheet withdrew, then the outflow of Arctic Ocean water would have been impeded and a warm West Greenland Current could have penetrated the inshore zone along Baffin Island (Aksu, this volume). This would occur under a North Atantic circulation similar to the present. Analyses of deep-sea cores suggests that the circulation and temperature of the North Atlantic remained essentially interglacial (i.e. like today) throughout isotope stage 5. Consequently the most plausible interpretation for the age of the Kogalu aminozone is post-isotope stage 5e and pre-stage 5/4 boundary. From Table 14.11, a reasonable EDT for a mid stage 5 event would be -9°C, associated with an averaged mean annual air temperature of -11 to -15°C, suggesting a glacial-age temperature 0 to 6°C lower than present. The amino acid data do not have sufficient resolving power to discriminate between the various substages of stage 5, but it is implied that the glacial event would have occurred during substage 5d or 5b.

Applying an EDT of -9°C to the older aminozones results in a striking 100 ka signal in the amino-acid-derived ages for major glacial events. These speculative age assignments suggest that major glaciations in the Eastern Canadian Arctic occur shortly after global interglacial events, presumably during periods of high-latitude insolation minima and relatively warm oceans with strong meridional transport into the Arctic region.

The age given to the unit bearing the oldest measured aIle/Ile ratios is conjectural due to the non-linearity of racemization kinetics and limited data on the early Quaternary thermal history, but does suggest the possibility of a Pliocene age for some of the raised marine units in Arctic Canada and Greenland (e.g. Feyling-Hanssen et al., 1982; Feyling-Hanssen, this volume).

SUMMARY

The aminostratigraphy of the Clyde Foreland Formation and available radiometric and amino-acid-derived ages are summarized in Table 14.12, along with the glacial, marine and climatic interpretations derived from deposits of the various aminozones. The calculated amino acid ages must be considered speculative until independent age determinations become available. At present, they represent the best age estimates based on all available information. Because of the inferred antiquity of the pre-Cape Christian members, the remanent magnetism in the older units should be reversely oriented if the sediment has not experienced a water-saturated freeze-thaw cycle. Such information could be used to test the validity of the amino acid ages (see Chapter 2).

ACKNOWLEDGEMENTS

This paper represents the composite efforts of many individuals working in or cooperating with the research program of the Amino Acid Geochronology Laboratory at the University of Colorado. I am greatly indebted to Dr. P.E. Hare, Carnegie Institution of Washington, for his long-standing collaboration on the application of amino acid geochronology to Arctic regions and to Drs. M. Stuiver (University of Washington), B. Szabo (U.S.G.S. - Denver) and W. Blake, Jr. (GSC - Ottawa) for their participation in our program to refine the absolute chronology of Quaternary events on Baffin Island. Dr. J.T. Andrews has actively participated in all aspects of this program.

Specific acknowledgement goes to Drs. J.K. Grette (nee Brigham) and P.C. Clark for their contributions to the sections on the sample preparation procedures and Clyde Foreland Formation revised aminostratigraphy, to Dr. W.N. Mode for the pyrolysis data in Mya, to Dr. J.K. Grette for unpublished amino acid analyses of samples from Alaska and the USSR, to S. Forman and S. Lehman for similar unpublished data on Spitsbergen samples, and to Drs. W.D. McCoy, J.K. Grette, W.N. Mode and Mr. D. Goter for assistance with laboratory analyses.

TABLE 14.12. Aminostratigraphy and chronological interpretation of the Clyde Foreland Formation, Eastern Baffin Island.

Clyde Foreland Formation Aminozones	Mean aIle/Ile		C-14 age (Ka)	U-series age[1] (Ka)	Amino Acid age[2] (Ka)	Comments
	Total	Free				
Modern	0.011	N.D.	-	-	-	--
Eglinton	0.013	N.D.	8-11	≥ 6	13	Limited glacial advance, panarctic fauna overlain by non-glacial sediment with non-diagnostic fauna
Loks Land	0.020	0.14	≥ 40	-	44	deglacial, sediment, panarctic fauna
late	0.027	0.22	≥ 54	-	69	Non-glacial sediment, warm subarctic fauna
Kogalu ice-proximal	0.029	0.30	≥ 54	> 70	110	Ice-proximal to de-glacial sediment with pan-arctic to subarctic fauna Over-lies inter-glacial soil, over-lain by marine subarctic sediement with warm fauna
Cape Broughton	0.045	0.43	≥ 40	≥ 140	210	Ice-proximal sediment, warm subarctic fauna
Kuvinilk	0.060	0.54	-	≥ 140	305	Includes ice-distal: ice-proximal: ice-distal sediment warm subarctic fauna throughout
Cape Christian	0.092	0.62	≥ 50	≥ 190	500	Includes a till overlain by warm subarctic fauna in marine sands on which is developed an inter-glacial soil
youngest pre-Cape Christian	0.116	0.72	-	-	650	Undivided multiple glacial and marine events (e.g. Nelson,
oldest pre-Cape Christian	0.55	1.02	-	> 300	3500	

[1] From Szabo et al., 1981
[2] From Table 14.16, -9°C EDT column.

Radiocarbon-dated calibration samples were supplied by Drs. T.O. Vorren (Norway), C. Hjort (Iceland), M. Stuiver (Maine), J. Mangerud (Norway), H. Krog and K-S. Petersen (Denmark), J.D. Peacock (Scotland), and J.T. Andrews (Vancouver).

Financial support has been provided by several grants from the National Science Foundation. Foremost of these are GA-41562 that initiated our amino acid geochronology program, EAR-79-26061 that supported the new analyses to derive standard ratios for the Baffin Island aminozones, DPP-78-26000 for the pyrolysis data and ATM-81-09013 that supported the C-14 dated calibration program and derivation of Arrhenius parameters for isoleucine epimerization in Mya.

J.T. Andrews, J. Mangerud and J.K. Grette provided input through discussions and critical review of earlier versions of this manuscript.

Appendix

TABLE 14.A. Comparison of aIle/Ile determinations in *Mya* *truncata* under various sample preparation procedures.

Locality	T[1] (°C)	aIle/Ile (Total)[2] A	B	C	Lab ID[3] AAL-
Baffin Is. (modern)	-10	0.011 (3)	0.016		3411/ -
USSR	-12	0.015 (1)	0.025 (1)		2595
Spitsbergen	-6	0.016	0.020		Episode A
USSR	-12	0.019 (3)		0.016 (3)	1479
USSR	-12	0.020	0.038 (1)		2591
USSR	-12	0.021 (2)		0.014 (2)	1485
USSR		0.022 (3)		0.017 (3)	1480
USSR	-12	0.022 (2)		0.017 (2)	1488
USSR		0.022 (1)	0.038 (1)		2586
USSR	-12	0.024 (2)		0.022 (2)	1486
USSR		0.024 (1)	0.043 (1)		2584
USSR	-12	0.024 (2)		0.022 (2)	1490
Greenland	-8	0.025 (3)	0.035 (3)		3144/BL-44
USSR		0.026 (1)	0.044 (1)		2585
USSR		0.026 (1)	0.046 (1)		2587
USSR		0.026 (1)		0.026 (1)	1453
USSR		0.026 (3)		0.018 (3)	1459
Spitsgergen	-6	0.027	0.032		Episode B
USSR		0.028 (1)	0.045 (1)		2590
USSR	-12	0.028 (3)		0.020 (2)	1498
Baffin Is.	-10	0.029 (18)	0.045 (25)		Table 14.7 Miller et al., 1977
USSR	-12	0.030 (2)		0.022 (2)	1489
Greenland	-8	0.030 (3)	0.057 (3)		3143/328
USSR		0.031 (1)	0.053 (1)		2593
USSR		0.035 (1)	0.062 (1)		2592
USSR		0.035 (1)		0.025 (1)	1452
Alaska	-12	0.08 (3)	0.102 (3)		2853/2060
Spitsbergen	-6	0.040	0.055		Episode C
Alaska	-12	0.043 (3)	0.109 (2)		2957/1684
Baffin Is.	-10	0.045	0.061 (14)		3329/107, 108
USSR	-12	0.045 (2)		0.036 (2)	1497
USSR	-12	0.046 (3)		0.029 (3)	1483
USSR	-12	0.048 (2)		0.052 (2)	1495
Spitsbergen	-6	0.048 (2)	0.080		Episode D
Baffin Is.	-10	0.056 (4)	0.081 (6)		3331/
Alaska	-12	0.076 (3)	0.157 (3)		2956/1688
Norway	7.6	0.090 (4)	0.14 (26)		3001/1698, 1699, BL86-89
Baffin Is.	-10	0.092 (8)	0.16 (16)		2845, 3332/147
USSR		0.114 (3)		0.066 (3)	1454
Baffin Is.	-10	0.115 (3)	0.20		3010/2844-
Alaska	-12	0.0119 (3)	0.120 (3)		2965/2735
Greenland	-8	0.140 (3)	0.25 (2)		3146/BL-35
Alaska	-12		0.190 (2)	0.123 (3)	2727/1309
Norway	7.6	0.192 (2)	0.27 (4)		2864/381, 2181

Table 14.A.   (continued)

Locality	T[1]	alle/Ile (Total)[2]			Lab ID[3]
Baffin Is. (modern)	-10	0.011 (3)	0.016		3411/ -
USSR		0.218 (3)		0.160 (3)	1455
Norway	7.6	0.218 (2)	0.28 (3)		2865/381
Norway	7.6	0.224 (4)	0.32 (6)		2866, 2867/385, BL-7
Norway	7.6	0.268 (3)	0.37 (7)		2868-2870/ 386, 2184
USSR		0.58 (2)		0.51 (2)	1456
USSR		0.70 (1)		0.64 (1)	1457
USSR		1.04 (1)		0.99 (1)	1458

[1] Current mean annual air temperature.
[2] Mean value, number of individual shell preparations given in parentheses.
[3] Most samples designated by AAL ID; BL-prefix are samples analyzed at the University of Bergen, 1979-1980. Spitsbergen ratios refer to aminozones defined in Miller, 1982, and all are derived from 4 or more analyses. Slashes separate IDs for different preparations (same order as data are listed).

TABLE 14.B.   Comparison of alle/Ile determinations in _Hiatella_ under various sample preparation procedures.

Locality	T[1] (°C)	alle/Ile (Total)[2]			Lab ID[3] AAL-
		A	B	C	
USSR	-12	0.021 (3)		0.013 (3)	1487
USSR	-12	0.023 (2)		0.022 (1)	1462
USSR	-12	0.024 (2)		0.019 (2)	1485
USSR	-12	0.028 (3)		0.022 (3)	1464
Greenland	-8	0.028 (3)	0.043 (3)		3148/1690
USSR		0.031 (1)		0.024 (1)	1452
USSR	-12	0.034 (2)		0.034 (2)	1486
USSR	-12	0.037 (3)		0.024 (3)	1461
USSR		0.038 (1)		0.026 (1)	1453
Alaska	-12	0.041 (3)	0.053 (3)		2840/1954
USSR	-12	0.042 (2)		0.037 (2)	1490
USSR	-12	0.044 (2)		0.033 (2)	1489
Greenland	-8	0.049 (2)	0.040 (3)		3145/BL-40
USSR		0.051 (3)		0.036 (3)	1463
USSR	-12	0.052 (3)		0.035 (3)	1494
USSR	-12	0.056 (3)		0.047 (3)	1460
USSR	-12	0.064 (3)		0.044 (3)	1493
USSR	-12	0.065 (2)		0.048 (2)	1495
USSR	-12	0.066 (2)		0.034 (2)	1497
Alaska	-12	0.084 (3)	0.144 (2)	0.056 (3)	3037/2728/1336
Greenland	-8	0.118 (1)	0.135 (3)	0.078 (3)	2963/2731/1310
Alaska	-12	0.136 (6)	0.181 (3)	0.112 (3)	2962 & 3124/ 2732/1301

TABLE 14.B. (continued)

Locality	T[1] (°C)	aIle/Ile (Total)[2]			Lab ID[3] AAL-
		A	B	C	
Alaska	-12	0.157((3)	0.181 (3)	0.121 (3)	3036/2727/1309
Alaska	-12	0.223 (3)	0.255 (3)	0.148 (3)	2964/2730/1338
USSR		1.00 (1)		1.07 (1)	1458
USSR		1.09 (1)		0.994 (1)	1457

[1]Current mean annual air temperature.
[2]Mean value; number of individual shell preparations given in
 parentheses.   Preparations A, B, and C as discussed in text.
[3]Most samples designated by AAL ID; BL-prefix are samples analyzed at
 the University of Bergen, 1979-1980. Slashes separate IDs for
 different preparation and are in the same order as data are presented.

TABLE 14.C. Comparison of aIle/Ile determinations in _Arctica islandica_ under various sample preparation procedures.

Locality	T[1] (°C)	aIle/Ile (Total)[2]			Lab ID AAL-
		A	B	C	
USSR		0.018 (1)	0.033 (1)		2588
USSR		0.022 (1)	0.038 (1)		2586
USSR		0.024 (1)	0.043 (1)		2584
USSR		0.026 (1)	0.044 (1)		2585
USSR		0.026 (1)	0.046 (1)		2587
USSR		0.028 (1)	0.045 (1)		2590
USSR		0.034 (1)	0.046 (1)		2589
USSR		0.035 (1)	0.060 (1)		2596
Scotland	9	0.050 (3)	0.098 (3)		2826/2160
Scotland	9	0.057 (3)	0.116 (3)		2827/2161
Denmark	8	0.166 (3)	0.248 (3)		2609/2795
Denmark	8	0.174 (3)	0.251 (3)		2608
Denmark	8	0.209 (3)	0.29 (3)		2607
Denmark	8	0.32 (3)	0.41 (3)		2606

[1]Current mean annual air temperature.
[2]Mean values; number of individual shell preparations given in
 parentheses. Preparations A, B, and C as discussed in text.

TABLE 14.D. High-temperature pyrolysis data, _Mya truncata_.

Lab ID	Temp (°C)	Time (yr)	D/L	k_1	avg k_1
AAL-1113 E1	80	2.74 E-1	0.037	9.44 E-1	
E2			0.031	7.27 E-2	
E3			0.033	7.99 E-2	
					7.80 E-2
AAL-1113 B1	80	7.01 E-1	0.061	7.08 E-2	
B2			0.066	7.79 E-2	
B3			0.062	7.22 E-2	

TABLE 14.D. High-temperature pyrolysis data, _Mya truncata_.

Lab ID	Temp (°C)	Time (yr)	D/L	k_1	avg k_1
AAL-1114 A1	100	4.11 E-2	0.034	5.57 E-1	
A2			0.038	6.54 E-1	
A3			0.049	9.19 E-1	
AAL-1114 B1	100	8.22 E-2	0.08	8.32 E-1	
AAL-1114 C1	100	1.64 E-1	0.12	6.57 E-1	7.42 E-1
C2			0.11	5.97 E-1	
C3			0.13	7.17 E-1	
AAL-1114 D2	100	2.47 E-1	0.16	5.96 E-1	
D3			0.22	8.35 E-1	
AAL-1114 E1	100	3.29 E-1	0.36	1.05	
AAL-1124 A1	120	5.7 E-3	0.043	5.57	
A2			0.045	5.92	
A3			0.044	5.57	
AAL-1124 B1	120	1.40 E-2	0.12	7.70	
B2			0.22	14.70	
B3			0.13	8.40	7.83
AAL-1124 C1	120	3.07 E-2	0.27	8.33	
C2			0.25	7.68	
C3			0.33	10.29	
AAL-1124 D1	120	4.85 E-2	0.26	5.07	
D2			0.48	9.72	
D3			0.44	8.84	
AAL-1124 E1	120	6.88 E-2	0.51	7.32	
E2			0.53	7.64	
E3			0.59	8.62	
AAL-1124 F1	120	9.03 E-2	0.64	7.22	
F2			0.58	6.44	
F3			0.54	5.94	
AAL-3548 1	130	2.74 E-3	0.051	14.6	
2	130	5.48 E-3	0.078	12.1	
3	130	8.21 E-3	0.151	16.8	
4	130	1.10 E-2	0.246	21.2	16.2
AAL-1084 A2	140	2.28 E-3	0.18	73.2	
B2			0.10	38.7	
C2			0.17	68.9	53.4
D2			0.11	43.0	
E2			0.11	43.0	
AAL-1404 A1	160	5.71 E-5	0.018	122	
A2			0.018	122	
A3			0.018	122	
AAL-1404 B2	160	1.14 E-4	0.023	105	
AAL-1404 C2	160	2.28 E-4	0.058	205	

TABLE 14.D. (continued).

Lab ID		Temp (°C)	Time (yr)	D/L	k_1	avg k_1
AAL-1404	D1	160	4.57 E-4	0.092	176	
	D3			0.064	115	
						171
AAL-1404	E1	160	9.13 E-4	0.20	204	
	E2			0.20	203	
	E3			0.20	204	
AAL-1404	F1	160	1.37 E-3	0.28	194	
	F2			0.28	194	
	F3			0.36	253	
AAL-1404	G1	160	1.83 E-3	0.41	217	
	G2			0.39	206	
AAL-1084	B3	160	2.74 E-3	0.60	221	
	D3			0.63	234	
	E3			0.66	247	
	F3			0.57	208	
AAL-1404	I1	160	4.57 E-3	0.68	153	
	I2			0.72	165	
	I3			0.88	216	
AAL-1404	J1	160	6.28 E-3	0.76	128	
	J2			0.96	180	
	J3			0.84	147	
AAL-1404	K1	160	1.14 E-2	1.08	124	
AAL-1404	H1	160	2.53 E-2	1.16	67	
	H2			1.19	72	
	L2			1.31	130	

TABLE 14.E. Radiocarbon-dated calibration sites for the genus *Mya*.

Locality	CMAT[1] (°C)	Age (yr)	Lab ID AAL-	aIle/Ile[2] (Total)	k_1
Qivitu, Baffin Is.	-9 (-11)	9300	3409	0.012 (3)	1.07 E-7
Cumb. Sound Baffin Is.	-7 (-9)	8700	3408	0.013 (3)	2.29 E-7
Cumb. Sound Baffin Is.	-7 (-9)	8800	3407	0.013 (3)	2.27 E-7
Frobisher Bay Baffin Is.	-6 (-8)	10 800	3410	0.014 (3)	2.77 E-7
Tromsø Norway[3]	3.2	12 000		0.031 (3)	1.66 E-6

TABLE 14.E.   (continued)

Locality	CMAT[1] (°C)	Age (yr)	Lab ID AAL-	aIle/Ile[2] (Total)	k_1
Iceland	5	12 470	3272	0.048 (3)	2.95 E-6
Melöy, Norway	5.2	11 610	2997	0.043 (3)	2.76 E-6
Maine	7.0	12 250	2780	0.062 (3)	4.13 E-6
Maine	7.0	12 625	2643	0.068 (3)	4.48 E-6
Maine	7.0	12 000	3065	0.060 (3)	4.05 E-6
Aalesund, Norway	7.5	11 090	3279	0.065 (3)	4.83 E-6
Aalesund, Norway	7.5	11 480	3274	0.059 (3)	4.15 E-6
Aalesund, Norway	7.5	12 330	3275	0.061 (3)	4.03 E-6
Denmark	7.8	12 000	2779	0.045 (3)	2.82 E-6
Denmark	7.8	12 000	2823	0.053 (3)	3.48 E-6
Denmark	7.8	13 010	3512	0.059 (4)	3.66 E-6
Scotland	9.1	11 000	2778	0.072 (3)	5.50 E-6
Vancouver, B.C.	10	12 625	2781	0.071 (3)	4.72 E-6

[1]CMAT = current mean annual air temperature except Arctic sites for which a value 2°C above the CMAT (given in parentheses) is used. See text for discussion.
[2]Number of individual shells prepared is given in parentheses.
[3]AIle/Ile ratio derived by multiplying original ratio by 0.61 (see Appendix 14.A).

REFERENCES

Andrews, J.T., Feyling-Hanssen, R.W., Hare, P.E., Miller, G.H., Schluechter, C., Stuiver, M. and Szabo, B.J., 1976: Alternative models of early and mid-Wisconsin events, Broughton Island, N.W.T., Towards a Quaternary chronology. International Geological Correlation Programme Report 3, Project 73/1/24, Bellingham WA, 28-61.

Andrews, J.T., and Miller, G.H., 1976: Quaternary glacial chronology of the eastern Canadian Arctic: a review and a contribution on amino acid dating of Quaternary molluscs from the Clyde Cliffs. In: Mahanney, W.C. (ed.), The Quaternary of North America. Stroudsberg, Dowden, Hutchinson and Ross, 2-32.

Andrews, J.T. and Miller, G.H., Quaternary glacial and nonglacial correlations in the eastern Canadian Arctic. Geological Survey of Canada Paper. In press.

Blake, W., Jr., 1980: Mid-Wisconsinan insterstadial deposits beneath Holocene beaches, Cape Storm, Ellesmere Island, Arctic Canada. Orono Maine: American Association for Quaternary Research, Abstracts with Programs, 261-27.

Brigham, J.K., 1984: Marine stratigraphy and amino acid geochronology of the Gubik Formation, western Arctic Coastal Plain, Alaska. Unpub. Ph.D. Thesis, University of Colorado, Department of Geological Sciences.

Brigham, J.K., 1983: Stratigraphy, amino acid geochronology and correlation of Quaternary sea-level and glacial events, Broughton Island, Arctic Canada. Canadian Journal Earth Sciences, 20:577-598.

Clark, P.U., 1984: Glacial history of northern Labrador, Canada, Unpublished Ph.D. Thesis, University of Colorado, Boulder, CO. 230 pp.

Dyke, A., 1979: Glacial and sea-level history of southwestern Cumberland Peninsula, Baffin Island, N.W.T., Canada. Arctic and Alpine Research, 11:179-202.

Dyke, A.S., Andrews, J.T., and Miller, G.H., 1982: Quaternary Geology of Cumberland Peninsula, Baffin Island, District of Franklin. Geological Survey Canada Memoir 403, 32 pp + map.

Feyling-Hanssen, R.W., 1967: The Clyde Foreland. Field Report North-Central Baffin Island 1966. In: Löken, O.H. (ed.), Ottawa: Department of Energy, Mines and Resources, Geographical Branch, 35-55.

Feyling-Hanssen, R.W., 1976a: A mid-Wisconsinan interstadial on Broughton Island, Arctic Canada, and its foraminifera. Arctic and Alpine Research, 8:161-182.

Feyling-Hanssen, R.W., 1976b: The stratigraphy of the Quaternary Clyde Foreland Formation, Baffin Island, illustrated by the distribution of benthic foraminifera. Boreas, 5:57-94.

Feyling-Hanssen, 1980: Microbiostratigraphy of young Cenozoic marine deposits of the Qivituq Peninsula, Baffin Island. Marine Micropalentology, 5:153-184.

Feyling-Hanssen, R.W., Funder, S. and Petersen, K.S., 1983: The Lodin Elv Formation; A ? Plio-Pleistocene occurrence in Greenland. Bulletin Geological Society Denmark, 31:81-106.

Kennedy, G.L., Lajoie, K.R. and Wehmiller, J.F., 1982: Aminostratigraphy and faunal correlations of late Quaternary marine terraces, Pacific coast, U.S.A. Nature, 299:545-547.

Klassen, R.A., 1981: Aspects of the glacial history of Bylot Island, District of Franklin. Current Research, Part A, Geological Survey of Canada, Paper 81-1A, 317-326.

Kriausakul, N., and Mitterer, R.M., 1978: Isoleucine epimerization in peptides and proteins: kinetic factors and application to fossil proteins. Science, 201:1011-1014.

Locke, W.W., 1980: The Quaternary geology of The Cape Dyer area southeasternmost Baffin Island, Canada. Unpublished Ph.D. Thesis, University of Colorado, Boulder, CO. 332 pp.

Löken, O.H., 1966: Baffin Island refugia older than 54,000 years. Science, 153:1378-1380.

Miller, G.H., 1980: Late Foxe glaciation of southern Baffin Island, N.W.T., Canada. Geological Society of America Bulletin, Part 1, 91:399-405.

Miller, G.H., 1982: Quaternary depositional episodes, western
   Spitsbergen, Norway: Aminostratigraphy and glacial history. Arctic
   Alpine Research, 14(4):321-340.
Miller, G.H., Andrews, J.T., and Short, S.K., 1977: The last
   interglacial/glacial cycle, Clyde Foreland, Baffin Island, N.W.T.:
   stratigraphy, biostratigraphy, and chronology. Canadian Journal of
   Earth Sciences, 14:2824-2857.
Miller, G.H. and Hare, P.E., 1980: Amino acid geochronology: Integrity
   of the carbonate matrix and potential of molluscan fossils. In: Hare,
   P.E., Hoering, T.C., and King, K., Jr. (eds.), Biogeochemistry of
   Amino Acids. New York: John Wiley and Sons, 415-444.
Miller, G.H., Brigham, J.K., and Clark, P., 1982: Alteration of the
   Total aIle/Ile ratio by different methods of sample preparation. In:
   Miller, G.H. (ed.), Amino Acid Geochronology Laboratory Report of
   Current Activities. Institute of Arctic and Alpine Research,
   University of Colorado, 9-20.
Mode, W.N., 1980: Quaternary stratigraphy and palynology of the Clyde
   Foreland, Baffin Island, N.W.T., Canada. Unpublished Ph.D. Thesis,
   University of Colorado, Department of Geological Sciences, 219 pp.

Nelson, A.R., 1978: Quaternary glacial and marine stratigraphy of the
   Qivtu Peninsula, northern Cumberland Peninsula, Baffin Island,
   Canada. Ph.D. thesis, University of Colorado, Boulder, CO. 297 pp.
Nelson, A.R., 1981: Quaternary glacial and marine stratigraphy of the
   Qivtu Peninsula, northern Cumberland Peninsula, Baffin Island.
   Geological Society of America Bulletin, 92, Part I, pp. 512-518, Part
   II, pp. 1143-1261.
Nelson, A.R., 1982: Aminostratigraphy of Quaternary marine and
   glaciomarine sediments, Qivitu Peninsula, Baffin Island. Canadian
   Journal of Earth Sciences, 19:945-961.

Schroeder, R.A., and Bada, J.L., 1976: A review of the geochemical
   applications of the amino acid racemization reaction. Earth and
   Planetary Science Letters, 12:347-391.
Szabo, B.J., Miller, G.H., Andrews, J.T., and Stuiver, M., 1981:
   Comparison of uranium-series, radiocarbon and amino acid data from
   marine molluscs, Baffin Island, Arctic Canada. Geology, 9:451-457.

Washburn, L., 1980: Geochronolgy. New York: John Wiley and Sons, 406
   pp.
Wehmiller, J.F., and Belknap, D.F., 1982: Amino acid age estimates,
   Quaternary Atlantic coastal plain: comparison with U-series dates,
   biostratigraphy, and paleomagnetic control. Quaternary Research,
   28:311-336.
Williams, K.M., and Smith, G.G., 1977: A critical evaluation of the
   application of amino acid racemization to geochronology and
   geothermometry. Origins of Life, 8:91-144.

15 An outline of glacial history of Bylot Island, District of Franklin, N.W.T.

R. A. Klassen

Bylot Island occupies a position that is key to the glacial history of much of the eastern Canadian Arctic. It is situated near the northeastern margins of former continental ice sheets, and lies along and in front of marine channels that have been main pathways of ice flow outwards from Arctic regions during past glaciations (Fig. 15.1). Consequently, the geologic record of glacier flow onto and around Bylot Island carries paleoclimatic implications that extend well beyond the area and links glacial events of Baffin Island-Foxe Basin with those of Arctic regions farther north.

The extent of ice sheets in the area of Bylot Island during the 'last glaciation' is not well known, although it has been speculated on by various workers, with major differences of interpretation evident among their accounts (cf. Craig and Fyles, 1960; Sim, 1960; Ives and Andrews, 1963; Prest et al., 1968; Hughes et al., 1977; Ives, 1978). In this chapter 'last glaciation' is used in a general sense to refer to glacial events that are thought to have occurred during the past 125,000 years. For the last glaciation, Craig and Fyles (1960) held that Baffin, Devon and Ellesmere islands were covered by a series of local ice caps named the Ellesmere-Baffin Glacier Complex. They concluded that the maximum extent of the Complex was achieved during late Wisconsinan time, and that the ice caps were coalescent and contiguous with both the Laurentide Ice Sheet to the southwest and the Greenland Ice Sheet to the northeast. Prest et al. (1968) placed the northern limits of the Laurentide Ice Sheet during late Wisconsinan time in Lancaster Sound to the north of Bylot Island, and both Prest (1970) and Blake (1970) postulated that continental ice merged there with ice flowing south from islands of the High Arctic. The interpretations of Prest and Blake are used as the basis for the hypothesis of extensive late Wisconsinan ice cover advanced by Hughes et al. (1977).

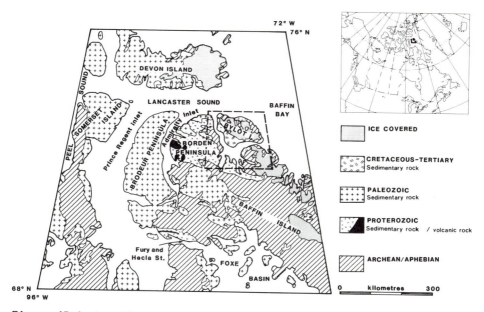

Figure 15.1 Location and generalized bedrock geology of study area and region of northern Baffin Island.

In contrast, Ives and Andrews (1963), and later Ives (1978), estimated that the maximum extent of ice cover was attained prior to late Wisconsinan time when continental ice was restricted on its northeastern margin to channels around Bylot Island, and highland areas of the eastern Arctic were covered by local ice caps that were thin and cold-based. The maximum extent of the Laurentide Ice Sheet during late Wisconsinan time was considered to be less extensive, and has been mapped by Andrews and Ives (1972) and by Miller and Dyke (1974) at the heads of eastward and northward facing inlets on Baffin Island, coincident with the Cockburn Moraine. England (1976) also suggests that late Wisconsinan glaciation was restricted, and portrays independent, possibly coalescent, ice caps among islands of the High Arctic, in contrast to the extensive single ice sheet described by Blake (1970, 1975). It is clear that direct geologic evidence of ice extent and age of glaciation is required for accurate paleoclimatic reconstructions of the northeastern sector of the Laurentide Ice Sheet.

The principal objective of this chapter is to develop an outline of the glacial history of Bylot Island and the geologic evidence upon which it is based. Emphasis has been placed on determining the extent of ice cover by 'foreign' and 'native' glaciers, patterns of regional ice movement, and the age and number of glacial events. The term 'foreign' is used here to describe either ice or debris that originated off Bylot Island, and the term 'native' is used to describe either ice or debris that originated on the island.

In common with many Arctic regions, evidence of continental glaciation on Bylot Island consists for the most part of erratics scattered over bedrock and of ice-marginal features such as meltwater channels and moraine ridges. Bedrock shows little evidence of glacial erosion and Quaternary sediments of any thickness are restricted to coastal areas. The stratigraphic sections of the coastal areas, however, are composed principally of marine sediments and of poorly sorted sediments that are genetically difficult to interpret.

The outline of glacial history presented here is based upon varied kinds of evidence obtained from widely separated locations. It has been interpreted through examination of lithologies and distributions of erratics, geomorphic relationships among landforms, and use of amino acid ratios. Initial discussion concerns the evidence of the last major foreign glaciation of Bylot Island, which has been named informally Eclipse glaciation (Klassen, 1982a). Because deposits and landforms of Eclipse glaciation are widespread they are used as a reference for interpretation of older and younger glacial events (Table 15.1). Eclipse moraines, formed at the edge of the continental ice sheet at the maximum of Eclipse glaciation, outline the extent of the ice sheet on Bylot Island and have provided the key for deducing the nature of interaction between glaciers impinging on the island and the island's local glaciers. Other glacial events are compared with Eclipse glaciation in terms of their relative ice extent and age.

The regional geologic setting of Bylot Island is well suited for recognition and separation of foreign and native glacial debris because lithologically distinct erratics occur that can be traced to specific source areas. Consequently, lithology of erratics is used to prove either a foreign or native origin of glacial deposits. In areas lying above former limits of marine submergence the presence of foreign debris is compelling evidence of glaciation by foreign ice, and where bedrock sources of this debris are known net flow paths within the ice can also be estimated. Development of a large ice sheet is required for foreign ice to fully occupy the deep (>500 m) channels around Bylot Island and to transport foreign debris onto it.

The term 'drift' is used here in a general sense to refer to deposits associated with a glaciation, and can include sediments of different origins. Drift is characterized generally by its content of abundant foreign erratics that demonstrate considerable activity by

TABLE 15.1. Summary of geologic evidence used to interpret the glacial history of Bylot Island.

EVENT	ROCK STRATIGRAPHIC	LANDFORM	RADIO-CARBON AGE (years BP)	AMINO ACID RATIO Free	Total	COMMENTS
Neoglaciation	Neoglacial drift	Neoglacial moraines	<120 ± 80 (GSC 3227)			Most glaciers are at or retreating from Neoglacial maximum positions.
Post-Eclipse foreign glaciations	Post-Eclipse marine sediments	Cape Hatt ice-contact delta	9510 ± 180 (GSC 3318)	N.D.[+]	N.D.	Cape Hatt ice-contact delta marks the approximate limits of continental ice during late Wisconsinan time, and lies north of (distal to) Cockburn Moraine
	Post-Eclipse drift	Minor moraine ridges		0.1-0.15	0.016-0.02	Foreign ice remained largely confined to channels around Bylot Island. Amino acid ratios based on detrital shells in glacial sediments.
Post-Eclipse native glaciation		Native ice-marginal deposits				This last major advance of native ice is associated with the withdrawl of Eclipse foreign ice from channels around Bylot Island.
Eclipse glaciation	Eclipse marine sediment	Raised marine deltas	>43,000 (GSC 3410) (GSC 2916)	0.20-0.32	0.016-0.072	Deltas bear no evidence of having been glaciated since their formation. They are located beside Eclipse Sound and extend to ≥60 m asl.
	Eclipse drift	Eclipse moraines				Surficial glacial deposits characterized by abundant foreign erratics, and distributed around outer Bylot Island.
						Maximum extent of foreign ice marked clearly by lateral moraines along the northeastern coast of Bylot Island.

TABLE 15.1. (continued)

EVENT	ROCK STRATIGRAPHIC	LANDFORM	RADIO-CARBON AGE (years BP)	AMINO ACID RATIO Free	Total	COMMENTS
Interglacial	Interglacial sediments					Buried organic detritus containing taxa of interglacial character underlying Eclipse drift.
Pre-Eclipse glaciation	Native erratics; Foreign erratics	Meltwater channels (native)				Evidence of Pre-Eclipse glaciation is based on native and foreign erratics found distal to (inland of) Eclipse drift, and on foreign erratics stratigraphically below interglacial sediments. On the northeastern coast, native meltwater channels are crossed by Eclipse moraines and formed during an earlier native glaciation.

+N.D. = Not detectable

foreign glaciers in the region, whether or not the sediments that the erratics are part of were deposited directly from ice.

The ratios of the amino acids D-alloisoleucine to L-isoleucine, based on the naturally hydrolyzed protein fraction (Free) and acid hydrolyzed fraction (Total), have been used in this chapter as a guide to the relative ages of glacial and marine deposits. Differences in the magnitudes of ratios of shells from deposits thought to be nearly contemporaneous suggest the possibility of alternate interpretations.

PREVIOUS WORK

The first scientific investigation of Bylot Island and northern Baffin Island was accomplished by members of the Danish Fifth Thule Expedition (1921-1924) who noted abundant evidence of glaciation on central northern Baffin Island and concluded, from geomorphic evidence, that ice moving northwards from there was deflected by, but did not override, Bylot Island (Freuchen and Mathiassen, 1925; p. 554). Later work by the Geological Survey of Canada resulted in identification of foreign glacial deposits, including moraines, on Bylot Island, and indicated that the last flow of foreign ice onto the island occurred prior to late Wisconsinan time (Hodgson and Hasleton, 1974; Klassen, 1981, 1982a). Evidence of two major expansions of native ice, one of which occurred subsequent to the last major foreign glaciation, was also reported. Glacial deposits that extend to the heads of north-facing inlets on Baffin Island, north of (distal to) the Cockburn Moraine, have been interpreted to be either of late Wisconsinan age (Hodgson and Hasleton, 1974), or much older (Miller and Dyke, 1974; Andrews, 1975).

Evidence of recent expansion of glaciers on Bylot Island has been described by various workers (Baird, 1955; Falconer, 1962; DiLabio and Shilts, 1978; Klassen, 1982b), and glacier margins appear now to be either at Neoglacial maximum positions or to be retreating from them (Falconer, 1962; Klassen, 1982b).

METHODS

This chapter is based almost entirely on fieldwork completed by the author during the summers of 1978, 1979, and 1981 for the Geological Survey of Canada. Most of the work was carried out on Bylot Island, although parts of adjacent Baffin Island were also examined. The area studied includes parts of map areas 38B, 38C, 48A, 48D (National

Topographic System of Canada; 1:250,000 scale), and is bounded approximately by latitudes 72° and 74°N, and longitudes 75° and 82°W (Fig. 15.1).

Bylot Island is about 170 km in length between its northwestern end near the Wollaston Islands and its southeastern end at Button Point, and it is about 110 km in maximum width between Eclipse Sound and Lancaster Sound. Full helicopter support, which was available throughout the course of fieldwork, allowed coverage of the large study area and access to many sites in diverse topographic settings. Despite this support, not all areas were examined with equal intensity, and many areas can be identified that require further study. During stops on traverses, sediment samples were collected and field examination of the lithologies of erratics made routinely. Over 300 sites located throughout the study area were sampled. Aerial photographs at 1:60,000 scale were used to supplement field observations and to extend mapping of surficial materials and landforms into areas not visited during field operations. The age of glacial and non-glacial events is based on radiocarbon and on amino acid analysis of marine shells and shell fragments, which were found to be widespread within both marine and glacial deposits. The use of amino acid ratios as a means of estimating relative age in the eastern Arctic has been discussed elsewhere (cf. Andrews and Miller, 1976; Miller et al., 1977; Miller and Hare, 1980; Szabo et al., 1981; Nelson, 1982; Miller, this volume).

GEOLOGIC SETTING

A variety of igneous and metamorphic rock types of Archean and Aphebian ages comprise a crystalline bedrock complex that underlies much of central Bylot Island and of northeastern and north-central Baffin Island (Fig. 15.1). The basement complex forms a rugged mountainous spine striking northwest-southeast across central Bylot Island where it supports an extensive system of mountain glaciers in a classical alpine setting. In contrast, crystalline terrane on Baffin Island forms high plateaus that support broad-domed ice caps. For purposes of boulder tracing, crystalline bedrock is considered to form a single lithologic unit, and crystalline erratics are used only to demonstrate former glacier cover. No attempt is made to distinguish between either a foreign or native provenance of crystalline erratics on Bylot Island.

Slightly deformed, unmetamorphosed sedimentary rocks of Proterozoic age crop out on western and northern Bylot Island and across much of Borden Peninsula (Fig. 15.1). They are composed chiefly of limestone, dolomite, mudstone and quartz sandstone. On western

Bylot Island the Proterozoic sediments form high tablelands, and on
northern Bylot Island they form broad lowlands sloping upwards from the
coast to the highland front of crystalline rock. Sedimentary rocks of
early Paleozoic age, some of which are fossiliferous, are mapped across
central northern Baffin Island and Brodeur Peninsula (Fig. 15.1). For
purposes of boulder tracing, two distinct lithologic units composed of
non-fossiliferous sedimentary rock (Proterozoic and Paleozoic) and of
fossiliferous sedimentary rock (Paleozoic) can be described.

Proterozoic volcanic bedrock occurs on Baffin Island near Fury and
Hecla Strait and along the eastern shore of Admiralty Inlet, and on
Bylot Island. It comprises a lithologic unit that is distinctive from
all others in the region. The largest areas of outcrop occur near
Admiralty Inlet. Volcanic bedrock on Baffin Island is fine-grained and
amygdaloidal, and differs in appearance from that on Bylot Island (G.
Jackson, personal communication, 1980).

Poorly consolidated sedimentary rocks of Tertiary age underlie
broad lowland areas of southwestern Bylot Island and northern Baffin
Island (Fig. 15.1). Tertiary sedimentary rocks, which include immature
to mature sandstone, mudstone and coal, are distinguished easily from
sedimentary rocks of Proterozoic and Paleozoic successions. Because
the Tertiary sedimentary rocks are poorly consolidated they are not
found as coarse clasts in glacial debris and consequently have not been
used in this study as indicators of ice flow directions.

SURFICIAL GEOLOGY

Eclipse Glaciation

Eclipse moraines. The maximum extent of foreign ice cover during
Eclipse glaciation is outlined clearly by lateral moraines that occur
along the northeastern and southern margins of Bylot Island, and on
northeastern Baffin Island (Fig. 15.2). They were formed beside large
glaciers lying within Lancaster Sound and Pond Inlet. The lateral
moraines lie along the upper outer edge of coastal cliffs and headlands
at elevations between 100 and 600 m asl, and loop inland to lower
elevations across the floors of valleys facing the coast (Figs 15.3 &
15.4). The moraines are composed largely of erratics of
unmetamorphosed sedimentary rock, and are well-defined geomorphically,
standing as ridges with crests 1 to 15 m above adjacent terrain (Fig.
15.5). Wherever they were seen in the field, the lateral moraines mark

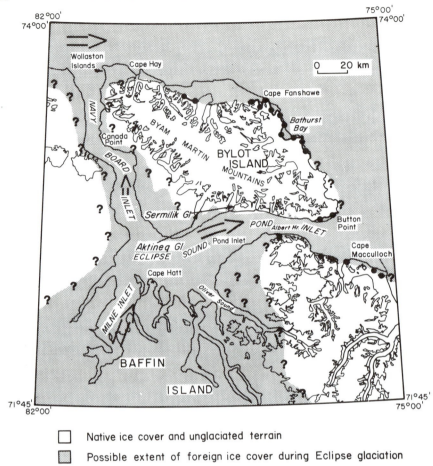

Figure 15.2 Extent of ice and major pathways of ice flow during
Eclipse glaciation. Eclipse moraines lying along the coasts of Bylot
Island and northeastern Baffin Island provide key evidence of the
extent of foreign ice cover then.

the inland limits of Eclipse drift on Bylot Island, and the rare
foreign erratics found inland of (distal to) the moraines are
considered to demonstrate an earlier foreign glaciation of greater
extent than Eclipse.

Ice sheet reconstructions based on Eclipse moraines. The profile of
the foreign ice surface facing the northeastern coast of Bylot Island
is similar within all valleys (Fig. 15.6a), and demonstrates the close

Figure 15.3 Position of Eclipse moraine on northeastern Bylot island
at Bathurst Bay, shown by dark arrows. The moraine lies along coastal
cliffs and loops inland at lower elevation across valley floors. Open
arrow gives view of Figure 15.5. White arrows shows location of
post-Eclipse moraines given in Figure 15.13 (GSC 203810-A).

relationship between the elevation of Eclipse moraines and their
distance inland from a prominent submarine escarpment lying directly
offshore. The position of Eclipse moraines, and the ice surface
profile derived from them, does not appear to have been influenced by
interaction with any of the native glaciers that presently occupy the
valley heads. Seaward extrapolation of the profile shown by Figure
15.6a indicates that ice surface elevations in Lancaster Sound over the
margin of the submarine escarpment were about 600 to 700 m asl.
Therefore, ice thicknesses in Lancaster Sound were about 1600 to 1700
m, and the ice was clearly grounded.

Within the Sound, ice surface gradients along the the margin of
the submarine escarpment were low (Fig. 15.6b), and eastward projection
of the surface profiles indicates that the ice in Lancaster Sound
during Eclipse glaciation may have remained grounded (i.e., >1000 m
thick) over 100 km out into the head of Baffin Bay. Reconstruction of
the ice profile westward of Cape Fanshawe indicates that foreign ice
moved against the mountain front on northern Bylot Island and extended
inland of the limits of some modern glaciers there (Fig. 15.2).

Figure 15.4 Eclipse moraine and the limits of Eclipse drift (marked by
arrows) outline clearly the extent of foreign ice cover on northeastern
Bylot Island near Cape Fanshawe. Eclipse deposits are light-toned due
to erratics of carbonate sedimentary rock within them. They lie at
about 600 m asl across headlands near Cape Fanshawe and about 225 m
asl across the valley floor in front of Glacier D-6. Open arrow gives
the view shown by Figure 15.9 (GSC 203803-Z).

Figure 15.5 Eclipse moraine lies along the outer edge of coastal
cliffs facing Lancaster Sound on northeastern Bylot Island. It is
commonly a single ridge and is composed almost entirely of sedimentary
rocks. The view of this photograph is given by Figure 15.3 (GSC
203099-H).

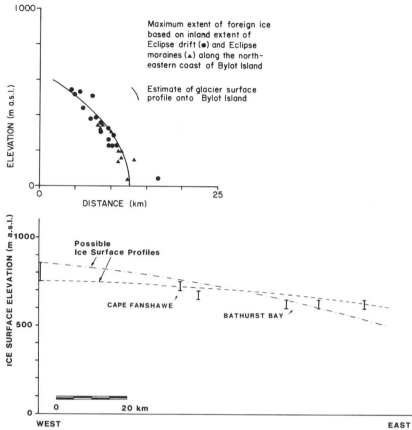

Figure 15.6 (a) Profile of ice surface facing northeastern Bylot Island during Eclipse glaciation. The positions of Eclipse moraines and the inland limits of Eclipse drift are shown relative to their distance inland from the margin of a major submarine escarpment and to their elevation above sea level. The information presented here is from seven separate valleys facing the northeastern coast. (b) Profile of the ice surface facing Baffin Bay, oriented along the northern coast of Bylot Island. The ice surface gradient is much less steep than shown in Figure 6a.

The elevation of Eclipse moraines beside Pond Inlet (Fig. 15.2) indicate that Eclipse Sound was also fully occupied by foreign ice during Eclipse glaciation and that ice surface elevations in the central part of Eclipse Sound could easily have been more than 1000 m asl. Across the southern lowland of Bylot Island the maximum extent of ice cover remains speculative because foreign moraines are not known in that area, although ice extended well up towards the mountain front based on the distribution of Eclipse drift. Because Eclipse drift is found inland of the present positions of Aktineq and Sermilik glaciers, foreign ice and native ice were in direct contact in some places on southern Bylot Island during Eclipse glaciation.

Foreign erratics in Eclipse drift and ice flow patterns. On southern Bylot Island, foreign erratics in Eclipse drift include non-fossiliferous carbonates, quartz sandstones, and fossiliferous carbonates of Ordovician and Silurian ages, listed in approximate decreasing order of abundance. The source areas of much of this debris lie either at or south of Paleozoic outcrops on central northern Baffin Island, and could include Foxe Basin (Fig. 15.1). The erratics demonstrate that regional ice flow impinging on southern Bylot Island was northward across north-central Baffin Island into Eclipse Sound then eastward into Baffin Bay via Pond Inlet during Eclipse glaciation (Fig. 15.2).

On northern Bylot Island, foreign erratics representative of all regional lithologic groups occur in Eclipse drift. They include Proterozoic and Paleozoic sedimentary rocks, some of which are fossiliferous, and volcanic rocks. The erratics of sedimentary rocks have potential bedrock source areas on Baffin Island that are located either on or south of Brodeur and Borden peninsulas (Fig. 15.1). No erratics of sedimentary rock-bearing fossils younger than Silurian age were found that could be used to demonstrate a provenance in regions lying west of Baffin Island.

Volcanic erratics found on northern Bylot Island are fine-grained and amygdaloidal, and they are similar to volcanic rocks in outcrop along Admiralty Inlet (G. Jackson, personal communication, 1980). The volcanic erratics are of particular interest because they are most likely derived from the eastern margin of Admiralty Inlet and they can be used to demonstrate a glacier flow path that was northward through Admiralty Inlet and eastward along Lancaster Sound during Eclipse glaciation (Fig. 15.2). Because the glacier that transported the erratics along Admiralty Inlet would have been insufficient to fill Lancaster Sound, it seems likely that it merged in the Sound as a tributary stream with a larger glacier flowing eastward toward Baffin Bay. The presence of volcanic rocks on Bylot Island indicates that the tributary stream from Admiralty Inlet remained against the southern coast of Lancaster Sound and was not significantly displaced by ice flowing northward off Baffin and Bylot Islands. This observation supports other evidence, cited below, that indicates glaciers on northern Baffin island as well as on Bylot Island were not well nourished at the maximum of Eclipse glaciation.

Source areas of the main glacier occupying Lancaster Sound could have included regions of the Arctic Archipelago to the north of the Sound as well as areas lying south of it, west of Baffin Island. Because the glacier appears to have extended well beyond the eastern mouth of Lancaster Sound, it is suggested that it could have merged in the head of Baffin Bay with other equally large glaciers moving southward between Greenland and Ellesmere Island.

Age of Eclipse glaciation. Eclipse glaciation occurred prior to 43,000 years ago, based on radiocarbon analysis of shells from raised marine deposits that postdate the last occupation of Eclipse Sound by grounded foreign ice (Table 15.2, Fig. 15.7). The marine deposits neither appear to have been eroded by glacial ice since their formation nor do they contain stratigraphic evidence of ice cover. Shell collections were made at Canada Point from within post-Eclipse deltaic sediments. The shells were paired, commonly with their siphon sheath attached, and were well preserved without evidence of pitting or chalkiness. Consequently, they are considered to be contemporaneous with enclosing sediments. Radiocarbon ages of the shells were >35,000 years (GSC 2916) and >43,000 years (GSC 3410), and are greater than estimates of about 33,000 years reported by Hodgson and Hasleton (1974) from shells collected at the Pond Inlet airstrip.

Shells from post-Eclipse marine deposits lying beside Eclipse Sound have amino acid ratios that vary between 0.15 and 0.32 (Free), and 0.015 and 0.04 (Total) (Fig. 15.8). Although amino acid ratios of detrital shells in Eclipse drift are generally larger, they vary widely and the population of ratios from Eclipse drift and from post-Eclipse marine sediments overlap (Fig. 15.8). The reasons for this overlap are not clear and may include a difference in the diagenetic histories of shells found in surficial deposits and in sections. Alternatively, deposits of post-Eclipse glacial events may have been incorrectly identified as Eclipse.

Eclipse glaciation appears to be correlative with glacial events interpreted from sediments of the Kogalu Member of the Clyde Foreland Formation (Miller et al., 1977) and of the Qaviq Member of the Qivitu Formation (Nelson, 1982). In the areas of the Clyde Forelands and Qivitu Peninsula, glaciers extended along inlets to the outer eastern coast during these events, and shells within associated lithostratigraphic units have amino acid ratios between about 0.3 and 0.25, (Free) which are similar in magnitude to those asociated with deposits of Eclipse glaciation. On Qivitu Peninsula ratios of this magnitude would correspond to 'Aminozone 2' of Nelson (1982), which is

TABLE 15.2. Summary of amino acid ratios and radiocarbon ages of shells from Eclipse marine deposits formed during submergence following Eclipse glaciation.

Amino Acid Lab. No.	Amino Acid Ratio		Shell Species	Comments
	Free	Total		
AAL 1218	0.23	0.022	M.t.*	Shells occurred as intact,
	0.26		"	paired valves within the
	0.13	0.016	"	deltaic complex at Canada
	0.24	0.03	"	Point. 25 m asl. Radiocarbon
	0.22	0.02	"	age was >35,000 years BP (GSC 2916).
--	0.27	0.036	H.a.+	Shells collected as intact valves near surface of the deltaic complex at Pond Inlet by D.A. Hodgson.
AAL 2623	0.15	0.023	M.t.	Collected from the deltaic
	0.15	0.027		complex at Canada Point in a bed 2 m higher in section than AAL 1218, 27 m asl. Radiocarbon age was >43,000 years BP (GSC 3410)
AAL 2626	0.32	0.059	M.t.	Shells collected as large
	0.29	0.035	"	fragments from sands of the
	0.38	0.053	"	deltaic complex at Pond Inlet, 66 m asl.
AAL 2471	0.30	0.035	M.t.	Shells collected as intact
	0.25	0.072	"	valves in a coarsening
	0.32	0.038	"	upward sequence of muddy sand-gravel covering valley floor 44 m asl near Janes Creek, 10 km east of Pond Inlet town site.

+ Hiatella
* Mya truncata

defined to include sediments with shells having minimum amino acid ratios of between 0.35 and 0.15 (Free), and 0.055 and 0.035 (Total).

Extent of native ice during Eclipse glaciation. Where Eclipse moraines lie across valley floors on northeastern Bylot Island there is no evidence that the foreign moraines either formed in contact with native glaciers or have ever been glaciated by native glaciers since their formation (Fig. 15.9). This is supported by the observation that the elevations of Eclipse moraines across valley floors are related only to their distance inland from the offshore escarpment (Fig. 15.6a), indicating that Eclipse ice was not apparently interfered with by

Figure 15.7 Deltaic deposit at Canada Point. Paired marine shells contemporaneous.with enclosing sediments of the deposit were >43,000 radiocarbon years old (GSC 3410). The Canada Point deposit and similar deposits elsewhere beside Eclipse Sound do not appear to have been glaciated since their formation and are considered to post-date Eclipse glaciation (GSC 203803-V).

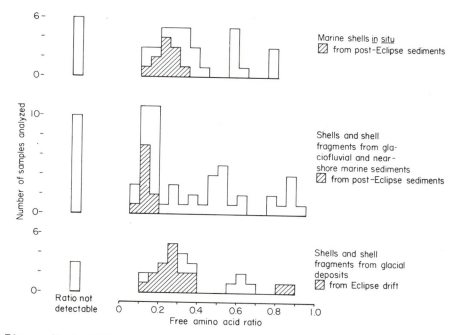

Figure 15.8 Histograms of amino acid ratios (Free) of shells analyzed in this study.

Figure 15.9 Eclipse moraine lying across the valley floor in front of glacier D-6 on Bylot Island. The view given here is indicated in Figure 15.4. The moraine has not been glaciated by native ice since its formation and, consequently, demonstrates that there has been no major advance of native glaciers on northeastern Bylot Island since Eclipse glaciation (GSC 203803-X).

expanded native glaciers. The native glaciers facing the northeastern coast of Bylot Island do not appear to have been in direct contact with foreign ice during Eclipse glaciation and have not advanced far, if at all, since.

On northeastern Baffin Island, there is evidence that glacial lakes were held in ice-free valleys during Eclipse glaciation by foreign ice lying in Pond Inlet and blocking free drainage to the sea. In valleys between Albert Harbour and Cape MacCullough (Fig. 15.2), glaciolacustrine deposits extend up to Neoglacial moraines of modern glaciers at altitudes of 200-300 m asl at valley heads. In some locations the position of the foreign ice dam across the valley mouth is marked by piles of drift (Fig. 15.10). The glaciolacustrine deposits demonstrate that the northward-sloping valleys of northeastern Baffin Island were substantially ice-free during Eclipse glaciation, like valleys on northeastern Bylot Island.

The limited extent of native ice on northern Baffin and northeastern Bylot Islands suggests that there also could have been

Figure 15.10 Ice-contact sediments, shown by arrows, mark a foreign
ice front position across a valley mouth on northeastern Baffin
Island. Deposits shown here extend to about 280 m asl. Foreign ice
held glacial lakes on Baffin Island during Eclipse glaciation, and
glaciolacustrine deposits extend to Neoglacial moraines in some
valleys. Valleys on northeastern Baffin Island were substantially
ice-free during Eclipse glaciation and have not been occupied since by
local glaciers (GSC 203803-W).

limited ice cover on Devon Island during Eclipse glaciation. Unless
glaciers on Devon Island and Bylot Island were nourished by different
sources, it would appear unlikely to have restricted local ice cover on
Bylot Island south of Lancaster Sound and greatly expanded ice cover on
Devon Island immediately to the north.

Pre-Eclipse Glaciation

Pre-Eclipse native glaciation. Evidence of pre-Eclipse glacial and
non-glacial events is found in surficial deposits inland of (distal to)
the maximum limits of Eclipse glaciation and in a few stratigraphic
sections located near the modern coast. The distribution and
lithologies of erratics are not well known in the highland interior of
Bylot Island, where field work was restricted due to the rugged
character of the terrain and absence of snow-free sites suitable for

landing an aircraft. No erratics were recognized on mountain peaks in the interior of Bylot Island, although on some peaks lying at lower overall altitude and located near the outer margins of the island, crystalline erratics of unknown provenance were found that demonstrate extensive glaciation. At lower elevations, and inland of Eclipse glacial limits, crystalline erratics are widespread. The crystalline erratics indicate that there was nearly complete glaciation of Bylot Island, likely by native glaciers, during at least one pre-Eclipse event.

Geomorphic evidence for major pre-Eclipse expansion of native glaciers is also found in valleys facing the northeastern coast. There, meltwater channels developed in bedrock on valley walls outline ice-marginal positions of native glaciers that completely filled the valleys and extended at least 10 km beyond their present positions to the coast. Because Eclipse moraines lie across some of the meltwater channels, the native glaciation to which the meltwater channels are related occurred prior to Eclipse glaciation.

Although stratigraphic evidence of pre-Eclipse foreign glaciation is known in a number of sections within the study area, there is no comparable stratigraphic evidence of pre-Eclipse native glaciation. This is suprising because at least three sections examined are located within several hundred meters of the modern fronts of large glaciers, and one of the sections visited is overlain directly by a Neoglacial moraine.

Pre-Eclipse foreign glaciation. The highest sites where foreign erratics are known are located beside Sermilik glacier on southern Bylot Island where erratics of quartz sandstone were found at 1136 m asl, and non-fossiliferous carbonate erratics were found at 820 m asl (Fig. 15.11). Elsewhere on Bylot Island, a few scattered foreign erratics have been found inland of Eclipse moraines, although at altitudes lower than reported above. Although the maximum limits of Eclipse glaciation in the area of Sermilik glacier are not clearly defined, the quartz sandstone erratics are interpreted to lie inland of Eclipse limits and to demonstrate a pre-Eclipse foreign glaciation of considerable magnitude. Transport of foreign erratics onto Bylot Island, against outflow directions of modern glaciers, requires that foreign ice surface elevations over channels were greater than over sites where erratics were deposited. Regional ice cover during such a pre-Eclipse glaciation would have been complete, or nearly so (Fig. 15.11).

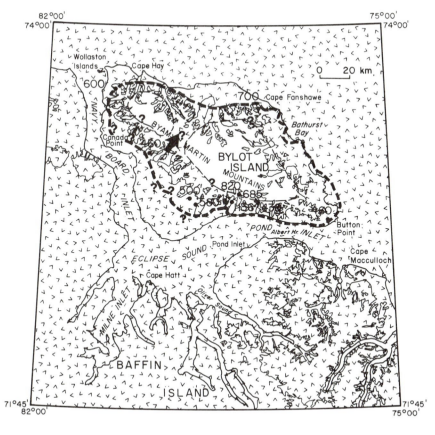

Figure 15.11 Possible extent of foreign ice cover during pre-Eclipse glaciation. Although largely interpretative, this map is based in part on foreign erratics found inland of Eclipse drift.

That pre-Eclipse foreign glaciation occurred prior to a period of interglacial climate is demonstrated by a section lying at the coast near the townsite of Pond Inlet where foreign erratics of carbonate sedimentary rocks are contained within marine sediments capped by buried organic beds. Disjunct taxa found within the organic beds include Betula, Rumex and Potamogeton Richardsonii, among plant macrofossils, and Dyschirius spp., among fossil arthropods (J.V.

Matthews, Jr., Geological Survey of Canada, Plan Macrofossil Report 81-9; Fossil Arthropod Report 81-10). Some of the disjunct taxa have modern distributions extending only to near treeline in northern Quebec, and indicate that the buried organic beds formed during climatic conditions much warmer than present. Because carbonate rocks are not present in local bedrock, the carbonate rocks found within the marine unit below the organic beds document foreign glacial activity prior to the period of interglacial climate. The marine unit and organic beds of the section are capped in turn by Eclipse drift and by marine sediment of probable Holocene age.

Post-Eclipse Glaciation

Post-Eclipse native glaciation. In front of many large glaciers on Bylot Island, particularly those facing the southwestern lowland, ice-marginal landforms, including ice-contact drift and meltwater channels, outline clearly the positions of native glaciers that extended 10 to 20 km beyond their present positions well into regions of the island that were covered by foreign ice during Eclipse glaciation (Fig. 15.12). In front of Aktineq and Sermilik glaciers, the landforms of the last major native advance lie almost wholly within the area of Eclipse drift and are characterized by foreign erratics at their surface, and stratigraphic sections in front of the glaciers are composed only of drift of foreign appearance. Native glacial deposits similar to those of modern native glaciers are not known. Only one section, located at the coast in front of Aktineq glacier, contains fold structures demonstrating movement of glacial deposits offshore subsequent to a foreign ice advance (Klassen, 1982a, p. 73). Such structures, however, could have been caused either by glacier movement or by mass movement processes not necessarily related to ice movement.

Two interpretations can be made of the timing and origin of the last major advance of native glaciers relative to Eclipse glaciation. One interpretation is that the last major native advance preceded Eclipse glaciation and that landforms related to the last major native advance (Fig. 15.12) are mantled by Eclipse drift. This interpretation is consistent with the history proposed for native glaciers facing the northeastern coast, where native glaciers have not eroded Eclipse moraines lying within a few kilometers of modern glacier margins.

Because the native glacial landforms on southern Bylot Island do not appear geomorphically to have been glaciated by foreign ice a second interpretation, which is favored by the author, is that the last

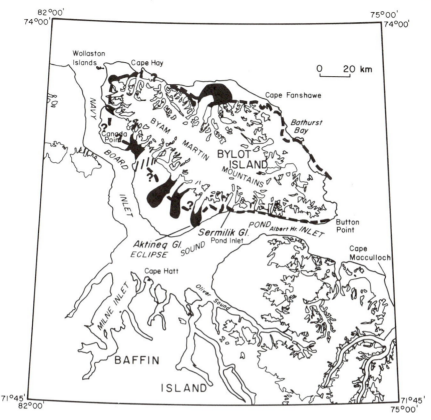

 Limits of foreign ice cover during Eclipse glaciation

Areas covered by native ice subsequent to Eclipse
foreign ice cover

Figure 15.12 Areas covered by native ice subsequent to foreign ice
cover of Eclipse glaciation. Glaciers facing northeastern Bylot Island
are not known to have advanced then.

major expansion of some native glaciers occurred following Eclipse
glaciation. For this interpretation, questions arise as to why the
deposits covered by native ice during its last major advance differ
from modern glacial deposits, and why only some native glaciers
advanced while others, for example those facing the northeastern coast,
are thought to have remained restricted, despite all glaciers on Bylot
Island having contiguous areas of highland accumulation.

     Because it appears geomorphically that only the native glaciers
were likely to have been in direct contact with foreign ice expanded
subsequent to Eclipse glaciation, it is proposed that this last advance

was connected closely with the movement of foreign ice. Foreign
erratics demonstrate clearly that at the maximum of Eclipse glaciation
ice flow was directed onto Bylot Island, and consequently native
glaciers contacted by foreign ice would have been 'held' or 'dammed'
within the highlands. During deglaciation, as foreign ice left
channels around Bylot Island and ice flow off the island resumed, it is
proposed that the native glaciers expanded and occupied valleys far
beyond modern limits. This linkage of foreign and native events could
explain why glacial deposits in front of some native glaciers retain
foreign lithologic and textural characteristics despite the geomorphic
evidence that associates them with native ice advance; the phase of
native ice cover following retreat of foreign ice was probably of short
duration and deposits typical of modern glaciers had little opportunity
to form. If this model is correct, the last major expansion of native
ice did not result from improvement in conditions for glacierization of
the island subsequent to Eclipse glaciation.

Post-Eclipse foreign glaciation. Evidence of post-Eclipse foreign
glaciation is known at coastal sites on Bylot Island and along
north-facing inlets on Baffin Island (Figs 15.13 & 15.14). Glacial
deposits lying along the coast contain detrital shells having amino
acid ratios much less than those from post-Eclipse marine sediments
(Table 15.3) and are considered to be the product of glaciation younger
than Eclipse. It is not known whether the deposits represent more than
one foreign glacial event, although their distribution indicates that
foreign ice was restricted almost entirely to channels around Bylot
Island during post-Eclipse time. A possible exception is found in the
area of northwestern Bylot Island, along Navy Board Inlet, where post-
Eclipse deposits may extend up to 100 m asl and a kilometer inland of
the coast (Table 15.3, AAL 906).

    Amino acid ratios of shells in post-Eclipse deposits correspond to
those of 'Aminozone 1" of Nelson (1982), which is associated
tentatively with glacial advance extending to mid-inlet positions on
eastern Baffin Island. However, correlative events in Arctic regions
are difficult to establish because reports of shells having amino acid
ratios much less than those of Eclipse marine sediments (~0.25 Free)
are not common. None are listed by Andrews and Miller (1976), Miller
et al. (1977), and Szabo et al. (1981), although some amino acid ratios
of magnitude 0.1 (Free) are given by Nelson (1982). Post-Eclipse
glaciation represented by the amino acid ratios in Table 15.3 could be
either middle- or late-Wisconsinan in age, although there is no direct
evidence for such an interpretation.

TABLE 15.3. Summary of amino acid ratios and radiocarbon ages of
shells from glacial and marine deposits post-dating Eclipse
glaciation.

Amino Acid Lab. No.	Amino Acid Ratio		Shell Species	Comments
	Free	Total		
906	0.14	0.016	H.a.+	Shells fragments collected at
	0.15	0.018	"	surface of a moraine ridge 100
	0.14	0.024	"	m asl and 1 km inland of the
				coast along northern Navy Board
				Inlet.
932	0.18	0.017	H.a.	Intact valves collected near
	0.13	0.017		Button Point from silts 22 m
	0.10	0.020		asl and underlying surface
				glacial deposits.
1472	0.12	0.023	H.a.	Shell fragments from coarse
	0.67	0.063		sand bed within foreign glacial
	0.60	0.072		deposits on northeast coast
				near Bathurst Bay. Analysis
				indicates a population of mixed
				ages.
1639	0.09	0.024	M.t.*	Sample from same unit as AAL
	0.12	0.028	"	932-see above.
	0.13	0.027	"	
--	--	--	--	In situ shells collected 75 m asl from Capa Hatt delta, which is graded to 80 m asl 9510±180 (GSC 3318).

+ Hiatella
* Mya truncata

On western Somerset Island glacial deposits formed by a large
glacier grounded in Peel Sound contain shells that have amino acid
ratios of about 0.15 (Free) (Dyke, in press). The deposits, which are
interpreted by Dyke to be of Late Wisconsinan age, could be correlative
with the morainic landforms lying along the northern and northeastern
coasts of Bylot Island (Fig. 15.13) that contain shells with similar
amino acid ratios.

At Cape Hatt, which is located at the northern end of Milne Inlet
on Baffin Island, a collection of articulated marine shells was made
from a large ice-contact delta that marks the northern terminus of a
glacier flowing northward from Baffin Island (Fig. 15.14). Based on
radiocarbon analysis of the shells, the delta formed about 9500 years
ago (Table 15.3). Ice-marginal deposits and moraines leading to this
deposit rise in elevation southward along the sides of Milne Inlet to
over 500 m asl near its head, and they outline the extent of a large

Figure 15.13 Morainic deposits and landforms related to post-Eclipse glaciation occur at the coast of northeastern Bylot Island. The view of this photograph is given by Figure 15.3 (GSC 203803-B).

Figure 15.14 Ice-contact marine delta at Cape Hatt, Baffin Island. The delta contains shells estimated to be about 9500 radiocarbon years old (GSC 3318) and approximates the northern extent of ice on Baffin Island during late Wisconsinan time (GSC 203803-B).

glacier that fully occupied Milne Inlet. The glacier extended more
than 50 km beyond the Cockburn Moraine, which has been mapped at the
southern end of the Inlet (Fig. 15.15). An advance of similar
proportions, and which may be contemporaneous, can be recognized by
moraines mapped by Hodgson and Hasleton (1974) along the margins of
Oliver Sound.

The ice-contact delta at Cape Hatt and ice marginal landforms
leading to it, provide a clear demonstration that moraines of the
Cockburn stage (cf. Andrews and Ives, 1978) do not mark the late
Wisconsinan glacial maximum on northern Baffin Island. The Cape Hatt
delta is interpreted to approximate the late Wisconsinan glacial
maximum on northern Baffin Island because no contemporaneous glacial
deposits inland of the coast of Eclipse Sound are known in areas lying
to the north. This interpretation of age has been offered earlier by
Hodgson and Hasleton (1974), although without supporting evidence.
Lateral moraines along Oliver Sound outline a glacier of similar extent
to that in Milne Inlet and they may be also of late Wisconsinan age.

In Frobisher Bay, moraines lying distal to moraines of the
Cockburn stage formed immediately prior to 10,760 years BP and they are
considered to outline a late Wisconsinan glacial maximum (Miller,
1980). Miller's evidence from southern Baffin Island and the evidence
presented here from Cape Hatt indicate that ice cover of late
Wisconsinan age was more extensive in parts of Baffin Island than
interpreted previously (cf. Andrews and Ives, 1978; Miller and Dyke,
1974). Because eastward and northward facing inlets on Baffin Island
were probably fed from a common ice sheet situated in the area of Foxe
Basin, mid-fiord moraines of unknown age on eastern Baffin Island could
be contemporaneous either with late Wisconsinan ice in Milne Inlet or
with an earlier and separate post-Eclipse glaciation.

Neoglaciation

Near the margins of most modern glaciers on Bylot Island terminal
moraines appear to have formed recently (Figs 15.16a & b). These
native moraines are composed typically of angular boulders of fresh
appearance, and support little, if any, colonization by lichens and
plants, in sharp contrast to surficial deposits lying beyond them which
are vegetated. Some glaciers end in direct contact with their terminal
moraines while others have diminished and end over a kilometer from
them. Generally, native glaciers terminating at higher elevations lie
closer to their terminal moraines than glaciers terminating at lower
elevations. A few modern glaciers front directly on tundra and have no

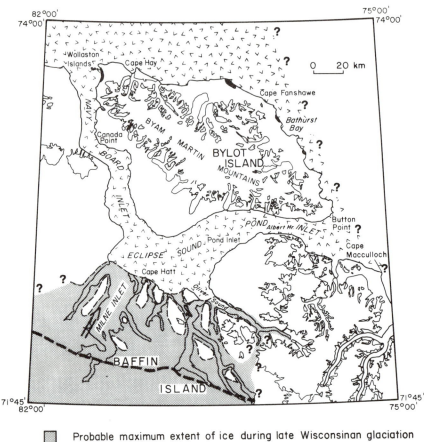

□ Probable maximum extent of ice during late Wisconsinan glaciation

□ Extent of ice during post-Eclipse glaciation; age and number of events uncertain

– – – Approximate limit of Baffinland Drift

⌐ Post-Eclipse moraines

Figure 15.15 Possible extent of ice during post-Eclipse glaciation. Deposits of more than one glacial event may have been used to make this reconstruction.

end moraine associated with them. Based on radiocarbon analysis of peats from a stratigraphic section overlain by a terminal moraine near the front of Aktineq glacier (Fig. 15.16b), DiLabio and Shilts (1978) concluded that Aktineq glacier was as far advanced during the formation of the moraine as it had been in the past 7000 to 8000 years, or longer, and that ice limits marked by the abandoned terminal moraine were attained within the past 400 years. For another glacier on Bylot

(a)

(b)

Figure 15.16 (a) Modern and Neoglacial moraines on Bylot Island. The moraines are thought to have been formed within the last hundred years, and to outline the maximum extent of ice on Bylot Island since Eclipse glaciation. (b) The section overlain by the Neoglacial moraine consists of till (?) (Eclipse drift?) and interbedded peat, sand and gravel. The lowermost peat is estimated to be about 7500 radiocarbon years old (GSC 10638-M [a] 203099-V [b]).

Island, Klassen (1982b) concluded that Neoglacial maximum limits were attained within the past 120 years.

Currently, most glaciers appear to be either stationary, at, or retreating from, their Neoglacial maximum positions. Based on examination of historical documents and photographs, Faconer (1962) concluded that there has been no marked variation in the position of ice margins within the few decades prior to his study. Comparison of modern ice front locations with those shown on the 1949 air photographs examined by Falconer (1962) indicates that there has been recent retreat of up to several hundred meters in some glaciers (e.g. Sermilik, C78), although no change is evident in others. None of the native glaciers, however, are thought to have advanced since the early 1960s including those glaciers without Neoglacial moraines.

The last major advance of native glaciers on Bylot Island has been linked closely in this chapter to retreat of foreign ice from surrounding channels late during Eclipse glaciation, although there is no direct geologic evidence of age of the post-Eclipse native advance. If this interpretation is correct, native glaciers have been no farther advanced than Neoglacial maximum positions since Eclipse glaciation more than 43,000 years ago. During late Wisconsinan time, native glaciers could have been less extensive than at present.

On northern Cumberland Peninsula the most recent advance of local glaciers, which began about 350 years BP and terminated about 65 BP, resulted in the most extensive ice coverage there during the past 5000 to 6000 years (Miller, 1973; Davis, this volume). At least some native glaciers of Bylot Island may have achieved Neoglacial maximum positions more recently than 65 years ago because they either stand in direct contact with their Neoglacial marines or front directly on tundra without an intervening moraine.

CONCLUSIONS

The principal conclusions of this chapter can be summarized as follows:

(1) Based on evidence of foreign erratics and foreign glacial landforms, at least three separate foreign glaciations are recognized on Bylot Island. During the earliest glaciation, glacier ice was greatest in extent, and it became successively less extensive during the two later events.

(2) During the last major foreign glaciation of Bylot Island, which has been named Eclipse glaciation, all channels around the island were occupied by grounded glacier ice that extended into the head of Baffin Bay. There were two principal pathways of ice flow then in the area of Bylot Island. One was northward across north-central Baffin Island into Eclipse Sound, then eastward into Baffin Bay via Pond Inlet. The other was northward along Admiralty Inlet, then eastward along Lancaster Sound. Surface elevations of the foreign ice were more than 600 m asl over modern coast positions, and decreased inland on Bylot Island. Moraines mark the maximum extent of ice along the coast of northeastern Bylot Island and along the coast of Pond Inlet, on Bylot and on Baffin islands. Elsewhere, a minimum estimate of the extent of foreign ice cover is shown by the distribution of Eclipse drift. The age of Eclipse glaciation is estimated to be more than 43,000 radiocarbon years.

(3) During the maximum of Eclipse glaciation most native glaciers on Bylot Island remained near their present positions. The northeastern margin of the continental ice sheet during Eclipse glaciation could not have had significant annual snowfall accumulation, and glacier mass balance on Bylot Island would appear to have been not much different than present. Subsequent to foreign ice cover of Eclipse glaciation some native glaciers experienced major advance 10 to 20 km beyond present positions. It is proposed that this advance is linked to the withdrawal of foreign ice from channels, when native glaciers dammed during the course of Eclipse glaciation re-established flow directions away from the highlands. This native advance concerned only those glaciers that were in direct contact with foreign ice and was likely short-lived.

(4) During post-Eclipse glaciation foreign ice was confined largely to channels around Bylot Island and created minor morainic landforms near modern coasts. The age of post-Eclipse glaciation is not known; it could include events of middle or late Wisconsinan age. About 9500 years ago a glacier that moved northward from an ice sheet on Baffin Island fully occupied Milne Inlet at least to the position of Cape Hatt, extending more than 50 km beyond the mapped extent of moraines of Cockburn age. The Cockburn Moraines do not mark the maximum extent of an ice sheet on northern Baffin Island during the late Wisconsinan.

(5) Native glaciers advanced to Neoglacial maximum positions within the last hundred years, and most are either at those limits, or are

retreating from them. The Neoglacial limits mark the greatest
extent of native ice on Bylot Island since Eclipse glaciation.

ACKNOWLEDGEMENTS

This work is a project of the Geological Survey of Canada (Project No.
780018), and formed the basis of a Ph.D. thesis by the author at the
University of Illinois at Urbana, Illinois (Dr. W.H. Johnson,
advisor). Logistic and aircraft support was provided generously by the
Polar Continental Shelf Project, directed by Mr. G. Hobson. Mr. F. Alt
and Mr. F. Hunt of P.C.S.P. are thanked for their significant
contributions to the success of field work.

    Foremost among those acknowledged and thanked for critical
discussion of this work are Dr. W.W. Shilts and Dr. R.N.W. DiLabio,
Terrain Sciences Division, G.S.C. In addition I thank Dr. J.T. Andrews
for early assistance with amino acid analysis; Dr. W. Blake, Jr. for
radiocarbon analysis; Dr. J.V. Matthews, Jr. for identification of
plant and insect taxa; and Dr. G.H. Miller for discussion in the
field. Assistance in the field was provided capably by Mr. S. Lavender
(1978), Mr. B. Needham and Ms. G.E. Iannuccilli (1979), Mr. J. Hornsby
(1981), and Mr. R. Muktar (1978, 1979).

REFERENCES

Andrews, J.T., 1975: Support for a stable late Wisconsin ice margin
    (14,000 to 9000 B.P.): a test based on glacial rebound. Geology,
    3:617-620.
Andrews, J.T. and Ives, J.D., 1978: "Cockburn" nomenclature and the
    Late Quaternary history of the eastern Canadian Arctic. Arctic and
    Alpine Research, 10:617-623.
Andrews, J.T. and Miller, G.H., 1976: Quaternary glacial chronology of
    the eastern Canadian Arctic: a review and a contribution on amino
    acid dating of Quaternary molluscs from the Clyde Cliffs; In: W.C.
    Mahaney (ed.), Quaternary Stratigraphy of North America, Dowden and
    Hutchinson, Stroudsburg, Pa. U.S.A., 1-32.

Baird, P.D., 1955: Glaciological research in the Canadian Arctic.
    Arctic, 8:96-108.
Blake, W. Jr., 1970: Studies of glacial history in Arctic Canada. I.
    Pumice, radiocarbon dates, and differential postglacial uplift in the
    eastern Queen Elizabeth Islands. Canadian Journal of Earth Sciences,
    7:634-664.
Blake, W. Jr., 1975: Radiocarbon age determinations and postglacial
    emergence at Cape Storm, southern Ellesmere Island, Arctic Canada.
    Geografiska Annaler, Series A, Physical Geography, 1-71.

Craig, B.G. and Fyles, J.G., 1960: Pleistocene geology of Arctic
   Canada. Geological Survey of Canada, Paper 60-10, 21 p.

DiLabio, R.N.W. and Shilts, W.W., 1978: Compositional variation of
   debris in glaciers, Bylot Island, District of Franklin; In: Current
   Research, Part B; Geological Survey of Canada, Paper 78-1B, 91-94.
Dyke, A.S., in press: Quaternary geology of Somerset Island, District
   of Franklin. Geological Survey of Canada, Memoir 404.

England, J.H., 1976: Late Quaternary glaciation of the eastern Queen
   Elizabeth Islands, N.W.T., Canada: alternative models. Quaternary
   Research, 6:185-202.

Falconer, G., 1962: Glaciers of northern Baffin and Bylot Islands,
   N.W.T. Department of Mines and Technical Surveys, Geographical
   Branch, Paper 33, 31 pp.
Freuchen, P. and Mathiassen, T., 1925: Contributions to the physical
   geography of the region north of Hudson Bay; In: The Danish
   Ethnographic and Geographic Expedition to Arctic America. Preliminary
   report of the Fifth Thule Expedition. Geographical Review,
   15:549-562.

Hodgson, D.R. and Hasleton, G.M., 1974: Reconnaissance glacial geology,
   northeastern Baffin Island. Geological Survey of Canada, Paper 74-20,
   10 p.
Hughes, T., Denton, G.H., and Grosswald, M.G., 1977: Was there a
   Late-Würm arctic ice sheet? Nature, 266:596-602.

Ives, J.D., 1978: The maximum extent of the Laurentide Ice Sheet along
   the east coast of North America during the last glaciation. Arctic,
   31:24-53.
Ives, J.D. and Andrews, J.T., 1963: Studies in the physical geography
   of North-Central Baffin Island, N.W.T. Geographical Bulletin,
   19:5-58.

Klassen, R.A., 1981: Aspects of the glacial history of Bylot Island,
   District of Franklin. In: Current Research, Part A, Geological Survey
   of Canada, Paper 81-1A, 317-326.
Klassen, R.A., 1982a: Quaternary stratigraphy and glacial history of
   Bylot Island, N.W.T., Canada. Ph.D. thesis, Department of Geology,
   University of Illinois at Urbana. 163 p.
Klassen, R.A., 1982b: Glaciotectonic thrust plates, Bylot Island,
   District of Franklin. In: Current Research, Part A, Geological Survey
   of Canada, Paper 82-1A, 369-373.

Miller, G.H., 1973: Late Quaternary glacial and climatic history of
   northern Cumberland Peninsula, Baffin Island, N.W.T., Canada.
   Quaternary Research, 3:561-583.
Miller, G.H., 1980: Late Foxe Glaciation of southern Baffin Island,
   N.W.T., Canada. Geological Society of America Bulletin, Part 1,
   91:399-405.
Miller, G.J., Andres, J.T., and Short, S.K., 1977: The last
   interglacial glacial cycle, Clyde Foreland, Baffin Island, N.W.T.:
   Stratigraphy, biostratigraphy, and chronology. Canadian Journal of
   Earth Sciences, 14:2824-2857.
Miller, G.H. and Dyke, A.S., 1974: Proposed extent of Late Wisconsin
   Laurentide Ice on eastern Baffin Island. Geology, 2:125-130.
Miller, G.H. and Hare, P.E., 1980: Amino acid geochronology: Integrity
   of the carbonate matrix and potential of molluscan fossils. In: Hare,
   P.E., Hoering, T.C., and King, K. Jr. (eds.), Biogeochemistry of
   Amino Acids. J. Wiley and Sons Ltd., New York, U.S.A. 558 pp.

Nelson, A.R., 1982: Aminostratigraphy of Quaternary marine and glaciomarine sediments, Qivitu Peninsula, Baffin Island. Canadian Journal of Earth Sciences, 19:945-961.

Prest, V.K., 1970: Quaternary geology in Canada. In: Douglas, R.J.W. (ed.), Economic Minerals of Canada. Geological Survey of Canada, Economic Geology Report 1, 5th ed., 675-764.
Prest, V.K., Grant, D.R., and Rampton, V.N., 1968: Glacial map of Canada. In: Douglas, P.J.W. (ed.), Geology and Ecomonic Minerals of Canada. Geological Survey of Canada, Economic Geology Report 1, 5th ed., 675-764.

Sim, V.W., 1960: A preliminary account of late 'Wisconsin' glaciation in Melville Peninsula, N.W.T. Canadian Geographer, 17:21-34.
Szabo, B.J., Miller, G.H., Andrews, J.T., and Stuiver, M., 1981: Comparison of Uranium-series, radiocarbon, and amino acid data from marine molluscs, Baffin Island, Arctic Canada. Geology, 9:451-457.

16 A review of the Quaternary geology of western Greenland

Michael Kelly

Important reviews of the Quaternary geology of Greenland, including a review map at 1:2.5 m, have been published by Weidick (1971, 1975a, 1976a). The Geological Survey of Greenland (G.G.U.) also has published two of the five Quaternary geology map sheets at 1:500,000 which will cover western Greenland (Weidick, 1974a, 1978a). In addition an extensive literature dealing with various aspects of the Quaternary of Greenland has resulted from the long history of research in the area, the scope of which is summarized by Funder (Chapter 1, this volume).

For the purpose of the present review western Greenland is taken as including the administrative districts of West Greenland and the southwestern part of North Greenland, from Kap Farvel at 59° 46'N, 43° 50'W to Kap Alexander at 78° 11'N, 73° 05'W. This large area is subdivided for convenience into the five sectors shown in Figure 16.1. The distribution of the geological evidence and its relationship to the topography in these sectors is illustrated by the series of cross sections in Figure 16.2. In the following discussion a local stratigraphic terminology of various ranks is used, except that, following local tradition, the chronostratigraphic nomenclature is taken from N. Europe (Mangerud et al., 1974, 1978). All dates used are in radiocarbon years BP and are quoted either as corrected for fractionation (normalised) and an assumed seawater error of 400 years, or are in an original form which is considered to be analogous. The stratigraphy is based on the recognition of major glacial and marine events, which are themselves related through the glacio-isostatic response of the area. A proposed succession of these informally named glacial and marine events is given in Table 16.1.

Over the major part of western Greenland the Quaternary deposits and minor landforms apparently belong to a single glacial episode, including its deglaciation phases, and to the associated

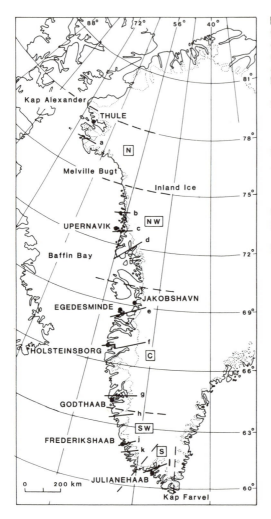

Figure 16.1 Location map of western Greenland showing geographical subdivisions used and locations of cross-sections.
Data Source: (a) Krinsley (in Davies et al., 1963); Weidick (1978b); Kelly (1980a) (b) & (c) Kelly (unpub.) (d) Laursen (1944), Kelly (unpub.) (e) Weidick (1968); Donner & Jungner (1975) (f) Kelly (1979, unpub.) (g) Weidick (1974a); Ten Brink & Weidick (1974); Kelly (unpub.) (h) Weidick (1975c, 1978a); Fredskild (1983) (j) Weidick (1975b, 1978a)(k) & (l) Kelly (unpub.) (m) Weidick (1963); Funder (1979); Kelly (unpub.)

subsequent glacio-isostatic marine transgression and regression. These are herein termed the Sisimiut glacial event, and the Disko Bugt marine event (Table 16.1). The definition and dating of this glacial event are consequently key points in the understanding of the stratigraphy of the area, and they in turn critically depend on the interpretation of the relationship between the marine (i.e. transgression) limit and the glacial deposits in the outer coastal regions of the C and SW sectors (Fig. 16.1).

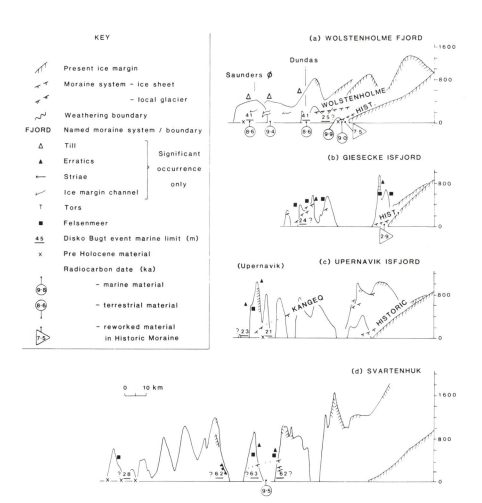

Figure 16.2 Representative geological cross-sections, western
Greenland (see Fig. 16.1 for locations).

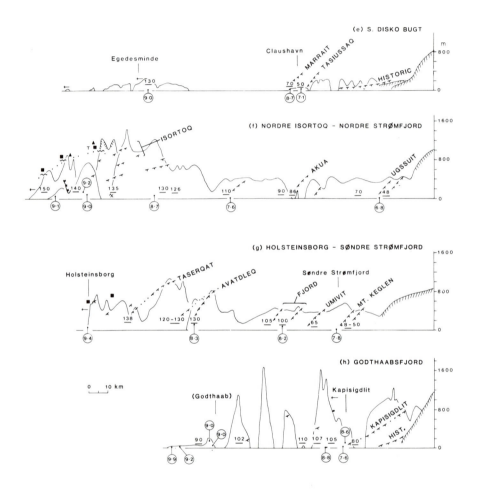

Figure 16.2 (continued)

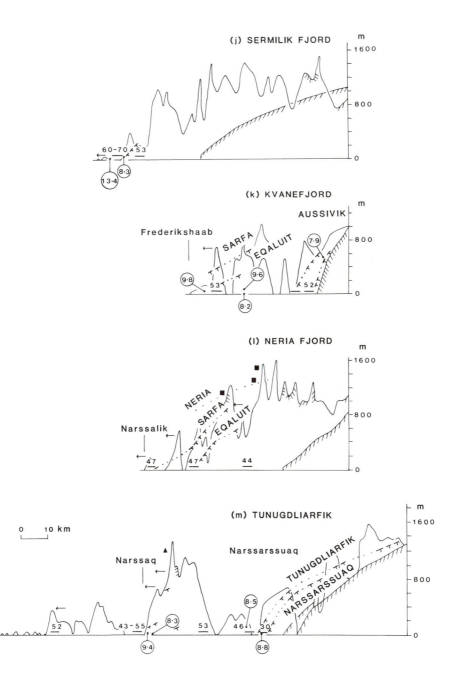

Figure 16.2   (continued)

Table 16.1. Stratigraphy of marine and glacial events in western
Greenland and their provisional ages.
==

	REGIONS	
60-69°N (CENTRAL & SOUTH)	SHELF	69-78°N (NORTH & NORTHWEST)
VESTERBYGD GLACIAL (0 - 3.5 ka)		as Central
DISKO BUGT MARINE (0 - 13.5 ka)		
SISIMIUT GLACIAL (3.5 - > 13.5 ka)	(SISIMIUT GLACIAL)	SISIMIUT GLACIAL ? GLACIAL
		SVARTENHUK MARINE (40-50 ka?)
LAKSEBUGT MARINE (130 ka?)	SEVERAL GLACIALS OR DEGLACIATION PHASES ?	SEVERAL POSSIBLE — KAFFEHAVN MARINE (70-120 ka?)
FISKEBANKE GLACIAL		GLACIALS
MUDDERBUGT MARINE		? — METEORBUGT MARINE (> 200 ka?)
	HELLEFISK GLACIAL	(Other marine events?)
		PATORFIK MARINE (> 500 ka?)

DISKO BUGT MARINE EVENT (LATE-WEICHSELIAN TO MID-HOLOCENE)

The deposits of the Disko Bugt event

Deposits of the Disko Bugt marine event are widespread in central
western Greenland, and again in the extreme north, where they are
represented by lithofacies ranging from muds to boulder gravels.
Although some descriptions of their sedimentology or stratigraphy have

been given (e.g. Laursen, 1950), a detailed study of both is long
overdue. Elsewhere in western Greenland deposits of this episode are
sparse, mainly consisting of coarse littoral gravels and sands.

Over 240 C-14 dates, mostly from bivalves, have been obtained for
the Disko Bugt event. Their geographical distribution strongly
reflects the regional difference in the frequency and lithology of the
deposits (Fig. 16.3). Evidence that these deposits also continue
beneath the present ice sheet is provided by the common occurrence of
shell fragments of this age in the glacial deposits at the ice margin
(see Fig. 16.15). Their age distribution is also highly significant,
being mainly restricted to the early and mid-Holocene (4000-10,000
BP). A few older dates indicate that, locally, the formation may go
back to at least 13,400 BP. The pattern of the early dates is
controlled by the early history of the Sisimiut event deglaciation,
whereas the absence of late Holocene dates is interpreted as being due
to the onset of the Neoglacial Vesterbygd glacial event, both of which
are discussed further below.

Emergence/Regression

The sea regressed during the early and mid-Holocene. This is shown by
the altitudinal relationships of radiocarbon dates on marine shells and
by the dates for marine regression horizons in the sediments of
present-day lakes. No evidence of any interruption of this regression
has been reported from western Greenland. Graphs illustrating the
history of emergence have been given for many localities, usually
interpolated and extrapolated into an "emergence curve" (Donner, 1978;
Donner and Jungner, 1975; Fredskild and Møller, 1981; Funder, 1979;
Kelly, 1973, 1979; Ten Brink, 1974, 1975; Weidick, 1972a). Figure 16.4
illustrates the emergence characteristics of selected areas. As can be
seen, in no area do the data really justify expressing the emergence by
a single well-defined curve; although some of the uncertainty may arise
from the need to include data from over relatively large areas, most of
it is inherent in the data themselves. Thus the majority of shell
dates provide only a minimum estimate of the height of the
contemporaneous sea level, being either from offshore sediments, or
less commonly, older shells incorporated by intraformational
reworking. In contrast, dates for lacustrine regression horizons are
normally minimum estimates of the sea level age, because the dated
sediment interval is above the actual horizon. However, they can also
be maximum estimates because of hard-water error or addition of old
organic matter. Obviously dates on terrestrial deposits give a maximum
value for the age of the contemporaneous sea level.

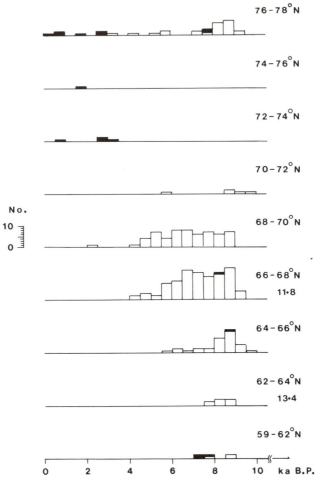

Figure 16.3   Distribution of radiocarbon dates from deposits of the Disko Bugt marine event (solid ornament indicates reworked material). Data Source: Blake (1975, 1977); Krinsley (in Davies et al., 1963); Donner (1978); Donner & Jungner (1975); Fredskild (1973, pers. comm.); Funder (pers. comm.); Kelly (1973, 1975, 1979, 1980a, 1980b, unpub.); Simonarson (1981); Sugden (1972); Ten Brink (1975); Weidick (1972a, 1972c, 1973, 1974b, 1975e, 1976b, 1977, 1978c).

Despite the limitations of the data certain features are apparent. In the central sector the earlier claimed difference in emergence rate between the coastal and interior sectors (Kelly, 1973; Ten Brink, 1974) is not substantiated. Instead emergence history over this area could have been very similar with maximum emergence rates of 0.05 m a^{-1}. In contrast, emergence in the N, SW, and S sectors appears to have been both slower and earlier achieved. In all areas sea level appears to have been close to the present by 2000-4000 BP.

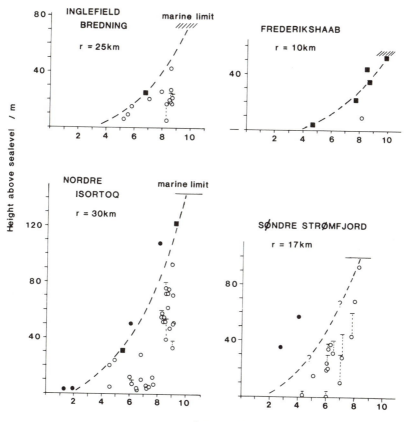

Figure 16.4 Representative emergence curves based on radiocarbon
data from defined areas (of radius r): open circles - marine shells,
with associated younger shorelines if present (bar); squares - lake
sediment regression horizons; solid circles - terrestrial peat.
Data Source: (a) Fredskild (pers. comm.); Weidick (1976b, 1978c).b)
Kelly (unpub.); Weidick (1972a) (c) Kelly (1973, 1979, 1981, unpub.);
Weidick 1972a, 1972c, 1973) (d) Ten Brink (1975); Weidick (1972a,
1972c, 1973).

Marine limit

The position of the maximum height of the Disko Bugt event
transgression in western Greenland, i.e. of the marine limit, is shown
in Figure 16.5a. The use of a variety of criteria to define the marine
limit is responsible for some of the spatial irregularity in the data.
Thus perched erratic and washing limits give a maximum estimate of the
height of the contemporaneous mean sea level, and marine sediments a
minimum one, whereas beach ridges and terrace breaks of slopes vary

between these in their significance. Older data are excluded from
Figure 16.5a because they often include non-marine features.

Figure 16.5b gives the age of the marine limit estimated by
extrapolation from the emergence curves, or the minimum age based on
individual dates. In only a few areas (Dundas, Holsteinsborg,
Frederikshaab) do the dated levels lie high enough to reasonably
closely date the oldest, highest marine limits of the Disko Bugt event
transgression. However, the marine limit everywhere in the C, SW, and
S sectors is considered to be of this age because it is spatially
related to the well-dated localities, and because there is as yet no
evidence to the contrary.

The maximum height, 140-150 m, occurs in the coastal areas of the
central sector (Figs 16.2f & g), in which sector there is also a strong
E-W gradient. The idea that the transverse profile of the marine limit
was dome-shaped, with the highest elevations east of the coast (Weidick
1972a, 1975a; Ten Brink, Weidick, 1974), does not appear to apply in

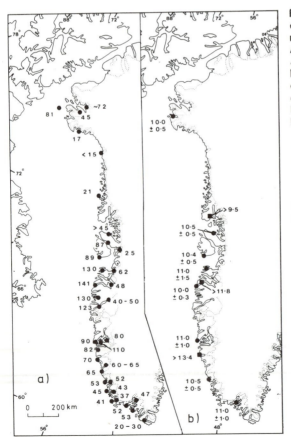

Figure 16.5 (a) Height
of the Disko Bugt event
marine limit (m). (b)
Age of the marine limit
at the outer coast (ka
BP): circles - date
extrapolated from
emergence curve; square -
based on C-14 date.
Data Source: (a) Blake
(1977); Donner & Jungner
(1975); Fredskild (1973,
pers. comm.); Funder
(pers. comm.); Kelly
(1980a, unpub.); Krinsley
(in Davies et al., 1973);
Laursen (1944); Sugden
(1972); Weidick (1974a,
1975b, 1975c, 1978a,
1978b) (b) Redrawn from
sources listed for Fig.
16.3.

this area. Maximum heights decrease to the south, and in the SW and S
sectors the limit becomes remarkably uniform, 40-55 m, with negligible
transverse gradients (Figs 16.2 k-m).

Much less is known about the Disko Bugt event marine limit in the
N and NW sectors where it is difficult to identify. In the area north
of Kap York (76°N), Disko Bugt event deposits have been dated up to 24
m and the marine limit, put at 45 m in Wolstenholme (Fig. 16.2a) and
Olrik fjords (Krinsley in Davis et al., 1963; Weidick, 1978b) appears
to be of Disko Bugt age. On the offshore Cary Øer raised beaches up to
84 m have been considered as Holocene, although the highest dated level
is 20 m (Bendix- Almgreen et al., 1967; Blake, 1977).

Between Umanaq Bugt and Melville Bugt (71°-76°N) there is only one
dated Disko Bugt event deposit (Fig. 16.3), and in the southern half of
this area there are no other known marine deposits even potentially of
this age. In western Svartenhuk (Fig. 16.2d) the marine limit may be
from an earlier marine event, or at least formed extra-marginally at
the maximum of the preceding glacial event (Sisimiut). Around
Upernavik (Figs 16.2b & c), reworking of moraines up to 21-25 m may be
of early Disko Bugt event age, but to the east the marine limit is
close to present sea level. Further north in Melville Bugt there are
undated, well-developed cobble beaches up to 17 m which can be assigned
to this event. Higher morphological features, up to 30-40 m, may date
from an earlier marine episode (Kelly, 1980a).

As Figure 16.5b shows, the oldest marine limits in the west date
from the end of the Weichselain, 10,000-13,500 BP. The initial
transgression over most of the outer coastal area probably occurred
contemporaneously with deglaciation. However, there may be important
exceptions to this where the marine limit was formed in the
extra-marginal zone of depression (see below).

Marine Faunas

The fauna of the Disko Bugt event has a long history of study, with
records of invertebrate collecting going back to Rink (1852), and of
vertebrates to 1765 (Bendix-Almgreen, 1976). Most attention has been
paid to the Mollusca, Cirripedia, Echinodermata and Brachiopoda, and
over 96 species have been recorded (Laursen 1944, 1950). In the
interpretation of these fossil faunas some consideration has been given
to their composition as representatives of life and death assemblages
of different habitats (Donner and Jungner, 1975; Kelly, 1973). Earlier
work concentrated on the recognition of sequences of faunas supposed to

have paleoclimatic significance (Harder, Jensen and Laursen, 1949; Laursen, 1950), and these were correlated with the N. European Late Weichselian and Holocene chronostratigraphy. Subsequent radiocarbon dating has shown that this scheme is no longer valid in its entirety, although Laursen (1976) has claimed that the basic sequence is correct. However, it is now known that supposedly key species overlap one another in time (Fig. 16.6), even in the type locality at Orpigsôq (Donner and Jungner, 1975).

Two features of the original scheme remain of interest, both initially described by Jensen (1942). Firstly, there is the anomalous behavior of <u>Portlandia</u> <u>arctica</u>. This species, which inhabits muddy sediments, was widespread in the central sector in the early Holocene where it was considered to be indicative of a cold water environment, as it is now extinct in W. Greenland south of 78°N. However,

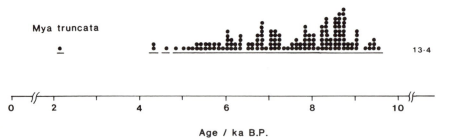

Figure 16.6 Distribution of C-14 dates of characteristic bivalve mollusc species: dot - individual date in 50 year increments, bar - range of dates including counting errors.
Data Source: From relevant sources listed for Figure 16.3.

historically it was not only generally synchronous with the relatively warmth-demanding Mytilus edulis (Fig. 16.6), but in the central sector it continued into the middle Holocene and was the contemporary of very thermophilous species (Donner and Jungner, 1975). It is known to be especially tolerant of the high turbidity associated with meltwater discharges (Ockelmann, 1958) and it therefore appears to have behaved at least partly as a facies fauna, but with its distribution becoming increasingly restricted until the point where communities could not be maintained in the remaining isolated suitable habitats, which still exist today. In contrast Jensen's concept of a period of optimum marine climate still seems valid as it can be defined by seven thermophilous mollusc species which have boreal distributions and which are absent from West Greenland today. The critical thresholds for these apparently were exceeeded only in the central sector, to which the thirteen known occurrences are restricted (Fig. 16.7). All appear to date from the Atlantic chronozone, and five occurrences of Zirphaea crispata and Arctica islandica are directly dated to between 4900-7320 BP (Fig. 16.6).

The records for two other species, Mytilus edulis and Chlamys islandica (Fig. 16.6), provide information about the development of the West Greenland Current. Both species occur today in western Greenland to 78°N, but have limited distributions in E. Greenland, where they are regarded as thermophilous elements (Funder & Hjort, 1974). Radiocarbon dates between 8500-8800 BP show that Mytilus occurred up to 67°N. A similar date from the extreme north (76°N) (Krinsley in Davies et al., 1963) needs confirmation because it was obtained for whale bone in the Mytilus horizon. (An early date of 9070 BP (K-1377) for Mytilus edulis (Weidick, 1972a) is a misprint for Mya truncata.) The first dates for Chlamys are earlier, 9400 BP at 67°N, which is in keeping with its more northerly distribution. The occurrence in dated faunas of a less demanding species, Mya truncata, is given in Figure 16.6 for comparison.

SISIMIUT GLACIAL EVENT - LATE WEICHSELIAN

Age and extent

The relatively fresh glacial deposits and landforms which occur sporadically over most of the central and southern sectors out to the outer coast are considered to be the products of the Sisimiut glacial event. The deglaciation of these outer coasts is dated to the end of the Late Weichselian, 10,000-14,000 BP, on the evidence of the age of

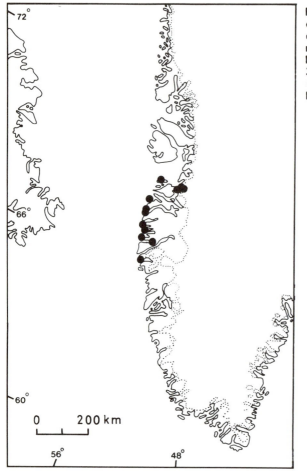

Figure 16.7 Occurrence of exotic mollusc species in deposits of the Disko Bugt marine event.
Data Source: Donner & Jungner (1975); Jensen (1942); Laursen (1950); Kelly (1979, unpub.)

the Disko Bugt event marine limit (Fig. 16.5b). This approximate age for deglaciation is supported by the distribution of radiocarbon dates between 8950 and 25,000 BP obtained from terrestrial and marine sediments not covered by tills (Fig. 16.8a). This shows that dates between 8950-10,200 are restricted to a relatively narrow coastal zone, and that there are only two dates older than this: 13,800 (Weidick, 1975b) and 11,800 BP (Kelly, unpub.). There is no reason to doubt the validity of the oldest date, except on the grounds of its uniqueness, but the other may be too old because the material dated both included shells and contained sediment. Since both of these sites are thought to be within the boundary of the Sisimiut glacial event (Figs 16.2f & j) and are not covered by tills, the Sisimiut glacial maximum could be

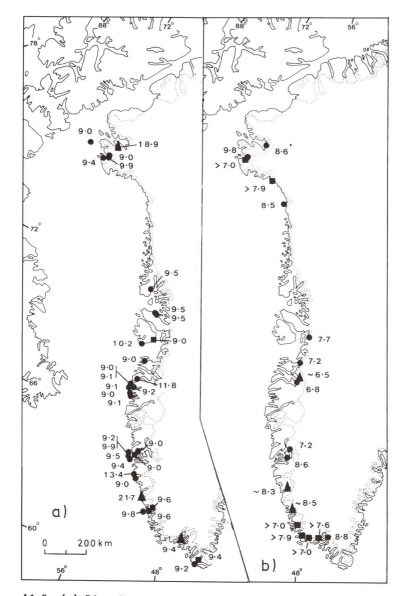

Figure 16.8   (a) Distribution of radiocarbon dates between 8950–
25,000 BP (ka BP): circles - marine shells and lake sediments below
marine limit; squares - lake sediments above marine limit; triangles -
material out of context. (b) date of attainment of the present level of
ice cover (ka BP): circles - dated material outside ice margin; squares
- glacially reworked material at ice margin; traingles - estimated from
marine shorelines at ice margin.
Data Source: From relevant sources listed for Figure 16.3.

dated to before c. 13,500 BP. However, these early dates may have only local significance since extrapolation of emergence curves generally suggests a younger date for the marine limit at c. 11,000 BP.

In places in the mountainous coastal areas of parts of the C and SW sectors two weathering zones can be recognized (Kelly, unpub.), with the upper zone characterized by predominately autochthonous felsenmeer, tors and a high degree of surface weathering of exposed clasts and bedrock; whereas the lower weathering zone is distinguished by the lack of these features, by more abundant glacial deposits and by the preservation of glacially striated and polished rock surfaces. In the area around Holsteinsborg (Fig. 16.2f) the boundary between the two zones, which in places is relatively sharply defined, correlates approximately with the upper limit of the Isortoq Moraine; a massive nunatak moraine system lying between 700-1000 m.

Although there is considerable discussion about the significance of such weathering features as tors and felsenmeer (e.g. Boyer & Pheasant, 1974; Dyke, 1976; Sugden & Watts, 1977; Watts, 1979) the boundary is assumed to represent a former ice margin which provisionally is considered as defining the maximum of the Sisimiut event ice sheet in the Holsteinsborg region. This would locate its frontal margin on the inner part of the continental shelf. Similar evidence in the SW sector (Arsuk & Neria Fjord; Fig. 16.2) also puts the ice margin just offshore and it seems reasonable to suppose that this was the general location of the Sisimiut event margin at its maximum in the C, SW, and S sectors. Its age therefore is not likely to be older than the Late Weichselian, although possibly > 13,500 BP.

The history of the Sisimiut glacial event in the northern half of western Greenland is more problematical. Around Upernavik (72-74°N; Figs 16.2b & c) moraines and weathering evidence suggest that an ice sheet presumed to be of this age reached just offshore. Further south, around Svartenhuk (71-72°N) (Fig. 16.2d), where extensive pre-Sisimiut event marine deposits are preserved (see below), this ice sheet may not have reached the outer coast. If so, its limit may be represented by the Umiarfik Moraine (Kelly, unpub.). However, the stratigraphy of this area is not yet fully established.

A similar situation exists in the N sector (76-78°), where there exists frequent pre-Sisimiut event marine sediments, and a poorly understood stratigraphy. The Wolstenholme Fjord Moraine (Fig. 16.2a) which relates to an ice margin at the mouth of this fjord, is of Late Weichselian age if the early Holocene marine sediments in the fjord are in situ. Hence it belongs to the Sisimiut event. A more extensive

stage of glaciation is indicated by tills overlying marine sediments at
Saunders Ø (Blake, 1975)and Melville Bugt (Kelly, 1980a), and by
glacial deposits and features over the whole area, including the Carey
Øer (Krinsley, in Davis et al., 1963; Blake, 1977; Kelly, 1980a).
However, whether these are also of Sisimiut event age is not known.

Deglaciation phases

Moraine systems are widespread in the central sector, less so
elsewhere, and in general they are more fragmentary in the outer
coastal areas than the interior. Weidick (1968) originally classified
them on the basis of their spatial distribution (nunatak, outer and
inner zones); a scheme which has been modified by later work. The
current nomenclature is based on morphostratigraphic units, i.e.
moraine systems, which are used to define glaciation phases or stages.
Figure 16.9 shows the major named moraine systems from western
Greenland.

The central sector has been the classic area for the study of
these deglaciation phases (Figs 16.2f & g; Weidick, 1968, 1972 a, b;
Ten Brink, 1975; Ten Brink and Weidick, 1974; Kelly, unpub.), during
which there has been some evolution of the nomenclature to the versions
given in Figure 16.9.

A major uncertainty exists about the origins of these moraine
systems. They may represent either significant readvances of the ice
sheet margin in response to regional positive changes in ice sheet mass
balance, albeit during a period of overall deglaciation, or
alternatively, local stillstands or readvances caused by topographic
influences or by shifts in ice sheet streamline patterns during
deglaciation. Ten Brink and Weidick (1974) consider that the Boreal
and early Atlantic moraine systems in the central sector were
climatically induced, pointing as evidence to their continuity in
upland and lowland areas, and also to the overall similarity in ages of
systems from different areas. According to them, changes in summer
temperatures were responsible since, as Weidick (1982) has pointed out,
the margin will respond rapidly to decreases in ablation because of the
differences in the dynamics of the ablation zone and of the whole ice
sheet. Weidick (1972a, 1976a) consequently has attempted to correlate
the moraine systems with fluctuations in the oxygen isotope ratios in
the Camp Century Ice core. On the other hand, it has been claimed that
the moraine system continuity is more apparent than real and that they
could be explained as metachronous local stillstands during a period of
slower retreat.

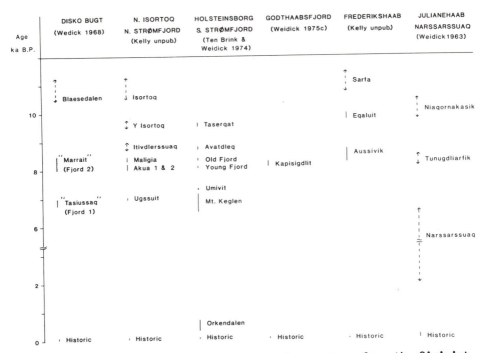

Figure 16.9 Ages and correlation of moraine systems from the Sisimiut event deglaciation and the Vesterbygd event.

This conflict is not made easier to resolve by the imprecision of the dating of the moraines, which is generally indirect, and commonly based on the dating of associated marine shorelines through the extrapolation of emergence curves. Furthermore, most of the moraines are not closely related to a contemporaneous sea level by an outwash feature but instead can only be said to be younger than a cross-cutting shoreline.

Another limitation is the lack of stratigraphic evidence of readvances. Only at Claushavn (69°N) (Fig. 16.2e) does this apparently exist. There the older "Marrait Moraine System" surrounds the younger "Tasiussaq Moraine System" which is clearly related by its sandur to a sea level at 50 m. According to the description in Weidick (1968) the Tasiussaq Moraine overlies marine sediments which post-date the Marrait Moraine. (The occurrence of fluvioglacial sediments alone lying on marine sediments is itself not evidence of a readvance.)

Despite these uncertainties, it does appear that there were two periods in the Holocene suited to moraine construction, especially in the central sector, broadly between 8000-8500 BP and around 7000 BP

(Fig. 16.9). These correspond to the Fjord and Mt. Keglen phases of Weidick (1972 a, b), terms which can still usefully be retained for chronostratigraphic purposes without implying any climatic changes. More precise dating of moraine systems is required to demonstrate whether other phases have a local or regional significance.

By the middle Holocene ($\sim$ 6000 BP) the Sisimiut event ice sheet had retreated to a size comparable to the present, probably reaching this stage considerably earlier in both the north and south ($\sim$ 8000 BP). The evidence for this, summarized in Kelly (1980b), is the age of the marine limit at the present ice margin and the age of reworked Disko Bugt event shells in the present glacial sediments (Fig. 16.8b).

Local glaciers

Little is known of the histories of the ice caps and valley and corrie glaciers which must have existed in suitable localities at all stages of the Sisimiut glacial event. At the maximum stage it is likely that more or less confluent local glacier complexes occurred on Disko, Nugssuaq, Svartenhuk and on the peninsulas of the extreme north and south. During the deglaciation, these became independent, though in a reduced state, whereas others came into existence, e.g. Sukkertoppen Ice Cap. Moraine systems belonging to these local glaciers are generally only poorly dated, but are thought to be of Late Weichselian-Early Holocene age (Kelly, 1980b). Although indicating periods of greater extent for local glaciers, it is not possible to correlate these phases with those of the ice sheet deglaciation. Not all the local glaciers present today came into existence during the deglaciation phase (or, if they did, they had a smaller extent then than at present). The present glaciers of this type cut across moraine systems belonging to the ice sheet glaciation. Examples are known from the NW and SW sectors (Kelly, 1980b) and the Nordre Isortoq area of the central sector (Kelly, unpub.).

PRE-HOLOCENE MARINE EVENTS

Evidence of pre-Holocene marine events is now known from 23 sites concentrated in the northern half of western Greenland (Fig. 16.10; Funder and Simonarson, 1983; Kelly, 1983). However, much of it derives not from in situ marine sediments but from reworked material incorporated in glacial sediments in a variety of forms: large slices of marine sediments, abundant lightly abraded shells, sparse abraded shell fragments, or concretions. Although most of these deposits were

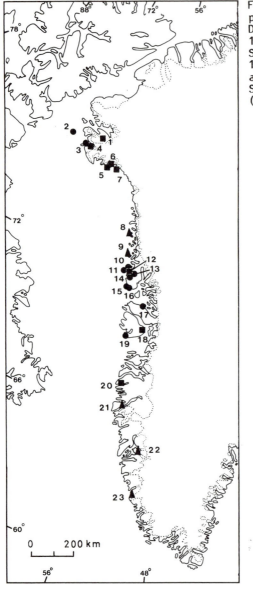

Figure 16.10 Occurrence of pre-Holocene marine material. Data Source: Blake (1975, 1977); Bryan (1954); Funder and Simonarson, 1983; Kelly (1980a, 1983); Krinsley (in Davies et al., 1973); Laursen (1944) Sugden & Miller (1976) Weidick (1976b, 1978b).

discovered recently, several have been known about since the beginnings of Quaternary research in Greenland, e.g. Pátorfik (Rink, 1852), Kugssineq (Steenstrup, 1883). Although Steenstrup (1883) considered that the Pátorfik fauna might be pre-glacial, the first pre-Holocene marine sediments to be positively identified as such by radiocarbon dating were those on Saunders Ø (Kingsley, in Davies et al., 1963). The age and sequence of marine events represented by this material are

still poorly understood. Very few individual successions contain
evidence of more than one marine episode (e.g. Saunders Ø [Blake,
1975]) so that reconstruction of the stratigraphy requires correlation
using faunal or amino acid information.

Radiocarbon dates for most sites are infinite > 32,000 - 42,000
BP. Of the four exceptions, two (Olrik Fjord and Frederikshaabs
Isblink) are thought to be due to contamination by Holocene material,
and two (Isbjörnö and Kingigtoq) to be finite only as a statistical
artifact of the counting error. Amino acid values, however, indicate
that the material covers a wide time span and probably represents a
number of distinct marine events.

The amino acid data, provided by the Geochronology Laboratory,
University of Colorado, are of two categories: those obtained by
current analytical methods (Miller et al., 1982), and those from
earlier versions which have been corrected to the same basis using
factors established by re-analyzing representatives of each set of West
Greenland samples (see Miller, this volume). Figure 16.11 shows the
geographical distribution of the total D-alloisoleucine/L-isoleucine
(aIle:Ile) ratios obtained.

Effective diagenetic temperatures (EDT) experienced by the
deposits are presumed to have varied geographically, as well as
temporally, because of the wide latitudinal spread of the samples
(67-77°N), which corresponds to a wide range in present temperature
climate (e.g. -2.7 to -10.5°C MAT [Lysgaard, 1969]). Consequently,
local aminozones have been defined for limited geographical regions
which are assumed to have experienced similar EDT regimes. Table 16.2
shows the aminozones proposed so far, based only on the recent analyses
(Funder and Símonarson, 1983; Kelly, 1983). Their correlation remains
problematical, depending on the emphasis placed on the Free and Total
ratios and on the weight given to possible latitudinal variation in
EDT. The scheme suggested is the simplest possible compatible with the
geological evidence.

From the faunal evidence, the Pátorfik deposits (Fig. 16.10, No.
17) clearly belongs to an interglacial marine event not represented
elsewhere in West Greenland. This, the most intensively studied of all
the pre-Holocene marine deposits (Laursen, 1944; Símonarson, 1981;
Funder and Simonarson, 1983), has a unique content of thermophilous
mollusc species absent from Greenland today, in addition to an extinct
gastropod Alvania patorfikensis. The majority of the faunas from other
sites are characterized by the presence of species typical today of the
subarctic water mass associated with the northward flowing West

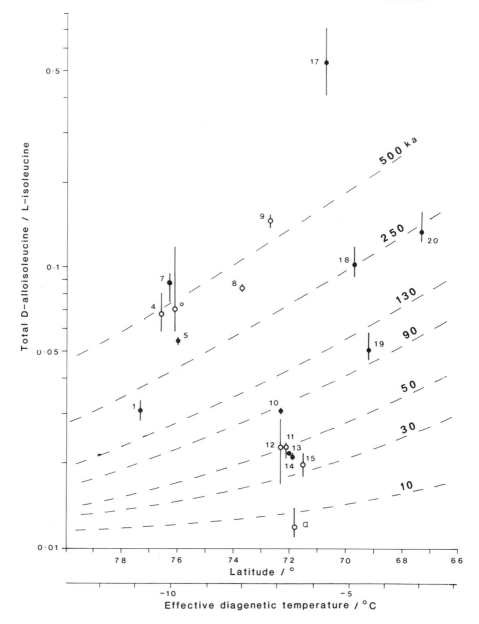

Figure 16.11 Latitudinal distribution of total D-alloisoleucine/
L-isoleucine ratios (means and ranges): solid circles - 1982/3
analyses, open circles - calculated from correction factors.
Isochrones are given for one alternative effective diagenetic
temperature model.
Data Source: Funder and Simonarson, 1983; Kelly (1980a, 1983);
Sugden & Miller (1976).

Table 16.2. Local aminozones with mean free (F) and total (T) alle:Ile ratios (site numbers in brackets).

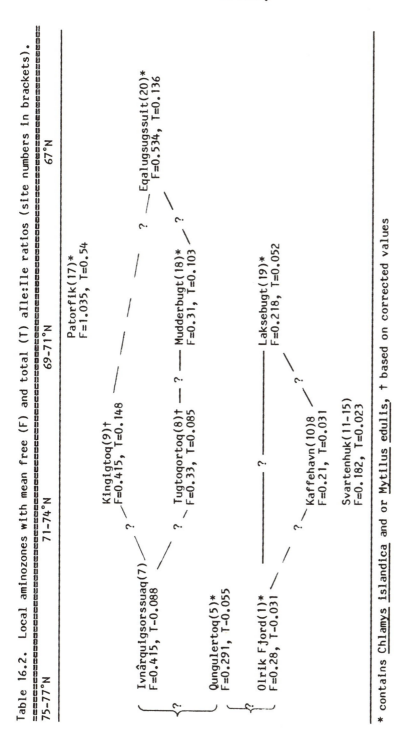

* contains Chlamys islandica and or Mytilus edulis, † based on corrected values

Greenland Current, i.e. <u>Mytilus edulis</u> and <u>Chlamys islandica</u>, including
sites near to the present northern limits of distribution of the two
species (Fig. 16.12). These species occur in deposits from at least
two aminozones in the northern region and at least three from further
south (Funder and Símonarson, 1983; Kelly, 1983). This indicates that
water mass distributions comparable to the present, or more favorable,
were established during 3-4 of the 4-6 marine events recognized. The
correlation of two important deposits from northern Greenland which
have this warm fauna (Carey Øer, No. 2, and Saunders Ø, No. 3) is
uncertain, because of the absence of amino acid data. In contrast,
none of the five deposits from the youngest event belonging to the
Svartenhuk aminozone (Fig. 16.10, Nos. 10-15), and its correlative

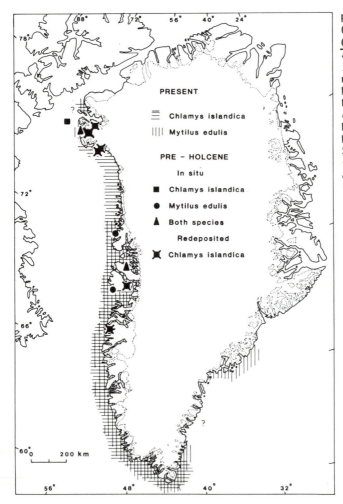

Figure 16.12
Occurrence of
<u>Chlamys islandica</u>
and <u>Mytilus edulis</u>
in pre-Holocene
marine material.
Data Source:
Modern
distributions from
Funder (1978) &
Funder &
Simonarson (1983);
fossil from Funder
& Simonarson
(1983) & Kelly
(1983).

Kugssineq (No. 16), include these species in their faunas (Kelly, 1983; Laursen, 1944). Although their absence may be due only to local habitat differences, it is compatible with the indication that this marine event was in part contemporaneous with a glacial event.

Overall, therefore, 4-6 pre-Holocene marine events are recognized provisionally; however, their ages remain problematical. The isochrones on Figure 16.11 show one model for their age distribution, based on the assumption that the estimated diagenetic temperature, EDT (for all time intervals), is the same as the present mean annual temperature, MAT (the temperature scale is linked to the latitude scale at four points for which MATs are known [Lysgaard, 1969]). The ages are calculated according to the Arrhenius equation for the total aIle:Ile ratio (Miller, pers. comm. 1983). Although this assumption is probably not valid, the model provides a baseline for further discussion.

The minimum age of the youngest marine event, Svartenhuk, is set at 40 ka by the C-14 dates. This puts the EDT for the interval at $\leq$ MAT. The low aIle:Ile ratios, however, suggest a young age and it is provisionally dated to 40-50 ka. The deposits from this event are covered by, or are associated with, glacial sediments which must derive either from the Sisimiut glacial event or an earlier one (see below).

The deposits belonging to the Olrik Fjord, Kaffehavn and Lakesbugt local aminozones may be from one or two marine events (Kaffenhavn/ Lakesbugt). The last was assigned to the oxygen isotope state 5e/6 transition, 130 ka, on the basis of its mixed warm and cold fauna (Funder and Símonarson, 1983). A similar relative EDT (3.5°C < MAT) puts the Kaffehavn deposit in the same interval (stage 5e), although one or both may equally well be Early Weichselian in age. The Kaffehavn and Lakesbugt marine event/s therefore may or may not be the same event. Correlation of the Olrik Fjord deposit with one or other of these requires a different EDT model and is therefore hypothetical.

On the basis of the aminozones one or two additional marine events may be represented between the Kaffehavn/Lakesbugt and the Pátorfik. Although the evidence is entirely from glacially reworked material, only at one site do the amino acid data suggest that the fauna may derive from more than one event. Provisionally one marine event is defined for each of the two sub-regions - Meteorbugt event and Mudderbugt event, based on the Ivnarquigsorssuaq and Mudderbugt aminozones, respectively. A Middle Quaternary age is possible for this event.

The Pátorfik marine event, the oldest known in Western Greenland, is now dated to the Early-Mid Quaternary on the basis of its fauna and high amino acid ratios (Funder and Simonarson, 1983), although it was intially referred to the Eemian (Simonarson, 1981).

PRE-SISIMIUT GLACIAL EVENTS

Glacial episodes older than the Sisimiut event include episodes of extensive ice cover. Evidence for these events comes from high-level erratics, glacial sediments associated with pre-Sisimiut event marine material, and from deposits and features on the continental shelf.

Onshore evidence

Erratics that occur on all the outlying islands and the highest summits, including summits lying above the level of the Sisimiut event ice sheet (Fig. 16.2) indicate that the whole of Greenland has been glaciated at one time. The margin of the ice sheet which deposited them must have reached well out onto the shelf, at least 50 km offshore at Nordre Isortoq (67°N). This is defined as the Fiskebanke glacial event.

Only at two of the sites where glacial sediments are found together with pre-Sisimiut marine sediments do they underlie this marine material and hence prove the existence of more than one glacial event, i.e. Saunders Ø (Blake, 1975) and Nerutussoq (Kelly, unpub.). These indicate the possible occurrence of pre-Kaffehavn and pre-Svartenhuk glacial events, respectively. The tills overlying the marine deposits may or may not belong to the Sisimiut event, since in the N and NW sectors where they mainly occur, the margin of the Sisimiut event ice sheet has not yet been defined. However, tills overlying Meteorbugt marine sediments in the N. sector were provisionally assigned to an older glacial event - the "Kay Melville glacial episode" by Kelly (1980a). Particularly important is the question of whether post-Svartenhuk event tills, which are < 40,000 BP, represent a Middle Weichselian glacial event, or whether they are of Late Weichselian (Sisimiut event) age, with a major hiatus occuring in the successions. If the former, it would imply an ice sheet more extensive than that of the Sisimiut event, and hence might belong to the main episode of shelf glaciation.

Offshore evidence

For some time the continental shelf of western Greenland has been
considered to bear signs of glaciation (Rvachev, 1964; Holtedahl, 1970;
Weidick, 1971). The shelf is everywhere broader than the adjacent
deglaciated land area, reaching 230 km at 68°N, although narrowing to
c. 50 km in the southwest, with the shelf break varying from c. 400 m
over the northern half to c. 200 m over the southern. The latter area,
south of 69°N, has extensive banks at < 100 m depth which are separated
on the landward side by a marginal channel from the submerged extension
of the coastal lowlands, and from each other by deep, broad troughs or
transverse channels (deeps) (Fig. 16.13). Much detailed information
about the morphology and shallow structure of this southern section has
recently become available from the geophysical surveys and drilling
undertaken for hydrocarbon exploration. This has substantiated the
general conclusions of the early work and shown that the southern shelf
has been completely glaciated at least once.

Much of the shelf is occupied by a sedimentary basin containing up
to 6000 m of Cretaceous-Tertiary sediments (Henderson et al., 1981).
From the seismic profiling records, this is considered to be covered
largely by a veneer of Quaternary sediments of variable thickness,
including both tills and fluvioglacial sediments (Brett and Zaruzki,
1979; Roksandic, 1979). Coarse clastic sediments also form the upper
units recorded from the exploratory wells, e.g. Hellefisk 1 at the
shelf edge had 200 m of "probable glacial moraine" (Risum et al.,
1980).

Of particular importance is the evidence of moraine ridge systems
from the sparker and boomer surveys. Between 64°-69°30'N four
categories of moraine systems have been identified: at the inner shelf
margin; along the flanks of the transverse deeps; on the surface of the
banks; and along the outer edge of the banks (Brett and Zarudzki, 1979;
Zarudzki, 1980). In the area from here south to 61°15'N, poorer
quality data have allowed only the patchy recognition of moraines of
the first two categories (Roksandic, 1979). These two types of moraine
systems also correspond with those proposed for various localities from
bathymetric evidence by Holtedahl (1970) and Weidick (1971).

Most notable are the moraines at the outer edge of the banks,
traceable over most of the region between 64°-69°30'N, but especially
well developed betwen 64°20' - 66°45'N. Brett and Zarudzki (1979)
describe them as a 150 km long and 4 km wide terminal moraine system of
5-7 narrow ridges up to 50 m high, with a second less well defined
system on its distal side (Fig. 16.14). This Hellefisk Moraine System,

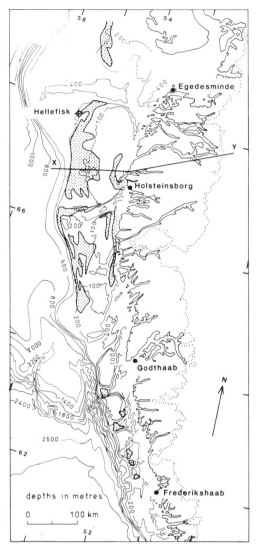

Figure 16.13 Morphology of the southern shelf and the distribution of morainic material (dotted) inferred from geophysical data. Data Source: Brett & Zarudzki (1979); Roksandic 1979; 1:500000 Bathymetric Maps, Greenland Geological Survey, 1975.

named here after the well sited on it, defines the Hellefisk glacial event. It clearly records the development of an ice sheet over the whole southern shelf out to its margin (Fig. 16.15). Brett and Zarudzki (1979) suggest that stratification in the moraine may indicate subaqueous deposition. However, because of the shallow depth of the shelf, most of the ice sheet must have been grounded.

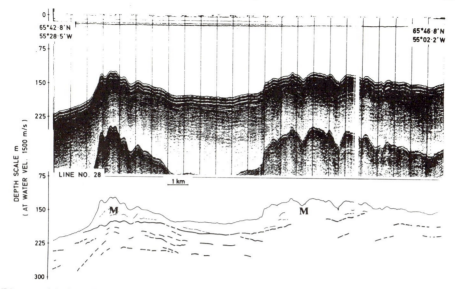

Figure 16.14   Sparker record of the "Hellefisk Moraine System" at the
shelfbreak (M-morainic deposits) (from Brett and Zarudzki, 1979).

Figure 16.15   Cross-section showing the relationship of the Hellfisk
and younger ice sheets. (See Fig. 16.13 for line of section.)

        Virtually nothing is known about the Quaternary deposits of the
northern half of the western Greenland shelf. However, with its
greater depth the ice cover, when it existed, must have occurred
largely as a floating shelf.

Correlation

The age of the offshore features and their correlation with the onshore
data is problematical. If the Hellefisk Moraine is older than the
Eemian, then a case for its preservation during subsequent regression
and transgression has to be made. If it is younger, i.e. Weichselian,
then there is a wide range of possible dates for it.

The easternmost shelf moraines can be considered provisionally as the terminal moraines of the Sisimiut glacial event, and the moraine systems on the banks therefore can be correlated loosely with the Fiskebanke event (Fig. 16.15). One or both of these last two phases saw tongues of ice extending westwards along the deep channels, resulting in the observed lateral moraine systems.

The offshore evidence also does not allow the relative significance of the different moraine systems to be determined, i.e. whether they represent independent advances or only successive retreat phases. This is of course linked to the question of their individual ages.

VESTERBYGD GLACIAL EVENT (MID-LATE HOLOCENE)

The latter part of the Holocene was a period of major advance of the ice sheet and of local glaciers in western Greenland. This constitutes the Neoglacial or Vesterbygd glacial event, whose history has been reviewed by Kelly (1980b).

As described above, the Sisimiut event ice sheet had retreated to its present size by the Middle Holocene, considerably earlier in the north and south (c. 8000 BP) than in the center (c. 6000 BP). That it subsequently continued to retreat beyond its present margins is demonstrated amply by the occurrence of reworked Disko Bugt event marine sediments and shells in the Veterbygd event moraine at 30 sites. (Fig. 16.16; Kelly, 1980b; Weidick, 1972b). The existence of a climate favorable to retreat in this ensuing interval is indicated by the thermophilous marine molluscs present for at least part of the time, and by the paleobotanical evidence (Fredskild, this vol.). With a period of possibly 3000-5000 years available, the extent of the retreat behind the present margin could have been considerable, especially in the northern and southern regions of the ice sheet.

Information about the early history of the Vesterbygd glacial event comes mainly from the paleobotanical evidence for climatic deterioration. Vegetation changes provide widespread evidence in the S, SW, C, and NW sectors for a general marked climatic deterioration to cooler and/or moister conditions between 2400-1800 BP (Fredskild, this vol.). In places, an earlier phase (c. 3600-3200 BP) of moister conditions is recognized also. Changes in the marine environment are less precisely dated, with the extinction of the boreal species datable only to < 4800 BP.

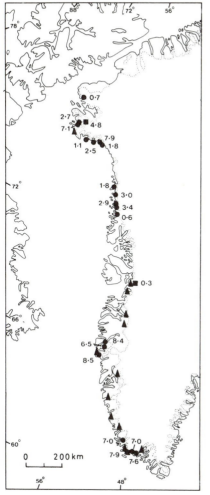

Figure 16.16   Age and occurrence of
reworked biogenic material in the
Historical Moraine System (ka BP):
circles - dated marine material;
squares - dated terrestrial;
triangles - undated marine.
Data Source: Dawes (in Weidick,
1978b); Goldthwait (1960);
Kelly(1980b); Weidick (1972a).

Thus the onset of the Vesterbygd glacial event can be put at c.
3500 BP, with a likely more marked response of the ice sheet to the
climatic change around 2000 BP. However, the probable advances of the
ice sheets and local glaciers which resulted from these early changes
in climate do not seem to have been great enough to bring their margins
to their present positions, since moraines from this period have not
been shown definitely to exist. The much quoted reference to
lichenometrically dated local glacier moraines at 2500 and 4000 BP
(Beschel, 1961) is believed to rest on a misunderstanding of that
author's conclusions, and the moraines are probably Early Holocene in
age (Kelly, 1980b). For the ice sheet, only two of the moraine systems
for which an early Vesterbygd episode age have been suggested, remain
as potential candidates for an early Neoglacial advance: the

Narssarssuaq Moraine (Fig. 16.2m; Weidick, 1963) and the Drygalski
Moraines (Weidick, 1968). However, the evidence available does not
preclude either moraine having an older age. Even if the Narssarssuaq
moraine belongs to the Vesterbygd episode, this sector is known to
behave in an anomalous fashion (Weidick, 1982). Thus, while Kelly
(1980b) considered that there was some evidence for an age < 2500 BP
for the Narssarssuaq Moraine System, Dawson's (1983) lichenometrical
analysis suggests a date older than this.

The implication of the lack of early Vesterbygd event moraines is
that this glacial event reached its maximum during the historically
recorded advances of the last few hundred years, which constructed the
moraine system termed the Historical Moraine by Weidick (1972b). From
a detailed study of the records of the frontal position of 135 local
glaciers and ice sheet outlet lobes, supplemented by some
lichenometrical data (Beschel, 1961), Weidick (1968, 1982) has dated
the attainment of this maximum, and the subsequent deglaciation
stages. He shows that, whereas the maximum was reached in the middle
18th century in the south, it did not occur until AD 1850-1930 in the
north. However, it has been proposed that in at least one area, the
maximum of the late Vesterbygd event occurred considerably earlier in
the historical period, i.e. at Söndre Strömfjord (Fig. 16.2g) where the
Ørkendalen Moraine System has been dated to 350-700 BP (Ten Brink,
1975).

The behavior of the ice margin in the 19th and 20th centuries can
be correlated with historical meteorological data (Fig. 16.17)
(Weidick, loc. cit.). The deglaciation following the maximum was
interrupted by readvances, which in the south date to AD 1850, 1890,
and 1920. The magnitude of the retreat varies widely between adjacent
sectors of the ice sheet, and with altitude, with calving outlet lobes
showing the most marked responses.

The paleoclimatic interpretation of the oxygen isotope data from
the Greenland Ice Sheet cores (Dansgaard et al., 1971, 1975) is in
general agreement with this history of the Vesterbygd glacial event,
with the prolonged Holocene warm interval (8000-3500 BP) ending with
two early Neoglacial cold intervals 3500-3000 and 2800-2400 BP),
followed by a complex sequence of cold intervals after 900 BP (c. AD
1000). Although there is no evidence that the early cold intervals
were any less severe than the later intervals, it is considered that
the successive readvances would need to be cumulative in effect to
offset the major mid-Holocene retreat of the ice margin, and hence
later readvances would override earlier ones. However, smaller ice

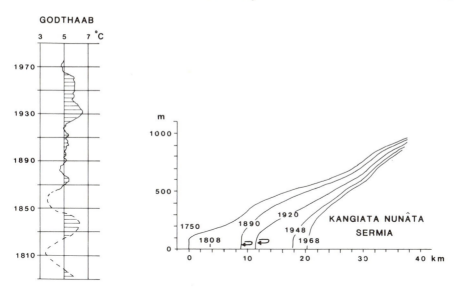

Figure 16.17   Correlation of climatic data and historical records of
glacier behavior in the Godthaabsfjord area (based on Weidick, 1982).
(a) 10 year running mean of summer temperatures from Gothaab, data
incomplete prior to 1870 (Vibe, 1967); (b) Ice surface profiles of
Kangiata nunata sermia(Weidick, 1982).

masses and certain sectors of the ice sheet may not show this effect
and moraines from early advances may be preserved (Kelly, 1980b).

The increased ice cover of the Vesterbygd glacial episode appears
to have had an isostatic response resulting in a marine transgression
(Kelly, 1980b; Weidick, 1968), here called the Sandnes transgression.
The evidence for this comprises early tidal observations; the
submergence of Norse and Paleo-Eskimo ruins, including the Norse church
at Sandnes, Godthaasbsfjord, and the apparent coincidence of present
sea level with past sea levels older than 1000-2000 BP in an area where
uplift could be expected to be still continuing. In many places the
transgression might have reached the position of yet older sea levels,
explaining the general lack of dated Disko Bugt event deposits younger
than 4000 BP (see Fig. 16.3). The early history of the transgression
is not known but it presumably began with the early Vesterbygd event
advances. The archeological data suggest that it was still in progress
in AD 1600-1860, and the tidal observations put its peak in the early
20th century, after which there has been a period of uplift.

DISCUSSION AND CONCLUSIONS

Despite increased evidence, there are still major uncertainties about the Weichselian glacial history in western Greenland. A particular problem that has important implications for the general modelling of the Weichselian ice sheets (Boulton, 1979; Denton and Hughes, 1981) is the extent and timing of the glaciation of the continental shelf.

The ice sheet of the Late Weichselian Sisimiut glacial event had a relatively restricted extent, reaching at most only to the inner shelf over much of its length, up to c. 74°N. Further north, the location of the margin depends upon the outcome of the current dispute about the date of the extensive phase of glaciation which covered the outer islands in this and adjacent areas to the north (e.g. Weidick, 1976, 1978b; Blake, 1977; England et al., 1981). The age of the Sisimiut event maximum is not closely defined but probably lies somewhere between 11-14,000 BP, the conflicting evidence about it suggesting perhaps that it was not a single spatially uniform event. The view that it was separated by a significant period of time from an earlier, more extensive glaciation depends on the interpretation of the evidence of the greater degree of weathering above the suggested ice sheet limit.

From the earlier Weichselian (> 40 ka) are known a number of marine deposits which are assigned to the Svartenhuk marine event. Their association with glacial deposits indicates the existence of ice sheets at least as large as that of the Sisimiut event. However, the current stratigraphic data do not allow either the number of independent events or their age to be established.

This leaves a large part of the Middle and Late Weichselian about which even less is known. The apparent lack of marine deposits of this age implies either that the area was completely glaciated, or that the eustatic/glacio-isostatic sea level was below the present, or a combination of both.

Evidence from the shelf itself indicates that one or more undated ice sheets extended onto it, ultimately reaching the shelf break in the Hellefisk glacial event. However, whether this represents the glaciation referred to above, or whether it should be correlated with an Early or pre-Weichselian event is not known.

The information from adjacent areas does not confirm any of the alternatives, although elements of the western Greenland stratigraphy can be recognized. One or more periods of extensive shelf glaciation is/are proposed for Baffin Island (Miller et al., 1977; Andrews, 1980) and Ellesmere Island (England et al., 1981), followed by a period of reduced glaciation and low sea level. In contrast, Funder (1982) considers that the eastern margin of the Greenland Ice Sheet extended onto the shelf after the Early Weichselian marine event (Flakkerhuk Stade), although the maximum glaciation in E. Greenland he considers to be Saalian. There are also more limited, later readvances in these areas, analogous to the Sisimiut, although not necessarily synchronous; e.g. Milne Land advance, E. Greenland, 10,400 - 9400 BP (Funder, 1978); N.E. Greenland, 10,300-9500 BP (Hjort, 1979); Late Foxe advance, Baffin Island 10,000-11,000 BP (Miller, 1980).

The amino acid data for Kaffehavn and Lakesbugt marine events, which may or may not be distinct events, allow either an Eemian (oxygen isotope stage 5) or Early Weichselian age. Deposits from both events contain a warm fauna which indicates the presence of the subarctic West Greenland Current. Although such evidence has been used to identify Eemian deposits in East Greenland (Funder, 1982), the same faunas exist in glacial marine deposits on Baffin Island for which an Early Weischselian age have been proposed (Miller et al., 1977), although this date is also being questioned (Andrews and Miller, 1983).

At least two older marine events have been distinguished on the basis of their amino acid ratios: Meteorbugt/Mudderbugt and Patorfik. These probably date from the Middle and Early Quaternary and both are probably of interglacial character.

Several features of the more recent geological history, after the Sisimiut glacial event maximum, are of interest. The development of the present 'interglacial' marine climate in the Disko Bugt episode appears to have lagged behind deglaciation and the initial transgression. Not until 9400 BP is there evidence of the existence of the subarctic West Greenland Current, but from 9100 BP it appears to have been permanently established at least to 67°N. This is generally compatible with the history of the N. Atlantic circulation reconstructed by Ruddiman et al. (1981), and more particularly for Baffin Bay and Davis Strait by Miller (1980), based on evidence from the Canadian Arctic Islands. However, the latter records an earlier date for the incursion of subarctic water to S. Baffin Island at 9725 BP. In western Greenland this was followed by the intrusion of water warmer than at present to the central coastal area (66°-69°N) by 7300 BP, lasting at least sporadically until 4800 BP.

Deglaciation during this interval eventually took the margin of the Sisimiut event ice sheet behind its present position. The extent to which this was interrupted by stillstands or readvances is debatable. However, there does seem to be evidence for several broadly defined intervals in the Holocene Atlantic and Boreal when there was a regional tendency for the margin to respond in this way.

Depending on the criterion used, the onset of the subsequent Neoglacial climate deterioration, which resulted in the readvances of the Vesterbygd glacial event, can be variously dated to 3200-4800 BP. Although it can be demonstrated clearly in places that the maximum position of the readvance was reached in AD 1700-1920, the idea that this was the general situation along the ice margin depends largely on the negative evidence of the absence of earlier dated moraines. Only one exception to this is definitely known, with the maximum dated to 350-700 BP (Orkendalen Moraine). The change in mass of the ice sheet during the Vesterbygd event had an isostatic effect, resulting in the Sandnes marine transgression. One result of this is that the records of Holocene marine faunas are artifically terminated around 4000 BP.

Weidick (1975d) has provided a conceptual model for the mass balance history of the Greenland Ice Sheet which shows a change from a large, low specific activity ice sheet in the Late Weichselian to a smaller, high activity ice sheet in the Neoglacial. The deglaciation is considered to result from the increase in accumulation lagging behind the increase in ablation. This model is compatible with our current knowledge of the glacial geology of this period.

Despite the long history of research into the Quaternary geology of western Greenland, many areas of uncertainty remain, even about the last glaciation, the Weichselian. The oxygen isotope profiles from the Greenland Ice Sheet (Dansgaard et al. 1971, 1975), which to various degrees are related to the history of the Greenland Ice Sheet, demonstrate in their complexity the gross simplification of the geological models. Thus, a considerable amount of further work remains to be done on the Quaternary geology of Western Greenland. Important areas, also, for future consideration are the correlations between the response of the Greenland Ice Sheet and nearby events in the Canadian Arctic which document glacial and marine events associated with the Laurentide Ice Sheet complex (e.g. Klassen, this volume; Miller, this volume).

ACKNOWLEDGEMENTS

I am grateful to B. Fredskild and S. Funder for the use of unpublished data, and to J. Andrews, S. Funder, H. Pinkerton, S. Watt and A. Weidick for their helpful criticism of the manuscript.

The results of work conducted under the auspices of the Geological Survey of Greenland are published by permission of the Director.

REFERENCES

Andrews, J.T., 1980: Progress in relative sea level and ice sheet reconstructions, Baffin Island, N.W.T., for the last 125,000 years. In: Morner, N.A., (ed.), Earth Rheology, Isostasy and Eustasy. New York: John Wiley and Sons, 175-200.

Andrews, J.T. and Miller, G.H., 1983: Quaternary glacial and nonglacial correlations for the eastern Canadian Arctic. Geological Survey of Canada, in press.

Bendix-Almgreen, S.E., 1976: Palaeovertebrate faunas of Greenland. In: Escher, A. and Watt, W.S. (eds.), Geology of Greenland. Copenhagen: Geological Survey of Greenland, 536-573.

Bendix-Almgreen, S.E., Fristrup, B. and Nichols, R.L., 1967: Notes on the geology and geomorphology of the Carey Øer, North-west Greenland. Meddelelser om Grönland, 164,8, 19 pp.

Beschel, R.E., 1961: Dating rock surfaces by lichen growth and itsapplication to glaciology and physiography (lichenometry). In: Raasch, G.O. (ed.), Geology of the Arctic. Toronto: Toronto University Press, 1044-1062.

Blake, W., 1975: Glacial geological investigations in northwesternGreenland. Geological Survey of Canada Paper 75-1A, 435-439. .

Blake, W., 1977: Radiocarbon age determinations fromthe Carey Islands, Northwest Greenland. Geological Survey ofCanada Paper 77-1A, 445-454.

Boulton, G., 1979: A model of Weichselian glacier variation in the North Atlantic region. Boreas, 8:373-395.

Boyer, S.J. and Pheasant, D.R., 1974: Delimitation of weatheringzones in the fjord area of eastern Baffin Island, Canada. Bulletin Geological Society of America, 85:805-810.

Brett, C.P. and Zarudzki, E.F.K., 1979: Project Westmar. A shallow marine geophysical survey on the West Greenland continental shelf. Rapport Grönlands Geologiske Undersögelse 87, 27 pp.

Bryan, M.S., 1954: Interglacial pollen spectra from Greenland. Danmarks Geologiske Undersögelse IIRk, 80:65-72.

Dansgaard, W., Johnson, S.J. Clausen, H.B. and Langway, C.C., Jr., 1971: Climatic record revealed by the Camp Century ice core. In: Turekian, K.K.(ed.), The Late Cenozoic Glacial Ages. London and New Haven: Yale University Press, 37-56.

Dansgaard, W., Johnsen, S. J., Reeh, N., Gundestrup, N. Clausen, H.B. and Hammer, C.U., 1975: Climatic changes, Norseman and modern man. Nature, 255:24-28.

Davies, W.E., Krinsley, D.B. and Nicol, A.H., 1963: Geology of theNorth Star Bugt area, northwest Greenland. Meddelelser om Grönland, 162:12. 68 p.

Dawson, A.G. 1983: Glacier dammed lake investigations in the lake area, South Greenland. Meddelelser om Gronland, Geoscience, 11, 22 p.

Denton, G.H. and Hughes, T.J. (eds.), 1981: The Last Great IceSheets. New York: John Wiley and Sons. 484 p.

Donner, J., 1978: Holocene history of the west coast of Disko, central West Greenland. Geografiska Annaler, 60A, 63-72.

Donner, J. and Junger, H., 1975: Radiocarbon dating of shells from marine Holocene deposits in the Disko Bugt area, West Greenland. Boreas, 4:25-45.

Dyke, A.S., 1976: Tors and associated weathering phenomena, Somerset Island, District of Franklin. Geological Survey of Canada Paper 76-1B, 209-216.

England, J. Bradley, R. S. and Stuckenrath, R., 1981: Multiple glaciations and marine transgressions, western Kennedy Channel, Northwest Territories, Canada. Boreas, 10:71-89.

Fredskild, B., 1973: Studies in the vegetation history of Greenland. Palaeobotanical investigations of some Holocene lake and bog deposits. Meddelelser om Grönland 198(4), 245 pp.
Fredskild, B., 1983: The Holocene vegetational development of theGodthabsfjord area, West Greenland. Meddelelser öm Grönland, Geoscience, 10, 28 p.
Fredskild, B. and Möller, M., 1981: Grönlands botaniske undersögelese 1980 og 1981. Rapport Grönlands Botaniske Undersögelse, 1981.
Funder, S., 1978: Holocene stratigraphy and vegetation history inthe Scoresby Sund area, East Greenland. Bulletin Grönlands Geologiske Undersögelese, 129, 66 pp.
Funder, S., 1979: The Quaternary geology of the Narssaq area, South Greenland. Rapport Grönlands Geologiske Undersögelse, 86, 24 pp.
Funder, S., 1982: Planterefugierne i Grönland. Naturens Verden, 7:241-255.
Funder, S. and Hjort, C., 1974: The subfossil occurrence of Mytilus edulis L. in central East Greenland. Boreas, 3:23-33.
Funder, S. and Símonarson, L.A., 1983: Bio- and aminostratigraphy of some marine Quaternary deposits in western Greenland (submitted).

Goldthwait, R.P., 1960: Study of an ice cliff in Nunatarssuaq, Greenland. U.S. Army CRREL, Technical Report 39, 108 pp.

Harder, P., Jensen, A.D. and Laursen, D., 1949: The marine Quaternary sediments in Disko Bugt. Meddelelser om Grönland, 149(1), 85 pp.
Henderson, G., Scheiner, E.J., Risum, J.B., Croxton, C.A. and Andersen, B.B., 1981: The West Greenland Basin. Memoirs Canadian Society of Petroleum Geologists, 7:399-428.
Hjort, C., 1979: Glaciation in northern East Greenland during the Late Weichselian and early Flandrian. Boreas, 8:281-296.
Holtedahl, O., 1970: On the morphology of the West Greenland shelfwith general remarks on the 'marginal channel' problem. Marine Geology, 8:155-172.

Jensen, A.S., 1942: Two new West Greenland localities for deposits from the ice age and the post-glacial warm period. Kongelige Danske Videnskabernes Selskabs Skrifter, 17(4), 35 pp.

Kelly, M., 1973: Radiocarbon dated shell samples from Nordre Strömfjord, West Greenland, with comments on models of glacio-isostatic uplift. Rapport Grönlands Geologiske Undersögelse, 59, 20 pp.
Kelly, M., 1979: Comments on the implications of new radiocarbondates from the Holsteinsborg region, West Greenland. Rapport Grönlands Geologiske Undersögelse, 95:35-42.
Kelly, M., 1980a: Preliminary investigations of the Quaternary ofMelville Bugt and Dundas, North-West Greenland. Rapport Grönlands Geologiske Undersögelse, 100:33-38.
Kelly, M., 1980b: The status of the Neoglacial in western Greenland. Rapport Grönlands Geologiske Undersögelse, 96, 24 pp.

Kelly, M., 1981: Permafrost related features in Holsteinborg district, West Greenland. Bulletin Geological Society of Denmark, 30:51-56.
Kelly, M., 1983: Review of Quaternary marine deposits in western Greenland older than the Holocene (submitted).

Laursen, D., 1944: Contributions to the Quaternary geology of northern West Greenland, especially the raised marine deposits. Meddelelser om Grønland, 135(8), 125 pp.
Laursen, D., 1950: The stratigraphy of the marine Quaternary deposits in West Greenland. Bulletin Grønlands Geologiske Undersøgelse, 2 (also Meddelelser om Grønland, 151(1), 142 pp.
Laursen, D., 1976: New contributions to the stratigraphy of the marine Pleistocene and Holocene of West Greenland. Abstracts American Quaternary Association, 4th biennial meeting, Tempe, Ariz. 144.
Lysgaard, L., 1969: Foreløbig oversigt over Grønlands klima i perioderne 1920-50, 1951-60 og 1961-65. Meddelelser Danske Meteorologiske Institut, 21, 35 p.

Mangerud, J., Andersen, S.T., Berglund, B.E. and Donner, J.J., 1974: Quaternary stratigraphy of Norden, a proposal for terminology and classification. Boreas, 3:109-128.
Mangerud, J. and Berglund, B.E., 1978: The subdivision of the Quaternary of Norden: a discussion. Boreas, 7:179-181.
Miller, G.H., 1980: Late Foxe glaciation of southern Baffin Island, N.W.T., Canada. Bulletin Geological Society of America, 91:399-405.
Miller, G.H., Andrews, J.T. and Short, S.K., 1977: The last interglacial-glacial cycle, Clyde foreland, Baffin Island, N.W.T.: stratigraphy, biostratigraphy, and chronology. Canadian Journal of Earth Science, 14:2824-2857.
Miller, G. and Hare, P.E., 1980: Amino acid geochronology: Integrity of the carbonate matrix and potential of molluscan fossils, In: Hare, P.E., Hoering, T.C. and King, K., Jr. (eds.), Biogeochemistry of amino acids, New York: John Wiley and Sons, 415-443.
Miller, G.H., Brigham, J. and Clark, P., 1982: Alteration of the total aIle/Ile ratio by different methods of sample preparation. In: Miller, G.H. (ed.), Report of current activities from the amino and geochronology laboratory, January 1981 through May 1982. Institute of Arctic and Alpine Research and Department of Geological Sciences, University of Colorado, 9-21.

Ockelmann, W.K., 1958: The zoology of East Greenland. Marine lamellibranchiata. Meddelelser om Grønland, 122, 256 pp.

Rink, H., 1852: Om den geographiske Beskaffenhed af de danske Handelsdisgrikter i Nordgrønland tilligemed En Udsigt over Nordgrønlands Geognosi. Kongelige Danske Videnskabernes Selskabs Skrifter, 5(3):37-98.
Roksandic, M.M. 1979: Geology of the continental shelf off West Greenland between 61°15'N and 64°00'N: an interpretation of sparker seismic and echosounder data. Rapport Grønlands Geologiske Undersøgelse, 92, 15 p.
Ruddiman, W.F., and McIntyre, A., 1981: The North Atlantic Ocean during the last deglaciation. Palaeogeography, Paleoclimatology, Paleoecology, 35:145-214.
Rvachev, V.D., 1964: Relief and bottom deposits of the shelf of southwestern Greenland. Deep Sea Research, 11:646-653.

Simonarson, L., 1981: Upper Pleistocene and Holocene marine deposits and faunas on the north coast of Nugssuaq, West Greenland. Bulletin Grønlands Geologiske Undersølgelse, 140, 107 pp.

Steenstrup, K.J.V., 1883: Bidrag til Kjendskab till de geognostikeog geographiske Forhold i en Del af Nord-Grönland. Meddelelser om Grönland, 4(5):173-242.
Sugden, D., 1972: Deglaciation and isostasy in the Sukkertoppen Ice Cap area, West Greenland. Arctic Alpine Research, 4:97-117.
Sugden, D. and Miller, G., 1976: Interglacial or early Wisconsinshell fragments in till on the flanks of Söndre Strömfjord, West Greenland. Arctic Alpine Research, 8:399-401.
Sugden, O.E. and Watts, S.H., 1977: Tors, felsenmeer and glaciation in northern Cumberland Peninsula, Baffin Island. Canadian Journal of Earch Science, 14:2817-2823.

Ten Brink, N.W., 1974: Glacio-isostasy: new data from West Greenland and geophysical implications. Bulletin Geological Society of America, 85:219-228.
Ten Brink, N.W., 1975: Holocene history of the Greenland Ice Sheetbased on radiocarbon-dated moraines in West Greenland. Bulletin Grönlands Geologiske Undersögelse, 113 (also Meddeleser om Grönland, 201(4), 44 pp.
Ten Brink, N.W. and Weidick, A., 1974: Greenland Ice Sheet history since the last glaciation. Quaternary Research,4:429-440.

Vibe, C., 1967: Arctic animals in relation to climtic fluctuations. Meddeleser om Grönland, 170, 227 pp.

Watts, S., 1979: Some observations on rock weathering, Cumberland Peninsula, Baffin Island. Canadian Journal of Earth Science, 16:977-983.
Weidick, A., 1963: Ice margin features in the Julianehaab district, South Greenland. Bulletin Grönlands Geologiske Undersögelse, 35 (also Meddeleser om Grönland, 165(3), 133 pp.
Weidick, A., 1968: Observations on some Holocene glacier fluctuations in West Greenland. Bulletin Grönlands Geologiske Undersögelse, 73 (also Meddeleser om Grönland, 165(6), 202 pp.
Weidick, A., 1971: Short explanation to the Quaternary Map of Greenland. Rapport Grönlands Geologiske Undersögelse, 36, 15 pp.
Weidick, A., 1972a: Holocene shore-lines and glacial stages in Greenland - an attempt at correlation. Rapport Grönlands Geologiske Undersölgelse, 41, 39 pp.
Weidick, A., 1972b: Notes on Holocene glacial events in Greenland. In: Vasari, Y., Hyärinen, H., and Hicks, S., Climatic changes in arctic areas during the last ten-thousand years, Acta Universitatis Ouluensis Series A3 geological 1), 177-204.
Weidick, A. 1972c: C-14 dating of Survey material performed in 1971. Rapport Grönlands Geologiske Undersölgelse, 45:58-67.
Weidick, A., 1973: C-14 dating of Survey material performed in1972: Rapport Grönlands Geologiske Undersögelse, 55:66-75.
Weidick, A., 1974a: Quaternary map of Greenland, 1:500000, SondreStrömfjord-Nugssuaq. Copenhagen: Geological Survey of Greenland.
Weidick, A., 1975a: A review of Quaternary investigations in Greenland. Ohio State University Institute of Polar Studies Report, 55, 161 pp.
Weidick, A., 1975b: Quaternary geology of the area between Frederikshabs Isblink and Ameralik. Rapport Grönlands Geologiske Undersögelse, 70, 22 pp.
Weidick, A., 1975c: Investigations on the Quaternary deposits inthe Godthaabsfjord area, southern West Greenland. Rapport Grönlands Geologiske Undersögelse, 75:63-65.

Weidick, A., 1975d: Estimates on the mass balance changes of theInland
   Ice since Wisconsin-Weichsel. Rapport Grönlands Geologiske
   Undersögelse, 68, 21 pp.
Weidick, A., 1975e: C-14 dating of Survey material performed in1974.
   Rapport Grönlands Geologiske Undersögelse, 75:19-20.
Weidick, A., 1976a: Glaciation and the Quaternary of Greenland. In:
   Escher, A. and Watt, W.S. (eds.), Geology of Greenland. Copenhagen:
   Geological Survey of Greenland, 430-460.
Weidick, A., 1976b: C-14 dating of Survey material carried out in1975.
   Rapport Grönlands Geologiske Undersögelse, 80:136-144.
Weidick, A., 1976c: Glaciations of northern Greenland- new
   evidence. Polarforschung, 46:26-33.
Weidick, A., 1977: C-14 dating of Survey material performed in
   1976. Rapport Grönlands Geologiske Undersögelse, 85:127-129.
Weidick, A., 1978a: Quaternary Map of Greenland Sheet 2 Frederikshabs
   Isblink - Söndre Strömfjord. Copenhagen: Geological Survey of
   Greenland.
Weidick, A., 1978b: Comments on radiocarbon dates from
   northernGreenland made during 1977. Rapport Grönlands Geologiske
   Undersölgelse, 90, 124-128.
Weidick, A., 1978c: C-14 dating of Survey material performed in 1977.
   Rapport Grönlands Geologiske Undersögelse, 90:119-124.
Weidick, A., 1982: Klima-og gletscheraendringer i det Sydlige
   Vestgronland i de sidste 1000 ar. Grönland, Nr. 5-7:235-251.

Zarudzki, E.F.K., 1980: Interpretation of shallow seismic profiles over
   the continental shelf in West Greenland between latitudes 64° and
   69°30'N Rapport Grönlands Geologiske Undersögelse, 100:58-61.

17 Pre-Holocene pollen and molluscan records from eastern Baffin Island

William N. Mode

The Quaternary history of eastern Baffin Island has consisted of glacial periods when relative sea level was higher than present (the marine limit is around 80 m asl; Andrews, 1980) separated by non-glacial (interglacial) intervals when relative sea level was near its present position. The record of the glacial intervals is found in raised, nearshore marine and glacial marine sediments which are well-exposed in coastal cliffs and which have been studied in detail on the Clyde Foreland (Feyling-Hanssen 1967, 1976, 1977; Miller et al., 1977; Mode, 1980; Mode et al., 1983), the Qivitu Peninsula (Andrews et al., 1975; Pheasant and Andrews, 1973; Nelson, 1978, 1981, 1982; Feyling-Hanssen, 1980), and Broughton Island (England and Andrews, 1973; Andrews et al., 1976; Brigham, 1980, 1983). Little sediment recording the non-glacial intervals prior to the Holocene is preserved above present sea level although lakes exist which lie beyond the limit of the last glaciation and which could contain sediment >40,000 years old.

The pollen and mollusc data reviewed here are derived primarily from sediments which were deposited during glacial isostatic submergence of coastal lowlands; the character of sedimentation in this setting has had an important influence on the fossils contained in the sediments.

The elevations of raised shorelines and the character of the sediments indicate that the coastal cliff sequences were deposited in less than 100 m of sea water (Mode et al., 1983). Disconformities, stone lines, channels, and cross-bedding are indicative of high wave and current energy in shoreface and ice-proximal environments. In these settings it is not surprising to find a predominance of disarticulated, broken, and abraded bivalves. Furthermore, such processes tend to winnow rather than concentrate the fine-grained sediments which commonly contain pollen (Stanley, 1969). Successive glacial isostatic transgressions covered similar areas, and hence there is a strong possibility that fossils have been reworked and transported stratigraphically upward.

Given the above constraints, the goals of this paper are to catalog fossil collections and assemblages and to use them to infer regional marine and terrestrial climatic conditions based primarily on comparison to the present biogeography of the same plants and animals. Such regional interpretation eliminates local transportation of fossils as a serious source of error. Reworked fossils can be distinguished by their abraded or poorly preserved appearance, but in the nearshore, shallow water sedimentary environments described above, even fossils which are contemporaneous with the sediment are typically abraded and fractured. Only in sublittoral muddy sand is there little likelihood of reworked bivalves occurring.

AMINOSTRATIGRAPHIC UNITS

Correlation of stratigraphic units in the coastal cliff sequences is difficult for several reasons. Glacial isostatic sedimentary events are represented by similar suites of lithofacies (Mode et al., 1983). Individual stratigraphic sections usually contain only two or three glacial marine diamictons, and because of numerous unconformities and the laterally discontinuous nature of the lithofacies, the diamictons are difficult to correlate between sections.

Amino acid ratios, in particular the ratios of D-alloisoleucene to L- isoleucene in the free and total amino acid fractions (Miller and Hare, 1975, 1980), have been measured in fossil bivalves from the glacial marine sequences. The appearance of discrete clusters in the frequency distributions of these ratios supports the hypothesis that sedimentation was episodic, not continuous. The modes in the frequency distributions have been informally designated as aminozones (Fig. 17.1; see Miller, this volume). Local correlation of stratigraphic units using amino acid ratios is possible because the integrated thermal history is virtually the same for all Pleistocene deposits (Miller et al., 1977; Nelson, 1978, 1982; Mode, 1980; Brigham, 1980, 1983). Regional correlation along the eastern coast of Baffin Island may produce aminozones which are somewhat time-transgressive but not greatly so, because temperatures do not range widely across the region (Andrews et al., 1981; Miller, this volume).

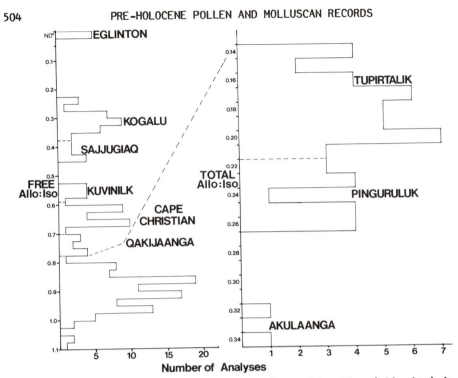

Figure 17.1 Frequency distribution of amino acid ratios (allo:iso) in
the free and total (free plus peptide-bound) amino acid fractions of
shells from the Clyde region (Mode, 1980).

In this paper, I will present the molluscan faunas and pollen
floras of these regional aminozones (Table 17.1) except for the most
recent one, the Eglinton aminozone (Holocene), which has been discussed
in some detail elsewhere (Andrews, 1972; Short et al., this volume).

PALYNOLOGY

Organic-rich sediment samples were collected from measured
stratigraphic sections near Clyde (Mode, 1980), on the Qivitu Peninsula
(Nelson, 1981), and on Broughton Island (Brigham, 1980) (Fig. 17.2).
Samples were analyzed using standard laboratory techniques (Mode, 1980)
and counted at 500x magnification.

Organic-rich strata were formed by two different proceses. A few
were in situ accumulations of organic matter in soil O- and A-horizons
formed during non-glacial or interglacial periods of emergence. They
became buried paleosols during succeeding periods of glacial isostatic

Table 17.1. Correlation of aminozones in the Clyde (left column; Mode, 1980), Qivitu (Nelson, 1982), and Broughton Island (Brigham, 1983) areas. Foraminiferal biostratigraphy is for the Clyde cliffs (Feyling-Hanssen, 1977), and the estimated ages are from Szabo et al. (1981).

AMINOZONES	MEAN ALLO/ISO (CLYDE) FREE TOTAL		FORAMINIFERAL BIOSTRATIGRAPHY	AMINOZONES QIVITU	AMINOZONES BROUGHTON	ESTIMATED AGE
EGLINTON	N.D.*	0.021		1	1	≤12,000
KOGALU	0.30	0.036		2	2	≥70,000
SAJJUGIAQ	0.42	0.051	Islandiella islandica zone	3	3	≥90,000
KUVINILK	0.55	0.065			4	≥136,000
CAPE CHRISTIAN	0.63	0.12	Cassidulina teretis zone	4	5	≥190,000
QAKIJAANGA	0.74	0.13		5		
TUPIRTALIK	0.90	0.18	Cibicides grossa zone	6	6	≥300,000
PINGURULUK	0.89	0.24				
AKULAANGA	1.00	0.33	Nonion tallahattensis zone	7		

*ALLOISOLEUCENE IS NOT DETECTABLE

submergence. The second sample type is composed of detrital organic matter, transported and winnowed before deposition.

The buried organic samples are difficult to interpret, because precise dating does not exist and sedimentation rates are unknown. Thus the pollen concentration data are not comparable, especially between buried soils and detrital organics, and interpretation must be based on the percentage data. Vegetational and climatic interpretation of fossil pollen spectra is done by comparison to modern pollen spectra (Short et al., this volume).

Twenty-nine pre-Holocene samples from the Clyde cliffs, nine from the Qivitu cliffs, and two from Broughton Island have been examined. In the pollen diagram (Fig. 17.3), the samples have been arranged in order of increasing age based on bracketing amino acid ratios (Table 17.1); this is the sequence in which they will be discussed.

Kogalu aminozone

The youngest samples (GRL-343-0 and 344-0) come from the upper sediments of the Kogalu aminozone (Qaviq Member) on the Qivitu Peninsula, and they are detrital organics. Both contain enough Betula pollen (6% and 12%, respectively) to suggest the presence of birch shrubs on the peninsula during the marine regression represented by the upper Kogalu sediments (Nelson, 1981). The present pollen rain on the

Figure 17.2 Location map. Samples discussed in this paper come from the Clyde, Qivitu Peninsula, and Broughton Island areas of eastern Baffin Island.

Qivitu Peninsula contains less than 2% Betula (Mode, 1980), and Betula percentages consistently greater than 5% occur only within the present range of shrub birch plants in the Low Arctic tundra of southern Baffin Island (Andrews et al., 1980).

Laminated detrital organic-rich sediment from the Kogalu aminozone (Harbour Member) near Broughton Island (Or 02) (Fig. 17.2) yielded a pollen assemblage dominated by Salix (willow; 89%), which is similar to

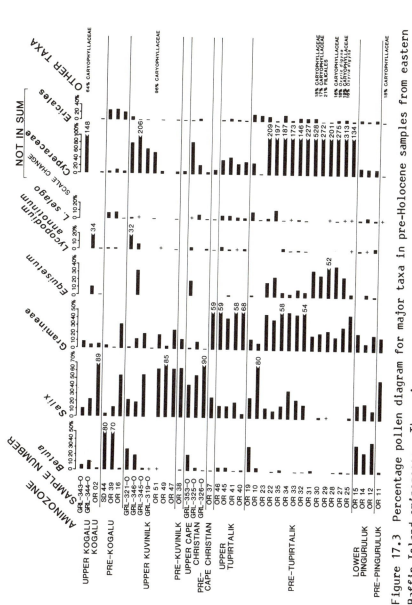

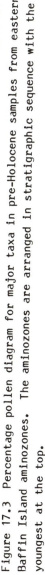

Figure 17.3 Percentage pollen diagram for major taxa in pre-Holocene samples from eastern Baffin Island aminozones. The aminozones are arranged in stratigraphic sequence with the youngest at the top.

the present pollen rain on Broughton Island (c. 50% Salix; Mode,
1980). The precise position of this sample within the Kogalu aminozone
is uncertain (Brigham, 1980 p. 137), but the well-sorted sands in which
it was contained are probably regressive deposits. Therefore, the
sample would be from the upper sediments of the Kogalu aminozone.

Last (pre-Kogalu) interglaciation

Three samples are buried A-horizons of paleosols separating Kogalu
aminozone sediments from underlying sediments. Two of these contain
very high Betula percentages (SD44 from Broughton Island and Or 39 from
Clyde) suggesting that the last (pre-Kogalu) interglaciation was
substantially warmer than present throughout eastern Baffin Island.
Similar, Betula-dominated pollen assemblages have been reported for
supposed last (Flitaway) interglacial beds near the northwest margin of
the Barnes Ice Cap in north-central Baffin Island (Terasmae et al.,
1966).

Sajjugiaq aminozone

No pollen-bearing sediments have been collected from the Sajjugiaq
(Cape Broughton of Miller, this volume) aminozone.

Kuvinilk aminozone

Four detrital organic samples (Or 47, 49, 51; GRL-345-0) from the upper
beds of the Kuvinilk aminozone have pollen assemblages rich in Salix
and similar to those of the modern pollen rain. These assemblages are
similar to those reported by Miller et al. (1977) for Kuvinilk
aminozone samples, though one sample (0-7; p. 2844) was dominated by
Gramineae (grass). Two samples (GRL-321-0 and -346-0) have high Betula
content (28% and 20% respectively). Both are from the Oivitu Peninsula
(Uivaruluk Member) and suggest that terrestrial conditions were again
warmer than present during the deposition of regressive marine
sediments, as was the case for the Kogalu aminozone.

Pre-Kuvinilk interglaciation

Two buried soil samples represent the pre-Kuvinilk interglaciation.
The pollen assemblage of Or 38 (from Clyde) is dominated by Salix (65%)
and Gramineae (13%), which is comparable to modern pollen samples from
Clyde (Salix averages 30% and Gramineae 28%; Mode, 1980). The Cape

Christian buried soil (Miller et al., 1977) also represents the
pre-kuvinilk interglacial at Clyde (sample O-8, p. 2833-2834), and it
contains over 50% Betula. Miller et al. (1977) correlated this sample
with the Flitaway Interglacial samples of Terasmae et al. (1966).

Cape Christian aminozone:

Of three samples from the upper sediments of the Cape Christian
aminozone (all from the Qivitu Peninsula), one contains 19% Betula
(GRL- 353-0) and may correlate with the Cape Christian buried organics
at Clyde (Miller et al., 1977, sample O-5, p. 2834) which contain over
50% Betula (p. 2844). This correlation may not be exact because the
samples from the Qivitu Peninsula are detrital organic-rich marine
sediments, whereas the Cape Christian organics may be from a buried
soil. However, the actual age difference is probably minimal. Betula
reached Qivitu and Clyde before the sea level had dropped to its
interglacial level near present sea level.

Pre-Cape Christian aminozones:

Betula apparently did not reach Clyde during the pre-Cape Christian
interglaciation (Or 37) or the period of deposition of the upper beds
of the Tupirtalik aminozone. No pollen samples of this or older age
have been recovered from Broughton Island or the Qivitu Peninsula.
Betula did reach Clyde during the pre-Tupirtalik interglacial, because
sample Or 19 contains 33% Betula (present Betula content of surface
samples around Clyde averages <1%; Mode, 1980). Samples Or 22-35 are
also from the pre-Tupirtalik interglaciation at Clyde, but their
stratigraphic position relative to Or 19 is uncertain. They were
collected from a single peat section and the sequence of their pollen
spectra shows a succession from pioneer plants which dominate the lower
horizons (Equisetum, Oxyria digyna, and Caryophyllaceae [pinks]) to
plants typical of somewhat stable sites (Gramineae and Salix) in the
upper horizons. Thus, the peat section probably pre-dates Or 19, which
probably represents vegetation in an advanced successional stage.

The only pollen samples from the lower part of any aminozone are
those from the lower Pinguruluk aminozone at Clyde (Or 12-15). Betula
percentages in those three detrital organic samples range from 27-41%
and suggest the presence of birch shrubs at Clyde during the
glacial-marine transgression of the Pinguruluk aminozone. Thus, this
lowest sequence of pollen samples apparently forms a different pattern
from those it underlies, because in those (beginning with the upper

Cape Christian aminozone sediments), <u>Betula</u> is present during marine regressions and succeeding interglacials.

MOLLUSCS

The molluscs in raised marine deposits provided the material needed for the aminostratigraphy. Whenever possible, large shell collections were made in order to permit characterization of mollusc faunal assemblages. Holocene mollusc assemblages of eastern Baffin Island have been discussed by Andrews (1972), and Andrews et al. (1981) characterized the assemblages of several pre-Holocene aminozones of eastern Baffin Island. Miller et al. (1977) reported mollusc assemblages and molluscan biostratigraphy from the Clyde cliffs. Following Andrews (1972), the main interpretation of these assemblages was in terms of the presence or absence of subarctic marine surface water. The repeated association of these subarctic mollusc-bearing deposits with glacial marine deposits suggests a marine climate warmer than present during glaciations of eastern Baffin Island. The warm water events are termed aquatherms (Andrews et al., 1981).

The molluscs are categorized by their present distributions as arctic, subarctic, or panarctic (distribution spans arctic and subarctic waters) as reported by several workers (Lubinsky 1980; Clarke 1974; MacPherson 1971; and Ellis 1960).

<u>Hiatella arctica</u>, a panarctic species, is by far the most common species, as it was present in 65 of the 66 collections at Clyde (Table 17.2). <u>Mya truncata</u>, <u>Balanus</u> spp., <u>Astarte</u> spp., and <u>Serripes groenlandicus</u> were also common and all are considered panarctic in distribution. <u>Astarte striata</u>, <u>Mya pseudoarenarica</u>, and <u>Chlamys islandica</u> are the three most common subarctic mollusks. Deposits of all aminozones except the Qakijaanga and Akulaanga (Fig. 17.1) contain at least one subarctic species. Because each aminozone is related to glaciation of the foreland, and because Subarctic molluscs were often found within glacial marine deposits, it is apparent that glaciers reached the outer coast during periods when surface waters warmer than present occupied western Baffin Bay (Andrews et al., 1981).

Because of the variable number of shell collections representing different aminozones it is not easy to make comparisons between faunal assemblages. In addition, it is uncertain whether all possible biofacies are represented. Given past water depths and sediment types the number of possible biofacies is somewhat limited, but when only a

TABLE 17.2. Frequency of fossils in 66 shell collections, Clyde
region.
===

Species*	Raw Frequency	Percentage Frequency
1. Hiatella arctica	65	98
2. Mya truncata	25	38
3. Balanus spp.	20	30
4. Astarte spp.	19	29
5. Serripes groenlandicus	19	29
6. Macoma calcarea	11	17
7. Astarte borealis	8	12
8. Astarte striata	8	12
9. Clinocardium ciliatum	7	11
10. Mya pseudoarenaria	6	9
11. Chlamys islandica	5	8
12. Bryozoa	4	6
13. Acmaea testudinalis	4	6
14. Astarte elliptica	3	5
15. Astarte montagui	3	5
16. Musculus discors	3	5
17. Hemithyris psittacae	3	5
18. Mytilus edulis	2	3
19. Delectopecten groelandicus	2	3
20. Echinoid plates	2	3
21. Macoma balthica	2	3
22. Bathyarca glacialis	1	2
23. Cyrtodaria siliqua	1	2
24. Portlandia arctica	1	2
25. Yoldia hyperboria	1	2
26. Yoldiella lenticula	1	2
27. Colus cf. pubsecens	1	2

*Identifications were made with the assistance of Drs. J.T. Andrews and
G.H. Miller, and Dr. F. Wagner of Geological Survey of Canada
generously contributed several identifications.

few shell collections comprise the data base for an aminozone it is
quite probable that all potential biofacies have not been sampled.

The faunal diversity of each aminozone is indicated by the number
of taxa (of various levels of identification) (Table 17.3). The upper
beds of the Kogalu aminozone have the greatest diversity with 27 taxa.
Other units with more than 10 taxa are the lower Kogalu, the middle
Kavinilk, the entire Cape Christian, and the middle Pinguruluk.

The number of subarctic taxa correlates with the diversity as
shown by the occurrence of nine subarctic species in the upper Kogalu
aminozone beds and four in the middle Sajjugiaq and the lower Cape
Christian beds.

TABLE 17.3. Aminostratigraphic distribution of marine macrofossils, eastern Baffin Island.

Zoogeographic Zone* AMINOZONES	Hiatella arctica (P)	Mya truncata (P)	Balanus spp. (P)	Serripes groenlandicus (P)	Macoma calcarea (P)	Astarte borealis (P)	Clinocardium ciliatum (P)	Astarte striata (S)	A. subaequilatera (S)	A. montagui (S)	A. elliptica (S)	Macoma balthica (S)	Chlamys islandica (S)	Mytilus edulis (S)	Mya pseudoarenarea (S)	Colus pubescens (S)	C. spitzbergensis (S)	Number of other taxa	Total number of taxa
KOGALU upper	X	X	X	X	X	X	X	X	X	X	X	X	X		X	X	X	11	27
KOGALU middle	X	X	X	X	X	X							X					3	10
KOGALU lower	X	X	X	X	X	X		X					X					10	18
SAJJUGIAQ upper	X	X		X														0	3
SAJJUGIAQ middle	X	X		X	X	X	X	X		X			X		X		X	11	22
SAJJUGIAQ lower	X		X															1	3
KUVINILK upper	X	X			X			X	X				X					0	6
KUVINILK middle	X	X	X		X	X		X				X	X					4	12
KUVINILK lower	X																	0	1
CAPE CHRISTIAN upper	X	X	X	X			X	X		X			X		X			2	11
CAPE CHRISTIAN middle	X	X	X	X		X	X	X						X				2	10
CAPE CHRISTIAN lower	X	X	X	X	X	X	X						X	X	X			3	13
QAKIJAANGA upper	X	X	X															2	5
QAKIJAANGA middle	X	X																1	3
QAKIJAANGA lower	X	X		X														2	5
TUPIRTALIK upper	X	X																1	4
TUPIRTALIK middle	X	X	X					X				X						4	9
TUPIRTALIK lower	X																	0	1
PINGURULUK upper	X	X	X	X	X	X												3	9
PINGURULUK middle	X	X	X	X	X	X	X	X							X			8	15
PINGURULUK lower	X	X	X		X	X		X										1	7
AKULAANGA	X						X			X	X							1	4

*P = Panarctic (Arctic-Subarctic), S = Subarctic, X = species present

Feyling-Hanssen's (1976, 1977) Islandiella islandica subzone, which has amino acid ratios spanning the Kogalu and Kuninilk aminozones (Table 17.1), has very high foraminifera faunal diversity and has been interpreted as interstadial or interglacial in character. This subzone probably represents surface water warmer than present in western Baffin bay. This supports the warm water implications of the macrofossils of the Kogalu, Sajjugiaq, and Kuvinilk aminozones.

The upper unit of the Kogalu aminozone has a mollusc assemblage indicative of conditions considerably warmer than present. This unit may correlate with foraminiferal zone c of Aksu and Piper (1979) which is associated with an ash zone dated at 65,000 years BP (Ruddiman and Glover, 1972, 1975).

The Cassidulina teretis subzone also has very high foraminiferal diversity indicative of conditions warmer than present (Feyling-Hanssen, 1976, 1977). This subzone has amino acid ratios which would place it within the Cape Christian aminozone (Table 17.1), and thus the foraminiferal data support the macrofossil data for this aminozone.

Foraminiferal zones underlying the Cassidulina teretis subzone have also been interpreted as representing interglacial or interstadial conditions, but the overall faunal diversity of these zones is lower than the overlying two subzones. This diversity pattern is mirrored in the macrofossils, as the aminozones underlying the Cape Christian have somewhat lower diversity and far fewer Subarctic species.

No molluscan biostratigraphic zonation is suggested in this paper, but Miller et al. (1977) suggested a Mytilus edulis zone of mid-Holocene age and a Chlamys islandica zone approximately coeval with the Islandiella islandica subzone of Feyling-Hanssen (1976, 1977). Mytilus edulis is now known from the Cape Christian aminozone on Broughton Island (Brigham, 1983). Chlamys islandica has also been found in Cape Christian aminozone sediments at Clyde (Mode, 1980) which correlate with Feyling-Hanssen's (1977) Cassidulina teretis zone. Therefore, the molluscan biostratigraphic zones in the Clyde cliffs (Miller et al., 1977) do not include the complete ranges of the species involved. Because the aminozones have most macrofossils in common, presence/absence data are not very useful for biostratigraphic zonation. Quantitative studies of mollusc assemblages might provide a better basis for biostratigraphic zonation, but these have not yet been undertaken.

PALEOCLIMATE

The climatic conditions associated with glaciation of eastern Baffin
Island during the Quaternary are an important unknown. Andrews et al.
(1981) suggested that Baffin Bay was filled with relatively warm
subarctic surface water during past glaciations. This hypothesis was
supported by the marine mollusc assemblages from Broughton Island, the
Qivitu Peninsula, and Clyde. In addition, each of Feyling-Hanssen's
foraminiferal zones from Clyde cliffs suggested episodes of marine
conditions warmer than present (Feyling-Hanssen, 1976, 1977, 1980).   In
cores from Baffin Bay, Aksu and Piper (1979) described two zones (b and
c) of planktonic foraminiferal assemblages indicative of subarctic
surface water.

     Figure 17.4 is a plot of mollusc assemblage and pollen data for
the eight pre-Holocene aminozones of eastern Baffin Island. Each
aminozone has been divided into lower, middle, and upper parts based on
lithofacies; the lower unit is transgressive, the middle represents a
glacial maximum, and the upper is regressive (Mode et al., 1983).
Between each aminozone, pollen data from interglacial soils are
plotted. The data are plotted so that points close to the vertical
axis represent faunas and floras representative of cool conditions,
approximately like the present; whereas points farther to the right
signify conditions warmer than present.

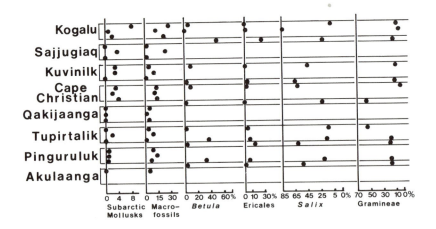

Figure 17.4 Marine and terrestrial paleoclimate indicators.
Relative warmth indicated by points increases (within each column)
toward the right. Each pollen spectrum represents the average
pollen percentages for all samples of each stratigraphic unit
(Fig. 17.3).

Four different interglacial pollen assemblages are shown from soils developed prior to deposition of the Kogalu, Kuvinilk, Cape Christian, and Pinguruluk aminozones. The presence of birch shrubs at least as far north as Clyde indicates a warm interglacial (terrestrial) climate during the pre-Kogalu and pre-Kuvinilk (Miller et al., 1977) intervals. Evidence for such conditions has not been found in the other two interglacial soils. As glaciers advanced, it is probable that the terrestrial climate remained warm (at least during the early phase of Pinguruluk aminozone glacial isostatic submergence). At the same time the marine climate during the glacial isostatically controlled transgressions was also warmer than present in at least the three aminozones which contain subarctic molluscs. Only one pollen sample comes from supposed glacial maximum sediments (Kogalu), and it suggests conditions comparable to present. The marine climate was usually warmer than present during the glacial maxima because six of the eight aminozones contain subarctic species in association with ice-proximal facies. As glaciers retreated, glacial isostatic rebound occurred, the sea regressed, and the marine climate apparently remained warmer than present, at least during the sedimentation of four aminozones. This is particularly true of the interval represented by the upper beds of the Kogalu aminozone which contain nine subarctic species. Terrestrial temperatures increased during regressions to permit birch to extend its range northward, at least as far as the Qivitu Peninsula, as shown by high proportions of Betula pollen in the upper beds of the Cape Christian, Kuvinilk, and Kogalu aminozones.

These diverse pieces of paleoclimatic information are difficult to assemble into a coherent picture, and the absence of pollen data from several aminozones limits generalization. However, comparison of these data with the better-known Holocene sequence may be helpful. Andrews (1972) showed that subarctic molluscs were present along eastern Baffin Island by about 8500 BP and they did not retreat until after 3500 BP. However, the terrestrial warming apparently occurred by at least 9500-9000 BP (Short et al., this volume), which is before the late glacial Baffinland moraine system of the Cockburn Substage had begun to form (9000-8000 BP; Andrews and Ives, 1978). The length and climatic record of the interval between the time of deposition of the Kogalu aminozone and the earliest Holocene are unknown. No buried organics have yet been reported from the middle Loks Land Member (Miller, this volume), and this unit has yet to be described micro- or macro-paleontologically. Apparently, relative sea level was similar to or lower than present (Miller et al., 1977), and climate was probably colder than present, at least during the interval 60,000-13,000 yr BP (Dansgaard et al., 1971). Holocene terrestrial climatic cooling preceded the marine cooling by at least 1000 years, with the initiation of deposition of the Scott Inlet Eolian Sands about 4500 BP (Miller et al., 1977), and possibly by as much as 2000 years with the retreat of _Betula_ from the Clyde region (ca. 5600 BP; Short et al., this volume). Furthermore, until 2500 BP, subarctic molluscs remained along southeastern Baffin Island (Andrews, 1972), but on the other island early Neoglacial advances had already occurred (ca. 3500 BP, Miller, 1973). A similar pattern of marine climate along West Greenland occurred with subarctic molluscs present by at least 8500 years BP, followed by an undated marine hypsithermal during which Boreal molluscs reached the area (Laursen, 1950; Weidick, 1975; see Kelly, this volume).

During the Holocene, terrestrial conditions became warmer than present earlier than did marine conditions, and post-Hypsithermal cooling is also recorded earlier in the terrestrial record. The first occurrence of marine waters warmer than present is at about the same time as the termination of the Cockburn glacial advance. This is a relationship seen in several pre-Holocene aminozones where the lower beds lack subarctic molluscs, but the middle and/or upper beds contain these warm water indicators. Post-Cockburn deglaciation was accompanied by warmer than present terrestrial and marine conditions; this pattern is also seen in several aminozones. Finally, there were long intervals when sea level was lower than present, and climate was colder than present, but for which there is no known record above sea level. These are periods for which there is no Holocene analog.

ACKNOWLEDGEMENTS

This paper comprises a portion of my dissertation, supervised by J.T. Andrews, to whom I am greatly indebted. Cathy Mode provided field assistance, and the generosity of the people of Clyde River, Baffin Island, helped make field seasons enjoyable. G.H. Miller, H. Nichols, and S.K. Short advised me on many topics. I also benefitted from discussions with a number of colleagues, including A.R. Nelson, P.T. Davis, W.D. McCoy, and J.K. Brigham. R. Stuckenrath, W. Blake, Jr., and R. Kihl generously provided laboratory analyses. B.K. McKnight read a draft of the paper and provided helpful suggestions. Support was from NSF Grants GA-41562, EAR-74-01857, and EAR-77-24555, and the Arctic Institute of North America, the Explorer's Club, and Sigma Xi.

REFERENCES

Aksu, A.E. and D.J.W. Piper, 1979: Baffin Bay in the past 100,000 yr. _Geology_, 7:245-248.

Andrews, J.T., 1972: Recent and fossil growth rates of marine bivalves, Canadian Arctic, and Late Quaternary arctic marine environments. Palaeogeography, Palaeoclimatology, Palaeoecology, 11:157-176.

Andrews, J.T., 1980: Progress in relative sea level and ice sheet reconstructions, Baffin Island, N.W.T., for the last 125,000 years, In: Mörner, N.A. (ed.), Earth rheology, isostasy, and eustasy: Wiley, New York, 175-200.

Andrews, J.T. and J.D. Ives, 1978: "Cockburn" nomenclature and the late Quaternary history of the eastern Canadian Arctic. Arctic and Alpine Research, 10:617-633.

Andrews, J.T., Szabo, B.J., and B. Isherwood, 1975: Multiple tills, radiometric ages, and assessment of the Wisconsin glaciation in eastern Baffin Island, N.W.T, Canada: a progress report. Arctic and Alpine Research, 7:39-59.

Andrews, J.T., Feyling-Hanssen, R.W., Hare, P.E., Miller, G.H., Schluchter, C., Stuiver, M., and B.J. Szabo, 1976: Alternative models of early and mid-Wisconsin events, Broughton Island, N.W.T.: toward a Quaternary chronology. International Union of Geological Sciences/UNESCO, International Geological Correlation Program, Project 73/1/24, Bellingham, Prague, 12-61.

Andrews, J.T., Mode, W.N., Webber, P.J., Miller, G.H., and J.D. Jacobs, 1980: Report on the distribution of dwarf birches and present pollen rain, Baffin Island, N.W.T., Canada. Arctic, 33:50-58.

Andrews, J.T., Miller, G.H., Nelson, A.R., Mode, W.N., and W.W. Locke, III, 1981: Quaternary near-shore environments on eastern Baffin Island, N.W.T., In: Mahaney, W.C. (ed.), Quaternary paleoclimate. Geo Books, Norwich, England, 13-44.

Brigham, J.K., 1980: Stratigraphy, amino acid geochronology, and genesis of Quaternary sediments, Broughton Island, E. Baffin Island, Canada. M.S. thesis, University of Colorado, Boulder. 199 pp.

Brigham, J.K., 1983: Stratigraphy, amino acid geochronology and correlation of Quaternary sea-level and glacial events, Broughton Island, arctic Canada. Canadian Journal of Earth Sciences, 20:577-598.

Clarke, A.H., 1974: Molluscs from Baffin Bay and the northern North Atlantic Ocean. National Museums of Canada, Publication in Biological Oceanography, No. 7, Ottawa, 23 pp.

Dansgaard, W., Johnson, S.J., Clausen, H.B., and C.C. Langway, Jr., 1971: Climatic record revealed by the Camp Century ice core. In: Turekian, K.K. (ed.), Late Cenozoic glacial ages. Yale University Press, New Haven, Connecticut, 37-56.

Ellis, D.V., 1960: Marine infaunal benthos in Arctic North America: Arctic Institute of North America Annual Report, Technical Paper No. 5, 53 pp.

England, J.H., and J.T. Andrews, 1973: Broughton Island--a reference area for Wisconsin and Holocene chronology and sea level changes on eastern Baffin Island. Boreas, 2:17-32.

Feyling-Hanssen, R.W., 1967: The Clyde foreland. In: Löken, O.H. (ed.), Field report, north-central Baffin Island (mimeo). Department of Energy, Mines, and Resources, Ottawa, 35-55.

Feyling-Hanssen, R.W., 1976: The stratigraphy of the Quaternary Clyde Foreland Formation, Baffin Island, illustrated by the distribution of benthic foraminifera. Boreas, 5:77-94.

Feyling-Hanssen, R.W., 1977: The Clyde Foreland Formation; a
    micropaleontologic study of Quaternary stratigraphy. Maritime
    Sediments, Special Publication No. 10, 315-377.
Feyling-Hanssen, R.W., 1980: Microbiostratigraphy of young Cenozoic
    marine deposits of the Qivitu Peninsula, Baffin Island. Marine
    Micropaleontology, 5:153-184.

Laursen, D., 1950: The stratigraphy of the marine Quaternary deposits
    in West Greenland. Meddelelser om Grönland, Bd. 151, nr. 1, 142 p.
Lubinski, I., 1980: Marine bivalve molluscs of the Canadian central and
    eastern Arctic: faunal composition and zoogeography. Canadian
    Bulletin of Fisheries and Aquatic Sciences, Bulletin 207, 111 p.

MacPherson, E., 1971: The marine molluscs of arctic Canada: National
    Museum of Natural Science Publications in Biological Oceanography,
    No. 3, National Museums of Canada, Ottawa, 120 pp.
Miller, G.H., 1973: Late Quaternary glacial and climatic history of
    northern Cumberland Peninsula, Baffin Island, N.W.T., Canada:
    Quaternary Research, 3:561-583.
Miller, G.H. and P.E. Hare, 1975: Use of amino acid reactions in some
    Arctic marine fossils as stratigraphic and geochronological
    indicators. Carnegie Institute of Washington Yearbook, 74:612-617.
Miller, G.H. and P.E. Hare, 1980: Amino acid geochronology: integrity
    of the carbonate matrix and potential of molluscan fossils, In: Hare,
    P.E., Hoering, T.C., and K. King, Jr. (eds.), Biogeochemistry of
    amino acids. John Wiley and Sons, New York, 415-443.
Miller, G.H., Andrews, J.T., and S.K. Short, 1977: The last
    interglacial/glacial cycle, Clyde foreland, Baffin Island, N.W.T.:
    stratigraphy, biostratigraphy, and chronology. Canadian Journal of
    Earth Sciences, 14:2824-2857.
Mode, W.N., 1980: Quaternary stratigraphy and palynology of the Clyde
    foreland, Baffin Island, N.W.T, Canada. Ph.D. thesis, University of
    Colorado, Boulder. 219 pp.
Mode, W.N., Nelson, A.R., and J.K. Brigham, 1983: A facies model of
    Quaternary glacial-marine cyclic sedimentation along eastern Baffin
    Island, Canada. In: Molnia, B.F. (ed.), Glacial-marine
    Sedimentation. Plenum, New York, 495-534.

Nelson, A.R., 1978: Quaternary glacial and marine stratigraphy of the
    Qivitu Peninsula, northern Cumberland Peninsula, Baffin Island,
    Canada. Ph.D. thesis, University of Colorado, Boulder. 298 pp.
Nelson, A.R., 1981: Quaternary glacial and marine stratigraphy of the
    Qivitu Peninsula, northern Cumberland Peninsula, Baffin Island,
    Canada. Geological Society of America Bulletin, Part II,
    92:1143-1261.
Nelson, A.R., 1982: Aminostratigraphy of Quaternary marine and
    glaciomarine sediments, Qivitu Peninisula, Baffin Island. Canadian
    Journal of Earth Sciences, 19:945-961.

Pheasant, D.R. and J.T. Andrews, 1973: Wisconsin glacial chronology and
    relative sea level movements, Narpaing Fiord-Broughton Island area,
    eastern Baffin Island, N.W.T. Canadian Journal of Earth Sciences,
    10:1621-1641.

Stanley, E.A., 1969: Marine palynology. Annual Review of Oceanography
    and Marine Biology, 7:277-292.
Szabo, B.J., Miller, G.H., Andrews, J.T., and M. Stuiver, 1981:
    Comparison of uranium-series, radiocarbon, and amino acid data from
    marine molluscs, Baffin Island, arctic Canada. Geology, 9:451-457.

Terasmae, J., Webber, P.J., and J.T. Andrews, 1966: A study of late
   Quaternary plant bearing beds in north-central Baffin Island,
   Canada. Arctic, 19:296-318.

Weidick, A., 1975: A review of Quaternary investigations in Greenland.
   Institute of Polar Studies Report No. 55, Ohio State University,
   Columbus, 161 pp.

18 Late and mid-Foxe glaciation of southern Baffin Island

L. E. Osterman, G. H. Miller, and J. A. Stravers

Southern Baffin Island consists of Frobisher Bay and the two adjacent land masses, Hall Peninsula to the north and Meta Incognita Peninsula to the south. Frobisher Bay is a half-graben with the fault escarpment lying along the coast of Meta Incognita Peninsula. Water depths in excess of 600 m occur along the narrow fault-escarpment trough, but shallower water depths (< 200 m) occur over the Calanus Shelf off Hall Peninsula (Fig. 18.1). Inner Frobisher Bay, less than 100 m deep in most places, is separated from outer Frobisher Bay by a group of islands. A sill with a water depth of 300 m separates outer Frobisher Bay from the continental shelf of the Labrador Sea.

The present tidal range of 10 to 13 m in Frobisher Bay is one of the highest in the world. As a result, the water of Frobisher Bay is well-mixed, and lacks fiord-type water stratification. The dominant water mass is an offshoot of the coastal, cold-water Baffin Current. This frigid (T < 0°C), low-salinity (S < 34°/₀₀ current originates as the surface water of the Arctic Ocean (Coachman and Aagard, 1974), which flows through the Queen Elizabeth Islands and southward along the coast of Baffin Island eventually forming the Labrador Current.

The bedrock of southern Baffin Island consists of widely occurring and diverse groups of migmitites, gneissic granites and quartz-feldspar gneisses of Precambrian age (Blackadar, 1967). Several outcrops of Paleozoic (Upper Ordovician?) limestone occur in the Foxe Basin lowland, and as small outliers at the head of Frobisher Bay. Extensive submarine limestone outcrops have also been mapped by MacLean et al. 1977) and MacLean and Falconer (1979) in Frobisher Bay and on the continental shelf of southern Baffin Island (MacLean, this volume, Chapter 6).

Reconstruction of the glacial history of southern Baffin Island reveals the complex interaction between three ice sources: a local Meta Incognita ice center, a Frobisher Bay outlet glacier from a regional Amadajuak Lake center, and a foreign Labrador/Ungava-based ice center. The late glacial history of adjacent Hall Peninsula is treated by Miller (this volume, Chapter 19).

The marine record of the glaciation of Frobisher Bay has been derived through the study of foraminifers and sediments in three piston cores (Fig. 18.1); cores HU77-159 (9.69 m) and HU77-157 (1.3 m) are located in the deep trough at 570 m and 497 m water depth, whereas core HU77-156 (2.6 m) is located on the continental shelf near Resolution Island at 487 m water depth. Radiocarbon dating of the cores indicates they span the last 20-30 ka in HU77-156, > 78 to 12 ka in HU77-159 (Table 18.1).

PRE-LATE FOXE GLACIATIONS

The record of glacial events prior to the last glacial maximum comes from the terrestrial depositional record in regions not subsequently glaciated, and from the more continuous record of marine cores in the outer reaches of the bay. Elevational zonations based on differences in diagnostic parameters related to rock weathering (bedrock and erratic clast weathering and soil development) have been noted by several researchers working on the peninsulas fronting the outer half of Frobisher Bay (Mercer, 1956; Miller, 1980; Müller, 1980; Stravers, unpublished data), whereas the inner reaches are characterized by only limited weathering features at all elevations (Colvill, 1982; Lind, 1983). Ascribing absolute ages to the weathering zones mapped in the mountainous regions fronting outer Frobisher Bay is hampered by the lack of an independent dating scheme. Nevertheless, the mature periglacial landscape mantling the highest summits of the Everett Mountains has led to the conclusion that these upland surfaces have not been inundated by actively eroding glacial ice during most of the Quaternary (i.e. at least 10^6 years). In addition, the presence of exceptionally mature cirque forms in the Everett Mountains, including many with floors below present sea level, argues that the glacial history of the outer half of Frobisher Bay has been dominated by local glacial and periglacial processes rather than by an ice stream draining a continental ice sheet centered to the west.

A very strong weathering break, separating highly weathered till with a deep soil profile from substantially less-weathered till, has been noted by several researchers occurring between 350 and 450 m aht

on the outer portions of Hall and Meta Incognita Peninsulas (Grinnell
Glacier area, Stravers unpublished; Terra Nivea Ice Cap, Müller, 1980;
Gold Cove area, Miller, 1980; inner Cornelius Grinnell Bay, Miller,
1980). This weathering break may separate deposits of the Foxe
Glaciation from the higher and older deposits of pre-Foxe glaciations.
Deposits of the Foxe Glaciation, differentiated from older deposits by
their limited weathering and soil characteristics, contain evidence for
at least three discrete glacial advances of early, middle, and late
Foxe ages.

Evidence for an early Foxe advance is found on Allen Island,
eastern Hall Peninsula (Fig. 18.1) where a recessional ice-contact
delta 37 m aht and lying 25 km beyond the late Foxe terminal position
contains a diverse mollusc assemblage (Miller, 1980) including Chlamys
islandicus (Müller). Amino acid analyses of three Hiatella arctica
(L.) valves from this deposit gave mean aIle/Ile ratios of 0.043
(Total) and 0.30 (Free) (AAL-650). These ratios suggest a correlation
with the Kogalu aminozone, and that the associated glacial event is
correlative with the early Foxe Ayr Lake till at Clyde River (Miller,
et al., 1977; Miller, this volume, Chapter 14).

A younger, mid Foxe event, is recorded at several sites generally
as a limestone-bearing till above or beyond the limits of late Foxe
ice. The oldest radiocarbon dates that constrain the timing of this
glacial event on southeastern Baffin Island come from the Queen
Elizabeth Foreland, eastern Loks Land (Fig. 18.2; Table 18.1), where a
broad strandflat is mantled by carbonate-rich glacial-marine sediment
deposited during deglaciation from the last glacial advance to reach
this area. In one locality a small sandy delta is superimposed over
the glacial-marine sediment. The delta retains its original morphology
and structure, hence has probably not been overridden by a subsequent
ice advance. In situ, paired valves of Macoma calcarea (Gmelin)
collected from this delta gave a finite radiocarbon date 41,900
+7100/-3700 (QC-446, Fig. 18.2, Table 18.1). Although the date is
finite at one standard deviation, it is infinite at the 95% confidence
level and is probably a minimum age. The entire collection was
consumed in the dating process and no amino acid data are available.
However, a second sample consisting exclusively of whole, unabraded but
not paired valves of H. arctica collected from the same delta was
submitted for both C-14 dating and amino acid analyses. The C-14 date
on this sample is also finite (32,500 +2400/-1800 (GX-8951; Table
18.1); but again at the 95% confidence level the date is infinite.
Amino acid analyses (Table 18.2) (aIle/Ile) on three separate valves of
H. arctica averaged 0.027 (Total) and 0.14 (Free) substantially lower
than the early Foxe Kogalu member of north-central Baffin Island and

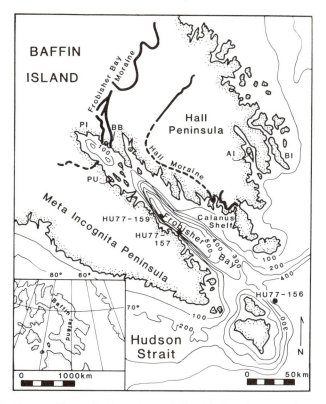

Figure 18.1 Generalized bathymetry of Frobisher Bay, and surrounding
Continental Shelf, along with the location of the marine cores
(HU77-156, 157, and 159), and the Hall and Frobisher Bay moraines.
Also shown are the locations of Peterhead Inlet (PI), Burton Bay (BB),
Pugh Island (PI), Allen Island (AI), and Breevort Island (BI).

have been used to define the Loks Land Meadow of the Clyde Foreland
Formation. Amino acid age estimates are between 30 and 70 ka (Miller,
this volume).

     Additional data from Meta Incognita Peninsula help reconstruct the
mid-Foxe glacial events. A sample of shell fragments collected from
till 170 m aht inland from Nannuk Harbour on southeastern Meta
Incognita Peninsula (Fig. 18.2) was C-14 dated at greater than 37,000
years (GX-8942, Table 18.1). Isoleucine epimerization ratios from
three H. arctica valves in the same collection (Table 18.2) are similar
to those in the dated Loks Land collection, suggesting an ice-free
Hudson Strait during a mid-Foxe interstadial. Blake (1966) also
reports several "old" shell dates from the south coast of Meta
Incognita Peninsula, all but one of which were collected from till

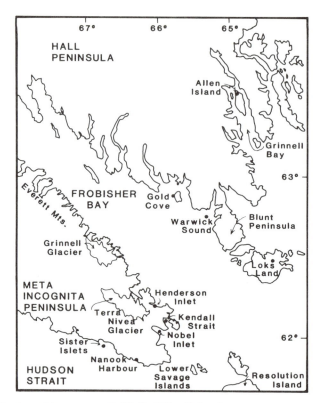

Figure 18.2   Location map of SE Baffin Island.

units and cannot be used to constrain the age of the glacial advance
that deposited the till.

Along the Frobisher Bay coast of Meta Incognita Peninsula Stravers
(unpublished) has identified a weathered, limestone-rich till that lies
above and/or beyond the limits of the late Foxe till. Fragmented
mollusc shells from this till yield aIle/Ile ratios similar to those
from Loks Land (Table 18.2), suggesting a mid Foxe advance of the
Frobisher Bay outlet glacier of greater extent than the late Foxe
maximum. Furthermore, striation evidence suggests that an independent
Meta Incognita ice center advanced after withdrawal of regional
mid-Foxe ice.

Insufficient evidence is yet available to specify the relative
extent of Frobisher Bay and foreign, Labradorean-based ice during the
mid-Foxe interval. Large-scale geomorphic features such as roches
moutonées, more than 200 m x 20 m in granitic bedrock, and the
distribution of cirque forms indicate that southeasternmost Meta

TABLE 18.1. Radiocarbon Dates Quoted in Text or Figures.
===

2,745±145	GX 7881	HU77-159 130-150 cm; pretreated organic carbon from core (Osterman, 1982).
4,560±180	GX 7091	HU77-159 26-290 cm; pretreated organic carbon from core (Osterman, 1982).
7080±175	GX-8160	Oldest date on deglaciation of head Frobisher Bay (Peterhead Inlet)(Squires, 1982).
7800±150	QC-905	Oldest date proximal to Frobisher Bay moraine (Covill, 1982).
8425±375	(AA-191)	Tandem Accelerator date on _Portlandia arctica_ fragment from 932 cm depth in core HU77-159. This date provides an alternative date for the onset of glacial conditions in core HU77-159.
8660±110	GSC-3468	Kendall Strait, SE Meta Incognita Peninsula. Shells from distal glacial-marine facies associated with ice-contact delta formed during ice withdrawl of Hudson Strait ice.
8580±150	GSC-3469	Noble Inlet, SE Meta Incognita Peninsula. Shells in ice-proximal glacial-marine sediment associated with retreat of Hudson Strait ice from Noble Inlet.
8600±110 9190±195 (8790)[+]	GSC-3648 GX-8194	S of Noble Inlet, SE Meta Incognita Peninsula. Shell from ice-proximal delta associated with retreat of Labradorean ice.
9725±120	QC-450	Gold Cove, outer Frobisher Bay. Fauna contains abundant subarctic watermass indicator species (Miller, 1980).
9735±235	GX-8670	Henderson Inlet, outer Frobisher Bay. Shells collected from stoney silt 2 m bht. Minimum date on deglaciation.
9,875±130	QC-903	Pugh Island. Dates retreat from Hall advance (Covill, 1982).
8590±100	GSC-3666	Pugh Island. Redate different sample.
10,025±225	GX 7882	HU77-159 410-430 cm dates post glacial marine conditions in Frobisher Bay; pretreated organic carbon in core (Osterman, 1982)
10,100±110	GSC-2725	Gold Cove, outer Frobisher Bay. Massive _M. truncata_ from marine limit (Miller, 1980).

TABLE 18.1. (continued)

10,200±210	GSC-2778	Gold Cave, outer Frobisher Bay. Single whole valve of M. truncata collected from till immediately above marine limit (Miller, 1980).
10,000±200	GSC-2813	Gold Cove, outer Frobisher Bay. Single 8 g fragment of M. truncata
41,900 + 7100 − 3700	QC-446	Loks Land. Paired Macoma calcarea in deltaic forests overlying glacial-marine drift.
32,500 + 2600 − 3700	GX-8591	Loks Land. Same site as QC-446. Whole valves H. arctica.
> 37,000	GX-8942	Nannuk Harbour, outer Meta Incognita Peninsula. Fragmented H. arctica in till ca. 180 m aht.
10,685±385	GX 6352	HU77-159 500-553 cm dates post-proximal glacial marine conditions in Frobisher Bay (Osterman, 1982).
11,910±380	GX 7119	HU77-159 860-920 cm near Base of piston core. Fauna indicating nonglacial conditions, thereby provides limiting date on beginning of Hall advance; pretreated organic carbon (Osterman, 1982).
27,255±1250	GX 7883	HU77-156 190-205 cm near Basal date; on pretreated organic carbon (Osterman, 1982).

[+] dates in brackets represent subtraction of 400 yrs to equal GSC and GX dates

Incognita Peninsula has been dominantly under the influence of northeasterly flowing ice derived from a southern source. At present, we prefer the concept that the limestone-rich glacial-marine sediment on Loks Land dated at greater than 40 ka was deposited by a major mid Foxe expansion of the Labradorean ice dome (Fig. 18.3), but no survey of ice-directional features has been made on Loks Land and the possibility of a Frobisher Bay outlet glacier reaching Loks Land cannot be excluded. High resolution HUNTEC seismic profiles suggest the presence of a deeply buried till (20-30 m) at the location of core HU77-159. This till may be related to a mid-Foxe glacier grounded in Frobisher Bay whose terminal position is uncertain. Regardless of ice source, the highlands of Loks Land bear no evidence of having been overrun at this time, and ice was confined to the low-relief southern half of the island. Similarly, the northeastern margin of Blunt

TABLE 18.2. Isoleucine epimerization ratios in molluscs and
foraminifera from southern Baffin Bay.

Lab ID	Locality	species[1]	C-14-age	alle/Ile Free	Total
AAL-3411	Cumberland Peninsula	M.t.	modern	N.D.[+]	0.011±0.001
AAL-3410	Warwick Sd., Frobisher Bay	M.t	10,760	N.D.	0.014±0.001
AAL-3000	Nannuk, Hbr, Meta Incognita Pen. Shells in till	M.t.	>37,000	0.14	0.029±0.004
AAL-2462	Loks Land, delta forests	H.a.	>40,000	0.14	--
AAL-2461	Loks Land, delta forests	M.t.	>40,000	0.14	--
AAL-3521	Loks Land, delta forests	H.a.	>40,000	0.12	0.021
AAL-3522	Loks Land, delta forests	M.t.	>40,000	0.12	0.024
AAL-650	Allen Is., Hall Peninsula	H.a.	--	0.30	--
AAL-3523	Allen Is., Hall Peninsula	H.a.	--	0.32	--
AAL-2738	HU77-156; 250 cm	C.l.	--	--	0.025

[1]Species abreviations: M.t.; Mya truncata
                          H.a.; Hiatella arctica
                          C.l.; Cibicides lobatulus
[+] N.D. = not detectable

Peninsula (Fig. 18.2) is essentially a non-glacial landscape, as are
the outer islands (such as Brevoort Is.) to the north (Fig. 18.1). In
contrast, the southwestern portion of Hall Peninsula contains abundant
glacial landforms, reaffirming the concept of actively eroding glacial
ice occupying outer Frobisher Bay but not sufficiently thick to overtop
the confining land (250 to 400 m height).

LATE FOXE GLACIATION

Field and marine core evidence reveals a complex interaction between
Labradorean (foreign), Amadjuak Lake (regional), and Meta
Incognita-based (local) ice-dispersal centers during the late Foxe

Figure 18.3 (See facing page.) Summary of the glacial history of
southern Baffin Island at (A) >>40,000 BP showing expansion of the Meta
Incognita Ice Cap, (B) >>40,000 BP Kogalu equivalent advance of
Labradorean ice to Loks Land, with the possible advance of grounded ice
in Frobisher Bay, (C) 28,000 to 11,000 BP showing withdrawal of ice from
both the Labradorean and Amadajuak Lake centers, (D) 11,000 BP Late Foxe
expansion of Labradorean center to Resolution Island, and advance of
Hall ice in Frobisher Bay as an ice shelf/ice rise complex, (E)
8000-9000 BP During the Cockburn substage ice retreated across the
southeastern tip of Meta Incognita Peninsula, with final retreat across
Hudson Strait occurring after 8000 BP. During this time, Frobisher Bay
ice is experiencing a standstill with slight fluctuations at the
Frobisher Bay moraine position, and ice caps on Meta Incognita and Hall
Peninsulas are advancing slightly, (F) 5000 BP Region largely ice-free,
reestablishment of the Labrador Current along the coast of southern
Baffin Island signifies the return of interglacial marine conditions.

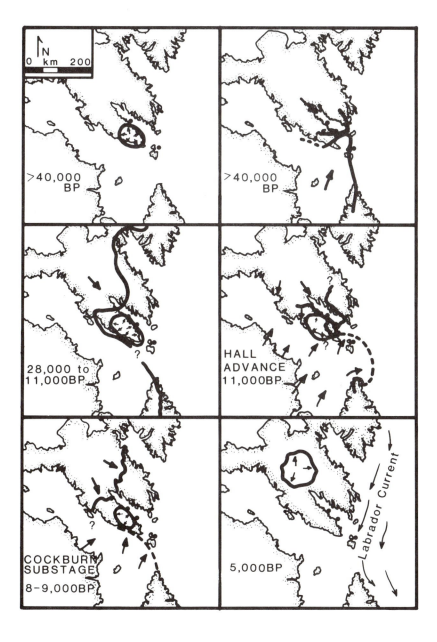

glacial stade. The timing of the onset of the last glacial stade is
dated only in a single marine core whereas a relatively well-dated
record of deglacial events is now available from land and marine
records.

Marine Record

Three continuous cores raised from the deeper basins of Frobisher Bay
(Fig. 18.1) provide evidence of glacial events affecting the southern
Baffin Island region since mid-Foxe time. Osterman and Andrews (1983)
discuss in detail the sedimentologic and biologic conditions associated
with three distinct glacial-marine sediment types in core HU77-159.
The detailed analysis of this core can be used to interpret the glacial
history in the older core HU77-156. Based on foraminiferal zonation,
core HU77-157 only records events of the last 10,000 years which are
covered in the discussion of HU77-159.

Core HU77-159 (Fig. 18.4) contains a record of proximal
glacial-marine sedimentation consisting of a rapidly deposited clay,
rich in detrital carbonate and other clay-sized particles that suggest
glacial erosion of bedrock, a low rate of sand influx, but with a high
frequency of angular quartz grain, and a characteristic low-diversity
Elphidium excavatum (Terquem) forma clavata (Cushman) benthic
foraminiferal assemblage. Distal glacial-marine sediments are found
directly above and below the proximal sediments, and these record
conditions prior to and after the glacial advance. Ice-distal
glacial-marine sediments are typically silty clays with moderate
$CaCO_3$ percentages, sand influx and grain angularity, and contain a more
abundant and diverse benthic foraminiferal assemblage including
Cassidulina reniforma (Nörvang) and Islandiella helenae
(Feyling-Hanssen and Buzas). In the upper portion of HU77-159,
ice-rafted glacial-marine sediments have been deposited since the
retreat of glacier ice from Frobisher Bay. Ice-rafted sediments are
slowly deposited, noticeably sandier, and contain maxima of sand influx
and grain angularity. Benthic foraminifera in the ice-rafted sediments
comprise the most abundant and diverse assemblages in core HU77-159,
and include I. helenae, C. reniforme and E. excavatum f. clavata (Fig.
18.4).

The chronology of these glacial-nonglacial events in HU77-159 is
uncertain at this time. A radiocarbon date of 11,900±380 BP (GX-7119)
on pretreated organics, with the carbonate removed, indicates the onset
of glacial conditions occurred soon after that time. However, a recent

Depth (cm)	^{14}C Dates on HU77-159	Foraminiferal Zonation	Other Biota	Sediments	Clay Zones	Stratigraphic Units	Glacial History Miller(1980)
100		C. lobatulus/ I. helenae		Silty Clay	I High chlorite and mica	Burton Unit	
200	2745±145	Nonion labradoricum			II High kaolinite High quartz High feldspar	Labrador Unit (Ice-Rafting)	
300	4560±180		Shell fragments more common	Clayey Silt			All ice gone
400		Immigration Zone	First occurrence of spicules	Clayey Silt	III		Frobisher mor. 10,100 BP Gold Cove readvance
500	10,025±450		Shell layer	Silty Clay	High mica Low chlorite Low feldspar	Frobisher Bay Unit (Ice Distal)	Ending before —10,760—
600	10,685±385	I. helenae					Hall Advance
700		C. reniforme		Clay	IV	Hall Clay (Ice Proximal)	
800		Low diversity E. excavatum f. clavata	Lowest diversity		High chlorite High feldspar High kaolinite		
900	11,910±380	C. reniforme		Silty Clay	V	Griffin Unit (Ice Distal)	
970		I. helenae			High mica		

Figure 18.4 Summary of radiocarbon dates, foraminiferal zonation, grain-size, clay mineralogy, glacial-marine sediment type and glacial history (from Miller, 1980) of core HU77-159 from Frobisher Bay.

tanderm accelerator datum of 8,425±375 (AA-191; Table 18.1) near the same depth suggests the chronology may have to be revised. The uncertainty over the timing of events does not, however, alter the interpretation of the events or their correlation between cores. It is our belief that radiocarbon dates on organic carbon in cores provide necessary intra-core correlations that may or may not be chronologically correct. This paper follows the chronology established by the earlier data and the correlation of the Hall marine clay with the Hall advance, when in actuality it may correlate instead with the later Cockburn advance.

The results of core HU77-159 can be used to interpret the longer record of core HU77-156. Figure 18.5 shows a summary of the foraminiferal zones of cores HU77-156 and 159, principal C-14 dates and implied terrestrial and marine history. A C-14 date of 27,255±1250 BP (GX-7883) on pretreated and concentrated detrital organic matter (Kihl, 1975) obtained near the base of the core suggests the core spans the last 30 ka (Table 18.1), although aIle/Ile ratios in Cibicides lobatulus from 260 cm depth in the core (Table 18.2) suggest the core may only span the last 18 ka. Sandy sediments (Fig. 18.6) containing a diverse Cibicides lobatulus (Dawson) assemblage occur at the top and bottom of core 156. This suggests interstadial conditions similar to the present occurred possibly prior to 27 ka ago. This interstadial

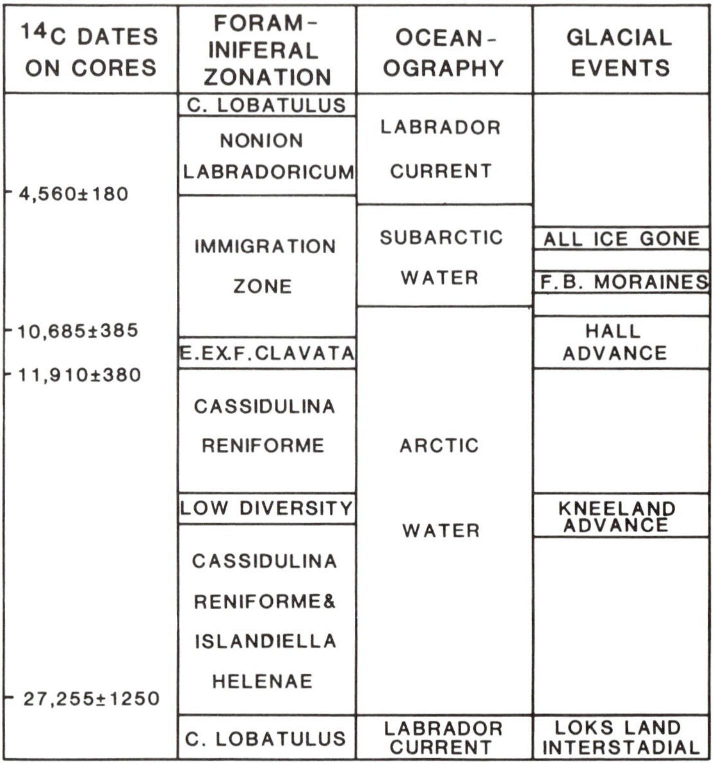

14C DATES ON CORES	FORAM- INIFERAL ZONATION	OCEAN- OGRAPHY	GLACIAL EVENTS
4,560±180	C. LOBATULUS NONION LABRADORICUM	LABRADOR CURRENT	
	IMMIGRATION ZONE	SUBARCTIC WATER	ALL ICE GONE F.B. MORAINES
10,685±385 11,910±380	E.EX.F.CLAVATA	ARCTIC WATER	HALL ADVANCE
	CASSIDULINA RENIFORME		
	LOW DIVERSITY		KNEELAND ADVANCE
	CASSIDULINA RENIFORME& ISLANDIELLA HELENAE		
27,255±1250	C. LOBATULUS	LABRADOR CURRENT	LOKS LAND INTERSTADIAL

Figure 18.5 Summary of foraminiferal zonation, and radiocarbon dates in core HU77-156 and 159, and the interpretation of the glacial and oceanographic history. The figures for the chronology of events here may have to be revised, but the stratigraphic interpretation of event is believed to be correct.

occurred after deglaciation of Loks Land and may be correlative with a part of the Loks Land Member of the Clyde Forland Formation (Miller, Chapter 14). The similar nature of foraminifers and sediments at the top and bottom of core 156 implies that the Baffin Current was in place, similar to today, at this time. Emplacement of the Baffin Current necessitates free exchange of water from the Arctic Ocean into Baffin Bay via the straits of the Queen Elizabeth Islands. This suggests those straits were ice-free and therefore, this interstadial appears to have a widespread occurrence. Blake (1980) also reports a Mid-Foxe (35-42 ka) interstadial with M. truncata in the Queen Elizabeth Islands,and Klassen (this volume, Chapter 15) identifies marine units correlative with the Loks Land Member in the Bylot Island, northern Baffin Bay.

HU77-156

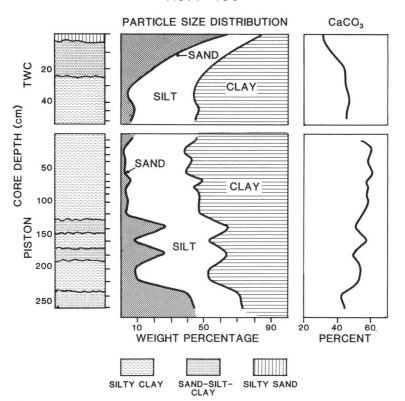

Figure 18.6   Results of the grain-size analysis of core HU77-156, showing the increase of sand at the tip and bottom of the core in response to sedimentation beneath the strong Baffin-Labrador Current. Non-glacial conditions are observed from 230 to 125 cm, and glacial conditions are found from 125 to 50 cm.

In core #156, the change from interstadial to glacial circulation and sedimentation is evidenced by decreased grain size and a change to a less diverse benthic foraminiferal assemblage of I. helenae and C. reniforme. Several sand peaks may signal increased ice-rafting triggered by slightly warmer temperatures (Fig. 18.6). From 125 to 50 cm depth, roughly 19,500 to 10,700 BP, the sediment in core 156 becomes finer grained (> 50% clay; Fig. 18.6) and contains two low-diversity benthic foraminiferal assemblage zones. The upper E. excavatum f. clavata zone from 50 to 80 cm in core 156 is correlated to the Hall glacial advance about 11 ka BP or the Cockburn advance from 8-9 ka BP). The second low diversity zone from 110 to 125 cm is interpreted to be a 18 ka BP glacial advance of regional ice in Frobisher Bay (or alternately the Hall advance at 11 ka BP). However, the possibility that this low diversity zone is related to the Labrador

center can not be excluded. Evidence for this glacial advance, here
termed the Kneeland advance, is preserved only in core 156 (Fig.
18.5). Moraines associated with the Kneeland advance have not been
identified and they are presumed to have been destroyed during the
maximum late Foxe event, the Hall advance. The sediments in core #156
associated with the Kneeland advance are slightly coarser than those
deposited during the Hall advance, which implies a more distal
glacial-marine environment, a conclusion also supported by the benthic
foraminifera. It appears therefore the Kneeland advance (18 ka BP) was
less extensive than the later Hall advance. A similar two-fold glacial
sequence has recently been suggested for east Greenland (Hjort and
Bjorg, submitted).

Hudson Strait Coast of Meta Incognita Peninsula

Blake (1966) noted the occurrence of a narrow zone of carbonate-rich
drift along the south coast of Meta Incognita Peninsula. He attributed
this drift to Hudson Strait ice impinging on the peninsula. Blake also
suggested (1966; p. 5, Fig. 2) that ice flowed northeasterly across the
tip of the peninsula. Marine shells in the carbonate drift yielded
radiocarbon dates in excess of 30 ka, and cannot be used to constrain
the age of the associated advance.

Ice-flow directional indicators are exceptionally well preserved
over large areas of SE Meta Incognita Peninsula as both macro- and
micro-morphologic features. Delicate polish, striae and even facets on
individual mineral grains are protected from chemical weathering by the
buffering influence of the overlying carbonate-rich till that mantles
much of this area. In contrast, such features are absent in areas of
locally derived acidic till. In most cases where striae are preserved,
both the orientation and sense of ice movement can be readily
discerned; Figure 18.7 is a map of such observations at sites on local
summits or broad uplands for which topographic channeling of ice flow
was minimal.

Across the broad dissected upland (100-200 m asl) inland and west
from Nannuk Harbour, at least three generations of striae are
preserved. The oldest set, oriented between S 5°W and S 15°W, are
preserved only on north-facing hillslopes. They reflect an episode of
local ice advance off the highlands of the peninsula, probably from an
expanded Terra Nivea Ice Cap. The striae are crosscut by a younger
suite that are the dominant directional features in the area. These
striae trend between N 30°E and N 55°E in this region (Fig. 18.7).
Associated till contains abundant limestone clasts, other dark

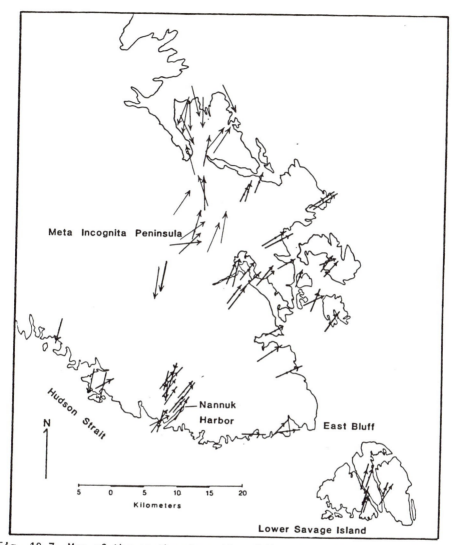

Fig. 18.7 Map of the southeastern tip of Meta Incognita Peninsula
showing the striation evidence for the dominant source of glacier
ice--the Hudson Strait Labradorean center.

sedimentary erratic clasts from source rocks on the Labrador-Ungava
Peninsula (Andrews et al., in press), scattered marine shell fragments,
and a matrix carbonate content of 20 to 50%. These observations
require an ice dispersal center, at some time, south of Baffin Island
feeding an ice tongue that crossed Hudson Strait, overwhelmed SE Meta
Incognita Peninsula and flowed into outer Frobisher Bay. Although
limestone erratics outcrop in Hudson Strait just off the coast of Meta

Incognita Peninsula, the dark sedimentary erratic suite does not, and must have been transported across Hudson Strait from the south and southwest. Thin section analysis of clasts from the till revealed that slates, metabasalts, and pyroxene-bearing gneiss (granulite facies) are present. The gneiss most likely comes from northern Ungava Peninsula whereas the slates and metabasalts originated from the Labrador Trough or the Cape Smith foldbelt (Hynes and Francis, in prep.; S.A. Morse and George Albino pers. comm.; Westra, 1978). The dispersal center feeding this ice tongue is referred to as the Labradorean center. A more recent set of striae oriented S 5°W found northwest of the Sister Islets (Fig. 18.2), indicates that an expansion of Meta Incognita-based ice followed recession of the Labradorean ice, but was not of sufficient magnitude to reoccupy the peninsula southeast of Sister Islets.

The timing of these Late Foxe events is defined by C-14 dates on molluscs from associated marine deposits. Although no material is available to date the oldest set of striae, numerous radiocarbon dates are available from ice-proximal raised marine sediments associated with the deglaciation of the outer east coast of Meta Incognita Peninsula. Shells from the intertidal zone in Henderson Inlet (Fig. 18.2) yielded a date of 9735±235 BP (GX-8670; Table 18.1) which indicates deglaciation had begun prior to that time and that the coast was ice-free shortly after 10,000 years ago. Shells in a distal silt facies related to a prominant ice-contact delta 40 m asl in Kendall Strait (Fig. 18.2) are dated to 8660±110 BP (GSC-3468; Table 18.1). Note that GSC dates are c. 400 yrs younger than dates reported by GX due to differences in the way shell dates are calculated. Deglaciation proceeded rapidly, and ice-proximal deposits near the head of Noble Inlet contain shells dated 8580±150 BP (GSC-3469; Table 18.1). No raised-marine deposits with in situ molluscs have yet been located along the Hudson Strait coast of outer Meta Incognita Peninsula, but Fillon (1978) reports C-14 dates on molluscs in cores taken from the floor of the strait that are c. 9100 BP and suggest very rapid deglaciation of NE-flowing ice from the peninsula. Dates from the Lake Harbour area average close to 8000 BP (Blake, 1966) but may be related to recession of local ice, as ice-flow indicators in that area are oriented toward the south.

Frobisher Bay

Moraines in Frobisher Bay have been mapped by Mercer (1956), Blake (1966), Miller (1980), Müller (1980), Colvill (1982), Squires (1982), and Lind (1983). Two prominent regional moraines are recognized. The

older of these, the Hall moraine (Miller, 1980; this volume, Chapter 9)
marks the most extensive recognized late Foxe advance of continental
ice onto Hall Peninsula and down Frobisher Bay. The moraine system
terminates in the Countess of Warwick Sound region (Fig. 18.2) on the
north side of outer Frobisher Bay where it consists of a limestone-rich
till, and a series of disjunct lateral moraines on gently sloping
hillsides, commonly associated with kame terraces and other
glacio-fluvial deposits.

Radiocarbon dates on the retreat from the Hall Moraine come from
paired valves of Mya truncata (L.) collected from a delta in Warwick
Sound (Fig. 18.2). Three C-14 dates, averaging 10,760±40 BP (Table
18.1), identify the time of delta formation when a moraine-dammed lake
was breached following deglaciation (Miller, 1980). Radiometric
evidence indicates a readvance in the Gold Cove area (Fig. 18.2) during
the period of general deglaciation. A single large (8g) fragment of
M. truncata collected from till between 219 and 232 m asl provides a
maximum date on the readvance of 10,000±200 BP (GSC-2813; Table 18.1)
and is supported by a date on shells in till of 10,200±210 BP
(GSC-2778; Table 18.1) from 80 m asl in the same area. Paired valves
in marine sediment 73 m asl dated 10,100±110 BP (GSC-2725; Table 18.1)
also indicate the area was deglaciated at that time.

Although there is no evidence of the Hall moraine on Meta
Incognita Peninsula at a similar elevation to those on Hall Peninsula,
Stravers (unpublished) noted the presence of flat-lying moraines 30 to
50 m above the marine limit, deposited by Hall-aged ice advancing onto
Meta Incognita Peninsula from Frobisher Bay. Osterman (1982) proposed
that these flat-lying Hall-equivalent moraines were deposited by an ice
shelf along the coast of Meta Incognita Peninsula. During the Hall
advance it is postulated that an ice shelf in Frobisher Bay was
ungrounded over the deep trough of the bay, but was grounded on the
Calanus Shelf to form an ice rise along the coast of Hall Peninsula
(Fig. 18.8). The ice shelf/ice rise model for the Hall advance is
supported by two lines of evidence. First is the moraine evidence
presented by Miller (1980) and Stravers (unpublished). Second, and
more problematically, is the evidence from the marine record of core
HU77-159, which indicates continuous sedimentation through the Hall
advance (if the organic dates are accepted), and no grounded ice at the
core location (Osterman, 1982).

The retreat of the Frobisher Bay outlet glacier after the Gold
Cove readvance must have been extremely rapid. Radiocarbon-dated
mollusc shells on Pugh Island (Fig. 18.1) indicate the island was
deglaciated by 9,875±130 (QC-903; Table 18.1). Another shell sample

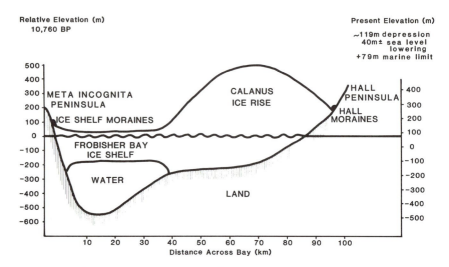

Figure 18.8 Cross sectional view of the proposed Frobisher Bay ice shelf/Calanus ice rise model from Osterman (1982). This model shows the position of the Hall moraine and the ice shelf moraine observed by Stravers (in prep).

from Pugh Island has been redated to 8590±100 (GSC-3666) or about 9000 BP for direct comparison with GX dates. Accepting this date requires 120 km of glacier retreat within 200-1000 yrs. Rapid removal of the Frobisher Bay outlet glacier may have been augmented by the eustatic sea level rise decoupling the ice from its bed, by the influx of warmer subarctic water into the bay and the instability of the ice shelf. Miller (1980) notes a shift in mollusc fauna from robust valves of the pan-arctic species Mya truncata (L.) in deposits dated older than 9900 BP, to a mollusc assemblage containing a diverse subarctic fauna at 9,725±120 BP (QC-450; Table 18.1). The change in the fauna implies the incursion of warmer water at that time.

A younger and more continuous moraine system is present near the head of Frobisher Bay and can be mapped across Hall Peninsula to the head of Cumberland Sound (Blake, 1966; Miller, 1980; this volume, Chapter 19) where it is correlated with the Ranger moraine (Dyke 1979). Radiocarbon dates on shells in marine deposits directly associated with the Frobisher Bay moraine range between 8200 and 8800 BP (Table 18.1). Lind (1983) presents evidence from facies relationships at Cape Rammelsberg that the Frobisher Bay moraine represents a readvance rather than a still-stand. The relationship of the massive left lateral moraine at the head of the bay, directly abutting stripped bedrock, also supports this interpretation.

A series of nested moraines proximal to the Frobisher Bay moraine indicates a period of minor oscillations occurred before final deglaciation from the bay. The sharp drop in the marine limit from 120 m on the distal side to 38 m on the proximal side of these moraines suggests they represent up to 1000 years of quasi-stability. Lind (1983) argues that during the final retreat ice was confined to the Frobisher Bay, and had separated from ice masses in paralleling valleys and on the upland surface of Meta Incognita Peninsula. Kame terraces and ice-contact deltas along the Bay mark the ice front during its final retreat.

Incursion of sea water behind the Frobisher Bay moraines occurred as the ice was removed. This ranged from 7800±150 BP (QC-905; Table 18.1) in Burton Bay to 7,080±175 BP (GX-8160; Table 18.1) near the head of the Bay in Peterhead Inlet. Final wastage of ice is believed to have been centered near Amadjuak Lake where basal organic dates indicate complete disappearance of the ice prior to 4550±220 BP (GSC-498; Blake, 1966).

Local Meta Incognita Ice

The relationship between the expansion of local ice and the impingement of foreign ice on Meta Incognita Peninsula has been studied by Mercer (1956), Blake (1966), Müller (1980), Lind (1983) and Stravers (in prep.). The current consensus is that the high plateau of the peninsula supported a series of inter-connected, semi-independent dispersal centers forming a broad ridge paralleling the axis of the peninsula, but displaced well to the southwest from the height of land and current ice divides. Large-scale landforms and striae observed along the Hudson Strait coast indicate dominant flow to the S-SW off the peninsula. The inner reaches of the peninsula would have been overrun by regional ice during the Hall maximum, whereas the outer half of the peninsula, shielded from an outlet glacier in the bay by the Everett Mountains (600 to 750 m asl) maintained an independent ice cap throughout late Foxe time.

Detailed stratigraphic studies in the Grinnell Glacier area fiords has revealed a complex interaction between local and Frobisher Bay ice (Stravers, unpublished data) during the interval from 12,000 to 9500 BP (Fig. 18.3). An extensive Late Foxe local advance occurred prior to the Hall Advance in Frobisher Bay. This is seen only in fresh, well-preserved striae that are morphostratigraphically older than the Hall moraine correlatives. They may represent a yet undated Late Foxe advance or alternatively could have occurred prior to the Hall Advance but postdating the initial ice-free interval (11,910±380?) recorded in core HU77-159.

Limestone-bearing tills and glacial-marine sediments deposited during the Hall advance are found along the inner parts of three fiords near Grinnell Glacier, indicating by their location that local ice had retreated prior to the Hall advance. Stratigraphic data show a subsequent merging of the two ice masses just prior to 10,530±110 (SI-5758). This was followed by a brief separation and a final merging during the Gold Cove readvance.

Cross-cutting striations along Hudson Strait indicate ice advanced off Meta Incognita Peninsula after the recession of Labradorean ice. This suggests an out-of-phase relationship that is also observed in Frobisher Bay, where radiocarbon dates indicate a delayed deglaciation of the Meta Incognita Ice Cap relative to the Frobisher Bay outlet glacier.

Deglaciation of the Meta Incognita Ice Cap began by 10,000 BP and was interrupted by a readvance of Grinnell Glacier and Terra Nivea outlet glaciers between 8500 to 8000 BP. Müller (1980) theorized that the incursion of subarctic water into Frobisher Bay at 9800 BP may have increased local precipitation and aided in the late preservation of the expanded Meta Incognita Ice Cap. Both the Meta Incognita local readvance and the regional Frobisher Bay moraine occurred during the Cockburn substage (8-9 ka BP), a time of general glacier readvance across eastern Baffin Island (Andrews and Ives, 1978). Separation of the Meta Incognita Ice Cap into the two smaller present-day ice caps occurred prior to the deposition of the York Sound delta about 8800 BP (Blake, 1966; Table 18.1).

During deglaciation of the peninsula the ice divide apparently migrated toward the northeast. The till directly abutting Neoglacial moraine surrounding outlet glaciers of the Terra Nivea Ice Cap in Jackman Sound (Fig. 18.2) contains abundant limestone erratics and is associated with ice-flow indicators oriented NW to NNW, requiring that the till was deposited by ice from the Hudson Strait. Deglaciation of Jackson Sound occurred prior to 9700 BP, hence the northeast-flowing Terra Nivea outlet glaciers are currently as extensive as at any interval since ca. 10,000 BP, whereas the southwest margin has receded over 30 km during the same time interval. Such asymetric retreat and ice-divide migration also characterized the deglaciation of the Foxe Basin dome, of which the Barnes Ice Cap is the residual remnant (Andrews and Falconer, 1969).

Present-day steady state equilibrium line and glacierization
threshold elevations on southeastern Meta Incognita Peninsula are as
low as anywhere on Baffin Island despite their southerly location and
relatively warm thermal regime (Andrews and Miller, 1972). The
favorable conditions for glacierization in this region may be related
to the strong tidal currents along the foot of the Everett Mountains
that extend the open water period and enhance local precipitation. The
summits of Terra Nivea (880 m asl) and Grinnell (850 m asl) ice caps
are at present the highest points on Meta Incognita Peninsula.
Although both ice masses have probably persisted throughout the
Holocene, Mercer (1954) speculated that they would not reform under
present climatic conditions. The turn-of-the-century Little Ice Age
expansion was the most extensive Neoglacial advance for both ice caps.
Recession from the Little Ice Age maximum has been greatest for outlet
glaciers that terminated in the sea (up to 2 km), whereas the
southwestern margin has remained relatively unchanged.

CONCLUSIONS

A complete understanding of the Quaternary history requires an
interpretation of both the terrestrial and marine records. This paper
includes the results of terrestrial (Miller, 1980; Stravers, in prep.),
and marine core investigations (Osterman, 1982; Osterman and Andrews,
1983) in Frobisher Bay. These studies record the complex interaction
of three separate ice masses: (1) a regional Amadjuak Lake centered
Frobisher Bay lobe; (2) a foreign Hudson Strait (Ungava centered -
Labradorean lobe?); and (3) a local Meta Incognita Peninsula ice cap.

     Investigation of weathering shows that only the highest summits of
the Everett Mountains of Meta Incognita Peninsula have escaped active
glaciation (Müller, 1980). Inner Frobisher Bay has been completely
inundated by the local Frobisher Bay ice lobe, but in the outer part of
the bay, weathering phenomena indicate a pre Foxe glacial event above
450 m, and provide evidence for three (early, middle, and late) Foxe
glaciations below 450 m.

     Amino acid data suggest that the early Foxe glaciation is possibly
equivalent to the Kogalu member of the Clyde Foreland Formation (Miller
et al., 1977). Evidence for the early Foxe advance on southern Baffin
Island has so far only been found on Allen Island (Fig. 18.1). The mid
Foxe glaciation is recorded at several sites on southern Baffin Island
and defines the Loks Land member of the same formation. During mid
Foxe time all three ice centers are believed to have been active, and
possibly contiguous.

There is evidence from marine cores (Osterman, 1982) that the mid and late Foxe glaciations of southern Baffin Island were separated by a regional interstadial. Such an interstadial has also been reported from Ellesmere Island (Blake, 1980) and Bylot Island (Klassen, this volume).

The late Foxe glaciation of southern Baffin Island is the longest and best documented record for Baffin Island. The onset of glaciation of the three ice centers is poorly constrained, but the recession is well dated. Marine cores suggest a two-fold advance, with the second advance the maximum. Extrapolation of the radiocarbon date on core HU77-156 places the early Kneeland advance at c. 18,000 BP (Fig. 18.5); the later Hall advance has been C-14 dated (Table 18.1) at shortly before 10,760 ± BP, but there are reasons not to accept these dates fully. It is uncertain, at this time, if the Kneeland advance involved only the foreign Hudson Strait ice, or the regional Frobisher Bay ice, or both.

The development of the Frobisher Bay ice shelf and Calanus ice rise during the Hall advance is not believed to have been characteristic of previous glaciations. HUNTEC seismic profiles reveal till beneath the sediment of core HU77-159, which indicates that grounded ice did indeed fill Frobisher Bay during a previous glaciation. However, during the Hall advance a floating ice shelf covered the deep trough of outer Frobisher Bay, and the Calanus ice rise developed on the shallow shelf off Hall Peninsula (Fig. 18.3).

Retreat of the Frobisher Bay lobe from the Hall moraine occurred shortly before 10,760 ± BP, but a minor readvance occurred into Gold Cove at 10,100 ± BP (Table 18.1). The Gold Cove readvance is not observed in the marine cores, and may have been an advance associated only with the Calanus ice rise. The retreat of the Frobisher Bay lobe continued up the bay to a position behind the Frobisher Bay moraines (Fig. 18.1). Lind (1983) believes that the Frobisher Bay moraines were formed by an advance after a period of a fluctuating ice margin.

The maximum late Foxe extent of the Hudson Strait Ungava-based ice across eastern Meta Incognita Peninsula is uncertain. However, the late Foxe ice is not believed to have been grounded at the location of HU77-156, 40 km to the northeast of Resolution Island. Radiocarbon dates indicate the Hudson Strait Ungava-based ice covered the southeast tip of Meta Incognita Peninsula, and Hudson Strait until sometime after 8,600 ± BP (Table 18.1) although Fillon's (1978) Hudson Strait core dates are c. 500 yr older. Retreat of the ice off Meta Incognita

Peninsula was followed by deglaciation of Hudson Strait and Hudson Bay at 8000 ± BP (Andrews and Falconer, 1969).

Expansion of the local Meta Incognita ice cap preceded and succeeded the Hudson Strait ice advance over Meta Incognita Peninsula. The later expansion of the Meta Incognita ice cap is dated at c. 8000-8500 BP (Müller, 1980), closely following the separation of the ice cap into two masses by 8800 ± BP (Lowden and Blake, 1973).

ACKNOWLEDGEMENTS

Field work and laboratory research has been supported by a number of NSF grants to Andrews, Miller and Osterman. We specifically wish to acknowledge grants EAR-79-26061, EAR-81-21296, and DPP-81-116948.

REFERENCES

Andrews, J.T. and Falconer, G., 1969: Late glacial and post glacial history and emergence of the Ottowa Islands, Hudson Bay, N.W.T.: Evidence on deglaciation of Hudson Bay. Canadian Journal of Earth Science, 6:1263-1276.

Andrews, J.T. and Ives, J.D., 1978: "Cockburn" Nomenclature and the late Quaternary history of the eastern Canadian arctic. Arctic and Alpine Research, 10:617-633.

Andrews, J.T. and Miller, G.H., 1972: The Quaternary history of Northern Cumberland Peninsula East Baffin Island, N.W.T. Part X: Radiocarbon date list. Arctic and Alpine Research, 4:261-277.

Andrews, J.T., Stravers, J.A., and Miller, G.H., in press: Patterns of glacial erosion and deposition around Cumberland Sound, Frobisher Bay, and Hudson Strait, and the location of ice streams in teh eastern Canadian Arctic. In: Woldenburg, M. (ed.), Models in Geomorphology. London: Allen & Unwin.

Blackadar, R.G., 1967: Geological reconnaissance, southern Baffin Island District of Franklin, Geological Survey of Canada Paper,66-47, 32 p.

Blake, W., Jr., 1966: End moraines and deglaciation chronology in Northern Canada with special reference to southern Baffin Island. Geological Survey of Canada Paper 66-26, 31 p.

Blake, W., Jr., 1980: Mid-Wisconsin interstadial deposits beneath Holocene beaches, Cape Storm, Ellesmere Island, Arctic Canada. American Quaternary Association Sixth Biennial meeting Abstracts and Program, Orono ME., 26-27.

Coachman, L.K. and Aagard, K., 1974: Physical Oceanography of Arctic and Subarctic Seas In: Herman, Y. (ed.), Marine Geology and Oceanography of the Arctic Seas. Springer-Verlag, New York, 1-72

Colvill, A.J., 1982: Glacial landforms at the head of Frobisher Bay, Baffin Island, Canada, M.S. thesis, University of Colorado, Boulder. 202 pp.

Dyke, A.S., 1979: Glacial and sea level history of southwestern Cumberland Peninsula, Baffin Island, N.W.T., Canada. Arctic and Alpine Research, 11:179-202.

Fillon, R.H., 1978: Glacier termini in eastern Hudson Strait. Geological Society of America, Abstracts, 10(7):401.

Hynes, A.Y. and Francis, D.M., in prep: A transect of the early Proterozoic Cape Smith fold belt, New Quebec.

Kihl, R., 1975: Physical preparation of organic matter samples for C-14 dating: In: Andrews, J.T., Radiocarbon date list II from Cumberland Peninsula, Baffin Island, N.W.T., Canada. Arctic and Alpine Research, 7:90-91.

Lind, E.K., 1983: Environmental conditions and history of the Holocene deglaciation for the Cape Rammelsberg area, southern Baffin Island, N.W.T., Canada, M.S. thesis, University of Colorado, Boulder. 219 pp.
Lowdon, J.A. and Blake, W., Jr., 1973: Radiocarbon dates VIII. Geological Survey of Canada Paper, 73(7), 37 pp.

MacLean B., Jansa, L.F., Falconer, R.K.H., and Srivastava, S.P., 1977: Ordovician strata on the southeastern Baffin Island shelf revealed by shallow drilling. Canadian Journal of Earth Sciences, 14:19-25-1939.
MacLean, B. and Falconer, R.K.H., 1979: Geological/geophysical studies in Baffin Bay and Scott Inlet-Buchan Gulf and Cape Dyer-Cumberland Sound areas of the Baffin Shelf. Geological Survey of Canada Paper, 79-1B, 231-244.
Mercer, J.H., 1954: The physiography and glaciology of southernmost Baffin Island, Ph.D. thesis, McGill Univ., Montreal, Quebec, Canada. 150 pp.
Mercer, J.H., 1956: Geomorphology and glacial history of Southernmost Baffin Island. Bulletin of the Geological Society of America, 67:553-570.
Miller, G.H., 1980: Late Foxe glaciation of southern Baffin Island, N.W.T., Canada. Geological Society of America Bulletin, 91:399-405.
Miller, G.H., 1982: Dynamics of the Laurentide ice sheet based on field evidence from northeastern Canada. INQUA, XI Congress, Moscow, Abstracts, 1:222.
Miller, G.H., Andrews, J.T., and Short, S.K., 1977: The last interglacial-glacial cycle, Clyde foreland, Baffin Island, N.W.T.: stratigraphy, biostratigraphy, and chronology. Canadian Journal of Earth Science, 14(12):2824-2857.
Müller, D.S., 1980: Glacial geology and Quaternary history of southeastern Meta Incognita Peninsula, Baffin Island, Canada, M.S. thesis, University of Colorado, Boulder. 211 pp.

Osterman, L.E., 1982: Late Quaternary history of southern Baffin Island, Canada: a study of foraminifera and sediments from Frobisher Bay, Ph.D. thesis, University of Colorado, Boulder. 380 pp.
Osterman, L.E., Miller, G.H., and Stravers, J.A., 1982: Late Quaternary history of southeastern Baffin Island, N.W.T., Canada. Geological Society of America Abstracts, 14:581.
Osterman, L.E. and Andrews, J.T., 1983: Changes in glacial-marine sedimentation in core HU77-159, Frobisher Bay, Baffin Island, N.W.T.: a record of proximal, distal and ice-rafting glacial-marine environments In: Molnia, B.F. (ed.), Glacial Marine Sedimentation. Pergamon Press, Oxford.

Squires, C.A., 1982: The late Foxe deglaciation of the Burton Bay area, southeastern Baffin Island, M.A. thesis, University of Windsor, Ontario. 115 p.

Stravers, J.A., in prep., Glacial and relative sea level history of outer Meta Incognita Peninsula, Baffin Island, Canada [unpublished Ph.D. thesis]: University of Colorado, Boulder.

Westra, L., 1978: Metamorphism in the Cape Smith-Wakeman Bay area north of 61°N, New Quebec. Geological Survey of Canada Paper, 78-10: 237-244.

19 Moraines and proglacial lake shorelines, Hall Peninsula, Baffin Island

Gifford H. Miller

Hall Peninsula is the central of the three great peninsulas of eastern Baffin Island. It is bounded by Frobisher Bay to the south and Cumberland Sound to the north. The land surface rises rather uniformly from 30 m elevation at Nettilling Lake on the western edge of the peninsula to 600-700 m over the gently rolling central plateau, rising again to the northwest to maximum elevations in excess of 1100 m along Cumberland Sound and the Davis Strait coast (Fig. 19.1). Most of the central and eastern portion of the peninsula is drained by the McKeand River that flows northwesterly along the axis of the peninsula for 200 km before turning abruptly to the northeast and emptying into the head of Cumberland Sound. Fiords, cut into the height of land along much of the perimeter of the peninsula, have captured the drainage of these marginal areas.

The physiography of Hall Peninsula can be approximated by a gently inclined surface dipping to the northwest and dissected around its perimeter (Fig. 19.1). Consequently, continental ice advancing onto the peninsula from a dispersal center to the west would flow against the regional drainage, creating a complex series of proglacial lakes, most of which would eventually drain to the southeast, into Frobisher Bay. During periods of advance or retreat, rapidly changing lake levels would have left little geomorphic evidence. In contrast, shore features of proglacial lakes associated with periods of ice-margin stability may be preserved in the geomorphic record. Proglacial lake deposits have been previously recognized in the eastern Canadian Arctic (e.g. Ives, 1960; Ives and Andrews, 1963; Barnett and Peterson, 1964; Lauriol and Gray, 1983).

A combination of aerial photograph interpretation and field studies allows the definition of two prominent moraine systems deposited by continental ice impinging onto Hall Peninsula from the

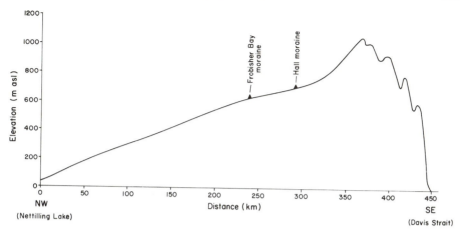

Figure 19.1 Generalized physiographic cross-section along the axis of
Hall Peninsula from Nettilling Lake at the NW end of the peninsula to
the Davis Strait coast, showing the steady rise of the land surface
toward the southeast. The rugged and dissected southeast margin of the
peninsula is only qualitatively represented. The location of the
Frobisher Bay and Hall Moraine systems are indicated.

west and northwest against the regional drainage of the peninsula.
During periods of ice-margin stabilization, proglacial lakes existed
long enough to create mappable shoreline deposits, overflow channels,
and glacio-lacustrine deltas. The relationships between moraines and
lake levels allow a reconstruction of the deglacial events across Hall
Peninsula and indicate a complex interaction between local ice centered
over the northeastern margin of the peninsula and fluctuations of the
continental ice margin. The moraine and shoreline deposits allow the
reconstruction of glacial events across the uplands where datable
deposits are essentially absent.

MORAINE SYSTEMS AND LAKE SHORELINES ON HALL PENINSULA

Field mapping has been carried out along much of the perimeter of Hall
Peninsula accessible from the sea, but no field parties have worked
across the interior. Mapping of the interior portions of the peninsula
has been from the interpretation of stereo aerial photography
(1:60,000) mapped onto 1:250,000 NTS topographic maps. Figure 19.2 is
a map of selected glacial and lacustrine features compiled from the
aerial photograph interpretation and field mapping.

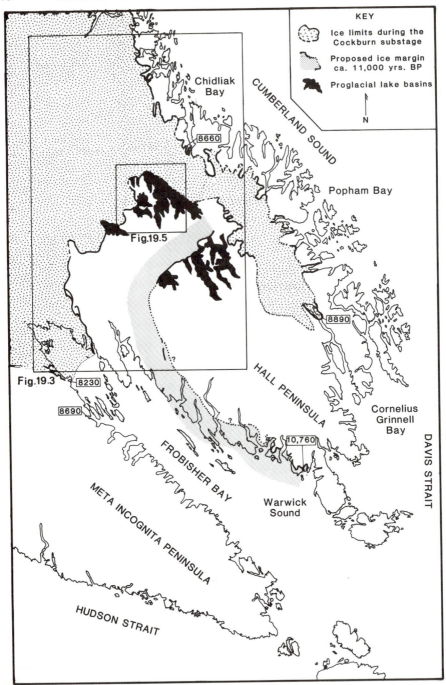

Figure 19.2 Generalized map of important morainal deposits, proglacial lake basins, and approximate glacial limits during the Cockburn substage (8-9 ka ago) and during formation of the Hall Moraine (11 ka ago). The locations and radiocarbon ages of samples that constrain the timing of these events are given in rectangular boxes.

Moraines

Moraines have been recognized previously on the peninsula. Blake
(1966) mapped a large segment of what has been subsequently called the
Frobisher Bay moraine system and provided a radiocarbon date on shells
from silt in juxtaposition with the moraine system in Frobisher Bay
(8230±240; GSC-462). He traced the moraine system most of the way
across Hall Peninsula and speculated that it may have continued across
Cumberland Sound. Blake's map (1966: Fig. 2) also shows moraine
segments farther east on Hall Peninsula but their significance is not
discussed nor is there any mention of lake shorelines on the peninsula.

A generalized outline of two major moraine systems is given in
Miller (1980: Fig. 1) including a more detailed map of moraine
segments in the Warwick Sound region of Hall Peninsula, outer Frobisher
Bay. Colvill (1982) and Lind (1983) treat the glacial history of inner
Frobisher Bay, including a new date on shells in sediment related to
the time when ice stood at the Frobisher Bay moraine (8690±120
[GSC-3157]; Lind, 1983).

Moraines identified in this study are outlined in Figure 19.2 and
shown in detail in Figure 19.3. Several long, nearly continuous
moraine segments are readily apparent.

Frobisher Bay moraine system. The outer moraine of this system can be
traced from its terminal position in Lewis Bay, inner Frobisher Bay, to
the northwest where it gradually ascends until its trace is lost at
about 500 m elevation. Drumlinized till proximal to the moraine
indicates ice flow paralleled the axis of the bay, and a complex
sequence of younger recessional moraines, some in cross-cutting
relationships, suggests numerous oscillations of the ice front between
deposition of the outermost terminal moraine in Lewis Bay and
deglaciation of the inner reaches of Frobisher Bay. Slightly north of
the last trace of the moraine system terminating in Lewis Bay, another
major moraine segment can be mapped as a prominent multi-ridged moraine
crossing the McKeand River valley and looping back to the north. The
morainal form is clearly controlled by relatively subtle topographic
obstacles, suggesting the ice was very thin at this time. Farther to
the northeast the trace of the moraine is lost.

Chidliak moraine. A second major moraine system can be traced with
only minor interpolation from the mouth of Irvine Inlet near the head
of Cumberland Sound, SSE along the coast of Hall Peninsula to Chidliak

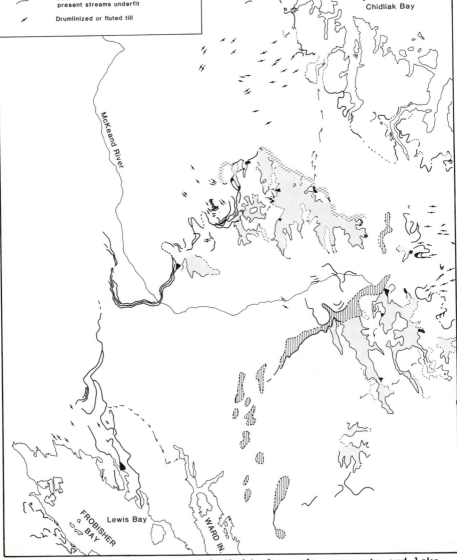

Figure 19.3 Detailed map of individual moraine segments and lake
shorelines as mapped from aerial photographs of central Hall
Peninsula.

Bay (Fig. 19.3). This moraine system was deposited by ice flowing off
the plateau of Hall Peninsula and funneled into the fiords and bays
fronting the sound. Numerous outlet glaciers reached tidewater,
depositing ice-proximal deltaic systems that are now 40-70 m asl. For
the most part the emerged delta complexes are composed of
coarse-grained, non-fossiliferous sediment, but the finer-grained
distal facies of one such delta in Chidliak Bay yielded shells of
Hiatella arctica that gave a date of 8660±160 (GSC-2466), providing a
close approximation of the time ice stood at the moraine system, and
indicating the correlation with the Frobisher Bay moraine farther
south. Drumlinized till east of Chidliak Bay trends northeasterly, in
contrast to the southeasterly trend of similar features in the center
of the peninsula (Fig. 19.3). Such changes in ice-directional
indicators suggests both thin ice and the possibility that multiple
domes characterized the margin of the continental ice sheet at this
time.

Additional moraine segments can be mapped along the coast
southeast of Chidliak Bay, but these cannot be reliably traced back to
those terminating in Chidliak Bay.

Local moraines. A broad morainic belt occurs west of the extant ice
caps inland from Popham Bay (Fig. 19.2), and requires ice flowing from
east to west, presumably from an expansion of the extant ice caps on
the peninsula. Drumlinized till indicates ice movement between west
and west-southwest. The terminal moraine formed during the expansion
of the local Hall Peninsula ice cap has been dated in inner Cornelius
Grinnell Bay. H. arctica shells collected from an ice-proximal raised
marine delta associated with an outlet glacier flowing down the bay
gave an age of 8890±100 (GSC-2568), thus indicating the broad
correlation of local ice advance, the inner continental moraine system
in Chidliak Bay (8660 BP) and the Frobisher Bay moraine (8230 BP, 8690
BP). A similar pattern of local and continental glaciers attaining a
maximum position during the Cockburn Substage (Andrews and Ives, 1978)
was documented on Cumberland Peninsula (Miller, 1973).

Hall moraine. The oldest mappable moraine system lies 50 km beyond the
inner moraine on Hall Peninsula. It is mapped as a broad morainic
belt, in places with discrete morainal form, that extends for 55 km
trending NE/SW just above 700 m elevation across the high plateau of
the peninsula (Fig. 19.2). The trace is lost completely to the
northwest; to the southwest a series of isolated patches of thick till
that swing to the southeast along the high land bordering Ward Inlet

are tentatively considered correlatives.   Miller (1980) provides three radiocarbon dates averaging 10,760±30 BP on deglaciation from a terminal moraine complex in Warwick Sound, outer Frobisher Bay. Collectively, these deposits constitute the Hall moraine system, representing the probable outer limit of late Foxe continental ice on Hall Peninsula and in Frobisher Bay.

Proglacial Lakes

In addition to the moraine systems, extensive lacustrine shorelines, deltas and overflow channels can be recognized in the aerial photographs of Hall Peninsula (Fig. 19.4). These deposits are evidence of proglacial lake systems, many mappable over several tens of kilometers with only intermittent breaks. Shorelines can be readily interpolated across areas of non-deposition by following the contours of the land surface which are at 60 m (200 ft) intervals for the NTS map sheets of this area.

Two main suites of lakes have been identified (Fig. 19.2). One series of lakes can be traced directly to the outer continental moraine system (Hall moraine) and a younger series is directly associated with the Frobisher Bay moraine system. Both series of lakes display well developed ice-contact/proximal deltas where the lakes abut the moraines. The older lakes lie at c. 630 m asl and drained to the south. Traces of a short-lived higher lake phase exist in disjunct segments, c. 20 m above the main lake shoreline. Lake shorelines are generally best developed along the southwestern perimeters of the former lake basins.

The younger lakes also lie close to 630 m asl and can be traced to the Frobisher Bay moraine along their western margins. However, the trace of the Frobisher Bay moraine disappears at the northwestern edge of the largest lake. The topography of the present-day land surface indicates that this lake could not have existed without a dam across much of the northeastern lake margin. Lake shorelines across this segment are found only in the isolated hills that rise above 650 m asl (Fig. 19.3). I postulate, therefore, that a thin residual ice cap must have occupied the region between the lake and Cumberland Sound at the time of formation of the largest lake in the younger lake series, although the relationship of such an ice dome and the ice that formed the Frobisher Bay moraine cannot be deciphered from the moraine evidence. That the two ice masses existed contemporaneously can be demonstrated by mapping lake overflow channels and associated deltaic systems. A detailed map of such relationships for two of the lakes in

(a)

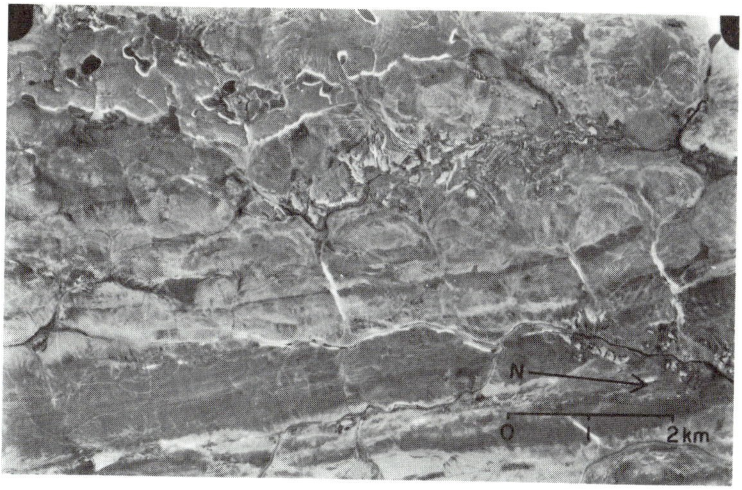

(b)

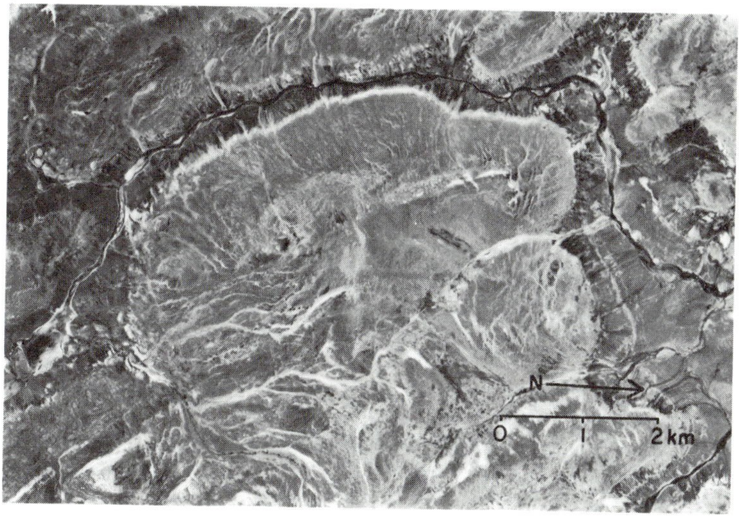

Figure 19.4 Two vertical aerial photograph of the interior of Hall
Peninsula showing proglacial lake shorelines and deltaic systems as
discussed in the text. The well defined proglacial lake shoreline in
the upper image (Fig. 4a) delimits a lake dammed by continental ice
lying at the Hall moraine. A segment of the Frobisher Bay moraine can
be seen at the bottom of the lower image (Fig. 4b) with an ice-contact
delta that was deposited in a small pro-glacial lake immediately distal
to the moraine (Lake B, Fig. 5) and the deltaic deposits (upper
left-hand corner) formed by overflow from this lake into a larger lake
(Lake C, Fig. 5) also dammed by the ice at this time.

Photos provided by the Department of Energy, Mines and Resources,
Government of Canada. Figure 4a is image A16744-116; Figure 4b is
image A16238-12.

the younger lake system is shown in Figure 19.5. The critical
relationships are the overflow channels from the isolated lake B that
was dammed solely by ice at the Frobisher Bay moraine, and the
associated deltas in the larger lake C. The overflow-delta
relationship clearly establishes that the two lakes were contemporary,
and, by implication, the damming ice masses must have also coexisted.

The geomorphic preservation of shorelines and deltas is similar
for both the inner and outer lake systems, suggesting no major age
differences.

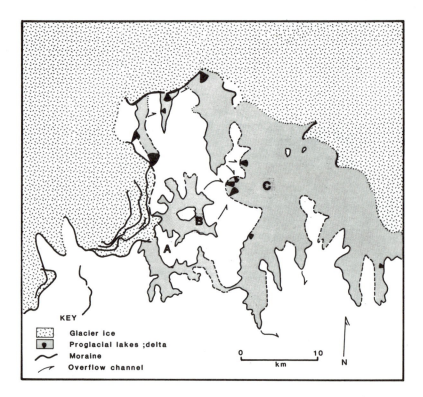

Figure 19.5 Detailed map of moraines, lake shorelines (both solid
where mapable, dashed where interpolated, dotted where presumed),
overflow channels and lacustrine deltas. The isolated lake B, dammed
by the Frobisher Bay Moraine drained to the northeast forming deltas in
Lake C, thereby proving the synchrony of the two lakes' existence.
Lake C can only exist if dammed along its northeast margin by a
residual ice mass, which, by inference, must have been contemporary
with continental ice at the Frobisher Bay moraine.

SUMMARY AND CLIMATIC IMPLICATIONS

From the evidence presented above the following sequence of events is
proposed:

(1) Advance of continental ice onto Hall Peninsula reaching the outer
moraine system, possibly coalescent with a local independent
ice-dispersal system centered over NE Hall Peninsula, was followed by
stabilization of the continental ice margin and the development of
prominent shorelines around the associated proglacial lakes. The
correlation of this event with the Hall moraine in outer Frobisher Bay
would suggest ice recession and lake drainage occurred shortly after
11,000 BP.

(2) Substantial ice recession without a major still-stand to a position
behind the Frobisher Bay moraine system. At this time local ice
dispersal centers became independent of the continental ice sheet and
did not undergo major recession. In addition, the ice dome that would
later form the northeast marginal dam for lake C in Figure 19.5 must
have persisted while the central plateau was deglaciated. The low
glacierization level that is maintained over this portion of the
peninsula at present is due to enhanced precipitation and decreased
ablation (local cloudiness) related to persistent ice-free conditions
along the outer Cumberland Sound coast of Hall Peninsula (Andrews and
Miller, 1972). A similar mechanism is postulated to maintain the ice
dome damming lake C (Fig. 19.5) even though the adjacent plateau was
deglaciated.

(3) A readvance of continental ice from the west to the Frobisher
Bay/Chidliak/Ranger moraine systems across southern Baffin Island,
damming the younger series of proglacial lakes, with the water in Lake
C (Fig. 19.5) blocked from draining into Cumberland Sound by the
residual ice mass that persisted northeast of the lake. During this
advance the continental ice sheet is postulated to have been thin along
its perimeter with numerous peripheral dispersal centers. At about the
same time the local ice cap system over the northeast uplands of the
peninsula expanded. There is no trace of an older, Hall moraine
correlative for the local ice cap system, hence the Frobisher Bay
correlative moraines are considered to represent the local late Foxe
maximum. Radiocarbon dates on molluscs in ice-proximal sediments
associated with these moraines indicate that all were being formed in
the time range of 8200 to 8900 BP.

    Lind (1983) presented geomorphic and sedimentary biofacies
evidence that the Frobisher Bay moraine was formed during a readvance

of continental ice into Frobisher Bay after a period of substantial deglaciation. Muller (1980) found evidence and obtained supporting C-14 dates that outlet glaciers of the Terra Nivea Ice Cap readvanced off Meta Incognita Peninsula at about the same time.

(4) Slow deglaciation. The residual ice cap forming the dam along the northeastern margin of lake C (Fig. 19.5) collapsed before continental ice withdrew far enough to the west to open the drainage into the McKeand River system. Massive fluvial deposits with large-scale bars indicate that lake C drained catastrophically into Cumberland Sound via outer Chidliak Bay as indicated in Figure 19.3. Small, less continuous lakes mapped proximal to the Frobisher Bay moraine suggest at least minor still-stands during regional deglaciation.

The primary uncertainties in this reconstruction are the correlation between the outer continental moraines on Hall Peninsula, with the dated Hall moraine in Warwick Sound, and the contact between the local ice cap and the residual ice dome damming the large proglacial lake at the Frobisher Bay moraine. Much of the upland area of the peninsula is characterized by a thin till veneer without distinctive morphology. Field mapping across the interior will be required to confirm the proposed correlations.

A certain amount of climatic inference is contained in the scenario outlined above. The well-developed shoreline deposits defining the proglacial lake systems require open-water conditions during most summers while ice stood at the adjacent damming moraines. Even under present climatic conditions, lakes above 600 m elevation occasionally remain ice-covered throughout the summer. Consequently the mean summer temperature prevailing during formation of both the Hall and Frobisher Bay moraine systems must have been as high or slightly higher than that of the last few decades. The abundance of glacio-fluvial features associated with the Hall moraine in the Warwick Sound region also indicates summer warmth and abundant meltwater.

The stronger development of shore features along the southern and western perimeters of the lakes suggests prevailing summer/fall winds from the NE. The modern-day distribution of wind-drifted perennial snowbanks indicates the dominant wind direction at present is also from the NE. Thus, in all likelihood, summer temperatures and wind regime between 11,000 and 8000 BP were similar to, and temperatures perhaps slightly warmer than, present conditions. The terrestrial climate appears to have warmed earlier than the inshore marine environment where near-modern conditions were attained in Frobisher Bay between 10 and 9.7 ka ago, after initial deglaciation.

ACKNOWLEDGEMENTS

Field work along the northern Hall Peninsula coastline was supported by
NSF grant GA-41562, and along the southern and eastern coasts by
contract from the Arctic Petroleum Operators' Association (APOA project
138). I am indebted to Malcolm Anderson and Harvey Moulton for
assistance in the field and to Dr. J.D. Jacobs (University of Windsor)
and D. Gardner (University of Calgary) for sharing logistical support.
I would also like to acknowledge support from the National Air Photo
Library in Ottawa, Dr. W. Blake, Jr. (Geological Survey of Canada) for
radiocarbon dating and the Inuit of Frobisher Bay and Pangnirtung for
their assistance with field logistics.

REFERENCES CITED

Andrews, J.T. and Miller, G.H., 1972: Quaternary history of northern
    Cumberland Peninsula, Baffin Island, N.W.T., Canada: Part IV:
    Maps of the present glaciation units and lowest equilibrium line
    altitudes for north and south Baffin Island. Arctic and Alpine
    Research, 4:45-59.
Andrews, J.T., and Ives, J.D., 1978: "Cockburn" nomenclature and the
    late Quaternary history of the Eastern Canadian Arctic. Arctic and
    Alpine Research, 10:617-633.

Barnett, D.M. and Peterson, J.A., 1964: The significance of glacial
    lake Naskaupi 2 in the deglaciation of Labrador-Ungava. Canadian
    Geographer, VIII: 173-181.
Blake, W., Jr., 1966: End moraines and deglaciation chronology in
    northern Canada with special reference to southern Baffin Island.
    Geological Survey Canada Paper 66-26, 31 pp.

Colvill, A.J., 1982: Glacial landforms at the head of Frobisher Bay,
    Baffin Island, Canada. Unpublished M.A. thesis, University of
    Colorado, Boulder. 202 pp.

Dyke, A.S., 1979: Glacial and sea-level history of southwestern
    Cumberland Peninsula, Baffin Island, N.W.T., Canada. Arctic and
    Alpine Research, 11:79-202.

Ives, J.D., 1960: Former ice-dammed lakes and the deglaciation of the
    middle reaches of the George River, Labrador-Ungava. Geographical
    Bulletin Canada, 4:44-70.

Ives, J.D. and Andrews, J.T., 1963: Studies in the physical geography
    of north-central Baffin Island, N.W.T. Geographical Bulletin,
    19:5-48.

Lauriol, B. and Gray, J.T., 1983: Un lac glaciare dans la region du lac
    Minto - Nouveau - Quebec. Canadian Journal of Earth Sciences,
    20:1488-1492.
Lind, E., 1983: Holocene paleoecology and deglacial history of the Cape
    Rammelsburg area, southern Baffin Island, N.W.T., Canada.
    Unpublished M.A. thesis, Unviversity of Colorado. 219 pp.

Miller, G.H., 1980: Late Foxe glaciation of southern Baffin Island,
    N.W.T., Canada. Geological Society of America Bulletin, Part I,
    91:399-405.
Müller, D.S., 1980: Glacial geology and Quaternary history of southeast
    Meta Incognita Peninsula, Baffin Island, Canada. Unpublished M.Sc.
    thesis, University of Colorado, Boulder. 211 pp.

Survey — Part IV

The chapters in Part IV covered a variety of topics on a variety of timescales but largely concentrated their attention on delimiting the glacial and marine events that characterized the coasts of Baffin Island, West Greenland, and Bylot Island. Emphasis was mainly placed on biostratigraphy, aminostratigraphy, and both relative and absolute chronologies. Locke (Chapter 12) considers some of the processes that operate in the arctic and control the manner and rate of soils formation and rock weathering. The remaining chapters are more concerned with stratigraphy rather than the environment of deposition, although all the chapters discuss the origin(s) of the units they are using. In a paper by Mode et al. (1983) a more detailed sedimentological description is given of the raised marine/glacial marine sediments that are discussed in other terms in Chapters 13, 14, and 18.

An important consideration at this juncture is how can the glacial, sea level, and climate events that are described in these 8 chapters be tied into the deep-sea or ice core records (see Part III)? The micropaleontological research of Feyling-Hanssen (Chapter 13) establishes that the exposed Quaternary record along eastern Baffin Island is surprisingly extensive in terms of time, and may include sediments of Pliocene age. This suggestion cannot be refuted by Miller's analysis (Chapter 14) of the rates of racemization of arctic molluscs. The foraminiferal assemblages described by Feyling-Hanssen represent intervals when sea level was above present and can thus record higher sea levels associated with glacial isostatic loading during the onset of a glaciation (= regression) (Andrews et al., 1981). The results of Feyling-Hanssen and those of Mode (Chapter 18) on the molluscs and pollen indicate that during those periods when sea level was above present, and sediment was being deposited against the outer coast of Baffin Island that the marine environment was more favorable than it is today. These are important results and must be considered in models of the inception of Northern Hemisphere Glaciation. Thus the initial suggestion that might be made is that the correlation between the land record and the marine record is not a simple: Cold = Glaciation equation but certainly involves a Warm = Glaciation element. Further refinement of the correlation between Baffin Bay and the land surrounding the Bay may be gleaned from the chapters that consider the glacial chronology of West Greenland (Chapter 16), Bylot Island (Chapter 15), and Baffin Island (Chapter 12, 14, 18, and 19). Klassen's work on Bylot Island may prove to be the pivotol research for several reasons: (1) It forms a potential link between the West Greenland and Baffin Island studies; and (2) the

island stands adjacent to deep water at the northern end of Baffin
Bay. Of particular note is Klassen's observations that the Eclipse
Glaciation represents a large ice sheet that eroded Paleozoic
limestones and which terminated at the northern end of Baffin Bay. On
the basis of amino acid ratios and minimum C-14 dates it is probable
that the Eclipse Glaciation is correlative with the Ayr Lake stadial of
Baffin Island and that both are included within the Kogalu aminozone.
On West Greenland, Kelly's data (Chapter 16) indicates that the amino
acid ratios on in situ and glacially transported marine shells may form
the platform for a relative aminostratigraphy (also, Funder and
Simonarson, 1984). On West Greenland, southern Ellesemere Island
(Blake, 1980) Bylot Island, and eastern Baffin Island, and northern
Labrador (Clark, 1984) (i.e. circum Baffin Bay!) there is growing
evidence for a middle Wisconsin (Foxe) glacial/deglacial event.
However, in all areas considered the suggestion has been made that the
most extensive glaciation during the last 120 ka BP was an early
Wisconsin event. What signal might exist to correlate from the land
into the deep sea? If there is one, it might be the link between the
erosion of Paleozoic carbonates and the development of distinctive
carbonate rich facies in Baffin Bay. Aksu (Chapter 7 this volume)
calls this Facies A. In cores where Facies A has been delimited it
occurs between the middle of isotope stage 5 and continues into stage 4
(Aksu, 1981). This is not the place to further develop these
correlations in detail, but...I suggest that we are close to obtaining
a significant level of real correlation between deep sea, ice core, and
terrestrial glacial/glacial marine records. This in itself is a major
acomplishement of this volume.

REFERENCES

Andrews, J.T., Miller, G.H., Nelson, A.R., Mode, W.N., and W.W. Locke, III, 1981: Quaternary near-shore environments on eastern Baffin Island, N.W.T., In: Mahaney, W.C. (ed.), Quaternary paleoclimate. Geo Books, Norwich, England, 13-44.

Blake, W., Jr., 1980: Mid-Wisconsinan interstadial deposits beneath Holocene beaches, Cape Storm, Ellesmere Island, Arctic Canada. Orono, Maine: American Association for Quaternary Research, Abstracts with Programs. 261-27.

Clark, P.U., 1984: Glacial history of northern Labrador, Canada. Unpublished Ph.D. Thesis, University of Colorado, Boulder, CO. 230 pp.

Funder, S. and Simonarson, L.A., 1984: Bio- and aminostratigraphy of some Quaternary marine deposits in West Greenland. Canadian Journal of Earth Sciences, 21:843-852.

Mode, W.N., Nelson, A.R., and J.K. Brigham, 1983: A facies model of Quaternary glacial-marine cyclic sedimentation along eastern Baffin Island, Canada. In: Molnia, B.F. (ed.), Glacial-marine Sedimentation. Plenum, New York, 495-534.

Part V

HOLOCENE SEA LEVELS AND CLIMATE

20 A numerical model of postglacial relative sea level change near Baffin Island

Garry Quinlan

It is intuitive that some form of causal relationship must exist between the ablation of late Pleistocene ice sheets and the variations in RSL (relative sea level) seen preserved in the geologic record of coastal areas. The form of this relationship has been quantified in numerical models which successfully explain the major features of the global postglacial RSL record (Farrell and Clark, 1976; Peltier and Andrews, 1976). By increasing the spatial resolution of these models, Quinlan and Beaumont (1981, 1982) examined details of the RSL record within Atlantic Canada and proposed a compatible late Wisconsin ice reconstruction for that area. The current paper uses the Quinlan and Beaumont (1982) approach to analyse the RSL record of the region around Baffin Island. There are two principal objectives: first, to derive an ice reconstruction consistent with the known RSL record; and second, to use this reconstruction to estimate the form of the RSL record for sites where direct RSL indicators are either absent or ambiguous in their implications.

The necessary calculations are discussed in detail by Farrell and Clark (1976) and will only be summarized in this paper in order to make clear the nature of the modifications to the original model. The fundamental equation relating ice ablation to RSL is of the form

$$[20.1] \quad RSL(\underline{r},t) = g^{-1}\{\iint_{ice} G^{E}(\underline{r}-\underline{r}')\rho_{I} I(\underline{r}',t)dA$$

$$+ \iint_{ocean} G^{E}(\underline{r}-\underline{r}')\rho_{w} RSL(\underline{r}',t)dA$$

$$+ \int_{t_o}^{t} dt'[\iint_{ice} G^{v}(\underline{r}-\underline{r}', t-t')\rho_{I} I(\underline{r}', t')dA$$

$$+ \iint_{ocean} G^V(\underline{r}-\underline{r}', t-t') \rho_w RSL(\underline{r}', t')dA]\}$$

$$- M_I(t)/A_o \rho_w -K$$

where   RSL $(\underline{r}, t)$ is RSL change at vector location $\underline{r}$ between time t_o

and time t.

$I(\underline{r},t)$ is ice thickness change at location $\underline{r}$ between time t_o

and time t.

$G^E(\underline{r}-r')$ is elastic Green function giving instantaneous RSL
change at location $\underline{r}$ due to a 1 kg point load change at
location $\underline{r}'$.

$G^V(\underline{r}-\underline{r}', t-t')$ is viscous Green function giving RSL change at
location $\underline{r}$ and time t due to a 1 kg point load change at
location $\underline{r}'$ and time t'.

g is acceleration due to gravity.

ρ_w is density of water.

ρ_I is density of ice.

t_o is time at which ablation commences.

$M_I(t)$ is total mass of ice melted between time t_o and time
t.

A_o is total area of world's oceans.

K is mass-conserving constant.

$\underline{r}-\underline{r}'$ is arc length distance between locations $\underline{r}$ and $\underline{r}'$.

dA is element of surface area.

Equation [20.1] is an integral equation in which the unknown, RSL,
appears in the second and fourth integral expressions of the right hand
side. This requires an iterative solution which is time-consuming and
which must proceed numerically because of the non-analytic form of the
Green functions, G^E and G^V, and of the ice function $I(\underline{r},t)$ (e.g.
Peltier and Andrews, 1976).

The basic idea is that one divides the world's oceans and
glaciated regions into discrete grids. Any particular ice
reconstruction is defined by specifying an average ice thickness value
for each grid of the glaciated region at each of a number of discrete
1000 year timesteps. These discrete values are meant to be
approximations to the continuous ice function $I(\underline{r},t)$ of equation [20.1]
and may be arrived at by informed geological judgement, by random
guessing or by any other process. What is important is how well the
RSL consequences of the reconstruction agree with RSL observations in
the field. Equation [20.1] must be solved to determine these

consequences.   In order to evaluate the second and
fourth integrals of equation [20.1] numerically, RSL must be defined as
an average value over each of the oceanic grids at each discrete
timestep.   The calculated RSL values are, therefore, only discrete
approximations to the continuous function RSL ($\underline{r}$,t).   This imposes two
restrictions: first, field observations of RSL must be compared with
calculated results from the nearest oceanic grid which may be up to c.
100 km distant; and second, the model is unable to resolve RSL
variations occurring on a spatial scale smaller than the oceanic grid
size.

There are ways to circumvent these limitations if one wishes to
make spatially detailed calculations of RSL.   Quinlan and Beaumont
(1981, 1982) used more and smaller grids to describe their area of
interest around Atlantic Canada than they used to describe the rest of
the world's ice sheets and oceans.   This method increases the time
required to solve equation [20.1] because the number of grid points is
increased.   An alternative approach is to solve an approximation to
equation [20.1] as discussed by Clark (1980).

This approximation is

$$[20.2] \quad RSL(\underline{r},t) = g^{-1} \left\{ \iint\limits_{ice} G^{E} (\underline{r}-\underline{r}')\rho_{I} I(\underline{r}', t)dA \right.$$

$$+ \int\limits_{t_{o}}^{t} dt' \iint\limits_{ice} \dot{G} (\underline{r}-\underline{r}', t-t')\rho_{I} I(\underline{r}', t')dA \Bigg\}$$

$$-M_{I}(t)/A_{o}\rho_{w}$$

where all symbols are as in equation [20.1].

This approximation makes use of the fact that water depth (RSL)
changes are typically an order of magnitude less than changes in ice
thickness for sites in formerly glaciated areas.   The integral
expressions in equation [20.1] which involve ice thickness dominate
those involving RSL and so the latter may be ignored to a first
approximation.   This is true only for sites in or near to formerly
glaciated areas but where it is applicable, equation [20.2] has
distinct advantages.   It is not an integral equation and so does not
require an iterative solution; this makes it much faster to solve than
equation [20.1].   Also it is not necessary to partition the oceans into
discrete grids in order to describe the water load because the effects
of the water load are ignored.   This means that RSL changes can be

calculated (approximately) for comparison with observations at any
arbitrary site and not just at the centers of discrete oceanic grids.
The current paper follows Quinlan and Beaumont (1981, 1982) in the use
of many small grids to describe the ice reconstruction around Baffin
Island but uses equation [20.2] to calculate the RSL consequences of
this ice reconstruction. As will be seen below, there are sufficient
ambiguities in the model results that the expense of a fully iterative
and slightly more accurate solution to equation [20.1] is probably not
warranted.

DERIVING THE ICE RECONSTRUCTION

The ice reconstruction for the Baffin region resulted from the Quinlan
and Beaumont (1982) approach of making systematic trial-and-error
modifications to a starting or "first guess" ice reconstruction. In
this case the starting ice model was the Peltier and Andrews (1976)
reconstruction, ICE-1, as modified by Quinlan and Beaumont (1982). The
ice within the heavy outline area of Figure 20.1 was defined on 1°x1°
grids whereas all global ice outside this area was defined more
coarsely on 5°x5° grids. Quinlan and Beaumont (1982) have shown that
the distinctive features of any calculated RSL curve can often be
associated with specific regions of the starting ice reconstruction.
By identifying those portions of the starting reconstruction which are
responsible for misfit between calculated and observed RSL curves, it
becomes possible to modify this reconstruction in a selective manner in
order to reduce the misfit. Successive trial-and-error adjustments to
the starting reconstruction are made until a point is reached at which
the calculated and observed RSL curves are similar to within likely
limits of observational error. At this point the ice reconstruction is
reasonable in the sense that it is compatible with the RSL record.

A PROPOSED ICE RECONSTRUCTION

Figure 20.2 shows contours of the proposed ice reconstruction at four
stages of deglaciation. It is assumed that the present-day ice
configuration over Baffin Island was attained 6 ky BP (6000 years
before present). Figures 20.3, 20.4 and 20.5 show that the RSL values
calculated from this reconstruction are generally in excellent
agreement with observations at the sites identified in Figure 20.1.
RSL indicators, however, often have large uncertainties both in their
absolute ages and in the elevation they imply for RSL. The size of
these uncertainties is indicated by the stippled envelopes of Figure
20.3. Obviously if the RSL data cannot define a unique RSL curve, then

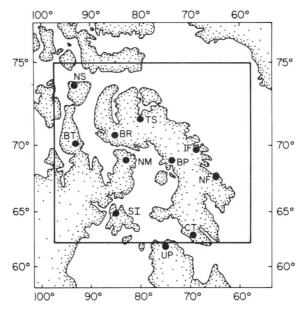

Figure 20.1 Location map showing the sites whose RSL records are used to constrain the proposed ice reconstruction. The ice reconstruction is defined on 1°x1° grids within the heavy border and on 5°x5° grids for other areas. BP = Baird Peninsula, BR = Brodeur Peninsula, BT = Boothia Peninsula, CT = Cape Tanfield, IF = Inugsuin Fiord, NF = Narpaing Fiord-Broughton Island, NM = Northern Melville Peninsula, NS = Northern Somerset Island, SI = Southhampton Island, TS = Tay Sound, UP = Ungava Peninsula.

there cannot be a unique ice reconstruction compatible with these data. Instead there is a range of acceptable reconstructions of which the proposed reconstruction of Figure 20.2 is only one. It is necessary to examine the extent to which departures from this reconstruction are allowed by the data. This is equivalent to asking whether specific features of the reconstruction are required by the data or are simply artifacts of the trial-and-error process used to generate the reconstruction.

Figure 20.2 shows a number of prominent features:

  (i)   The ice is basically stable over Baffin Island between 18
        ky BP and 8 ky BP with rapid deglaciation following this
        long period of stability.
  (ii) There is an elongated, bilobate ice ridge along western Baffin
        Island and over the Foxe Basin. This ridge shows most
        prominently after it begins to separate from the bulk of the
        Laurentide ice c. 10 ky BP.

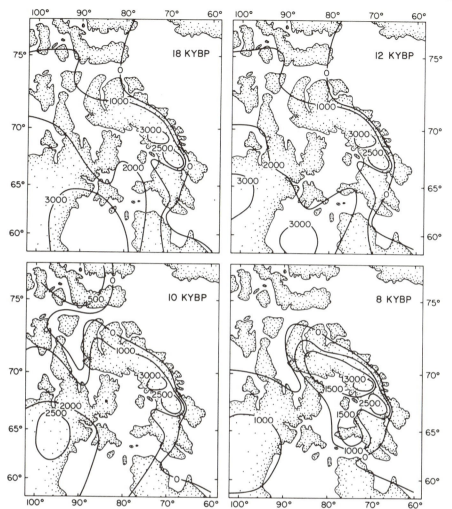

Figure 20.2. Contours of the proposed ice model for four stages of ablation. Contours are in metres (with permission).

(iii) The outer coast of eastern Baffin Island is ice-free throughout the time period of the model from 18 ky BP to the present.
(iv)   Within the Baffin Island ice cover there are local ice thickness highs in locations near the present day Barnes and Penny Ice Caps.

Only the first two of these features, as will be shown below, are well-constrained by observations. Let us examine these features individually.

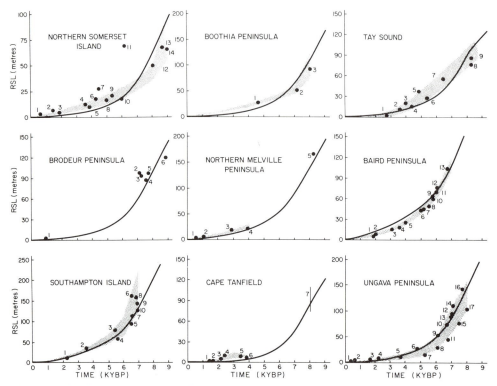

Figure 20.3.   Calculated RSL curves compared with observations from the
sites identified in Figure 20.1.   The stippled envelopes represent
likely limits on the position of past RSL consistent with the numbered
observations.   Northern Somerset Island: 1 = S1391, 2 = S1383,   3 =
S1405, 4 = GSC2233, 5 = S1393, 6 = 21389, 7 = S1386, 8 = GSC2081, 9 =
GSC2080, 10 = S1374, 11 = S1381, 12 = S1384, 13 = GSC450, 14 = GSC150.
(All data and envelope from Dyke, 1979.) Boothia Peninsula: 1 =
GSC45, 2 = I6212, 3 = I6213.   (All data and envelope from Walcott,
1972.) Tay Sound: 1 = GSC654, 2 = I1317, 3 = I1320, 4 = I1318, 5 =
I1245, 6 = L762C, 7 = GSC328, 8 = I1316, 9 = I724. (All data from
Andrews and Drapier, 1967.) Brodeur Peninsula: 1 = GSC239, 6 = GSC183
(Dyck et al., 1965); 2 = GSC307, 3 = GSC304, 4 = I1254, 5 = GSC306
(Dyck et al., 1966). Northern Melville Peninsula: 1 = K504 (Tauber,
1960); 2 = GSC691 (Lowdon and Blake, 1968); 3 = P213, 4 = P207 (Andrews
et al., 1971), 5 = Prest et al., 1967. Baird Peninsula: 1 = I1830, 2
= I489, 3 = GSC564, 4 = I2830, 5 = GSC557, 9 = I1831, 12 = I405, 13 =
I406 (Andrews and Drapier, 1967); 6 = F5, 7 = F4, 8 = F3, 10 = F2, 11 =
F1 (Andrews, 1970). Southampton Island: 1 = M1085 (Crane and Griffin,
1966); 2 = S12, 4 = S13 (Andrews and Drapier, 1967); 5 = GSC337, 6 =
GSC308, 7 = GSC309, 8 = GSC323, 9 = GSC334, 10 = GSC311 (Dyck et al.,
1965). Cape Tanfield: 1 = M1529, 2 = M1528, 4 = M1531, 5 = M1532
(Crane and Griffin, 1966); 3 = GSC596, 6 = Gif 2816, 7 = GSC425
(Andrews et al., 1970) Ungava Peninsula: 1 = Y1717, 4 = M1534
(Andrews et et al. 1971); 2 = Gak 1036, 3 = GSC537, 5 = NPL71, 7 =
N284, 8 = N285, 14 = L702A, 15 = L702B, 16 = I729, 17 = GSC672
(Mathews, 1967); 6 = GSC812 (Lowdon and Blake, 1968); 10 = NPL84, 11 =
NPL82, 13 = NPL85 (Callow et al., 1966); 12 = NPL58 (Callow et al.,
1965).

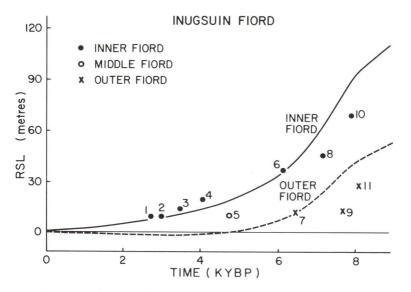

Figure 20.4   Calculated RSL curves for the inner and outer Inugsiun fiord are compared with observations.   1 = I1555, 2 = I1599, 3 = 1601, 4 = I1597, 5 = I2961, 6 = GSC631, 7 = I2962, 8 = I1554, 9 = I2831, 10 = I1673, 11 = Y1705 (Andrews and Drapier, 1967).

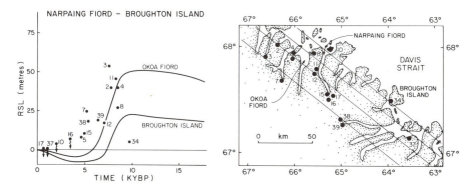

Figure 20.5   Calculated RSL curves for the Narpaing Fiord-Broughton Island area are compared with observational data. The data are all from Pheasant and Andrews (1973) as is the inset map. The numbered data were collected from the correspondingly numbered sites shown on this map.   There is a trend in the height of these features, with features lying along a common NW-SE line tending to define a different RSL curve than features lying along a different NW-SE line. Two calculated RSL curves are shown, one for the inner area near Okoa Fiord and one for the outer area near Broughton Island.

Late Melting Baffin Ice

Quinlan and Beaumont (1982) showed that the exact timing of ice
thinning was less important a factor in determining the RSL record of
Atlantic Canada than was the magnitude of ice thinning. This
conclusion is valid for Atlantic Canada because that area was largely
ice-free by c. 13 ky BP and, therefore, the model has at most four or
five 1000 year melting episodes in which to remove any ice present at
the 18 ky BP glacial maximum. With such little leeway for choice the
exact manner of thinning was relatively unimportant.

Baffin Island presents a much different problem in that it is
known to have been largely ice covered at least as recently as 8 ky
BP. Baffin ice cover must therefore be described in the model for more
time steps than was the case for Atlantic Canada ice and these extra
parameters of the Baffin ice reconstruction must be constrained by
available RSL data. It is only reasonable that broader perturbations
about the proposed Baffin reconstruction will be tolerated by a
comparable amount of data. In fact the situation is somewhat worse
than this in that the Baffin RSL record goes back only as far as 8-9 ky
BP rather than the 13 ky BP value in Atlantic Canada. This results in
a shorter time window of data to constrain more ice sheet parameters.
The effect is greater uncertainty in the early ice cover of Baffin
Island than was the case in Quinlan and Beaumont's (1982) analysis for
Atlantic Canada.

Figure 20.6 illustrates the effect of differences in the assumed
melting timetable for Baffin ice. Suppose that a basic 1°x1° ice grid
on Baffin Island is covered by 1000 m of ice, allowed to establish
isostatic equilibrium, and then deglaciated according to one of two
alternative timetables, both of which leave the grid ice-free by 8 ky
BP. In one case the ice remains intact until 9 ky BP and then melts
completely by 8 ky BP. In the second case the ice is thinned linearly
from 1000 m to zero between 18 and 8 ky BP. Curve 1 of Figure 20.6
represents the RSL consequences of the former case and curve 2 the
consequences of the latter. In both cases the RSL curve is calculated
for a site directly beneath the grid center. The linear melting
history produces a RSL curve which is consistently below that for the
late melting ice. In order to mimic the effects of late melting ice it
would be necessary to assume at least double the thickness of linearly
melting ice. This suggests that rather large trade-offs are possible
between the proposed ice thickness and the rate at which the ice is
assumed to have thinned. The timetable used in the proposed
reconstruction corresponds most closely to the late melting case for
Baffin ice and to the linearly melting case for other ice. The large

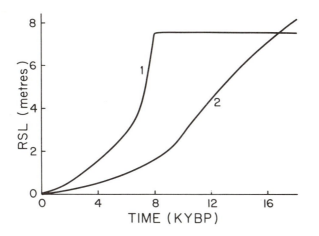

Figure 12.6 The effect of the timing of glacial melting on the
calculated values of RSL. A 1°x1° Baffin Island ice grid is covered by
1000 m of ice and deglaciated according to one of two alternative
timetables. RSL is calculated at the center of the grid and shown by
curves 1 and 2. Curve 1 corresponds to instantaneous melting of the
entire 1000 m at 8 ky BP. Curve 2 corresponds to linear reduction of
the ice thickness between 18 and 8 ky BP.

range of possible melting timetables precludes a detailed search
through all of them. It is clear, however, that the proposed ice
thickness for Baffin Island would need to be much greater, perhaps
twice as great, if it were to have undergone substantial thinning much
prior to 8 ky BP.

The 8 ky BP model ice ridge along Baffin Island has a half width
of c. 150 km and so a central thickness of c. 1700 m would be
consistent with simple ice mechanical theory (e.g. Paterson, 1969, p.
146). The proposed ice thickness is close to this value although it
should be emphasized that no form of glaciological theory was used in
the construction of the proposed ice model. This approximate agreement
could be considered fortuitous. It does suggest, however, that trading
the proposed late melting ice for substantially thicker and earlier
melting ice is not an attractive option.

Geological field work often implies the occurrence of glacial
readvances on a local scale; an example is Miller's (1980) work in
Frobisher Bay which documents an ice advance down Frobisher Bay between
10 - 10.7 ky BP. As Quinlan and Beaumont (1981) have shown, small
local advances need not leave highly discernible evidence in the RSL
record of surrounding sites. This suggests that a precise analysis of

the local RSL record will be needed to see the effects of such
readvances. It also means that the general predictions of the model
will not suffer drastically by failure to include small readvances in
the proposed ice reconstruction. In general, no attempt has been made
to deal with such readvances in specifying the timing of ice ablation.

Western Baffin Ice Ridge

The ice ridge over western Baffin Island and the Foxe Basin shown in
Figure 20.2 is similar to one proposed by Andrews and Miller (1979) to
explain the distribution of limestone erratics on Baffin Island. The
major difference between these two reconstructions is the bilobate form
of the ridge shown in Figure 20.2. Andrews and Miller (1979) saw this
ridge as a more uniform feature along its strike but conceded that
their data were also consistent with a ridge which swung onshore in
southern Baffin Island. The southern lobe of Figure 20.2 is similar to
the late melting ice mass proposed by Blake (1966) for the Amadjuak
Lake area. How well do RSL observations constrain the form of this ice
ridge and is it a feature required by the RSL record?

Figure 20.7 uses the uncertainty bounds on data from Tay Sound,
the Brodeur Peninsula, and Cape Tanfield-Ungava Peninsula to assess
allowable perturbations to the ice reconstruction in northern Baffin
Island, over the Foxe Basin, and in southern Baffin Island
respectively. In the graphs of Figure 20.7 the horizontal axis
represents the value of RSL as calculated from the ice reconstruction
of Figure 20.2. The stippled regions of these graphs depict the data
uncertainty drawn with respect to the calculated RSL values. For
example, the 4 ky BP RSL value for Tay Sound has been calculated as 12
m above present but the data would allow this value to be 8 m higher or
4 m lower. Notice that Figure 20.7 ignores the actual form of the RSL
curve and focuses instead on how much leeway exists for changes to the
calculated curve. This permissible leeway is a measure of allowable
perturbations to the proposed ice reconstruction. Any perturbations
will produce departures from the curves of Figure 20.3. These
departures brand the perturbation as unacceptable if they lie outside
the bounds of data uncertainty.

Suppose that ice in northern Baffin Island (areas 1 and 2 of Fig.
20.7) were taken to be either twice or half its proposed thickness.
The effects of such perturbations on the calculated RSL for Tay Sound
are shown by curves 1 and 2 respectively. It is clear that either of
these perturbations is too extreme because both curves 1 and 2 lie well
outside permissible limits. The ice thickness in northern Baffin

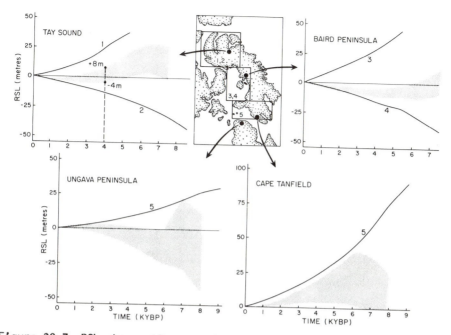

Figure 20.7 RSL observations at the sites shown are used to constrain
ice thickness within the numbered areas of the inset map. Each
numbered curve represents the effect of a perturbation to the proposed
ice reconstruction within the correspondingly numbered area. The
details of these perturbations are described in the text.

Island is constrained to within a factor of two and the northern ice
lobe is apparently required by the Tay Sound RSL data.

Suppose that ice cover over the Foxe Basin (areas 3 and 4 of Fig.
20.7) were either twice or half its proposed thickness. The effects of
such perturbations on the calculated RSL values for the Baird Peninsula
are shown by curves 3 and 4 respectively. Either of these
perturbations produces excessively large data misfit. It seems
unlikely that the bilobate form of the ridge could be modified to a
continuous, uniformly thick ridge without degrading the model's ability
to reproduce RSL observations from the Baird Peninsula. If anything, a
slight thinning of this ice is suggested.

The calculated RSL curve for Cape Tanfield slightly underestimates
the observed height of raised marine features (Fig. 20.3). For this
reason there is no point considering reductions to the proposed ice
thickness in southern Baffin Island (area 5 of Fig. 20.7); such
reductions would only aggravate this underestimate. Curve 5 of Figure

20.7 shows the effect on RSL at both Cape Tanfield and the Ungava Peninsula of doubling the ice in area 5. This modification is too extreme but something between a 25% and 50% increase cannot be ruled out on the basis of Cape Tanfield observations. Notice, however, that the proposed reconstruction almost overestimates the height of raised marine features at nearby Ungava Peninsula (Fig. 20.3). Increasing ice thickness in southern Baffin Island can be expected to cause problems in the Ungava Peninsula and any adjustments must take this into account. Recall that the Ungava data site lies outside the area whose ice cover is defined on 1°x1° grids. The coarser definition of ice in the Ungava region prevents "fine tuning" of the calculated Ungava RSL curve and doubtless has some secondary influence on the model's ability to fit the nearby Cape Tanfield data. Given the scale of spatial definition for this area, the data fit seems quite adequate and apparently requires an ice thickness high over southwestern Baffin Island as shown in Figure 20.2.

Ice-Free Eastern Coast?

Andrews (1980) argues against late Foxe glaciation of the outer coast of eastern Baffin Island in the area between the Clyde River Forelands and the southeastern Cumberland Peninsula. The argument is based on the vertical juxtaposition of raised marine indicators 30 ky BP with those younger than 10 ky BP. The absence of marine features of intervening age may be due to local ice cover (Feyling-Hansen, 1976) but the lack of an appropriately old till discredits this explanation. Farther to the north Klassen (1981) finds evidence that Bylot Island has been isolated from external ice sources for at least the last 35,000 years. The proposed ice reconstruction also has late Foxe age ice confined westward of the eastern Baffin Island highlands but it is not clear that the RSL record demands this conclusion. It is necessary to establish that the RSL record is incompatible with a glaciated outer coast and not simply compatible with an unglaciated one.

The sites most sensitive to this portion of the ice reconstruction are those at Inugsuin Fiord and Narpaing Fiord-Broughton Island. The RSL indicators at these sites form a gradational record along the outer coast with data from each area overlapping those of the intervening Home Bay region. In addition, there are signs of a distinct trend in the data from the mouths of the coastal fiords to their heads. It is difficult, therefore, to take the RSL curve calculated for a single point as being representative of, for example, the entire Inugsuin Fiord. A more correct evaluation of the ice reconstruction involves determining RSL values for a range of locations within the vicinity of the data sites and comparing the gradation in calculated values with that seen in the data. This is the procedure followed in Figures 20.4 and 20.5.

Although the calculated RSL curves for the inner and outer regions of the Inugsuin Fiord (Fig. 20.4) are very close to the observed values, there are difficulties apparent in the calculated trends for the Narpaing Fiord-Broughton Island area (Fig. 20.5). The inset map of Figure 20.5 is from Pheasant and Andrews (1973) and shows the location of each of the numbered data.

There is clearly a trend in the height of raised marine features normal to the superimposed NW-SE lines. Overlap between individual data points makes it difficult, however, to recognize distinct RSL curves associated with each of the NW-SE lines. The calculated RSL curves for Okoa Fiord and Broughton Island reproduce the observed trend but not the actual values of RSL. The heights of the most landward marine features are underestimated while the heights of the most seaward are overestimated. The seaward gradient of calculated RSL values is, therefore, not as steep as that observed and this misfit does not seem able to be corrected by modifying the ice reconstruction. Certainly having ice cover over the outer coast would only aggravate the overestimated height of Broughton Island raised marine features. On the other hand, lowering the assumed ice thickness in more landward areas cannot correct the Broughton Island misfit without aggravating that at Okoa Fiord.

It could be, as Quinlan (1981) has suggested, that the high level of seismic activity in this area (e.g. Basham et al., 1977) may have disrupted the RSL record. For whatever reason, however, the model is apparently unable to account for the gradient in the observed RSL data at these sites. This makes it very difficult to be dogmatic about the presence or absence of coastal ice based solely on examination of the RSL record.

Local Ice Domes

The proposed ice domes near the present-day Barnes and Penny Ice Caps were postulated largely to account for the height of raised marine features near the Narpaing and Okoa fiords. The reliability of these data and/or the model's ability to reproduce them has already been questioned. The necessity of these domes is, therefore, also open to question.

Ice Flow Directions

Although no attempt has been made to incorporate ice mechanics into the derivation of the reconstruction, it is worth considering the reconstruction's implications for ice flow directions. The ice thickness over northwestern Hudson Bay at 18 ky BP is a maximum value within the area of Figure 20.2. This suggests that ice should have flowed from Hudson Bay across the Foxe Basin and onto Baffin Island between 18 and 10 ky BP contrary to the field evidence cited by Andrews and Miller (1979).

Two points must be emphasized, however. First, the contours of Figure 20.2 are thickness, not topographic, contours. One must consider isostatic effects in order to convert these to the topographic contours that influence flow patterns. Second, the ice thickness high over Hudson Bay is not well constrained by the data of this study. The high is a feature inherited from the Peltier and Andrews (1976) ICE-1 reconstruction that served as the starting point for trial-and-error modifications. A detailed analysis of RSL indicators from the vicinity of this high would be needed to establish its reality. More work is needed, therefore, before reconstruction of Figure 20.2 can be used to determine the directions for comparison with field evidence.

SPECULATION ON THE FORM OF THE RSL RECORD

The second objective of this study is to use the proposed ice reconstruction to estimate the form of the RSL record for areas where direct observations are sparse. Given the level of uncertainty associated with some features of the reconstruction, one might question the wisdom of pursuing this objective. Figure 20.8, however, establishes that the model gives reasonable results not only for the sites used to derive the reconstruction (Fig. 20.1) but also for a wide range of other land-based data sites.

The left (a) panels of Figure 20.8 show calculated RSL isobases for 8, 6, 4, 2 and 1 ky BP while the right (b) panels are isobases as constructed by Dyke (1974) from RSL observations. Note that Dyke's (1974) figures were originally published as uplift isobases, his RSL data having been adjusted by Shepard's (1963) eustatic sea level estimates. These eustatic adjustments have been removed from Dyke's figures in order to produce RSL isobases for direct comparison with those calculated. The dashed contour lines of Figure 20.8b represent features of Dyke's (1974) isobase pattern which are significantly different from the calculated pattern but which do not seem to be well

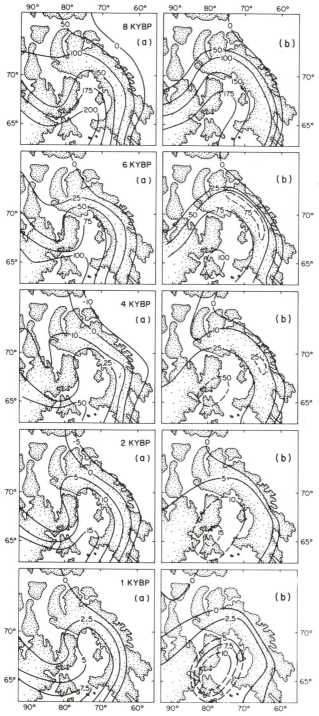

Figure 20.8
Calculated
isobases of RSL
change (a) are
compared with
observed isobases
(b). The contours
are in m with
positive values
being in excess of
present RSL.
Dashed features
are felt to be
poorly constrained
by observations
(with permission).

defined by observations. Except for these features, the data are in
general agreement with a single rebound center over western Baffin
Island and the Foxe Basin as predicted by the model. The short
wavelength uplift features resulting from the proposed Barnes and Penny
ice domes are not noticeable in the calculated isobase pattern, being
superimposed onto longer wavelength features associated with the
western Baffin ice ridge, which in turn are superimposed onto yet
longer wavelength features resulting from a Laurentide rebound centre
in the Keewatin region to the southwest.

The close correspondence of the two isobase patterns justifies
extending the model calculations into the offshore where virtually no
well-constrained data are available. Such an extrapolation from the
known RSL record is possible only if one has a physically reasonable
model to use as a guide. The predictive capability of such models is
considered to be one of their significant strengths.

In principle the model also permits extrapolation in time to
estimate the RSL record for periods prior to the formation of the
earliest RSL indicators. In the case of Baffin Island this would allow
estimates of RSL prior to 9-10 ky BP. Quinlan and Beaumont (1982) have
shown, however, that one should be cautious in accepting such
predictions. Spatial extrapolation within the time frame spanned by
the known RSL record is generally more reliable than temporal
extrapolation outside this time frame. The earth apparently responds
to load changes sufficiently quickly that any load changes which may
have affected the RSL record prior to c. 15 ky BP need not be expected
to have much expression in the RSL record formed within the last 9000
years. The ice reconstruction has been designed to explain only the
more recent RSL record and so may not contain all aspects of the ice
load necessary for proper reproduction of the earlier RSL record. The
same limitation does not apply to purely spatial extrapolations of the
known RSL record into, for example, the offshore. This is because any
RSL changes occurring offshore are in response to the same load changes
affecting onshore areas. Once the onshore RSL record is explained, the
offshore RSL record for the same period can be determined with some
confidence.

Figure 20.9 shows isobases of calculated RSL for the continental
shelf area bordering Baffin Island. These isobases are for times going
back to 18 ky BP, although the above restrictions should be kept in
mind regarding these early isobases. The calculations imply that RSL
was higher than present between 18,000 and c. 5000 years ago along most
of the eastern coast of Baffin Island with the exception of the
southern Cumberland Peninsula near Cape Dyer (cf. Locke, 1980).

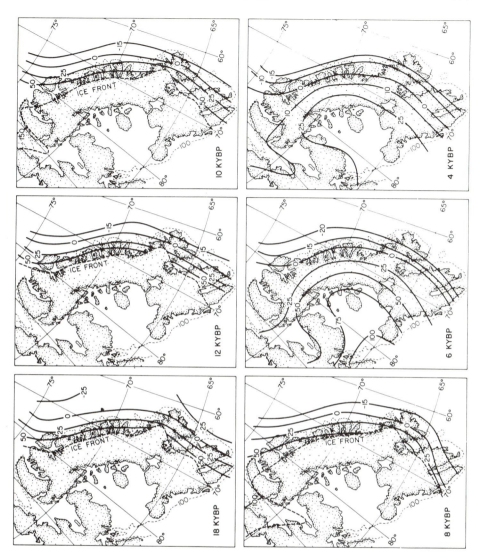

Figure 20.9 Isobases of RSL change calculated for the continental
shelf bordering eastern Baffin Island. Heavy dashed lines identify the
proposed ice front and light dashed lines the 100 m bathymetric
contour.

Andrews (1980), on the other hand, has interpreted the geological
record and results of a simple 2-D isostatic model (Brotchie and
Sylvester, 1969) to imply that RSL was lower than present along the
outer coast of Cumberland Peninsula and possibly the entire outer coast
between 10 and 30 ky BP. The potential problems with early model

calculations make it difficult to argue strongly against Andrews'
(1980) interpretation. It is nevertheless instructive to review the
evidence in order to see whether the model results can be reconciled
with it.

Miller et al. (1977) described a section from the Clyde River
Forelands in which Kogalu marine sands dated at 47.7 ky BP are
unconformably overlain by the Ravenscraig marine sand whose elevation
reaches 27 m above present sea level. Freshwater organisms and moss,
apparently in situ, are found 4 m below the top of the Ravenscraig sand
and have been dated at 9880 ± 200 BP (GSC-2201). This freshwater
material marks the Ravenscraig sand as a transgressive unit deposited
no earlier than 9-10 ky BP. Echo soundings from the Pigjoat Forelands,
Broughton Island and the Kangotokjuak Fiord, all to the southeast,
reveal submerged glacial-marine deltas and beach ridges at depths
ranging from 18 to 102 m. Those features, reasonably assigned Late
Foxe or Holocene ages, are found between 18 and 32 m depth (J.T.
Andrews, pers. comm., 1982). Andrews (1980) has interpreted these
observations to imply a RSL lower than present prior to 10 ky BP. This
lowstand was called the Pigjoat Regression and was associated with
formation of the presently submerged deltas and deposition of the
freshwater and terrestrial material presently found beneath the
Ravenscraig sand. The Pigjoat Regression was followed by the Cape
Adair Transgression during which the Broughton Island RSL indicator
(pt. 34, Fig. 20.5) and the Ravenscraig sand were deposited. This
transgression was in turn followed by the Inugsuin Regression during
which RSL fell to or below its present level, leaving behind the post-9
ky BP raised marine features which constitute the bulk of the Holocene
RSL record for this part of Baffin Island.

Figure 20.10 shows the RSL curve proposed by Andrews (1980) for
the east coast of Baffin Island along with curves calculated by the
model. Note that the model predicts a series of curves depending on
distance from the ice front. This gradational behavior is related to
migration of a glacial forebulge (e.g. Quinlan and Beaumont [1981]
among others). There is some doubt about the model's ability to
reproduce precisely the RSL record from the Broughton Island area as
has been discussed earlier. For this reason the calculated curves of
Figure 20.10 should not be taken as exact but rather as representative
of predicted RSL trends.

The model suggests that RSL was lower than present near Cape Dyer
for at least the last 18,000 years. Farther to the northeast, however,
RSL may have been higher than present and slowly transgressive until
some time between 8 and 10 ky BP depending on distance from the ice

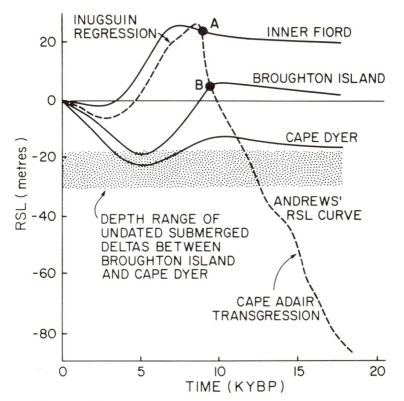

Figure 20.10 The RSL history of eastern Baffin Island as hypothesized by Andrews is contrasted with the history calculated from the model. A = freshwater material buried by Ravenscraig Sand. B = Broughton Island RSL indicator. Shaded region is depth of undated but probably late Foxe age submerged features.

margin. If these calculations are reasonable then one must explain why there are no raised marine features dating between 10 and 30 ky BP from this area (Andrews, 1980). Perhaps the absence of such features is due to erosion by the slow transgression calculated to have affected the coast. This explanation is not entirely satisfactory because the freshwater and terrestrial materials found beneath the Ravenscraig sand at the Clyde River (Miller et al., 1977) have survived the transgression. One is left to appeal to random chance or to argue why marine materials should be less robust.

Setting aside this potential difficulty there is another significant difference between the model calculations and Andrews' (1980) interpretation of the data. Andrews' RSL curve in Figure 20.10 associates the presently submerged deltas and beaches with the pre-10 ky BP Pigjoat Regression. If these features formed prior to the 18 ky BP model starting time, then clearly the model has nothing to say about them. If this is not the case, then the model calculations permit these features to be older than 10 ky BP only in the vicinity of Cape Dyer. For other sites along eastern Baffin Island RSL is calculated to have been lower than present only in the last 10,000 years.

It is important to realize that these submerged features need not be associated entirely with a single RSL lowstand. For this reason, even if one set of submerged features is demonstrably older than 10 ky BP, the model's prediction of a post-10 ky BP lowstand need not be incorrect. Although they are not presently dated, evidence that Pigjoat Forelands, submerged features formed 15 k BP for example, would certainly discredit the model's predictions of a higher-than- present RSL at that time. The model results over the last 9000 years would, however, still be valid because they are in good agreement with observations. I would contend that the calculated post-10 ky BP RSL lowstand is a real feature, whether or not it is the cause of all the submerged features seen offshore.

The maximum lowering of RSL during the well-constrained period from 10 ky BP to the present is c. 20 m (Fig. 20.9). The rapid deepening of shelf bathymetry away from the eastern Baffin coast makes it unlikely that such a small amount of shallowing would have exposed more than the most nearshore of areas. A corollary is that any RSL features presently found in more than c. 20 m of water are likely to be older than 10,000 years.

SUMMARY AND DISCUSSION

A reconstruction of Late Foxe age ice has been presented that accounts for many relative sea level observations from the Baffin Island area during the past 10,000 years. This reconstruction, however, is not unique. The lack of observations from the interval 10,000 - 30,000 years BP and the inherent uncertainty in the existing observations make a range of ice reconstructions equally valid. Despite this unavoidable limitation, there are some features that are common to all acceptable reconstructions and, therefore, required if the relative sea level record is to be explained.

The best defined of these features is a bilobate ice ridge over western Baffin Island and the Foxe Basin that is suggested to have separated from the bulk of the Laurentide Ice Sheet c. 10,000 years ago. This ridge could have been a late glacial remnant of ice originating to the southwest over Hudson Bay and Keewatin or it could have functioned as a separate dispersal centre throughout Late Foxe time. The calculations of this paper are more sensitive to changes in loads than to the origin of such loads and so the present calculations do not resolve this question. Further refinement of the ice reconstruction near Hudson Bay and examination of the flow directions implied by the refined reconstruction may be able to shed some light on this problem.

Ice is suggested to have been essentially stable over Baffin Island between 18,000 and 8000 years BP. It is perhaps possible to satisfy the relative sea level constraints by assuming a more gradual thinning of Baffin Island ice. Such gradual thinning would, however, require a maximum ice thickness over Baffin Island that is up to twice the value shown in Figure 20.2. This increased thickness is substantially greater than simple ice mechanics calculations would allow either for a stable ice sheet whose basal area is equal to that of Baffin Island or for the marginal regions of a more extensive ice sheet. For this reason, the late melting ice is believed to be a well-constrained feature of the proposed reconstruction.

The reconstruction assumes that the outer coast of eastern Baffin Island has been ice-free for at least the last 18,000 years but it is difficult to verify this assumption. The model fails to reproduce the observed seaward gradient of relative sea level change near Narpaing Fiord and Broughton Island whether or not ice cover is assumed on the outer coast. This is presently an area of high seismic activity and it is possible that the original relative sea level record has been seismically disrupted. In such a case the preserved record is no longer simply related to the distribution of glacial ice loads and cannot be used to infer ice conditions on the outer coast. The potential difficulty with the record and/or with the model also prevents any firm statement on the reality of the postulated ice domes near the present-day Barnes and Penny Ice Caps.

Whether or not the outer coast was glaciated, there is a gap in the record of raised marine deposits along this coast which must be explained. No such deposits have been found that date between c. 10,000 and 30,000 years BP. The model implies that this gap may be due neither to coastal ice cover (Feyling-Hansen, 1976) nor to a lower-than-present relative sea level (Andrews, 1980) but rather to the erosive influence of a slow transgression affecting the eastern coast of Baffin Island north of Cape Dyer between 18,000 and 10,000 years ago. This suggestion is tentative because it relies on extension of the known relative sea level record back in time, a process which is more prone to error than is spatial extension of the known record within its own time frame. The suggestion might be tested by coring of nearshore lakes and examining the recovered sediment for indicators of marine conditions between 10,000 and 18,000 years ago.

ACKNOWLEDGEMENTS

This work formed part of my PhD research done at Dalhousie University under the direction of Christopher Beaumont whom I wish to thank for many valuable discussions. W.R. Peltier of the Univesity of Toronto provided the Green functions used in the calculations. Brian MacLean and P.J. Mudie read and commented on an early version of this paper. Drafting, typing, and photographic services were supplied by the Atlantic Geoscience Centre during my tenure of a Visiting Fellowship there. Funding for the research came from the Natural Sciences and Engineering Research Council of Canada.

REFERENCES

Andrews, J.T., 1970: A geomorphological study of post-glacial uplift with particular reference to Arctic Canada. Institute of British Geographers Special Publication 2. 156 pp.

Andrews, J.T., 1980: Progress in relative sea level and ice sheet reconstructions Baffin Island N.W.T. for the last 125,000 years. In: N.-A. Mörner (ed.), Earth rheology, isostasy and eustasy. John Wiley and Sons, London. 175-200.

Andrews, J.T. and L. Drapier, 1967: Radiocarbon dates obtained through Geographical Branch field observations. Geographical Bulletin, 9:115-162.

Andrews, J.T. and G.H. Miller, 1979: Glacial erosion and ice sheet divides northeastern Laurentide Ice Sheet, on the basis of the distribution of limestone erratics. Geology, 7:592-596.

Andrews, J.T., J.T. Buckley, and J.H. England, 1970: Late glacial and glacio-isostatic recovery chronology, Home Bay, east Baffin Island, Canada. Geological Society of America Bulletin, 81:1123-1148.

Andrews, J.T., R. McGhee, and L. McKenzie-Pollack, 1971: Comparisons of elevations of archaeological sites and calculated sea levels in arctic Canada. Arctic, 24:210-228.

Basham, P.W., D.A. Forsyth, and R.J. Wetmiller, 1977: The seismicity of northern Canada. Canadian Journal of Earth Sciences, 14:1646-1667.

Blake, W., Jr., 1966: End moraines and deglaciation chronology in northern Canada, with special reference to southern Baffin Island. Geological Survey of Canada, Paper 66-21, 1-31.

Brotchie, J.F. and R. Sylvester, 1969: On crustal flexure. Journal of Geophysical Research, 74:5240-5252.

Callow, W.J., M.J. Baker, and G.I. Hassall, 1965: National Physical Laboratory radiocarbon measurements III. Radiocarbon, 7:156-161.

Callow, W.T., M.J. Baker, and G.I. Hassal, 1966: National Physical Laboratory radiocarbon measurements IV. Radiocarbon, 8:340-347.

Clark, J.A., 1980: The reconstruction of the Laurentide Ice Sheet of North America from sea-level data: Method and preliminary results. Journal of Geophysical Research, 85:4307-4323.

Crane, H.R. and J.B. Griffin, 1966: University of Michigan radiocarbon date XI. Radiocarbon, 8:256-285.

Dyck, W., J.G. Fyles, and W. Blake, Jr., 1965: Geological Survey of Canada radiocarbon dates IV. Radiocarbon, 7:24-46.

Dyck, W., J.A. Lowdon, J.G. Fyles, and W. Blake Jr., 1966: Geological Survey of Canada radiocarbon dates V. Radiocarbon, 8: 96-127.

Dyke, A.S., 1974: Deglacial chronology and uplift history: northeastern sector, Laurentide Ice Sheet. Institute of Arctic and Alpine Research, University of Colorado, Occasional Paper No. 12, 73 pp.

Dyke, A.S., 1979: Radiocarbon-dated Holocene emergence of Somerset Island, central Canadian Arctic. Current Research, Part B, Geological Survey of Canada, Paper 79-1B, 307-318.

Farrell, W.E. and J.A. Clark, 1976: On postglacial sea level. Geophysical Journal of the Royal Astronomical Society, 46:647-667.

Feyling-Hansen, R.W., 1976: The stratigraphy of the Quaternary Clyde Foreland Formation, Baffin Island, illustrated by the distribution of benthic foraminifera. Boreas, 5:77-94.

Klassen, R.A., 1981. Aspects of the glacial history of Bylot Island, District of Franklin. Current Research, Part A, Geological Survey of Canada, Paper 81-1A, 317-326.

Locke, C.W., III, 1980: The Quaternary geology of Cape Dyer area, southeasternmost Baffin Island, Canada. PhD thesis, University of Colorado, Boulder. 331 pp.

Lowdon, J.A. and Blake, W., Jr., 1968: Geological Survey of Canada radiocarbon dates VII. Radiocarbon, 10:207-245.

Mathews, B., 1967: Late Quaternary land emergence in northern Ungava, Quebec. Arctic, 20:176-201.

Miller, G.H., 1980: Late Foxe glaciation of Baffin Island, N.W.T., Canada. Geological Society of America Bulletin, 91:399-405.

Miller, G.H., Andrews, J.T., and Short, S.K., 1977: The last interglacial/glacial cycle, Clyde Foreland, Baffin Island, N.W.T.: stratigraphy, biostratigraphy and chronology. Canadian Journal of Earth Sciences, 14:2824-2857.

Paterson, W.S.B., 1969: The physics of glaciers. Pergamon Press, Oxford, 250 pp.

Peltier, W.R. and Andrews, J.T., 1976: Glacial-isostatic adjustment-I. The forward problem. Geophysical Journal of the Royal Astronomical Society, 46:605-646.

Pheasant, D.R. and Andrews, J.T., 1973: Wisconsin glacial chronology and relative sea level movements, Narpaing Fiord- Broughton Island area, eastern Baffin Island, N.W.T. Canadian Journal of Earth Sciences, 10:1621-1641.

Prest, V.K., Grant, D.R., and Rampton, V.N., 1967: Glacial Map of Canada, Geological Survey of Canada Map 1253A.

Quinlan, G., 1981: Numerical models of postglacial relative sea level change in Atlantic Canada and the eastern Canadian Arctic. PhD thesis, Department of Oceanography, Dalhousie University, Halifax, Nova Scotia, Canada. 499 pp.

Quinlan, G. and Beaumont, C. 1981: A comparison of observed and
    theoretical postglacial relative sea level in Atlantic Canada.
    Canadian Journal of Earth Sciences, 18:1146-1163.
Quinlan, G. and Beaumont, C. 1982: The deglaciation of Atlantic Canada
    as reconstructed from the postglacial relative sea level record.
    Submitted to Canadian Journal of Earth Sciences.

Shepard, F.P., 1963: 35,000 years of sea level. In: Essays in marine
    geology. University of Southern California Press, 1-10.

Tauber, H., 1960: Copenhagen radiocarbon dates IV. American Journal of
    Science Radiocarbon Supplement, 2:12-25.

Walcott, R.I., 1972: Late Quaternary vertical movements in eastern
    North America. Reviews of Geophysics and Space Physics, 10:849-884.

21 Holocene sea level variations within Frobisher Bay

J. T. Andrews and G. H. Miller

A feature of Quaternary studies on Baffin Island for the last two decades, if not earlier, has been the attention paid to raised marine beaches. These features were commented on by nearly all workers (e.g. Blackadar, 1958; Mercer, 1956; Ives, 1964; Loken, 1965; Andrews, 1966; Miller, 1980) and they are a conspicuous element of many coastal localities around Baffin Island. With the advent of radiocarbon dating there was a nearly immediate application of the method to date marine shells, whalebone, and driftwood, particularly on the uppermost beaches. Indeed these dates on or close to the local marine limit constituted the main evidence for reconstructing the pattern and timing of ice retreat throughout much of Arctic Canada (Craig and Fyles, 1960; Prest, 1969; Bryson et al., 1969). In addition, the combination of altimetric observations and C-14 dated levels brought to the fore the study of glacial isostatic rebound (e.g. Matthews, 1967a; Andrews, 1966, 1970a; Walcott, 1972; Blake, 1975) and with it, a focus on the distribution and thickness of late Quaternary ice sheets (Peltier, 1976; Peltier and Andrews, 1976, 1983; Clark, 1980; Quinlan, this volume).

Frobisher Bay has been the site for several brief studies of raised marine shorelines and the subject is not without controversy (Wengerd, 1951; Mercer, 1956; Ward, 1952; Blake, 1966). However, despite the relative accessibility of the area suprisingly little Quaternary research had been carried out within the Bay (see, however, Matthews, 1967a; Blake, 1966) until INSTAAR developed a major research interest in southern Baffin Island from 1977 onwards. Prior to 1977 only 4 or 5 C-14 dates had been reported around the shores of this large bay (Blake, 1966; Matthews, 1967a). Since then several research parties have greatly expanded the base of radiocarbon-dated shorelines (Figs 21.1 & 21.2; Andrews, 1978; Miller, 1979; Andrews and Short,

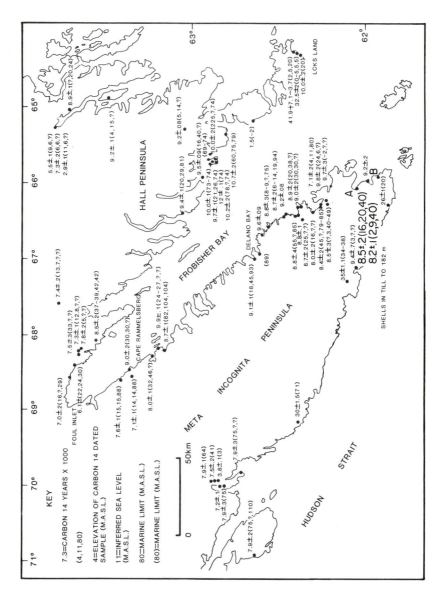

Figure 21.1 Map of dated shorelines in and around Frobisher Bay. The
map shows the locations of the sites, the elevation the sample was
collected from, the probable sea level at the time, and the height of
the local marine limit (from: Miller, 1980; Muller, 1980; Blake, 1966;
Colvill, 1982; Lind, 1983; Miller, Stravers and Osterman, Unpubl.; data
from Jacobs and Mode, Pers. commun.) Sites A and B, Figure 21.6, are
also located.

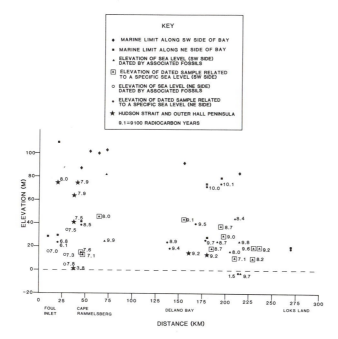

Figure 21.2   Shoreline diagram for Frobisher Bay showing the elevation and ages of radiocarbon dated sites (see Fig. 21.1) projected onto a plane running parallel with the axis of the Bay.

1983) although there are still some notable gaps in the distribution of dated sites especially along the northern shore of Frobisher Bay, as well as along the north coast of Hudson Strait between Lake Harbour and the Resolution Island to the east.

The purpose of this paper is to outline the sea level history of Frobisher Bay during the last 10 to 11 ky and compare it with other studied areas, such as Cumberland Peninsula to the north (Dyke, 1979; Andrews, 1980; Clark, 1980; Fig. 21.3). In addition, we compare the observed data, assembled as part of our study, with the predictions of a Holocene sea level history developed by Quinlan in Chapter 20 of this volume. In terms of the Clark et al. (1978) specifications, Frobisher Bay provides a transect from the Zone I region of continuous Holocene emergence to the zone I/II transition which delimits the Earth's response to unloading in the vicinity of the ice margin.

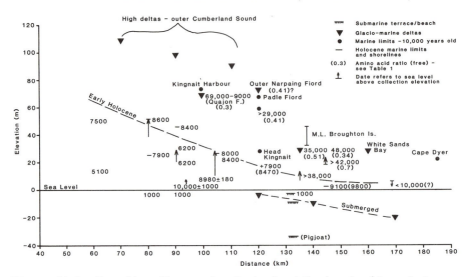

Figure 21.3 Shoreline diagram for Cumberland Peninsula (from Andrews, 1980) showing elevation/distance relationships for shorelines of Holocene age and older (with permission).

Structural and Tectonic setting:

Previous work on sea level changes along Baffin Island has ignored possible tectonic influences on postglacial recovery. Nearly all studies to date have shown that postglacial emergence curves and shoreline diagrams can be reconstructed with a minimum of data points especially where those data points represent situations identified unambiguously (Löken, 1965; Andrews, 1970b; Andrews et al., 1970; Dyke, 1979). However, in recent years the seismicity of Baffin Island, especially of the outer fiords, is becoming understood, and in addition there is some evidence of Holocene faulting offshore (McLean and Falconer, 1979). Thus it appears that a possible tectonic influence on glacial-isostatic recovery cannot be dismissed out-of-hand.

     Frobisher Bay is a half-graben with the southerly coastline clearly fault-controlled. Paleozoic limestone crops out over much of the floor of the outer bay and outliers occur at the head of Frobisher Bay. On the seismic map of the eastern Canadian Arctic (Balsham et al., 1977; Reid and Falconer, 1982), few epicenters are located around Frobisher Bay. Reid and Falconer (1982, p. 1530) note that " . . . The activity is shallow and near the coast." They further go on to remark that " . . . The seismicity may be controlled by old lines of weakness related to the rifting process". Thus, present indications are that seismicity is low around Frobisher Bay. However, in reading C.F.

Hall's (1865) account of his travels within Frobisher Bay during the middle of the 19th century it is striking that his description of why the Inuit left the region of Loks Land (outer Frobisher Bay) calls to mind the effect of a major earthquake. The description is as follows: ". . . . There was no sea, no wind, no ice; . . . weather good yet the dreadful noise continued . . . . The earth cracked and rumbled, and seemed as if breaking up in all directions."

Interpretation of C-14 shell dates:

Over the last few years we have had a number of shell samples redated by two or more laboratories. We did this partly as a check on the consistency of C-14 dates on shells but also because it became apparant that different laboratories report dates on marine shells in somewhat different ways. Stuiver and Polach (1977) have indeed suggested that all radiocarbon laboratories should report their shell dates in a consistent fashion. Mangerud (1972) has specifically pointed out some of the problems in the calculations of shell ages and discussed the application of a "reservoir" correction. In Table 21.1 we briefly list the laboratories that have provided C-14 dates for us (Fig. 21.2) with a statement as to how we judge their quoted ages in association with dates from the Geological Survey of Canada. In Table 21.2 we list some of the inter-laboratory comparisons with some comments where judged useful. In comparing the dates with those of the Geological Survey of Canada we are correcting a "conventional radiocarbon age" to a "reservoir corrected age." Thus in practice we could add c. 410 years to all GSC dates or subtract 410 years from dates with QC, GX, SI, or QL designations. As noted by Stuiver and Polach (1977), p. 357):

>"The corrections for C-13 fractionation and reservoir C-14 deficiency cancel each other more or less for shells from non-polar regions. . . ."

COMPARATIVE SEA LEVEL STUDIES, EASTERN BAFFIN ISLAND

Andrews (1980) presented two shoreline diagrams for the coast of eastern Baffin Island north of Frobisher Bay. The first included an area from the outer coast to the fiord heads in the vicinity of Clyde River whereas the second area was from Cumberland Peninsula. This second diagram is presented as Figure 21.3. It is based on a wide regional survey of beach elevations and their association with glacial deposits. Dating is provided by radiocarbon assays, and some U-series dates (cf. Szabo et al., 1981), and other correlations are based on

TABLE 21.1.   Radiocarbon laboratory procedures used in reporting dates
on marine shells.
==

Lab. Designation	Reporting Procedure
QC	conventional C-14 years
SI	conventional C-14 years
QL	conventional C-14 years
Beta	conventional C-14 years
DICAR	conventional C-14 years but normalized to $-19°_o$ Sc^{13}
GSC	quoted date normalized to $0°/_o$*

*NB. For dates to be compared between these laboratories, 410 years
would have to be added to the GSC reported ages. The shell dates from
the other facilities will be too old by an amount equal to the
"reservoir effect".

TABLE 21.2.   Comparison of C-14 dates on shells from different
laboratories.
==

Site	C-14 Date	Lab #	Comment
Noble Inlet	9190 ± 195	GX-8194	Shells from the same site. Initial date was GX but this
Noble Inlet	8600 ± 110	GSC-3648	was followed by GSC date because other dates from
Noble Inlet	(8780 ± 195) reservoir corrected	GX-8194	area significantly younger than GX dates. Discussions suggest difference is associated with different ways in which dates are reported.
Pugh Island	9875 ± 130	QC-903	Shells may be from the same site although this is not absolutely
Pugh Island	8560 ± 100	GSC-3666	certain. The reservoir correction in this case
Pugh Island	(9465 ± 130) reservoir corrected	QC-903	certainly lessens the difference between the QC and GSC dates. At the extremes of the 95% confidence levels the dates still differ by 515 years.
Cape Hooper	7960 ± 140	Y-1833	Shells from a single site dated
Cape Hooper	8050 ± 115	QC-457	by 3 laboratories. All dates
Cape Hooper	7640 ± 125	Beta-2362	are in conventional C-14 years according to laboratory reports although Beta-2362 is approximately 400 years younger than the other two.
Countess of Warwick Sound	10700 ± 140 10600 ± 150 10900 ± 70 10510 ± 70	QC-480A QC-480B QL-1173 QL-1174	Multiple dates obtained on robust paired valves of Mya truncata. QL 1174 was on the periostracum. All dates conventional C-14 years.

amino acid ratios (Nelson, 1978; Brigham, 1983; Locke, 1980; Hawkins, 1980; Dyke, 1977, 1979; Dyke et al., 1982). Many of the major features of the shoreline diagram noted for the Clyde River transect were also apparent in the data from Cumberland Peninsula. These features can be itemized as follows (from youngest to oldest):

1. a series of peats and buried soils indicating a 0.5 to 1 m rise in sea level over the last 1000 to 3000 years;

2. an early Holocene raised marine shoreline that is dated between 8000 and 8600 BP and which slopes seaward from a maximum present elevation of about 70 m asl in the inner fjords to very near or below present sea level at Cape Dyer where no Holocene raised beaches have been recognized (Locke, 1980); there is also evidence that sea level changed little between 10,000 and 8000 BP and there may have been one or more transgressions (Nelson, 1978);

3. a series of major but undated, submerged glacial-marine deltas occur (Miller, 1975) that may predate this 8000-8600 BP shoreline;

4. a series of older and higher glacial-marine deltas and shorelines that generally lie along the outer coast of Cumberland Peninsula (Löken, 1966; Ives and Bradley, 1969; Nelson, 1980; Brigham, 1983; Dyke, 1979; Locke, 1980) which have associated radiocarbon dates on shells of >35,000 BP, and U-series dates greater than 45,000 BP (Szabo et al., 1981). There is a notable <u>total</u> absence of C-14 dated sediments in the range between 10,000 and 35,000 BP which has been one major thread in the argument that there was no extensive 18,000 BP glacial advance along the northeastern margin of the Laurentide Ice Sheet.

Curves showing changes in relative sea level throughout the Holocene have been presented for parts of Cumberland Peninsula by Pheasant and Andrews (1973), Dyke (1979), and Clark (1980). Andrews (1980) employed a simple two-layer earth model to predict the sea level history of Cumberland Peninsula over the last 125,000 years with results not too dissimilar from those proposed by Dyke (1979) and Clark (1980). Thus a major implication from these studies was that sea level changes during the Holocene were predictable and generally followed an exponential decrease in uplift as a function of time.

In Quinlan's analysis (Chapter 20, this volume) of the late Quaternary sea level variations of the eastern Canadian Arctic, he used a series of radiocarbon dated relative sea level curves to develop a time history of ice thickness and extent. In addition his analysis produced a series of maps showing the predicted variations in sea level (Chapter 19). Our Figure 21.4A is taken from Quinlan's analysis and consists of a shoreline diagram constructed normal to the isobases running from Foul Inlet at the head of Frobisher Bay to Loks Land at

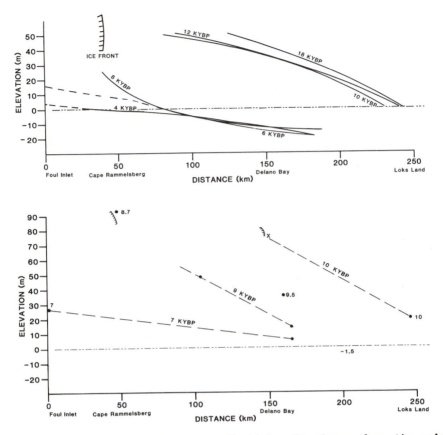

Figure 21.4A Shoreline diagram for Frobisher Bay drawn along the axis of the Bay (from Quinlan, this volume). 21.4B Shoreline diagram for Frobisher Bay based on reconstruction of Holocene isobases for 10,000, 9000, and 7000 BP.

the outer eastern limit. This shoreline diagram is based on Quinlan's ice sheet reconstruction which in turn is based on an analysis of radiocarbon dated sea levels. However, at the time that Quinlan commenced his research there were few data points for Frobisher Bay and the nearest series of dated shoreline elevations was around Lake Harbour, northern Hudson Strait. Figure 21.4A shows that high shorelines are predicted to occur in the outer 150 km of Frobisher Bay and date between 10,000 and 18,000 BP. Thereafter, sea level falls rapidly as the ice across Baffin Island thins and disintegrates. It is predicted that the 8000 BP shoreline would now have an elevation of about 30 m near Cape Rammelsberg and that it would be below sea level 100 km from Foul Inlet (Figs 21.1 & 21.4). The outer 150 km of Frobisher Bay is predicted to have experienced a 10-15 m rise in sea

level over the last 4000 years, whereas the very head of the Bay has
seen a very slight fall of sea level over the same interval.

Quinlan's analysis of the pattern of Holocene level variations
differs from other reconstructions (Andrews, 1980) (e.g. Fig. 21.3) in:
(i) the presence of suggested shorelines 12,000 years or older; (ii)
the amount of middle and late Holocene sea level rise along the outer
coast; and (iii) the magnitude of the recovery in the vicinity of the
margin of the ice sheet. In Quinlan's inverse reconstruction (Chapter
20) the ice sheet is essentially constant (cf. Andrews, 1975) in extent
and thickness between 18,000 and 8000 BP.

CHANGES IN THE ELEVATION AND AGE OF THE MARINE LIMIT

Changes in the elevation of the marine limit along a transect normal,
or approximately so, to local isobases is a powerful method of
examining the interactions between ice load, rate of glacial retreat,
and the time taken for deglaciation (Andrews et al., 1970). Miller
(1980), Müller (1980), and Colvill (1982) all commented on the
elevation/distance/age relationship of marine limits in outer and inner
Frobisher Bay.

The marine limit on Resolution Island and the Savage Islands is
present sea level. In contrast, on the eastern tip of Meta Incognita
Peninsula the marine limit lies at about 40 m asl and dates from about
8600 BP. From this point westward there are differences in the trends
of marine limits depending on whether we consider the north coast of
Hudson Strait or the south or north coasts of Frobisher Bay.

There has been no detailed study of the marine limit and lower
beaches along northern Hudson Strait, although Blake (1966) presents
useful initial data. The marine limit is slightly lower (around 90
m). Radiocarbon dates indicate deglaciation of Big Island occured
prior to 7900 BP, whereas toward the outer coast the 20 m marine limit
at Nannuk Harbour is associated with a C-14 date of 26,000 BP (Blake,
1966; Fig. 21.1). The meaning of this date is currently unclear.

Along the south side of Frobisher Bay marine limit determinations
have increased significantly since the early work of Mercer (1956) and
Blake (1966). The glacial history of this coast is complex as sites
were alternately under the influence of local ice flowing toward
Frobisher Bay from the uplands of Meta Incognita Peninsula or of
regional ice moving eastward along the main axis of Frobisher Bay
(Osterman, 1982; Miller, 1980; Colvill, 1982: Lind, 1983; Stravers,

unpubl.). The marine limit is relatively low southeast of Jackman Sound but from there to Cape Rammelsberg marine limits on the outer coast vary between 80 m and 122 m asl with a general tendency to increase inland toward the distal zone of the Frobisher Bay moraine (Miller, 1980; Müller, 1980; Colvill, 1982). The presence of late-lying, local ice is probably reflected in a variety of lower marine limit heights, but even these only vary between 67-73 m asl. Marine limits fall dramatically on the proximal side of the Frobisher Bay moraine (Blake, 1966; Colvill, 1982; Lind, 1983; Miller, 1980) to altitudes of 29-32 m asl (Fig. 21.1). Radiocarbon dates (Miller, 1979; Andrews and Short, 1983) indicate that whereas the outer sections of Frobisher Bay were deglaciated between 10,700 and 8700 BP (Miller, 1980; Müller, 1980), the inner bay became ice free by around 7000 BP.

Marine limits are lower along the north side of Frobisher Bay than the corresponding southern shores by 10-20 m. This may reflect in part the trend of the regional isobases as well as differences in the style of deglaciation. On Loks Land, the marine limit is low, and in situ molluscs have been collected that date the middle Foxe Loks Land member (Miller, this volume) as well as earliest Holocene sea levels. At the margin of the Hall moraine the marine limit is 79 m and dates from 10,700 BP (Miller, 1980; Fig. 21.1). Further west along the north side of Frobisher Bay a date of 9400 BP relates to a sea level of 29 m asl whereas the local marine limit is 81 m. This suggests both rapid deglaciation and rapid rebound of the coast immediately after deglaciation.

HOLOCENE RELATIVE SEA LEVELS

Observations on the elevation and age of raised shorelines around Frobisher Bay (Figs 21.1 & 21.2) include publications or theses by Mercer (1954, 1956), Blake (1966), Matthews (1967b), Müller (1980), Miller (1980 and unpubl.), Colvill (1982), Lind (1983), plus observations by Mode, Jacobs and Osterman (in Andrews and Short, 1983). Figure 21.1 shows the distribution of radiocarbon-dated marine deposits. Where appropriate (in most cases) each date is associated with 2 to 3 numbers, such as (4, 11, 80). The explanation for this sequence is given in the legend to the figure. A shoreline diagram (Fig. 21.2) runs along the main axis of the Bay and sites have been projected onto the plane of the diagram (Løken, 1962; Andrews, 1970a). Compared with the amount of data available prior to 1977, Figures 21.1 and 21.2 represent a substantial increase in information. There are now unpublished observations (J. Stravers, in Andrews and Short, 1983) for the coast between Delano Bay and Cape Rammelsberg. The two major areas requiring field survey are the north coast of Frobisher Bay between Ward Inlet and the outer coast (Fig. 21.1) and in particular, there is a large and critical gap in field observations and dated sequences between Lake Harbour and the Savage Islands. The dates along this coast which are 26,000 BP are in part on transported shells, but the 26,000 BP date (Blake, 1966) from Nanook Harbour may be in situ.

The glacial history of Frobisher Bay is discussed in a companion chapter (Osterman, Miller, and Stravers) and will only be briefly outlined here. The deglaciation of Frobisher Bay and Hudson Strait is currently far from clear--the problem that we face is that the region of Figure 21.1 was influenced by several glacial sources including local ice caps, ice flowing northward from Hudson Strait, ice on the shelf, and ice flowing seaward down Frobisher Bay.

Löken (1962) showed how a series of 3-point solutions can be used to determine the direction of dip of isostatically deformed shorelines. We have followed this procedure for the dates and elevations shown on Figure 21.1. Although there are a considerable number of C-14 dates for the region several of these represent shell collections that cannot be specifically tied to a former sea level. In several cases all we can say with certainty is that sea level was higher than the collecting site. However, there are enough control points to develop 10 triangular solutions. The dates on the surfaces range between 10,000 and 7000 years old. The average directions of the isobases is 011°-191° true with a standard deviation of ±10°. There is a suggestion that the isobases are swinging toward a more north-south orientation throughout the early Holocene as the average isobase strike for the 10,000 BP shorelines is 24°-204°, it changes to 17°-197° for 9000 year deposits, and has dropped to 10°-19° by 7000 BP. There are, however, too few sites for each dated interval to provide a firm check on this suggestion. Isolines drawn normal to the strike of Frobisher Bay would have a direction of 045°-225°. The implications of isobases striking 011°-191° are that the ice thickness contours also run approximately north-south.

The isostatic data cannot be easily reconciled with the idea of a late Foxe ice flow directed across southern Baffin Island to elevations of 400 m+ (see Chapter 18, this volume) nor do the low marine limits along the extreme tip of Labrador (Löken, 1962, 1964) fit within this concept. At the moment it is fair to state that the glacial and post-glacial isostatic interactions for northernmost Labrador and southeast Baffin Island are not well resolved! It must, however, be noted that for a grounded ice sheet flowing over substantial bodies of water, the uncompensated isostatic depression is approximately the mass of ice that extended above sea level.

The shoreline diagram (Fig. 21.2) is not directed normal to the
isobases. The figure does show that there is a general increase in the
height of the marine limit toward the head of the Bay, specifically to
Cape Rammelsberg (Blake, 1966; Miller, 1980; Colvill, 1982; Lind, 1983)
and thereafter the marine limit falls to 20-24 m near the head of Foul
Inlet. Figure 21.1 shows the elevation of shell collections. Many
samples have been taken from proximal and distal glacial-marine
sediments that delimit ice proximity but do not provide good control on

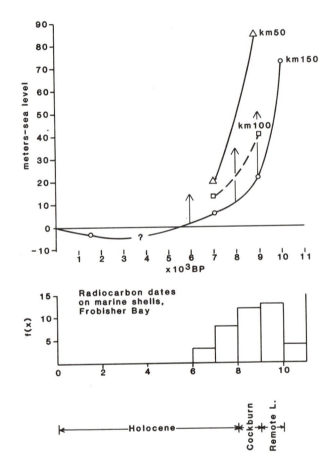

Figure 21.5 Upper: Emergence curves for Frobisher Bay at 50, 100 and
150 km from the 0 km mark of Figure 21.4B. The arrows on the 150 km
curve show the predicted elevation of the sites based on the Andrews'
(1968) model for glacial isostatic recovery. Lower: Histogram of
radiocarbon dates accumulated in 1000 yr class intervals showing the
apparent absence of dated shorelines younger than 6000 BP.

former sea levels. Figure 21.5 illustrates the wealth of C-14 dated
sites with ages 7000 BP and present. Note that there is good evidence
for a recent rise in sea level of at least 2 m in the last 1500 years
(Miller et al., 1981).

Figure 21.4B is a simplified diagram normal to the isobases showing
the predicted shoreline elevations--it can thus be compared directly
with the shoreline model proposed by Quinlan (Fig. 21.4A). This
comparison suggests that the overall changes in gradients of the
shorelines are comparable but that there has been significantly more
emergence during the early and middle Holocene than predicted by
Quinlan's model. Thus the 10,000 BP shoreline (10 ky BP shoreline on
Fig. 21.4A and B) indicates between 10 and 30 m more emergence than is
predicted and at 8000 BP the difference between the observed and the
predicted is nearly 70 m. It must be remembered that Quinlan's
analysis was carried out without the benefit of most of the data
portrayed on Figures 21.1 and 21.2 and, prior to this present study,
the glacial isostatic recovery of southeast Baffin Island was hardly
known; hence any modeling was poorly constrained throughout the region
from Hall Peninsula southward to the east side of Ungava Bay and the
fiords of the Torngat Mountains. The differences between Figures 21.4A
and B might suggest that the ice sheet reconstruction of Quinlan (this
volume) should be changed in the vicinity of Hudson Strait and
Frobisher Bay to take into account these new data. Of course this
interactive process between field data-- modelling-- field data merely
serves to highlight the importance of models in developing questions
that can be researched by the field scientists. In a similar manner
the shoreline diagram of Figure 21.4B provides some limits on expected
elevations of marine deposits and their heights above present sea level
in areas that are presently not surveyed, such as the northern coast of
Frobisher Bay (cf. Miller, 1980).

Examination of Figure 21.1 indicates where the next major series
of field studies should concentrate and that is along the northern
shore of Hudson Strait between Lake Harbour in the west and the islands
off the eastern tip of Meta Incognita Peninsula. At the moment, the
very limited observational base is restricted to the work of Blake
(1966) and unpublished observations. Transported "old" shells exist in
tills along large sections of this coast (Blake, 1966). Blake
suggested that the marine limit in the vicinity of a 26,000 years date
at Nanook Harbour was only 20 m which would be compatible with a
deglaciation of the outer coast of Hudson Strait between 10,000 and
9000 BP (Fig. 21.4B). There is in fact a 9400±700 BP date from
Pritzler Harbour about 40 km west along the coast that does not
contradict this idea and a shell from a marine core in outer Hudson has

a C-14 date of 9100± BP (Fillon et al., 1982). However, Miller has
C-14 dates from the very outer coast of Meta Incognita which indicate
ice was immediately offshore between 9200 and 8600 BP. These
contradictions are yet another indication of the deglacial and glacial
complexity that affected this area. Judging from field measurements
toward Lake Harbour the marine limit reaches 110 m on Big Island
(Blake, 1966) whereas for the same date of deglaciation the marine
limit is nearly 40 m lower (Blake, 1966; Clark, in Andrews and Short,
1983) inland at Lake Harbour.

A significant weakness in the glacial isostatic reconstructions
has been, and still is, the extremely limited number of dated
elevations from a single locality. On the coasts of west and east
Baffin Island earlier workers had relatively little difficulty in
finding sequences of raised marine deposits that could be dated to
provide emergence curves (e.g. Ives, 1964; Andrews, 1966; Löken, 1966;
Andrews et al., 1970). However, as the research areas moved southward,
first to Cumberland Peninsula and then to the Frobisher Bay region, the
number of sites which have been discovered to produce such a curve is
extremely small and indeed as of 1981 no studied site within Frobisher
Bay had the necessary ingredients. This in itself is a weakness but
one that cannot be readily overcome, although additional field work may
come up with suitable sites.

The form of the postglacial (Holocene) emergence of Frobisher Bay
can be gained from a study of the shoreline diagram (Fig. 21.4B) and a
comparison with the form of other emergence curves from western and
eastern Baffin Island (Andrews, 1970b). Emergence curves are
reconstructed from the Shoreline Diagram (Fig. 21.5) at 150, 100, and
50 km from the head of Frobisher Bay. The resulting emergence takes
place in 5000 to 7000 years. In the outer part of Frobisher Bay
present sea level was attained by 6000 BP, and our data and Quinlan's
model (Fig. 21.4A) suggests that a marine transgression affected the
outer coast over the last 4000 years or so. Miller et al. (1981)
report a date on submerged peat at -2 m of about 1500 BP. This site
projects onto Figure 21.4B at about the 200 km distance.

Nearer the head of the Bay it is likely that slow emergence
characterized the middle and late Holocene. This assertion is partly
supported by the emergence curve at km 50 (Fig. 21.5) but it is more
specifically supported by Jacobs' analysis (this volume) of the
elevational relationships of the various paleo-eskimo archeological
sites in the inner Bay.

The difference in the rate of emergence within Frobisher Bay and
other sites from western and eastern Baffin Island (Andrews, 1966;

Löken, 1965; Andrews et al., 1970) is shown on Figure 21.5 by the
arrows that extend vertically from the km 150 emergence curve. The
heads of these arrows mark the <u>predicted</u> sea level based on the Andrews
(1968) model for glacial isostasy in Arctic Canada. This model has
proved useful in other studies (e.g. Andrews et al., 1970) and it even
provides a very close approximation to the detailed emergence curve
from Cape Storm, Ellesemere Island which is drawn on the basis of over
50 radiocarbon dates (Blake, 1975)! There are however, several
theoretical reasons why the relaxation of the Earth's crust from the
ice load should not be constant (see papers <u>in</u> Mörner, 1980 [ed.]);
particularly toward the edge of the ice load, the rebound should be
forced by the shorter wave lengths and predictably would occur faster
than toward the interior areas of the former ice sheet. Whether there
may be a discernible tectonic imprint on the rate and pattern of
emergence (e.g. Mörner, 1980; Sissons and Cornish, 1982) is possible
but cannot be determined from the data we have at present. A histogram
of available C-14 dates from Frobisher Bay (Fig. 21.5) imply that large
sections of the present coast were close to or above sea level in the
period 6000 BP to the present.

MARINE REGRESSION AND TRANSGRESSION BETWEEN 8000 AND 9000 BP

Andrews (1980) proposed that during the global glacial maximum at about
18,000 BP sea level off the eastern coast of Baffin Island was below
present. Quinlan (1981, this volume) has argued against that
interpretation. In Andrews' (1980) scenario, sea level rises between
18,000 and c. 10,000 BP against an isostatically depressed coastline.
This event is the Cape Adair transgression and is recorded along much
of the outer east coast of Baffin Island (Andrews, 1975; Miller et al.,
1977; Nelson, 1978, 1981). During the next 2000 years there is
evidence from the Clyde River area (King, 1969; Andrews, 1980) that sea
level did not simply fall during the subsequent Inugsuin regression,
but rather stayed high relative to the present sea level.

    Over the last few years there have been indications that
significant oscillations of sea level occurred during the Remote Lake
and Cockburn chrons (Fig. 21.5). On Qivitu Peninsula, Nelson (1978,
pp. 209-210; 1981) noted radiometric and stratigraphic evidence for a
complicated history of sea level which included: (1) a rise of sea
level to about 25 m (?) close to 10,000 BP; (2) a fall of sea level to
no more than 10 m above present shortly thereafter; (3) a rise of sea
level to at least 12 m between 9000 and 10,000 BP; (4) a fall of sea
level to about 4 m about 8200 BP; and (5) a transgression extending to
15 m asl. Following this last transgression sea level fell, and more

recently, gave way to a late Holocene marine transgression, named the
Canso Channel transgression (Pheasant and Andrews, 1973; Andrews,
1980).

Within Frobisher Bay, Lind (1983) has postulated that ice
retreated toward the head of Frobisher Bay after 9800 BP and then
readvanced to the Frobisher Bay moraines. Recognition of proximal and
distal glacial-marine litho- and biofacies in sections near Cape
Rammelsberg led Lind (1983) to propose that a significant oscillation
of sea level (50 m or so) was associated with the glacial retreat and
subsequent readvance.

Evidence of a major sea level oscillation during the early
Holocene is preserved in the geomorphic and stratigraphic records
across SE Meta Incognita Peninsula. The marine limit phase is
generally recorded as a washing limit (erosional evidence) in till
(Blake, 1966; Miller, unpublished observation). Along gently sloping
coastlines the marine limit may lie 1 or more km inland from the
present coast. In such areas there are few or no marine deposits
between this limit and about half the elevation drop to sea level. The
highest marine deposits often dam small local drainages and are thicker
than are the marine deposits at lower elevations.

One possible scenario to explain this geomorphologic evidence is
that relative sea level remained at the marine limit phase briefly,
then fell rapidly, followed by a new transgression that attained an
elevation well below the marine limit but for a larger duration.
Marine deposits laid down during this later transgression dammed local
drainages. Subsequently the sea regressed leaving scattered marine
deposits down to present sea level.

Such a reconstruction explains the geomorphic evidence but is not
unique in doing so. However, stratigraphic evidence from two sites in
the area strongly argue that the model is correct. In Noble Inlet
(Site A, Fig. 21.1), ice-proximal glacial-marine sediment, containing
abundant striated limestone erratics in a fine-grained matrix, extends
to c. 40 m aht. In situ and whole valves of molluscs collected from
the distal facies of this deposit were radiocarbon dated at 8580±150
(GSC-3469). The deposit was subsequently incised by a small stream;
the eroded gully extends below present-day high tide. The gully was
later back-filled with non-glacial normal marine sediment, in marked
contrast to the heterogeneous, carbonate-rich, ice-proximal sediment
(Fig. 21.6). This deposit forms a prominent terrace at 9 m aht.
Paired valves of Hiatella arctica from just above the erosional
unconformity separating the two marine units gave a C-14 age of 8220±90

(GSC-3404). The two dates are surprisingly close. The geologic and chronologic evidence indicate that ice flowing out Noble Inlet from the west carrying abundant limestone erratics receded across the site about 8600 years ago when relative sea level stood 40 m above present. Sea level dropped rapidly by more than 40 m and the ice-proximal sediments were dissected below present high tide. Relative sea level rose again to 9 m aht about 8200 years ago, then regressed a final time. Relative sea level is presently transgressing at the site and across the region.

A similar situation is found in a small inlet 15 km to the south (site B, Fig. 21.1). At the head of this bay a major pro-glacial delta was formed as ice withdrew to the west during deglaciation. Large (>30 cm); erractic lithologies are common, although the matrix is coarser grained than at the previous site. The delta reaches c. 28 m aht at the head of the present bay. A cross-section of the raised delta, exposed by present day wave erosion, reveals a massive cut and fill cycle. The entire central portion of the delta has been eroded by later fluvial activity. The erosional unconformity extends below present high tide and has been infilled again with a better-sorted marine sediment that is capped by a storm beach at 20 m aht and which dams a small lake in the drainage behind the delta. The main river was diverted at this time and now occupies a channel 1 km to the south. Only abraded shell fragments were found in either sediment body and due to the possible admixture of populations of various ages, the shells have not been radiocarbon dated. Nevertheless, the proglacial delta is clearly of late-Foxe age and the cut and fill evidence requires a major regression-transgression cycle of the sea similar to the Noble Inlet site.

Such fluctuations of relative sea level are difficult to explain as eustatic. Our present working hypothesis is that much of the glacial loading and unloading of this region may have been accommodated along preexisting fault systems. In such cases isostatic adjustment could have occurred rapidly, even episodically, resulting in a complex relative sea-level history unlike the predictable pattern of postglacial emergence documented elsewhere on the island.

Across outer Frobisher Bay, the rate of emergence was initially rapid. In most areas at least 80% of the net emergence was accomplished prior to 8000 years ago, and no emerged deposits <7000 years old have yet been dated (Fig. 21.4B). Relative sea level probably reached modern level before 6000 years ago. The sea has been transgressing for at least the last 1500 years(Miller et al., 1980).

The observations outlined above can not be reconciled by simple exponentially decelerating uplift after glacial unloading. A relative

sea level oscillation of between 10 and 30 m or more apparently
occurred between c. 8200 and 8600 years ago, immediately subsequent to
regional deglaciation. The most plausible explanation is that
isostatic adjustment of the peninsula to changes in the ice mass
involved rapid and possibly episodic movement of the crust along planes
of structural weakness remnant from early Tertiary rift processes.
Additional field evidence is required to confirm the oscillation of
relative sea level whereas detailed seismic profiling may provide
evidence to test the proposed explanation. Of particular inportance
would be the offset of Holocene marine sediment in the vicinity of
projected fault planes.

Discussion

The evidence presented above for a rapid regression, transgression, and
final regression of sea level on outer Meta Incognita Peninsula during
the Cockburn substage is broadly matched by evidence presented by
Nelson (1978) and to a lesser extent by King (1969). Based on simple
radiocarbon dated sea level indicators, graphs of the change in
relative sea level during the early Holocene (Fig. 21.5) show a rapid
but monotonic decrease in sea level. Conversely, the stratigraphy of
sites on the outer Meta Incognita coast (Fig. 21.6) and their
associated C-14 dates demonstrate that the sea level history is more
complex than first realized.

There are three possible explanations for the sea level
oscillations. The **first** would invoke a significant increase in the
iceload during the deglacial cycle; the **second** would suggest a possible
neotectonic influence; whereas the **third** postulates these sea level
oscillations may be in response to the migration of a forebulge.

Before attempting briefly to evaluate these explanations it is
worth noting Løken's (1962) Holocene sea level history of northernmost
Labrador. Løken's shoreline diagram indicates that:

"Strandline S1-4 is of special interest as it is postglacial
and associated with a major transgression." (p. 52)

He further noted that the strandline was apparently horizontal over
extended distances, and that the magnitude of the transgression was at
least 15 m. A C-14 date of 8190±710 (I-1322) (in Andrews and Drapier,
1967, p. 146) is: " . . . below surface of a delta assigned to the
S1-4 strandline. . ." Løken's (1962) strandline diagram suggested to
him that the S1-4 shoreline was younger than c. 8000 BP because a C-14
date of 9000±200 (L-642) was obtained from an altitude of 29 m and

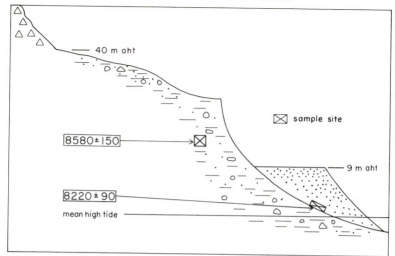

40 m aht

⊠ sample site

8580±150

9 m aht

8220±90

mean high tide

Figure 21.6 Stratigraphy and C-14 dates at site A (see Fig. 21.1) on the east coast of Meta Incognita Peninsula.

provided a minimum date for shoreline S1-3 which sloped down toward N 25° E at 0.6 m km^{-1}.

The data presented by Løken (1962, 1964), Nelson (1978), and Miller (this chapter) imply that significant changes of sea level occurred during the Cockburn substage, and that these changes affected a significant portion of the outer coast of northernmost Labrador and eastern Baffin Island. Whether neotectonics can be invoked to explain such a widespread phenomenon is perhaps questionable, but it is equally difficult to account for it via glacial isostatic mechanisms. Although glacial advances were common during the Cockburn substage (Falconer et al., 1965; Andrews et al., 1970; Miller, 1973) along the eastern seaboard, it is difficult to envision that the build up of mass was sufficient to result in a glacial isostatically induced marine transgression of 15 m or so on coasts some tens of kilometers removed from the ice margin. Further, the oscillations of sea level appear too dramatic to be explained by forebulge migration. The nature and origins of these sea level oscillations are worthy of more directed study.

ACKNOWLEDGEMENTS

This chapter is a contribution to NSF grant EAR-81-21296. In particular we wish to acknowledge the importance of field observations collected by D. Müller, A. Colvill, L. Osterman, E. Lind, and J.

Stravers as parts of this grant. J. Jacobs and W. Mode have also
shared with us information on the height and age of marine features
from the innner part of Frobisher Bay.

REFERENCES

Andrews, J.T., 1980: Progress in relative sea level and ice sheet
   reconstructions Baffin Island N.W.T. for the last 125,000 years. In:
   Mörner, N.-A. (ed.), Earth Rheology, Isostasy, and Eustasy. London,
   John Wiley and Sons, 175-200.
Andrews, J.T., 1976: Radiocarbon Date List III, Baffin Island, N.W.T.,
   Canada. University of Colorado, Institute of Arctic and Alpine
   Research, Occasional Paper 21. 50 pp.
Andrews, J.T., 1975: Support for a stable late Wisconsin ice margin
   (14,000 to c. 9000 BP); a test based on glacial rebound. Geology,
   4:617-620.
Andrews, J.T., 1970: A geomorphological study of post-glacial uplift
   with particular reference to Arctic Canada. Institute of British
   Geographers Special Publication 2. 156 pp.
Andrews, J.T., 1970: Differential crustal recovery and glacial
   chronology (6,700 to 0 BP), west Baffin Island, N.W.T., Canada.
   Arctic and Alpine Research, 2:115-134.
Andrews, J.T., 1966: Pattern of coastal uplift and deglacierization,
   west Baffin Island, N.W.T. Geographical Bulletin, 8:174-193.
Andrews, J.T., Buckley, J.T., and England, J.H., 1970: Late- glacial
   chronology and glacio-isostatic recovery, Home Bay, east Baffin
   Island. Geological Society of America Bulletin, 81:1123-1148.
Andrews, J.T. and Drapier, L., 1967: Radiocarbon dates obtained through
   Geographical Branch field observations. Geographical Bulletin,
   9:115-162.
Andrews, J.T. and Short, S.K., 1983: Radiocarbon Date List V from
   Baffin Island; Radiocarbon Date List II from Labrador. University of
   Colorado, Boulder, Institute of Arctic and Alpine Research,
   Occasional Paper 40. 71 p.

Basham, P.W., Forsyth, D.A. and Wetmiller, R.J., 1977: The seismicity
   of northern Canada. Canadian Journal Earth Sciences, 14:1646-1667.
Blackadar, R.B., 1958: Patterns resulting from glacier movements north
   of Foxe Basin, N.W.T. Arctic, 11:157-165.
Blake, W., Jr., 1966: End moraines and deglaciation chronology in
   northern Canada, with special reference to southern Baffin Island.
   Geological Survey of Canada, Paper 66-21, 1-31.
Blake, W. Jr., 1975: Radiocarbon age determinations and postglacial
   emergence at Cape Storm, southern Ellesmere Island. Geografiska
   Annaler, 62A:1-71.

Brigham, J.K. 1983: Stratigraphy, amino acid geochronology, and correlation of Quaternary sea-level and glacial events, Broughton Island, arctic Canada. Canadian Journal Earth Sciences, 20:577-598.

Bryson, R.A., Wendland, W.M., Ives, J.D., and Andrews, J.T., 1969: Radiocarbon isochrones on the disintegration of the Laurentide Ice Sheet. Arctic and Alpine Research, 1:1-14.

Clark, J.A., 1980: A numerical model of world wide sea level changes on a visco-elastic earth. In: Mörner, N.-A. (ed.), Earth Rheology, Isostasy, and Eustasy. London, John Wiley, 525-534.

Clark, J.A., Farrell, W.E., and Peltier, W.R., 1978: Global changes in postglacial sea level: a numerical calculation. Quaternary Research, 9:265-287.

Colvill, A., 1982: Glacial landforms at the head of Frobisher Bay, Baffin Island, Canada, MA thesis, University of Colorado, Boulder. 202 pp.

Craig, B.G. and Fyles, J.C., 1960: Pleistocene geology of Arctic Canada. Geological Survey of Canada, Paper 60-10. 21 pp.

Dyke, A.S., 1979: Glacial and sea level history of the southwestern Cumberland Peninsula, Baffin Island, N.W.T., Canada. Arctic and Alpine Research, 11:179-202.

Dyke, A.S., 1974: Deglacial chronology and uplift history: northeastern sector, Laurentide Ice Sheet. University of Colorado, Institute of Arctic and Alpine Research, Occasional Paper, 12. 73 pp.

Dyke, A.S., 1977: Quaternary geomorphology, glacial chronology, and climatic and sea-level history of southwestern Cumberland Peninsula, Baffin Island, Northwest Territories, Canada. Unpublished Ph.D. thesis, University of Colorado, Boulder, Colorado. 185 pp.

Dyke, A.S., Andrews, J.T., Miller, G.H., 1982: Quaternary geology of Cumberland Peninsula, Baffin Island, District of Franklin. Geological Survey of Canada Memoir 403. 32 pp.

Dyke, A.S., Dredge, L.A. and Vincent, J.-S., 1982: Configuration of the Laurentide ice sheet during the Late Wisconsin Maximum. Geographie physique et Quaternaire, 26:5-14.

Fillon, R.H. et al., 1981: Labrador shelf: Shell and total organic matter C-14 date discrepancies. Geological Survey of Canada Paper, 81-1B, 105-111.

Hall, C.F., 1865: Arctic Researches and Life among the Esquimaux. New York: Harper and Brothers. 595 pp.

Hawkins, F.F., 1980: Glacial geology and late Quaternary paleoenvironment in the Merchants Bay area, Baffin Island, N.W.T., Canada. M.Sc. thesis, University of Colorado, Boulder. 145 pp.

Ives, J.D., Deglaciation and land emergence in northern Foxe Basin. Geographical Bulletin, 21:54-65.

Ives, J.D. and Buckley, J.T., 1969: Glacial geomorphology of Remote Peninsula, Baffin Island, N.W.T. Arctic and Alpine Research, 1:83-96.

King, C.A.M., 1969: Glacial geomorphology and chronology of Henry Kater Peninsula, east Baffin Island, N.W.T. Arctic and Alpine Research, 1:195-212.

Lind, E.K., 1983: Sedimentology and paleoecology of the Cape Rammelsberg area, Baffin Island, Canada. M.Sc. thesis, University of Colorado, Boulder.

Locke, W.W. III, 1980: The Quaternary geology of the Cape Dyer area, southeasternmost Baffin Island, Canada. Ph.D. thesis, University of Colorado, Boulder, 331 pp.

Løken, O.H., 1962: The late glacial and postglacial emergence and deglaciation of northernmost Labrador. Geographical Bulletin, 17:23-56.

Løken, O.H., 1964: A study of the late and postglacial changes of sea level in northernmost Labrador. Unpublished report to the Arctic Institute of North America (mimeo). 80 pp.

Løken, O.H., 1965: Postglacial emergence at the south end of Inugsuin Fiord, Baffin Island, N.W.T. Geographical Bulletin, 7:243-258.

Løken, O.H., 1966: Baffin Island refugia older than 54,000 years. Science, 153:1378-1380.

Mangerud, J., 1972: Radiocarbon dating of marine shells, including a discussion of apparent age of recent shells from Norway. Boreas, 1:143-172.

Matthews, B., 1967: Late Quaternary land emergence in northern Ungava, Quebec. Arctic, 20:176-201.

Matthews, B., 1967: Late Quaternary marine fossils from Frobisher Bay (Baffin Island, N.W.T., Canada). Palaeogeography, Palaeoclimatology, Palaeoecology, 3:243-263.

Mercer, J.H., 1956: Geomorphology and glacial history of southernmost Baffin Island. Geological Society of America Bulletin, 67:553-570.

Mercer, J.H., 1954: The physiography and glaciology of southernmost Baffin Island. PhD thesis, McGill University, Montreal. 150 pp.

Miller, G.H., 1980: Late Foxe glaciation of southern Baffin Island, N.W.T., Canada. Geological Society of America Bulletin, Part I. 91:399-405.

Miller, G.H., 1979: Radiocarbon Date List IV: Baffin Island, N.W.T., Canada. University of Colorado, Institute of Arctic and Alpine Research, Occasional Paper 29. 61 p.

Miller, G.H., Andrews, J.T., and Short, S.K., 1977: The last interglacial-glacial cycle, Clyde Foreland, Baffin Island, N.W.T.: stratigraphy, biostratigraphy and chronology. Canadian Journal of Earth Sciences, 14:2824-2857.

Miller, G.H., Locke, W.W. III, and Locke, C., 1980: Physical characteristics of the southeastern Baffin Island coastal zone. Geological Survey of Canada, Paper 80-10, 251-265.

Mörner, N.-A. (ed), 1980: Earth Rheology, Isostasy and Eustasy. London, John Wiley and Sons, 599 pp.

Müller, D.S., 1980: Glacial geology and Quaternary history of southeast Meta Incognita Peninsula, Baffin Island, Canada. MSc thesis, University of Colorado, Boulder. 211 pp.

Nelson, A.R., 1978: Quaternary glacial and marine stratigraphy of the Qivitu Peninsula, northern Cumberland Peninsula, Baffin Island, Canada. Ph.D. dissertation, University of Colorado, Boulder. 215 pp.

Nelson, A.R., 1980: Chronology of Quaternary landforms, Qivitu Peninsula, northern Cumberland Peninsula, N.W.T, Canada. Arctic and Alpine Research, 12:265-286.

Nelson, A.R., 1981: Quaternary glacial and marine stratigraphy of the Qivitu Peninsula, northern Cumberland Peninsula, Baffin Island. Geological Society of America Bulletin, 92 Part I 1:512-518, Part II, 1143-1281.

Osterman, L.E., 1982: Late Quaternary history of southern Baffin
   Island, Canada: A study of foraminifera and sediments from Frobisher
   Bay. Ph.D. dissertation, University of Colorado, Boulder. 380 pp.

Peltier, W.R., 1976: Glacial isostatic adjustment--II: the inverse
   problem. Geophysical Journal Royal Astronomical Society, 46:669-706.
Peltier, W.R. and Andrews, J.T., 1976: Glacial-isostatic
   adjustment--I: the forward problem. Geophysical Journal Royal
   Astronomical Society, 46:605-646.
Peltier, W.R. and Andrews, J.T., 1983: Glacial geology and glacial
   isostasy, Hudson Bay, Canada. In: Smith, D.I. (ed.), Glacial
   Isostasy. Academic Press.
Pheasant, D.R., and Andrews, J.T., 1973: Wisconsin glacial chronology
   and relative sea level movements, Narpaing Fiord Broughton Island
   area, eastern Baffin Island N.W.T. Canadian Journal Earth Sciences,
   10:1621-1641.
Prest, V.K., 1969: Retreat of Wisconsin and Recent ice in North
   America. Geological Survey of Canada Map 1257A.

Quinlan, G., 1981: Numerical models of postglacial relative sea level
   change in Atlantic Canada and the eastern Canadian Arctic.
   Unpublished Ph.D. thesis, Department of Oceanography, Dalhousie
   University, Halifax, Nova Scotia, Canada. 499 pp.

Reid, I. and Falconer, R.K.H., 1982: A seismicity study in northern
   Baffin Bay. Canadian Journal Earth Sciences, 19:1581-1631.

Stuiver, M. and Pollach, H., 1977: Discussion reporting of C-14 data
   from marine molluscs, Baffin Island, Arctic Canada. Geology,
   9:451-457.
Szabo, B.J., Miller, G.H., Andrews, J.T., and Stuiver, M., 1981:
   Comparison of uranium-series, radiocarbon, and amino acid data from
   marine molluscs, Baffin Island, Arctic Canada. Geology, 9:451-457.

Walcott, R.I., 1972: Late Quaternary vertical movements in eastern
   North America. Reviews of Geophysics and Space Physics, 10:849-884.
Ward, W.H., 1952: A note on elevated strandlines of Frobisher Bay,
   Baffin Island. Geographical Review, 42:651.
Wengerd, S.A., 1951: Elevated strandlines of Frobisher Bay, Baffin
   Island. Geographical Review, 41:622-637.

22 The Holocene record from Baffin Island: modern and fossil pollen studies

S. K. Short, W. N. Mode, and P. T. Davis

Palynology in Baffin Island is in its infancy, and there is a paucity of data with which we can assess the Holocene changes of pollen accumulation over a complex mosaic of tundra. Most of the collections are from short, truncated, organic-rich sediment sections typically exposed in coastal or fluvial cliffs. Longer peat sections are rare, thus a program to locate and core pond and lake sediments for continuous stratigrapic sections began in 1975; results from two lake cores will be reported in this paper. This limited pollen evidence contrasts with that of Greenland (Fredskild, this volume) which has seen more systematic and extensive research in Holocene palynology; the need for a more detailed record of Holocene environmental history on Baffin Island is also stressed by other authors in this volume (Andrews and Funder; Andrews et al.; Davis; Bradley and Williams).

Buried organic layers are common features in late Holocene sediments from Baffin Island, and the problems inherent in dating and interpreting episodes of presumed environmental stability have been discussed by Dyke (1977) and Stuckenrath et al. (1979). These problems include small sample size (i.e. the organic nature of the sample commonly appears more evident in the field than in the lab), ambiguous C-14 dates on buried soils especially those soils developed on eolian sediments (Goh et al., 1977), wind erosion and redeposition of older materials, the well-drained nature of the sediments which facilitates the passage of modern groundwater with high C-14 levels due to post-World War II bomb tests, and possible contamination of the finer organic fraction with decayed rootlets. Stuckenrath et al. (1979) concluded that consistent C-14 dates could be obtained by dating the NaOH insoluble residue of the <125 μm fraction. However, dating problems with Arctic sediments remain a potential block to understanding the climate of the Holocene. For example, although fifteen radiocarbon dates provide a framework for the Windy Lake peat

section, dating reversals are common and require the analyst to make
decisions regarding the reliability of certain dates. Nevertheless,
progress toward understanding the Holocene environmental history of
Baffin Island has occurred through study of modern pollen deposition
and fossil pollen in lake sediment cores and peat sections. The
palynological methods employed here are described in other publications
(cf. Short and Andrews, 1980).

CONTEMPORARY POLLEN DEPOSITION

When fossil pollen spectra are compared to modern pollen assemblages
and associated vegetation, palynology becomes a powerful tool for the
interpretation of Holocene environmental and climatic changes.
However, little quantitative study of the relationship between the
plant ecology (Polunin, 1948; Webber, 1971) and palynology has been
made on Baffin Island. One approach that circumvents the lack of
information on pollen-vegetation relationships on Baffin Island is the
use of transfer functions. These equations disregard the complicated
relationship between pollen and vegetation and proceed directly to an
empirical relationship between pollen and climate (Imbrie and Kipp,
1971; Webb and Bryson, 1972; Webb and Clark, 1977). This method and
its application to Arctic pollen data have been discussed in a series
of articles (Andrews et al., 1980a; Andrews and Nichols, 1981; Andrews
et al., 1981; Elliott-Fisk et al., 1982).

    Modern pollen data from two transects have been used to construct
transfer function equations; the first transect runs SW-NE from south
of treeline in Keewatin to Clyde River, Baffin Island (KWA transect)
whereas the second runs S-N from Fort Chimo, Labrador to Clyde River
(ECA transect) (Andrews et al., 1981: Fig. 1). The Imbrie/Kipp and
multiple stepwise linear regression models were used to show the
statistical association between pollen percentages (13 taxa and 19
taxa, respectively for KWA and ECA) and climatic variables: January,
June, July, and summer (June, July, August) temperatures, Young's
(1971) index of summer warmth (i.e. the sum of monthly average
temperatures above 0°C), and summer precipitation. The results suggest
that transfer functions provide useful _relative_ measures of climatic
parameters. The coherence between records strongly suggests that the
initial modern pollen counts provide good estimates of the true pollen
frequencies. The two different approaches yield different absolute
predictions, but relative differences appear to be validated by both
approaches. The limited number of weather stations in the eastern
Canadian Arctic is seen as a primary limitation in the use of transfer
function equations in the Arctic.

Modern pollen fallout studies in the Arctic are important not only for understanding modern tundra vegetation but also for the interpretation of tundra zones in late-glacial pollen diagrams of temperate regions (Birks, 1973). Before reviewing the data on contemporary pollen rain, the processes of pollen production, dispersal, and deposition must be briefly discussed.

Pollen Production and Dispersal

The relative abundance of plants usually is not accurately reflected in the pollen rain within the same vegetation. To test this relationship in the Broughton Island area (Fig. 22.1), moss and lichen polsters were

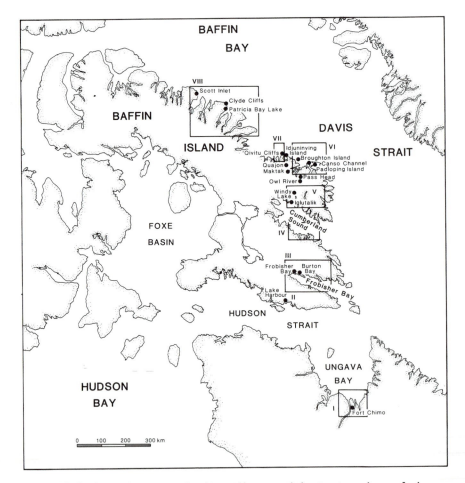

Figure 22.1 Location map of sites discussed in text and sampled regions of the eastern Canadian Arctic modern pollen transect.

collected and analyzed from vegetation quadrats which were quantified
in terms of cover and frequency of plant species (Webber, 1971).
Comparison of the relative importance of five major taxa (Salix
[willow], Gramineae [grass family], Cyperaceae [sedge family], Ericales
[heath group], and Lycopodium [clubmoss]) in the vegetation analysis
versus the pollen analyses shows that pollen percentages are generally
representative of plant importance in the local vegetation (Mode,
1980). Lycopodium and Salix pollen tend to be slightly more common
than clubmoss and willow plants, whereas the ericaceous plants are
somewhat under-represented in the pollen rain. The under-
representation of Ericales pollen in surface samples has been noted in
other tundra pollen studies (Iversen, 1945; Fredskild, 1967; Rymer,
1973), though Empetrum (crowberry), which is not separated from the
Ericales in this study, is commonly well represented in areas where it
has been distinguished (Pennington, 1980).

The major difference between the Broughton Island pollen rain and
that of other Arctic tundra areas (Birks, 1973; Pennington, 1980) is
that the pollen from the over-productive shrub Betula (birch) does not
dominate the Broughton Island spectra because the Broughton Island
site is just north of the northern limit of Betula nana (dwarf birch)
(Andrews et al., 1980b; see also Fig. 22.2). However, Salix pollen,
which as noted is slightly over-represented, fills the "void" left by
Betula.

The concept of a regional pollen rain implies that forest pollen
is dispersed at a regional scale because it is released into the
turbulent or mixed layer of the atmosphere. However, in the Arctic,
where most shrubs are prostrate, little pollen is transported on a
regional scale. Gregory (1973) suggested that airborne particles are
trapped in the laminar boundary of the atmosphere, which is next to the
ground. The thickness of this layer varies with wind velocity and
local surface roughness and surface heating, but clearly a large
portion of pollen released in the High and Middle Arctic is probably
released in or very near the laminar boundary layer and transported
only very short distances upon release. Thus, limited pollen transport
in the Arctic may inhibit a regional pollen rain from developing
through atmospheric mixing. Instead, the vegetational mosaic of the
tundra is probably characterized by a mosaic of pollen assemblages
(Bartley, 1967; Pennington, 1980). We have adopted a method to mix
local pollen spectra in order to produce a regional pollen rain (e.g.
Adam and Mehringer, 1975). The method entails analysis of a number of
moss polsters from each of the regional vegetation units and then use
of the median values for each pollen type as a representation of the
regional pollen rain.

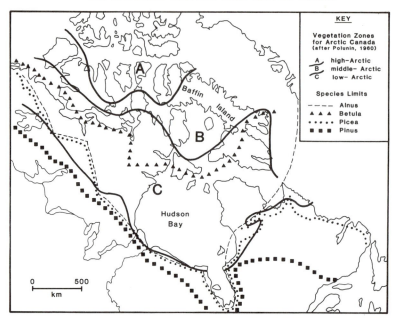

Figure 22.2. Map of vegetation zones of the eastern Canadian Arctic according to Polunin (1960). The northern limits of Alnus, Betula, Picea, and Pinus are also indicated.

Regional Pollen Rain

Most authors divide the vegetation of Baffin Island into three phytogeographic regions. The largest region is classified as Middle Arctic (Polunin, 1960) or Stony Sedge-Moss-Lichen Tundra (Porsild, 1957). In southern Baffin Island, the area classified as Low Arctic or Dwarf Shrub-Sedge-Moss-Lichen Tundra roughly corresponds to the area inhabited by shrub birch (Betula nana and B. glandulosa, Andrews et al., 1980b). A number of other plant species, including Artemisia (sage), Rubus chamaemorus (cloudberry), and Lycopodium annotinum (bristly clubmoss), reach their northern limits in this zone. A small area in northern Baffin Island is characterized as High Arctic, Rock Desert, or Fell-Field. The classification of Polunin (1960) is illustrated in Figure 22.2 as are the northern limits of the important exotic genera Alnus (alder), Picea (spruce), and Pinus (pine). Young (1971) established four circumpolar floristic zones based on the sum of summer (June, July, August) degree-days above 0°C. The existence of these zones supports the concept that vegetation is a function of climate with summer warmth as the primary limiting factor on plant growth. Most of the sites discussed in our paper lie in Young's Zone 3 which is transitional between the High Arctic (Zones 1 and 2) and the

Low Arctic (Zone 4); however, the Frobisher Bay and Lake Harbour areas
lie in Zone 4.

Mode (1980; unpublished data) and Mode et al. (1981) analyzed 124
surface samples to define the regional pollen rain along a north-south
transect in the eastern Canadian Arctic. This transect (Fig. 22.1) is
tied to Fort Chimo, Quebec at its southern end, and to the Clyde River
area at its northern end. The samples were divided into eight regional
groups: (I) the Fort Chimo, Quebec area with 6 polsters; (II) the Lake
Harbour area with 8 polsters; (III) the Frobisher Bay area which
includes samples from the Cape Rammelsberg region, 30 km to the south
(Lind, 1983), with 34 polsters; (IV) the Cumberland Sound area with 7
polsters; (V) the Pangnirtung area with 14 polsters, 2 lake sediment
samples, and 6 Tauber trap samples (Davis, 1980b); (VI) the Broughton
Island and Canso Channel area with 36 polsters; (VII) the Qivitu
Peninsula area with 12 polsters; and (VIII) the Clyde area with 18
polsters and 3 lake sediment samples (Table 22.1; Fig. 22.3).

Pollen concentration data are presented in Elliott-Fisk et al.
(1982). The data can be summarized as follows:

(1) Fort Chimo, located in the forest-tundra ecotone, has high tree
    and shrub (Alnus, Betula, Larix [larch], Picea, Pinus)
    concentrations and thus contrasts with all of the Baffin Island
    region.
(2) The Lake Harbour region of southern Baffin Island has a pollen rain
    dominated by Ericales, Salix, and Betula. These pollen data
    accurately represent a vegetation in which areas of scrub and
    shrubby heath dominated by these taxa are common (Polunin, 1948).
(3) The Frobisher Bay region has Alnus and Betula concentrations lower
    than Fort Chimo but higher than at any other region on Baffin
    Island. Salix, Gramineae, and Cyperaceae pollen are more important
    here than at Fort Chimo or Lake Harbour, and reflect the greater
    importance of herbaceous vegetation versus shrubby and dwarf shrub
    vegetation types in the Frobisher Bay area. Decreasing shrubby
    vegetation is a trend which continues with distance northward along
    the transect (Polunin, 1948), and this trend is recognizable in the
    pollen rain. Exotic Pinus pollen is present in amounts comparable
    to the Fort Chimo area.
(4) Alnus and Betula concentration values in the Cumberland Sound
    region are lower than around Frobisher but higher than in regions
    further north. Pinus concentration values are low here but
    percentages reach a maximum for the transect due to the combination
    of low local tundra pollen production and the significant long-
    distance transport of Pinus pollen from the boreal forest, an

TABLE 22.1. Median pollen percentage data for seven key taxa of Figure 21.3 in each of the eight regions of the eastern Canadian Arctic. (No exclusions from pollen sum).

	Fort Chimo	Lake Harbour	Frobisher Bay	Cumberland Sound	Pangnirtung	Broughton Island	Qivitu	Clyde
Alnus %	26.4	1.4	7.2	3.4	2.6	0.4	0.3	1
Betula %	18.4	4.4	5.5	2.2	0.5	1	1.3	0.4
Pinus %	3.3	2.2	4.0	3.4	1.4	1	0.2	1
Salix %	6.6	5.5	10.9	8.3	10.0	14.5	24.0	22
Gramineae %	1.4	2.4	6.0	2.8	24.8	4	3.6	22
Cyperaceae %	4.1	2.8	8.1	14.3	11.0	21	9.4	6
Ericales %	10.6	9.4	4.8	24.3	12.0	17	3.2	7

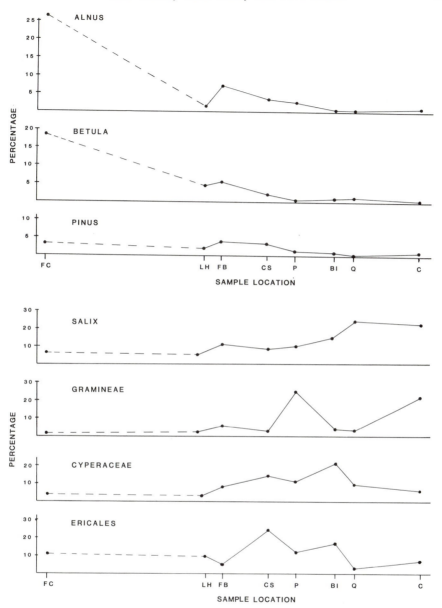

Figure 22.3 Median pollen percentages along the eastern Canadian
Arctic transect for <u>Alnus</u>, <u>Betula</u>, <u>Pinus</u>, <u>Salix</u>, Gramineae,
Cyperaceae, and Ericales. (FC = Fort Chimo, LH = Lake Harbour, FB =
Frobisher Bay, CS = Cumberland Sound, P = Pangnirtung, BI = Broughton
Island and Canso Channel, Q = Qivitu, C = Clyde).

effect first documented by Aario (1940) in Finland and noted
elsewhere since then (Ritchie and Lichti-Federovich, 1967; Birks,
1973). Ericales percentages are much greater in this region than
in the areas to the south.

(5) Gramineae percentages dominate the pollen spectra from the
Pangnirtung area. Betula percentages are very low, corresponding
to the region's position very near the northern limit of isolated
stands of shrub birch (Andrews et al., 1980b, Fig. 1; see also
Fig. 22.2, this paper). In all other respects, the median pollen
spectrum is similar to that of Cumberland Sound.

(6) The Broughton Island and Canso Channel region includes polster
samples from near the northernmost outliers of shrub birch on
Baffin Island (Andrews et al., 1980b, Fig. 1; see also Fig.22.2,
this paper). Pollen spectra are dominated by Cyperaceae, Ericales,
and Salix. The exotic taxa Alnus and Pinus occur in very small
percentages. Concentration values are low and similar to those
recorded in the Qivitu region.

(7) The pollen rain from the Qivitu foreland is dominated by Salix, and
the concentrations of exotic and local pollen taxa are lower than
those in any other region except for the Clyde area.

(8) Salix and Gramineae dominate the pollen rain in the Clyde region,
and the latter reaches its percentage maximum here. Pollen
concentrations are at their lowest for the transect.

Figure 22.3 summarizes the trends of the surface pollen percentage
data along the eastern Canadian Arctic transect. The Alnus
concentration data provide an example of the non-linear decrease in
pollen deposition with distance from the source plants (Faegri and
Iversen, 1975). The pollen rain at the northern limit of alder shrubs
(Fig. 22.2) is approximated by the six Fort Chimo polsters in which
Alnus pollen concentration exceeds that of all other pollen types. The
Alnus percentages north of this southern region decrease along a curve,
except for the Lake Harbour area where local pollen productivity is
great enough to minimize the effect of long-distance dispersal.

The southernmost region, the Fort Chimo area, is well to the north
of the northern limit of pine trees; therefore, the Pinus concentration
curve consists entirely of the asymptotic tail of the pollen deposition
curve, and it reflects the prodigious amounts of pine pollen released
as well as the ability of this pollen to remain suspended in the
atmosphere (Faegri and Iversen, 1975). The precise shape of the Pinus
percentage curve reflects the interaction of the long distance
transport of pine pollen and the low pollen productivity of the Baffin
Island tundra. The over-representation of pine pollen is an effect
noted in other Arctic areas (cf. Birks, 1973), but never confirmed in

the literature for a long transect into the Middle and High Arctic regions. Pinus pollen is also over-represented in the northern regions of the transect, but much less so than in the Frobisher Bay and Cumberland Sound regions.

Birch shrubs reach their northern limit in small outlier stands found in the southern part of the Broughton Island-Canso Channel region (Fig. 22.2). The Betula curve (Fig. 22.3) reflects the decreasing numbers of shrub birches with distance north of Frobisher Bay. Median Betula percentages of about 5% distinguish the Low Arctic zone of southern Baffin Island (Lake Harbour and Frobisher Bay) from the vegetation zones to the north where median Betula percentages range up to 2.2%.

The Cyperaceae and Ericales percentage curves both reach maxima in the Cumberland Sound, Pangnirtung, and Broughton Island regions. These maxima presumably represent the greater importance of these taxa in the vegetation of the Middle Arctic tundra compared to the forest-tundra ecotone (Fort Chimo) and the High Arctic (Qivitu and Clyde). With long-distance transport of Alnus, Betula, and Pinus declining, Cyperaceae and Ericales become proportionally more important in the pollen rain than they are in the Low Arctic zone, though the plants producing these pollen types are probably not more abundant than they are in the Low Arctic zone. Salix percentages increase with distance northward, but as is the case for all pollen types, Salix concentration values decrease. However, Salix concentration does not decrease northward as rapidly as that of other taxa, especially the Ericales and Cyperaceae groups. High Salix and low Ericales and Cyperaceae percentages in the Clyde and Qivitu regions are distinct from lower values in the other regions. Gramineae percentages peak at two separate places along the transect, Pangnirtung and Clyde. No trend is obvious for this family except that Gramineae never constitutes a large percentage of the Low Arctic pollen rain.

To summarize, three modern pollen assemblages can be identified in the eastern Canadian Arctic: (1) a Low Arctic assemblage typified by the Frobisher Bay and Cumberland Sound regions, characterized by significant amounts of pollen from the Low Arctic shrubs Alnus (wind-transported) and Betula; (2) a Middle Arctic assemblage, represented in the Broughton Island-Canso Channel regions and to a lesser extent by the Cumberland Sound region, where Cyperaceae and Ericales are the dominant pollen types; and (3) a High Arctic assemblage, typified by Qivitu and Clyde, dominated by Salix and Gramineae. This threefold zonation resembles Polunin's (1948) three vegetation zones discussed above (Fig. 22.2).

Mode's (1980) analyses, are supported by recent work by Elliott-Fisk et al. (1982) who have analyzed a suite of modern samples from 39 sites in eastern and central northern Canada, ranging from treeline to northern Baffin Island. Elliott-Fisk et al. (1982) present a series of isopoll maps of percentage and concentration data for eight major pollen types (Alnus, Betula, Picea, Pinus, Salix, Gramineae, Cyperaceae and Ericaceae [= Ericales]) which dominate the pollen spectra. Alnus percentages are significantly higher in Mode's (1980) Fort Chimo samples than in the Labrador-Ungava and Keewatin sites reported in Elliott-Fisk et al. (1982, Fig. 3A); Alnus values on Baffin Island, however, are similar, as are the concentration values (Elliott-Fisk et al., 1982, Fig. 5A). Betula values are similar in the two studies except for a greater importance in the Frobisher Bay region in Mode's (1980) study. The Pinus curves compare broadly, as do the Salix, Gramineae, and Ericales data. Although both studies record a decrease in Cyperaceae percentages northwards from a maximum on the central Baffin Island coast, the values are larger in Mode's (1980) study. Both studies record maximum Ericales concentration values in the Low and Middle Arctic sites.

A major difference in the two studies occurs in the comparison of the Cyperaceae concentration data. Elliott-Fisk et al. (1982, Fig. 6C) record a consistent decrease in values north of the central Labrador coast; Mode (1980), however, records a maximum in Cyperaceae values in the Cumberland Sound and Broughton Island regions. The former study correlated higher Cyperaceae values with "warm" wet habitats, and we believe that such a description best describes the Low Arctic regions analyzed by Mode (1980).

Cluster analysis (Elliott-Fisk et al., 1982) yields a set of six distinct pollen assemblages, with good correspondence to Mode's (1980) analysis. Cluster I, characteristic of the Transitional Middle to High Arctic tundra zones, is defined by the dominance of Gramineae, Salix, and Ericales; the latter probably reflects the "mix" of Middle Arctic sites in this cluster. Cluster 2, also described as Transitional, and therefore probably best described as Middle Arctic, is defined by Ericales, Cyperaceae, and Salix. Cluster 3, Low Arctic mesic shrub communities, is made up of Ericales, Betula, and Pinus, whereas Cluster 4, also Low to Transitional Arctic but believed to represent wetter sites, is defined by Salix, Cyperaceae, and Gramineae. The other two clusters delineate northern boreal assemblages and do not apply to this review.

Our studies of the modern pollen fallout on Baffin Island can also be compared to Mudie's (1982) analysis of 90 surface samples from

coastal and continental shelf areas off eastern Canada between 42 and
69°N (i.e. Nova Scotia to northern Baffin Island). Mudie (1982; see
also Mudie and Short, this volume) found that the latitudinal gradient
in the marine pollen assemblages is related to terrestrial climatic
conditions rather than ocean current circulation or sedimentation
rates. Q-mode factor analysis of the percentage pollen data produces
geographical groups of four pollen assemblages that correspond closely
to the distribution of vegetation zones. The Arctic Assemblage is
characterized by Gramineae, Rosaceae (rose family), and other NAP
(non-arboreal pollen) species, and Cyperaceae pollen with maximum
values off Baffin Island (60-69°N). Percentage contours for Rosaceae
pollen (Mudie, 1982, Fig. 7b) reach up to 20% in the core-top samples
off eastern Baffin Island; these values are generally higher than those
found in our terrestrial modern samples and may reflect differential
preservation of this taxon in the marine environment. Mudie (1982)
mentions that in the northern regions of the Arctic Assemblage, the NAP
species are dominated by more Arctic taxa, such as Saxifraga
(saxifrage) and Oxyria (alpine sorrel); thus, further analyses could
sub-divide this assemblage.

HOLOCENE ASSEMBLAGES

The Holocene pollen record from Baffin Island is limited in time,
depth, and areal extent. Organic-rich terrestrial sediment sections
suitable for palynology are frequently short, truncated, and cluster in
the last 2500-3000 years. However, buried soils in uplifted marine
sections have been recovered which date from 10,000 to 8000 BP. Lake
sediment cores that span a major part of the Holocene have been
analyzed from two regions, Cumberland Peninsula and the Clyde Foreland
(Fig. 22.1).

Clyde River - Lake Sediments

Mode (1980) collected and analyzed a sediment core from Patricia Bay
Lake on the Clyde foreland (Fig. 22.4). Four C-14 dates provide a
chronology for the core which spans the last 7000 years. A date of
6320±130 BP on moss fragments from 105 to 108 cm suggests emergence of
the lake basin above sea level about 6500 BP an age supported by other
dates from the foreland (Miller et al., 1977; Mode, 1980); therefore, a
date of 8810±205 BP on lake sediments from 84-90 cm depth is rejected.
Mode suggests that a possible source of contamination is old, reworked
organics in the fine fraction of that sample (in Andrews, 1983). The
core was sampled at 2.5 cm intervals, providing a sampling resolution

Figure 22.4a Pollen percentage (relative data) diagram, Patricia Bay Lake, based on a pollen sum excluding Cyperaceae, Ericales, Sphagnum and undeterminable grains.

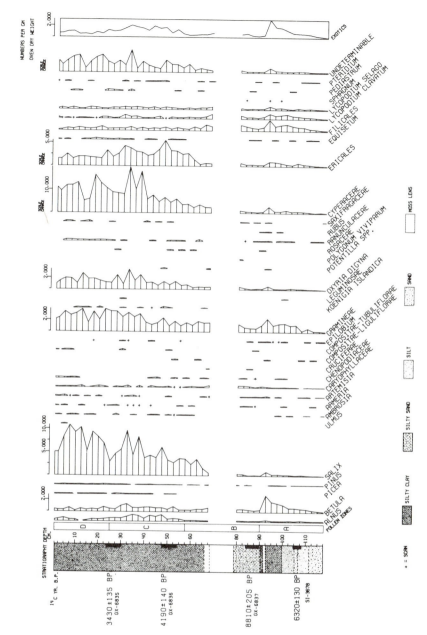

Figure 22.4b   Pollen concentration diagram, Patricia Bay Lake, based on numbers of grains per gram, oven-dry weight.

of 95 years for the basal section, 80 years for the middle section, and 320 years for the last 3400 years. The top sample is at 1 cm rather than zero because the surface sediment was lost during coring.

Four pollen zones were identified (Fig. 22.4a). The Betula zone (A), 6800-5600 BP, is characterized by peaks in Betula, Filicales (ferns), and Lycopodium clavatum/annotinum (clubmoss) percentages as well as moderate to maximum concentration values for these taxa. An analog for this assemblage has not been found in the modern samples analyzed; however, a similar assemblage commonly appears in Scandinavian, early post-glacial pollen diagrams (Vasari, 1974; Hyvärinen, 1975). Mode (1980) concluded that shrub birch grew at or quite near the Clyde area, some 450 km north of the present limit, and that this period represented the local Holocene climatic optimum.

The Caryophyllaceae zone (B), 5600-4500 BP, has the highest percentage of Caryophyllaceae (pink family) and the lowest concentration values of exotic pollen of the four zones. The rise in the Salix percentage curve occurs in the upper part of the zone as does a drop in spore percentages. During this period, a climatic deterioration occurred that was severe enough to depress influx of exotic pollen and cause birch shrubs to disappear from the Clyde region.

Alnus and Ericales percentages peak in the Alnus zone (C), 4500-3100 BP; Betula percentages recover to moderate values, and Picea pollen first appears in the diagram. Increased Ericales percentages suggest that local warming occurred, and this is supported by the increased exotic influx due either to a northward advance of Picea, Alnus, and Betula caused by ameliorated climatic conditions or increased frequency of advective transport of these pollen types northward.

The Salix zone (D) is characterized by maximum Salix percentages and concentration values over the last 3000 years. Cyperaceae is co-dominant with Salix and exotic pollen concentrations are lower than in zone C. This zone represents deteriorating climatic conditions. Several local Salix species (S. herbacea, S. cordifolia, and S. reticulata) are snow-tolerant (Porsild, 1957), and colder summers may have caused snowbanks to persist longer, enhancing Salix pollen concentrations. The higher concentrations of Cyperaceae pollen probably reflect a concomitant decrease in the depth of the active permafrost layer and increased surface ponding which would have increased areas of sedge vegetation.

Cumberland Peninsula - Lake Sediments

A 290-cm lake sediment core was collected by Davis (1980a, b; Davis et
al., 1980) at Iglutalik Lake on the east coast of Cumberland Sound.
Extrapolation from a date of 8815±275 BP centered at 269 cm, an age of
9200 BP is suggested for the base of the core (Fig. 22.5). Thus, this
site provides the longest and most continuous Holocene terrestrial
climatic record available from Baffin Island. The site was analyzed
for pollen at 2-cm intervals for the upper 90 cm and at 5- or 10-cm
intervals for the lower 2 m. Five C-14 dates (Fig. 22.5) suggest
sediment accumulation rates of 0.2 to 0.6 mm/year, thus the 69 pollen
levels range from about 110 to 260 years/sample. Percentage and
concentration pollen data for Iglutalik Lake are illustrated in Figures
22.6a & b.

Exotic tree and shrub pollen influx (Alnus, Picea, and Pinus) were
higher between 9200 and 8700 BP than at any time until 5800 BP,
suggesting the importance of strong summer airflows from the boreal
forest, at that time at least 2000 km to the south and/or west.
Interpretation of transfer function equations (Fig. 22.5) suggests that
this short climatic episode was warm and moist. Gramineae and
Cyperaceae dominated the local pollen spectra during this period, but

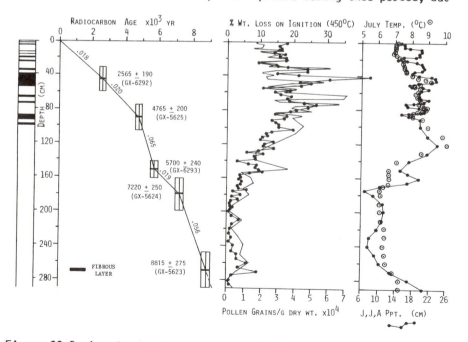

Figure 22.5 Age-depth, weight loss, and transfer functions curves,
Iglutalik Lake.

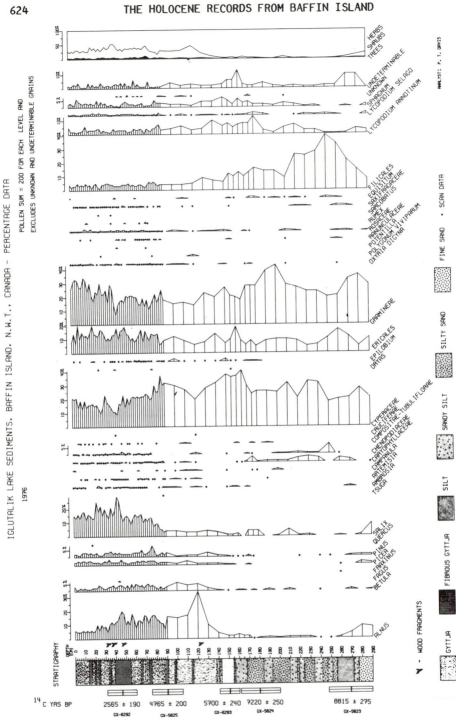

Figure 22.6a   Pollen percentage (relative data) diagram, Iglutalik
Lake, based on a total count of 200 grains for each level, excluding
unknown and undeterminable grains.

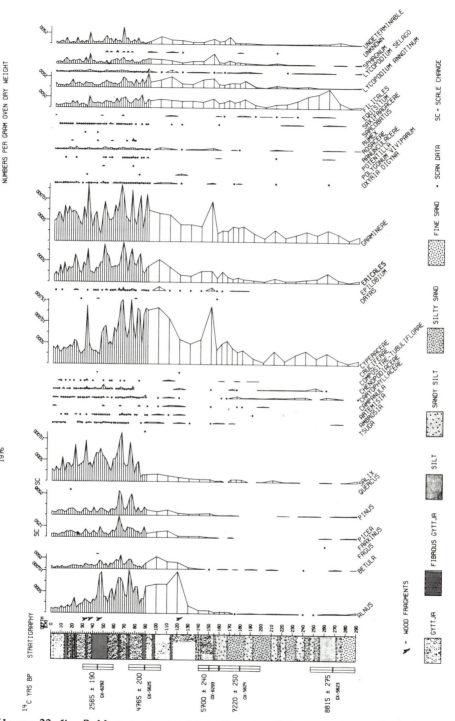

Figure 22.6b Pollen concentration diagram, Iglutalik Lake, based on numbers of grains per gram, oven-dry weight.

an episode of Filicales co-dominance was recorded from 8700 to 7800
BP. Subsequently, Filicales values declined and Cyperaceae and
Gramineae continued to dominate the spectrum; Salix was a minor element
of the vegetation until 5800 BP. Organic content and pollen
concentration values were low between 8700 and 5800 BP, suggesting a
dry, cool environment; predicted July temperatures for this time period
remained at their lowest values during the Holocene (Davis, 1980b;
Davis et al., 1980; Figs 22.6a, & 22.9). These climatic predictions
from the transfer functions may reflect local shelf ice conditions as
Laurentide ice was present at the head of Cumberland Sound until 5700
BP (Dyke, 1979; Davis, 1980b; Davis et al., 1980; Dyke et al., 1982).

At 5800 BP sediment accumulation rates again increased and organic
content and pollen concentration values rose dramatically as vegetation
became well established and the local Laurentide ice margin receded.
Cyperaceae and Gramineae continued to dominate until 5200 BP when
increased values were registered for both exotic tree and shrub pollen
and for local shrub taxa. Exotic Alnus pollen percentages and
concentration values increased to a Holocene maximum at about 5200 BP
and continued to contribute to the pollen spectra until about 2700-2600
BP. Exotic Picea and Pinus values also increased as treeline moved
northward in Labrador-Ungava (Short, 1978a, b) and Keewatin (Nichols,
1975, Fig. 26). Betula nana and/or B. glandulosa reached a Holocene
maximum between 5200 and 4700 BP suggesting the local growth of birch
shrub, possibly marking the local Hypsithermal. Predicted July
temperature and summer precipitation values peaked at 5200 BP. After a
decrease in these "warm" indicators at 4700 BP, the pollen spectra for
Iglutalik Lake register a recovery until about 2700 BP, although this
episode is characterized by fluctuating values both in the pollen
diagrams and in the transfer function data (see Figs 22.5 & 22.9).
Picea and Pinus concentration values increased to peak Holocene levels
between 4200 and 3300 BP, suggesting a prolonged influx of southerly
air concomitant with the northern extent of the forests in central
Labrador-Ungava (Short, 1978a, b). Salix, Cyperaceae, Ericales, and
Gramineae dominate the local pollen spectra at this time, suggesting a
local maximum in shrub development.

The pollen spectrum is dominated by local taxa to the top of the
core, and a strong Neoglacial cooling from 2700 BP onwards is
suggested. Picea and Pinus values fell after 3300 BP, following the
decrease in forest cover in Labrador-Ungava, but minor peaks in exotics
were recorded at 2700-2500 BP, 2200 BP, and 1800 BP. Alnus and Betula
also declined to low values, reflecting cold conditions. Predicted
July temperatures fell by 2-3°C during the last 3300 years.

Early Holocene Peats

The Clyde Cliffs form one of the major Quaternary exposures in North
America, with a complex sequence of glacial, glaciofluvial, and marine
sediments interbedded with buried soils and organic accumulations
(Miller et al., 1977; Mode, 1980; see Mode, this volume). These
organic sediments range in age from pre-Holocene to late Holocene;
seven short sections were analyzed for pollen and reported in Miller et
al. (1977, Fig. 10). The earliest Holocene record, the base of the
Ravenscraig marine sediments, is based on two samples dated 9880±200 BP
and 8210±50 B.P., and is characterized by very low pollen
accumulation. Moderately large amounts of Betula (about 50%)
contribute to the large percentage of shrub pollen, and mesic and dry
indicators (Miller et al., 1977, Table 9) are dominant. Exotic taxa
are very low or non-existent, and the lack of Alnus and Picea pollen
probably reflects the very long distances to source areas in southern
Canada or western North America in the early Holocene. The low
concentration values could be due to: (1) very low pollen productivity
due to a severe local climate, or (2) rapid organic growth and/or
washing-out during the subsequent marine phase. Whether the moderately
large Betula percentages represent small local clones or long-distance
transport is uncertain; however, Betula has also been reported for this
time period in east Greenland (Funder, 1978), western North America
(Ritchie, 1976, 1977), and along the northern Labrador-Ungava coast
(Short, 1978a, b).

Mode (1980) reported the pollen analysis of a 80-cm thick peat
monolith dated between 8000 and 6000 BP from the Qivitu Cliffs, 325 km
south of Clyde. The pollen assemblage (Mode, 1980, Figs 6.3 & 6.4) is
remarkably uniform and is dominated by Cyperaceae, Salix, and
Gramineae. The basal 15 cm of the diagram represents vegetational
succession on freshly exposed surfaces near the site, with
Caryophyllaceae and Filicales the first colonizers; the role of the
Caryophyllaceae as an early colonizer was also noted by Bartley and
Matthews (1969) at Sugluk, northwestern Ungava. Betula percentages are
high only in the two basal samples, probably because of the very low
local pollen production at this time. The apparent absence of birch
shrubs at this site conflicts with their inferred presence at Clyde
during this time. Possibly the shrubs at Clyde were an outlier well
north of the Betula "shrub" line or possibly birch shrubs were present
but not pollinating at Qivitu. Qivitu is about 40 km north of isolated
B. nana stands today (Andrews et al., 1980b).

Mode also analyzed several miscellaneous samples from the Qivitu
and Cumberland Peninsula regions; some of these are relevant to this

paper. Sample GRL-318-0, dating 9950±185 BP from the Qivitu region (Mode, 1980, Figs 6.5 & 6.6), contains higher Betula and Alnus percentages than modern samples from the region, suggesting either closer proximity to these shrub stands and/or more frequent advective transport of these grains from southerly or westerly sources. Sample GRL-318-0 compares palynologically to the Ravenscraig samples in the Clyde region. Sample #207 from southwestern Cumberland Peninsula is assumed to be the same age as a nearby sample (Usualuk) with a date of 6800±800 BP. The latter contained a pollen assemblage dominated by Betula (Short, in Andrews, 1976, p. 8). Sample #207 contains a diverse assemblage (Mode, 1980, Figs 6.5 & 6.6), with sufficiently high Betula values to suggest that birch was present in the flora. Birch shrubs grow at present within a few tens of kilometers of the site, and these samples suggest an antiquity of some 7000 years for the migration of this taxon into the region. This supports Schwarzenbach's (1975) earlier suggestion that the June/Padle/Kingnait valleys of Cumberland Peninsula were a plant refugium during the last glaciation. This record predates the amelioration at 5800 BP, as well as the increase in Betula pollen which occurred about 5200 BP, in the nearby Iglutalik Lake core.

Middle and Late Holocene Peats

The middle Holocene at Clyde Cliffs is represented by a short (13-cm) peat section at Scott Inlet (north of Clyde) with a basal date of 4260±475 BP (Miller et al., 1977, Fig. 10). This section records the transition from the local climatic optimum, a period of environmental stability, toward the more severe climate of the last 4000 years, marked by the initiation of eolian sand deposition. The pollen spectra are dominated by Gramineae and Cyperaceae with the addition of moderate amounts of Salix pollen and a small but consistent influx of exotic taxa. This contrasts with the modern spectra which contain less Salix pollen, more Caryophyllaceae pollen, and little exotic pollen. Indeed, the present-day pollen record is the most impoverished of the entire period represented in the Clyde Cliffs.

Short and Andrews (1980) reported the results of palynological investigations on six short peat sections from northern Cumberland Peninsula dating from the middle and late Holocene. The six sites were compared to two pollen spectra from Maktak Fiord (Boulton et al., 1976, Fig. 2) and Windy Lake (Nichols, 1975, Figs 24 & 25; Andrews et al., 1979, Figs 6 & 7; Davis, 1980b, Figs 5.3 & 5.4). The most important trend observed is the change from a mixed and relatively diverse shrub and heath assemblage that characterized the period between about 5500

BP (Pass Head site) and 1500 BP (Owl River, Windy Lake, Maktak Fiord, and the lower part of Padloping Island sites) to a predominantly graminoid (grass and sedge) phase that characterized the past 1500 years (Quajon Fiord, Idjuniving Island, Broughton Island, and the upper part of Padloping Island sites). The combined record also suggests that even the shrub assemblage from the period between 3000 and 2000 BP records a climatic deterioration from the local climatic optimum, registered at Pass Head, which is characterized by a much greater pollen productivity with higher Salix values.

A major peak of Salix pollen is recorded in the pollen diagrams from Padloping Island, Owl River, Maktak Fiord, and Windy Lake (Short and Andrews, 1980, Fig. 9). At Padloping Island, Windy Lake, and Maktak Fiord, interpolation between C-14 dates puts the willow peaks at 2200-2400 BP; however, at Owl River the willow spike dates closer to 2600-2800 BP. With the errors inherent in dating these sediments, Short and Andrews proposed that the age of the willow peak was bracketed between 2200 and 2600 BP, and it was correlative with a willow peak at 2570 BP recorded by Fredskild (1967) at Sermermiut, West Greenland.

The peat site near Windy Lake was resampled by Davis in 1976, who extended the record from 127.5 to 225 cm, or from about 2700 BP to 3700 BP (Davis, 1980b). The pollen spectra in this basal section are dominated by Gramineae; Cyperaceae and Ericales are also important at this time (Fig. 22.7). Exotic pollen taxa, Alnus, Picea, and Pinus, occur in low but consistent numbers throughout the lower section which contrasts with the more intermittent record in the upper part of the diagram. Thus, the extended record from Windy Lake supports the above trend of a climatic deterioration from the mid-Holocene. Nichols et al. (1978) suggest that Alnus, Picea, and Pinus peak concentrations at the Windy Lake site result from regional paleo-wind shifts, possibly due to periodic shifts of a trough in the atmosphere. Spectral analysis of the exotic taxa profiles in the upper section exhibited peaks with a periodicity of roughly 200 years between 2500 and 600 BP. Nichols et al. (1978) recognize the possibility of irregular sediment accumulation rates for peat, but Davis noted that three of the five highest Picea and Pinus peak concentrations correlate with the three highest organic matter percentages in the upper section (Davis, 1980b, Fig. 5.5). Thus, higher organic content could indicate slower sedimentation accumulation due to decreased eolian sand contribution, and a slower sediment accumulation rate could explain high exotic concentrations of exotic windblown pollen into the site. Moreover, Davis (1980b) found no evidence from Iglutalik Lake for correlation of exotic pollen peaks and paleo-wind shifts.

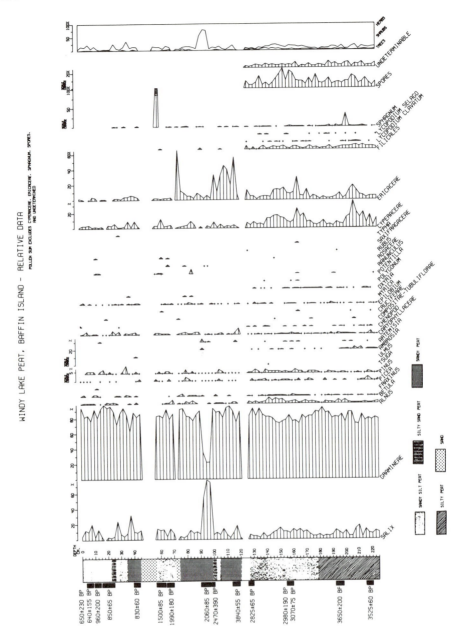

Figure 22.7a. Pollen percentage (relative data) diagram, Windy Lake Peat, based on a total count of 100 grains for the upper 120 cm and 200 grains for the lower 100 cm, excluding Cyperaceae, Ericaceae, Sphagnum, undifferentiated spores, and undeterminable grains. Sand layers (46-57 cm and 71-74 cm) were not collected.

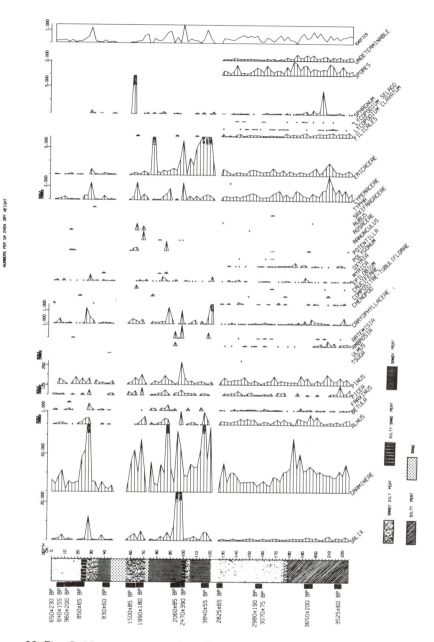

Figure 22.7b Pollen concentration ("absolute" data) diagram, Windy
Lake Peat, based on numbers of grains per gram, oven-dry weight. Sand
layers (46-57 cm and 71-74 cm) were not collected.

To the southwest of Cumberland Peninsula, Short and Jacobs (1982) reported a shallow peat section in the Frobisher Bay region which dates 2000 to 900 BP. The Burton Bay Cliffs peat section (1982, Figs 4 & 5) registers two phases: (1) an earlier pollen assemblage, 2000-1650 BP, which is more diverse, includes significant numbers of shrub (Betula, Salix, Ericales) pollen and records a large influx of exotic taxa, especially Alnus; and (2) an impoverished grass-dominated phase, from about 1650 BP to 900 BP. The assemblage from the earlier phase suggests warmer conditions, with possible increased advection of southerly air masses bringing in large numbers of exotic pollen and supporting a dense shrubland. The subsequent phase was certainly colder and drier than either the earlier period or the present. Modern pollen samples suggest that present-day climatic conditions are close to those before 1650 BP, although perhaps somewhat cooler due to the absence of any birch in the immediate vicinity of the cliffs today. The proposed colder and drier climatic conditions for the later phase are supported by eolian deposition at the site after 900 BP (Short and Jacobs, 1982, Table 4). Transfer function estimates of July temperature for the Burton Bay site (Short and Jacobs, 1982, Fig. 6) also suggest a shift to lower temperatures after 1600 BP.

Additional palynological studies are presently being carried out in the Frobisher Bay area. These studies include an expanded modern pollen rain analysis (Lind, 1983; Mode, unpublished) and the collection of lake sediment cores which will provide more continuous and extensive paleoclimatic records. Mudie and Short (this volume) discuss the palynology of core HU77-159 from marine sediments in Frobisher Bay.

DISCUSSION

The sites that have proved most useful for interpretation of Holocene paleovegetation and paleoclimate are part of a C-14 dated sequence and show fluctuations in the major taxa, Gramineae, Salix, Ericales, Cyperaceae, Betula, as well as the exotic taxa through time (Fig. 22.8). This figure illustrates the general paucity of data before about 4000 BP on Baffin Island and the need for continuous records, specifically for the period $\geq$ 10,000 to 5000 BP.

An early Holocene warm period, first reported from the Clyde area (Miller et al., 1977), is corroborated by sample #207 from the Qivitu region and by the basal sediment of Iglutalik Lake, and suggests a northern extension of birch shrubs along the east coast of Baffin Island and an increase of exotic pollen influx. Increased advection of southerly air in the early Holocene has been postulated as a mechanism

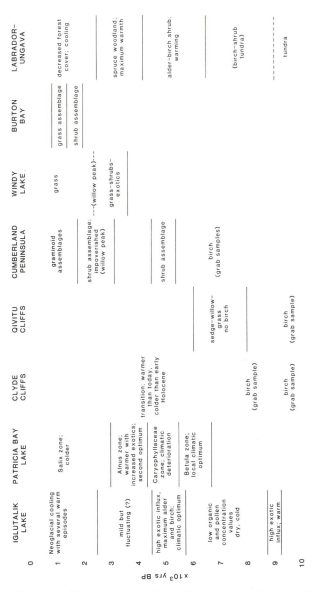

Figure 22.8 Summary diagram comparing the seven palynological records from Baffin Island discussed in the review with the published palynological record from northern Labrador-Ungava Peninsula (Short, 1978b).

for nourishing the northeastern part of the Laurentide Ice Sheet, causing the glacial advances of the Cockburn Substage dated between 9000 and 8000 BP on Baffin Island (Andrew and Ives, 1972, 1978).

However, this period may also have experienced heavy winter snowfall as
Andrews (1982) has pointed out that the rate of retreat of tidewater
glaciers along eastern Baffin Island was surprisingly slow.

A pollen record from Qivitu dating from 8000 to 6000 BP provides a
uniform assemblage dominated by Cyperaceae, Salix, and Gramineae; an
apparent lack of Betula in this region at that time conflicts with its
inferred presence at Patricia Lake Bay near Clyde 325 km to the north
in and in two samples dated about 6800 BP from southwestern Cumberland
Peninsula. Possibly Betula was present but not pollinating at Qivitu
(e.g. Matthews, 1975). The period from 7800 to 5800 BP was interpreted
by Davis (1980a) and Davis et al.(1980) as the most severe of the
Holocene at Iglutalik Lake. However, this period of climatic
deterioration at Iglutalik Lake was probably due to local sea-ice
conditions (Davis, 1980a), as grounded ice remained at the head of
Cumberland Sound until 5700 BP (Dyke, 1979). The presence of birch
shrubs represented in the basal zone of the Patricia Bay Lake core,
6800-5700 BP, was interpreted as a warm period, presumably the Holocene
climatic optimym in the Clyde region. Molluscs of subarctic affinity
are present in raised marine sediments from about 9700 to about 3000 BP
(Andrews, 1972; Miller, 1980); thus seasonal ice cover was less
restrictive along the coast of eastern Baffin Island, suggesting that
the climate was warmer and wetter than present during this interval.
Notable by its absence from the basal sediments of the oldest sequences
on Baffin Island is any evidence for the pioneer vegetation phase
recognized in Greenland (Fredskild, this volume) which has significant
proportions of Oxyria and Koenigia pollen.

Subsequently, the climate in the Clyde region deteriorated as
suggested by the disappearance of Betula from the area and by the very
low exotic pollen concentrations from 5700-4500 BP in the Patricia Bay
Lake sediments. This record again contrasts with that at Iglutalik
Lake which records maximum Alnus influx values about 5200 BP and a
local birch optimum between 5300 and 4700 BP, and with the record at
Pass Head, Cumberland Peninsula, which was interpreted as part of the
climatic optimum in that region.

Warmer climatic conditions are recorded at Patricia Bay Lake after
4500 BP; this amelioration is reflected locally by increased heath
percentages and also by increased influx of exotic alder, spruce, and
birch pollen caused either by a northward advance of these genera or
increased frequency of advective transport of the grains. Similarly,
100 km to the north, Miller et al. (1977) interpreted a thick organic
lens dated to about 4200 BP and covered by eolian sand of the Scott
Inlet phase as indicative of warmer and possibly wetter than present

conditions. At Iglutalik Lake, mild but possibly fluctuating climatic
conditions were proposed by Davis (1980b) until about 2700 BP,
following a brief but severe cold episode at 4700 BP.

Evidence for vegetational and climatic change during the last 4000
years is provided at ten sites; however, seven are short, intermittent
records with poor dating control common. In general, although
palynological records suggest various climate scenarios for the period
between 10,000 and 5000 BP, nearly all suggest that the climate started
to deteriorate sometime between 5000 and 2500 BP with a trend from a
more diverse, commonly shrub-dominated assemblage to a more
impoverished, graminoid spectrum. At Patricia Bay Lake, this
deterioration in climate is dated about 3100 BP with a change to a
sedge-willow dominated landscape. At Iglutalik Lake, the diversity of
local vegetation is even more limited after about 2700 BP. A similar
change from a more diverse shrub assemblage to a grass-willow-sedge
tundra at the Windy Lake peat site is dated about 2400 BP.

A major peak of Salix is an intriguing feature of pollen diagrams
from Windy Lake, Maktak Fiord, Owl River, and Padloping Island. Short
and Andrews (1980) bracketed the age of the willow peaks between 2200
and 2600 BP and correlated the peaks with a similar spike in West
Greenland. However, willow spikes are not as significant from the lake
sediment sections on Baffin Island (Figs 22.4a, b & 22.6a, b).

In several of the short peat sections from southern Baffin Island,
similar moderately rich and diverse pollen assemblages are recorded
between 2500 and 2000 BP. However, all the assemblages recorded in the
last 2000-1500 years are impoverished, dominated by pollen of the
graminoid group, mainly grass, but also sedge at the wetter sites. In
the Clyde region, the modern pollen spectra are the most impoverished
of the Holocene and are dominated by grass pollen; little exotic pollen
reaches the Clyde area today. However, in the Cumberland Sound region,
the modern pollen rain from Pangnirtung Pass (Andrews et al., 1979,
Table IV) suggests that modern conditions most closely approximate the
local climate between about 2500 and 2000 BP, or intermediate between
the conditions at 5000 BP and those after 1500 BP. This interpretation
is supported by the record in the Frobisher Bay region.

Transfer Function Results

Transfer function estimates have been developed for Windy Lake (Andrews
et al., 1981, Fig. 3), Maktak Fiord (Andrews et al., 1980a, Fig. 9),
Iglutalik Lake (Davis, 1980b, Fig. 6.3; Andrews et al., 1981, Fig. 3),

Burton Bay (Short and Jacobs, 1982, Fig. 6), Patricia Bay Lake, the
Qivitu Cliffs, Pass Head (Andrews et al., 1981, Fig. 3), and Padloping
Island (Andrews et al., 1980a, Fig. 8). Figure 22.9 summarizes the
estimated departures from the mean July temperature for all but the
Padloping Island section. All sites show a significant decrease in
average July temperature over the last 6000 years. The local thermal
maximum (+Δ 2°C) terminated about 5000 years ago on Baffin Island,
although warm summer temperatures are registered at several sites until

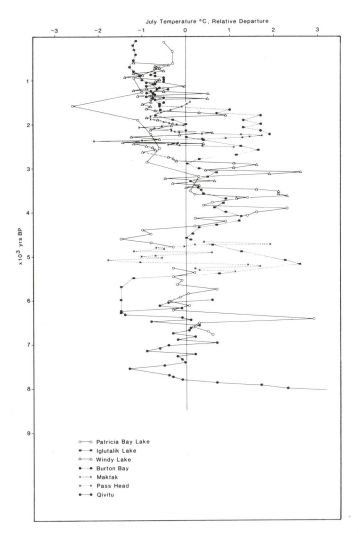

Figure 22.9 Estimated relative departures, July temperatures (°C),
based on the eastern Canadian Arctic transect (Andrews et al., 1980a).

3000 to 2500 years BP. Generally cool summer conditions are indicated
after about 2500 BP, and the lowest July departures during the past few
thousand years occurred about 1500 BP at Patricia Bay Lake (-2.6°C from
the present) and later between 600 and 200 years ago at Windy Lake and
Iglutalik Lake (-1 to -1.5°C from the present) (Andrews et al., 1981,
Fig. 3).

Superimposed on this general trend are a number of temperature
oscillations. Three sand layers and cessation of peat growth at Windy
Lake at about 2050-1950 BP, 1700-1450 BP, and 600-0 BP probably
indicate greater cold, but the transfer function equations from Davis
(1980b, Fig. 5.5) suggest that the sand layers may not reflect
extremely dry conditions as previously asserted (Davis in Andrews,
1976, p. 17). At Iglutalik Lake, the dominant pollen taxa fluctuate
synchronously and exhibit six peaks over the past 4000 years: 3950 BP,
3350-3450 BP, 2650 BP, 1800 BP, 1200-1100 BP, and 550 BP; these peaks
generally correlate with periods of glacier retreat (Davis, 1980b,
Fig. 6-7) and have been interpreted as periods of warmer climate
superimposed on the general cooling trend.

COMPARISONS WITH ADJACENT REGIONS

A generalized vegetational/climatic sequence for northern Labrador-
Ungava is included in Figure 22.8. Birch-shrub tundra (9000-6700 BP)
and alder-birch shrub tundra zones (6700/6500-4000 BP) recognized in a
series of diagrams from this region (Short and Nichols, 1977; Short,
1978a, b) can be interpreted as a warm period characterized by a
productive vegetation. The delay in tree migration into the
Labrador-Ungava region is not well understoood but may be due to
topographic controls, climatic fluctuations, and/or the presence of
late-lying ice in the center of the peninsula (Ives et al., 1975). The
inferred presence of shrub birch in limited areas of the northern
Labrador coast may provide a source for the birch recorded in early
Holocene pollen spectra in north-central Baffin Island. The climatic
optimum recorded by maximum Betula values at Patricia Bay Lake between
6800 and 5700 BP matches the beginning of the very productive
alder-birch shrub tundra zone at the Labrador sites. However, the
climatic optimum recorded by maximum Alnus, Betula, Picea, and Pinus
values at Iglutalik Lake between 5700 and 4700 BP matches the peak of
the alder-birch shrub tundra zone in Labrador-Ungava. In addition, B.
nana arrived in West Greenland about 6500 BP (Fredskild, this volume).
The climatic deterioration at Patricia Bay Lake at 5700-4300 BP is not
recognized in the Labrador data; although a short period of
deteriorating climate is noted at Iglutalik Lake about 4700 BP.

The peaks in exotic alder and spruce pollen after 4500 BP at Patricia Bay Lake are coeval with the northernmost extent of the boreal forest in Labrador-Ungava and with a secondary advance of the forest in Keewatin (Nichols, 1975, Fig. 26). Local climatic warming is also seen in the short peat records from Clyde Cliffs and Pass Head, in addition to Patricia Bay Lake, and a period of maximum exotic influx is recorded in the Iglutalik Lake sediments at 5200 BP and again from 4200-3300 BP.

The climatic deterioration which began as early as about 3000 BP at Patricia Bay Lake and continued through to about 2000 BP is correlative with the decline in lake productivity and the interpretation of a reduction in forest extent in pollen diagrams from northern Labrador-Ungava; there the timing of this event ranges from about 3000 BP to 2400 BP at several sites (Short, 1978a, Table VIII). This period is also considered the end of the Hypsithermal in West Greenland (Fredskild, this volume). Transfer function estimates for a number of the Labrador pollen sites (Andrews et al., 1981, Fig. 3) show a decrease in July temperature between 4000 and 3000 BP which continues through the late Holocene.

CONCLUSIONS

This review of the Holocene palynological record from Baffin Island is based on the analyses of two lake sediment sections and 16 peat sections. Dating control is provided by over 50 C-14 dates. Modern pollen deposition data are recorded in a suite of over 175 polster samples plus six trap and five surface lake sediment samples. Longer sediment cores from lakes, fiords, and from the continental shelves in the eastern Canadian Arctic are needed so that a more complete chronology of Holocene paleoclimate can be constructed. In particular, deep lakes near the outer coasts and fiords can be expected to provide these necessary long records. A broad geographic collection of surface lake sediment samples would allow the comparison of these pollen data to polster data and thus allow improved application of pollen-climate transfer function equations for interpretation of palynological data from lake sediments. "Gaps" in the modern record include the Frobisher Bay area and regions north of Clyde; recent field collections in these areas will aid in our analysis.

Future work must emphasize filling in the gaps in the fossil record, such as the poorly understood early Holocene, and the analysis of recently collected fossil sites from southern and southwestern Baffin Island, as well as from the High Arctic vegetation regions of the northern parts of Baffin Island.

ACKNOWLEDGEMENTS

This work represents a contribution to NSF grants DPP-8121774 and
ATM-8208677 to Short and Andrews, which focus on the question of
synchronology of Holocene pollen changes across arctic North America
and reconstructed paleoclimates respectively. This chapter was
developed in part from NSF ATM-77-17549 which supported
paleoenvironmental research in the eastern Canadian Arctic. Specific
field collections were supported by several grants-in-aids from the
Arctic Institute of North America, the American Alpine Club, the
Explorers Club, Sigma Xi, the Geological Society of America, the
Natural Sciences and Engineering Research Council of Canada, and the
University of Windsor Research Board.

The authors wish to thank R. Kihl of the INSTAAR Sedimentology
Laboratory who prepared samples for radiocarbon dating and Drs. R.
Stuckenrath, Smithsonian Institution, and W. Blake, Jr., Geological
Survey of Canada, for providing many of the dates used in this paper.
The authors benefited from discussions with many colleagues, and we
thank Dr. J.T. Andrews for his extensive review of this manuscript.

REFERENCES

Aario, L., 1940: Waldgrenzen und subrezente pollenspektren in Petsamo
    Lappland. Annales Academiae Scientarum Fennicae, Ser. A., 54, 120
    pp.
Adam, D.P. and Mehringer, P.J., Jr., 1975: Modern pollen surface
    samples - an analysis of subsamples. U.S. Geological Survey, Journal
    of Research, 3:733-736.
Andrews, J.T., 1972: Recent and fossil growth rates of marine bivalves,
    Canadian Arctic and Late-Quaternary arctic marine environments.
    Palaeogeography, Palaeoclimatology, Palaeoecology, 11:187-192.
Andrews, J.T., 1976: Radiocarbon date list III, Baffin Island,
    N.W.T.,Canada. Institute of Arctic and Alpine Research, Occasional
    Paper No. 21, 47 pp.
Andrews, J.T., 1982: Holocene glacier variations in the eastern
    Canadian Arctic: A review. Striae, 18:9-18.
Andrews, J.T., 1983: Radiocarbon date list V: Baffin Island, N.W.T.,
    Canada. Institute of Arctic and Alpine Research, Occasional Paper
    No. 40:1-53.
Andrews, J.T. and Ives, J.D., 1972: Late- and postglacial events
    (<10,000 BP) in the eastern Canadian Arctic with particular reference
    to the Cockburn moraines and breakup of the Laurentide ice sheet. In:
    Vasari, U., Hyvärinen, H., and Hicks, S. (eds.), Climate Changes in
    Arctic Areas during the Last Ten-Thousand Years. Acta Universitatis
    Ouluensis, Series A, No. 3, 149-174.
Andrews, J.T. and Ives, J.D., 1978: "Cockburn" nomenclature and the
    late Quaternary history of the eastern Canadian Arctic. Arctic and
    Alpine Research, 10:617-633.
Andrews, J.T., Webber, P.J., and Nichols, H., 1979: A late Holocene
    pollen diagram from Pangnirtung Pass, Baffin Island, N.W.T., Canada.
    Review of Palaeobotany and Palynology, 27:1-28.
Andrews, J.T., Mode, W.N. and Davis, P.T., 1980a: Holocene climate
    based on pollen transfer functions, eastern Canadian Arctic. Arctic
    and Alpine Research, 12:4-64.

Andrews, J.T., Mode, W.N., Webber, P.J., Miller, G.H., and Jacobs, J.D., 1980b: Report on the distribution of dwarf birches and present pollen rain, Baffin Island, N.W.T., Canada. Arctic, 33:50-58.

Andrews, J.T. and Nichols, H., 1981a: Modern pollen deposition and Holocene paleotemperature reconstructions, central northern Canada. Arctic and Alpine Research, 13:387-408.

Andrews, J.T., Davis, P.T., Mode, W.N., Nichols, H., and Short, S.K., 1981b: Relative departures in July temperatures in northern Canada for the past 6,000 yr. Nature, 289:164-167.

Bartley, D.D., 1967: Pollen analysis of surface samples of vegetation from Arctic Quebec. Pollen et Spores, 9:101-105.

Bartley, D.D. and Matthews, B., 1969: A paleobotanical investigation of post-glacial deposits in the Sugluk area of northern Ungava, Quebec. Review of Palaeobotany and Palynology, 9:45-61.

Birks, H.J.B., 1973: Modern pollen rain studies in some arctic and alpine environments. In: Birks, H.J.B. and R.J. West (eds.), Quaternary Plant Ecology. (Wiley, New York), 143-168.

Boulton, G.S., Dickinson, J.H., Nichols, H., Nichols, M., and Short, S.K., 1976: Late Holocene glacier fluctuations and vegetation changes at Maktak Fiord, Baffin Island, N.W.T., Canada. Arctic and Alpine Research, 8:343-354.

Davis, P.T., 1980a: Holocene vegetation and climate record from Iglutalik Lake, Cumberland Sound, Baffin Island, N.W.T., Canada. Abstracts and Programs, Sixth Biennial Meeting, American Quaternary Association, Orono, Maine, p. 61

Davis, P.T., 1980b: Late Holocene glacial, vegetational, and climatic history of Pangnirtung and Kingnait Fiord area, Baffin Island, Canada. Ph.D. thesis, Department of Geological Sciences, University of Colorado. 366 pp.

Davis, P.T., Nichols, H., and Andrews, J.T., 1980: Holocene vegetation and climate record from Iglutalik Lake, Baffin Island. Abstracts, Fifth International Palynological Conference, London, p. 105.

Dyke, A.S., 1977: Quaternary geomorphology, glacial chronology, and climatic and sea-level history of southwestern Cumberland Peninsula, Baffin Island, Northwest Territories, Canada. Ph.D. thesis, Department of Geography, University of Colorado. 185 pp.

Dyke, A.S., 1979: Glacial and sea-level history of southwestern Cumberland Peninsula, Baffin Island, N.W.T., Canada. Arctic and Alpine Research, 11:179-202.

Dyke, A.S., Andrews, J.T., and Miller, G.H., 1982: Quaternary geology of Cumberland Peninsula, Baffin Island, District of Franklin. Geological Survey of Canada Memoir, 403, 32 pp.

Elliot-Elliott-Fisk, D.L., Andrews, J.T., Short, S.K., and Mode, W.N., 1982: Isopoll maps and analysis of the distribution of the modern pollen rain, eastern and central northern Canada. Geographie physique et Quaternaire, 36:91-108.

Faegri, K. and Iversen, J., 1975: Textbook of Pollen Analysis. (New York:Hafner, 295 pp.

Fredskild, B., 1967. Paleobotanical investigations at Sermermiut, Jakobshavn, West Greenland. Meddelelser om Grønland, 278, 54 pp.

Funder, S., 1978: Holocene stratigraphy and vegetation history in the Scoresby Sund area, East Greenland. Grønlands Geologiske Undersøgelse Bulletin No. 129, 66 pp.

Goh, K.M., B.P. Molloy, Jr., and T.A. Rafter, 1977: Radiocarbon dating of Quaternary loess deposits, Banks Peninsula, Canterbury, New Zealand. Quarternary Research, 7:177-196.

Gregory, P.H., 1973: The Microbiology of the Atmosphere (second edition). London: Hill. 377 pp.

Hyvärinen, H., 1975: Absolute and relative pollen diagrams from northernmost Fennoscandia. Fennia, 142, 23 pp.

Imbrie, J. and Kipp, N.G., 1971: A new micropaleontological method for quantitative paleoclimatology: application to a late Pleistocene Caribbean core. In: Turekian, K.K. (ed.), The Late Cenozoic Glacial Ages. New Haven: Yale University Press, 71-181.

Iversen, J., 1945: Origin of the flora of western Greenland in the light of pollen analysis. Oikos, 4(2):85-103.

Ives, J.D., Andrews, J.T., and Barry, R.G., 1975: Growth and decay of the Laurentide Ice Sheet and comparisons with Fennoscandia. Naturwissenschaften, 62:118-125.

Lind, E.K., 1983: Holocene paleoecology and deglacial history of the Cape Rammelsberg area, southern Baffin Island, N.W.T., Canada. M.S. thesis, Department of Geological Sciences, University of Colorado.

Miller, G.H., Andrews, J.T., and Short, S.K., 1977: The last interglacial/glacial cycle, Clyde foreland, Baffin Island, N.W.T.: stratigraphy, biostratigraphy, and chronology. Canadian Journal of Earth Sciences, 14:2824-2857.

Mode, W.N., 1980: Quaternary stratigraphy and palynology of the Clyde Foreland, Baffin Island, N.W.T., Canada. Ph.D. thesis, Department of Geological Sciences, University of Colorado. 219 pp.

Mode, W.N., Davis, P.T., and Andrews, J.T., 1981: Pollen floristic zonation of the eastern Canadian arctic tundra. Association of American Geographers Annual Meeting, Los Angeles, Program Abstracts, 51.

Mudie, P.J., 1982: Pollen distribution in recent marine sediments, eastern Canada. Canadian Journal of Earth Sciences, 19:720-747.

Nichols, H., 1975: Palynological and paleoclimatological study of the late Quaternary displacement of the boreal forest-tundra ecotone in Keewatin and Mackenzie, N.W.T., Canada. Institute of Arctic and Alpine Research, Occasional Paper No. 15, 87 pp.

Nichols, H., Kelly, P.M., and Andrews, J.T., 1978: Holocene paleo-wind evidence from palynology in Baffin Island. Nature, 273:140-142.

Pennington, W., 1980: Modern pollen samples from West Greenland and the interpretation of pollen data from the British late-glacial (Late Devensian). New Phytologist, 84:171-201.

Polunin, N., 1948: Botany of the Canadian eastern Arctic, part III, vegetation and ecology. National Museums of Canada, Bulletin No. 104, 304 pp.

Polunin, N., 1960: Introduction to plant geography and some related sciences. London: Longmans, 640 pp.

Porsild, A.E., 1957: Illustrated flora of the Canadian Arctic archipelago. National Museums of Canada, Bulletin No. 146, 218 pp.

Ritchie, J.C., 1976: The late-Quaternary vegetational history of the Western Interior of Canada. Canadian Journal of Botany, 54:1793-1818.

Ritchie, J.C., 1977: The modern and late Quaternary vegetation of the Campbell-Dolomite uplands, near Inuvik, N.W.T., Canada. Ecological Monographs, 47:401-423.

Ritchie, J.C. and Lichti-Federovich, S., 1967: Pollen dispersal phenomena in arctic-subarctic Canada. Review of Palaeobotany and Palynology, 3:255-266.

Rymer, L., 1973: Modern pollen rain studies in Iceland. New Phytologist, 72:1367-1373.

Schwarzenbach, F.H., 1975: Botanical observations on the Penny Highlands of Baffin Island. Translated from German manuscript by Dr. D. Love, available from INSTAAR, University of Colorado, Boulder. 164 pp.

Short, S.K., 1978a: Holocene palynology in Labrador-Ungava: climatic history and culture change on the central coast. Ph.D. thesis, Department of Anthropology, University of Colorado. 231 pp.

Short, S.K., 1978b: Palynology: a Holocene environmental perspective for archaeology in Labrador-Ungava. Arctic Anthropology, 25:9-35.

Short, S.K. and Nichols, H., 1977: Holocene pollen diagrams from subarctic Labrador-Ungava: vegetational history and climatic change. Arctic and Alpine Research, 9:265-290.

Short, S.K., and Andrews, J.T., 1980: Palynology of six middle and late Holocene peat sections, Baffin Island. Geographie physique et Quaternaire, 34:61-75.

Short, S.K. and Jacobs, J.D., 1982: A 1100 year paleoclimatic record from Burton Bay - Tarr Inlet, Baffin Island. Canadian Journal of Earth Sciences, 29:398-409.

Stuckenrath, R., Miller, G.H., and Andrews, J.T., 1979: Problems of radiocarbon dating Holocene organic-bearing sediments, Cumberland Peninsula, Baffin Island, Arctic Canada. Arctic and Alpine Research, 22:109-120.

Vasari, Y., 1974: The vegetation of northern Finland-past and present. Inter-Nord, 13-14:98-108.

Webb, T., III and Bryson, R.A., 1972: Late- and postglacial climatic change in the northern Midwest, U.S.A.: quantitative estimates derived from fossil pollen spectra by multivariate statistical analysis. Quaternary Research, 2:70-115.

Webb, T., III and Clark, D.R., 1977: Calibrating micropaleontological data in climatic terms: a critical review. Annals of the New York Academy of Science, 288:93-118.

Webber, P.J., 1971: Gradient analysis of the vegetation around the Lewis Valley, north-central Baffin Island, Northwest Territories, Canada. Ph.D. thesis, Department of Biology, Queens University. 366 pp.

Young, S.B., 1971: The vascular flora of St. Lawrence Island with special reference to the floristic zonation of the arctic regions. Contributions to the Gray Herbarium, Harvard University, No. 201, 115 pp.

23 Holocene pollen records from west Greenland

Bent Fredskild

Ever since Johs. Iversen first published pollen diagrams from peat deposits at some West Greenland Norse farmsteads, and 20 years later from lakes nearby (Iversen, 1934, 1954), a significant number of pollen diagrams have been produced and published (Fig. 23.1). Despite this wealth of material the Holocene vegetation and climatic history is still not fully known. Yet, especially as far as West Greenland is concerned, a picture begins to emerge of a metachronous succession of plant communities, somewhat different from site to site regarding species composition, but physiognomically similar and which reflect the changing environmental conditions. This will be elucidated by examples from different parts of the area before trying to identify the common trends. The reader is urgently warned against using the C-14 dates in this chapter too literally. Suffice it to say that: (1) often too few radiocarbon dates on lake sediments are obtained; (2) the dating of many pollen zone borders are interpolated over long time spans, on the assumption of constant sedimentation rates, and 3) a hard water effect in certain areas and an outwashing of terrestrial, possibly older organic material, in certain periods cannot be excluded.

SOUTH GREENLAND

Kap Farvel Area (c. 60° N)

The outer coast is marked by the wide zone of pack ice which follows the East Greenland Current. At Kap Farvel this current turns northwestward along the coast, causing cool, cloudy, often foggy summers with a mean temperature (June - August) of only 5-6° C. Apart from windswept ridges, the snow cover in winter is deep and constant, and the overall dominating vegetation is dwarf-shrub heath, rich in

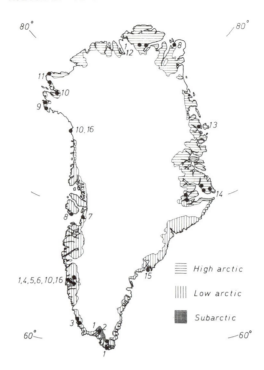

Figure 23.1 Sites of Greenland pollen diagrams from lakes and bogs
covering the Holocene or part of it. 1: Fredskild, 1973; 2: Fredskild,
1978; 3: Kelly and Funder, 1974; 4: Iversen, 1934; 5: Iversen, 1954; 6:
Fredskild, 1983b; 7: Fredskild, 1967a,b; 8: Funder, unpublished; 9:
Hyvärinen, 1972a; 10: this paper; 11: Vasari, 1972; 12: Fredskild,
1969; 13: Björck and Persson, 1981; 14: Funder, 1978a; 15: Bick, 1978;
16: Fredskild, unpublished.

mosses and lichens. Empetrum hermaphroditum is the most frequent of
the ericaceous plants, sometimes almost alone, but usually with
Loiseleuria procumbens, Phyllodoce coerulea, Ledum spp., and Vaccinium
spp., besides Betula glandulosa and dwarfish Salix glauca.
Snow-patches and fens are frequent. The deglaciation began c. 10,000
BP, and the oldest dating of a gyttja is 9,440±140 BP. The following
phases can be separated (Fredskild, 1973):

Oxyria-Koenigia phase (c. 9600-9100 BP). The fresh soil was invaded by
a pioneer vegetation of herbs, club-mosses and ferns. The herbs are
widespread, arctic or arctic-alpine species, ubiquitous but low
competitive species being restricted to unstable soil, frost boils
etc. in the lowland today, but ascending to high altitudes. Saxifraga
is copiously represented (pollen of S. caespitosa type, S.

<u>oppositifolia</u> <u>t</u>., <u>S</u>. <u>nivalis</u> t., <u>S</u>. <u>foliolosa</u> t.) as are <u>Oxyria</u> <u>digyna</u>,
<u>Sagina</u>, <u>Koenigia</u> <u>islandica</u>, <u>Minuartia-Silene</u> t., Cruciferae, Gramineae
and Cyperaceae. Reflecting the sparse vegetation cover, the sediment
is always a clay-gyttja. In some diagrams the curve of fern and
club-moss spores, being more resistant to corrosion than most pollen,
parallels the curve of minerogenic content, indicating the existence of
secondary spores (and pollen). This has been confirmed by analyses of
the pollen from clay in soil profiles. Nevertheless, the early
existence of <u>Huperzia</u> <u>selago</u>, <u>Diphasium</u> (presumably <u>D</u>. <u>alpinum</u>) and
some ferns is most probable. Climatically the lack of any
warmth-demanding plants in this early assemblage indicates arctic
conditions.

<u>Oxyria-Loiseleuria</u> phase (9,100-8,400 BP). After half a millennium a
number of ericaceous dwarf-shrubs immigrate: <u>Loiseleuria</u>, <u>Harrimanella</u>
<u>hypnoides</u>, <u>Empetrum</u> <u>hermaphroditum</u>, <u>Vaccinium</u> <u>uliginosum</u>, <u>Phyllodoce</u>
<u>coerulea</u>. Pollen of <u>Armeria</u> and <u>Chamaenerion</u> (presumably <u>A</u>. <u>scabra</u> and
<u>C</u>. <u>latifolium</u>), otherwise rarely found in pollen analyses, show that
these plants were common among the pioneers. Important immigrants are
<u>Cerastium</u> <u>cerastoides</u> (to 73° in W. Greenland), <u>Angelica</u> <u>archangelica</u>
(north to 70°) and <u>Alchemilla</u> <u>alpina</u> (to 67°), showing that although
the vegetation was still rather open with extensive snow-patches and
moist, sandy-clayey spots, conditions were improving, being now low
arctic. <u>Thalictrum</u> <u>alpinum</u> (to 72° in W. Greenland) possibly
immigrated during this phase. For want of sufficient climatic
observations, the terms high, low and subarctic are here defined on a
phytogeographical basis (Fig. 23.1), subarctic areas having low birch
forest (<u>Betula</u> <u>pubescens</u>, only in interior S. Greenland), low arctic
having willow copses of <u>Salix</u> <u>glauca</u>, in the south up to 4 m high,
towards the north only knee-deep, and high arctic areas being north
hereof.

<u>Sedum-Lycopodiaceae</u> phase (c. 8400-c. 7500 BP). The dwarf-shrub
vegetation covered a still greater part of the area, but the numerous
spores of <u>Diphasium</u> <u>alpinum</u> (the only reasonable explanation of the
many <u>Diphasium</u> spores), <u>Huperzia</u>, and <u>Lycopodium</u> <u>dubium</u> also show the
existence of chinophilous vegetations, rich in mesophytes and mosses.
The maximum in <u>Sedum</u> pollen may originate from <u>S</u>. <u>villosum</u>, growing on
rocks with seeping water, or <u>S</u>. <u>rosea</u>. The entomostrac <u>Lepidurus</u>
<u>arcticus</u>, which today is not found south of 67° (apart from a high
altitude relict occurrence at Narssaq, 61°), was living in some of the
lakes, during the early part of this phase but it is not recorded after
7800 BP (Fredskild et al., 1975).

Salix-Gramineae phase (c. 7500-5300 BP). The indications of the
immigration of Salix are not unambiguous in the three lakes dealt
with. A few specimens of S. glauca may well have been growing in the
dwarf-shrub heaths of the preceding phases, but with the opening of
this phase, Salix (S. glauca and S. herbacea) are spreading. Absolute
pollen counts show that shortly after, Salix was as common as it is
today. Most of the pioneer plants disappeared, and a common vegetation
type must have been dwarf- shrub heaths, yet still with fairly few
Empetrum. If Salix had already been in the area for some time, its
extension must be a response to an increase in temperature. Other new
incomers are Coptis trifolia, Elymus, and Juniperus.

Salix-Juniperus-Empetrum phase (c. 5300-3800 PB). The most significant
changes are registered in the lake vegetation i.e. the contemporaneous
immigration of Isoëtes lacustris, the southernmost of the Greenland
hydrophytes, which now grows only in the Kap Farvel area. If this late
immigration is not fortuitous, an increase in temperature might be
implied, but indications, based on the terrestrial plants, are weak and
contradictory, and any inferred climatic change must have been small.

Betula glandulosa-Empetrum phase (c. 3800-2200 BP). The opening of
this phase is defined as the, indeed very late, immigration of Betula
glandulosa. The appearance of this prolific pollen producer tends to
prevent any climatic change being traced in the pollen diagrams.

Empetrum-Betula glandulosa phase (c. 2200-0 BP). Empetrum and Betula
dominate the pollen spectra. A marked decrease in Juniperus at the
border and the re-occurrence of some of the pioneer plants point to a
marked change to more cool, humid conditions. This may be corroborated
by the indications in one of the lakes of a thicker ice-cover, which
eroded older sediments.

Tunugdliarfik (c. 61° N)

The vegetation of the subarctic interior S. Greenland is dominated by
extensive Salix glauca copses, with Betula pubescens, up to 4-5 m, on
protected sites. Heath communities, likewise dominated by Salix glauca
with abundant Betula glandulosa, contain only few ericaceous plants
among which Empetrum is extremely rare. Herb-slopes and fens, rich in
species, and grassy communities on more xeric areas, are frequent,
whereas snowbed communities are sparse in the lowland. One of the

cores (Comarum Sø, Fredskild, 1973) penetrates the late-glacial clay, and the date on the oldest gyttja is 8530±140 BP.

Cyperaceae-Plantago maritima-Empetrum phase (c. 8700-8000 BP). A pioneer vegetation dominated by Chamaenerion (presumably C. latifolium), Silene acaulis, Minuartia rubella, Plantago maritima, and with Oxyria, Empetrum, Saxifraga nivalis t., Elymus, Dryas and Angelica more or less covered the soil. No decidedly southern plants were found, but there are several indications that suggest conditions were analagous to those found today between 67° and 69° in W. Greenland. Lepidurus arcticus was found in the oldest samples.

Salix-Thalictrum phase (c. 8000-6900 BP). Salix immigrates around 8,000 BP and immediately spreads, attaining the same absolute values as today. Its immigration therefore seems to have been delayed, due to lack of dispersal.

Juniperus-Gramineae phase (c. 6900-4500 BP). Juniperus immigrates around 6900 BP. The occurrence of Eleocharis palustris and Selaginella selaginoides indicates favorable climatic conditions, not much different from those of today, but presumably slightly warmer. An equilibrium between vegetation and environmental factors came about during 500 yrs. In the following period low, open, xerophytic heath types dominated. At 5500-5300 BP an increase in Juniperus pollen with a maximum of 29% in Galium Kaer indicates a change to even more dry/warm conditions, whereas no major change is found in the diagrams from Comarum Sø and Comarum Mose.

Juniperus-Betula glandulosa phase (c. 4500-3600 BP). The spreading of Betula glandulosa is very gradual, indicating that it had to oust other plants. Juniperus, Gramineae, Botrychium and Dryopteris types decrease, while Angelica, Gymnocarpium dryopteris and also, to a slight extent, Salix glauca increase. In all, this points to slightly more humid conditions, but it is not possible to trace any connected change in temperature, unless the blooming of Menyanthes trifoliata and Hippuris in Comarum Sø is taken as an indication of increasing temperature, and not only as a reflection of the filling-in of the basin.

Betula pubescens-Betula glandulosa-Salix phase (c. 3600-1900 BP).
Betula pubescens-Salix scrubs with an undergrowth of Angelica,
Gymnocarpium, grasses etc. covered protected valleys and slopes, with
low, shrubby Betula glandulosa-Salix-Juniperus heaths occupying exposed
areas. Climatic conditions still seem favorable, but towards the end
of the phase the ratio B. pubescens:B. glandulosa changes in favor of
the latter, a reflection of the deterioration which characterizes the
coming phase. At 3400 BP a drop in Juniperus influx to one-third may
have climatic implications and may reflect more humid conditions, and
thereby competition with the shrubs which occupied the lower part of
the slopes.

Betula glandulosa-Salix phase (c. 1900-1000 BP). The decrease in
Juniperus is most marked at c. 1900 BP. Together with the decreasing
Betula pubescens this indicates decreasing temperatures. In small
depressions between some raised beaches the micro- and macrofossil
content show that some time prior to the Norse landnam c. AD 982, a
highly humified peat was covered by a less humified peat which formed
under a willow scrub with Angelica, Selaginella, and ferns, and with
temporary puddles (Fredskild, 1978).

Rumex acetosella phase (c. 1000-0 BP). The effect on the vegetation of
the Norse landnam was overwhelming: clearance of the birch "forest",
grazing of the area, and introduction of many annual as well as
perennial weeds. Among the latter Rumex acetosella, of which native
strains had long been growing in more northern latitudes, and R.
acetosa are most pronounced in the diagrams. Soil erosion is traced in
the increasing wash of sand into lake basins, and on the newly opened
ground many of the late-glacial pioneer plants turn up as apophytes.

During the Norse era the peat deposits in Comarum Mose reflect a
dry-wet-dry-wet cycle. The landnam took place during the first dry
stage, the extinction during the second wet stage. The obvious
climatic explanation is only valid if the Norsemen themselves did not
affect the drainage system of the bog.

SOUTHWEST GREENLAND

Frederikshaab (c. 62° N)

The vegetation is a mosaic of scrub, heath, herb and bog communities,
belonging to the low arctic oceanic to sub-oceanic vegetation region.

There is the usual difference between the outer coast, dominated by
Empetrum heaths, and the warmer and drier interior. The difference
includes a more varied vegetation cover with low shrubs, including
isolated stands of Alnus crispa. This difference can be detected over
most of the period in the pollen diagrams from three lakes, placed on a
line from the outer coast to close to the present ice margin. The
diagrams can be divided into six zones, which are not quite identical
with those proposed by Kelly and Funder (1974).

Oxyria-Gramineae-Saxifraga-Caryophyllaceae zone (>c. 9600 BP). This is
the pioneer zone with very low pollen concentration but great diversity
in species. Other plants occurring are Sedum (?villosum or rosea),
Chamaenerion cf. latifolium, Potentilla and Campanula. Apart from
Angelica, no thermophilous plants are found. The ericaceous
dwarf-shrubs are gradually spreading, and the next zone is not clearly
delineated.

Cyperaceae-Empetrum-Harrimanella zone (c. 9600-8950 BP). The pollen
spectra are dominated by Cyperaceae, Gramineae, Empetrum and ferns, but
the pollen concentration data show that the pollen deposition of these
species was the same as today. Most significant are the fairly
frequent pollen of Harrimanella type, indicating snowbeds. As in the
early Kap Farvel samples the interpretation of the occurrence of
scattered Salix pollen is not obvious, but the occurrence of a few
specimens can not be excluded. The most warmth-demanding plants found
in this zone are Botrychium (to 72° N) and Angelica.

Salix-Cyperaceae-Polypodiaceae zone (c. 8950-7000 BP). In this zone
Salix attains its maximum, relatively as well as absolutely. In the
upper part of the zone, pioneer plants decrease significantly, while
Lycopodium dubium and Diphasium alpinum decrease, indicating heaths
with good snow-cover.

Lycopodiaceae-Selaginella-Juniperus zone (c. 7000-5750 BP).
Pteridophyte spores dominate the spectra. Significant is the
immigration of Selaginella selaginoides (to 65° N) and Juniperus, and
the increase in Thalictrum and Artemisia, indicating warm and dry
conditions. This zone was defined by Kelly and Funder (1974) because
of the marked increase in Alnus (at 7600 BP) which was interpreted as a
result of its early immigration to the area. As will be discussed in

connection with long-distance pollen under the heading Exotic Pollen in
the Greenland Lakes, (Fig. 23.14), this is far from proved.

Betula glandulosa-Juniperus-Gramineae zone (c. 5750-3200 BP). Pollen
concentrations are highest in this zone. Thymus drucei and Coptis
trifolia (both north to 66°) first emerge in this zone, which possibly
was the period of optimum temperatures.

Betula-Empetrum-Cyperaceae zone (c. 3200-0 BP). The zone border is
marked by a decrease in Juniperus, Alnus, Gramineae and
Lycopodium/Diphasium, an increase in Cyperaceae, and to a still less
extent in Selaginella, and in the Empetrum + Vaccinium curve. The date
of 3200 BP marks the end of the climatic optimum.

WEST GREENLAND

Godthaabsfjord (64-65° N)

The outer coast area is dominated by Empetrum-Rhacomitrium-lichen
heaths with only a few Betula nana, Ledum ssp. and Vaccinium
uliginosum. Snowbeds with Salix herbacea and Harrimanella hypnoides,
and fens with Scirpus caespitosus, Eriophorum spp. and Salix arctophila
are frequent. Godthab has a summer mean temperature of 6.7°C and a
yearly precipitation of 515 mm, with corresponding figures for
Kapisigdlit, at the head of the fjord, being 9.7°C and 255 mm. This is
clearly reflected in the inland vegetation, dominated by Betula
nana-lichen dwarf-shrub heaths, often with Salix glauca, which on
protected sites forms 3 m high copses together with Alnus crispa.
Ledum groenlandicum is frequent in heaths on more mesic sites.

Ten lakes have been cored (Iversen 1954; Fredskild 1973, 1983b).
Radiocarbon dated micro- and macrofossil diagrams are at hand from four
of these, two of which are from the interior, two from the outer coast
on the lowland peninsula Nordlandet. Figure 23.2 represents a pollen
diagram from the interior. This lake was cored in 1935, and the
results published in Iversen (1954, "Lake 100 m s.m."). I cored it
again in 1973, and the official name is now Johannes Iversen Sø.

In spite of the differences caused by local topography and
climate, the diagrams show that the vegetation has passed through
similar stages, especially in the first period after deglaciation: a
pioneer stage, an Ericales stage, a Salix-Cyperaceae stage, a Betula

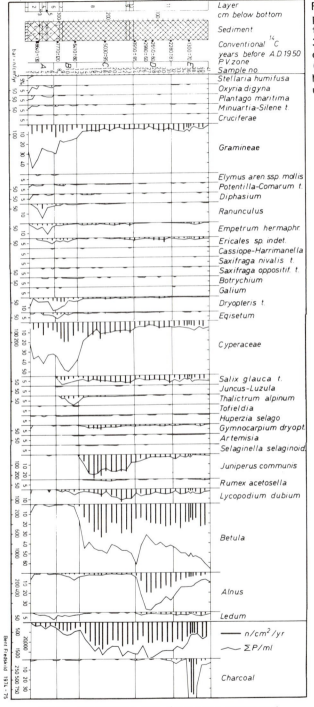

Figure 23.2 Selected
pollen curves of
terrestric plants from
Johs. Iversen Sø, head
of Godthaabsfjord
(64°24'N 50°12'W).
Note change in influx
unit (with permission).

JOHS. IVERSEN SØ (64°24'N 50°12'W)
Godthåbsfjord   W.Greenland   Microfossils

nana-Juniperus stage and a Betula-Empetrum stage can be recognized. An Alnus-Betula stage is inserted in the inland diagrams between the two last mentioned ones. The following stages can be distinguished (Fig. 23.3):

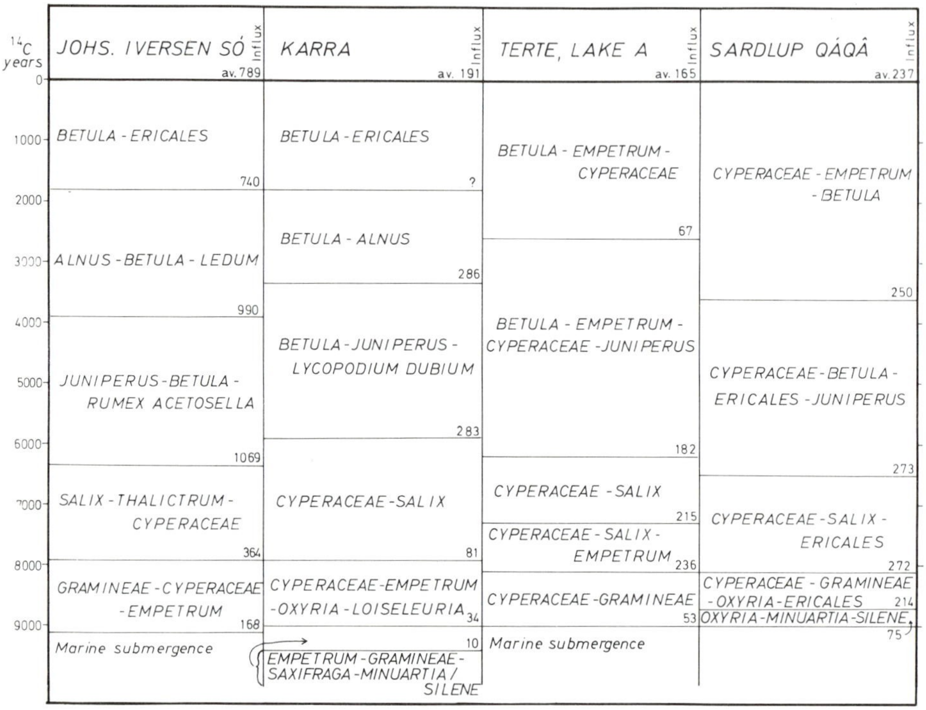

^{14}C years	JOHS. IVERSEN SÓ (av. 789) Influx	KARRA (av. 191) Influx	TERTE, LAKE A (av. 165) Influx	SARDLUP QÁQÁ (av. 237) Influx
0				
1000	BETULA - ERICALES	BETULA - ERICALES	BETULA - EMPETRUM - CYPERACEAE	CYPERACEAE - EMPETRUM - BETULA
2000	740	?		
3000	ALNUS - BETULA - LEDUM	BETULA - ALNUS 286	67	
4000	990	BETULA - JUNIPERUS - LYCOPODIUM DUBIUM	BETULA - EMPETRUM - CYPERACEAE - JUNIPERUS	250
5000	JUNIPERUS - BETULA - RUMEX ACETOSELLA			CYPERACEAE - BETULA - ERICALES - JUNIPERUS
6000	1069	283	182	273
7000	SALIX - THALICTRUM - CYPERACEAE	CYPERACEAE - SALIX	CYPERACEAE - SALIX 215	CYPERACEAE - SALIX - ERICALES
8000	364	81	CYPERACEAE - SALIX - EMPETRUM 236	272
9000	GRAMINEAE - CYPERACEAE - EMPETRUM 168	CYPERACEAE - EMPETRUM - OXYRIA - LOISELEURIA 34	CYPERACEAE - GRAMINEAE 53	CYPERACEAE - GRAMINEAE - OXYRIA - ERICALES 214 OXYRIA - MINUARTIA - SILENE 75
	Marine submergence	10 EMPETRUM - GRAMINEAE - SAXIFRAGA - MINUARTIA / SILENE	Marine submergence	

Figure 23.3 Holocene Palaeo vegetation zones from four Godthaabsfjord lakes, two from the interior (left), two from the outer coast (right). The mean pollen influx (n cm^{-2} yr^{-1}) for each zone is given (with permission).

The pioneer stage (c. 9400-c. 8000 BP). The fresh, unstable, minerogenous soil, still with permanent snow-drifts and dead-ice, was invaded by the widespread, ubiquitous arctic pioneer plants, but even in the first samples, pollen and seeds of some ericaceous dwarf-shrubs have been found, and after the initial phase they slowly spread. A detailed study, especially of the macrofosssils, have yielded a list of no less than 64 plants, determined to the species or generic level, besides those only determinable to the family level. There are plants from exposed soil, frostboils, outwash plains etc. (e.g. Oxyria, Tofieldia, Silene acaulis, Minuartia groenlandica, M. rubella, Phippsia algida), from seashore or moist, ion-rich soils (Carex ursina, C. bicolor, C. glareosa, Juncus ranarius, J. arcticus), from rocks,

screes, fell-fields (Poa glauca, Cystopteris fragilis ssp. fragilis and
ssp. dickieana, Carex nardina), from lake shores (Carex rufina,
Limosella aquatica, Subularia aquatica), from fens (Carex saxatilis,
Galium brandegei), snowbeds (Harrimanella, Carex lachenalii) and from
heaths (besides Ericales also Dryas, Carex bigelowii, C. norvegica, C.
arctogena, Huperzia). Of climatic significance are Galium brandegei
(which today ranges to 68° N), Menyanthes (to 69° N), Subularia (to
70°), Limosella and Myriophyllum spicatum (to 71°), indicating that
soon after the deglaciation the climate was low arctic, corresponding
to that of today a few latitudes further north. This conclusion is
supported by, for example, Lepidurus arcticus, which lived in the lakes
during this period.

Salix-Cyperaceae stage (c. 8000-6300 BP). Judging from the pollen
diagrams, the immigration of Salix glauca and S. herbacea does not have
much influence on the vegetational development. Salix glauca entered
the dwarf-shrub heaths, while S. herbacea found its place in the
snowbeds and in snow-protected heaths. Alone, the emergence of Salix
does not indicate a climatic change, and no obvious climatic indicator
species immigrates during this stage. But towards the end, a decrease
in those Ericaceae taxa which characterize snow-protected heaths, and a
decrease in the pioneer plants, both indicate less snow. Furthermore,
the extinction of Lepidurus between 8000 and 7500 BP points to an
increase in temperature. The increase in Alnus in the second half of
the stage is interpreted as a result of long-distance wind
transportation.

Betula nana-Juniperus stage (c. 6300-? BP). Once the European Betula
nana had passed the large Greenland ice cap, its spreading within the
comparatively small area around Godthaabsfjord expectedly should be
very fast. The dating of the Betula increase varies between 6500 and
5900 BP, with the youngest date in a lake at 265 m asl. The frequency
of Juniperus and Rumex acetosella indicate warm, dry conditions.

Alnus immigrates into the interior around 3800 BP (there are
indications that the upper radiocarbon dates in Johs. Iversen Sø are
c. 300 years too old (Fig. 23.3)). Alnus hardly ever grew on
Nordlandet. As the late immigration of Alnus may be caused simply by
lack of dispersal at an earlier time, this event has no implication
for climatic change. However, the high influx of its pollen shows it
to have been much more frequent in the first part of its era, thereby
indicating warmer, but possibly also more humid, conditions, as its
preferred habitats are along streams and in ravines.

On Nordlandet, a subtle vegetational change is reflected around 3600 BP at Sardlup qaqa, based on an increase in Diphasium alpinum, whereas in Terte, Lake A, the decrease in Juniperus and Rumex acetosella is gradual. However, an increase in Diphasium and Ledum may also be traced here at the middle of the millennium, indicating a slight change towards more humid and/or cooler conditions.

Betula nana-Ericales stage in the interior (c. 1800-0 BP) and Empetrum-Cyperaceae-Betula nana stage at the outer coast (c. 3500-0 BP). In the interior an Alnus decrease and an increase in Betula and Ericales are interpreted as reflecting a change toward cooler and presumably also more humid conditions, with Betula nana dominating the dwarf-shrub heaths. Contrary to this, the change to more humid conditions at the outer coast is disadvantageous to Betula, and here Ericales, Cyperaceae, and snowbed plants profit by the change. In the Sardlup diagram minor changes, i.e. an increase in Harrimanella/ Cassiope indicate that cooler, more humid conditions set in around 2500-2000 BP. It must be admitted, however, that lake diagrams are not as useful in registering climatic changes as are peat profiles. In Godthaabsfjord area, peat growth formation started under a dwarf-shrub heath on a raised beach at Itivnera c. 3200 BP (Fredskild 1973), and c. 630 BP this peat changed to a peat formed under a moist, fen-like community.

Diskobugt (68-70° N)

Corings in a number of lakes on Disko and around Disko Bugt are currently being worked up by Funder and Fredskild, but are not yet ready for publication. The early immigration to this area of Betula nana (before 7850 BP), published in Kelly and Funder (1974) and repeated in Funder (1979, Fig. 5), seems invalid and may be due to reworked material. Five lakes of which two are fairly close to Kelly's locality, have shown no Betula nana pollen in c. 7000 year old samples (Funder, personal communication and my own investigation).

Sermermiut (69° N)

A raised beach next to Jakobshavn Isfjord has repeatedly been settled by seal-hunting Eskimoes. This first occured around 3400 BP when Elymus and Stellaria humifusa covered the beach (Fredskild 1967 a,b). Alopecurus alpinus, a species highly favored by human activity, dominated the habitation period. After the abandonment a short period

with an Empetrum-Ledum decumbens heath was succeeded by the driest
plant community represented in the profile: a Betula nana heath.
Presumably as a result of moister conditions, the area was overgrown by
a Salix glauca scrub c. 2600 BP, which in turn was replaced by a very
wet stage, a Sphagnum squarrosum bog with Montia, Ranunculus
hyperboreus, R. lapponicus and Carex rariflora (2350 BP) This moist
period was very short, and an Empetrum-Ledum heath, later with many
Gramineae and Stellaria longipes, covered the area until a new
habitation c. 2000-1900 BP. Some time after the disappearance of this
group of hunters a new Sphagnum bog, resembling the first one, spread.
A dating of the middle of the Sphagnum peat is 1540±100 BP; a new heath
community then recaptured the area.

NORTHWEST GREENLAND

Melville Bugt (75°22' N)

A 25 km long, narrow peninsula, Tugtuligssuaq, is surrounded by calving
glaciers, the Greenland Ice Sheet, and the ice-filled Melville Bugt.
The climate is high arctic maritime, which is reflected in a very open,
Cassiope tetragona dominated vegetation with many Salix herbacea-
Huperzia-Carex bigelowii snowbeds. Salix arctica is fairly frequent,
whereas Empetrum and Vaccinium uliginosum are restricted to the most
protected south slopes. Upland sites and north-exposed slopes are
almost devoid of vegetation. Fragmentary Carex stans meadows can be
found in valley bottoms (Fredskild and Bay, 1980). Pollen diagrams are
at hand from two lakes, of which one is shown in Figure 23.4. This
oblong lake, situated just above the marine limit at c. 16 m asl,
receives meltwater only from the 20-30 m high ridges on both sides,
whereas the other lake, Rundesø, receives meltwater from a 400 m
slope. Four pollen zones can be separated:

Gramineae-Oxyria zone (c. 8600-7700 BP). Gramineae, Oxyria and other
pioneer herbs dominated this zone. During this interval Armeria
scabra, now extinct on this peninsula, was growing. Judging from the
pollen content per ml, grasses must have been more frequent than today,
even after considering the slow sedimentation rate.

Oxyria-Minuartia/Silene-Huperzia zone (c. 7700-6600 BP). Empetrum and
Vaccinium immigrated at the beginning, Cassiope in the middle of the
zone, spreading seemingly at the expense of Gramineae. Papaver

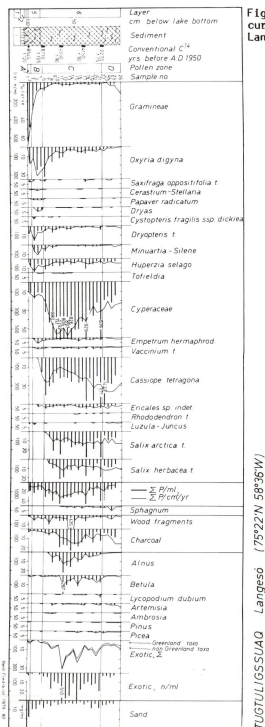

Figure 23.4 Selected pollen curves of terrestrial plants from Langesö (75°22'N 58°36'W).

radicatum, which is very rare today, growing only on some sandy, south-exposed screes, was frequent.

Cyperaceae-Salix-Cassiope zone (c. 6600-2000 BP). _Salix arctica_ and _S. herbacea_ immigrated around 6600 BP, and before a millennium had passed, they both found a level in the percentage diagram, which has been almost unchanged until today. _Cassiope_ seems to have ceded part of its area, most likely to _Salix arctica_ and _Carex bigelowii_. Cyperaceae increase in the first half of the zone, whereas _Oxyria_, _Empetrum_ and pioneer plants decrease. In both diagrams _Tofieldia_ pollen are only found in the first half of the zone. Neither the southern _T. pusilla_ (north to 74°30') nor the more northerly, rare _T. coccinea_ (between 67° and 78° in W. Greenland), are found on the peninsula today. From the middle of the zone _Cassiope_ increases and Cyperaceae decrease, but as the macrofossil samples have not been investigated, the implication for the vegetation cannot be determined. Everything points to a period with greater precipitation, beginning around 5500-5000 BP, caused by more frequent winds from the southwest. Whereas in the Langesø diagram the pollen content per ml from now on exhibits a slight decrease, almost paralleling the Cyperaceae curve, it steadily increases in the Rundesø diagram, reaching a maximum toward the end of the zone ten times as great as in the beginning of the zone. Parallel to this is an increase in sand and in rebedded pollen and spores, mainly consisting of _Betula_ and _Dryopteris_ type. This was caused by a greater supply of meltwater washing down contemporary pollen and reworked material. The reworked pollen had been brought to the surface partly by a deeper erosion and partly as a result of increasing solifluction and patterned ground activity. The influx of long-distance pollen, especially _Alnus_ and _Betula_, has a maximum in the middle of the zone, and shows a drastic reduction at the transition to the next zone.

Cassipe-Salix-Cyperaceae zone (c. 2000-0 BP). The increase in _Cassiope_, which began c. 5300 BP, peaks at the opening of the zone, making this plant the greatest pollen producer, but then it decreases. _Luzula-Juncus_ pollen, most likely originating from _L. confusa_, another of the dominating plants today shows few fluctuations and Gramineae increase. The influx of exotic pollen, apart from that of _Artemisia_, is highly reduced, best illustrated by the influx of _Alnus_, which drops from 1.50 (mean of samples 7-17) via 0.26 (sample 18-21) to 0.05 (samples 22-26), a reduction by a factor of 30. An explanation might be, that from c. 2000 BP moisture (and long-distance pollen) carrying winds from the southwest became less frequent.

Thule area (76-78° N)

Vasari (1972) and Hyvärinen (1972a) each presents two pollen diagrams
from peat monoliths from the Thule area. One of these (Igdlolorssuit)
consists of an extremely rapidly sedimented moss-grass peat, formed
under a bird cliff. Gramineae make up 85-95% of all spectra. With the
exception of one sample, Gramineae make up 65-95% in another diagram
(Ivsugissoq). The third one, from the settlement of Thule/Dundas,
shows at c. 750 BP a marked change from a Cyperaceae dominated, medium
humified peat to fresh moss peat, dominated by Gramineae. This
probably does not date a climatically caused change but rather the
effect of human habitation.

Only the fourth of the diagrams, from Etah (Vasari 1972), does not
seem to be influenced by man or bird. The peat formation on a raised
beach began at 1860±150 BP. The plant community was dominated by
grasses with some Salix arctica, Cassiope and Cerastium/Stellaria. A
fast peat formation, roughly 1 cm in 10 years for the major part of the
monolith, ends with the 12-15 cm sample, the dating of which is 1200
BP. The vegetation changes towards a more dry Salix arctica heath,
still with many grasses, but also with Potentilla and Saxifraga.
Cassiope becomes rarer. The peat formation under this type of
vegetation is small, as illustrated by the dating of the next sample,
viz. 12-9 cm below the surface (330 BP) and 9-6 cm (175 BP) This
points to a climatic change, the dating of which is uncertain.

On Carey Island the formation of a 2.6 m thick peat deposit began
c. 6500 BP and ceased by 4500-4000 BP, thus bracketing the Hypsithermal
(Brassard and Blake 1978).

Qeqertat (77°30' N)

Judging from the vegetation, these islands at the head of Inglefield
Bredning have a climate intermediate between the high arctic maritime
climate on Tugtuligssuaq, with optimal conditions to Cassiope, and the
high arctic desert in central Peary Land, where Dryas is dominating.
The vegetation on dry rocks is dominated by Carex rupestris with Dryas
integrifolia as subdominant, and with Carex nardina, Hierochloë alpina,
Melandrium triflorum and Saxifraga tricuspidata. Cassiope-moss heaths
are mainly restricted to north-exposed slopes. Vaccinium uliginosum,
usually sterile, and Empetrum are frequent in heaths, not quite as dry,
often with Salix arctica which is the species with the widest
ecological amplitude, ranging from hummocky fens to dry rocks. The
fens are otherwise dominated by Cyperaceae: Eriophorum triste, E.

scheuchzeri, Carex stans, C. misandra, and C. atrofusca, besides
Kobresia simpliciuscula, Arctagrostis latifolia, and Juncus biglumis
and J. triglumis.

At the time of deglaciation the western half of Qeqertat,
consisting only of some tiny skerries, emerged by a couple of meters.
The marine limit is 72 m, and by means of dating shells at 43 m
(8620±125 BP, K-3504) and 22.5 m (7930±120 BP, K 3503) and a date on 4
cm gyttja overlying marine clay in the cored lake at 22 m asl (6800±85
BP, K-3502), an uplift curve can be constructed which estimates the
time of deglaciation at 9500-9000 BP. Unfortunately, the first
millennia are missing in the pollen diagram, and the plant communities
were well established, when the limnic sedimentation started. The
gyttja throughout the core is extremely well preserved, full of Nostoc
balls, which are still green and often with the cell content intact.

Dryopteris-Cyperaceae zone (c. 6900-6300 BP). It is difficult to
imagine Cystopteris fragilis or Woodsia glabella--the only possible
contributors to the high frequency of Dryopteris type--being that
common on the rocks of Qeqertat as suggested by their frequencies
(Fig. 23.5), and some of these spores may well have been embedded in
the clay, washed out in the lake from the surrounding, recently emerged
clay deposits. Nevertheless, four out of 45 spores were still
surrounded by their perine, by means of which they could be determined
as Cystopteris fragilis ssp. dickieana. The perine would hardly be
preserved in a rebedded spore.

Cyperaceae zone (c. 6300-3300 BP). After some centuries a fairly wide,
almost horizontal area to the south and east of the lake was covered by
fens. The pollen influx is relatively high: 7.3 grains cm^{-2} yr^{-1}.
Towards the end of the period, a slight decrease in Cyperaceae
corresponds with a slight increase in Cassiope and Gramineae. Without
the help of macrofossils, the interpretation is difficult, but for the
whole zone a milder, possibly more humid, climate is suggested. The
high influx of long-distance pollen, especially Alnus, brought in by
southerly winds, supports this interpretation.

Salix arctica-Cassiope-Cyperaceae zone (c. 3300-1000 BP). The increase
in Salix arctica, Ranunculus and Dryas and the corresponding decrease
in Cyperaceae can be explained as reflecting more dry conditions with
northerly winds. But at the same time the sudden occurrence of marine
diatoms and "Hystrix", likewise of marine origin, combined with a

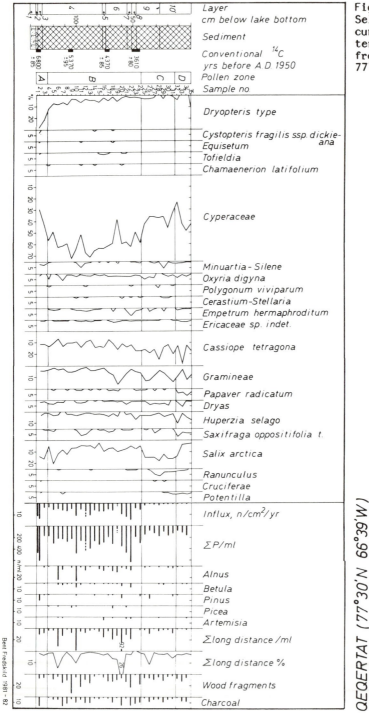

Figure 23.5
Selected pollen
curves of
terrestrial plants
from Qeqertat
77°30'N 66°39'W.

QEQERTAT (77°30'N 66°39'W)
Inglefield Bredning NW. Greenland

Bent Fredskild 1981-82

drastic increase in Pediastrum indicate that the many frostboils in the
fen area were formed at the opening of the zone, as a result of a
climatic change towards colder conditions. Marine organisms as well as
more nutrients were brought into the lake by the meltwater. The pollen
influx dropped, now averaging 1.6 gr. cm^{-2} yr^{-1}.

Cyperaceae-Papaver-Saxifraga zone (c. 1000-0 BP). The increase in
Papaver, Potentilla, and Saxifraga oppositifolia types (including at
this latitude S. oppositifolia and S. tricuspidata) may point to even
drier conditions. The pollen influx is still low.

NORTH GREENLAND

Peary Land, Sølejren (82°13' N)

A 30 cm deep hole was dug through alternating moss and swamp peat
layers in a Carex stans meadow at Sølejren in the heart of Peary Land.
A pollen diagram (Fredskild, 1973, Plate 21) supplemented by
macrofossil analyses, indicate that the local vegetation has passed
through four stages:

A. A Carex stans meadow with scattered Salix arctica and Pedicularis
    hirsuta was overgrown by a Carex stans meadow with Eutrema
    edwardsii. Conditions resembled those of today.
B. After some time the vegetation was totally dominated by mosses and
    Cyperaceae, Carex stans still being the most important among
    these. This period was moist.
C. A somewhat drier mossy vegetation, in which Cyperaceae had to share
    the place with Papaver, Dryas, and Draba bellii.
D. At the transition to this zone, Salix arctica spread, the peat
    became more humified, yet macrofossils of Draba bellii, Melandrium
    apetalum, and Ranunculus sulphureus show that it was not formed
    under a dry heath community. Permafrost is found at a depth of
    13-15 cm.

The age of a sample 23.5-30 cm below the surface (1,520±100 BP)
may date or post-date the beginning of peat formation at the site.
However, a date on one recognizable layer, 10-11.5 cm (the bottom of
Stage D) in three peat monoliths, only 10 cm apart, gave ages of 1170,
770 and 360, all ± 100 BP. Cases of inverted radiocarbon dates on peat
are numerous (Andrews et al., 1979; Hyvärinen, 1972a; Richard, 1981;
Short and Nichols, 1977; and Vasari, 1972), and it goes without saying

that the use of such dates in high arctic peat profiles, subject to
solifluction and periodical ceasing of growth, is not recommended
without a degree of caution.

A hint of the age of zone C is given in the maximum of microscopic
charcoal, most likely resulting from Neo-Eskimo camp fires at
Sølejren. Charred Salix arctica from a fire-place at the opposite side
of the lake has been dated at 770±100 BP.

Klaresø (82°10' N)

102 cm of lake marl and lime gyttja overlie marine clay in lake
Klaresø, 45 m asl, south of Jørgen Brønlund Fjord. This northernmost
pollen diagram from Greenland (Fredskild, 1969, 1973) can be divided
into five zones:

A. Gramineae-Cyperaceae-Oxyria zone, reflecting a pioneer vegetation
    on the surrounding raised marine beds, when Klaresø was still a
    brackish water bay of the fiord. Rebedded pollen is abundant.
B. Saxifraga oppositifolisa-Gramineae zone. During this zone the lake
    is isolated, and the pollen content increases.
C. Salix arctica immigrates either to the surroundings of the lake or
    most likely to the area south of Brønlund Fjord. During the zone
    its pollen increases, reaching 50-60% of the total pollen towards
    the end. A comparison between its recent area cover around the
    lake and its frequency in the topmost sediment and in two pollen
    traps shows Salix to be overrepresented in pollen spectra in this
    high arctic area by a factor of 5.
D. Salix arctica zone. The number of pollen per g sediment is at the
    highest in this zone. The productivity of the lake, as expressed
    in number of Botryococcus reaches its maximum. The high pollen
    influx must be caused by a more dense vegetation cover which in
    this desert (yearly precipitation about 25 mm) can be caused only
    by higher precipitation both in winter and during the growing
    season. This can be connected with the opening of the today
    permanently ice-covered North Greenland fiords and the occurrence
    of more open water in the Polar Basin, reflected in the driftwood
    on raised beaches along the fiords, i.e. Brønlund Fjord. This is
    in accordance with the greatest abundance of driftwood between 6000
    and 4200 BP in Markham Inlet, Ellesmere Island (Stewart and England
    1983).
E. Gramineae-Salix zone. Relatively speaking, Gramineae, Saxifraga,
    Cyperaceae, and Oxyria are more frequent than in zone D, but a very

marked drop in pollen/g and in sedimentation rate does not permit
detailed conclusions to be made.

As a result of the lime content, the dating of the isolation, the
immigration of _Salix_, and of the zone borders and thereby the implied
climatic changes, are subject to some uncertainty. Nine samples have
been dated, of which the lowermost 3 cm limnic gyttja gave an age of
6850±140 BP, whereas a sample 3-6 cm below the present lake bottom was
found to be 2610±120 years old. Based on the assumption that the
supply of "old" carbonate to the lake has been constant, all dates were
calibrated by 1900 years in the original papers, which is the "date" of
the gyttja-water interface according to a best-fit C-14 age versus
depth curve. However, recent investigations by Funder (pers. comm.) in
this area has questioned the tenability of this assumption. A new
uplift curve, less steep than the one previously published (Fredskild,
1969, Fig. 3), indicates a much smaller correction of the oldest date.
A gradual increase in the hardwater effect throughout the history of
the lake, possibly supported by the upward increasing content of
carbonate + soluble matters and corresponding decreasing loss on
ignition (Fredskild, 1973, Plate 20) may be more correct.

If the climatic interpretation of the highly productive zones, C
and D, is correct, dates on them would be most important. As to the
opening of zone C the pollen diagram reflects an immigration and rapid
spreading of _Salix_ which does not exclude the climate from being
favorable right from the immigration or even earlier. The driftwood on
the highest terraces in these fiords has been dated at 5780±100 BP.
Musk oxen profited from a more luxuriant vegetation cover, and from c.
4500 BP the people of the Independence I culture hunted them for about
a millennium (the youngest date being c. 3600 BP). After some
centuries without habitation the Independence II people, who also
hunted seals, settled c. 3300-2500 BP. Everything considered, a major
climatic change closed the Peary Land fiords some time after 2700 BP,
and this also influenced the vegetation and the game, and thereby
banished the inhabitants.

EAST GREENLAND

The pollen analysis of cores from five lakes at Scoresby Sund (70-71°,
Funder, 1978a) and from two on Hochstetter Forland (75° N, Björck and
Persson, 1981), as well as from a 3 m peat deposit at Angmagssalik (66°
N, Bick, 1978), show that the Holocene stratigraphy and climatic
history is different from that of W. Greenland, being connected with
the North Atlantic-Scandinavian development (Funder, 1978b). The

Hypsithermal and the following deterioration are earlier than in W. Greenland, as seen for example in the coastal area at Scoresby Sund, where a pioneer vegetation spread c. 10,000 BP. Right from the beginning, such thermophilous plants as Thalictrum, Botrychium, Salix herbacea, Lycopodium dubium, and Diphasium alpinum were included, showing that by this time the summer temperature was already similar to the present. At 8000 BP Betula nana immigrated to the area and expanded rapidly. The very high frequencies attained by this species in the period from 8000 to 5000 BP indicate a climate warmer than the present. At c. 5700 BP Salix arctica immigrated and later on expanded together with the other high arctic species Cassiope tetragona. At the same time an increase in the pioneer species is also indicative of a temperature decline, causing more unstable soil, and this trend seems accentuated, as from c. 2800 BP a change to minerogenous sedimentation in the lakes reflects more open ground.

PALEOLIMNOLOGY

In the preceding, only the terrestric vegetation has been considered, owing to the fact that the development within the lakes often proceeds more or less independent of this, at least as far as the climatic implications are concerned. This can be exemplified by the development of the four lakes at Godthaabsfjord (Fig. 23.6).

Between 8500 and 8000 BP the climatic threshold for all the macrophytes found seems to have been passed, and two species only, Isoëtes setacea and Myriophyllum alterniflorum, had not emerged. Of course the flowering of a plant, approaching its northern limit, is dependent on the water temperature, but with this reservation it seems likely to relate the succession, the flourishing, and in many cases eventually the extinction, of a limnic organism to the trophic state of the lake.

Every lake cored so far in S. and W. Greenland runs through a succession of stages, starting fairly eutrophic, passing through a mesotrophic and ending in an oligotrophic stage. As to the macrophytes this is illustrated in Figure 23.6, which is parallel to the diagrams from S. Greenland (Fredskild, 1973, Figs 10 and 21). Species preferring lakes with a high conductivity, like Myriophyllum spicatum and Potamogeton filiformis as well as Chara and Nitella, serve as pioneer plants in the ion-rich lakes, which are still supplied with clay and ions, washed out from the surroundings, not yet covered by humus-accumulating vegetation. The primary production in this initial stage is very high, with up to 50,000 Pediastrum boryanum coenobia

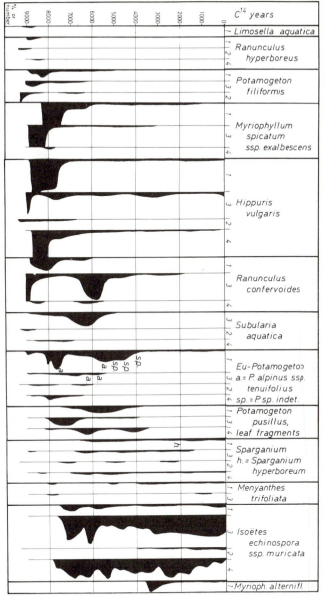

Figure 23.6   The development of macrophytic vegetation in four Godthaabsfjord lakes (1: Johs. Iversen Sø, 2: Karra, 3: Terte, Lake A, 4: Sardlup qaqa) (with permission).

being sedimented per cm^2 per year in a rich lake (and only 1,500 in a poor one). To this must be added a similar amount of Botryococcus braunii plus Scenedesmus, Nostoc, diatoms and other algae not counted, plus the basal food for the myriads of entomostracs.

Sooner or later the content of nutrients in the run-off water from the surroundings diminished, partly because of leaching, partly because the nutrients were used by the terrestrial plants and, to a certain degree, bound in the slowly decomposed litter. Besides this, the humus formation made the acidity of the lakes increase. Contemporarily with the often drastic reduction in the species characterizing the eutrophic stage is the immigration of Potamogeton pusillus and Isoëtes setacea, and the more frequent occurrence of Sparganium hyperboreum. This often happens at the change from a clay gyttja to an almost pure organic gyttja. Other species of Pediastrum have successive, but much smaller maxima after the reduction in P. boryanum (Fredskild, 1983a). The oligotrophication and acidification are also reflected in the diatom flora in the Godthaabsfjord lakes (Foged, 1972). In the ultimate oligotrophic stage the number of species and individuals of macrophytes and of animals is severely reduced. Besides being reflected in the vegetation, chemical analyses have also shown the decrease in yearly deposition of a number of chemical components, often by a factor of 10 from the late-glacial clay-gyttja to the present loose, watery gyttja (Fredskild, 1983b).

Some high arctic lakes do not follow this scheme: for example, the lake on Qeqertat and some of the Scoresby Sund lakes. In the Hochstetter Forland diagrams the curves of Pediastrum and Botryococcus are based on total pollen and therefore not immediately interpretable.

The history of a few animals in the same lakes is illustrated in Figure 23.7. The stickleback, Gasterosteus aculeatus, is frequent in nutrient rich Greenland lakes, northwards to 74°. Daphnia pulex is common everywhere, whereas Simocephalus vetulus has its northern limit at 74°. Climatically significant is Lepidurus arcticus, which is common in the beginning in all lakes as well as in the earliest stages in some S. Greenland lakes (Fredskild et al., 1975). It became extinct between 8000 and 7500 BP, possibly a little earlier in some of the S. Greenland lakes. It is striking that in these four lakes the macrofossil sample following the one with the youngest find of Lepidurus still contains the oldest find of Cristatella mucedo statoblasts. So far, this species has not been found alive in Greenland, but the finding of its statoblasts in a recent gyttja sample in Johs. Iversen Sø indicates that it is still living in the area. Spongilla spicules, dissolved by the standard preparation of pollen same samples, have only been searched for in one lake, where they emerge in the sample as Cristatella. Eurycercus glacialis is distributed from S. Greenland to 72° N.

For each lake, the length of each stage, its diversity as regards species, and its productivity are highly dependent on the soil of the

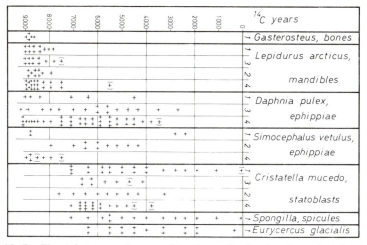

Figure 23.7   The history of some limnic animals in four Godthaabsfjord
lakes (1-4: see Fig. 23.6) (with permission).

catchment area, as is the present vegetation in the lakes and ponds.
This is convincingly illustrated by the former and present distribution
of Myriophyllum spicatum. At Godthaabsfjord it occurred at an early
stage in eight out of ten cored lakes, but since then it has been
extinct in all of them. The present-day distribution, mainly
restricted to the interior, continental area of W. Greenland between
61° and 70° N, may lead to the conclusion that it is related to summer
temperature, but the broad valleys in these areas are rich in till of
Neoglacial ge and of raised marine beds, and the lakes consequently
have a high conductivity, contrary to the nutrient poor, outer coast
lakes, often found in depressions in a polished gneissic bedrock.

SEDIMENTATION RATES

In most unglaciated areas of Greenland it is fairly easy to find a lake
suitable for coring, i.e. some hundreds of meters wide, fairly shallow
and without inlet. This makes the results of the corings comparable,
as regards both sedimentation rates and pollen influx. The sediment in
many of the cored lakes starts with a clay-gyttja, which is succeeded
by a jelly-like algae-gyttja, gradually becoming more loose and watery
upwards, and often the upper part of the core, consisting mainly of
precipitated humic substances washed out from the surroundings, is so
loose that it cannot keep shape when extruded. Following the local
"late-glacial" phase, in which only clay and sand is sedimented, about
half of the lakes exhibit a fast sedimentation rate (3-5 cm/100 yr) in

the beginning, due to a combination of clay influx and high algae production. Contemporarily with the covering of the surroundings by dwarf-shrub heaths, a change in the sediment is often seen. From that time the sedimentation is dependent on two major factors: (1) the autochthonous algae production, depending mainly on available nutrients and temperature, and (2) allochthonous organic matter. All calculations on sedimentation rates and pollen influx are based on best-fit curves on calibrated C-14 dates. From 6542 BP, I have used the results of Damon et al. (1973), whereas a curve, based on Stuiver (1971) has been used for older samples.

In the gneissic area of Greenland oligotrophication often causes a decrease in autochthonous matter, and the rate in the topmost layer is c. 1 cm/100 yr. This type is illustrated by the first two lakes in Figure 23.8. Other lakes have only a slightly changing rate through time, sometimes showing a tendency of a slightly higher rate in the middle, which in the two shown examples are not connected with the maximum in Pediastrum or Botryococcus. Finally, some lakes exhibit an increasing rate towards the top. These are shallow lakes being filled in with macrophytes spreading all over the bottom, illustrated by Comarum Sø (Fig. 23.8). This is the Greenland lake with the fastest but one average Holocene rate (3.91 cm/100 yr).

In some high arctic lakes the rate in the upper part is extremely low. The date of 6120±90 BP on a sample only 9-15 cm below the bottom in Ailsa Sø (75°19' N) made Björck and Persson (1981) suggest that the sedimentation ceased about 5000 years ago, which is close to the end of the East Greenland Hypsithermal. They find support for this in the similarity of the uppermost pollen spectra with a nearby mid-Holocene pollen zone in the Peters Bugt Sø. However, above where a radiocarbon sample (9540 BP) was obtained, only five pollen samples have been analysed, and furthermore two deeper samples give inverse datings. Regardless, the mean rate for the whole series is extremely low, c. 0.6 cm/100 yr, if no erosion of the lake bed has taken place. An even lower rate is seen in the upper part of the Klaresø core (Fig. 23.8) with a mean of 0.4, the same as that of a 6430 years old core from Station Nord (c. 81°30' N) in NE Greenland (Abrahamsen, 1982).

The Greenland mean sedimentation rates range between 1 and 5 cm/100 yr with 20 out of 37 being less than 2.0 cm/100 yr. The highest numbers are found at the head of the S and SW Greenland fiords, with the exception of Morten Sø, Scoresby Sund (Funder, 1978a).

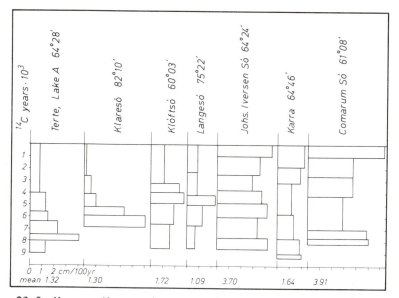

Figure 23.8 Mean sedimentation rates in different periods for some
Greenland lakes.

IMMIGRATION AND SPREADING OF SOME IMPORTANT PLANT SPECIES

The isolated position of Greenland and the many topographic obstacles
within the country makes the spreading of plant species without
effective dispersal propagules time-consuming, and this makes climatic
interpretations, based on first appearance of indicator species
hazardous. Very often an immigrating species attains its maximum,
relatively as well as absolutely, shortly after is first appearance,
e.g. Betula nana and Alnus crispa in Fig. 23.2. The only climatic
implication of this is that at the time of immigration conditions for
some time had been favorable to the species.

     Among the earliest immigrants in low arctic Greenland are the
ericaceous dwarf-shrubs, usually represented by several species in the
earliest pollen sample. Some fairly thermophilous species are early
immigrants too, e.g. Angelica archangelica, Botrychium, Thalictrum
alpinum, whereas others are late, reflecting a less effective dispersal
capacity, e.g. Betula spp. and Alnus crispa. The high arctic Salix
arctica spread southward reaching Melville Bugt by 6700 BP (Fig.
23.9). Its advance to the present south limit at 69° in W Greenland is
not known, but in E Greenland this did not happen until c. 5700 BP, 2-3
millennia later than the immigration of S. herbacea. This species was
also an early immigrant in W Greenland. The immigration of Salix to

South Greenland is still ambiguous. No pollen older than 8000 BP were
found in the interior, while scattered grains occur even back to 9400
BP in the Kap Farvel area. The European <u>Betula nana</u> first reached E
Greenland by 8000 BP (Fig. 23.10), but the ice cap was an effective
barrier. It arrived at Godthaabsfjord c. 6500 BP, whereas its arrival
at Disko Bugt has not yet been dated precisely. It happened later than
7000 BP, which makes its spreading to Eastern Baffin Island earlier
than 6000 BP difficult to explain (Miller et al., 1977). The N
American counterpart, <u>B. glandulosa</u>, did not arrive in SW Greenland
until 5,700 BP, and the spreading to S. Greenland took two millennia.
The European tree birch, <u>B. pubescens</u>, is a very late immigrant.
<u>Juniperus</u> reached S Greenland by 7000 BP and Godthaabsfjord by 6400 BP
(Fig. 23.11).

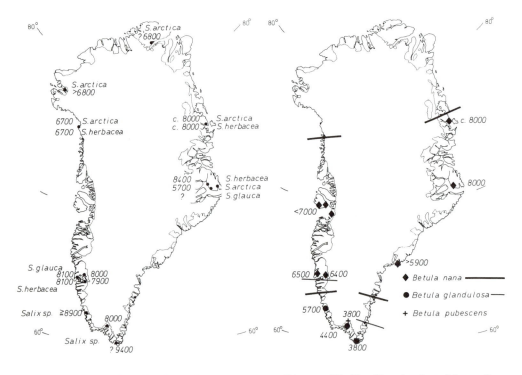

Figure 23.9 The immigration of
some <u>Salix</u> species to different
parts of Greenland.

Figure 23.10 The immigration of
three Greenland species of
<u>Betula</u>. The present north and
south limits <u>Betula glandulosa</u>
(thin line) are given.

POLLEN INFLUX

Considering the variation in pollen influx both within and between
lakes (Davis and Brubaker, 1973), the results from Greenland give a
fairly consistent picture which demonstrates the usefulness of
employing pollen influx in reconstructing past vegetation. The average
influx for three S Greenland outer coast lakes is 177-327 against
1710-2610 gr cm^{-2} yr^{-1} from the interior (Fig. 23.12). Northward the
number decreases. Based on the number of pollen per mg in the
Nigerdleq diagram (Kelly and Funder, 1974) a cautious estimate gives an
average influx for Holocene of 150-200 gr cm^{-2} yr^{-1}. The numbers for
the outer coast diagrams from Godthaabsfjord are 67 and 250 against 286
(the penultimate zone) and 740 gr cm^{-2} yr^{-1} in the interior. The
results from four lakes at Sdr. Strömfjord (393-828 gr cm^{-2} yr^{-1}) and
five lakes on Disko (172-251 gr cm^{-2} yr^{-1}) are tentative estimates,
based on the

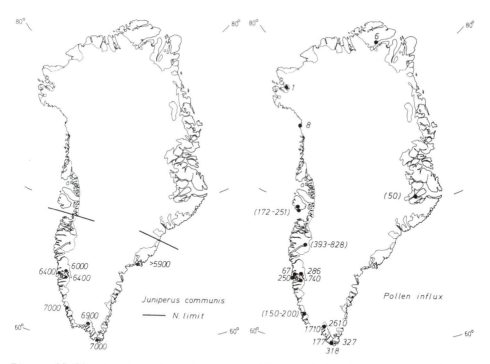

Figure 23.11 The immigration of
Juniperus communis to different
parts of Greenland.

Figure 23.12 Pollen influx (n
cm^{-2} yr^{-1}) to different parts of
Greenland. Unbracketed numbers
are mean of the recent pollen
influx (or subrecent: 286 in
Godthaabsfjord) pollen zone.
Bracketed numbers: see text.

assumption of a sedimentation rate of 1 cm in 30 years for near shore surface mud (Pennington, 1980). In E Greenland the average influx for the past millennium in Bramgaassø is c. 50 (Funder, 1978a, Fig. 14). In the high arctic lakes the influx is less than 10. As mentioned above, a lake on Tugtuligssuaq receiving meltwater from a 400 m high ridge had an influx ten times as high as Langesø, (Fig. 23.12, influx = 8 gr cm^{-2} yr^{-1}).

From low values in the pioneer phase the influx increases to reach maximum values during Hypsithermal (Fig. 23.3; Fredskild, 1973, Tables 16-21), and then decreases parallel to the climatic deterioration. With the exception of Comarum Sø, which is being filled in, this decrease, in many cases by a factor of 1.5-2, is seen in every lake investigated so far. In some high arctic lakes the decrease is even more drastic: Langesø 20 to 8 gr cm^{-2} yr^{-1}, Qeqertat c. 7 to 1 gr cm^{-2} yr^{-1}, besides Ellesmere Island, where Hyvärinen and Blake (1981) found an influx of 5-15 gr cm^{-2} yr^{-1} in early Holocene, decreasing to c. 1 gr cm^{-2} yr^{-1} in the subrecent sample. The post-hypsithermal decrease is not caused by a major change in the composition of vegetation to less pollen producing species, but rather by a reduced viability caused by the climatic deterioration, or a reduced number of individuals of pollen producing plants. Lichens and mosses may thus have conquered their area.

The Greenland results compare well with those from North America, where values for the initial herb zone following deglaciation are 500-1,000 for Rogers Lake, Conn. (Davis, 1969), 300 gr cm^{-2} yr^{-1} in Quebec (Richard, 1971) or even lower (Mott and Farley-Gill, 1981). Average pollen influx over the past 6-7 millennia for three lakes on Ile du Diana in the northwestern part of Ungava Bay, today covered by herb tundra, ranges between 75 and 415 gr cm^{-2} yr^{-1}. The corresponding number for 4-5 millennia for a lake west of Ungava Bay in the shrub tundra is 730, and for two lakes in the forest tundra 50 km further south 1320 and 2120 gr cm^{-2} yr^{-1} (Richard, 1981).

HOLOCENE CLIMATIC CHANGES

The inferred climatic changes can be summarized as follows: In the low arctic S and W Greenland the temperature increased from deglaciation 10,000 to 9000 BP. It increased during the following millennia which was the warmest and driest period, but at what time the temperature curve peaked cannot be deduced on the basis of the diagrams, mainly because of a delay in immigration of some of the most important plants. In S Greenland northward to 62° some minor changes occurred around 5700-5300 BP, and the following two millennia may have been the warmest. A first deterioration sets in at 3400-3200 at Frederikshaab and the interior further south, where the next, more marked change to cool and humid conditions is registered 2200-1900 BP. At Godthaabsfjord a more moist, but still fairly warm period is registered in the interior from c. 3900 BP or some centuries later, and from 1800 BP conditions were more humid and cool. At the outer coast more moist and presumably also colder conditions set in c. 3600 BP, followed by a new deterioration between 2500 and 2000 BP. Peat profiles show more humid periods to begin around 3200, 2400, 1500 and 600 BP.

In the Melville Bugt area higher precipitation begins between 5500 and 5000 BP. A change to more high arctic conditions is seen around 2000 BP. Further north, at Qeqertat, colder, possibly drier conditions set in around 3300, and another change, to even drier conditions, is registered at 1000 BP. In Peary Land evidence of a different origin, but ultimately bound to climate, indicate a change to harsh, high arctic conditions between 2500 and 2100 BP, and peat profiles have registered later, minor fluctuations very roughly dated around 1500 and 1000 BP. The expansion in N Greenland, during the 15th and 16th centuries AD (Knuth 1981), of Neo-Eskimo whalers must be connected with a climatic change resulting in more open water. Having sailed north of the island, they left an umiaq on the beach of Koln s, NE Greenland.

In discussing the Neoglacial history of W Greenland, Kelly (1980) concludes that glacierization began to increase about 3500-3000 BP. Although there is evidence of a more severe climate around 2000 BP, the glacierization did not reach its maximum until another cold phase after 1000 BP. This is in accordance with the results obtained from the Camp Century core (Dansgaard et al., 1973) showing a prolonged duration of a climatic optimum from 8000-3500 BP followed by two early Neoglacial cold intervals (3,500-3,000 and 2800-2400 BP) and a complex one in the late Neoglacial.

Broadly, these climatic events show an overall similarity with the Holocene climatic history of the low and subarctic Canada (Ritchie and Hare, 1971; Nichols, 1975; Jordan, 1975; Andrews and Nichols, 1981; Richard, 1981; Macpherson, 1982), where a marked temperature decline is registered c. 4000-3500 BP and another one around 2000 BP. The climatic history of E Greenland is different (Funder, 1978b), with an early Hypsithermal and a major deterioration commencing around 5000 BP.

EXOTIC POLLEN IN THE GREENLAND LAKES

Long-distance transported pollen are very important in subfossil arctic
samples, and their occurrence have been used to correlate local
diagrams with changes in distant areas (e.g. Hyvärinen, 1972b) or to
reconstruct paleo-wind patterns (Nichols et al., 1978; Andrews et al.,
1979). For S. Greenland the feasibility of the method in tracing
source area for the exotic pollen has been investigated (Fredskild,
1973), and now with more material at hand, it seems reasonable to make
an up-to-date review.

Ambrosia

Ambrosia pollen are far from frequent in Greenland samples, and yet
their occurrence varies with time, best visualized by the mean Ambrosia
pollen per sample for each millennium of calibrated dates. The picture
for the four S Greenland lakes and the four Godthaabsfjord lakes seems
clear (Fig. 23.13): summer winds from the Ambrosia region in the
Midwest were frequent in early Holocene, peaking in the period
7000-6000 C-14 years BP and then decreased until the period c. 3200-
2400, followed by a slight increase in the following millennia. The
increase in the past 500 years must be a reflection of forest clearance
(most pollen are found in the recent gyttja sample). For the two NW
Greenland lakes the picture is similar, yet the lack in the first two
millennia might indicate that not until 7000-6000 BP did the pollen-
carrying southwesterly winds penetrate that far north to a greater
extent. This is confirmed by other exotic pollen taxa.

Artemisia

The present distribution within Greenland of Artemisia borealis, the
pollen of which cannot be distinguished from other Artemisia pollen, is
restricted to sunny slopes, especially in the interior, between 62° and
72° in W Greenland. Figure 23.13 shows the mean pollen per sample for
four S Greenland lakes and two lakes from NW Greenland, well outside
this area. Artemisia is unlikely to have lived in S Greenland during
Holocene, as the highest frequency (2.33 per sample) is found in
Isoëtes Sø at the outer coast, whereas it is much more sparse in
Comarum Sø (0.28) in the interior where Artemisia no doubt would have
preferred to grow. The content of windblown pollen is fairly constant
up to c. 4,000 BP, then decreasing slightly and, as with Ambrosia,
increasing in the recent centuries. The picture in the NW Greenland
lakes is quite different; from low values in early Holocene the

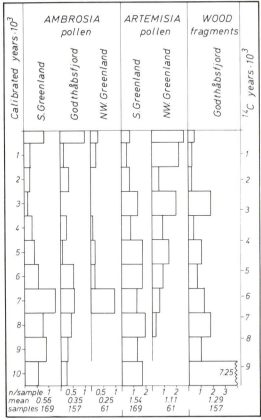

Figure 23.13 Mean influx per millennium of Ambrosia and Artemisia pollen and of microscopic wood fragments to some Greenland lakes.

frequency increases to reach a maximum at the same time as the minimum in Ambrosia. This seems to indicate that the major part of Artemisia pollen comes from the arctic tundra regions of N America, where the southward retreat of the timberline provided places for extended open communities with Artemisia frigida on xeric sites (Ritchie and Hare, 1971; Nichols, 1975).

Alnus

The eagerly discussed question of an early immigration of Alnus to SW Greenland around 7600 BP (Iversen, 1954; Kelly and Funder, 1974) or a late one shortly after 4000 BP (Fredskild, 1973, 1983b) will not be referred in detail, but the evidence for a late immigration, shown here in Figure 23.14, can be briefly explained. Only the two lakes to the right of Figure 23.14 are within today's Alnus area, with Johs. Iversen Sø right in the heart of it. All lakes but the two most northerly

ALNUS

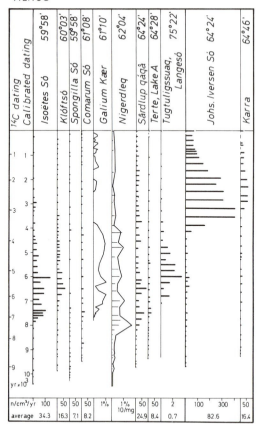

Figure 23.14 Influx or percentages of <u>Alnus</u> pollen in 12 south, west, and northwest Greenland lakes, of which 10 (left) are outside the present area of <u>Alnus crispa</u>. Note difference in unit (with permission).

show an increase in influx between 8000 and 7000 BP, and a marked decrease 4000-3500 BP. This seems to reflect conditions in Labrador, where <u>Alnus</u> spread along the coast around 7200 BP (Jordan, 1975), showing metachronous maxima (Morrison, 1970; Lamb, 1980; Short and Nichols, 1977), until a synchronous decline is registered west of Ungava Bay c. 3500 BP (Richard, 1981). Winds from this area must have been frequent during springtime, at least in the said period. Whether the <u>Alnus</u> decline in the Greenland diagrams simply reflects the Labrador decline, or whether changing wind directions were involved, cannot be decided on this basis alone. Incidently <u>Alnus</u> spread to W Greenland about the same time, almost immediately obtaining its maximum.

Conifers

Conifer pollen are found in virtually all S and W Greenland pollen

23.15). Pinus is fairly frequent throughout, yet often with higher
frequencies between 7000 and 3000 BP, thus resembling Alnus. It is
rare in the northernmost lakes but higher frequencies tend to occur
between 5000 and 3000 BP. Picea, rare in early Holocene, shows a
marked increase in frequency between 5000 and 4000 BP, having a maximum
in the first two millennia in the two Kap Farvel lakes (in the third,
Isoëtes Sö, conifers were not separated during counting), whereas it
continues with fairly constant frequencies in the W Greenland lakes.
In many lakes the Pinus:Picea ratio changed from 7-10:1 to 1:2 around
5,000 BP.

Unfortunately the Holocene isopoll maps of northeastern USA
(Bernabo and Webb, 1977) have not yet been extended to cover Canada,
but in summarizing, it seems justified to assume paleo-winds in S and W
Greenland prior to c. 7000 BP in spring and summer came mainly from the
area south of the Laurentide ice shield, where the Picea area was
compressed between the Pinus area to the south and the ice to the

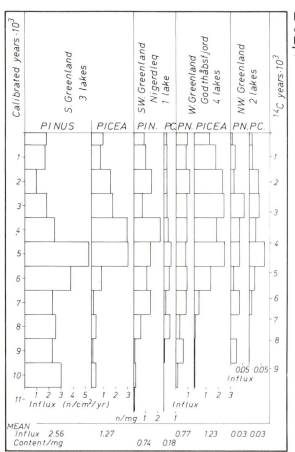

Figure 13.15 Mean influx
or content per mg of
Pinus and Picea pollen to
some Greenland lakes.

BP the source area for most exotics was probably the Labrador area.
The change in the Pinus:Picea ratio around 5000 BP indicates that from
that time winds from the Canadian taiga became more frequent. In the
period 3000-2000 BP winds mainly from north of the taiga reached the
Melville Bugt area.

Long-distance pollen transportation tells nothing about the winds
in the winter half-year, but part of the wood flour, resulting from the
abrasive effect of the ice-crystal-carrying winter storms at the
timberline (Savile, 1972) may end in the gyttja or in the ice cap
(Fredskild and Wagner, 1974). It is made up almost exclusively from
microscopic fragments of tracheids with bordered pits or of pith rays,
easily recognizable as originating from Conifers. The maximum in the
first millennium (Fig. 23.13) is ascribed to rebedded interglacial
material, originating from supposed Picea mariana in W Greenland
(Bryan, 1954). The maximum in the period 3500-2500 BP is mainly due to
one sample, containing 14 fragments. Apart from this, the histogram,
as well as the Langesø diagram (Fig. 23.4) suggest a decreasing number
of winter storms from that part of N. America, especially from 3000 BP
and onwards. The curves of microscopic charcoal, originating from
forest fires (or local Eskimo or Norse camp fires), do not give any
consistent picture. In some W Greenland lakes there is an increase
after c. 5000 BP, in some S Greenland lakes there is a decrease after
c. 3000 BP, whereas in others the curve is irregular.

REFERENCES

Abrahamsen, N., 1982: C-14-dating of samples collected during the 1979
    expedition to North Greenland. In: Funder, S. (ed.), The Geological
    Survey of Greenland. Report No. 110:9-14.
Andrews, J.T., Webber, P.J., and Nichols, H., 1979: A late Holocene
    pollen diagram from Pangnirtung Pass, Baffin Island, N.W.T., Canada.
    Review of Palaeobotany and Palynology, 27:1-28.
Andrews, J.T. and Nichols, H., 1981: Modern pollen deposition and
    Holocene paleotemperature reconstructions, Central Northern Canada.
    Arctic and Alpine Research, 13(4):387-408.

Bernabo, J.C. and Thompson Webb III, 1977: Changing patterns in the
    Holocene record of Northeastern North America: A mapped summary.
    Quaternary Research, 8:64-96.
Bick, H., 1978: A postglacial pollen diagram from Angmagssalik, East
    Greenland. Meddelelser om Grønland, 204(1):1-22.
Björck, S. and Persson, T., 1981: Late Weichselian and Flandrian
    biostratigraphy and chronology from Hochstetter Forland, Northeast
    Greenland. Meddelelser om Grønland, Geoscience, 5:1-19.
Brassard, G.R. and Blake, W. Jr., 1978: An extensive subfossil deposit
    of the arctic moss Aplodon workmskioldii. Canadian Journal of Botany
    56(16):1852-1859.
Bryan, M.B., 1954: Interglacial pollen spectra from Greenland. Danmarks
    Geologiske Undersøgelse II, 80:65-73.

Damon, P.E., Long, A. and Wallick, E.I., 1973: Dendrochronologic
    calibration of the carbon-14 time scale. Proceedings of the 8th
    International Conference on radio carbon dating, Wellington, New
    Zealand October 1972. Contribution No. 57, Department of Geoscience,
    University of Arizona.
Dansgaard, W., Johnson, S.J., Clausen, H.B., and Gundestrup, N., 1973:
    Stable isotope glaciology. Meddelelser om Grønland, 197(2):1-53.
Davis, M.B., 1969: Climatic changes in southern Connecticut recorded by
    pollen deposition at Rogers Lake. Ecology, 50(3): 409-422.
Davis, M.B. and Brubaker, L.B., 1973: Differential sedimentation of
    pollen grains in lakes. Limnology and Oceanography, 18(4): 635-646.

Foged, N., 1972: The diatoms in four postglacial deposits in
    Greenland. Meddelelser om Grøland, 194(4):1-66.
Foged, N., 1977: The diatoms in four postglacial deposits at
    Godthaabsfjord, West Greenland. Meddelelser om Grønland, 199(4):
    1-64.
Fredskild, B., 1967a: Palaebotanical investigations at Sermermiut,
    Jakobshavn, West Greenland. Meddelelser om Grønland, 178(4):1-54.
Fredskild, B., 1967b: Postglacial plant succession and climatic changes
    in a West Greenland bog. Review of Palaeobotany and Palynology,
    4:113-127.
Fredskild, B., 1969: A postglacial standard pollen diagram from Peary
    Land, North Greenland. Pollen et Spores XI, 3:573-583.
Fredskild, B., 1973: Studies in the vegetational history of Greenland.
    Palaeobotanical investigations of some Holocene lake and bog
    deposits. Meddelelser om Grønland, 198(4):1-245.
Fredskild, B., 1978: Palaeobotanical investigations of some peat
    deposits of Norse age at Qugssiarssuk, South Greenland. Meddelelser
    om Grønland, 204(5):1-41.
Fredskild, B., 1983a: The Holocene development of some low and high
    arctic Greenland lakes. Developments in Hydrobiology, in print.
Fredskild, B., 1983b: The Holocene vegetational development of
    Godthaabsfjord area, West Greenland. Meddelelser om Grønland,
    Geoscience, 10:1-28.
Fredskild, B. and Wagner, P., 1974: Pollen and fragments of plant
    tissue in core samples from the Greenland Ice Cap. Boreas, 3:105-108.
Fredskild, B., Jacobsen, N., and Røen, U., 1975: Remains of mosses and
    freshwater animals in some Holocene lake and bog sediments from
    Greenland. Meddelelser om Grønland, 198(5):1-44.
Fredskild, B. and Bay, C., 1980: Botanical investigations. In:
    Jacobsen, N.K., The Knud Rasmussen Memorial Expedition. Geografisk
    Tidsskrift, 80:29-44, 39-41.
Funder, S., 1978a: Holocene stratigraphy and vegetation history in the
    Scoresby Sund area, East Greenland. Grønlands Geologiske
    Undersøgelse, Bulletin 129:1-66.
Funder, S., 1978b: Holocene (10,000-0 years BP) climates in Greenland,
    and North Atlantic atmospheric circulation. Danish Meteorological
    Institute Climatological Papers, 4:175-180.
Funder, S., 1979: Ice-age plant refugia in East Greenland.
    Palaeogeography, Palaeoclimatology, Palaeoecology, 28:279-295.

Hyvärinen, H., 1972a: Pollen analyses of three peat sections from NW
    Greenland: Savigssivik, Thule and Ivssugissoq. Acta Universitatis
    Ouluensis, A 3, Geologica, 1:127-129.
Hyvärinen, H. 1972b: Pollen-analytic evidence for Flandrian climatic
    change in Svalbard. Acta Universitatis Oulensis, A 3, Geologica,
    1:225-236.

Hyvärinen, H. and Blake, W. Jr., 1981: Lake sediments from Baird Inlet, East-Central Ellesmere Island, Arctic Canada; radiocarbon and pollen data. Third International Symposium on Paleolimnology, Joensuu, Finland, September, 1981. Abstracts, 35.

Iversen, J., 1934: Moorgeologische Untersuchungen auf Grönland. Meddelser fra Dansk Geologisk Forening, 8:341-358.
Iversen, J., 1954: Origin of the flora of Western Greenland in the light of pollen analysis. Oikos, 4: II: 85-103.

Jordan, R., 1975: Pollen diagrams from Hamilton Inlet, Central Labrador, and their environmental implications for the northern Maritime Archaic. Arctic Anthropology, XII-2:92-116.

Kelly, M., 1980: The status of the Neoglacial in western Greenland. The Geological Survey of Greenland, 96:1-24.
Kelly, M. and Funder, S., 1974: The pollen stratigraphy of late Quaternary lake sediments of South-West Greenland. The Geological Survey of Greenland, 64:1-26.
Knuth, E., 1981: Greenland News from between 81° and 83° North. Folk, 23:91-111.

Lamb, H.H., 1980: Late Quaternary vegetational history of south-eastern Labrador. Arctic and Alpine Research, 12(2):117-135.

Macpherson, J.B., 1982: Postglacial vegetational history of the eastern Avalon Peninsula, Newfoundland, and Holocene climatic change along the eastern Canadian seaboard. Geographie physique et Quaternaire, XIII:175-196.
Miller, G.H., Andrews, J.T., and Short, S.K., 1977: The last interglacial-glacial cycle, Clyde Foreland, Baffin Island, N.W.T.: stratigraphy, biostratigraphy, and chronology. Canadian Journal of Earth Sciences, 14:2824-2857.
Morrison, A., 1970: Pollen diagrams from interior Labrador. Canadian Journal of Botany, 48:1957-1975.
Mott, R.J. and Farley-Gill, L.D., 1981: Two Late Quaternary pollen profiles from Gatineau Park, Quebec. Geological Survey of Canada, 80-31:1-10.

Nichols, H., 1975: Palynological and paleoclimatic study of the Late Quaternary displacement of the boreal forest-tundra ecotone in Keewatin and Mackenzie, N.W.T., Canada. Institute of Arctic and Alpine Research, Occasional Paper, 15:1-87.
Nichols, H., Kelly, P.M., and Andrews, J.T., 1978: Holocene palaeo-wind evidence from palynology in Baffin Island. Nature, 273(5658):140-142.

Pennington, W., 1980: Modern pollen samples from West Greenland and the interpretation of pollen data from the British Late Glacial (Late Devensian). New Phytologist, 84:171-201.

Richard, P.J.H., 1971: Two pollen diagrams from the Quebec city area, Canada. Pollen et Spores XIII, 4:523-559.
Richard, P.J.H., 1981: Paleophytogeographie postglaciaire en Ungava par l'analyse pollinique. Paleo-Quebec, 13:1-153.
Ritchie, J.C., and Hare, F.K., 1971: Late-Quaternary vegetation and climate near the Arctic tree line of Northwestern North America. Quaternary Research 1(3):331-342.

Savile, D.B.O., 1972: Arctic adaptations in plants. Monograph No. 6: 1-81. Canada Department of Agriculture.

Short, S.K. and Nichols, H., 1977: Holocene pollen diagrams from
 subarctic Labrador-Ungava: Vegetational history and climatic change.
 Arctic and Alpine Research, 9(3):265-290.
Stewart, T.G. and England, J., 1983: Holocene sea-ice variations and
 paleoenvironmental change, northernmost Ellesmere Island, N.W.T.,
 Canada. Arctic and Alpine Research, 15(1):1-17.
Stuiver, M., 1971: Evidence for the variation of atmospheric C^{14}
 content in the Late Quaternary. In: Turekian, K.K. (ed.), Late
 Cenozoic Glacial Ages. Yale University Press.

Vasari, Y., 1972: Pollen-analytical observations: Ita and
 Idglolorssuit. Acta Universitatis Ouluensis, A 3, Geologica 1:
 124-127.

24 Neoglacial moraines on Baffin Island

P. T. Davis

In conjunction with the development of lichenometry in the 1960s, a
renewed effort has been made to determine the ages of Neoglacial
moraines in alpine areas throughout the world. Sequences of Neoglacial
moraines with a variety of ages were recognized, although more commonly
only moraines dating to the Little Ice Age (100-300 yrs BP) were
identified. The occurrence of Neoglacial moraines that predate the
Little Ice Age in a few areas led to the following questions:
(1) If the Little Ice Age, with its associated glacier advances, was
not unique during the Holocene, when did the earlier "Little Ice Ages"
or Neoglacial advances occur, and were these glacier advances periodic
and/or globally synchronous?
(2) When will the next "Little Ice Age" occur, and will it be a
global-wide phenomenon?
(3) When will the present interglacial (the Holocene) end?

Opinions differ concerning the answer to question (1). Denton and
Karlén (1973, 1977) feel that major glacial advances occurred
throughout the world at roughly 2500 yr intervals during the Holocene,
as well as during the late Wisconsin. On the other hand, Benedict
(1973) built a case for non-synchronous global behavior of glaciers
during the Holocene, based primarily on his own research in the
Colorado Front Range. Also, two Neoglacial moraine chronologies from
Baffin Island (Miller, 1973a, 1975; Andrews and Barnett, 1979) are at
variance with the Denton and Karlén model. Although Denton and Karlén
(1977, p. 109) recognize that some moraine chronologies do not fit
their model, they argue that global synchrony of climatic change is the
best interpretation of alpine glacier fluctuations during the
Holocene. Finally, answers to questions (2) and (3) above will be
dependent, at least in part, on an answer to question (1).

There are two broad approaches for resolving the conflict concerning synchrony of global climatic change. One approach, used by Wendland and Bryson (1974), involves critical review of all the literature available on late Holocene climatic chronologies, and generalizing temporal patterns of climatic change on a global scale. However, problems persist in the critical evaluation and comparison of variable types of climatic data sources, such as historical, glacial, palynological, and dendroclimatological records (Grove, 1979). Another means for evaluation of proxy climatic records requires return to a well-studied field area to replicate the original results (e.g. moraine ages), as well as to provide additional data sources (e.g. pollen or tree rings). For instance, recent radiocarbon ages (Davis and Waterman, 1979; Davis, 1982) from Arapaho Cirque in the Colorado Front Range necessitate a revision of at least part of Benedict's (1973) well-known Holocene climate reconstruction, for it is now known that the type Triple Lakes Advance occurred thousands of years prior to 3000-5000 BP. This latter method may be costly; however, such results are necessary before further modeling of proxy data be undertaken.

Previous results from Baffin Island led Miller (1975) to suggest that a 1600 yr periodicity of glacial advance may be appropriate for glaciers on northern Cumberland Peninsula, whereas Andrews and Barnett (1979) feel a 600 yr return period is applicable for advances of lobes of the Barnes Ice Cap (Fig. 24.1); thus both areas are postulated not to follow a simple 2500 yr recurrence interval. The aim of this chapter is comparison of Neoglacial moraine records from southern Cumberland Peninsula (Fig. 24.1; Davis, 1980) with moraine records from northern Cumberland Peninsula and from around the Barnes Ice Cap. Thus, a composite Baffin Island Holocene glacial chronology may be compared with other Neoglacial moraine records from around the globe and serve as a test for the Denton and Karlén (1973, 1977) model for global synchrony of Holocene climatic change.

Exclusive use of moraine ages as records of climatic change is a problem because moraines represent only end points of glacier responses to threshold equilibrium conditions. Because moraines only record the terminations of events within a changing climatic continuum, more continuous records of climatic change are desirable (Davis, 1980; Karlén, 1976, 1981). However, long-term, continuous palynological records from sediment sections on Baffin Island are few (Short et al., this volume), especially if compared with the extensive Holocene palynological studies to the west in Keewatin (Nichols, 1975), to the east in Greenland (Fredskild, 1973, this volume), and to the south in Labrador (Short and Nichols, 1977). This area of research is important because of our lack of knowledge regarding arctic glacier/vegetation/

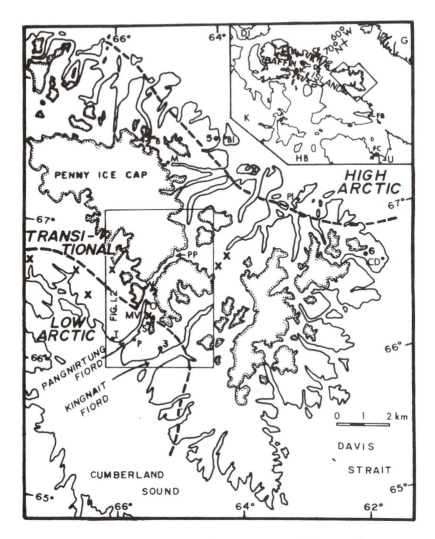

Figure 24.1 Index map of Baffin Island, emphasizing southern
Cumberland Peninsula. Glacier areas generalized with stippled
boundaries and shown in more detail in Figs 24.2, 24.3, 24.4, & 24.5.
Sites of glaciological studies labeled O (Ookalik Glacier) and B (Boas
Glacier). Site of lacustrine sedimentological study labeled S (Shadow
Lake). Sites of 5 weather stations labeled P (Pangnirtung), CD (Cape
Dyer), PI (Padloping Island), BI (Broughton Island), and DL (Dewar
Lakes). Sites of dwarf birch observations marked by X and sites for
radiocarbon dates on lichen-kill (Table 24.2) numbered 1-6. Vegetation
regions shown by dashed line and labeled after plates 45-46, The
National Atlas of Canada (1973). Sites of palynological and
sedimentological studies referenced in this chapter and Chapter 22
marked I (Iglutalik Lake), S (Shadow Lake), W (Windy Lake peat), and M
(Maktak peat). Other place names mentioned in this and other chapters
marked PP (Pangnirtung Pass), MV (Moon Valley), CF (Clyde Foreland,
including Pat Bay Lake), FB (Frobisher Bay), FC (Fort Chimo), G
(Greenland), DI (Devon Island), K (Keewatin), HB (Hudson Bay), and L-U
(Labrador-Ungava).

climate dynamic interactions; however, few studies to date have
produced a variety of proxy data from one geographic area on these
interrelationships. Least understood are the leads and lags between
various natural responses to climatic change. Leads and lags are
important if comparisons of climatic chronologies derived from a
variety of proxy data sources are used from different areas. Reliable
proxy data are especially important when considering Baffin Island's
sensitivity to climatic change (Bradley and Miller, 1972). Moreover,
Keen (1979) suggests that Baffin Island shows the best correlations
with the general temperature trends of the Northern Hemisphere above
55°N latitude over the past few decades. The above questions (2) "When
will the next 'Little Ice Age' occur?" and (3) "When will the present
interglacial end?", are highly relevant for climatic studies in
sensitive areas such as Baffin Island, because Holocene climatic data
are a prerequisite for further development of numerical models of
glacial-climatic interactions for the eastern Canadian Arctic (Williams
and Bradley, this volume).

LOCATIONS AND GENERAL DESCRIPTIONS OF FIELD AREAS

The Pangnirtung and Kingnait Fiord area of southern Cumberland
Peninsula is an area of spectacular beauty, with 300-1200 m high fiord
and valley walls, hundreds of cirque and valley glaciers, and the 6000
km^2 Penny Ice Cap centered on a plateau just to the north (Fig. 24.1).
Three local areas were selected for lichenometric studies of moraine
systems fronting 22 glaciers (Fig. 24.1). Nearby and within these
three areas, lacustrine sediment cores and peat monoliths were
collected for palynological, sedimentological, and paleomagnetic
analyses (Fig. 24.1; Davis, 1980).

Field study area (1) is located east of the northern end of
Pangnirtung Fiord, between 12 and 25 km from the settlement of
Pangnirtung (Figs 24.1 & 24.2). This rarely visited area of rugged
alpine terrain is part of the Fretted Uplands Province of Dyke (1977),
with over 25 cirques separated by horns, cols, and arêtes. Altitudes
range from sea level to about 600 m on cirque and valley floors, and up
to 1500 m on surrounding summits.

Field study area (2) includes 40 km of Pangnirtung Pass between
the head of Pangnirtung Fiord and Summit Lake (Figs 24.1 & 24.3). This
scenically magnificent through-valley with hanging glaciers and 1200 m
high side walls has become the most visited part of Auyuittuq National
Park. All of the surrounding summits, which reach up to 2100 m asl,
and most of the side valley glaciers in Pangnirtung Pass have been

Figure 24.2 Part of Canadian Government vertical air photograph
A-16255-37 (1958), Neoglacial moraine study area number 1, between 12
and 25 km northeast of Pangnirtung (Fig. 24.1). Pangnirtung Fiord at
lower left. Shadow Lake marked S, just down-valley from Glacier D135
(Fig. 24.11). Soil profile and boulder weathering study sites are
numbered and referenced in Davis (1980). Glaciers numbered according to
<u>Glacier Atlas of Canada.</u>

Figure 24.3 Part of Canadian Government vertical air photograph
A-16817-29 (1959), Neoglacial moraine study area number 2, Pangnirtung
Pass (southern section). Head of Pangnirtung Fiord at lower left.
Turnweather Glacier moraines (Fig. 24.12) in shadow marked 1, Tete de
Cirques Glacier moraines (Fig. 23-13) marked 2, and Tumbling Glacier
moraines (Fig. 23-14) marked 3. Glaciers numbered according to <u>Glacier
Atlas of Canada.</u>

given common names, which for glaciers are included along with their glacier inventory numbers in this chapter. The valley glaciers in Pangnirtung Pass, generally less than 10 km long, commonly drain upland icecaps with altitudes up to 1200 m.

Field study area (3) includes the remote "South America Lake Valley" (Barton and Harben, 1974) located northwest of the head of Kingnait Fiord (Figs 24.1 & 24.4). In this splendid valley, glaciers up to 6 km long originate in cirques with floors about 1000-1200 m asl and terminate on the main valley floor at about 500 m asl. Surrounding summits reach 1800 m asl and valley wall relief is up to 800 m.

PRESENT-DAY ICE MASSES ON CUMBERLAND PENINSULA

Glaciers on Baffin Island's Cumberland Peninsula represent a wide variety of glacier morphologies, with the Penny Ice Cap the dominant glaciological feature. The northwestern portion of this 6000 km^2 ice body is situated at altitudes up to 2000 m on the upland plateau of Baffin Island, whereas the southern part of the ice cap is superimposed on alpine topography and feeds numerous valley glaciers which commonly radiate outward and terminate below 500 m asl. Due to the continental location, the glaciation level and ELA (equilibrium line altitude) of the Penny Ice Cap lie up to a couple of hundred meters above equivalent levels on valley and cirque glaciers in the field study areas (Fig. 24.5; Andrews and Miller, 1972a). Besides the Penny Ice Cap, local ice caps, which vary in size from <1 km^2 to 50 km^2, cover many of the broader and higher summits. Outlet glaciers of variable lengths radiate from these ice caps, as well, and commonly drop over walls up to 1000 m high into cirques and troughs. Hundreds of individual cirques are occupied with glaciers, from <1 km^2 to 15 km^2 in size, whereas numerous other cirques are empty of ice (Williams, 1975).

CLIMATE, GLACIATION LIMITS, AND ELAs

Because proxy climatic records based on lichenometrically dated moraines are the major emphasis of this chapter, comparison of climatic data between the northern and southern sides of the Cumberland Peninsula is relevant. Climatic data are scarce for the Pangnirtung and Kingnait Fiord area, and the only published continuous climatic data from Pangnirtung span the period from 1931 to 1940 (Bradley, 1973, p. 232). These data suggest that Pangnirtung has higher mean annual, mean summer, and mean winter temperatures than do three climate stations situated on northern Cumberland Peninsula. However, Bradley

Figure 24.4 Part of Canadian Government vertical air photograph
A-16817-137 (1959), Neoglacial moraine study area number 3 (South
America Lake Valley), between 20 and 30 km northwest of the head of
Kingnait Fiord (Fig. 24.1). Soil profile and boulder weathering study
sites are numbered and referenced in Davis (1980). Glaciers numbered
according to Glacier Atlas of Canada.

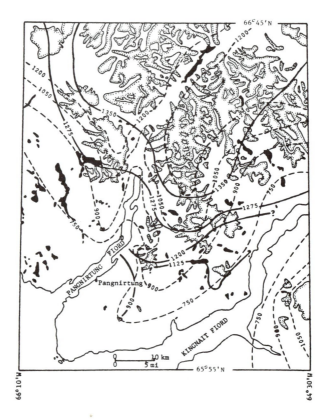

Figure 24.5 Generalized location of glaciers (stippled) and lakes
(solid), Pangnirtung and Kingnait Fiord area, southern Cumberland
Peninsula. Contours of the steady state equilibrium line altitudes
(ELA's) shown by dashed lines (in meters) and contours of the
glaciation limit shown by solid lines (in meters). Expanded from
Anderson (1976).

(1973, p. 234) suggests that ablation season temperatures (June-August)
for Baffin Island were cooler during the late 1960s than they had been
for the previous 30 to 40 yrs. Because the recording periods, as well
as the station altitudes, are not similar between Pangnirtung and
northern Cumberland Peninsula, comparison of climatic data is
difficult, and one cannot prove that there is a pronounced difference
in temperatures. Moreover, although the precipitation data for
Pangnirtung, Broughton Island, and Padloping Island (Fig. 24.1) appear
to be similar, Bradley (1973, p. 235, 238) notes that generally there
has been an increase in precipitation over Baffin Island during the

1960s; thus, the Pangnirtung and Padloping Island precipitation records
from the 1930s and 1940s may be comparatively too low. It is clear
also that precipitation data are strongly influenced by seasonal open
water (vs. sea ice) conditions, as evidenced by twice the mean annual
precipitation at Cape Dyer as at Broughton Island, 150 km to the north
(Fig. 24.1). Although the longer-term climatic data do not demonstrate
conclusively that a climatic gradient exists across the Peninsula,
local variability in temperature was noted during the 1976 summer
season; Parks Canada data also suggest that Pangnirtung averaged about
3°C warmer than Broughton Island in 1976 (E. Sieber, oral
communication, 1977).

Numerous workers have recognized that large areas of Baffin Island
are near full glacial conditions (Tarr, 1897; Ives, 1962; Andrews et
al., 1972). Baffin Island is situated beneath one of two major troughs
in the upper westerlies, thus changes in the amplitude and number of
Rossby waves in the upper level flow may have a major effect on the
climate of Baffin Island (Barry et al., 1977). During the Little Ice
Age (100-300 BP) 30-40 percent of northern Baffin Island was covered by
thin snowbanks, whereas only 2 percent is covered at present (Ives,
1962; Davis and Wright, 1975; Locke and Locke, 1977). Bradley and
Miller (1972) suggest that climatic shifts spanning less than a decade
have recognizable effects on the amount of permanent ice and snow on
Baffin Island. Thus, Baffin Island is a "sensitive" area for studies
of climatic change (Andrews et al., 1972).

Small-scale (1:1,000,000) maps of present glaciation limits and
lowest equilibrium line altitudes (ELAs) for Baffin Island have been
constructed by Andrews and Miller (1972a). A larger-scale map
(1:250,000) showing the present glaciation limits and ELAs for the
Pangnirtung Fiord area also has been constructed (Anderson, 1976, p.
10). This later map has been extended to include the Kingnait Fiord
and Pangnirtung pass areas (Fig. 24.5). The maritime influence of
Pangnirtung and Kingnait Fiords on lowering present glaciation limits
and especially ELAs is apparent. Thus, by comparison with topographic
maps of the Cumberland Peninsula, it appears that lowering the
glaciation limit by only 200-300 m would significantly increase the
area of glacierization.

PREVIOUS INVESTIGATIONS

In 1884 Franz Boas visited Pangnritung Pass and recorded the terminal
positions of several glaciers. Some of these glaciers were
photographed by J.D. Soper when he visited the area in 1924-25. Not

until 1953, when the Arctic Institute of North America sent a
scientific expedition to the Penny Ice Cap, did Thompson (1954, 1957)
describe the landforms and propose a glacial chronology for Pangnirtung
Pass. Thompson (1957) recognized a major phase of through-valley
glaciation of Pangnirtung Pass, followed by local glacier advances
which deposited the Little Ice Age moraines recognized by Boas (1884)
and Soper (1925).

Since 1967 workers from the Institute of Arctic and Alpine
Research (University of Colorado) have been active on the Cumberland
Peninsula. The late-glacial and Neoglacial chronologies provided by
Carrara (1972), Carrara and Andrews (1972), Boyer (1972), Mears (1972),
Miller (1973a, 1975), Dyke (1977, 1979), Birkeland (1978), Bockheim
(1979), Locke (1979, 1980), and Hawkins (1980) are most relevant to
this region. However, lichenometric studies of moraines from northern
Baffin Island (Andrews and Barnett, 1979; Løken and Andrews, 1966;
Harrison 1964, 1966), from southern Baffin Island (Müller, 1980), and
from northern Labrador (McCoy, 1983) also are useful for comparison
with Neoglacial moraine records from the Cumberland Peninsula.
Glaciological studies have been made on the "Boas" Glacier (Fig. 24.1)
since 1969 (Andrews and Barry, 1972; Jacobs et al., 1972; Weaver,
1975), and on Ookalik Glacier (Fig. 24.1) since 1976 (Andrews, 1977,
1978; Davis, 1977). Recent climatic trends for Baffin Island have been
discussed by Bradley and Miller (1972). Little Ice Age snow cover on
Baffin Island has been mapped and subjected to climate modeling
(Andrews et al., 1976; Davis and Wright, 1975; Locke and Locke, 1977;
Williams, 1978a, b). Erosion rates for Cirque glaciers on Cumberland
Peninsula (Andrews, 1972; Anderson, 1978) also are relevant to Holocene
climatic chronologies.

STRATIGRAPHIC NOMENCLATURE

In this chapter late Holocene refers to about the last 4000 C-14 BP.
All ages reported in this chapter are in radiocarbon (C-14) BP, before
1950 AD. The chronostratigraphic term Holocene has the rank of a
series, and thus may be subdivided into stages. Mangerud et al. (1974,
p. 120) suggested that the Holocene and Flandrian Stage are identical
and have the same lower boundary, and that the Flandrian may be broken
up into late, middle, and early substages. Later, Mangerud and
Berglund (1978) recommended that the terms Flandrian and Holocene
should not be used interchangeably in Norden, although they suggested
that the terms late, middle, and early may be retained as subdivisions
of the Holocene in a similar manner to that proposed for the Flandrian.

Such a subdivision would place the late/middle and middle/early
Holocene boundaries at 2500 and 5000 BP (Fig. 24.6), thus this chapter
would concern the later 1500 of the middle Holocene, as well as the
late Holocene. However, as Hopkins (1975) stated, "a complex
time-stratigraphic nomenclature for an episode as brief and
geochronologically well documented as the Holocene Epoch obscures more
than it clarifies."

Although the local stage is unnamed on northern Baffin Island,
Andrews and Ives (1978, p. 625) named substages: Remote Lake

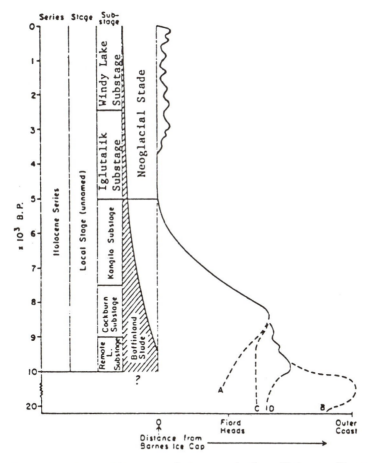

Figure 24.6 Stratigraphic nomenclature and time-distance diagram
of possible models for fluctuations of northeastern margin of the
Laurentide ice sheet and cirque glaciers on Cumberland Peninsula during
the Baffinland and Neoglacial Stades. Modified from Andrews and Barnett
(1979). Cirque glacier variations during Neoglacial Stade based on
estimated relative distances beyond present ice marginal positions.

(10,000-9000 BP), Cockburn (9000-7500 BP), and Kangilo (7500-5000 BP);
these parallel the Fennoscandian chronozones Preboreal, Boreal, and
Atlantic (Godwin, 1966), respectively. Cockburn age moraines are also
recognized on both the northern side (Miller, 1973a, 1975) and the
southern side of the Cumberland Peninsula (Dyke, 1977, 1979).
Substages equivalent to the Sub-boreal and Sub-atlantic Fennoscandian
chronozones spanning the last 5000 are named on Baffin Island as the
Iglutalik and Windy Lake substages (Andrews, 1982).

For the purpose of this chapter, geologic-climatic and lithostratigraphic terms are more useful than time-stratigraphic
nomenclature. The terms Drift and Glacial Drift Sheet apply to surface
litho-morphostratigraphic units (Porter, 1976; Karlén, 1973; Denton,
1974; Denton and Karlén, 1977). Drift units may be differentiated by
surface weathering of boulders, rind thickness of clasts, soil profile
development, and lichen diameters, as well as morphology and geographic
position. The Code of Stratigraphic Nomenclature (1970) makes no
provision for the use of physical characteristics that result from
postdepositional modification to define litho-morphostratigraphic units
(Birkeland, personal communication, 1979), although Ruhe (1974)
proposed use of the soil-geomorphic unit, which is in part a time unit
and in part a litho-morphostratigraphic unit, but more importantly a
mapping unit. On northern Baffin Island, Andrews and Ives (1978)
defined Baffinland Drift as an extensive, mappable litho-
morphostratigraphic unit composed of moraines, till, outwash, and
glaciomarine deposits. The Baffinland Drift is the basis for the
Baffinland Stade, a geologic-climatic unit with a retreat phase that is
radiocarbon dated between 9000 and 5000 yrs BP (Andrews and Ives,
1978). Although Andrews and Ives (1978, p. 625, Fig. 4) show the
Baffinland Stade extending to the present, I prefer to use the term
Neoglacial Stade for the last 5000 of geologic-climatic history on
Baffin Island (Fig. 24.6). In fact, Andrews and Miller (in press) have
proposed that the Baffinland Stade which is equivalent to the end of
the Foxe Glaciation, terminate 5000 BP.

Neoglacial (Moss, 1951) is a geologic-climatic term defined as
"the climatic episode characterized by rebirth and/or regrowth of
glaciers following maximum shrinkage during the Hypsithermal interval"
and "may fall anywhere within the last five millenia" (Porter and
Denton, 1967, p. 205). Deevey and Flint (1957, p. 12) defined the
Hypsithermal as the period of time allowing formation of pollen zones V
through VIII in the Danish System, beginning about 9000 ago and ending
about 2500 ago in Europe (Godwin, 1966). Heusser (1966, p. 124),
however, suggested that the transition to Neoglacial climate was
time-transgressive with latitude in northwestern North America.

Indeed, on Baffin Island the Hypsithermal ended and the Neoglacial began 4000 to 5000 ago.

In most parts of the world Holocene glacier fluctuations were minor compared to ice advances of the Wisconsin glacial age, and thus may not be worthy of stade designation. Porter (1974) suggested that such events should be referred to informally as advances. The term "Advance" is adopted in this chapter for individual periods of glacier fluctuation during the Neoglacial Stade on southern Cumberland Peninsula, as well as across northern Cumberland Peninsula (Miller, 1973a, 1975) and around the Barnes Ice Cap (Andrews and Barnett, 1978). Although Porter and Denton (1967, p. 181) rejected the phrase, Little Ice Age has become firmly entrenched in recent literature in reference to the climatic deterioration of the last several centuries, or the late Neoglacial. On the Cumberland Peninsula of Baffin Island, moraines, about 100 old suggest that many glaciers reached their maximum Holocene positions during the Little Ice Age.

MORAINES ON SOUTHERN CUMBERLAND PENINSULA

General Description

Moraines fronting 22 glaciers on southern Cumberland Peninsula were investigated. These glaciers ranged from 0.10 to 14.55 km^2 in area, from 0.20 to 9.75 km in length, from 815 to 1110 m asl in steady state ELA, from 71 to 920 m asl in snout altitude, from 2.5 to 30.0 km distance to coast (fiord), and through all quadrants of the azimuth. Histograms for the ranges of four of the physical parameters appear in Figure 24.7 and the data for these histograms, as well as for other physical parameters, are summarized in Table 24.1. Similar data for 49 glaciers studied by Miller (1973a, 1975) on northern Cumberland Peninsula are also summarized in Figure 24.7. The advantage of studying moraines from a large number of glaciers with a wide range of characteristics is the probable cancellation of any specific local effects on individual glaciers which might respond differently to climatic changes (Meier, 1965).

Neoglacial moraines that lie at the margins of cirque and valley glaciers on Baffin Island are usually large, steep, high ridges containing ice cores, whereas end moraines of the Baffinland Stade of the Foxe Glaciation are generally low, broad features that apparently lack an ice-core. Miller (1973a) suggested that moraines of Cockburn age (9000-8000 BP) lost their ice cores during the mid-Holocene warm

TABLE 24.1. Size, orientation, estimated ELA, Snout Altitude, and
Distance to Coast for the Glaciers Investigated.

Glacier Name and/or Inventory number (area 46204-)	Area[1] (km^2)	Length (km)	Orien-tation[2]	Esti-mated ELA[3] (m asl)	Snout Alti-tude (m asl)	Distance to Coast[4] (km)
C15	0.25	0.20	N59E	845	725	7.0
C16	2.10	2.10	N55E	900	755	6.0
C18	1.95	3.05	N63E	1010	790	6.0
D121	1.80	2.40	N47E	870	755	6.0
D124	2.55	3.30	N10W	945	455	4.5
D128	0.40	0.90	S70W	1030	900	3.0
D129	0.65	1.70	N72W	885	665	2.5
D135	1.30	2.10	N5E	845	665	5.0
DI35b	0.10	0.10	S76W	870	800	4.5
D136	1.65	2.75	N47W	870	575	5.0
D137	0.75	1.40	S67W	1030	900	3.0
D137b	0.15	0.75	N80W	1060	920	3.5
D138	1.50	3.70	S71W	970	545	4.5
Turnweather (D78)	14.75	9.75	N86W	1000	500	9.0
Tete de Cirques (69)	8.15	8.25	S57W	1030	150	10.0
Tumbling (D178)	12.95	5.50	N67E	1110	71	10.0
Windy (D7, 8, 9)	10.50	5.00	S11E	1000	240	17.5
Caribou (D24)	10.00	7.25	S42E	1050	635	30.0
Spire (C147)	4.95	5.57	N82E	815	575	16.0
Crown (C192)	2.10	3.50	S39E	1050	760	21.0
Throne (C156)	1.15	2.20	N46W	950	665	21.0
Odlid (C145)	0.65	2.30	N86E	880	580	14.0

[1] Accuracy of the area measurements is ±2%.
[2] Orientation defined by a line trending parallel to the major
   directions of flow in the accumulation area.
[3] Transient ELAs estimated by air photograph (1956 series) and
   topographic map interpretation, and by field observation of
   snowline during July-August 1975 and 1976.
[4] Distance to coast measured from glacier Ela to nearest fiord
   coast in km.

interval, although he recognized that some moraines of Cockburn age
receiving little solar radiation have retained their ice cores to the
present. Carrara (1975) recognized that the outermost moraine fronting
the Akudnirmuit glacier on northern Cumberland Peninsula does not
retain an ice core, whereas all three inner moraines are ice-cored
within 40 cm of the debris surface. The presence of ice-cores is
important for understanding moraine stabilization and lag times in
lichen colonization.

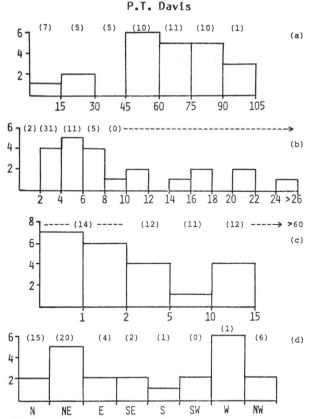

Figure 24.7 Histograms for the range of physical parameters of 22 glaciers investigated on southern Cumberland Peninsula for delimiting a Neoglacial chronology: (a) altitude of the glacier snouts (x 10 m), (b) distance from the coast (glacier ELA to fiord, x 10 km), (c) glacier area (km²), and (d) orientation. The vertical scale indicates the number of glaciers within each class. See Table 24.1 for data by glacier name and/or inventory number. Numbers of glaciers studied in each category on northern Cumberland Peninsula by Miller (1975) shown in parentheses.

Lichenometry

General theory. Absolute dating techniques such as dendrochronology, radiocarbon dating of organic material within and underlying till, and historical records are of limited use in the eastern Canadian Arctic. Organic material has not been found within or underlying Neoglacial till on the Cumberland Peninsula, although Stuckenrath et al. (1979) discuss a series of radiocarbon ages from organic sands that underlie possible early Neoglacial outwash from Thompson Glacier (No. 4620M-119) in central Pangnirtung Pass (Fig. 24.1). The coarse fraction (>125 μm) radiocarbon ages (2920±100 and 2715±75 BP, SI-2071) are about 1000 yrs younger than the fine fraction (<125 μm) radiocarbon ages (3825±85 and 3810±85 BP, SI-2071); however, all four ages on the possible advance of Thompson Glacier are compatible with ages of early Neoglacial advances determined by lichenometry elsewhere on Cumberland Peninsula (Miller, 1973a; Davis, 1980). Andrews et al. (1976) summarized four radiocarbon ages on dead mosses and lichens from former snowfield sites. Extensive areas that appear as light tones on LANDSAT-1 satellite imagery have been mapped and interpreted as areas of expanded, thin snow cover on Baffin Island during the Little Ice Age (300-0 BP; Table 24.2). However, an additional radiocarbon age (1025±100 BP, SI-2550) obtained by Andrews (1976) indicates that some expanded snowbanks also predate the Little Ice Age. Although relative dating techniques, including boulder weathering, soil profile development, and moraine slope angle, were used by most workers on Baffin Island, lichenometry proves to be far more useful for distinguishing one Neoglacial moraine's age from another. However, relative dating data are useful for distinguishing Neoglacial moraines from Cockburn age deposits.

Moraines with heavy lichen cover, and larger lichen diameters, are recognized as darker toned features on black and white air photographs. These multicrested moraines with variable lichen cover and sizes are the focal point of Neoglacial lichenometric moraine studies on Baffin Island (Fig. 24.8). Assumptions and restrictions concerning lichenometry for dating of rock substrates such as moraines have been reviewed on numerous occasions (Beschel, 1950; Andrews and

TABLE 24.2. Radiocarbon dates on dead willow, moss and lichens from former snowfield sites on Baffin Island.

Site[a]	Radio-carbon Date (Yrs BP)[b]	Range (±2σ) (Yrs BP)	Laboratory Number	Reference
1	330±75	180–480	I-1204	Falconer, 1966
2	330±90	150–510	GaK-3099	Carrara and Andrews, 1972
3	120±70	0–260	GaK-4835	Andrews, 1975
4	0±90	0–180	GaK-3100	Andrews and Miller, 1972b
5	1025±100	825–1225	SI-2550	Andrews, 1976
6	0±100	0–200	SI-2548	Andrews, 1976

[a] See Figure 24.1 for locations.
[b] Radiocarbon dates of 0 yrs BP may be as old as 500 yrs BP, because Stuiver (1978) demonstrated that radiocarbon dates 0-500 yrs old can have multiple ages.

Figure 24.8 Looking south at moraines fronting Spire Glacier (C147) in South America Lake Valley, moraine study area number 3 (Figs 24.4 & 24.15). Large, steep, high ice-cored Neoglacial moraines overlie small, subdued, vegetated Cockburn age moraines at left and right (out of view) sides of the photograph. The Neoglacial moraines represent at least eight advances of Spire Glacier and are conducive to lichenometric study.

Webber, 1964; Benedict, 1967; Locke et al., 1979; and numerous workers in Arctic and Alpine Research, 1973, v. 4, no. 3). The largest lichen thalli or percentage cover of lichens on boulders may be used as relative age indicators for moraine substrates. Only the diameters of circular lichens or the shortest diameters of non-circular or intergrown lichens are measured (Fig. 24.9). A critical assumption is that the largest lichen on a moraine began to grow during moraine stabilization immediately following a glacier advance. To use lichen diameters as absolute age indicators for substrates, a growth curve for the lichen species must be established for each geographic region.

Lichen growth curves. Preferably a slow-growing species such as Rhizocarpon geographicum is chosen for the lichen growth curve. Miller and Andrews (1972) developed lichen growth curves for Rhizocarpon geographicum sensu lato and Alectoria minuscula for northern Cumberland Peninsula. R. geographicum s. l. is used in the general sense because taxonomy of the group is too complex to distinguish to species level in

Figure 24.9 Rhizocarpon geographicum sensu lato on an outer Neoglacial
moraine fronting Throne Glacier (C156) in South America Lake Valley,
moraine study area number 3 (Fig. 24.4). The maximum R. geographicum
s. l. thallus diameter on the boulder in the photograph is 95 mm, thus
the moraine stabilized about 3250 ^{14}C yrs BP (Fig. 24.10). Either
circular lichens or the shortest dimension of non-circular or
intergrown lichens provide maximum thallus diameters in this study, as
exhibited above.

the field. Construction of the R. geographicum s. l. growth curve for
northern Cumberland Peninsula was based on three independent approaches
(Miller and Andrews, 1972): (1) historical records (whalers' graves),
(2) interspecific ratios between R. geographicum s. l. and direct
measurements of the faster growing Alectoria minuscula, and (3)
radiocarbon ages related to initial R. geographicum s. l. colonization
of rock substrates. Miller (1973b) further supported curve of R.
geographicum s. l. with additional interspecific ratio data and another
radiocarbon age from northern Cumberland Peninsula (Table 24.3).

    Andrews and Barnett (1979) revised a growth curve for R.
geographicum s. l. from the area around the Barnes Ice Cap through the
addition of three radiocarbon ages (Table 24.3). The growth curves
from the Barnes Ice Cap area and northern Cumberland Peninsula are
remarkably similar, thus one general growth curve for R. geographicum
s. l. on Baffin Island is used (Fig. 24.10). This lichen growth curve
is unique because it includes five radiocarbon ages as control points

TABLE 24.3. Radiocarbon dates used as control points for growth curve
of Rhyzocarpon geographicum sensu lato on Baffin Island (see Fig.
24.10).
==

Laboratory Number	Radiocarbon Date (C-14) BP	Location and Interpretation	Reference
GaK-3722	680±80	Date on seal bone from floor of house ruins from Thule culture. Thule houses were sodded, prohibiting lichen growth until abandoned; therefore, maximum R. geographicum diameter of 43 mm began growing about 680 years ago. Although Miller (1973a) included this date to modify the lichen growth curve, he now feels that the date is about 500 years too young and that the original curve should be retained (oral comm., 1977). Possibly the site was reoccupied at the later time (680 yrs BP) by seal hunters.	Miller (1973a)
GaK-1992	2400±90	Date on seal bone from floor of house ruins of Dorset culture. Maximum R. geographicum diameter of 80 mm from an abandoned house wall began growing about 2400 years ago, a date consistent with other archeological evidence for the Dorset culture in the eastern Canadian Arctic.	Miller and Andrews (1972)
GSC-1304	2520±150	Date on detrital organics collected from deltaic sediments related to raised shorelines of Generator Lake at southeast margin of Barnes Ice Cap. Maximum R. geographicum diameter on shorline is 84 mm.	Andrews and Barnett (1979)
GSC-1315	2620±150	Date on detrital organics collected from deltaic sediments related to raised shoreline of Generator Lake at southeast margin of Barnes Ice Cap. Maximum R. geographicum diameter on shoreline is 71 mm.	Andrews and Barnett (1979)
GSC-1276	3090±170	Date on detrital organics collected from deltaic sediments related to raised shoreline of Generator Lake at southeast margin of Barnes Ice Cap. Maximum R. geographicum diameter on shoreline is 110 mm.	Andrews and Barnett (1979)

TABLE 24.3.   (continued)

Laboratory Number	Radiocarbon Date (^{14}C Yrs BP)	Location and Interpretation	Reference
GaK-5479	8980±180	Date on shells within marine sands related to moraine at head of Quajon Fiord. Maximum R. geographicum diameter on moraine is 280 mm. Although the relationship between the marine sands and the moraine is inferential, an extrapolation of the lichen growth curve based on the above data points yields an age of 9800 C-14 BP for a 280 mm diameter lichen.	Andrews and Barnett (1979)

GaK = Gakushuin University, Tokyo, Japan. GSC = Geological Survey of Canada, Ottawa, Canada.

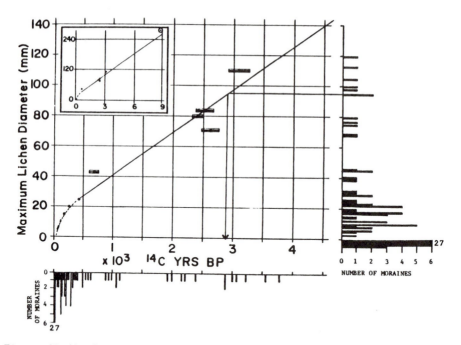

Figure 24.10   Growth curve for Rhizocarpon geographicum sensu lato on Baffin Island. Radiocarbon dates used as control points for growth curve shown by rectangles and circled dot on insert, and described in Table 24.3. Other control points for younger part of growth curve discussed in Miller (1973a, 1973b). Ages of Neoglacial moraines on southern Cumberland Peninsula also shown. Thallus diameters and nomenclature for Neoglacial advances given in Table 24.5.

for the older part of the curve, rather than the usual one or two
control ages. Moreover, the linear part of the Baffin Island curve
represents a R. geographicum s. l. growth rate of about 0.03 mm/yr,
similar to the linear parts of R. geographicum growth curves developed
for Swedish Lappland (Karlén, 1973; Denton and Karlén, 1973; 0.04
mm/yr) and the St. Elias Mountains in Alaska/Yukon (Denton and Karlén,
1977; 0.035 mm/yr).

MAPPING AND DATING NEOGLACIAL MORAINES ON SOUTHERN CUMBERLAND PENINSULA

Method

Neoglacial moraines fronting 22 glaciers were mapped and
lichenometrically dated in three field areas on southern Cumberland
Peninsula. Errors in measuring the maximum diameter of R. geographicum
on a moraine were ±1.5 mm, or ±50 yrs for the linear part of the R.
geographicum growth curve for Baffin Island (Fig. 24.10). Total lichen
cover has not been calibrated on an absolute time scale for moraines on
Baffin Island, thus the maximum lichen cover on a boulder surface was
recorded only as a relative age check between moraines. In only one
case was there an inconsistency between the largest diameter
R.geographicum and maximum percentage cover. There were far more
inconsistencies between the diameters of R. geographicim and A.
minuscula, therefore, only R. geographicum diameters were used for
absolute dating of moraines. Although searches for largest lichen
diameters were carried out over entire moraine surfaces, the largest
R. geographicum was usually found on a large boulder close to a moraine
crest, probably because those areas are: (1) well-drained, (2)
relatively more stable than the moraine slopes, and (3) less influenced
by late-season snowbanks and microclimates. Table 24.4 summarizes
between-worker comparison of lichen data from Turnweather and Tete de
Cirques glaciers in moraine study area 92 (Figs 24.1 & 24.3). Although
comparison of total lichen cover is difficult because different
measuring methods were employed, Davis (1980) universally found larger
R. geographicum. This may be the result of different techniques for
measuring lichen diameters, but more probably a result of differential
person/hours used in search for the largest lichens; the author and an
assistant invariably spent a full day or more on a moraine, whereas
Birkeland and Locke spent only a few hours searching for the largest
lichen (Birkeland, oral communication, 1975). Of the 22 moraine
systems studied on southern Cumberland Peninsula, five are discussed in
detail in this chapter.

TABLE 24.4. Between-worker comparison of lichen data from Turnweather and Tete de Cirques Glaciers,[a], [b].

	Maximum thallus diameter (mm)		Total Lichen Cover (%)
	R. geographicum	A. minuscula	
Turnweather Glacier (46204-D78)			
Crest #1 (right lateral)[a], [b]			
PWB/WWL	15	52	3.9 (aver.)
PTD	22	77	30 (max.)
Crest #2 (right lateral)[a], [b]			
PWB/WWL	30, 35	115	45-60 (range)
PTD	39	155	70 (max.)
Crest #3[c] (left lateral)			
PTD	10	54	20 (max.)
Crest #4[c] (left lateral)			
PTD	18	55	20 (max.)
Tete de Cirques Glacier (46204-D69)			
Crest #1			
PWB/WWL	0	0	0
PTD	0	0	0
Crest #2			
PWB/WWL	13, 14	75, 64	2.0, 14.8 (aver.)
PTD	17	68	20 (max.)
Crest #3[d]			
PTD	12	50	10 (max.)

[a] PWB/WWL = 1974 study by Peter W. Birkeland and William W. Locke (Birkeland, P.W., 1978).
[b] PTD = 1975-1976 study by author.
[c] Two additional moraine crests on north side of valley studied by author. Neither of these two moraine crests could be traced stratigraphically to the two moraine crests on the south side of the valley, and all four moraine crests yielded different maximum R. geographicum thallus diameters, therefore all four moraine crests could represent glacier expansions of different ages.
[d] Additional moraine crest studied by author just inside outermost moraine crest yielded maximum R. geographicum thallus diameters different than lichens measured on the other two moraine crests.

Moraine Study Area (1)

All 13 glaciers in moraine study area (1) (Fig. 24.2) were visited during the 1975 field season, and one glacier (D135) was revisited in 1976 (Fig. 24.11). The largest R. geographicum found in this area was 40 mm diameter (900 yrs BP), and occurs on the outermost of seven distinct moraine crests fronting Glacier D137. Six of the 13 glaciers in this area have largest R. geographicum between 24 and 31 mm diameter (350-625 yrs BP). However, the majority of moraines in this area have largest R. geographicum <12 mm diameter, indicative of moraine

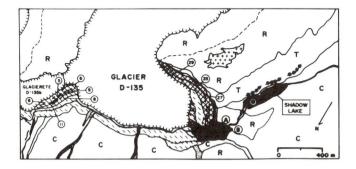

EXPLANATION

GLACIAL DRIFT SHEETS (Maximum *R. geographicum* diameters in mm)		AGE ^{14}C YRS BP
0	Cumberland Advance I	≤100
≤12		
17-26	Cumberland Advance II	200-400
28-32	Cumberland Advance III	500-650
39-46	Pangnirtung Advance	900-1150
68-69	Kingnait Advance I	1900-2000
75-80	Kingnait Advance II	2200-2400
no. 95-100, 105	Snow Creek Advance	2900-3100, 3250
no. 120 (?)	Earliest Neoglacial	3800 (?)
soil data	Baffinland	5000-10,000

R   Bedrock

⌒⌒⌒   Glacier Margin

⌒⌒⌒   Moraine Ridge

⌒⌒⌒   Protalus Ridge

⌒⌒⌒   Mudflow Levee

T   Talus

C   Colluvium

⊙   Alluvium

╱   Geologic Contact (commonly located approximately)

⟨69⟩   Maximum *R. geographicum* diameter (mm)

Ⓐ   Soil Pit (see text for profile description)

497   Altitude (m)

⟍   Stream, permanent or intermittent

+ +
+ + + Snowfield

Figure 24.11. Part of Canadian Government vertical air photograph A-16255-37 (1958) and map of Neoglacial deposits fronting Glacier #46204-D135 and D135b, moraine study area number 1 (Fig. 24.2), southern Cumberland Peninsula, Baffin Island. Location of Shadow Lake sediment coring site marked by X. Soil profiles labeled A and B referenced in Davis (1980). Drainage divide and ridge crest marked by dashed line.

stabilization only within the last 100 years. The glaciers in area (1) are generally smaller and closer to the coast than glaciers in the other two moraine study areas (Table 24.1). These two factors may result in more snowfall, higher mass turnover rates, and smaller moraines. Thus, Little Ice Age glacier advances (<100 yrs BP) in this area more easily obliterated the small moraines of earlier Neoglacial and Cockburn age advances.

Moraine Study Area (2)

In Pangnirtung Pass (Fig. 24.3) four glaciers were visited during both the 1975 and 1976 field seasons, and one (Tumbling Glacier) was studied only during the 1975 field season. In addition, P.W. Birkeland and W.W. Locke also studied moraines fronting Turnweather and Tete de Cirques Glaciers in 1974 (Birkeland, 1978; Locke, 1979). Two major moraine crests were mapped on each side of Turnweather Valley (Figs 24.3 & 24.12), but the four moraine ridges could not be traced laterally into one another because of stream dissection. The largest R. geographicum diameters were different on all four deposits, with the largest R. geographicum occurring on the outer left lateral moraine crest (numerous 39 mm lichens; 900 BP). R. geographicum (10, 18, 22 mm diameters) on the other three moraines all suggest moraine stabilization during the last 300 years. A fifth minor moraine without lichen cover lies close to the glacier margin. Turnweather Glacier's marginal position is 2 km distance and 1000 km above its major moraines on the floor of Pangnirtung Pass, unlike most present-day glacier terminal positions which generally lie within a few 100 m of their Little Ice Age limits on Baffin Island. Of three continuous moraines fronting Tete de Cirques Glacier (Figs 24.3 & 24.13) only two are colonized with lichens, and probably both of these stabilized during the last 200 years (12 and 17 mm diameter R. geographicum).

    Tumbling Glacier (Figs 24.3 & 24.14) calves into Crater Lake which is enclosed by a complex system of moraines. Lichenometrically, the oldest Neoglacial moraines lie on the down-valley side of the lake and moraine complex (23 mm R. geographicum, 350 BP). This relative position on the valley floor is opposite to older Neoglacial moraines on the up-valley sides of glaciers on northern Cumberland Peninsula (Miller, 1975). Small R. geographicum diameters (6 mm maximum) suggest that all other Tumbling Glacier moraines have stabilized very recently (<50 BP); however, strong prevailing winds laden with sand and silt blowing down-valley through Pangnirtung Pass could cause slower lichen colonization on the windward (up-valley) moraines than on the leeward (down-valley) moraines.

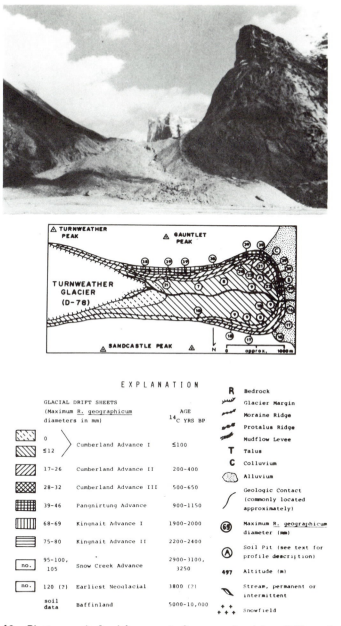

EXPLANATION

GLACIAL DRIFT SHEETS
(Maximum R. geographicum diameters in mm)

		AGE 14C YRS BP
0	Cumberland Advance I	≤100
≤12		
17-26	Cumberland Advance II	200-400
28-32	Cumberland Advance III	500-650
39-46	Pangnirtung Advance	900-1150
68-69	Kingnait Advance I	1900-2000
75-80	Kingnait Advance II	2200-2400
95-100, 105	Snow Creek Advance	2900-3100, 3250
120 (?)	Earliest Neoglacial	3800 (?)
soil data	Baffinland	5000-10,000

R   Bedrock
     Glacier Margin
     Moraine Ridge
     Protalus Ridge
     Mudflow Levee
T   Talus
C   Colluvium
     Alluvium
     Geologic Contact (commonly located approximately)
(65) Maximum R. geographicum diameter (mm)
(A) Soil Pit (see text for profile description)
497 Altitude (m)
     Stream, permanent or intermittent
+ + +
+ + + Snowfield

Figure 24.12 Photograph looking east from west side of Weasel River and map of Neoglacial deposits fronting Turnweather Glacier (#46204-D78), moraine study area number 2 (Fig. 24.3), Pangnirtung Pass, southern Cumberland Peninsula, Baffin Island. Soil profiles at sites labeled A, B, and C referenced in Davis (1980). Turnweather Peak's 1500 m vertical southwest face in background. Relevant part of Canadian Government air photograph A-16817-29 (1959) in shadow (Fig. 24.3).

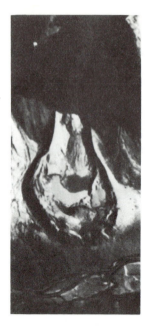
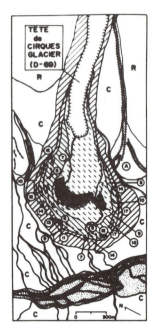

EXPLANATION

GLACIAL DRIFT SHEETS
(Maximum *R. geographicum*
diameters in mm)

AGE
^{14}C YRS BP

Drift	Diameter	Advance	Age
	0	Cumberland Advance I	≤100
	≤12		
	17-26	Cumberland Advance II	200-400
	28-32	Cumberland Advance III	500-650
	39-46	Pangnirtung Advance	900-1150
	68-69	Kingnait Advance I	1900-2000
	75-80	Kingnait Advance II	2200-2400
no.	95-100, 105	Snow Creek Advance	2900-3100, 3250
no.	120 (?)	Earliest Neoglacial	3800 (?)
soil data		Baffinland	5000-10,000

R — Bedrock
— Glacier Margin
— Moraine Ridge
— Protalus Ridge
— Mudflow Levee
T — Talus
C — Colluvium
— Alluvium
/ — Geologic Contact (commonly located approximately)
(69) — Maximum *R. geographicum* diameter (mm)
Ⓐ — Soil Pit (see text for profile description)
497 — Altitude (m)
— Stream, permanent or intermittent
+ + + — Snowfield

Figure 24.13   Part of Canadian Government vertical air photograph
A-16817-29 (1959) and map of Neoglacial deposits fronting Tete de
Cirques Glacier (#46204-D69), moraine study area number 2 (Fig. 24.3),
Pangnirtung Pass, southern Cumberland Peninsula, Baffin Island. Fan
deposit labeled A studied by Birkeland and Locke, and believed to be
late-Foxe age (Birkeland, 1978).

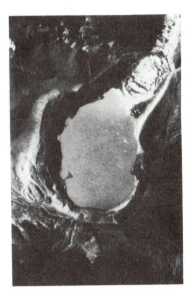
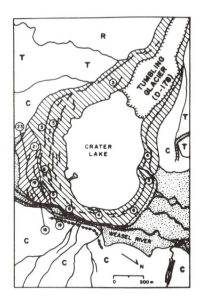

EXPLANATION

GLACIAL DRIFT SHEETS

(Maximum <u>R. geographicum</u> diameters in mm)

		AGE ^{14}C YRS BP
0 ≤12	Cumberland Advance I	≤100
17-26	Cumberland Advance II	200-400
28-32	Cumberland Advance III	500-650
39-46	Pangnirtung Advance	900-1150
68-69	Kingnait Advance I	1900-2000
75-80	Kingnait Advance II	2200-2400
no.	95-100, 105	Snow Creek Advance	2900-3100, 3250
no.	120-(?)	Earliest Neoglacial	3800 (?)
soil data	Baffinland	5000-10,000

R — Bedrock

Glacier Margin

Moraine Ridge

Protalus Ridge

Mudflow Levee

T — Talus

C — Colluvium

Alluvium

Geologic Contact (commonly located approximately)

69 — Maximum <u>R. geographicum</u> diameter (mm)

A — Soil Pit (see text for profile description)

497 — Altitude (m)

Stream, permanent or intermittent

Snowfield

Figure 24.14 Part of Canadian Government vertical air photograph A-16817-29 (1959) and map of Neoglacial deposits fronting Tumbling Glacier (#46204-D178), moraine study area number 2 (Fig. 24.3), Pangnirtung Pass. southern Cumberland Peninsula, Baffin Island.

Moraine Study Area (3)

Four complex moraine sequences fronting Spire, Throne, Crown, and Odlid
Glaciers in South America Valley (Fig. 24.4) were mapped and
lichenometrically dated during a three-week period in 1976. The outer
ice-cored moraines of Spire Glacier are in places more than 80 m high.
H. Bleuer and J. Amatt in 1975 described the moraines around Spire
Glacier as the largest they had ever seen during the course of numerous
mountaineering expeditions throughout the world. P.S. Marshall (oral
communication, 1976) likened the Spire Glacier moraines to those
fronting the Kaskawalsh Glacier in Alaska. The simplest explanation
for preservation of earlier Neoglacial moraines in South America Lake
Valley is that the presence of large moraines of Cockburn age, without
ice-cores, confined subsequent glacier advances. Thus, in South
America Lake Valley three of the four glaciers studied (not Odlid
Glacier) exhibit both large, ice-cored Neoglacial moraines, as well as
low, non-ice-cored moraines formed during the Baffinland Stade.

The lichenometric ages of moraines in the South America Lake
Valley represent the most complete record of Neoglacial activity on
southern Cumberland Peninsula studied to date. Moraines fronting Spire
Glacier (Fig. 24.15) stabilized about 3000-2900, 2300, 1900, 1150, 650,
400, and <100 yrs BP (largest R. geographicum 98-95, 77, 68, 45, 32,
25, and 10-0 mm diameters, respectively). In South America Lake
Valley, the relative extents of glacier expansion during the Neoglacial
were of similar magnitude, probably due to the inability of the
glaciers to break through large moraines formed during previous glacier
advances.

Summary of Neoglacial Moraine Record on Southern Cumberland Peninsula

The age groupings of Neoglacial moraines on southern Cumberland
Peninsula are derived from the growth curve for Rhizocarpon
geographicum sensu lato in Figure 24.10. Ages of Neoglacial moraines
studied on southern Cumberland Peninsula are summarized in Figure
24.10. One controversial moraine in Moon Peak Valley (Fig. 24.1), with
a largest R. geographicum 113 mm diameter mapped by Birkeland and
Locke, is also included in Figure 24.10. Although Birkeland (1978)
noted a 65+ cm soil oxidation depth suggesting an early-Foxe age for
the moraine, Locke (oral communication, 1978) found that etching on
hornblende grains from the soil profile was consistent with a
Neoglacial age for the deposit.

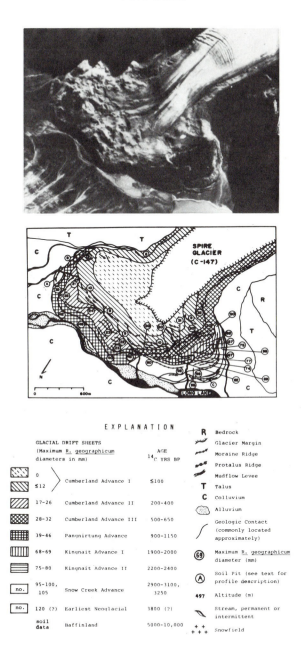

Figure 24.15 Part of Canadian Government vertical air photograph
A-16817-137 (1959) and map of Neoglacial and Baffinland Stade deposits
fronting Spire Glacier (#46204-C147), moraine study area number 3
(Fig. 24.4), southern Cumberland Peninsula, Baffin Island. Soil
profiles labeled A, B, and C referenced in Davis (1980).

NEOGLACIAL CHRONOLOGY IN THE EASTERN CANADIAN ARCTIC

Neoglacial moraines mapped and lichenometrically dated in this study
are compared with moraines similarly studied on northern Cumberland
Peninsula (Miller, 1973a, 1975), around the Barnes Ice Cap (Andrews and
Barnett, 1979), and in the Torngat Mountains of northern Labrador
(McCoy, 1983. A schematic time/distance diagram of late Holocene
glacier margins (Fig. 24.16) suggests potential time lags between
initiation of a glacier advance and stabilization of moraines caused by
glacier stillstand or retreat. Neoglacial moraines from the four areas
studied in the eastern Canadian Arctic are all dated by lichenometry
and thus incorporate an unknown time lag following advance.

    Broad synchrony for moraine stabilization occurs in the three
regions of Baffin Island and in one area of northern Labrador (Fig.
24.17). Although the Dorset Advance of northern Cumberland Peninsula
(Table 24.5) is not recognized on southern Cumberland Peninsula,
moraines from many glaciers in these two areas represent synchronous
glacier behavior during the Neoglacial. Fewer Neoglacial moraines were
examined around the Barnes Ice Cap and in northern Labrador; however,
most lichenometric data from these two areas suggest synchronous
Neoglacial fluctuations.

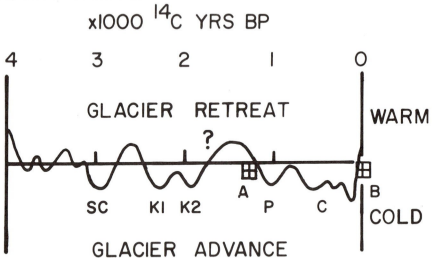

Figure 24.16 Schematic time-distance diagram for Neoglacial
fluctuations on southern Cumberland Peninsula. Late Holocene glacier
advances labeled SC (Snow Creek), K1 and K2 (Kingnait), P
(Pangnirtung), and C (Cumberland). Hypothetical radiocarbon dated soil,
wood, or moss buried by advancing ice such as for situations in
Switzerland (Schneebeli and Röthlisberger, 1976; Röthlisberger and
Schneebeli, 1979) labeled A; hypothetical lichenometrically dated
ice-cored moraine surface such as for situations on Baffin Island
labeled B. The lacking moraines of the Dorset advance on northern
Cumberland Peninsula (Miller, 1975) labeled "?".

TABLE 24.5. Comparison of Minimum Thallus Diameters of Largest Rhizocarpon geographicum and associated moraine ages from three areas of Baffin Island.

		Southern Cumberland Peninsula (this study)			Northern Cumberland Peninsula (Miller, 1973a, 1975)			Barnes Ice Cap (Andrews and Barnett, 1979)		
		R. geog. mm	#[a]	yrs BP	R. geog. mm	#	yrs BP	R. geog. mm	#	yrs BP
CUMBERLAND ADVANCE	I	0-12	43	<100	0-15	60	<150	10	2	100
	II	17-26	19	200-400	20-25	15	250-400	18-20	6	200-250
	III	28-32	6	500-650	31-35	5	625-800	30-32	2	600-650
					37-38	2	825-875	35-39	2	800-900
PANGNIRTUNG ADVANCE		39-46	6	900-1150	41-44	3	950-1050			
					48[b]	1	1200			
DORSET ADVANCE					55-58	9	1500-1600	53-54	2	1400-1450
					60[b]	1	1650	60	1	1650
KINGNAIT ADVANCE	I	68-69	2	1900-2000	67-70	3	1850-2050	70	2	2050
	II	75-80	3	2200-2400	80[c]	1	2400	85	1	2550
SNOW CREEK ADVANCE		95-100	4	2900-3100	95-100	3	2900-3100	90	2	2750
		105	1	3250				100	1	3100
		113?[d]	1	3600	1152?[b]	1	3650	133	1	4300
		120	1	3800						

[a] Number of moraines.
[b] From Carrara and Andrews (1972).
[c] From Mears (1972).
[d] From W.W. Locke (oral communication, 1978).

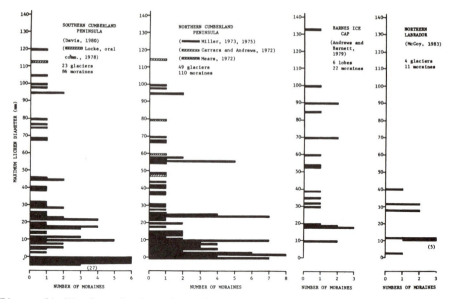

Figure 24.17 Correlation chart for Neoglacial moraine records from three areas on Baffin Island and another area in northern Labrador. Nomenclature for Neoglacial advances given in Table 24.5.

The names Kingnait and Pangnirtung Advances are proposed for lichenometrically dated moraines on southern Cumberland Peninsula not previously emphasized on northern Cumberland Peninsula (Table 24.5). The Kingnait Advances (2400-2200 and 2000-1900 BP) and Pangnirtung Advance (1150-900 BP) include representative moraines in all four study areas in the eastern Canadian Arctic, with the exception of moraines dating to the Kingnait Advances in northern Labrador and the Pangnirtung Advance around the Barnes Ice Cap.

The Neoglacial moraine chronology for Baffin Island resembles glacier records from Swedish Lappland and Alaska-Yukon, although the Baffin Island moraines show many more events than accommodated by a simple 2500-yr periodicity.

REFERENCES

American Commission on Stratigraphic Nomenclature, 1970: Code of stratigraphic nomenclature, 2nd edition. American Association of Petroleum Geologists Bulletin, 21 pp.

Anderson, L.W., 1976: Rates of cirque glacier erosion and source of glacial debris, Pangnirtung Fiord area, Baffin Island, N.W.T., Canada. Unpublished M. Sc. thesis, University of Colorado, Boulder. 78 pp.

Anderson, L.W., 1978: Cirque glacier erosion rates and characteristics
   of Neoglacial tills, Pangnirtung Fiord area, Baffin Island. N.W.T.,
   Canada. Arctic and Alpine Research, 10(4):749-760.
Andrews, J.T., 1972: Glacier power, mass balance, velocities, and
   erosion potential. Zeitschrift für Geomorphologie, 13:1-17.
Andrews, J.T., 1975: Radiocarbon date list II from Cumberland
   Peninsula, Baffin Island, N.W.T., Canada. Arctic and Alpine Research,
   7(1):77-92.
Andrews, J.T., 1976: Radiocarbon date list III, Baffin Island, N.W.T.,
   Canada. University of Colorado, Institute of Arctic and Alpine
   Research, Occasional Paper, 21, 47 pp.
Andrews, J.T., 1977: Installation of a glaciology mass balance stations
   network in Auyuittuq National Park. Unpublished final report to Parks
   Canada, Ottawa, 59 pp.
Andrews, J.T., 1978: Glaciology mass balance data, Boas and Ookalik
   glaciers, Auyuittuq National Park. Unpublished report to Parks
   Canada, Ottawa.
Andrews, J.T., 1982: Chronostratigraphic division of the Holocene,
   Arctic Canada. Striae, 16:56-64.
Andrews, J.T. and Barnett, D.M., 1979: Holocene (Neoglacial) moraine
   and proglacial lake chronology, Barnes Ice Cap, Canada. Boreas,
   8(3):341-358.
Andrews, J.T. and Barry, R.G., 1972: Present and paleo-climatic
   influences on the glacierization and deglacierization of Cumberland
   Peninsula, Baffin Island, N.W.T., Canada. University of Colorado,
   Institute of Arctic and Alpine Research, Occasional Paper, 2, 160 pp.
Andrews, J.T., Barry, R.G., Bradley, R.S., Miller, G.H., and Williams,
   L.D., 1972: Past and present glaciological response to climate in
   eastern Baffin Island. Quaternary Research, 2(3):303-314.
Andrews, J.T., Davis, P.T., and Wright, C., 1976: Little Ice Age
   permanent snowcover in the eastern Canadian Arctic: Extent mapped
   from LANDSAT-1 satellite imagery. Geografiska Annaler, 58A:71-82.
Andrews, J.T. and Ives, J.D., 1972: Late- and postglacial events
   (<10,000 B.P.) in the eastern Canadian Arctic with particular
   reference to the Cockburn moraines and break-up of the Laurentide ice
   sheet. In: Vasari, Y., Hyvärinen, H., and Hicks, S. (eds.), Climatic
   changes in Arctic areas during the lastten-thousand years. Finland:
   Acta University Oulensis, Oulu, Series A(3):149-174.
Andrews, J.T. and Miller, G.H., 1972a: Quaternary history of northern
   Cumberland Peninsula, Baffin Island, N.W.T., Canada: Part IV: Maps of
   the present glaciation limits and lowest equilibrium line altitude
   for north and south Baffin Island. Arctic and Alpine Research,
   4(1):45-59.
Andrews, J.T. and Miller, G.H., 1972b: Quaternary history of northern
   Cumberland Peninsula, east Baffin Island, N.W.T., Part X, Radiocarbon
   date list. Arctic and Alpine Research, 4(3)261-277.
Andrews, J.T. and Miller, G.H., in press: Quaternary glacial and
   nonglacial correlations for the eastern Canadian Arctic. Geological
   Survey of Canada, Paper.
Andrews, J.T. and Webber, P.J., 1964: A lichenometrical study of the
   northwestern margin of the Barnes Ice Cap: A geomorphological
   technique. Geographical Bulletin, 22:80-104.
Arctic and Alpine Research, 1973: Webber, P.J. and Andrews, J.T.
   (eds.), Lichenometry: Dedicated to the memory of the late Roland E.
   Beschel 5(4)-2983-424.

Barry, R.G., Arundale, W.H., Andrews, J.T., Bradley, R.S., and Nichols,
   H., 1977: Environmental change and cultural change in the eastern
   Canadian Arctic during the last 5000 years. Arctic and Alpine
   Research, 9(2):193-210.

Barton, R. and Harben, C., 1974: Arctic Canada 74, The Clifton College Expedition to Baffin Island. Unpublished manuscript, (available through Clifton College, Bristol, England), 72 pp.

Benedict, J.B., 1967: Recent glacial history of an alpine area in the Colorado Front Range, U.S.A., I: Establishing a lichen-growth curve. Journal of Glaciology, 6:817-832.

Benedict, J.B., 1973: Chronology of cirque glaciation, Colorado Front Range. Quaternary Research, 3(4):584-599.

Beschel, R.E., 1950: Flechten als Altersmasstab rezenter Moränen. Zeitschrift für Gletscherkunde und Glazialgeologie, 1:152-161. Translated 1973 as "Lichens as a measure of the age of recent moraines." Arctic and Alpine Research, 5(4):303-309.

Birkeland, P.W., 1978: Soil development as an indicator of relative age of Quaternary deposits, Baffin Island, N.W.T., Canada. Arctic and Alpine Research, 10(4):733-747.

Boas, F., 1884: A journey in Cumberland Sound and on the west shore of Davis Strait in 1883 and 1884. Journal of the American Geographical Society, 16:242-272.

Bockheim, J.G., 1979: Properties and relative age of soils of southwestern Cumberland Peninsula, Baffin Island, N.W.T., Canada. Arctic and Alpine Research, 11(3):289-306.

Boyer, S.J., 1972: Pre-Wisconsin, Wisconsin, and Neoglacial ice limits in Maktak Fiord, Baffin Island: A statistical analysis. Unpublished M. Sc. thesis, University of Colorado, Boulder. 117 pp.

Bradley, R.S., 1973: Seasonal fluctuations in Baffin Island, N.W.T. during the period of instrumental records (1914-1970). Arctic, 26(3):230-243.

Bradley, R.S. and Miller, G.H., 1972: Recent climatic change and increased glacierization in the eastern Canadian Arctic. Nature, 237:385-387.

Carrara, P.E., 1972: Late- and Neoglacial history in Smirling and Sulung Valleys, eastern Baffin Island, N.W.T., Canada. Unpublished M.Sc. thesis, University of Colorado, Boulder. 50 pp.

Carrara, P.E., 1975: The ice-cored moraines of Akudnirmuit Glacier, Cumberland Peninsula, Baffin Island, N.W.T., Canada. Arctic and Alpine Research, 7(1):61-68.

Carrara, P.E. and Andrews, J.T., 1972: Quaternary history of northern Cumberland Peninsula, Baffin Island, N.W.T., Canada, Part I: The late- and Neoglacial deposits of Akudlermuit and Boas glaciers. Canadian Journal of Earth Sciences, 9(4):403-414.

Davis, P.T., 1977: The first Inuit glacio-meteorologists. Canadian Alpine Journal, 60:75-77.

Davis, P.T., 1980: Late Holocene glacial, vegetational, and climatic history of Pangnirtung and Kingnait Fiord area, Baffin Island, N.W.T., Canada. Ph.D. dissertation, University of Colorado, Boulder. 366 pp.

Davis, P.T., 1982: Chronology of Holocene Glaciation, Arapaho Cirque, Colorado Front Range. XI INQUA Congress Abstracts, II, p. 54.

Davis, P.T. and Waterman, S.E., 1979: New radiocarbon ages for type Triple Lakes moraines, Arapaho Cirque, Colorado Front Range. Geological Society of America, Abstracts with Program, 11(6):270.

Davis, P.T. and Wright, C., 1975: Extent of Little Ice Age snow cover in the eastern Canadian Arctic: An example of an abortive glaciation. Geological Society of America, Abstracts with Program, 7(7):1046-1047.

Deevey, E.S. and Flint, R.F., 1957: Postglacial hypsithermal interval. Science, 125:182-184.

Denton, G.H., 1974: Quaternary glaciations of the White River Valley,
    Alaska, with a regional synthesis for the northern St. Elias
    Mountains, Alaska and Yukon Territory. Geological Society of America
    Bulletin, 85:871-892.
Denton, G. H. and Karlen, W., 1973: Holocene climatic variations--their
    pattern and possible cause. Quaternary Research, 3(2):155-205.
Denton, G.H. and Karlen, W., 1977: Holocene glacial and tree-line
    variations in the White River Valley and Skolai Pass, Alaska and
    Yukon Territory. Quaternary Research, 7(1):63-111.
Dyke, A.S., 1977: Quaternary geomorphology, glacial chronology, and
    climatic and sea-level history of southwestern Cumberland Peninsula,
    Baffin Island, Northwest Territories, Canada. Ph.D. dissertation,
    University of Colorado, Boulder, 184 pp.
Dyke, A.S., 1979: Glacial and sea-level history of the southwestern
    Cumberland Peninsula, Baffin Island, N.W.T., Canada. Arctic and
    Alpine Research, 11(2):179-202.

Falconer, G., 1966: Preservation of vegetation and patterned ground
    under a thin ice body in northern Baffin Island, N.W.T. Geographical
    Bulletin, 8:194-200.
Fredskild, B., 1973: Studies in the vegetational history of Greenland.
    Meddeleser om Grønland, 198(4):245.

Godwin, H., 1966: Introductory address. In: World Climate from 8,000 to
    0 B.C. Royal Meteorological Society, London, 3-14.
Grove, J.M., 1979: The glacial history of the Holocene. Progress in
    Physical Geography, 3:1-53.

Harrison, D.A., 1964: A reconnaissance glacier and geomorphological
    survey of the Duart Lake area, Bruce Mountains, Baffin Island,
    N.W.T. Geographical Bulletin, 22:57-71.
Harrison, D.A., 1966: Recent fluctuations of the snout of a glacier at
    McBeth Fiord, Baffin Island, N.W.T. Geographical Bulletin, 8:48-58.
Hawkins, F.F., 1980: Glacial geology and late Quaternary
    paleoenvironment in the Merchants Bay area, Baffin Island, N.W.T.,
    Canada. Unpublished M.Sc. thesis. University of Colorado, Boulder.
    146 pp.
Heusser, C.J., 1966: Polar hemispheric correlation: Palynologic
    evidence from Chile and the Pacific Northwest of America. In: World
    Climate from 8,000 to 0 B.C. Royal Meteorological Society, London,
    124-141.
Hopkins, D.M., 1975: Time-stratigraphic nomenclature for the Holocene
    Epoch. Geology, 3:10.

Ives, J.D., 1962: Indications of recent extensive glacierization in
    north-central Baffin Island. Journal of Glaciology, 4:197-205.

Jacobs, J.D., Andrews, J.T., Barry, R.G., Bradley, R.S., Weaver, R.,
    and Williams, L.D., 1972: Glaciological and meteorological studies on
    the Boas Glacier, Baffin Island, for two contrasting seasons (1969-70
    and 1970-71). In: The Role of Snow and Ice in Hydrology, (Proceedings
    of the Banff Symposia, September, 1972), 1:371-381. UNESCO-WMO-IASH.

Karlen, W., 1973: Holocene glacier fluctuations and climatic
    variations, Kebnekaise Mountains, Swedish Lapland. Geografiska
    Annaler, 55A:29-63.
Karlen, W., 1976: Lacustrine sediments and tree-limit variations as
    indicators of Holocene climatic fluctuations in Lappland: Northern
    Sweden. Geografiska Annaler, 58A:1-34.

Karlen, W., 1981: Lacustrine sediment studies, a technique to obtain a
continuous record of Holocene glacier variations. Geografiska
Annaler, 63A:273-281.
Keen, R.A., 1979: Temperature and circulation anomalies in the eastern
Canadian Arctic, Summer 1946-76. Ph.D. dissertation, University of
Colorado, Boulder. 158 pp.

Locke, W.W., III, 1979: Etching of hornblende grains in arctic soils:
An indication of relative age and paleoclimate. Quaternary Research,
11(2):197-212.
Locke, W.W., III, 1980: The Quaternary geology of the Cape Dyer area,
southeasternmost Baffin Island, Canada. Ph.D. dissertation,
University of Colorado, Boulder. 331 pp.
Locke, W.W., III, Andrews, J.T., and Webber, P.J., 1979: A Manual for
Lichenometry. Technical Bulletin 26, Geo. Abstracts Ltd., University
of East Anglia, Norwich, 47 pp.
Locke, C.W. and Locke, W.W., III, 1977: Little Ice Age snow-cover
extent and paleoglaciation thresholds: North-central Baffin Island,
N.W.T., Canada. Arctic and Alpine Research, 9(3):291-300.
Løken, O.H. and Andrews, J.T., 1966: Glaciology and chronology of
fluctuations of the margin at the south end of the Barnes Ice Cap,
Baffin Island, N.W.T. Geographical Bulletin, 8:341-359.

Mangerud, J., Andersen, S.T., Berglund, B.E., and Donner, J.J., 1974:
Quaternary stratigraphy of Norden, a proposal for terminology and
classification. Boreas, 3(3):109-128.
Mangerud, J. and Berglund, B.E., 1978: The subdivision of the
Quaternary of Norden: A discussion. Boreas, 7(3):179-181.
McCoy, W.D., 1983: Holocene glacier fluctuations in the Torngat
Mountains, northern Labrador. Geographie physique et Quaternaire, 37:
211-216.
Mears, A.I., 1972: Glacial geology and crustal properties in the
Nedlukseak Fiord region, eastern Baffin Island, Canada. Unpublished
M.Sc. thesis, University of Colorado, Boulder. 60 pp.
Meier, M.F., 1965: Glaciers and Climate. In: Wright, Jr., H.E. and
Frye, D.G. (eds.), The Quaternary of the United States. Princeton
University Press, N.J., 795-805.
Miller, G.H., 1973a: Late-Quaternary glacial and climatic history of
northern Cumberland Peninsula, Baffin Island, N.W.T., Canada.
Quaternary Research, 3(4)561-583.
Miller, G.H., 1973b: Variations in lichen growth from direct
measurements: Preliminary curves for Alectoria minuscula from eastern
Baffin Island, N.W.T., Canada. Arctic and Alpine Research,
5(4):333-339.
Miller, G.H., 1975: Glacial and climatic history of northern Cumberland
Peninsula, Baffin Island, during the last 10,000 years. Ph.D.
dissertation, University of Colorado, Boulder. 226 pp.
Miller, G.H. and Andrews, J.T., 1972: Quaternary history of northern
Cumberland Peninsula, east Baffin Island, N.W.T., Canada, Part VI:
Preliminary lichen growth curve for Rhizocarpon geographicum.
Geological Society of America Bulletin, 83(4):1133-1138.
Moss, J.A., 1951: Late glacial advances in the southern Wind River
Mountains, Wyoming. American Journal of Science, 249:865-883.
Müller, D.S., 1980: Glacial geology and Quaternary history of southeast
Meta Incognita Peninsula, Baffin Island, Canada. Unpublished M.Sc.
thesis, University of Colorado, Boulder. 211 pp.

National Atlas of Canada, 1973: Surveys and Mapping Branch, Department
of Energy, Mines, and Resources, Ottawa.

Nichols, H., 1975: Palynological and paleoclimatic study of the late Quaternary displacement of the boreal forest-tundra ecotone in Keewatin and Mackenzie, N.W.T., Canada. University of Colorado, Institute of Arctic and Alpine Research, Occasional Paper, 15, 87 pp.

Porter, S.C., 1974: Holocene glacier fluctuations. American Quaternary Association Abstracts, 3rd Biennial Meeting, Madison, Wisconsin, 68-72.

Porter, S.C., 1976: Pleistocene glaciation in the southern part of the North Cascade Range, Washington. Geological Society of America Bulletin, 87:61-75.

Porter, S.C. and Denton, G.H., 1967: Chronology of Neoglaciation in the North American Cordillera. American Journal of Science, 265:177-210.

Röthlisberger, F. and Schneebeli, W., 1979: Genesis of lateral moraine complexes, demonstrated by fossil soils and trunks; indicators of postglacial climatic fluctuations. In: Schlüchter, C. (ed.), Moraines and Varves: Origin/Genesis/Classification. Rotterdam, A.A. Balkema, 387-419.

Ruhe, R.V., 1974: Holocene terminology and the continental record. American Quaternary Association Abstracts, 3rd Biennial Meeting, Madison, Wisconsin, 97.

Schneebeli, W. and Röthlisberger, F., 1976: 8000 Jahre Walliser Gletschergeschichte ein Beitrag zür Erforschung des Klimaverlaufs in der Nacheis eit. Die Alpen, 52:1-152.

Short, S.K. and Nichols, H., 1977: Holocene pollen diagrams from subarctic Labrador-Ungava: Vegetational history and climatic change. Arctic and Alpine Research, 9(3):265-290.

Soper, J.D., 1925: Report on the Baffin Island Expedition of 1924-1925. Unpublished manuscript available from Arctic Institute of North America Library, Calgary, Alberta, 181 pp.

Stukenrath, R., Miller, G.H., and Andrews, J.T., 1979: Problems of radiocarbon dating Holocene organic-bearing sediments, Cumberland Peninsula, Baffin Island, N.W.T., Canada. Arctic and Alpine Research, 11(1):109-120.

Stuiver, M., 1978: Radiocarbon timescale tested against magnetic and other dating methods. Nature, 273:271-274.

Tarr, R.S., 1897: Difference in the climate of the Greenland and American sides of Davis' and Baffin's Bay. American Journal of Science, series 4(3):315-320.

Thompson, H.R., 1954: Pangnirtung Pass, Baffin Island; an exploratory regional geomorphology. Ph.D. dissertation, McGill University, Montreal. 227 pp.

Thompson, H.R., 1957: The old moraines of the Pangnirtung Pass, Baffin Island. Journal of Glaciology, 3:42-49.

Weaver, R.L., 1975: "Boas" Glacier (Baffin Island, N.W.T., Canada) mass balance for the five budget years 1969-1974. Arctic and Alpine Research, 7(3):279-284.

Wendland, W.M. and Bryson, R.A., 1974: Dating climatic episodes of the Holocene. Quaternary Research, 4(1):9-24.

Williams, L.D., 1975: The variation of corrie elevation and equilibrium line altitude with aspect in eastern Baffin Island, N.W.T., Canada. Arctic and Alpine Research, 7(2):169-181.

Williams, L.D., 1978a: Ice sheet initiation and climatic influences of expanded snow cover in arctic Canada. Quaternary Research, 10(2):141-149.

Williams, L.D., 1978b: The Little Ice Age glaciation level on Baffin
   Island, arctic Canada. Palaeogeography, Palaeoclimatology,
   Palaeoecology, 25:199-207.

25 Environment and prehistory, Baffin Island

J. D. Jacobs

In this chapter we briefly review current thought on the prehistory of the eastern Canadian Arctic against the background of changes in climate and faunal resource availability, before turning to a more detailed analysis of the situation in the Baffin Island region.

Recent works by Dekin (1978) and Maxwell (1980a) have reviewed progress and problems in eastern Arctic prehistory. In addition, symposia edited by Maxwell (1976), McCartney (1979) and Arundale and Schledermann (1980) have provided a wealth of new material, ranging from site-specific studies to attempts at broad synthesis. As Maxwell (1980a, p. 163) has pointed out for the Arctic and Subarctic in general, the proliferation of research has been such that half of all the literature ever published on the prehistory of the area has appeared in the last decade. It would be a difficult task to distill all that is essential from such a wealth of material. For present purposes, it must suffice to attempt a cursory summary while pointing to particular issues that are pertinent to the situation in Baffin Island.

The exact timing and number of successive waves of migration to the eastern Arctic of the first Eskimo peoples, the Arctic Small Tool tradition (ASTt) or Pre-Dorset peoples, is a matter of some debate (cf. Maxwell, 1980a, p. 167). However, as presaged by Collins (1950; in Taylor, 1966), it is now generally agreed that by about 4000 BP, the eastern Canadian Arctic and northwestern Greenland had become occupied by a culture of recent Alaskan or possibly Asian origin that was characterized by a flexible economy adapted to both coastal and inland environments. Over the next two millennia and in widely separated locations, from the Labrador coast to the High Arctic, there evolved the distinctive Dorset culture, first recognized by Jenness (1925) from artifact assemblages obtained from southwestern Baffin Island.

Considerable debate arose in the interpretation of subsequent Dorset culture finds as to whether it represented a direct and purely Eskimoid development from the Pre-Dorset, or a distinct transition resulting from the infusion of certain subarctic traits. While the issue has not been resolved to everyone's satisfaction (cf. Maxwell, 1976a, p. 3), the most recent evidence tends to support the former interpretation, that is, of a Pre-Dorset through Dorset continuum with the latter characterized by a greater reliance on marine mammals, as evidenced by the development of a more refined capability for hunting on the ice (Maxwell, 1980a: p. 169). For the most part, Dorset culture appears to have remained relatively stable in terms of cultural development over a period of nearly 1500 years (c. 500 BC - AD 1000) until it was displaced or assimilated by a new wave of migrants from the west, who reached the eastern Arctic between AD 1000 - AD 1200. Still, the demise of the Dorset culture was not an abrupt one: remnant populations persisted south of Hudson Strait in the area of the northern Labrador and Ungava coasts and in southeastern Hudson Bay as late as AD 1450.

This new culture, the Thule, had its origins on the north coast of Alaska and was superbly adapted to marine mammal hunting, including the bowhead whale. More than anything else, it is the remains of these large mammals and the associated hunting implements found at their sites throughout the Canadian Arctic that has come to characterize the Thule culture, although the relative importance of this one species to the economies of Thule culture populations certainly varied with time and place (cf. Freeman, 1979; McCartney, 1979b, 1980). As much as any previous Arctic culture, the Thule people as a whole had a varied subsistence strategy, out of which had developed a number of regional variants by the beginning of the Historic Inuit period.

ENVIRONMENTAL AND CULTURAL CHANGE

Over the past decade or more, several attempts have been made at meaningful synthesis of the parallel sequences of cultural and environmental change in the eastern Arctic. McGhee (1969/70) set the pattern with his explanation of the expansion and development of Thule culture, based on a scheme of Holocene climatic changes put forward by Bryson and Wendland (1967), and using interpretations of relationships between climate and faunal resources availability due in part to Vibe (1967). McGhee later extended this approach to the entire Arctic cultural sequence (McGhee, 1974). Among other things, he suggested that the entry of both Pre-Dorset and Thule cultures into the eastern Arctic occurred during warmer climatic intervals in their respective time periods. Subsequent cultural developments, including regional variants, represented adaptations to changing environmental conditions following these milder periods.

Another attempt at a systematic treatment of climate-cultural response relationships was that of Dekin (1972). Based on a number of sources, he distinguished seven different climatic stages encompassing the prehistoric period in the eastern Arctic, and developed regional examples of climate-culture interactions, particularly with the view of explaining Dorset culture development and movement.

A specific regional focus was developed in detail by Fitzhugh in his studies of the Labrador coast (Fitzhugh, 1972) and of Hudson Bay (Fitzhugh, 1976). In the latter study, he made a serious effort to explain differences in the distribution around Hudson Bay of sites representing successive Pre-Dorset and Dorset temporal stages in terms of contemporaneous climatic conditions and the associated constraints on faunal resources. He saw the southern Hudson Bay area as a marginal one for human occupation, in contrast to the northwestern part, and explained this in terms of the different coastal and sea ice environments of the two areas.

Barry et al. (1977) carried out a critical review of such studies, incorporating the paleoenvironmental knowledge of the eastern Arctic as it existed up to that time. Their results are summarized in Figure 25.1. As those authors point out, there is some disagreement in detail in the trend of the climatic intervals between regions. The problem is compounded by the fact that dissimilar paleoclimatic methods are used in each. Whether or not such episodes are actually time-transgressive between regions or whether the apparent lags are simply artifacts of the different methods used must await the results of further study (see also Chapters 22, 23, 24, and 26).

There does appear to be a correspondence between some significant cultural events and the reconstructed paleoclimatic record, as presented in Figure 25.1. However, the ambiguities in the latter record preclude extending the correlation beyond the kinds of general relationships already proposed by others (e.g. McGhee, 1976). The lack of a precise paleoenvironmental record makes it difficult to understand some critical aspects of the archeological record, such as the direction and causes of the development and decline of the Dorset culture (Barry et al. 1977, p. 205), while others, such as the retreat of Thule culture populations from the High Arctic during a cooler period c. 400 BP, are less problematic.

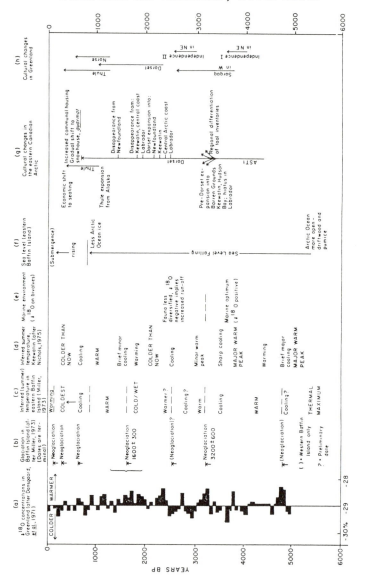

Figure 25.1 Overview of environmental and cultural changes in the eastern Arctic (Barry et al., 1977) (with permission).

ARCHEOLOGY IN BAFFIN ISLAND

As a contribution to the Inuit Land Use and Occupancy Project, M. Maxwell, R. McGhee, and W. Noble prepared maps showing the distribution of prehistoric sites and presumptive land use areas for the Canadian Arctic (Freeman, 1976; Vol. 2, pp. 117-122). Although the number of

known sites has increased since that report, the general pattern of
occupation in the various cultural stages remains basically the same.
Figure 25.2 is a composite map for the Baffin Island region derived
from that study.

The shaded areas are those for which archeological or ethnographic
evidence indicates occupation at least on a seasonal basis in the
past. Empty areas are those for which no such evidence exists. Since
not all of the region has been systematically surveyed, it is likely
that some of these areas will eventually be filled in; however, in many
others, local environmental conditions must have discouraged
settlement.

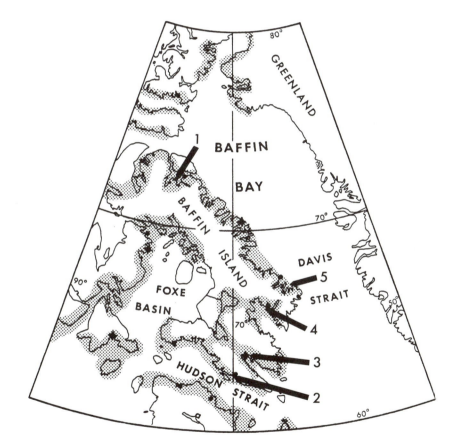

Figure 25.2 Prehistoric occupation areas in the eastern Canadian
Arctic. Shaded areas are those in which archeological and ethnographic
evidence shows past settlement and/or land use activities. Numbers
refer to particular areas in Baffin Island that are the subject of this
discussion. (1 - Northern Baffin, 2 - South Coast, 3 - Frobisher Bay,
4 - Cumberland Sound, 5 - East Coast).

For purposes of this discussion, the Baffin Island region is
divided into six archeological subregions: Northern Baffin, the South
Coast, the East Coast, Frobisher Bay, and Cumberland Sound (Fig. 25.2
and Table 25.1). This division is based in part on the physiography of
the region, which has historically tended to constrain settlement to
those areas, with movement between the subregions more difficult than
movement within. This is not to say, however, that movement between
subregions was uncommon (cf. Maxwell, 1979). Baffin Island cannot
really be viewed in isolation from prehistoric occupation areas to the
north and south; and, of course, the area around Foxe Basin must be
acknowledged as the source area of much of the material and ideas of
eastern Arctic prehistory (Mathiassen, 1927; Meldgaard, 1960; Campbell,
1966, p. 898). However, there is much that is distinctive about the
prehistory of Baffin Island that warrants our focussing on its
subregions.

Table 25.1. Selected sources on Baffin Island prehistory by subregion
(see Fig. 25.2).
==
1. Northern Baffin: Mathiassen, 1925, 1927; Mary-Rousseliere,
    1976, 1979; Hunston, 1979.

2. South Coast: Jenness, 1925; Mathiassen, 1927; Manning, 1950;
    Taylor, 1968; Maxwell, 1962, 1973, 1976, 1979, 1980b, 1981; Dekin,
    1975; Arundale, 1976a, 1976b, 1980; Sabo, 1981; Sabo and Sabo,
    1978; Sabo and Jacobs, 1980.

3. Frobisher Bay: Collins, 1950; Maxwell, 1973, 1980c; Sabo and
    Jacobs, 1979; Stenton, 1982, 1983; Jacobs, 1982.

4. Cumberland Sound: Boas, 1885, 1888; Schledermann, 1974, 1975,
    1976a; Neatby, 1977; Gardner, 1979; Sabo, 1979b.

5. East Coast: Boas, 1885; Mathiassen, 1927; McKenzie-Pollack, 1969;
    Schledermann, 1972, 1974, 1976a; Kemp, 1975; Sabo, 1979a, 1979b;
    Jacobs and Sabo, 1978; Gardner, 1979; Wenzel, 1981.

Northern Baffin Island

Initial discoveries by Mathiassen (1925, 1927) in the vicinity of Pond
Inlet and at Button Point on Bylot Island have been greatly advanced by
Mary-Rousselière, who has excavated numerous sites in this area. The
earliest finds there are Pre-Dorset and have been dated c. 4000 BP,
based on an adjusted radiocarbon date (Mary-Rousselière, 1976, p. 41;
cf. McGhee and Tuck, 1976, Arundale, 1981, p. 261). It is not clear
whether the entire Dorset sequence is represented: early and late

Dorset elements dominate, and Mary-Rousselière interprets the dearth of middle Dorset material to mean that the Pond Inlet area was a relatively marginal area compared to Foxe Basin (Mary-Rousselière, 1976, p. 53).

Of particular importance is the Nunguvik site on Navy Board Inlet, which contains, in addition to some 80 early to late Dorset house remains, about 50 Thule culture winter houses (Mary-Rousselière, 1979). At this site there is an overlap in dates for the earliest Thule occupation and the latest Dorset, which suggests contemporaneous occupation of the site by the two cultures (cf. Wenzel, 1979). Dorset traits soon disappeared, however, while the Thule occupation continued into the transitional Historical period, without apparent interruption.

Several additional significant points can be summarized from the work of Mary-Rousselière. While he regards northern Baffin Island as something of a marginal area, it appears that the Pond Inlet area was occupied to some degree over most of the last 4000 years. For part of this time, it served as a jumping off place for periodic movements northward across Lancaster Sound into the High Arctic, i.e. into even more "marginal" areas. The economies of all stages have both terrestrial and marine aspects but caribou bone is more abundant in most Dorset and Thule houses, except for the earliest Thule phase. Baleen appears in Dorset as well as Thule sites; this, coupled with evidence of their use of kayaks, leads Mary-Rousselière (1976, p. 53) to suggest that these Dorset culture people may have hunted the bowhead whale. Their earliest Thule successors at Nunguvik certainly did; and, although a decline in whaling seems to have occurred after the second century of the Thule occupation (c. 1200-1300 AD), this activity was never entirely abandoned in the Pond Inlet area (Mary-Rousselière, 1979, p. 59). Finally, there is the question of links between northern Baffin Island and the Igloolik area, northern Foxe Basin, particularly in the Pre-Dorset and Dorset stages. The close proximity of these two areas (Fig 25.2) and the physiography of the region suggest the possibility of close cultural ties. It might be argued on those grounds alone that northern Foxe Basin should be included as part of the northern Baffin subregion. This has not been done in the present review more from a lack of information than from anything else. The work of Meldgaard (1960) at Igloolik has not yet been developed to the point of providing evidence of such cultural affinities; however, Maxwell (pers. comm., 1982) has suggested that further analysis of the Igloolik materials might produce such evidence. On the other hand, Mary-Rousselière (1976, p. 53), in an analysis of Dorset lithic traits, has found what he believes is a closer stylistic association between the Pond Inlet area and the Lake Harbour area of southern Baffin Island than between Pond Inlet and the Igloolik area.

South Coast

The artifacts which first led to recognition of what has become known as the Dorset culture came to the National Museum of Canada from an unknown site or sites near Cape Dorset, on the southwest coast of Baffin Island (Jenness, 1925). While some limited work has been done in that area (Mathiassen, 1927; Manning, 1950; Arundale, 1980), systematic archeological investigations have yet to be undertaken there.

The principal focus of research on the south coast has been the Lake Harbour area (Fig. 25.2). Work begun there by Maxwell in 1960 has since continued with few interruptions. Initial excavations at Juet Island yielded material that ranged stylistically from Pre-Dorset through Dorset, revealing that a continuous cultural development had occurred there (Maxwell, 1962; 1973). Subsequent work (see Table 25.1) has demonstrated that the continuum extends from the earliest Pre-Dorset, dated at c. 4000-4200 BP at the Closure Site (Arundale, 1981, p. 261) through early Thule (Sabo, 1981; Maxwell, 1981) and up to the present time. There is even compelling evidence for early Thule contact with Vikings from West Greenland (Sabo and Sabo, 1978).

The collective results of these studies, combined with observations of modern Inuit populations there, have led Maxwell to conclude that "...despite technologic innovations such as gunpowder and gasoline, human population and resource utilization have followed persistent and relatively unchanging patterns from 2000 BC to the present" (Maxwell, 1979, p. 78). The pattern has been one of occupation by a stable population of up to 200 individuals distributed among a number of winter settlements along a 500 km long stretch of coast centered on Lake Harbour. There is a mixed and generally reliable faunal resource base there, which Maxwell estimates to have a potential carrying capacity of about three times the long-term average population.

The Pre-Dorset to Dorset transition at Lake Harbour occurred between about 2600 to 2400 BP (Arundale, 1981, p. 261), with a shift to greater emphasis on hunting on the sea ice, but still retaining the dual (land and sea) economic base of the Pre-Dorset. Early Dorset manifestations here coincide with a significant intensification of permafrost. Maxwell interprets aspects of Pre-Dorset and Dorset tool assemblages as indicative of a fair degree of technological conservatism in these cultures, possibly to the extent of discouraging both innovation within the system and the infusion of new ideas from without (Maxwell, 1976b, p. 70).

The subsequent Thule occupation presented a more generalized adaptation to the same ecological niche formerly occupied by the Dorset people. Here and in other areas, it may have been this difference in adaptive strategy that led to the demise or assimilation of the Dorset after the Thule people had arrived on the scene (Maxwell, 1979, p. 85).

Thule culture components in the Lake Harbour area include at least one example of a winter dwelling that Maxwell has identified as being contemporaneous with the earliest phases of Thule occupation in the eastern High Arctic. On the basis of this evidence and what is known of an early Thule occupation in Frobisher Bay (see below), he suggests that southern Baffin Island may have been occupied as an offshoot from the earliest Thule migration, which is generally regarded to have had its main thrust northeastward, toward northwestern Greenland (Maxwell, 1981).

From excavations at three sites in the area, Sabo (1981) has identified three phases in the Thule culture occupation, distinguished by artifact types and dwelling styles. In the first or Classical Thule phase, there is a high degree of cultural similarity with early sites elsewhere. This period lasted some 200-300 years at most. In the next or Developed Thule phase, the population has altered its subsistence adaptation, apparently in response to environmental changes. Finally, in the Historic Thule phase, Euro-Canadian materials make their appearance. This approach to distinguishing more-or-less distinct changes in Thule culture is a useful one that can be applied in principle to other areas as well (see Fig. 25.3).

The system of three cultural phases has been used by Sabo, in conjunction with data on recent Inuit land use in the area (Kemp, 1976), in the interpretation of temporal changes in faunal procurement systems (Sabo and Jacobs, 1980). Bowhead whale (bone and baleen) is present in all phases, but is more abundant in the Classical and again in the Historic phase, when European whalers had entered the scene. Considering all other faunal resources (op. cit., Table 3) it is evident that a significant shift from land to sea mammals occurred between the Classical and Developed Thule phases.

Frobisher Bay

Frobisher Bay assumed a place of prominence in Arctic archeology through the identification by Collins at the Crystal II site (see Fig. 25.4) of an early Thule occupation overlying a late Dorset one, the two being separated by a sterile horizon (Collins, 1950). More than two

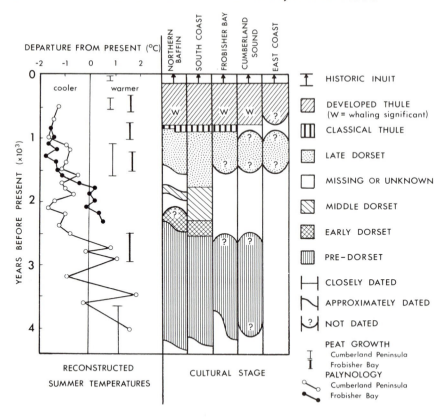

Figure 25.3.   Occupational sequences on Baffin Island in relation to summer paleoclimate.   (For explanation, see text.)

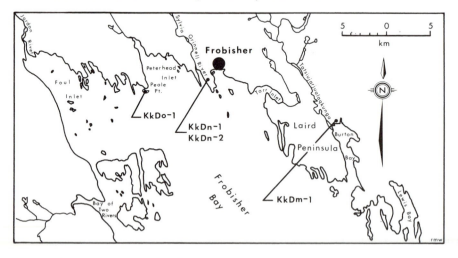

Figure 25.4.   Archeological sites at the head of Frobisher Bay.   Site designations are according to Borden number.   KkDo-1 = Peale Pt., KkDn-1 = Crystal II, KkDn -2 = Shaymarc, KkDm-1 = Burton Bay Paleoeskimo site.

decades elapsed before further work was done by Maxwell, who, while
adding to what was known about the Dorset component at Crystal II,
located a Pre-Dorset occupation site (Shaymarc site) at higher
elevation close by (Maxwell, 1973). This site yielded an adjusted
radiocarbon date of 3675±144 BP (Arundale, 1981, p. 261).

In 1979, Sabo and Jacobs located a major Thule winter settlement
at Peale Pt., about 10 km west of the Crystal II site and the village
of Frobisher Bay. Excavations begun there in 1980 by Maxwell (1980c)
and continued by Stenton (1982, 1983), have yielded an abundance of
materials ranging from Pre-Dorset through Historic Thule. The combined
evidence from the aforementioned sites and other finds southeast of the
village of Frobisher Bay, including the only reported Paleoeskimo
(probably Pre-Dorset) dwelling structure in southern Baffin Island
(KkDm-1) point to a significant Pre-Dorset and Dorset use of upper
Frobisher Bay (Sabo and Jacobs, 1979; Maxwell, 1980c; Jacobs, 1981).
This is not surprising in view of the fact that the area has
historically had an abundance of fish, waterfowl, and caribou, which
appear to have diminished only recently under pressures from the modern
settlement.

The remoteness of the area from the fast ice edge, some 100 km to
the southwest, at first makes it difficult to explain the presence of
Thule winter settlements at the head of Frobisher Bay. Bowhead whale
was undoubtedly available there during the open water season in the
early phases, as indicated by baleen and whalebone at the sites.
Maxwell (1980c) has suggested that during this earlier phase, a
combination of locally available terrestrial game, seals taken at
breathing holes, and stored whale blubber and meat allowed occupation
through the winter. Later, there appears to have been a decline in
bowhead whale and a greater reliance on hunting caribou inland from the
head of the bay. Interestingly, Stenton's (1983) analysis of ringed
seal age classes at this site indicates that winter breathing-hole
sealing was not practiced to any significant extent.

There was apparently no winter occupation at the head of Frobisher
Bay by 1861, when C.F. Hall arrived there (Hall, 1865). Although not
always consistent, Hall made a point of describing settlement sites,
including the abandoned Thule winter houses on the Sylvia Grinnell
River (Crystal II). He camped near the Peale Pt. site, though not
directly upon or in sight of it, but made no comment about the
existence there of either an active or abandoned settlement. In 1982,
another Thule-Historic winter settlement with some 14 semi-subterranean
houses was reported at Ward Inlet, 90 km from the head of the bay
(Jacobs, 1982). This site had been passed by Hall during a sledge

journey in 1862, also without comment. From the observations Hall did make, it seems that by that late period, winter settlement was confined mainly to snow house encampments near the floe edge in the lower part of the bay.

Cumberland Sound

This area was a focus for commercial whaling activities during the latter part of the 19th century and subsequently for the establishment of missions, trading posts, and government service functions among the Inuit. Several early researchers used these posts as bases for ethnographic and geographical studies in the area. Most prominent of these was Franz Boas who, starting from the whaling station at Kikerten Island, not far from the present-day settlement of Pangnirtung, traveled extensively along the Cumberland Sound shore before crossing Cumberland Peninsula to Davis Strait (Boas, 1884, 1885, 1888). In 1910, 25 years after Boas, Bernard Hantzsch arrived at Blacklead Island on the south side of Cumberland Sound. From there he began a journey northwestward, overland to the Foxe Basin shore. There, Hantzsch succumbed to illness, but his journals were brought back by his Inuit companions and have recently been published in translation (Neatby, 1977). Both Boas and Hantzsch commented on seeing ruins of habitations which were attributed by their Iniut companions to a vanished race, the Tunnit, a term usually interpreted as referring to people of the Dorset culture, though the structures in this case were most likely of the Thule culture.

Archeological studies were started in the area by Schledermann in 1971. His main concern was with Thule culture development and, in the course of three years, he located 23 Thule winter settlement sites in the upper part of Cumberland Sound (Schledermann, 1975). Selective testing and excavations at several of these sites yielded a wealth of Thule culture material. Paleoeskimo artifacts were also recovered, which Schledermann (1975, p. 278) believes to represent late Pre-Dorset and late Dorset phases. This result, though based on limited evidence, fits the pattern noted elsewhere in Baffin Island of an apparent occupational hiatus or population reduction in the middle Dorset phase.

Schledermann recovered cultural material typologically associated with early Thule phases, an inference supported by adjusted radiocarbon dates of about 700 to 800 BP. He feels that this earliest occupation was of limited proportions, but suggests that further investigations might yield other sites coeval with the early sites on Frobisher Bay (Schledermann, 1975, p. 244). Beginning with the first major

occupation of the area, he distinguishes three stages: (1) an early
(1220-1650 AD) "baleen period" which came to an abrupt end with the
onset of severe sea ice conditions, (2) a subsequent period of reduced
population that lasted about a century and is attributed by him to the
relocation of winter settlements, perhaps in the interior on Nettilling
Lake or possibly on the outer coast, and (3) a period of reoccupation
of inner Cumberland Sound that lasted into historic times, but saw a
major shift in emphasis away from bowhead whale to seal and caribou.

Here and elsewhere, Schledermann has noted an evolution in
dwelling styles away from smaller structures to larger, "communal"
winter houses in the later Thule phase as suggested earlier by McGhee
(1969/70, p. 181). While acknowledging that communal houses sometimes
occur in earlier Thule phases, a point made also by Mary-Rousselière
(1979: 57), Schledermann (1975, p. 267, 1976b) sees a distinct trend
toward communal living, with the advantages of shared resources and
reduced fuel consumption, coinciding with worsening climate.

East Coast

The east coast is topographically the most rugged and climatically the
most severe of the subregions (see Chapter 2). It is here that one
might expect to find the clearest indications of environmental
constraints on prehistoric occupation. In one respect, at least, we
have a good data base on which to proceed. These are the results of
extensive reconnaissance and careful site surveys carried out in the
course of the Thule Archeology Conservation Project (McCartney, 1979b)
and surveys done prior to the establishment of Auyuituq National Park
(Kemp, 1975; Schledermann, 1972, 1974, 1976a). However, actual
excavations have been few and these not necessarily at the potentially
most productive sites.

A few Paleoeskimo sites have been located, and artifact types and
radiocarbon dates on these place them in the late Dorset phase
(McKenzie-Pollack, 1969). Thule culture sites are numerous, but the
excavations to date have produced material confined mainly to the early
contact and subsequent Historic phases of the 19th and early 20th
centuries (Sabo, 1979b). On the other hand, a large proportion of the
Thule winter dwellings are of the single sleeping-platform type that is
characteristic of early Thule sites elsewhere in Baffin Island, and
whalebone is found at several sites; therefore, it is to be assumed
that the area was occupied by at least the early Developed Thule phase,
if not in the Classical phase.

Lacking significant prehistoric cultural and faunal evidence from the area, it is not yet possible to reconstruct the sequence of past procurement systems, as has been done for the south coast, for example. When such archeological data are forthcoming, investigators will find a basis for comparison with present-day Inuit of the subregion in a study of the cultural ecology of the Clyde Inuit by Wenzel (1981), as well as in the land use study edited by Freeman (1976).

Using the considerable survey data available for a 200 km stretch of coastline centered on Broughton Island, Jacobs and Sabo (1978) analyzed the site and situation characteristics of 63 Thule and Historic settlement sites, broken down into temporal and seasonal groupings. All categories showed a marked preference for site characteristics that combine to maximize protection from prevailing (northwesterly) winds and provide a restricted wave fetch from seaward. Four major clusters of sites were identified, separated by partial barriers in the form of headlands with perennially rough ice offshore. The maximum number of sites was found in a cluster centered in Merchant's Bay, the most southerly area and one that is connected through a low pass with Cumberland Sound. It remains for future excavations in this largely unstudied area to produce information on movements between these two subregions and the circumstances under which they occurred.

CORRELATION WITH THE PALEOENVIRONMENTAL RECORD

The archeological record for Baffin Island is summarized by subregion in Figure 25.3. Note that the onset of the Historic Inuit phase is presented as being more-or-less simultaneous across the region in the latter half of the 19th century. This follows the interpretation of Ross (1979), supported by Sabo's (1979a) results, which sees the first significant exotic cultural impact occurring with the start of overwintering by British and American whalers. Other temporal stages are according to the various authors whose work was discussed in the previous section.

Paleoenvironmental data relevant to the prehistoric sequence include climatic reconstructions from paleovegetation (peat and pollen) and the record of relative sea level. The latter is less significant to archeology in Baffin Island than in Foxe Basin, where there is a good temporal correspondence between archeological site elevations and an uncomplicated shoreline emergence (Meldgaard, 1960). Over the eastern Arctic in general these relationships are less clear and of

limited value in dating occupations (Andrews et al., 1971). Thule
winter settlements in Baffin Island are not constrained to a strict
site-elevation-to-contemporary-sea-level rule (cf. Mary-Rousselière,
1979), mainly as a consequence of the sharp vertical relief and complex
coastal topography in most areas. A separate problem is that of the
inundation and erosion of Thule sites in areas of eastern Baffin Island
where local coastal submergence is occurring (Andrews, et al., 1971;
Gardner, 1979). This may account for the apparent paucity here of
early Paleoeskimo material.

In areas where there has been substantial uplift during the
prehistoric period, changes in coastline configuration and the
shallowing up of bays and channels may have affected settlement
patterns in some minor way. An example is the isolated Paleoeskimo
site at the head of Burton Bay, now separated from the open sea by
extensive tidal flats (Fig. 25.4). In upper Frobisher Bay, where
emergence of some 5-10 meters has occurred in the last 4000 years,
there does appear to be a relationship between elevations at the few
known Pre-Dorset and Dorset sites and site components and relative sea
level. The Pre-Dorset components appear to cluster near 15 m above
present sea level and the (late) Dorset near 5 m, with nothing in
between. This distribution may be compared with Schledermann's (1978)
analysis of the elevations of some 66 Paleoeskimo sites in the eastern
Bathurst Island - MacDougall Sound area. His results showed a distinct
bimodal frequency distribution with a trough that he suggests might be
associated with a decline in population during the period of generally
cooler climate c. 3150-1750 BP identified by Nichols (1975).

The paleoclimatic evidence which has the most direct bearing on
prehistoric settlement in Baffin Island is the record of summer
temperatures derived from pollen stratigraphy using transfer function
methods, and the related evidence of buried peat horizons. Miller
(1973) used the latter to infer a sequence of alternating cool/dry and
warm/wet periods for Cumberland Peninsula, the assumption being that
peat growth is encouraged by relatively warm and wet summer climate.
Conversely, the cessation of peat growth is seen as an indication of a
return to colder, dryer summers. The palynological evidence, discussed
in detail in Chapter 22, relies on the correspondence between present
vegetation and climate to infer past climate from the pollen record.
Several pollen diagrams from southern and eastern Baffin Island have
provided the basis for reconstructing summer temperatures there
(Andrews et al., 1980; Short and Jacobs, 1982).

Shown in Figure 25.3 are estimated summer (July) temperatures for Cumberland Peninsula and upper Frobisher Bay, based on pollen records for those areas. Also shown are warm-summer intervals inferred from peat stratigraphy for the same areas. The methods used in the pollen-derived temperature estimates for both areas are the same. The qualitative temperature estimates of Miller (1973) for Cumberland Peninsula used both basal and upper radiocarbon dates for a peat horizon interval, whereas those for Frobisher Bay (Jacobs and Short, unpublished data) are single dates for a whole peat stratum, with the interval determined by one standard deviation either side of that average date.

Clearly evident in both sets of records is the onset of Neoglacial conditions around 1700 to 2500 BP. The difference in time of onset between the two areas is probably not real and more recently recovered sections are being analyzed in an attempt to resolve the uncertainty. The earlier date is more consistent with the cultural record, in which the Pre-Dorset to early Dorset transition is seen to occur at that time (Fig. 25.3). The next "warm" interval, seen in the peat record but less evident in the pollen record, commences c. 1500 BP and coincides roughly with the beginning of the late Dorset Phase.

Interestingly, although the medieval warm epoch appears in the peat growth record, there is no evidence for it in the pollen record, indicating that the impact of any such amelioration on composition of the terrestrial flora was not great in southern Baffin Island. Studies of modern pollen from the area suggest that it is the climate of the past few decades that bears the closest resemblence to conditions of 2000-2500 BP (Andrews et al., 1980; Short and Jacobs, 1982). Thus, it would appear to be unwarranted to draw too strong an analogy between conditions under which the Thule culture became established in the area and those experienced by the early Pre-Dorset.

The peat record confirms a trend to cooler conditions around 200-300 BP, which fits the end of Schledermann's "Baleen Period" and the movement of Thule winter habitations away from the head of Cumberland Sound and Frobisher Bay, presumably to sites closer to the fast ice margin. Whether or not the most recent period of milder climate would have seen the reoccupation of these and other more marginal areas on purely environmental grounds, apart from recent Euro-Canadian influences, cannot be said for certain, but the archeological and historical evidence would seem to suggest so.

The reconstructed temperature curves do not show any obvious periods of unstable conditions, such as Arundale (1976b) and others

have invoked to explain certain cultural and settlement changes. The
periods of major shift almost certainly contained intervals of
significant short-term fluctuations, but the palynological
investigations in the region are not sufficiently advanced to resolve
changes unequivocally on that time scale. In any case, it is not at
all certain that such fluctuations would have a major impact on the
total faunal resource base. Warmer conditions may cause reductions in
the extent of fast ice habitat for ringed seal, for example, but even
given occasional icing conditions in the accompanying milder winters,
the net effect on caribou numbers may be a positive one through
increases in terrestrial primary production. Therefore, a culture
whose adaptive strategy encompasses a dual land and sea-ice based
economy should have persisted in Baffin Island under these
circumstances.

CONCLUSION

The archeological record reveals certain similarities in major cultural
developments throughout the Baffin Island region. These include the
timing of the earliest arrivals of Pre-Dorset in northern Baffin, on
the south coast, and in Frobisher Bay, as well as of the earliest Thule
phase in those same areas. There is the apparent decline or hiatus in
the middle Dorset phase in most subregions, which may be due to either
a withdrawal from the region entirely or a substantial change in
seasonal settlement patterns, for example to winter snowhouses on the
fast ice (Maxwell, 1980b, p. 514), coinciding with the deepening of the
Neoglacial climatic interval. Then there is the reappearance of Dorset
in all subregions at the beginning of a period of moderating climate.

     Could it be that, with the possible exception of the south coast,
the Pre-Dorset and early Dorset populations of Baffin Island never
really adapted to neoglacial conditions there? If so, did the
terrestrial part of the Dorset procurement system continue to dominate
the maritime, which might explain the apparent resurgence in population
during the medieval warm period? Clearly, in order to address these
questions, more evidence is needed on the Paleoeskimo chronology over
all of Baffin Island, along with accurate reconstructions of
contemporaneous paleoenvironments.

     The Thule culture, from its earliest appearance, seemed very much
at home in Baffin Island. The intriguing questions here concern the
nature of local adaptive strategies, particularly in the Developed
Thule phase. Fitzhugh (1980), has found evidence for different
adaptations (procurement systems) among several contemporaneous Thule

populations along the Labrador coast, which he attributes to dissimilar local environments. In contrast to that, the record we have seen thus far suggests that there was considerable uniformity among Baffin Island Thule populations in their adjustments to a changing environment. Much work remains to be done to fill the empty areas in the Thule site surveys and to sample the Thule culture settlements adequately for both cultural and faunal materials as a basis for explaining these adaptations. At the same time it is necessary to improve the temporal and spatial resolution of the palynological record in order to remove the ambiguities in paleoclimatic reconstructions for various parts of the region. Beyond this, what are ultimately needed are good estimates, based on paleoecological considerations, of the relative abundances of faunal resources at each stage of the prehistoric period.

ACKNOWLEDGEMENTS

This chapter has been improved by the critical comments of Drs. M. Maxwell and G. Sabo III: their help and assistance is much appreciated.

REFERENCES

Andrews, J.T., McGhee, R., and McKenzie-Pollack, L., 1971: Comparison of elevations of archaeological sites and calculated sea levels in Arctic Canada. Arctic, 24(3):210-228.
Andrews, J.T., and Miller, G.H., 1979: Climatic change over the last 1000 years, Baffin Island, N.W.T. In: McCartney, A.P. (ed), Thule Eskimo Culture: An Anthropological Retrospective. Archaeological Survey of Canada, Mercury Series, No. 88, 541-554.
Andrews, J.T., Mode, W.N., and Davis, P.T., 1980: Holocene climate based on pollen transfer functions, Eastern Canadian Arctic. Arctic and Alpine Research, 12(1):41-64.
Arundale, W.H., 1976a: The archaeology of the Nanook Site: An explanatory approach. PhD. Dissertation, Michigan State University, East Lansing.
Arundale, W.H., 1976b: A discussion of two models related to climatic change in the eastern Arctic. Paper presented at the 41st annual meeting of the Society for American Archaeology, St. Louis, Mo. May 6-8, 1976.
Arundale, W.H., 1980: Functional analysis of three unusual assemblages from the Cape Dorset area, Baffin Island. Arctic, 33(3):464-486.
Arundale, W.H. and Schledermann, P. (eds.), 1980: Recent Research in Eskimo Archaeology. Arctic, 33(3), 670 pp.
Arundale, W.H., 1981: Radiocarbon dating in eastern Arctic archaeology: a flexible approach. American Antiquity, 46(2):244-271.

Barry, R.G., Arundale, W., Andrews, J., Bradley, R., and Nichols, H., 1977: Environmental change and cultural change in the eastern Canadian Arctic during the last 5000 years. Arctic and Alpine Research, 9(2):193-210.

Boas, Franz, 1884: A journey in Cumberland Sound and on the west shore of Davis Strait in 1883 and 1884. Bull. Amer. Geogr. Soc., 3:241-272.
Boas, Franz, 1885: Baffin-Land. Geographische Ergebnisse einer in den Jahrer 1883 und 1884 ausgeführten Forschungsreise. Petermanns Geogr. Mitteilungen, Erg. No. 80. (Gotha, Justus Perthes), 1-100.
Boas, Franz, 1888: The Central Eskimo. Sixth Annual Report of the Bureau of American Ethnology. Smithsonian Institution, Washington, 261 pp.
Bryson, R.A. and Wendland, W.M., 1967: Tentative climatic patterns for some Late Glacial and Post Glacial episodes in Central North America. In: W.J. Mayer-Oaks (ed.), Life, Land, and Water, Winnipeg, Canada, 271-298.

Campbell, J.M., 1966: Current research, Arctic. American Antiquity, 31:895-899.
Collins, H.B., 1950: Excavations at Frobisher Bay, Baffin Island, N.W.T. National Museum of Canada Bulletin No. 118, (Ottawa), 18-43.

Denkin, Albert A. Jr., 1972: Climatic change and cultural change; A correlative study from Eastern Arctic Prehistory. Polar Notes, 12:11-31.
Dekin, Albert A. Jr., 1975: Models of pre-Dorset culture: Towards an explicit methodology. Ph.D. dissertation, Michigan State University, East Lansing.
Dekin, Albert A. Jr., 1978: Arctic Archaeology: A Bibliography and History. New York: Garland Publishing, 279 pp.

Fitzhugh, W.W., 1972: Environmental archaeology and cultural systems in Hamilton Inlet, Labrador. Smithsonian Contributions to Anthropology, 16, 299 pp.
Fitzhugh, W.W., 1976: Environmental factors in the evolution of Dorset culture: a marginal proposal for Hudson Bay. In: Maxwell, M.S. (ed.), Eastern Arctic Prehistory: Paleoeskimo Problems. Memoirs of the Society for American Archaeology, 31:139-169. 170 pp.
Fitzhugh, W.W., 1980: Preliminary report on the Torngat Archaeological Project. Arctic, 33(3):585-606.
Freeman, M.R. (ed.), 1976: Inuit Land Use and Occupancy Project. Department of Indian and Northern Affairs, Ottawa, 3 vols.
Freeman, M.R., 1979: A critical view of Thule culture and ecologic adaptation. In: McCartney A.P. (ed.),Thule Eskimo culture: An Anthropological Retrospective. Archaeological Survey of Canada, Mercury Series, 88:278-285. 586 pp.

Gardner, D., 1979: Area III site surveys: Clyde Area. In: McCartney, A.P. (ed), Thule Eskimo Culture: An Anthropological Retrospective. Archaeological Survey of Canada, Mercury Series, No. 88.

Hall, C.F., 1865: Life with the Esquimaux. Reprinted in 1970 by M.G. Hurtig, Ltd., Edmonton, Canada, 547 pp.
Hunston, J., 1979: 1975 Area II Site Survey: Northern Baffin Island. In: McCartney, A.P. (ed), Thule Eskimo Culture: An Anthropological Retrospective. Archaeological Survey of Canada, Mercury Series, No. 88, 143-182.

Jacobs, J.D., 1981: Archaeological observations in Frobisher Bay, Baffin Island, Summer, 1981. Unpub. report to Arch. Survey of Canada, National Museum of Man., Ottawa.
Jacobs, J.D., 1982: Archaeological Observations in Frobisher Bay during 1982. Unpublished Report to Archaeological Survey of Canada, National Museum of Man., Ottawa.

Jacobs, J.D. and Sabo, G. III, 1978: Environments and adaptations of the Thule culture on the Davis Strait coast of Baffin Island. Arctic and Alpine Research, 10(3):595-615.

Jenness, Diamond, 1925: A new Eskimo culture in Hudson Bay. Geog. Rev., 15:428-437.

Kemp, W.B., 1975: Archaeological survey and excavation in Baffin Island National Park. Unpublished report to Parks Canada, Ottawa.

Kemp, W.B., 1976: Inuit land use in south and east Baffin Island. In: Freeman, M.R. (ed.), Inuit Land Use and Occupancy Report (Vol. I), 125-151.

McCartney, A.P. (ed.), 1979a: Thule Eskimo culture: An Anthropological Retrospective. Archaeological Survey of Canada, Mercury Series, No. 88, 586 pp.

McCartney, A.P. (ed.), 1979b: Archaeological Whale Bone: A Northern Resource. University of Arkansas Anthropological Papers, No. 1, 558 pp.

McCartney, A.P., 1980: The nature of Thule Eskimo whale use. Arctic, 33(3):517-541.

McGhee, R., 1969/70: Speculations on climatic change and Thule culture development. Folk, 11-12:173-184.

McGhee, R., 1974: The peopling of arctic North America. In: Ives and Barry, (eds.), Arctic and Alpine Environments. London, Methuen, 831-855.

McGhee, R., and J.A. Tuck, 1976: Un-dating the Canadian Arctic. In: Maxwell, M.S. (ed.), Memoirs of the Society for American Archaeology, No. 31, 6-14.

McKenzie-Pollack, L., 1969: Archaeological survey of Broughton Island and adjacent mainland in East Baffin Island, N.W.T. Unpublished report to Archaeological Survey of Canada.

Manning, T.H., 1950: Eskimo stone houses in Foxe Basin. Arctic, 3(2): 108-112.

Mary-Rousselière, Guy, 1976: The Palaeoeskimo in northern Baffinland. In: Maxwell (ed.), Memoirs of the Society for American Archaeology, No. 31, 40-57.

Mary-Rousselière, Guy, 1979: The Thule culture on north Baffin Island: early Thule characteristics and the survival of the Thule tradition. In: McCartney A.P. (ed), Thule Eskimo Culture: An Anthropological Retrospective. Archaeological Survey of Canada, Mercury Series, No. 88, 54-75.

Mathiassen, Therkel, 1925: Preliminary report of the Fifth Thule Expedition: Archaeology. Proc. 21st International Congress of Americanists, (Goteborg), 202-215.

Mathiassen, Therkel, 1927: Archaeology of the Central Eskimos, the Thule Culture, and its Position within the Eskimo Culture. Report of the Fifth Thule Expedition, 1921-1924, 4(1 & 2) (Copenhagen).

Maxwell, M.S., 1962: Pre-Dorset sites in the vicinity of Lake Harbour, Baffin Island, N.W.T., Canada. National Museum of Canada, Bulletin, 180:20-55.

Maxwell, M.S., 1973: Archaeology of the Lake Harbour District, Baffin Island. Archaeological Survey of Canada, Mercury Series, No. 6, 362 pp.

Maxwell, M.S. (ed.), 1976a: Eastern Arctic Prehistory: Paleoeskimo Problems. Memoirs of the Society for American Archaeology, No. 31, 170 pp.

Maxwell, M.S., 1976b: Pre-Dorset and Dorset artifacts: The view from Lake Harbour. In: Maxwell, M.S. (ed.), 1976a, 58-78.

Maxwell, M.S., 1979: The Lake Harbour Region-Ecological Equilibrium in
sea coast adaptation. In: McCartney, A.P. (ed), Thule Eskimo Culture:
An Anthropological Retrospective. Archaeological Survey of Canada,
Mercury Series, No. 88, 76-88.

Maxwell, M.S., 1980a: Archaeology of the Arctic and Subarctic Zones.
Ann. Rev. Anthropol., 9:161-185.

Maxwell, M.S., 1980b: Dorset site variation on the southeast coast of
Baffin Island, Arctic, 33(3):505-516.

Maxwell, M.S., 1980c: Preliminary report of the 1980 Field Season
Archaeological Investigations of sites Burton Bay (Kk Dm-1) and Peale
Point (Kk Do-1), Vicinity of Frobisher Bay, N.W.T., Canada.
Unpublished report to Archaeological Survey of Canada, Ottawa, 19 pp.

Maxwell, M.S., 1981: A southeastern Baffin Thule house with Ruin Island
characteristics. Arctic, 34(2):133-140.

Meldgaard, Jörgen, 1960: Prehistoric culture sequences in the Eastern
Arctic as elucidated by stratified sites at Igloolik. In Selected
Papers of the Fifth International Congress of Anthropological and
Ethnological Sciences, A.F.C. Wallace, ed. Philadelphia, 588-595.

Miller, G.H., 1973: Late Quaternary Glacial and Climatic history of
northern Cumberland Peninsula, Baffin Island, N.W.T., Canada.
Quaternary Research, 3(4):561-583.

Neatby, L.H., (Translator and editor), 1977: My Life among the Eskimos:
Baffinland Journeys in the Years 1909 to 1911 by Bernard Hantzsch.
Trans. from German. Inst. for Northern Studies, University of
Saskatchewan, Saskatoon, Sask., 396 pp.

Nichols, H., 1975: Palynological and Paleoclimatic study of the Late
Quaternary Displacements of the Boreal Forest - Tundra Ecotone in
Keewatin and Mackenzie, N.W.T., Canada. Institute of Arctic and
Alpine Research Occasional Paper, No. 15. 87 pp.

Ross, W.G., 1979: Commercial whaling and Eskimos in the eastern
Canadian Arctic, 1819-1920. In: McCartney, A.P. (ed), Thule Eskimo
Culture: An Anthropological Retrospective. Archaeological Survey of
Canada, Mercury Series, No. 88, 242-266.

Sabo, D., and Sabo G., III, 1978: A possible Thule carving of a Viking
from Baffin Island, N.W.T. Canadian Journal of Archaeology, 2:33-42.

Sabo, G., III, 1979a: Archaeological reconnaissance in eastern Baffin
Island. In: McCartney A.P. (ed), Thule Eskimo Culture: An
Anthropological Retrospective. Archaeological Survey of Canada,
Mercury Series, No. 88, 183-250; 329-364.

Sabo, G., III, 1979b: Development of the Thule culture in the historic
period: patterns of material culture change on the Davis Strait coast
of Baffin Island. In: McCartney, A.P. (ed.), Thule Eskimo Culture: An
Anthropological Retrospectice. Archaeological Survey of Canada,
Mercury Series, 88:212-231.

Sabo, G., III, 1981: Thule Culture Adaptations on the South Coast of
Baffin Island, N.W.T. Ph.D. Thesis, Department of Anthropology,
Michigan State University, East Lansing, Michigan. 516 pp.

Sabo, G., III and Jacobs, J.D., 1979: Unpublished report to the
Archaeological Survey of Canada on two archaeological sites near
Frobisher Bay, N.W.T.

Sabo, G., III and Jacobs, J.D., 1980: Aspects of Thule Culture
adaptations in southern Baffin Island. Arctic, 33(3):487-504.

Schledermann, P., 1972: An archaeological site survey of Baffin Island
National Park, Baffin Island, N.W.T. Manuscript Report No. 101,
National Historic Parks and Sites Branch, Parks Canada, Ottawa.

Schledermann, P., 1974: Salvage Archaeology and site survey in
  Cumberland Sound and Merchants Bay Area, Baffin Island, 1973.
  Unpublished Manuscript. Archaeological Survey of Canada, National
  Museum of Man, Ottawa.
Schledermann, P. 1975: Thule Eskimo Prehistory of Cumberland Scound,
  Baffin Island, Canada. Archaeological Survey of Canada, Mercury
  Series, No. 38, 297 pp.
Schledermann, P., 1976a: History of Human Occupation. In: Wilson,
  (ed.), 1976. The Land That Never Melts: Auyuittuq National Park.
  Peter Martin Associates, Toronto, 63-93.
Schledermann, P., 1976b: The effect of climatic/ecological changes on
  the style of Thule culture winter dwellings. Arctic and Alpine
  Research, 8(1):37-47.
Schledermann, P., 1978: Prehistoric demographic trends in the Canadian
  High Arctic. Canadian Journal of Archaeology, 2:43-58.
Schledermann, P., 1979: The "Baleen Period" of the Arctic whale hunting
  tradition. In: McCartney (ed.), Thule Eskimo culture: An
  Anthropological Retrospective. Archaeological Survey of Canada,
  Mercury Series, No. 88, 134-148.
Short, S.K. and Jacobs, J.D., 1982: A 1100 year paleoclimatic record
  from Burton Bay-Tarr Inlet, Baffin Island. Canadian Journal of Earth
  Sciences, 19(3):339-409.
Stenton, D., 1982: Archaeological investigations at the Peale Point
  Site (Kk Do-1) Baffin Island, N.W.T., 1982. Unpublished report to
  Archaeological Survey of Canada, Ottawa, 40 pp.
Stenton, D., 1983: An Analysis of Faunal Remains from the Peale Point
  Site (KkDo-1), Baffin Island, N.W.T. M.A. thesis, Department of
  Anthropology, Trent University Peterborough, Ontario, 204 pp.

Taylor, W.E., Jr., 1966: An archaeological perspective on Eskimo
  economy. Antiquity, 40:114-120.
Taylor, W.E., Jr., 1968: The Arnapik and Tyara Sites: An archaeological
  study of Dorset Culture origins. Society for American Archaeology,
  Memoirs, No. 22, 129 pp.

Vibe, C., 1967: Arctic animals in relation to climatic fluctuations.
  Meddler om Grønland, 170(5):1-227.

Wenzel, G., 1979: Analysis of a Dorset-Thule structure from
  northwestern Hudson Bay. In: McCartney, A.P. (ed), Thule Eskimo
  Culture: An Anthropological Retrospective. Archaeological Survey of
  Canada, Mercury Series, No. 88, 122-133.
Wenzel, G., 1981: Clyde Inuit Adaptation and Ecology. Canadian
  Ethnology Service, National Museum of Man, Mercury Series, No. 77,
  179 pp.

26 Paleoclimatology of the Baffin Bay region

L. D. Williams and R. S. Bradley

The paleoclimatology of the Baffin Bay region is based on a variety of
proxy data, some of which have been discussed in considerable detail in
previous chapters. Here, we compare the paleoclimatic record from
various sources and attempt to identify areas of agreement as well as
inconsistencies in the record, as presently understood. First,
however, we consider the value of paleoclimatic studies in this region.

THE SIGNIFICANCE OF BAFFIN BAY REGION PALEOCLIMATE

How important are climatic fluctuations of the Baffin Bay area (=
Baffin region)? What significance do Baffin region paleoclimatic data
have for other Arctic areas? Meteorological records, in the Arctic
are, unfortunately, quite brief (Chapter 2) but studies of data from
recent decades clearly indicate that temperature fluctuations in the
Baffin region are highly correlated with those elsewhere in the
Arctic. For example, Walsh (1977) showed that in all seasons, the
principal eigenvectors of Arctic surface air temperatures were centered
over Baffin Island, indicating temperature variations in the Baffin
region are typical of a very large part of the Arctic. Similarly, Keen
(1980) demonstrated that temperature fluctuations (1951-76) at c. 80°W
are more highly correlated with the 70°N zone average than any other
zone in the Arctic (Fig. 26.1). Consequently, Keen concluded that
"Baffin temperature [is] a sensitive indicator of summer conditions
across the Arctic as a whole." If such relationships are typical of
longer time scales (probably a reasonable assumption for periods when
surface boundary conditions were not drastically different from those
of recent decades) then the paleoclimatology of the Baffin region
assumes added importance. Paleoclimatic fluctuations of the Baffin
region are thus of significance to studies of the entire Arctic.

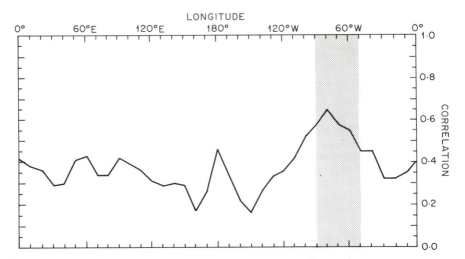

Figure 26.1   Correlation coefficient of summer (J,J,A) mean temperature
at different longitudes at 70°N with the zonal average (1951-76).
Shaded area signifies longitudinal zone from West Greenland to the west
coast of Baffin Island at 70°N. Temperatures in this zone are highly
correlated with the zonal mean (after Keen, 1980) (with permission).

It has often been said that the Baffin region is particularly
'sensitive' to climatic fluctuations (e.g. Tarr, 1897; Andrews et al.,
1972). This sensitivity results from two main factors: low summer
temperatures and the position of Baffin Island in relation to the mean
position of the mid-tropospheric trough over northeastern North
America. On a latitudinal basis (Fig. 26.2) summer temperatures on
Baffin Island are lower than in any other sector of the Arctic (except
for the Greenland Ice Cap) (Barry et al., 1977; Keen, 1980). Thus,
relatively small changes in summer temperature markedly affect snow
cover conditions and glacier mass balance (Bradley and Miller, 1972).
In particular, the upland regions of Baffin Island, which are
extensive, are near the threshold of glacierization; relatively small
decreases in ablation season temperature would initiate large scale
expansion of firn on these uplands (Williams, 1978). Indeed, there is
much evidence that this did occur during the Little Ice Age (17th-19th
centuries) (Locke and Locke 1977; Andrews et al., 1976).

The sensitivity of Baffin Island is also closely related to the
regional circulation patterns. A major feature of the northern
hemispheric general circulation is an upper level trough over Baffin
Bay. The precise position of this trough is very important for
large-scale advection of air across Baffin Island. If the trough is
displaced westward, relatively warm southerly and southeasterly winds

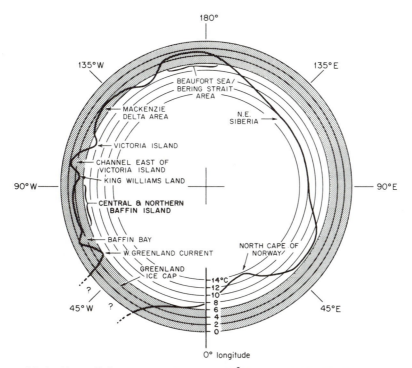

Figure 26.2 Mean July temperature at 70°N around the Northern
Hemisphere (from Barry et al., 1975) (with permission).

increase in frequency leading to above average summer temperatures.
Conversely, with an eastward displacement of the trough axis, airflow
from the north and northwest increases in frequency resulting in
colder, generally drier summer conditions (Fig. 26.3). Statistically,
the relationship between mean trough position in summer and mean summer
temperatures in the Baffin region is strong. For example, for the
period 1949-1976, the correlation coefficient (r) was 0.57 (significant
at >99% level) (Keen, 1980). Such changes in trough position should
not be viewed in isolation from changes in the general circulation of
which they are a part. In a study of the relationships between
hemispheric circulation indices and summer temperatures in the Baffin
area, Keen (1980) convincingly demonstrated that temperature changes on
a hemispheric scale are accentuated in the Baffin region due to the
effects of trough displacement (Fig. 26.4). As hemispheric cooling
occurs, meridional temperature gradients are enhanced, leading to
stronger mid-latitude westerly airflow. This, in turn, causes the
wavelength of upper level waves in the atmosphere to increase, with the
result that the Baffin trough is displaced eastward. The trough may
also be deepened by increased cyclogenesis along the arctic front

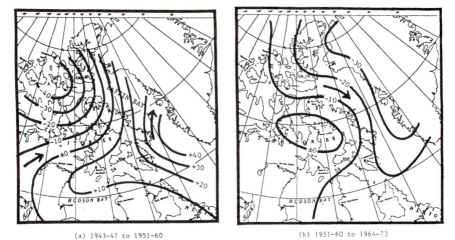

(a) 1943-47 to 1951-60   (b) 1951-60 to 1964-73

Figure 26.3   Changes in mean July 700 mb heights durng two contrasting
periods. From 1943-47 to 1951-60 the mean trough position migrated
westward, resulting in increased southerly and westerly airflow over
Baffin Island and consequently higher summer temperatures.   By
contrast, the mean trough position 1964-73 was further east, leading to
enhanced northerly and northeasterly airflow and significantly lower
summer temperatures (after Keen, 1980).

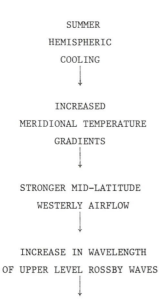

Figure 26.4   Schematic diagram
illustrating relationship between
hemispheric cooling and enhanced
temperature decline in the Baffin
Island area.

leading to more depressions tracking northward into Baffin Bay. As a
result, cold northerly airflow is drawn over Baffin Island,
accentuating the hemispheric coolong in progress, and leading to
reduced regional ablation and enhanced glacierization.

REGIONAL CLIMATIC GRADIENTS

Climatic conditions across Baffin Bay are by no means uniform (Figs
26.5, 26.6 & 26.7), and these sub-regional scale differences have
significance for paleoclimatic studies. Of particular importance is
the marked contrast in mean annual temperature on opposite sides of
Baffin Bay which is related to oceanic circulation patterns. In
eastern Baffin Bay/Davis Strait, the West Greenland Current carries
relatively warm water (and associated warm air) along the west coast of
Greenland. In western Baffin Bay, southward flowing, cold Arctic water
prevails. As a result, although latitudinal temperature <u>gradients</u> are
similar (a decrease of 0.65°C per 1° increase in latitude), mean annual
temperatures on the west Greenland coast are 8°C warmer than at the
same latitude on the east coast of Baffin Island (Fig. 26.5). South of

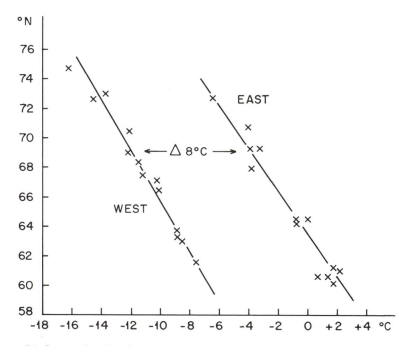

Figure 26.5 Latitudinal mean summer temperature gradients along the
western and eastern sides of Baffin Bay and Davis Strait.

64°N, along the Greenland coast, open water is present throughout the
year (Fig. 26.6(a)). However, on the western side of Baffin Bay,

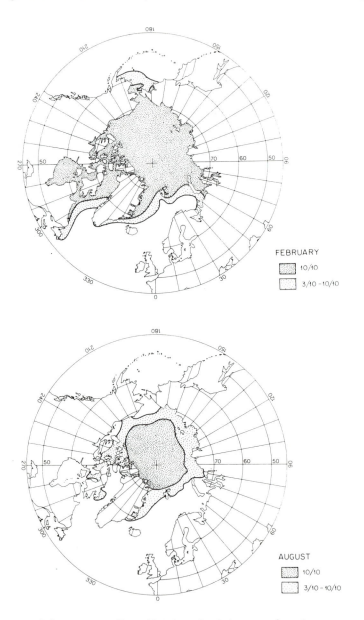

Figure 26.6 (a) Sea-ice distribution in February (maximum extent).
(b) Sea-ice distribution in August (minimum extent). Note contrast in
sea-ice extent in mid-winter between the western and eastern sides of
Davis Strait/Baffin Bay.

mid-winter ice cover extends southward along the Labrador coast as far
as 49°N and even in late summer, considerable ice may remain around the
Cumberland Peninsula and within Foxe Basin (Fig. 26.6(b)). Sea-ice
extent plays an important role in determining the amount of
precipitation which occurs in different parts of the region. As shown
in Figure 26.7, precipitation amounts increase significantly in
southernmost Greenland (south of c. 62°N) where open water occurs
throughout the year. On southeastern Baffin Island, heaviest
precipitation amounts occur where locally extensive open water occurs
year round, as, for example, near Cape Dyer (Fig. 26.6(a) & (b)).
Elsewhere, mean annual precipitation amounts are generally les than 400
mm.

From this consideration of contemporary climatic conditions, it is
apparent that oceanographic conditions play a very important role in
regional climatic differences. More extensive open water in the past
would have been associated with high temperature (though not
necessarily in the ablation season). At the same time, there would

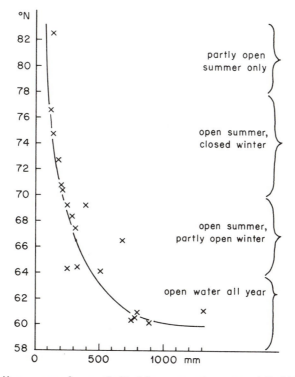

Figure 26.7 Mean annual precipitation amounts around Baffin Bay/Davis
Strait, in relation to seasonal sea-ice extent. Note marked increase
in precipitation towards zone with open water all year (southern
Greenland). Locations with locally persistent polynas year round
(e.g. Cape Dyer at c. 66°N) have anomalously heavy precipitation.

have been an even more pronounced increase in precipitation (since the
precipitation/proximity to open water relationship is non-linear).
Conversely, when the West Greenland current was absent due to more
extensive Arctic water, as for example from >20,000 to c. 10,000 BP
(Ruddiman and McIntyre, 1981), precipitation would have been greatly
reduced and the temperature decrease along the West Greenland coast
would have been much greater than on the coast of eastern Baffin
Island. In short, strong <u>differential</u> changes in climate across Baffin
Bay are likely to have occurred in the past as a result of changes in
large-scale oceanographic conditions.

EVIDENCE USED IN PALEOCLIMATIC STUDIES

The principal lines of evidence used in the reconstruction of
paleoclimates in the Baffin region pertain to temperatures during the
summer, which is the environmentally sensitive period; large changes in
winter temperatures <u>may</u> have taken place in the past, but there are few
means of detecting if such variations occurred. It is possible that
studies of periglacial features (e.g. ice and sand wedges) could
provide a valuable winter season perspective on paleoclimatic
conditions (cf. Hopkins, 1982).

    Proxy data (Table 26.1) from the Baffin area are, of course,
subject to the same limitations as similar data from other areas. The
data are often of dubious paleoclimatic significance and may be
discontinuous in time and space, providing only episodic information.
On Baffin Island, only peat and lake sediments have provided <u>continuous</u>
paleoclimatic records and pollen studies of these sediments have formed
the basis of several regional quantitative paleoclimatic studies (e.g.
Andrews et al., 1980; Andrews and Diaz, 1981; Diaz and Andrews, 1982).
Not all of the records used in these studies are well dated; in
particular, many lake sediment records from northern Labrador have
dates which are out of sequence stratigraphically (Short and Nichols,
1977; Short, 1978), possibly as a result of disturbance and/or
contamination during the coring operation (removal of lower core
sections). Further lake sediment studies are required to resolve the
many uncertainties in these records.

THE PALEOCLIMATIC RECORD

In this section we compare the various proxy data records which bear on
paleoclimatic conditions in the Baffin Island region. Particular
attention is given to events over a few decades (for the more recent

TABLE 26.1. Types of evidence used in paleoclimatic reconstructions in the Baffin Bay area.

Proxy Data	Continuous (C) or Discontinuous (D)	Method of Dating	Time Frame (year)	Dating Uncertainty Factor	Minimum Sampling Interval (yrs)*	Resolution (period length in years)	Paleoclimatic Inference	Remarks	Reference Example
Glacial Deposits (moraines, till, outwash)	D	C-14 lichens	0-10,000+ 5-5,000	±50-200 yrs. ±10%			(Advances): cooler summers and/or higher accumulation. (Recession): warmer summers and/or less accumulation.	Glacier advance or retreat reflects mass balance changes; though these may be complex, they are generally ascribed to changes in summer temperatures.	Carrara and Andrews, 1972. Miller, 1973. Davis, this volume.
Soils	D	C-14	0-5,000+	±50-200			Colder after date on top surface of soil.	Dates generally on uppermost soil profile when overlain by sand, outwash or till. May be dating problem (J.A. Mathews, 1980).	Miller, 1973.
Peat Growth	C or D	C-14	0-7,000+	±50-200	50-600		Summers warmer during peat growth interval, cooler after peat growth ceases.	Basal and/or uppermost dates generally supplemented by pollen data.	Nichols, 1974
Pollen in: a) lake sediment	C	C-14	0-9,200	±100-300		100-1200	Summers warmth; southerly airflow	Calibration with modern data possible to give quantitative	Davis, 1980. Short and Andrews, 1980.

	C or D	Dating	Range	Precision	Resolution		Climate signal	Comments	Reference
b) peat	C or D	C-14	0-7,000	±50-7,000			frequency.	paleoclimatic estimates. Isotopic studies (0-18) possible also (e.g. Gray 1982).	Andrews et al. 1979, 1980.
Ice: a) 0-18	C	Multivariate	0-10,000	±5% to 5,000, increasing to ±10% to 10,000	1-50+ increasing with time	2-100+	Snowfall event temperatures and/or proximity to open water and/or ice surface height.	Seasonal signals possible.	Koerner, 1977. Short and Andrews, 1980. Andrews et al. 1979 - 1980.
b) Melt Layers	C	Multivariate	0-2,000	±5%	1-10+	2-20	Summer temperature.	Percolation may cause meltwater to penetrate deeper layers, thereby "extending" period of apparent warmth.	Fisher and Koerner 1981. Herron et al. 1981.
Tree Rings	C	Annual ring counts	0-300	±1 yr.	1	2	Summer temperature.	Northern Labrador treeline only.	Cropper and Fritts, 1981.
Drift Wood	D	C-14	0-9,000+	±50-200 yrs.			Summer temperature.	Periods when driftwood frequency is highest are thought to reflect reduced sea-ice cover/ open water.	Stewart and England, 1983.

records) to a few centuries. Of course, dating uncertainty and differing responses of different proxy climatic indicators make comparisons on these time scales difficult. However, some knowledge of short-term paleoclimatic variations would be useful, for comparison with similar changes in other regions, and with the short-term fluctuations of atmospheric C-14 (revealed by the departure of radiocarbon ages from dendrochronological ages), which may have been caused by solar variations (Denton and Karlen, 1973).

In view of the scarcity of information and dating uncertainty prior to the postglacial period, consideration is confined to the last 12,000 C-14 years. This period subdivides naturally, on the basis of records of comparable detail, into three nested intervals: (1) the last 12,000 C-14 years BP*, (2) 4000 BC to present, and (3) AD 600 to present. It many not be possible to identify short-term climatic changes in the earlier part of this period, the glacial to interglacial transition, for the evidence must reflect such things as changes in oceanic circulation, rates of glacier calving, and floral/faunal succession, as well as climatic variations. Even later evidence does not necessarily permit a simple climatic interpretion. Changes in vegetation, as measured by pollen assemblages, are probably indicative of summer temperature changes, although moisture conditions may also have been important (Andrews et al., 1980). In this regard, it should be noted that temperature may also control surface-water availability, at least locally, by its effect on thawing of frozen ground and the depth of the active layer in summer. Glacier fluctuations may be related to summer temperature, but could also be caused by changes in snowfall or even by non-climatic factors. The amount of refrozen meltwater in ice cores (Fisher and Koerner, 1981; Herron et al, 1981) can be directly attributed to summer temperature changes, although even this may be complicated by local factors such as radiation climate.

In general, we can only assume that most of the proxy climatic evidence considered (at least in the two later intervals) are indicative of summer temperature variations, with the understanding that some of the discrepancies observed may be due to violations of the various assumptions. Oxygen isotope variations in ice cores are problematical as whatever climatic information they contain is related to conditions on precipitation days, most of which occur in the months of May and September to November (Bradley, 1983). However, two records

---

* 'BP' is to be understood to mean conventional C-14 years before present.

of 0-18 maxima (i.e. maxima in the annual cycles) are included from cores for which data on intra-annual variations are available, and also two mean annual 0-18 records for comparison with meltwater records from the same sites.

The locations from which the data were obtained are shown in Figure 26.8. Some information from locations in the Canadian Arctic outside the Baffin Bay region is included in this study for comparison.

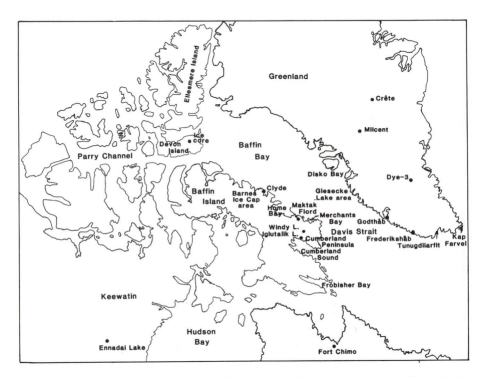

Figure 26.8   Location of places from which data have been used in this study.

THE LAST 12,000 C-14 YEARS

Figure 26.9 compares events during the last 12,000 C-14 years, arranged roughly from northwest at the top to southeast at the bottom. At the beginning of this period, it appears that there was a great difference in climatic conditions between the Baffin Bay region and the western Canadian Arctic. The first postglacial appearance of open water in western Parry Channel about 11,500 BP (Fig. 26.9a) coincided with the

advent of spruce on the Mackenzie River delta, which is north of the
present tree line (Ritchie and Hare, 1971; Ritchie et al., 1983). The
northwestern margin of the Laurentide ice sheet may have been
retreating at the rate of about 200 m per year between 12,000 and
10,000 BP (Bryson et al., 1969; Prest, 1970). In contrast, the retreat
rate of part of the ice sheet in southwestern Greenland during this
time was only 10 m per year (Fig. 26.9k) and the ice margin in eastern
Baffin Island appears to have been stable (Andrews, 1975).

By 10,000-9000 BP, there was open water throughout Parry Channel
and in northern Baffin Bay (Fig. 26.9a). The correspondence between
radiocarbon dates and actual (calendar) dates is not definitely known
before 7000 BP, but at that time the C-14 dates are about 1000 years
too young, and if such is also true of dates 10,000-9000 BP, then the
advent of open water in northern Baffin Bay roughly coincided with the
very abrupt rise of 0-18 values in the Camp Century ice core (Hammer et
al., 1978). Meanwhile, in the southern part of the Baffin Bay region,
there seems to have been a rapid retreat of the ice margin in Frobisher
Bay (Fig. 26.9i) between 10,500 and 8800 BP (and also after c. 10,300
BP in Merchants Bay, not shown in Fig. 26.9). This may have been due
partly to increased rates of calving as relative sea level rose (cf.
Fig. 2 in Andrews, 1982), but it may also have been partly climatic, as
suggested by exotic pollen influx on southwestern Cumberland Peninsula,
10,000-8700 BP (Fig. 26.9g; see Short et al., this volume), and the
intrusion of subarctic species of molluscs in Frobisher Bay about 9700
BP (Fig. 26.9j).

In contrast, there was an advance or stillstand of the ice margin
in southwestern Greenland between 10,000 and 9500 BP (Fig. 26.9k).
However, another period of advance of the southwestern Greenland ice
margin between 8800 and 8100 BP (although interrupted by a rapid
retreat) does coincide with an extensive advance of the Laurentide ice
margin on Baffin Island (the 'moraines of Cockburn age'; Andrews and

Figure 26.9 Some events of the last 12,000 C-14 years in the region.
(a) Inferences made by Blake (1972) on the basis of fossil molluscs and
driftwood in the Queen Elizabeth Island. (b,c,e,f,i,k) Relative rates
of ice margin retreat (up) or advance (down) estimated on Baffin Island
by Andrews (1982) and on southwestern Greenland by Ten Brink and
Weidick (1974). (d,g,h,k,l,m,n,p) Relative changes in summer
temperature inferred from pollen assemblages on Baffin Island (Short et
al., this volume) and in southwestern Greenland (Fredskild, this
volume). (j) Changes in mollusc assemblages on the coasts of Baffin
Island (Andrews, 1972; Miller, 1980) and West Greenland (Laursen,
1976).

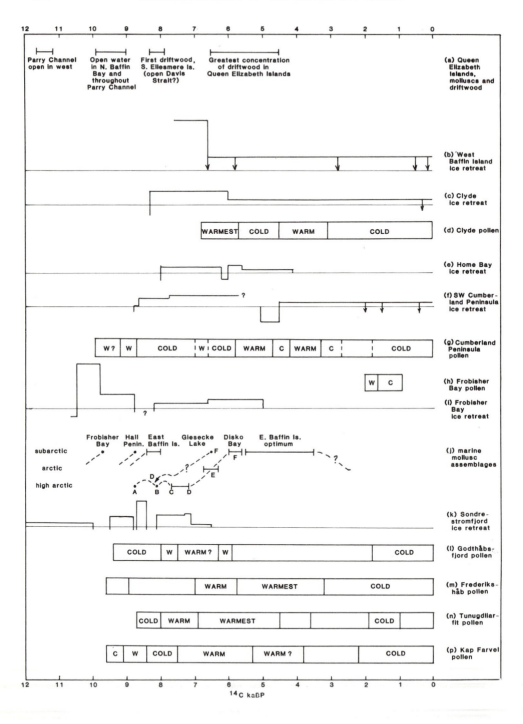

Ives, 1978). Pollen assemblages from Godthaabsfjord and Tunugliarfik (Figs 26.91 & n) indicate that this was a relatively cold period. The pollen record from Kap Farvel (Fig. 26.9p) disagrees in part, but would correlate well with the other evidence if its time scale was a few centuries older.

The glacial records from Clyde, Home Bay, Frobisher Bay, and Sondre Stromfjord (Figs 26.9c, e, i, & k) all agree on the termination date of this glacial stage at 8300-8000 BP. This coincides with the first appearance of driftwood on southern Ellesmere Island (Fig. 26.9a) (which suggested to Blake, 1972 the opening of Davis Strait) and with the (probably related) intrusion of subarctic species of molluscs on the east coast of Baffin Island (Fig. 26.9j). It also coincides with the beginning of a long period of warmth-indicating pollen assemblages in southwestern Greenland (again, with Kap Farvel slightly later).

The evidence from Cumberland Sound on glacier retreat (Fig. 26.9f) and pollen (Fig. 26.9g) appears to be contradictory, both with each other and with the other records. There, ice margin retreat seems to have commenced with the onset of a cold period about 8700 BP which lasted through most of the following three millenia. Another apparent contradiction to climatic amelioration about 8000 BP is the continued presence of high arctic mollusc species in the Disko Bay (West Greenland) area until about 7000 BP (horizons C and D, Fig. 26.9j). This may be a matter of succession rather than climate as both Laursen (1976) and Miller (1980) have noted the time-transgressive spread of subarctic molluscs on the coasts of West Greenland and Baffin Island, respectively (see Fig. 26.9j; see Fig. 26.8 for the locations). Curiously, the subarctic species seem to have arrived first on the Baffin Island coasts, although today they are restricted in this region to the warm West Greenland Current (Andrews, 1972).

Andrews and Ives (1972) suggested that ice margin advance (or stillstand) on Baffin Island in the Cockburn Stade might be attributed to increased snowfall due to the incursion of warmer water into Davis Strait while the land remained cold. Given the dating uncertainty, this is certainly a possibility, but an alternative explanation could be that ice margin advance (or stillstand) was associated with a cold period, which was terminated upon incursion of warmer water into Davis Strait about 8300-8000 BP. Although not necessarily associated, it should be noted that this was also about the time of the breakup of the Laurentide ice sheet over Hudson Bay (Andrews and Falconer, 1969).

The middle Holocene records from the Baffin Bay region are difficult to characterize. The greatest concentration of driftwood in the Queen Elizabeth Islands, presumably implying the most open water, occurred between 6500 and 4200 BP (Fig. 26.9a; Stewart and England, 1983). The climatic optimum in the Clyde (Patricia Bay) pollen record at 6800-5700 BP is in partial agreement (Fig. 26.9d), but also coincides with ice margin advance, or at least reduced retreat rate, in central Baffin Island (Figs 26.9b, c, & e). The latter may perhaps be explained as a response of the ice sheet to the cessation of calving when it became wholly land-based (Andrews, 1973, 1982).

A readvance of the ice on Cumberland Peninsula at 5000-4500 BP (Fig. 26.9f), and reduced retreat rate in the Home Bay area at 5600-4100 BP (Fig. 26.9e), may be related to the severe cold period (Short et al., this volume) indicated by the Patricia Bay pollen record between 5700 and 4500 BP (Fig. 26.9d) though there is considerable uncertainty in the dating of the lower part of this core. By contrast the pollen evidence from Iglutalik Lake (Fig. 26.9g) is entirely contrary to the Patricia Lake record during this period and other evidence points to relatively mild conditions in Baffin Bay and southwestern Greenland at this time. For example, subarctic species of molluscs flourished on the east coast of Baffin Island between 5500 and 3500 BP (Fig. 26.9j; Andrews, 1972), and the climatic optimum in the pollen record from Frederikshab covers nearly the same interval, 5800-3200 BP (Fig. 26.9m). Other pollen records from southwestern Greenland indicate a generally warm climate from c. 8000 BP to at least 4000 BP. Eigenvector analysis of pollen-based temperature records from Baffin sland, Labrador and Keewatin also indicate relatively warm conditions from 5500 to 3500 BP (Andrews and Diaz, 1981).

THE LAST 6000 YEARS

Figures 26.10 and 26.11 show records covering the last 6000 and 1400 calendar years, respectively. For comparison with ice-core and tree-ring records, radiocarbon dates and the time scales of C-14-dated records have been converted to calendar years according to the dendrochronological calibration of Stuiver (1982), which covers the last 2000 years, and Ralph et al. (1973) for older dates. The calibration of Stuiver (1982) is probably very accurate, but another recent calibration by Klein et al. (1982) indicates that some of the earlier BC dates used here (i.e. according to the Ralph et al. 1973 curve) are about a century too old at times. However, the Klein et al. (1982) data are presented in a form which is not easy to use, and since no ice-core records before 300 BC are considered anyway, the older calibration is used. The radiocarbon date list and conversions are given in Appendix A.

Eight different kinds of paleoclimatic evidence are presented in
Figures 26.10 and 26.11: (1) Wood (charcoal) north of present tree
line, (2) transfer function reconstructions of summer temperature based
on pollen assemblages, (3) lichenometric dates on moraines, (4)
radiocarbon dates on various deposits which suggest climatic change,
(5) proglacial sediment stratigraphy, (6) tree-ring width, (7) percent
refrozen meltwater in ice cores, and (8) oxygen isotope ratios (0-18)
in ice cores (cf. Table 26.1).

For the sake of comparability, and also for smoothing, the time
series in Figure 26.10 have been reduced to approximate 120-year
running means, 120 years being the approximate sampling interval of two
of the records (Figs 26.10c and the upper part of 26.10b). At the top
of Figure 26.10, two kinds of evidence from a location outside the
Baffin Bay region, the vicinity of Ennadai Lake in Keewatin, are shown
for comparison, both with each other and with the records from the
Baffin Bay region. Back to about 1000 BC there is very good agreement
between dates on fossil wood (charcoal) north of present tree line
(Fig. 26.10a) and the pollen transfer function reconstruction of July
temperature at Ennadai Lake (26.10b). However, the dates on the
farthest northward extension of forest, at around 2000 BC, coincide
with a relative low in the July temperature reconstruction, although it
was still warmer than most of the post-1000 BC part of the record. The
lifetime of this forest is unknown, but it may have existed through the
long warm period before about 2200 BC indicated by the pollen record,
and survived the subsequent cooling, until destroyed by fire. No wood
from north of present tree line in this region has been reported which
dates from later in the 2nd millennium BC (for which the transfer
function also gives high temperatures), but there are several dates on
spruce above present tree-line altitude in the Yukon from 3380-3050
C-14 years BP (c. 1900-1400 BC) (Rampton, 1971; Denton and Karlen,
1973).

The Keewatin records provide a well-dated and reasonably
consistent climatic history with which to compare the data from the
Baffin Bay region. The episode of relatively cool summers about
2100-1700 BC, which interrupted very warm intervals, is also found in
the Baffin Island pollen transfer function results from Iglutalik Lake
(Fig. 26.10c) and Windy Lake (Fig. 26.10d). However, the Baffin Island
transfer function results show little resemblance after 1000 BP, other
than long-term climatic deterioration. Their dating is uncertain to
some degree, but it seems impossible to reconcile them entirely by any
reasonable shift of their time scales. The part of the Iglutalik Lake
profile (Fig. 26.10c) which has been analyzed by transfer function is
not at all well-dated (Appendix A); the top of the section is assumed

to be modern (Davis, 1980), but the long climatic quiescence after 900 AD which this implies, makes the assumption seem questionable. The Windy Lake record (Fig. 26.10d) has breaks where sand layers in the section were not sampled, which are interpreted by Andrews et al. (1980) as representing cold intervals. This section has many C-14 dates, but there is a fair amount of scatter in the depth-age data, and we have derived a time scale simply by a linear least-squares fit (see Appendix A).

The Maktak Fiord record (Fig. 26.10e) is well-dated, with four very consistent dates within 1300 years. However, much of the variation in the Maktak series (even in the non-smoothed data) is about the same as the standard error of estimate (about 1°C) of the transfer function used to derive it (Table 7 in Andrews et al., 1980). Indeed, this is true of much of the post-1000 BC variation in all three Baffin Island transfer function temperature reconstructions (note the difference in scales on Figs 26.10b, c, d, & e).

Some indication of climatic change on Baffin Island may perhaps be gained from lichenometric dates on moraines (Fig. 26.10f) and radiocarbon dates on various kinds of deposits which suggest transition from warmer to colder conditions or the converse (Fig. 26.10g). these data are drawn as triangles with the points toward the colder episodes (except for one date marked W which indicates warm conditions, and the lichenometric dates after 1700 AD, which are too numerous to portray in this way). The lichenometric dates from the vicinity of the Barnes Ice Cap and from Cumberland Peninsula are plotted together in Figure 26.10f, for they fall in similar clusters, using the same lichen growth curve for both (as suggested by Andrews and Barnett, 1979).

Figure 26.10   Data (and derived data) from the last 6000 calendar years. (a) Radiocarbon dates on fossil wood (charcoal) north of present tree line in Keewatin (Bryson et al., 1965; Bender et al., 1965, 1966, 1967). (b,c,d,e) Transfer function reconstructions of mean July temperature based on pollen assemblages in Keewatin and Baffin Island (Andrews et al., 1980; Andrews and Nichols, 1981; data supplied by J.T. Andrews). Reduced to approximate 120-year running means, except where the sampling interval is about 120 years. Arrows denote C-14 dates. (f) Lichenometric dates on Baffin Island moraines (Miller, 1973; Andrews and Barnett, 1979; Davis, 1980). Minimum dates for glacier advance. (g) Radiocarbon dates on deposits suggestive of climatic change, Cumberland Peninsula, Baffin Island (Miller, 1973; Dyke, 1977). Triangles point toward colder episodes. See Appendix B. (h) Percent refrozen meltwater in an ice core from southern Greenland (Herron et al., 1981), 120-year running mean.

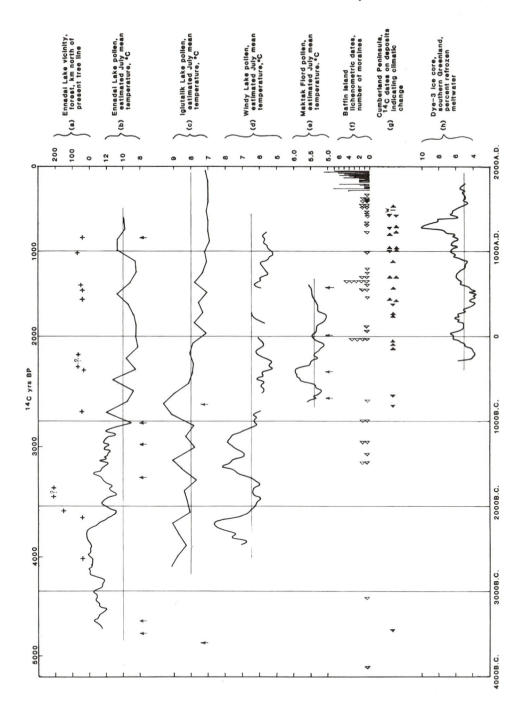

The oldest dates are widely scattered, and the lichenometric dates may be considerably in error, for the lichen growth curve is extrapolated from a control date at 500 BC. A cluster of lichenometric dates after 1500 BC indicates termination of a glacial advance which is probably related to the relatively cold period of 2100-1700 BC shown by the Windy Lake pollen data (the mean July temperature of about 6°C obtained for this interval by the transfer function is just below the 6.5°C it predicts from the modern pollen rain). Scattered lichenometric dates and C-14 dates on basal peat during the early 1st millennium BC are not clearly associated with the pollen transfer function results.

The cluster of lichenometric dates about 2000 years ago may be related to the cooling episode 400-100 BC in the Maktak (Fig. 26.10c) record (which amounts to 1.4°C in the non-smoothed data, and is therefore probably significant). The Dye-3 meltwater record (Fig. 26.10h) is also compatible with these lichenometric dates. The other Baffin Island data (Figs 26.10c, d, & g) are contradictory; however, C-14 dates (Fig. 26.10g) are on soils underlying eolian deposits (which Dyke, 1977, interpreted as a change to cold, dry conditions), and radiocarbon dates on soils can be centuries too old (Stuckenrath et al., 1979; Matthews, 1980). It is likely that the lichens grew on moraines deposited 200-100 BC, and the eolian sands were deposited during a later cold episode.

A large cluster of lichenometric dates after 500 AD (with a peak at 650 AD), together with two 5th century C-14 dates on glacial outwash (Fig. 26.10g; Appendix B), indicate a glacial advance sometime around the 5th century AD. This agrees very well with the low at that time in

Figure 26.11 The last 1400 years. (a) As Figure 26.10, but for individual samples at 40-year intervals. (b) As Figure 26.10f. (c) As Figure 26.10g. (d) Sediment stratigraphy in front of a Baffin Island glacier (Miller, 1973). Dashed - silt; fine stipple - fine sand; coarse stipple - coarse sand, gravel, cobbles; shaded - organic horizon. See text for interpretation. (e) Tree-ring width indices from vicinity of Fort Chimo, northern Quebec (Cropper and Fritts, 1981), 30-year running means. (f) Percent refrozen meltwater in two cores from the Devon Island ice cap (Fisher and Koerner, 1981), 30-year running means. (g) As Figure 26.10h, but 20-year means (up to AD 1300) and 30-year running means (after AD 1300). (h,i) 30-year running means of annual maxima of 0-18 in Greenland ice cores, from (h) Milcent, and (i) Crete (data supplied by World Data Center A for Glaciology). (j,k) 30-year running means of mean annual 0-18 in ice cores from (j) Devon Island (Paterson et al., 1977), and (k) Dye-3 Greenland (Herron et al., 1981).

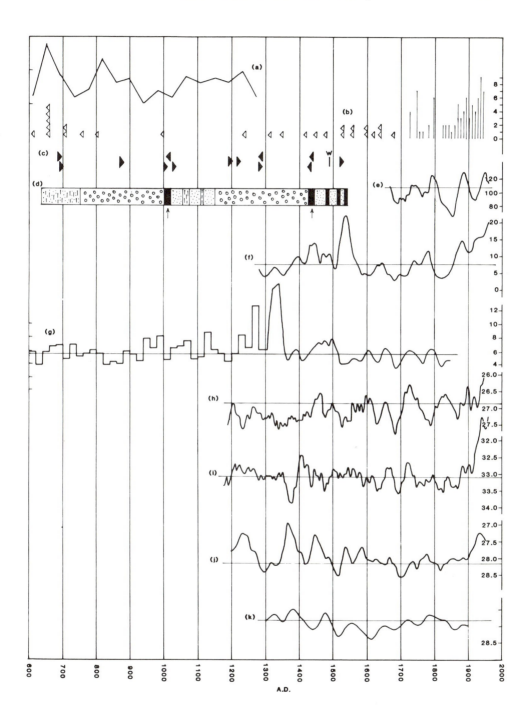

the Dye-3 meltwater record (Fig. 26.10h), and with the interpretation
that the contemporaneous sand layer in the Windy Lake section (Fig.
26.10d) represents a cold period. Variations in the Iglutalik Lake and
Maktak Fiord records (Fig. 26.10c, e) during this period are not
significant. The timing of this cold episode is worthy of some
consideration, for if it occurred in the 5th century in the Baffin Bay
region, then it was clearly opposite to climatic change in Keewatin
(Figs 26.10a & b).

600 AD TO PRESENT

The records for the last 1400 years (Fig. 26.10) may be considered in
somewhat more detail than the previous sets. The points on the Windy
Lake transfer function results (Fig. 26.11a) represent individual
samples at about 40-year intevals. As noted previously, this record is
ambiguously dated, and this is especially so in the upper part of the
section (Appendix A). However, the concentration of lichenometric
dates (Fig. 26.11b) around the 7th century AD does imply a warming at
that time, in agreement with the Windy Lake result (see previous
section).

     The set of C-14 dates on deposits indicative of climatic change
(Fig. 26.11c and Appendix B), together with the sediment stratigraphy
in front of a glacier (Fig. 26.11d), provide a consistent and fairly
detailed history of climatic change on the Cumberland Peninsula of
Baffin Island between the 7th and 16th centuries AD. The proglacial
sediment stratigraphy is interpreted by assuming that gravel was
deposited when the glacier was close to the site, and fine sediment or
organic material when it was withdrawn. This differs from the original
interpretaiton of Miller (1973), but agrees with most of the
radiocarbon-dated deposits (Fig. 26.11c), and follows the
interpretation of Patzelt (1974) for a similar proglacial sediment
section in the Alps. It also agrees very well with the meltwater
record from the Devon Island ice cores (Fig. 26.11f).

     Later, in the 17th to 20th centuries, the Devon Island meltwater
record is remarkably similar to the Fort Chimo (northern Quebec)
tree-ring series (Fig. 26.11e). The Baffin Island lichenometric dates
(Fig. 26.11b) are fairly consistent with the Devon Island record (Fig.
26.11f) after 1680. However, the scattered lichen dates in the
preceding few centuries show no clear association with the other
evidence from the eastern Canadian Arctic.

The 'Little Ice Age' minima in the Devon Island meltwater record
at around 1600 AD, the late 17th century, and the early 19th century,
are very similar in timing to those in many other proxy paleoclimate
records from North America and Europe (Lamb 1977; Williams and Wigley,
1983). Prior to the late 16th century, however, there appear to be
some important differences between climatic change in the eastern
Canadian Arctic and the rest of North America and Europe.

Many proxy climate records from North America and Europe show a
major transition from a cold episode in the 9th - 10th centuries (in
places comparable in severity to the Little Ice Age) to a very warm
episode in the 11th - 12th centuries or later (Williams and Wigley,
1983). In Europe, the latter is known as the 'Medieval Warm Period'
(Lamb, 1977). The Windy Lake temperature reconstruction (Fig. 26.11a)
and our interpretation of the proglacial sediment section (Fig. 26.11d)
do imply warming from the 10th to the 11th century. However, the
temperature increase at Windy Lake was less than the standard error of
th transfer function used to derive it. Also, interpretation of the
C-14-dated deposits (Fig. 26.11c and Appendix B) in the early 11th
century is ambiguous, and only one moraine on Baffin Island has been
lichenometrically dated to the 10th or 11th century. Thus, there may
have been a slight climatic amelioration on Baffin Island at the time,
but it appears to have been far less of a change than that implied by
records from elsewhere in North America and Europe. Similarly, there
was only a slight increase in snowmelt at Dye-3 (Greenland) at the time
(Figs 26.10h, & 26.11g).

Summer climate in the eastern Canadian Arctic seems to have been
even more anomalous in the 15th to early 16th centuries (Williams and
Wigley, 1983). In the western United States, ring-widths of
high-altitude, summer temperature-sensitive trees indicate a major cold
period in the late 15th century, locally perhaps the most severe of the
Little Ice Age (LaMarche, 1974; LaMarche and Stockton, 1974), and
well-dated glacier advances occurred at the same time in the Yukon
(Denton and Stuiver, 1966). Proxy climatic data from Europe also
indicate a period of relatively cold summers in the 15th century (e.g.
Schweingruber et al., 1978). In contrast, evidence from the esatern
Canadian Arctic (Fig. 26.11c, d, & f) indicates that summers were
relatively warm in the 15th and early 16th centuries, following a cold
period in the 13th - 14th centuries. However, it must be noted that
this evidence, although strong, does not receive support from
lichenometric dates (Fig. 26.11b).

The date (1490) marked 'W' in Figure 26.11c is on organic material
at the bottom of an ice wedge on Baffin Island, which implies seasonal

thawing of ground perennially frozen at present. This was also the
case in southern Greenland at the time, for material from burials of
Norse settlers dating up to at least the late 15th century have been
recovered from ground which was perennially frozen at the time of
excavation in the early 20th century (Nörlund, 1924; Hovgaard, 1925).
The good state of preservation of these materials suggested that the
ground became perennially frozen not long after burial, and remained so
until excavation.

These burials give general support to the meltwater record from
the nearby Dye-3 ice core (Fig. 26.11g), which has high values during
the time of Norse settlement from the late 10th century until the early
16th century, and generally low values thereafter. However, the Dye-3
meltwater record shows very little agreement with that from Devon
Island (Fig. 26.11f); in fact the differences between the two in the
14th and 16th centuries are astounding. The two sites are a fair
distance apart (Fig. 26.8) so it is possible that local climatic
changes were quite different. Some difference is even perhaps to be
expected, considering that melting on the Devon Island ice cap is
strongly dependent on synoptic climate, especially on the frequency of
Baffin Bay lows (Alt, 1978), which might have an opposite effect on
melting at Dye-3. Yet we found that the Devon Island meltwater series
bears a strong resemblance to the tree-ring series from distant Fort
Chimo (Fig. 26.8) during the last few centuries, and the minima in the
Dye-3 meltwater record (Fig. 26.10h) seem to be closely associated with
clusters of lichenometric dates on Baffin Island moraines (Fig.
26.10f). The question of synchrony of climatic change even within the
eastern Canadian Arctic - southern Greenland region remains unresolved,
much less that over the entire Northern Hemisphere or globally (cf.
Denton and Karlen, 1973).

An attempt has been made here to relate oxygen isotope data from
two Greenland ice cores to summer temperature, by taking only the
maximum 0-18 values in an annual cycle. This was done for the Milcent
(Fig. 26.11h) and Crete (Fig. 26.11i) cores, for the period of record
since about 1200 AD when data on annual variations are available (date
provided by P.K. MacKinnon, World Data Center A for Glaciology, 1981).
As seen in Figures 26.11h & i, the exercise was not particularly
informative, as the two records generally disagree with each other and
with the Dye-3 meltwater record, except perhaps for some similarity
between the Dye-3 and Milcent series (Figs 26.11g & h) from 1400 to
1700 AD.

Finally, the mean annual 0-18 records from the Devon Island and
Dye-3 ice cores (Figs 26.11j & k) are shown for comparison with the

summer melt records from the same locations (Figs 26.11f & g). In the
case of Devon Island, there is a fair correspondence between the two
curves after about 1650, and around 1300. However, between 1350 and
1600 the two curves show little resemblance. The mean annual 0-18
record and melt record from Dye-3 (Figs 26.11g & k) are even less in
agreement.

SUMMARY

There is little coherence in the pattern of early- to mid-Holocene
climatic change in the Baffin Bay region when various records of
ice-margin retreat, pollen, and mollusc assemblages are compared. The
period of ice margin readvance on Baffin Island (and to some extent, in
southwest Greenland) sometime between 8000 and 9000 C-14 years BP may
have terminated synchronously with the opening of Davis Strait, the
influx of subarctic species of molluscs on the east coast of Baffin
Island, and the start of a long period of relatively warm-climate
pollen assemblages in southwestern Greenland. Subsequently, a climatic
optimum seems to have prevailed in most of the Baffin Bay region until
about 3000 C-14 years BP, although interrupted in places by colder
episodes.

For the period 4000 BP to present, pollen transfer function
reconstructions of summer temperatures in Keewatin and Baffin Island
indicate generally warm conditions up to the first millennium BC, and
generally colder thereafter, although they differ greatly in detail.
They do agree on a relatively cold interval within the earlier period,
from about 2100-1700 BC, and this is supported by a cluster of
lichenometric dates on Baffin Island moraines. However, a
far-northward extension of treeline in Keewatin is also dated within
the period 2100-1700 BC. After 100 BC, dates on wood north of the
present treeline in Keewatin compare well with temperature peaks based
on palynological (transfer function) reconstruction for the same
vicinity at about 900 BC, 500-300 BC, 400-600 AD, and 1000-1200 AD. On
Baffin Island, two episodes of glacier expansion, dated
lichenometrically, compare very well with minima in a southern
Greenland ice-core meltwater record at around 200 BC and 400-500 AD.
The latter contrasts with the warm episode in Keewatin at that time.

Various kinds of evidence for climatic change in the eastern
Canadian Arctic during the last 1400 years are reasonably consistent,
but those from the Greenland ice cores disagree. The combined evidence
from the eastern Canadian Arctic suggests that after a relatively warm
7th century AD, the climate deteriorated into the 10th century. Warmer

conditions then prevailed in the 11th and 12th centuries (as in
Keewatin, and indeed elsewhere in North America and Europe), followed
by a cooling to the 14th century. The 15th and early 16th centuries
were warm; at times, at least as warm as at present. This contrasts
with evidence (tree-ring and glacial) for a cold episode in western
North America and Europe in the 15th century. The Devon Island
ice-core meltwater record indicates a sharp cooling in the late 16th
century. This record, the northern Quebec tree-ring record, and the
Baffin Island lichenometric dates on moraines all show very good
agreement from 1650 to the present. These indicate cold periods in the
late 17th century and early 19th century, with warming during the 18th
century and late 19th to 20th century (the latter interrupted by
cooling around 1900).

It would be naive to suppose that the few cases discussed, in
which two or more kinds of paleoclimatic evidence agree, provide a
definitive history of climatic change in the region. Some may be
fortuitous, and there are many other cases in which there is
disagreement among different records. The inconsistencies may be due
to misinterpretation, dating uncertainty, or differing responses to
climatic change, as well as the possibility of local anomalies
Paleoclimatic interpretation is only made possible by aggregation of
different kinds of supporting evidence, and this survey shows that
there is a need for much additional information to be collected in the
Baffin Bay region.

APPENDIX A

SOURCES OF DATA, AND CONVERSION OF RADIOCARBON DATES TO CALENDAR YEARS

Figure No.	Reference
26.9a	Blake (1972)
26.9b, c, e, f, i	Andrews (1982)
26.9d, g, h	Short et al. (this volume)
26.9j	Andrews (1972), Laursen (1976), Miller (1980)
26.9k	Ten Brink and Weidick (1974)
26.9l, m, n, p	Fredskild (this volume)

Fig. No.	Calendar AD/BC	C-14 age BP	Lab. No.	Reference
26.10a	AD 1160	880 180	WIS-5	Bender et al. (1965)
		1050 180		
	AD 980	1140 90	WIS-17	"
	AD 600	1450 90	WIS-15	"
	AD 540	1530 80	WIS-96	Bender et al. (1966)
	AD 430	1590 80	WIS-37	Bender et al. (1965)
	220-360 BC	2140 80	WIS-136	Bender et al. (1967)
	400 BC	2210 160	WIS-29	Bender et al. (1965)
	890 BC	2670 105	WIS-93	Bender et al. (1966)
	1790-1890 BC	3430 110	WIS-12	Bender et al. (1965)
	2060 BC	3540 110	WIS-52	Bender et al. (1966)
	2140 BC	3650 100	WIS-80	Bender et al. (1966)
	2620 BC	4000 160	WIS-7	Bender et al. (1965)
26.10b	AD 1160	870 60	Gak-5062	Nichols (1975)
	1020 BC	2790 100	Gak-5061	"
	1270 BC	2960 100	Gak-5060	"
	1660 BC	3340 120	Gak-5059	"
	3350 BC	4520 110	Gak-5057	"
	3500 BC	4690 140	GSC-1781	"
26.10c	795 BC	2565 190	GX-6292	Davis (1980)
	3610 BC	4765 200	GX-5625	"
26.10d	AD 1300	640 155	Gak-5449	Davis (1980)(mean depth 2.5 cm)
	AD 1030	960 200	Gak-5450	" (7.5 cm)
	AD 1200	850 65	DIC-327	" (15.5 cm)
	AD 560	1500 85	DIC-390	" (62 cm)
	AD 10	1990 180	Gak-5411	" (69 cm)
	130 BC	2060 85	Gak-5412	" (90 cm)
	730 BC	2470 390	DIC-515	" (99 cm)
	1060 BC	2825 65	SI-2950	" (128 cm)
	1410 BC	3070 75	DIC-402	" (157.5 cm)
	2140 BC	3650 200	SI-2556	" (205 cm)
	2050 BC	3525 60	SI-2951	" (217.5 cm)

Least-squares fit: Date (AD)   1273 = 16.545 x depth(r^2 = .97)

Fig. No.	Calendar AD/BC	C-14 age BP	Lab. No.	Reference
26.10e	AD 580	1480 160	Birm-370	Boulton et al. (1976)
	AD 25	1970 200	Birm-535	"
	410 BC	2240 190	Birm-536	"
	765 BC	2500 170	Birm-380	"
26.10f	Historical dates			Miller and Andrews (1972)
	AD 1290	680 80	Gak-3722	Miller (1973)
	500 BC	2400 90	Gak-1992	Miller and Andrews (1972)
26.10g	See Appendix B			
26.11a	See 26.10d			
26.1b	See 26.10f			
26.11d	AD 1435	450 130	Gak-3726	Miller (1973)
	AD 1010	1010 100	Gak-3725	"

Radiocarbon dates on deposits which indicate climatic change,
Cumberland Peninsula, Baffin Island

Lab. No.	C-14 yrs BP	AD/BC	Description of deposit	Interpre- tation
Gak-3099	330±90	AD 1520	Dead moss in area of lichen kill	Cold after
Gak-2983	350±100	1490	Organic debris in bottom of ice wedge	Warm
Gak-3357	430±90	1440	Soil overlain by peat	Warm after?
Gak-3726	450±130	1435	Organic horizon underlain by outwash	Cold before
Gak-3098	680±90	1290	Base of peat section	Warming?
Gak-2792	730±70	1280	Top of peat section	Cooling?
Qu-305	830±70	1215	Soil overlain by eolian sand	Cold after
Gak-3094	850±110	1190	Top of peat overlain by eolian sand	Cold after
BGS-267	970±80	1025	Soil overlain by eolian sand	Cold after
Gak-4839	970±70	1025	Organic loess overlain by eolian sand	Cold after
Gak-3725	1010±100	1020	Soil underlain by outwash	Cold before
SI-2550	1025±100	1000	Dead moss in area of lichen kill	Cold after
Qu-301	1170±150	870	Peaty soil overlain by outwash	Cold after
Gak-3160	1260±150	690	Soil buried between moraine crests	Cold after
Gak-4307	1290±100	685	Soil overlain by loess	Cold after
BGS-268	1500±80	560	Soil overlain by eolian sand	Cold after
Qu-307	1610±230	425	Peaty soil overlain by outwash	Cold after
Qu-303	1640±130	410	Peaty soil overlain by outwash	Cold after
Gak-2575	1670±90	390	Base of peat section	Warming?
SI-1703	1740±70	260	Soil overlain by eolian sand	Cold after
GSC-2084	1790±80	230	Soil overlain by eolian sand	Cold after
SI-1700	2015±60	60 BC	Soil overlain by eolian sand	Cold after
SI-1702A	2025±105	95 BC	Soil overlain by eolian sand	Cold after
Gx-3271	2080±190	150 BC	Soil overlain by eolian sand	Cold after
BGS-269	2450±90	660 BC	Base of peat	Warming?
SI-2555	2570±75	800 BC	Base of peat	Warming?
SI-1699	4660±90	3440 BC	Base of peat	Warming?

REFERENCES

Alt, B.T., 1978: Synoptic climatic controls of mass-balance variations on Devon Island ice cap. Arctic and Alpine Research, 10:61-80.

Andrews, J.T., 1972: Recent and fossil growth rates of marine bivalves, Canadian Arctic, and Late-Quaternary arctic marine environments. Palaeogeography, Palaeoclimatology, Palaeoecology, 11:157-176.

Andrews, J.T., 1973: The Wisconsin Laurentide ice sheet: dispersal centers, problems of rates of retreat and climatic implications. Arctic and Alpine Research, 5:185-199.

Andrews, J.T., 1975: Support for a stable late Wisconsin ice margin (14,000 to c. 9000 BP): a test based on glacial rebound. Geology, 3:617-620.

Andrews, J.T., 1982: Holocene glacier variations in the eastern Canadian Arctic: a review. Striae, 18:9-14.

Andrews, J.T. and Barnett, D.M., 1979: Holocene (Neoglacial) moraine and proglacial lake chronology, Barnes Ice Cap, Canada. Boreas, 8:341-358.

Andrews, J.T. and Diaz, H.F., 1981: Eigenvector analysis of reconstructed Holocene July temperature departures over Northern Canada. Quaternary Research, 16:373-389.

Andrews, J.T. and Falconer, G., 1969: Late glacial and postglacial history and emergence of the Ottawa Islands, Hudson Bay, N.W.T.: evidence on the deglaciation of Hudson Bay. Canadian Journal of Earth Sciences, 6:1263-1276.

Andrews, J.T. and Ives, J.D., 1972: Late-and postglacial events (<10,000 BP) in the eastern Canadian Arctic with particular reference to the Cockburn moraines and break-up of the Laurentide Ice Sheet. In: Y. Vasari, H. Hyvärinen and S. Hicks (eds.), Climatic change in Arctic areas during the last ten-thousand years. Oulu, Finland, University of Oulu, 149-174.

Andrews, J.T. and Ives, J.D., 1978: "Cockburn" nomenclature and the late Quaternary history of the eastern Canadian Arctic. Arctic and Alpine Research, 10:617-633.

Andrews, J.T. and Nichols, H., 1981: Modern pollen deposition and Holocene paleotemperature reconstruction, central northern Canada. Arctic and Alpine Research, 13:387-408.

Andrews, J.T., Mode, W.N., and Davis, P.T., 1980: Holocene climate based on pollen transfer functions, eastern Canadian Arctic. Arctic and Alpine Research, 12:41-64.

Andrews, J.T., Barry, R.G., Bradley, R.S., Miller, G.H., and Williams, L.D., 1972: Past and present glaciological responses to climate in eastern Baffin Island. Quaternary Research, 2:303-314.

Andrews, J.T., Davis, P.T., and Wright, C., 1976: Little Ice Age permanent snowcover in the eastern Canadian Arctic: extent mapped from Landsat-1 satellite imagery. Geografiska Annaler, 58A:71-81.

Andrews, J.T., Webber, P.J. & Nichols, H., 1979: A late Holocene pollen diagram from Pangnirtung Pass, Baffin Island, N.W.T., Canada. Review of Palaeobotany and Palynology, 27:1-18.

Barry, R.G., Arundale, W.H., Andrews, J.T., Bradley, R.S., and Nichols, H., 1977: Environmental change and cultural change in the eastern Canadian Arctic during the last 5000 years. Arctic and Alpine Research, 9:193-210.

Bender, M.M., Bryson, R.A., and Baerreis, D.A., 1965: University of Wisconsin radiocarbon dates I. Radiocarbon, 7:399-407.

Bender, M.M., Bryson, R.A., and Baerreis, D.A., 1966: University of Wisconsin radiocarbon dates II. Radiocarbon 8, 522-533.

Bender, M.M., Bryson, R.A., Baerreis, D.A., 1967: University of Wisconsin radiocarbon dates III. Radiocarbon 9, 530-544.

Blake, W., Jr., 1972: Climatic implications of radiocarbon-dated driftwood in the Queen Elizabeth Islands, arctic Canada. In: Vasari, Y., Hyvarinen, H., and Hicks S., (eds.), Climatic change in Arctic areas during the last ten-thousand years. Oulu, Finland, University of Oulu, 77-104.

Boulton, G.S., Dickson, J.H., Nichols, H., Nichols, M., and Short, S.K., 1976: Late Holocene glacier fluctuations and vegetation changes at Maktak Fiord, Baffin Island, N.W.T., Canada. Arctic and Alpine Research, 8:343-356.

Bradley, R.S., 1983: Arctic precipitation-temperature relationships and the interpretation of ice core isotopic records. Abstracts 12th Arctic Workshop, Contribution No. 44, Department of Geology and Geography, University of Massachusetts, Amherst, p. 16.

Bradley, R.S. and Miller, G.H., 1972: Recent climatic change and increased glacierization in the eastern Canadian Arctic. Nature, 237:385-387.

Bryson, R.A., Irving, W.N., and Larsen, J.A., 1965: Radiocarbon and soil evidence of former forest in the southern Canadian tundra. Science, 147:46-48.

Bryson, R.A., Wendland, W.M., Ives, J.D., and Andrews, J.T., 1969: Radiocarbon isochrones on the disintegration of the Laurentide Ice Sheet. Arctic and Alpine Research, 1:1-14.

Carrara, P. and Andrews, J.T., 1972: The Quaternary history of northern Cumberland Peninsula, Baffin Island, N.W.T.: Part I: The late and neoglacial deposits of the Akudlermuit and Boas Glaciers. Canadian Journal of Earth Sciences, 9(403):403-413.

Cropper, J.P. and Fritts, H.C., 1981: Tree-ring width chronologies from the North American Arctic. Arctic and Alpine Research, 13:245-260.

Davis, P.T., 1980: Late Holocene glacial, vegetational and climatic history of Pangnirtung and Kingnait Fiord area, Baffin Island, N.W.T., Canada. Ph.D. thesis, University of Colorado, Boulder. 366 pp.

Denton, G.H. and Karlen, W., 1973: Holocene climatic variations--their pattern and possible cause. Quaternary Research, 3:155-205.

Denton, G.H. and Stuiver, M., 1966: Neoglacial chronology, northeastern St. Elias Mountains, Canada. American Journal of Science, 264:577-599.

Diaz, H.F. and Andrews, J.T., 1982 Analysis of the spatial patterns of July temperature departures (1943-1972) over Canada and estimates of the 7000 mb mid-summer circulation during Middle and Late Holocene. Journal of Climatology, 2:251-265.

Dyke, A.S., 1977: Quaternary geomorphology, glacial chronology, and climatic and sea-level history of southwestern Cumberland Peninsula, Baffin Island, Northwest Territories, Canada. Ph.D thesis, University of Colorado, Boulder. 185 pp.

Fisher, D.A. and Koerner, R.M., 1981: Some aspects of climatic change in the high Arctic during the Holocene and deduced from ice cores. In: Mahaney, W.C. (ed.), Quaternary Paleoclimate, Norwich, Geo Abstracts, 249-271.

Hammer, C.U., Clausen, H.B., Dansgaard, W., Gundestrup, N., Johnsen, S.J., and Reeh, N., 1978: Dating of Greenland ice cores by flow models, isotopes, volcanic debris and continental dust. Journal of Glaciology, 20:3-26.

Herron, M.M., Herron, S.L., and Langway, C.C., Jr., 1981: Climatic signal of ice melt features in southern Greenland. Nature, 293:389-391.

Hopkins, D.M., 1982: Aspects of the paleogeography of Beringia during the late Pleistocene. In: Hopkins, D.M., Matthews, J.V., Jr., Schweger, C.E., and Young, S.B. (eds.), Paleoecology of Beringia, New York, Academic Press, 3-28.
Hovgaard, W., 1925: The Norsemen in Greenland. Recent discoveries at Herjolfnes. Geographical Review, 15:605-616.

Keen, R.A., 1980: Temperature and circulation anamalies in the eastern Canadian Arctic, Summer 1946-76. Occasional Paper No. 34. Institute of Arctic and Alpine Research, University of Colorado, Boulder.
Klein, J., Lerman, J.C., Damon, P.E., and Ralph, E.K., 1982: Calibration of radiocarbon dates: tables based on the consensus data of the Workshop on Calibrating the Radiocarbon Time Scale. Radiocarbon, 24:103-150.
Koerner, R.M., 1977: Devon Island ice cap: core stratigraphy and paleoclimate. Science, 196:15-18.

LaMarche, V.C., Jr., 1974: Paleoclimatic inferences from long tree-ring records. Science, 183:1043-1048.
LaMarche, V.C., Jr. and Stockton, C.W., 1974: Chronologies from temperature-sensitive bristlecone pines at upper treeline in western United States. Tree-ring Bulletin, 34:21-45.
Lamb, H.H., 1977: Climate: Present, Past and Future. Volume 2, London, Methuen. 835 pp.
Laursen, D., 1976: New contributions to the stratigraphy of the marine Pleistocene and Holocene of West Greenland. In: Abstracts of the fourth biennial meeting, American Quaternary Association.
Locke, C. and Locke, W., 1977: Little Ice Age snow-cover extent and paleoglaciation thresholds: North-central Baffin Island, N.W.T., Canada. Arctic and Alpine Research, 9:291-300.

Matthews, J.A., 1980: Some problems and implications of C-14 dates from a podzol buried beneath an end moraine at Haugabreen, southern Norway. Geografiska Annaler, 62A:185-208.
Miller, G.H., 1973: Late Quaternary glacial and climatic history of northern Cumberland Peninsula, N.W.T., Canada. Quaternary Research, 3:561-583.
Miller, G.H., 1980: Late Foxe glaciation of southern Baffin Island, N.W.T., Canada. Geological Society of America Bulletin, 91:399-405.
Miller, G.H. and Andrews, J.T., 1972: Quaternary history of northern Cumberland Peninsula, east Baffin Island, N.W.T., Canada. Part IV: preliminary lichen growth curve for Rhizocarpon geographicum. Geological Society of America Bulletin, 83:1133-1138.

Nichols, H., 1974: Arctic North American paleoecology: the recent history of vegetation and climate. In: Ives, J.D. and Barry, R.G. (eds.), Arctic and Alpine Environments. Methuen, London, 637-668.
Nichols, H., 1975: Palynological and paleoclimatic study of the late Quaternary displacement of the boreal forest-tundra ecotone in Keewatin and Mackenzie, N.W.T., Canada. Occasional Paper 15, Institute of Arctic and Alpine Research, University of Colorado, Boulder. 87 pp.
Nørlund, P., 1924: Buried Norsemen at Herjolfnes. Meddelelser om Grønland, 67(1):1-270.

Paterson, W.S.B., Koerner, R.M., Fisher, D., Johnsen, S.J., Clausen, H.B., Dansgaard, W., Bucher, P., and Oeschger, H., 1977: An oxygen-isotope record from the Devon Island ice cap, arctic Canada. Nature, 266, 508-511.

Patzelt, G., 1974: Holocene vartiations of glaciers in the Alps. In: Variations du Climat au cours du Pleistocene, 51-60, Paris, Colloques Internationaux du Centre de la Recherche Scientifique, 219.

Prest, V.K., 1970: Quaternary geology in Canada. In: Douglas, R.J. (ed.), Geology and Economic Minerals of Canada, 675-764, 5th Edition, Geological Survey of Canada, Economic Geology Report No. 1.

Ralph, E.K., Michael, H.N., and Han, M.C., 1973: Radiocarbon dates and reality. MASCA Newsletter, 9:1-19.

Rampton, V., 1971: Late Quaternary and climatic history of the Snag-Klutlan area, southwestern Yukon Territory, Canada. Geological Society of America Bulletin, 82:959-978.

Ritchie, J.C. and Hare, F.K., 1971: Late-Quaternary vegetation and climate near the Arctic treeline of northwestern North America. Quaternary Research, 1:331-342.

Ritchie, J.C., Cwynar, L.C., and Spear, R.W., 1983: Evidence from north-west Canada for an early Holocene Milankovitch thermal maximum. Nature, 305:126-128.

Ruddiman, W.F. and McIntyre, A., 1981: The North Atlantic Ocean during the last deglaciation. Palaeogeography, Palaeoclimatology, Palaeoecology, 35:145-214.

Schweingruber, F.H., Braeker, O.U., and Schaer, E., 1978: Dendroclimatic studies in Great Britain and the Alps. In: Evolution des Atmospheres Planetaires et Climatologie de la Terre, Toulouse, Centre National d'Etudes Spatiales, 369-372.

Short, S.K., 1978: Holocene palynology in Labrador-Ungava: climatic history and culture change on the central coast. Ph.D. thesis, Department of Anthropology, University of Colorado, Boulder.

Short, S.K. and Nichols, H., 1977: Holocene pollen diagrams from subarctic Labrador-Ungava: vegetational history and climatic change. Arctic and Alpine Research, 9:265-290.

Short, S.K. and Andrews, J.T., 1980: Palynology of six middle and late Holocene peat sections, Baffin Island. Geographie Physique et Quaternaire, 34:61-75.

Stewart, T.G. and England, J., 1983: Holocene sea-ice variations and paleoenvironmental change, northernmost Ellesmere Island, N.W.T., Canada. Arctic and Alpine Research, 15:1-17.

Stuckenrath, R., Miller, R.H., and Andrews, J.T., 1979: Problems of radiocarbon dating Holocene organic-bearing sediments, Cumberland Peninsula, Baffin Island, N.W.T., Canada. Arctic and Alpine Research, 11:109-120.

Stuiver, M., 1982: A high-precision calibration of the AD radiocarbon time scale. Radiocarbon, 24:1-26.

Tarr, R.S., 1897: Differences in the climate of the Greenland and American sides of Davis' and Baffin's Bay. American Journal of Science, 4:315-320.

Ten Brink, N.W. and Weidick, A., 1974: Greenland Ice Sheet history since the last glaciation. Quaternary Research, 4:429-440.

Walsh, J., 1977: The incorporation of ice station data into a study of recent arctic temperature fluctuations. Monthly Weather Review, 105:1527-1535.

Williams, L.D., 1978: The Little Ice Age glaciation level on Baffin Island, Arctic Canada. Palaeogeography, Palaeoclimatology, Palaeoecology, 25:199-207.

Williams, L.D. and Wigley, T.M.L., 1983: A comparison of evidence for late Holocene summer temperature variations in the Northern Hemisphere. Quaternary Research, 20:286-307.

Survey — Part V

The chapters in the final section of this volume deal largely with the history of events during the last 10,000 years. This period includes the retreat of ice sheets and glaciers from the eastern Canadian Arctic and West Greenalnd during early and middle Holocene time, but it also includes the readvances of local glaciers during the geologic-climate interval called the neoglacial. However, it should be noted that chapters in other sections of this volume deal with glacial, climatic, and oceanographic events during the Holocene. Thus chapters 15, 16, 18, and 19 contribute substantial information to the broad topic of Holocene glacial and climatic events in the area surrounding Baffin Bay.

The chapter by Quinlan (Chapter 20) represents a significant contribution to our attempts to portray the changing character of the late Quaternary North American ice sheet. The maps and figures provided by Quinlan represent an important series of hypotheses against which field observations can be tested. This in turn will allow a re-evaluation of the glacial isostatic model that Quinlan and others have employed. A preliminary test of the model(s) in Frobisher Bay indicates some areas of disagreement. However, the Frobisher Bay area has largely been worked on during large-scale reconnaissance surveys where the emphasis has been on mapping of glacial and glacial marine sediments (see Chapter 18).

Chapters 22, 23, 24, 25, and 26 deal with various aspects of Holocene climate. The detailed investigation of neoglacial moraines by Davis (Chapter 24) indicates that the fluctuation of glacier snouts in southern Cumberland Peninsula has a much higher frequency content than the proposed 2500 year cycle. However, the pollen records from both the eastern Canadian Arctic and West Greenland (Chapters 22 and 23) illustrate broader amplitude oscillations of climate.

A comparison of these two chapters indicates a substantial difference in the degree to which palynology has been used as a tool for reconstructing Holocene vegetation and climate histories. In Greenland there has been a long tradition of palynological studies and the reader must be impressed by the wealth of detail in Fredskild's chapter. This detail extends to a recognition of local pollen taxa that is virtually unparalleled in North American studies. In contrast, the beginning of palynological studies in the Eastern Canadian Arctic have tended to focus more on the climatic implication of the pollen record and less on the history of vegetation succession. However, a theme that might link and serve as a correlation between these areas

well north of the conifer tree line is the documentation of extensive
influx and accumulation of exotic pollen taxa in the Holocene
sediments.

Both West Greenland and the Eastern Canadian Arctic were peopled
during the middle/late Holocene when successive waves of immigrants
moved eastward from a possible "homeland" in Alaska. Of great interest
to the archeologist, paleoclimatologist, and even the current
populations that live in the area is the extent to which climate forced
adaptations in hunting and survival strategies. Chapter 25 explores
these questions and utilizes data of the kind discussed in Chapters 22,
23, and 24 as a framework for understanding changes in human occupation
of these northern lands.

In the last chapter of this volume Bradley and Williams (Chapter
26) discuss the nature of the present climate system in the area of
Baffin Bay. In some ways this chapter, or the first half of it, could
have been printed toward the beginning of the volume as the authors
clearly show the climatic sensitivity of the Baffin region. However,
their discussion then moves on and presents a detailed comparison of a
variety of climatic "proxy" records from Greenland, Baffin Island, and
adjacent sites. These records are examined on a variety of time-scales
and resolutions and, as might be expected, both similarities and marked
differences can be found between the various attempts to reconstruct
"climate." These differences might be real or they might reflect some
of the severe problems asociated with climatic reconstructions among
which the establishment of a valid chronology must loom as a major
goal.

DATE DUE